Annotated Instructor Edition for

# INVESTIGATING BIOLOGY
# Laboratory Manual

*Seventh Edition*

**Judith Giles Morgan**
*Emory University*

**M. Eloise Brown Carter**
*Oxford College of Emory University*

D0904741

**Benjamin Cummings**

Boston   Columbus   Indianapolis   New York   San Francisco   Upper Saddle River
Amsterdam   Cape Town   Dubai   London   Madrid   Milan   Munich   Paris   Montréal   Toronto
Delhi   Mexico City   São Paulo   Sydney   Hong Kong   Seoul   Singapore   Taipei   Tokyo

The authors and publisher believe that the lab experiments described in this publication, when conducted in conformity with the safety precautions described herein and according to the school's laboratory safety procedures, are reasonably safe for the students to whom the manual is directed. Nonetheless, many of the described experiments are accompanied by some degree of risk, including human error, the failure or misuse of laboratory or electrical equipment, mismeasurement, spills of chemicals, and exposure to sharp objects, heat, bodily fluids, blood, or other biologics. The authors and publisher disclaim any liability arising from such risks in connection with any of the experiments contained in this manual. If students have questions or problems with materials, procedures, or instructions on any experiment, they should always ask their instructor for help before proceeding.

Vice President/Editor-in-Chief: Beth Wilbur
Senior Editorial Manager: Ginnie Simione Jutson
Project Editors: Nina Lewallen Hufford, Susan Berge
Managing Editor, Production: Michael Early
Production Project Manager: Jane Brundage
Production Management and Composition: S4Carlisle, Carla Kipper
Art: Precision Graphics, S4Carlisle
Cover Design Production: Seventeenth Street Studios
Photo Researcher: Maureen Spuhler
Photo Editor: Donna Kalal
Image Rights and Permissions Manager: Zina Arabia
Image Permissions Coordinator: Elaine Soares
Manufacturing Buyer: Michael Penne
Executive Marketing Manager: Lauren Harp
Text, Cover and Color Insert Printer: Edwards Brothers
Cover Photo Credit: "Succulent I" ©2005 Amy Lamb, www.amylamb.com

**Library of Congress Cataloging-in-Publication Data**

Morgan, Judith Giles.
  Investigating biology laboratory manual / Judith Giles Morgan M. Eloise Brown Carter.—7th ed.
    p. cm.
  Includes bibliographical references and index.
  ISBN 978-0-321-66821-9 (student edition : alk. paper)—ISBN 978-0-321-67668-9
(annotated instructor edition : alk. paper) 1. Biology—Laboratory manuals. I. Carter, M.
Eloise Brown, 1950- II. Title.
  QH317.M74 2010
  570.78—dc22                                    2010030416

          Student Edition ISBN-13: 978-0-321-66821-9          ISBN-10: 0-321-66821-9
Annotated Instructor's Edition ISBN-13: 978-0-321-67668-9     ISBN-10: 0-321-67668-8

**Benjamin Cummings**
is an imprint of

**www.pearsonhighered.com**          2 3 4 5 6 7 8 9 10—EB—14 13 12 11

# Preface

*The scientist is not a person who gives the right answers,
he [she] is the one who asks the right questions.*

CLAUDE LEVI-STRAUSS

Our knowledge of the biological world is based on the scientific enterprise of asking questions and testing hypotheses. An important aspect of learning biology is participating in the process of science and developing creative and critical reasoning skills. Our goal in writing this laboratory manual is to present a laboratory program that encourages participation in the scientific process. We want students to experience the excitement of discovery and the satisfaction of solving problems and connecting concepts. For us, investigating biology is more than just doing experiments; it is an approach to teaching and learning.

The laboratory exercises are designed to encourage students to ask questions, to pose hypotheses, and to make predictions before they initiate laboratory work. Students are required to synthesize results from observations and experiments, then draw conclusions from evidence. Students are provided with many opportunities to design their own open-inquiry investigations as part of the laboratory. In addition, students can pursue independent investigations using the suggestions and extensions provided at the end of lab topics. Finally, whenever possible, students apply their results to new problems and case studies. Scientific writing and communication are emphasized throughout the laboratory manual and supported by an appendix on scientific writing and communication including suggestions for a laboratory writing program.

We are convinced that involving students in the process of science through investigating biological phenomena is the best way to teach. The organization of this laboratory manual complements this approach to teaching and learning.

## New in the Seventh Edition

The seventh edition of *Investigating Biology* offers enhanced opportunities for students to participate in science. You will find new and revised open-inquiry exercises, suggestions for extending lab topics with independent investigations, new case studies for practicing problem solving, a new section on student media, and a revised appendix to support scientific writing in the laboratory. Adopters of the seventh edition will notice an enhanced emphasis on recurring themes in biology, including structure and function, unity and diversity, and the overarching theme of evolution. These themes are developed across hierarchical levels from cells to organisms to ecosystems.

In the previous edition we introduced a new laboratory, Lab Topic 17 Bioinformatics: Molecular Phylogeny of Plants. This lab topic connects organismal biology, molecular genetics and evolution, using the techniques of computer science to analyze nucleotide sequences and develop phylogenetic trees. We have modified the organisms used in this laboratory, provided suggestions for teaching, additional analysis tools, and new questions that involve "tree thinking."

We have continued to revise and propose ideas for the student-designed investigations in Lab Topic 5 Cellular Respiration, Lab Topic 11 Population Genetics I, Lab Topic 14 Protists and Fungi, Lab Topic 21 Plant Growth, Lab Topic 26 Animal Behavior, and Lab Topic 28 Ecology II. These open-inquiry laboratory experiences allow students to independently investigate questions, thus providing team research opportunities in the introductory laboratory program. For almost all other lab topics, we have further developed the open-inquiry section: "Investigative Extensions." For programs interested in providing independent or team research opportunities, students may pursue these investigations. Additional support for the investigations is provided in the *Preparation Guide* available to Instructors. In Appendix A, Scientific Writing and Communication, we included a section on oral and poster presentations to introduce students to other formats for scientific communication.

Other changes are more subtle but represent fine-tuning based on our experiences and those of instructors and students who used the first six editions. We have reorganized and added to the questions at the end of each lab topic. Questions in "Reviewing Your Knowledge" allow students to test their knowledge and comprehension. Questions in "Applying Your Knowledge" allow students to apply their understanding to new problems that feature current research and societal issues. These questions reinforce the investigative approach of the laboratory manual in which students develop their skills in analyzing results, synthesizing, and communicating as they participate in the scientific process. Figures and photos have been revised and new ones added for visualization of procedures, structures, and organisms. We have updated taxonomic classifications to be consistent with *Campbell Biology,* 9th edition. However, we recognize that these will continue to change as new research reveals evolutionary relationships.

With increasing use of technology in the laboratory by instructors and students, we have provided all data tables in Excel format at *masteringbiology.com* under the Study Area. These tables can be modified and used in open-inquiry investigations, to teach data analysis, and as a model for developing independent projects. All websites have been updated and new Web sources and resources are included. The use of media and technology is further supported by inclusion of a new section in each lab topic called, "Student Media: BioFlix, Activities, Investigations and Videos," which coordinates with *Campbell Biology* student media materials.

The *Annotated Instructor's Edition (AIE)* and the *Preparation Guide* have also been revised to coordinate with and support the laboratory manual. For instructors we have modified the *AIE* Teaching Plans to provide clear objectives and support for teaching using inquiry driven approaches. We include recommendations for time, organization, and student development. No other laboratory manual is supported by a comprehensive laboratory preparation guide that includes detailed information on planning, ordering, materials preparation, organization, and care and culture of organisms. There is an online version of *Preparation Guide for Investigating Biology Laboratory Manual* that allows for downloading and customizing preparation lists each semester.

Writing and scientific communication continue to be a strong component of the laboratory manual. In the 7th edition we enhanced support for the writing program by reorganizing and revising Appendix A Scientific Writing and Communication with easily located information on each part of the scientific paper, notes for successful writing, and suggestions for locating appropriate references. We have added a table with a Plan for Writing a Scientific Paper. We provide tips for preparing and presenting oral papers and posters. The program is supported by materials in the *Preparation Guide* as well as coordinating with web resources and the latest editions of guides to scientific writing and presentations.

In all of these changes and modifications, our objective has been to provide laboratory experiences that will engage students in the scientific enterprise and prepare them to participate in science. We are keenly aware of the constraints on laboratory programs, including number of students and sections, preparation of laboratories and instructors, and the expense and mentoring that are required to teach inquiring young scientists. We hope we have provided choices and options that can meet the needs of a wide range of programs and instructors.

## Laboratory Topics

The laboratory topics build on information and techniques in previous exercises. Various laboratory exercises incorporate a combination of directed and open-ended procedures. There are basically three types of lab topics included in the manual.

1. **Directed-Inquiry investigations** in which exercises have been constructed to involve students in the process of science. We have organized these lab topics to include introductory information from which students develop hypotheses and then predict the results of their experiments. They collect their data and summarize it in tables and figures of their own construction. The students must then accept or reject their hypotheses, based on their results. Examples of these directed-inquiry investigations include Lab Topic 3 Diffusion and Osmosis, Lab Topic 4 Enzymes, and Lab Topic 6 Photosynthesis.

2. **Key Theme investigations** in which laboratory exercises have been designed and reorganized with a focus on key themes of biology, for example, the unity and diversity of life. In these thematic exercises, students summarize and synthesize their results and observations connecting to the key themes. They use their observations as evidence in support of these major concepts and apply their understanding to new problems. Examples of these laboratories (and their underlying themes) include Lab Topic 2 Microscopes and Cells (Unity and Diversity of Life); Lab Topics 15 and 16 Plant Diversity (Adaptation to the Land Environment); and Lab Topics 22 to 24 Vertebrate Anatomy (Structure and Function).

3. **Open-Inquiry investigations** in which students generate their own hypotheses and design their own experiments. These exercises begin with an introduction and a simple experiment that demonstrates procedures. Then students are given suggestions and encouraged to develop their own questions and methodologies for further investigation. Examples of these open-inquiry investigations include Lab Topic 5 Cellular Respiration and Fermentation, Lab Topic 11 Population Genetics I, Lab Topic 21 Plant Growth, Lab Topic 26 Animal Behavior, and Lab Topic 28 Ecology II.

Lab Topics are designated as *Directed-Inquiry, Key Theme,* or *Open-Inquiry Investigations* in the table of contents.

## Scientific Communication: Writing and Presenting

Scientists communicate their results in writing and in presentations to research groups and at meetings. Undergraduates need instruction in writing and an opportunity to practice these skills; however, instructors do not have the time to critique hundreds of student research reports for each exercise. Throughout this lab manual, teams of students work together on improving their skills. They are asked to organize and present their results to their peers during the discussion and summary sessions in the laboratory. Students are also required to write as part of each laboratory. They summarize and discuss their results and then apply information to new problems in the questions at the end of the laboratory.

We have also incorporated a scientific writing program into our lab manual in a stepwise fashion. Students must answer questions and summarize results within the context of the laboratory exercises. For directed-inquiry investigations, students are required to submit one section of a scientific paper. For example, they might submit the Results section for one experiment in Lab Topic 3, and the Discussion section for one experiment in Lab Topic 4. Once students have experience writing each section, they write at least one complete scientific paper for an open-inquiry investigation, for example, Lab Topics 21 and 26. Instructions for writing a scientific paper, developing an oral presentation, and creating a poster are included in Appendix A, which also contains suggestions for developing an organized writing program.

Instructors may also choose to organize a session for oral or poster presentations similar to scientific meetings. Appendix A includes advice and references for these other forms of scientific commmunication.

 See Appendix A and the Teaching Plans for additional information on scientific writing and presentations.

## Integration of Other Sciences and Mathematics

Students often view biology as a separate and isolated body of knowledge. We have attempted to integrate biology, chemistry, some physics, and geology whenever possible. For example, in Lab Top 17 Bioinformatics: Molecular Phylogeny of Plants, students must use bioinformatics to analyze their data from molecular biology. Students use computer models to understand relationships and test hypotheses in Lab Topic 11 Population Genetics I: The Hardy-Weinberg Theorem and Lab Topic 28 Ecology II: Computer Simulations of a Pond Ecosystems. We also provide opportunities for students to quantify observations, analyze and summarize results in tables and figures, and, ultimately, to use these data to construct arguments in support of their hypotheses.

# Special Features

**Reviewing Your Knowledge:** Students recall terminology and content by describing and explaining fundamental concepts. Students examine their results and then use evidence as they evaluate their understanding and knowledge. (Bloom's Taxonomy Level I: Knowledge and Comprehension).

**Applying Your Knowledge:** As instructors, we want our students to be challenged to think and to develop critical thinking skills. Throughout this manual, students are asked to work logically through problems, critique results, and modify hypotheses. To emphasize these skills further, we have developed a section in each laboratory topic called Applying Your Knowledge, in which students are asked to apply their knowledge to new problems and to make connections between topics. The questions encourage students to use their knowledge as they solve problems developed from current research, science, medicine and societal issues. (Bloom's Taxonomy Level II: Application and Analysis; Synthesis and Evaluation).

**Excel Tables:** All data tables used for recording and analyzing results are available in Excel format for modification and use in the laboratory. These tables are designated with an icon in the text and can be downloaded to laboratory computers for student use at www.masteringbiology.com. in the Study Area.

**Images and Data:** Lab Topic 17 Bioinformatics: Molecular Phylogeny of Plants, is supported by the masteringbiology.com website. Folders found under the Study Area contain images of the plants used in the investigation along with the edited and ready-to-use nucleotide sequences for each species. These can be downloaded to laptops or accessed on the website.

**Investigative Extensions and Case Studies:** Students and instructors have expressed interest in extending laboratory topics to an open-inquiry investigation or simply to pursue additional questions that connect the topic to current research and issues. At the end of most lab topics, we have provided questions to prompt student-designed investigations. For a few we provide case studies that build on the lab topic and require additional reading and research.

**Student Media and Web Resources:** Providing media and Web resources can enhance teaching and learning in the laboratory. For students we have included references to videos that connect to their laboratory activities. These are designated in the text by a laptop icon. Also at the end of each lab topic, we have included a section, **Student Media: BioFlix, Actvities, Investigations and Videos**, which directs students to the website for **activities** and **investigations** that can be used to prepare for laboratory or review and practice after the laboratory. These media resources are available at the website: www.masteringbiology.com. under the Study Area.

**Safety Considerations:** Safety concerns are noted in the text by the use of icons for general safety and for biohazards. Laboratory safety is also addressed in the teaching plans at the end of each lab topic in the Instructor's Edition. Note **Laboratory Safety Guidelines** printed on the inside front cover.

**Notes to Students:** To assure student success, cautionary reminders and notes of special interest are also highlighted in the text.

**Appendixes:** Information needed in several laboratory topics is included in the appendixes: scientific writing and communication, using chi-square analysis, and dissection terminology.

**Color Insert:** Color photographs are particularly helpful in the study of cells and organisms. Photos and illustrations in the color insert are cross-referenced to the text.

# Instructional Support

 The *Preparation Guide* and the Teaching Plans in the *Annotated Instructor's Edition* provide valuable suggestions and essential information for the successful implementation of the laboratory topics.

**Preparation Guide:** A detailed *Preparation Guide for Investigating Biology* accompanies the laboratory manual. It contains materials lists, suggested vendors, instructions for preparing solutions and constructing materials, schedules for planning advance preparation, and organization of materials in the lab. The *Preparation Guide is essential* for successfully preparing and teaching these investigative laboratories. It is now available in hard copy and electronically at masteringbiology.com. Instructors can download preparation and ordering lists as needed and customize these for their program.

**Annotated Instructor's Edition:** Teaching biology using an investigative approach requires that instructors guide students in posing questions and hypotheses from which they can predict the results of their experiments. We have included additional support for the instructor in the form of instructor's annotations in each lab topic. These annotations are intended to guide the instructor in responding to students, not to provide the right answers to every student question. We encourage instructors to become the guide to discovery rather than the repository of correct answers. Features of the *Annotated Instructor's Edition* include:

- **Margin notes** with simple suggestions, such as accepting hypotheses as long as they are testable, hints for success, and additional explanations appropriate for the instructor.
- **Suggested answers** to student questions.
- **Appendix D** with calculations and answers to questions in Lab Topic 12 Population Genetics II: Determining Genetic Variation.
- **Typical results**
- **Explanatory figures**
- **Teaching Plans**

**Teaching Plans:** The teaching plans are the instructor's guide to organizing and teaching each laboratory, and they reflect our objective to systematically develop more effective ways to engage students in the study of biology. The teaching plans have been particularly useful for instructors initiating investigative approaches to lab programs. Instructors should feel free to modify these plans to meet their specific needs. The teaching plan for each lab topic includes:

- **Detailed objectives**, both for content and for development of skills in problem solving and scientific methodology.
- **Suggested order of the lab.**
- **Estimated time requirements** for each portion of the lab.
- **Suggested options for organizing the activities for a 2-hour lab period.**
- **Hints on how to manage groups of students** and involve them in investigations that might otherwise become passive learning experiences.
- Information about **lab safety precautions**.

## Acknowledgments

The development of our ideas, the realization of those ideas in laboratory investigations, and the preparation of this laboratory manual are the result of collaborations with many colleagues over the years. We are indebted to our teaching assistants whose critical evaluations and insightful suggestions helped shape the exercises. Several colleagues made especially helpful or critical contributions to our efforts, including: Evelyn Bailey, Steve Baker, Chris Beck, Jim Brown, Joy Budensiek, Alex Escobar, Andrea Heisel, Nitya Jacob, Amanda Pendleton, and Theodosia Wade. We are grateful for the support and guidance provided by the editorial and production group at Benjamin/Cummings. Our thanks to Beth Wilbur, Susan Berge, Kim Wimpsett, Lauren Harp, Mike Early, Jane Brundage, Carla Kipper, and Yvo Riezebos. We are especially indebted to all the laboratory educators who have shared their ideas, hints for success, and philosophies of teaching with us, particularly our friends in the Association for Biology Laboratory Education (ABLE). We are grateful to our reviewers for sharing their words of encouragement and criticism, which were essential to the success of our work. We thank our students who, over the years, have provided the ultimate test for these investigations and our ideas. Finally, we extend special appreciation to our families for their good humor, patience, and encouragement.

## Reviewers for Previous Editions

Mitch Albers, *Minneapolis Community College*

Carol Alia, *Sarah Lawrence College*

Connie L. Allen, *Edison Community College*

Phyllis Baudoin, *Xavier University of Louisiana*

Mariette Baxendale, *University of Missouri, St Louis*

Carol Bernson, *University of Illinois*

Charles Biggers, *Memphis State University*

Sandra Bobick, *Community College of Allegheny County*

Jerry Bricker, *Cameron University*

Jeffrey T. Burkhart, *University of LaVerne*

Linda Burroughs, *Rider University*

Neil Campbell, *University of California, Riverside*

Nickie Cauthen, *Clayton College and State University*

Deborah Clark, *Oregon State University*

Sunita Cooke, *Montgomery College*

Charles Creutz, *University of Toledo*

Diana R. Cundell, *Philadelphia University*

Jean DeSaix, *University of North Carolina, Chapel Hill*

Rebecca Diladdo, *Suffolk University*

John Doherty, *Villanova University*

Susan Dunford, *University of Cincinnati*

April Ann Fong, *Portland Community College, Sylvania Campus*

Caitlin Gabor, *Southwest Texas State University*

William Garnett, *Raymond Walters College*

Patricia S. Glas, *The Citadel*

Michelle F. Guadette, *Tufts University*

Regina Guiliani, *Saint Peter's College*

Tracy Halward, *Colorado State University*

Teresa Hanlon, *Washington and Lee University*

John Hare, *Linfield College*

Laurie Henderson, *Sam Houston State University*

Daniel Hoffman, *Bucknell University*

Alan Jaworski, *University of Georgia*

Virginia A. Johnson, *Towson State University*

Tamas Kapros, *University of Missouri, Kansas City*

Nikolai Kirov, *New York University*

George Krasilovsky, *Rockland Community College*

Laura G. Leff, *Kent State University*

Brian T. Livingston, *University of Missouri, Kansas City*

Raymond Lynn, *Utah State University*

Presley Martin, *Drexel University*

Barbara J. Maynard, *Colorado State University*

Dawn McGill, *Virginia Western Community College*

Jacqueline McLaughlin, *Pennsylvania State University, Lehigh Campus*

Michael McVay, *Green River Community College*

Charles W. Mims, *The University of Georgia*

Alison Mostrom, *Philadelphia College of Pharmacy and Science*

Greg Paulson, *Washington State University*

Shelley A. Phelan, *Fairfield University*

Marcia Pierce, *Eastern Kentucky University*

Jane Rasco, *University of Alabama*

Carolyn Rasor, *Southern Methodist University*

John Rousseau, *Whatcom Community College*

Michael R. Schaefer, *University of Missouri, Kansas City*

Daniel C. Scheirer, *Northeastern University*

Bin Shuai, *Wichita State University*

Sandra Slivka, *San Diego Miramar College*

Nanci Smith, *University of Texas, San Antonio*

Gerald Summers, *University of Missouri*

Sheryl Swartz Soukup, *Illinois Wesleylan University*

Michael Toliver, *Eureka College*

Geraldine W. Twitty, *Howard University*

Roberta Williams, *University of Nevada, Las Vegas*

Dan Wivagg, *Baylor University*

Carol Yeager, *The University of Alabama*

*To Mary, Bill, Rob, Laura, Victoria, Kate, and Will, with love*
J.G.M.
*To Stefanie and Cyndi, with love*
M.E.B.C.

Judith Giles Morgan                                              M. Eloise Brown Carter

# About the Authors

**Judith Giles Morgan** received her M.A. degree from the University of Virginia and her Ph.D. from the University of Texas, Austin. She is Professor Emeritus of Biology at Emory University. She developed a general biology laboratory curriculum for majors to incorporate an investigative approach and a TA training program for multisection investigative laboratories. She served as an officer and board member of the Association for Biology Laboratory Education. Dr. Morgan has worked to improve science education in the community through a number of enrichment programs for high school students.

**M. Eloise Brown Carter** earned her Ph.D. from Emory University and is Professor of Biology at Oxford College of Emory University. With expertise in plant ecology, she has incorporated independent and team research into lecture and laboratory courses in introductory and advanced biology. Dr. Carter served as President of the Association of Southeastern Biologists. She has received several teaching awards including Oxford's Phi Theta Kappa and Fleming Awards, and Emory's Williams Award. Dr. Carter also teaches in the Oxford Institute for Environmental Education, a program for precollege teachers to improve science education through development of schoolyard investigations.

# Contents

Directed
Inquiry

## 6  Photosynthesis   131

Key Theme

## 7  Mitosis and Meiosis   151

Key Theme   **19  Animal Diversity II: Nematoda, Arthropoda, Echinodermata, and Chordata  485**

Key Theme   **20  Plant Anatomy  519**

# Credits

## Photographs

1.9: Barry Mansell/Nature Picture Library. 2.1a, b: Judith Morgan. 3.6a, b: Eloise M. Carter. p.102: Steve Shott/Dorling Kindersley. p.172: USDA/Agricultural Research Service. 7.3: Biophoto/Photo Researchers. 7.9a: Hans Pfletschinger/PhotoLibrary. 7.9c: B. A. Palevitz, Courtesy of E. H. Newcomb, University of Wisconsin. 7.13 left: George Barron, University of Guelph, Canada. 7.13 right: Michael Franklin (mfmicroimaging.com). 11.5a, b: David Fox. Oxford Scientific/Photo Library. 11.6a: Photo Researchers. 11.6b: Omikron/Photo Researchers. p.311a, b: Courtesy of Helena Laboratories, Beaumont, TX and Paul Herbert and Margaret Beaton, Universisty of Guelph. 13.2a, b: Centers for Disease Control and Prevention (CDC). 13.9: Courtesy of and © of Becton, Dickinson and Company. 14.1: John Mansfield. 14.2b: Dr. David Patterson/Photo Researchers. 14.3a: Commonwealth of Australia. 14.3b: D P Wilson/FLPA. 14.4a: Sidney Moulds/Photo Researchers. 14.4b: John BurbidgePhoto Researchers. 14.5a: Colin Bates/Coastal Imageworks (coastalimageworks.com). 14.5b: David Hall/Photo Researchers. 14.5c: Walter H. Hodge/Peter Arnold Images/PhotoLibrary. 14.6: Manfred Kage/Peter Arnold Images/PhotoLibrary. 14.7: Manfred Kage/Peter Arnold Images/PhotoLibrary. 14.8: Stephen Sharnoff (sharnoffphotos. com). 14.9a: Ed Reschke/Peter Arnold/Alamy. 14.9b: D. P. Wilson, Eric & David Hoskins/Photo Researchers. 14.9c: Wayne P. Armstrong, Palomar College. 14.10a: blickwinkel/Alamy. 14.10b: Laurie Campbell/ NHPA Limited/Photoshot. 14.10c: Lee W. Wilcox. 15.5b: Steven P. Lynch, Lynch Images. 15.7: The Field Museum, Neg #GEO85637C, Chicago. Photographer: John Weinstein. 20.1a: Judith Morgan. 20.1b: Eric Grave/Photo Researchers. 20.1c: Judith Morgan. 20.1d: Jim Wehtje/ Photodisc/Getty Images. 20.1e: Judith Morgan. 20.3: Radboud University Nijmegen, Virtual classroom Biology, (www.vcbio.science.ru.nl/ image-gallery/show/print/PL0070/). 20.5b: Ed Reschke. 20.5c: Richard Gross. 20.6b: Ed Reschke/Peter Arnold Images/PhotoLibrary. 20.6c: Natalie Bronstein. 20.8b: Professor Ray F. Evert. 20.8c: Lee W. Wilcox. 20.10b: Michael Clayton, University of Wisconsin-Madison. 21.7: John Innes Centre. 22.1a-d: Marian Rice. 22.2a: Marian Rice. 22.2b: Nina Zanetti. 22.2c, d: Marian Rice. 22.2e: Biophoto/Photo Researchers. 22.3a-c: Marian Rice. 22.4b: Victor Eroschenko. 22.5a: Judith Morgan. 24.7a, b: The Exploratorium, www.exploratorium.edu. Photograph by Ester Kutnick. 24.7c: Judith Morgan. 25.5a-d: Hans Pfletschinger/Peter Arnold Images/photolibrary.com. 25.5e: Photo courtesy Dr. Andrew Ewald. 25.5f: Hans Pfletschinger/Peter Arnold Images/photolibrary.com. 25.7a-f: Judith Morgan. 25.9: From Matthews and Schoenwolf, Atlas. 25.10: From Matthews and Schoenwolf, Atlas. 25.11: From Matthews and Schoenwolf, Atlas. 25.12: From Matthews and Schoenwolf, Atlas. 26.1: Bruce Taylor/Bruce Coleman/Photoshot. 26.2: Veer Marketplace. 26.3: Thinkstock.

## Color Plates

Plate 1: NHPA/Photoshot. Plate 2: Richard L. Howey. Plate 3: Image by Yuuji Tsukii, Laboratory of Biology, Science Research Center, Hosei University, Japan. Protist Information Server, (protist.i.hosei.ac.jp). Plate 4: Ralf Wagner (dr-ralf-wagner.de). Plate 5: Manfred Kage/Peter Arnold/PhotoLibrary. Plate 6: Perennou Nuridsany / Photo Researchers. Plate 7: Bruce Iverson/SPL/Photo Researchers. Plate 8: CNRI/SPL/Photo Researchers. Plate 10: Judith Morgan. Plates 11a-d: Elisabeth Pierson, FNWI-Radboud University Nijmegen, Pearson Science. Plate 11e: Ed Reschke/Peter Arnold Images/PhotoLibrary. Plate 11f: Elisabeth Pierson, FNWI-Radboud University Nijmegen, Pearson Science. Plates 12a-c: Ed Reschke/Peter Arnold Images/PhotoLibrary. Plate 12d: M. Abbey/Photo Researchers. Plate 12e: Ed Reschke/Peter Arnold Images/PhotoLibrary. Plate 13: Judith Morgan. Plate 14: Michael Clayton. Plate 15: JS Photo/Alamy. Plate 16: Eloise M. Carter. Plate 17: Christine Case. Plate 18: Centers for Disease Control. Plate 19: Barry L. Batzing. Plate 21: W. P. Dailey (http://micrap.selfip.com:81/micrapp/samples.htm). Plate 22: J. R. Waaland/Biological Photo Service. Plate 23: Michler Hanns-Frieder/AGE Fotostock. Plate 24: George Barron. Plate 25: William Crowder/National Geographic Collection. Plate 26: D P Wilson/FLPA. Plate 27: Laurie Campbell/NHPA/Photoshot. Plate 28: CAIP IMAGES Copyright © 2006 University of Florida (http://plants.ifas.ufl.edu/ slidecol.html). Plate 29: Michael Fogden/Oxford Scientific/PhotoLibrary. Plate 30: Tony Wharton/FLPA. Plate 31: K. G. Vock/Okapia/Photo Researchers. Plate 32: Pat Lynch/Photo Researchers. Plate 33: Jody Banks, Purdue University. Plate 34: Michael Viard/Peter Arnold Images/ PhotoLibrary. Plate 35: Shari L. Morris/AGE Fotostock/PhotoLibrary. Plate 36: Thinkstock. Plate 36 inset: Walter H. Hodge/Peter Arnold Images/PhotoLibrary. Plate 37: Goodshoot/Thinkstock. Plate 37 inset: Fred Spiegel. Plate 38: Eloise M. Carter. Plate 38 inset: Judith Morgan. Plate 39: Ken Brate/Photo Researchers. Plate 40: Eloise M. Carter. Plate 41: Biophoto Associates/Photo Researchers. Plate 42: Keith Gunnar/ Bruce Coleman/Photoshot . Plate 42 inset: Al Schneider, www .swcoloradowildflowers.com. Plate 43: M. E. Warren/Photo Researchers. Plate 44: Stephen Dalton/Photolibrary.com. Plate.45: D. Wilder/Wilder Nature Photography. Plate 46: Anthony Mercieca/Photo Researchers. Plate.47: Merlin D. Tuttle/Photo Researchers. Plate 48: Ed Reschke. Plate 49: Eloise M. Carter. Plate 50: Robert Brons/Biological Photo Service. Plate 51: Ryan Photographic (www.ryanphotographic.com). Plate 52: Biophoto Associates/Photo Researchers. Plate 53: Tom E. Adams/Peter Arnold Images/PhotoLibrary. Plate 54: Robert Dunne/Photo Researchers. Plate 55: Kjell Sndved/Photo Researchers. Plate 56: S. Stephanowicz/SPL/Photo Researchers. Plate 57: Kjell Sandved/Bruce Coleman/Photoshot . Plate 58: Luis Castaneda/Getty Images. Plate 59: Jeff Rotman/Nature Picture Library. Plate 60: Heather Angel/Natural Visions. Plate 61: Judith Morgan. Plate 62: Ed Reschke/Peter Arnold Images/PhotoLibrary. Plate 63: Richard Gross. Plate 64: LUMEN - Loyola University Medical Education Network. Plate 65: SPL/Photo Researchers. Plate 66: George von Dassow. Plate 67: George von Dassow. Plate 68: Nancy Hopkins. Plate 69: Oxford Scientific/PhotoLibrary. Plate 70: Jane Burton/Bruce Coleman/Photoshot.

## Trademarks

Clorox is a registered trademark of the Clorox Company.
FlyNap is a registered trademark of Carolina Biological Supply Company.
Kimwipes is a registered trademark of Kimberly Clark Corporation.
Listerine is a registered trademark of Warner-Lambert Company.
Lysol is a registered trademark of Sterling Drug Inc.
Parafilm is a registered trademark of American National Can Company.
Protoslo is a registered trademark of Carolina Biological Supply Company.
Roquefort is a registered trademark of Roquefort Brand.
Windows is either a registered trademark or a trademark of Microsoft Corporation in the United States and/or other countries.

# Art

**1.3:** Figure from *Science News,* June 21, 2008; Vol. 173, #19, p. 4, "Science Stats." © 2008 Society for Science and the Public. Data from Child and Family Statistics, 2009. **1.4:** Figure from *Science News,* July19, 2008; Vol. 174 #2, p. 4, "Science Stats." © 2008 Society for Science and the Public. Data from Centers for Disease Control, 2008. **1.5:** Figure from *The Return of DDT* by A. Opar from SEED © 2006 All Rights reserved. Used by permission and protected by the Copyright Laws of the United States. **1.6, 1.7:** Adapted from B. R. Grant and P. R. Grant. 2003. "What Darwin's Finches Can Teach us About the Evolutionary Origin and Regulation of Biodiversity," *BioScience,* vol. 53, pp. 965–975. **1.8:** Data from Centers for Disease Control and Prevention. Graph by Rodger Doyle, "Lifestyle Blues,"*Scientific American,* vol. 284, p. 30. Copyright © 2001. Rodger Doyle. Used with permission.
**1.2, 4.3a/b, 5.2, 5.5, 7.1, 7.2, 7.4, 7.5, 7.6, 7.7, 7.8, 7.9 (art), 7.10, 8.1, 8.2, 10.1, 10.2, 11.3, 14.11, 16.1, 16.2, 16.3, 18.2, 18.5, 18.7, 22.1d, 22.5, and 23.3:** From N. Campbell, J. Reece, and L. Mitchell, *Biology,* 8th ed. (San Francisco, CA: Benjamin Cummings, 2005), © 2008 Pearson Education, publishing as Benjamin Cummings Publishing. **2.4a/b:** Adapted from N. Campbell, *Biology,* 3rd ed. (Redwood City, CA: Benjamin/Cummings, 1993), © 1993 Pearson Education, publishing as Benjamin Cummings Publishing. **4.4:** Adapted from N. Campbell and J. Reece, *Biology,* 7th ed. (San Francisco, CA: Benjamin Cummings, 2005), © 2005 Pearson Education, publishing as Benjamin Cummings Publishing. **7.13:** From H. C. Bold et al., *Morphology of Plants and Fungi,* 5th ed., p. 677. Copyright © 1980 HarperCollins. Reprinted by permission of Pearson Education. **7.UN.03:** Adapted from Becker et al., *The World of the Cell,* 5th ed. (San Francisco, CA, 2003), © 2003 Pearson Education, publishing as Benjamin Cummings Publishing.
**8.3, 8.4, 8.5, 8.6, and 8.7:** Based on information from the *Wisconsin Fast Plants Manual,* published by Carolina Biological Supply Company. © 1989 by Wisconsin Alumni Research Foundation.
**9.2, 9.3:** From Demerec and Kaufman, *Drosophila Guide,* 9th ed. (Washington, DC: Carnegie Institution of Washington, 1986). Reprinted by permission. **11.6c:** From Starr, *Biology: Concepts and Applications,* 1e © 1991 Brooks/Cole, a part of Cengage Learning, Inc. Reproduced by permission. www.cengage.com/permissions. **11.UN.2:** Adapted from L. M. Cook. 2003. The rise and fall of the *Carbonria* form of the peppered moth. *The Quarterly Review of Biology,* 78(4), 399—417. The University of Chicago. All rights reserved. **12.1:** Adapted from F. Ayala, *Population and Evolutionary Genetics: A Primer* (Menlo Park, CA: Benjamin/Cummings, 1982), © 1982 Pearson Education, publishing as Benjamin Cummings Publishing. **13.6:** Adapted from E. Alcamo, *Fundamentals of Microbiology,* 5th ed., fig. 3.11 (Menlo Park, CA: Benjamin/Cummings, 1997), © 1997 Pearson Education, publishing as Benjamin Cummings Publishing.
**14.12, 14.13a/b, 14.14a, and 14.15:** Illustrated by Miguel Ulloa Sosa. Reprinted by permission. **15.3, 15.7b, c, d:** Adapted from James D. Mauseth, *Botany: An Introduction to Plant Biology.* Used by permission of the author. **17.2—10:** Screen shots from Biology Workbench, San Diego Supercomputer Center, University of California, San Diego. Used with permission. **18.3, 18.4, 19.1, 19.2, 19.3, 19.4, 19.7:** Adapted from L. Mitchell, J. Mutchmor, and W. Dolphin, *Zoology* (Menlo Park, CA: Benjamin/Cummings, 1988), © 1988 Pearson Education, publishing as Benjamin Cummings Publishing. **Un 19.1 (AIE only):** Illustration from Charles F. Lytle and J.E. Wodsedalek, *General Zoology Laboratory Guide,* complete version, 9th ed. (Dubuque, IA: Wm. C. Brown, 1987). © 1987 Wm. C. Brown Communications, Inc. Reproduced by permission of the McGraw-Hill Companies. **19.6c:** Campbell, Neil A., et al., *Biology,* 8th ed. © 2008, p. 693. Reprinted by Permission of Pearson Education Inc., Upper Saddle River, New Jersey. **20.4:** Adapted from J. Dickey, *Laboratory Investigations for General Biology* (Redwood City, CA: Benjamin Cummings, 1994) © 1994 Pearson Education, publishing as Benjamin Cummings Publishing.
**21.1, 21.2a/b, 21.3, 21.5:** Based on information from the *Wisconsin Fast Plants Manual,* published by Carolina Biological Supply Company. © 1989 by Wisconsin Alumni Research Foundation. **21.4:** Figure used with permission, courtesy of the Wisconsin Fast Plants Program, College of Agricultural and Life Sciences: University of Wisconsin-Madison: www.fastplants.org. **22.1 a, b, and c (art):** Adapted from E. Marieb, *Human Anatomy and Physiology Laboratory Manual,* Cat Version, 6th ed. (Menlo Park, CA: Benjamin/Cummings, 1999), © 1999 Pearson Education, publishing as Benjamin Cummings Publishing. **22.1d (art):** Adapted from N. Campbell, J. Reece, and L. Mitchell, *Biology,* 5th ed.

(Menlo Park, CA: Benjamin/Cummings, 1999), © 1999 Pearson Education, publishing as Benjamin Cummings Publishing. **22.2a (art):** E. Marieb, *Human Anatomy and Physiology Laboratory Manual,* Cat Version, 5th ed. (Menlo Park, CA: Benjamin/Cummings, 1996), © 1996 Pearson Education, publishing as Benjamin Cummings Publishing. **22.2d (art):** E. Marieb, *Human Anatomy and Physiology Laboratory Manual,* Cat Version, 3rd ed. (Menlo Park, CA: Benjamin/Cummings, 1989), © 1989 Pearson Education, publishing as Benjamin Cummings Publishing.
**22.2e (art):** E. Marieb, *Human Anatomy and Physiology Laboratory Manual,* Cat Version, 4th ed. (Menlo Park, CA: Benjamin/Cummings, 1993), © 1993 Pearson Education, publishing as Benjamin Cummings Publishing. **22.2c; 22.3a and c (art), 22.4a:** Adapted from E. Marieb, *Human Anatomy and Physiology Laboratory Manual,* Cat Version, 6th ed. (Menlo Park, CA: Benjamin/Cummings, 1999), © 1999 Pearson Education, publishing as Benjamin Cummings Publishing. **22.7:** Adapted from Stephen Gilbert, *Pictorial Anatomy of Fetal Pig,* 2nd edition. © 1966 University of Washington Press. Reproduced by permission of the University of Washington Press. **23.8:** Adapted from E. Marieb, *Human Anatomy and Physiology Laboratory Manual,* Fetal Pig Version, 6th ed. (Menlo Park, CA: Benjamin/Cummings, 1999), © 1999 Pearson Education, publishing as Benjamin Cummings Publishing. **24.5:** E. Marieb, *Human Anatomy and Physiology Laboratory Manual,* Cat Version, 4th ed. (Menlo Park, CA: Benjamin/Cummings 1993), © 1993 Pearson Education, publishing as Benjamin Cummings Publishing.
**25.1, 25.2, 25.3, 25.4:** L. Mitchell, J. Mutchmor, and W. Dolphin, *Zoology,* figure 16.2 (Menlo Park, CA: Benjamin/Cummings, 1988), © 1988 Pearson Education, publishing as Benjamin Cummings Publishing. **26.4:** From R. J. Greenspan. 1995. Understanding the genetic construction of behavior. *Scientific American,* April 1995, pp. 72-73, figures 1-6. Illustrations copyright © by Suzanne Barnes. **28.1:** From *Living in the Environment: Principles, Connections and Solutions,* 7/e by Miller, © 1992. Reprinted with permission.
**28.2 through 28.8, 1A 28.1, 1A 28.2:** E. C. Odum, H. T. Odum, and N. S. Peterson, *Environmental Decision Making:The BioQUEST Library Volume V.* San Diego, CA: Academic Press, 2002. Lab written to correspond to software by permission of publisher.

# Text

**Lab Topics 1, 5:** Portions adapted from J. Dickey, *Laboratory Investigations for Biology* (Menlo Park, CA: Benjamin Cummings, 1995) © 1995 Pearson Education, publishing as Benjamin Cummings Publishing.
**Lab Topic 3,Exercise 3.1, Experiment B:** Lab exercise from D. R. Helms and S. B. Miller, *Principles of Biology: A Laboratory Manual for Biology.* © 1978 Contemporary Publishing. Used with permission.
**Lab Topics 8, 21:** Based on information from the *Wisconsin Fast Plants Manual,* published by Carolina Biological Supply Company © 1989 by Wisconsin Alumni Research Foundation. Reprinted by permission.
**Lab Topic 9:** First published as J. G. Morgan and V. Finnerty, "Inheritance of aldehyde oxidase in *Drosophila melanogaster*" in *Tested Studies for Laboratory Teaching,* Volume 12, Proceedings of the 12th Workshop/Conference of the Association for Biology Laboratory Education (ABLE), Corey A. Goldman, Editor. Used by permission.
**Lab Topic 11:** Adapted from J. C. Glase, 1993. A laboratory on population genetics and evolution: A physical model and computer simulation. In *Tested Studies for Laboratory Teaching,* Volume 7/8, pp. 29-41, Proceedings of the 7th and 8th Workshop/Conference of the Association for Biology Laboratory Education (ABLE), C. A. Goldman and P. L. Hauta, Eds. Used by permission.
**Lab Topic 14:** Excerpt from William Crowder, "Marvels of Mycetozoa," *National Geographic Magazine,* April 1926. Copyright © 1926 National Geographic Society. Reprinted by permission.
**Lab Topic 17:** Coauthored with Dr. Nitya Jacob, Assistant Professor of Biology, Oxford College.
**Lab Topic 25: Exercise 25.3:** Adaptation of *Development in the Zebra Fish* by John Pilger. Reprinted by permission of the author.
**Lab Topic 28:** Adapted from E. C. Odum, H. T. Odum, and N. S. Peterson, "Environmental Decision Making."The BioQUEST Library: http://bioquest.org/BQLibrary. Reprinted by permission of the BioQuest Library.
**Appendix B, Table B.2:** Table B.2: From R. A. Fisher and F. Yates, *Statistical Tables for Biological, Agricultural and Medical Research,* 6th ed, Table IV. © 1974 Pearson Education Ltd. Used with permission.

## Lab Topic 1
# Scientific Investigation

 Before going to lab, read the Introduction and Exercises 1.1 and 1.2. Be prepared to answer all questions and contribute your ideas in a class discussion.

## Laboratory Objectives

After completing this lab topic, you should be able to:

1. Identify and characterize questions that can be answered through scientific investigation.
2. Define *hypothesis* and explain what characterizes a good scientific hypothesis.
3. Identify and describe the components of a scientific experiment.
4. Summarize and present results in tables and graphs.
5. Discuss results and critique experiments.
6. Design a scientific experiment.
7. Interpret and communicate results.

*For a 2-hour lab: Reduce the time for the introduction to the topic and lab discussion (Exercise 1.1). Students can critique the experiment outside the lab time and write a report (Exercise 1.5). See Teaching Plan.*

## Introduction

Biology is the study of the phenomena of life, and biological scientists—researchers, teachers, and students—observe living systems and organisms, ask questions, and propose explanations for those observations. Scientific investigation is a way of testing those explanations. Science assumes that biological systems are understandable and can be explained by fundamental rules or laws. Scientific investigations share some common elements and procedures, which are referred to as the *scientific method*. Not all scientists follow these procedures in a strict fashion, but each of the elements is usually present. Science is a creative human endeavor that involves asking questions, making observations, developing explanatory hypotheses, and testing those hypotheses. Scientists closely scrutinize investigations in their field, and each scientist must present his or her work at scientific meetings or in professional publications, providing evidence from observations and experiments that supports the scientist's explanations of biological phenomena.

*Instructors and/or students can download data tables in Excel format. Go to www.masteringbiology.com. Look in the Study Area under Lab Media.*

In this lab topic, you will not only review the process that scientists use to ask and answer questions about the living world, but you will develop the skills to conduct and critique scientific investigations. Like scientists, you will work in research teams in this laboratory and others, collaborating as you ask questions and solve problems. Throughout this laboratory manual, you will be investigating biology using the methodology of scientists, asking questions, proposing explanations, designing experiments, predicting results, collecting and analyzing data, and interpreting your results in light of your hypotheses.

---

EXERCISE 1.1

# Questions and Hypotheses

---

This exercise explores the nature of scientific questions and hypotheses. Before going to lab, read the explanatory paragraphs and then be prepared to present your ideas in the class discussion.

## Lab Study A. Asking Questions

Scientists are characteristically curious and creative individuals whose curiosity is directed toward understanding the natural world. They use their study of previous research or personal observations of natural phenomena as a basis for asking questions about the underlying causes or reasons for these phenomena. For a question to be pursued by scientists, the phenomenon must be well defined and testable. The elements must be measurable and controllable.

There are limits to the ability of science to answer questions. Science is only one of many ways of knowing about the world in which we live. Consider, for example, this question: Do excessively high temperatures cause people to behave immorally? Can a scientist investigate this question? Temperature is certainly a well-defined, measurable, and controllable factor, but morality of behavior is not scientifically measurable. We probably could not even reach a consensus on the definition. Thus, there is no experiment that can be performed to test the question. Which of the following questions do you think can be answered scientifically?

*The purpose of this activity is to lead students to begin to think about characteristics of good scientific questions. It is important that students discuss what is definable, measurable, and controllable, but it is not important that everyone agree. Scientists from various disciplines may have different opinions about the degree of precision required when measuring variables. Mention that positive correlation does not necessarily indicate cause and effect.*

1. Do kids who play violent video games commit more violence?
2. Did the consumption of seven cans of "energy drink" cause the heart attack of a motorcyclist in Australia?
3. Will increased levels of $CO_2$ in the atmosphere stimulate the growth of woody vines, such as poison ivy and kudzu?
4. How effective are extracts of marigold and rosemary as insect repellents?
5. Should it be illegal to sell organs, such as a kidney, for transplant purposes?

How did you decide which questions can be answered scientifically?

*Items investigated must be well defined, measurable, and controllable. The questions should be reasonable and consistent with existing bodies of knowledge. Students will have a variety of ways to exclude wild speculations.*

# Lab Study B. Developing Hypotheses

As questions are asked, scientists attempt to answer them by proposing possible explanations. Those proposed explanations are called **hypotheses.** A hypothesis tentatively explains something observed. It proposes an answer to a question. Consider question 4, preceding. One hypothesis based on this question might be "Marigold and rosemary extracts are more effective than DEET in repelling insects." The hypothesis has suggested a possible explanation that compares the difference in efficacy between these plant extracts and DEET.

A scientifically useful hypothesis must be testable and falsifiable (able to be proved false). To satisfy the requirement that a hypothesis be falsifiable, it must be possible that the test results do not support the explanation. In our example, the experiment might be to spray one arm with plant extracts and the other with DEET. Then place both arms in a chamber with mosquitoes. If mosquitoes bite both arms with equal frequency or the DEET arm has fewer bites, then the hypothesis has been falsified. *Even though the hypothesis can be falsified, it can never be proved true.* The evidence from an investigation can only provide support for the hypothesis. In our example, if there are fewer bites on the arm with plant extracts than on the DEET-treated arm, then the hypothesis has not been proved, but has been supported by the evidence. Other explanations still must be excluded, and new evidence from additional experiments and observations might falsify this hypothesis at a later date. In science seldom does a single test provide results that clearly support or falsify a hypothesis. In most cases, the evidence serves to modify the hypothesis or the conditions of the experiment.

Science is a way of knowing about the natural world (Moore, 1993) that involves testing hypotheses or explanations. The scientific method can be applied to the unusual and the commonplace. You use the scientific method when you investigate why your once-white socks are now blue. Your hypothesis might be that your blue jeans and socks were washed together, an assertion that can be tested through observations and experimentation.

Students often think that controlled experiments are the only way to test a hypothesis. The test of a hypothesis may include experimentation, additional observations, or the synthesis of information from a variety of sources. Many scientific advances have relied on other procedures and information to test hypotheses. For example, James Watson and Francis Crick developed a model that was their hypothesis for the structure of DNA. Their model could only be supported if the accumulated data from a number of other scientists were consistent with the model. Actually, their first model (hypothesis) was falsified by the work of Rosalind Franklin. Their final model was tested and supported not only by the ongoing work of Franklin and Maurice Wilkins but also by research previously published by Erwin Chargaff and others. Watson and Crick won the Nobel Prize for their scientific work. They did not perform a controlled experiment in the laboratory but tested their powerful hypothesis through the use of existing evidence from other research. Methods other than experimentation are acceptable in testing hypotheses. Think about other areas of science that require comparative observations and the accumulation of data from a variety of sources, all of which must be consistent with and support hypotheses or else be inconsistent and falsify hypotheses.

*To generate discussion, ask your students if the Nobel Prize should be taken away from Watson and Crick because they did not actually perform a controlled experiment to discover the structure of DNA.*

The information in your biology textbook is often thought of as a collection of facts, well understood and correct. It is true that much of the knowledge of biology has been derived through scientific investigations, has been thoroughly tested, and is supported by strong evidence. However, scientific knowledge is always subject to novel experiments and new technology, any aspect of which may result in modification of our ideas and a better understanding of biological phenomena. The structure of the cell membrane is an example of the self-correcting nature of science. Each model of the membrane has been modified as new results have negated one explanation and provided support for an alternative explanation.

## Application

Before scientific questions can be answered, they must first be converted to hypotheses, which can be tested. For each of the following questions, write an explanatory hypothesis. Recall that the hypothesis is a statement that explains the phenomenon you are interested in investigating.

*The hypothesis should be broad and more general. Students will come back to these hypotheses and state predictions.*

1. Does supplemental feeding of birds at backyard feeders affect their reproductive success?

   *Supplemental feeding of birds at backyard feeders increases their reproductive success.*

2. Do preschool boys in coed classes develop better verbal skills than boys in all-male classes?

   *Preschool boys in classes that include girls develop better verbal skills than boys in all-male classes.*

*Depending on time constraints, set limits for discussion of these hypotheses or omit one or two.*

Scientists often propose and reject a variety of hypotheses before they design a single test. Discuss with your class which of the following statements would be useful as scientific hypotheses and could be investigated using scientific procedures. Give the reason for each answer by stating whether it could possibly be falsified and what factors are measurable and controllable.

*Encourage students to discuss additional factors that need to be considered or defined to test these hypotheses. The study of evolution still uses the scientific method of hypothesis testing, but some hypotheses cannot be tested with controlled experiments. Tests for these involve the researcher's compilation of different sources of information, all of which must be consistent with the hypothesis.*

1. The use of pesticides in farming causes deformities in nearby frog populations.
2. Sea turtles are more likely to hatch during a new moon.
3. Drinking two or three cups of coffee a day reduces the risk of heart disease in women.
4. Manatees and elephants share a common ancestor.
5. Organic food is healthier than conventionally produced food.

## EXERCISE 1.2
# Designing Experiments to Test Hypotheses

The most creative aspect of science is designing a test of your hypothesis that will provide unambiguous evidence to falsify or support a particular explanation. Scientists often design, critique, and modify a variety of experiments and other tests before they commit the time and resources to perform a single experiment. In this exercise, you will follow the procedure for experimentally testing hypotheses, but it is important to remember that other methods, including observation and the synthesis of other sources of data, are acceptable in scientific investigations. An experiment involves defining variables, outlining a procedure, and determining controls to be used as the experiment is performed. Once the experiment is defined, the investigator predicts the outcome of the experiment based on the hypothesis.

Read the following description of a scientific investigation of the effects of sulfur dioxide on soybean reproduction. Then in Lab Study A you will determine the types of variables involved, and in Lab Study B, the experimental procedure for this experiment and for others.

### INVESTIGATION OF THE EFFECT OF SULFUR DIOXIDE ON SOYBEAN REPRODUCTION

Agricultural scientists were concerned about the effect of air pollution, sulfur dioxide in particular, on soybean production in fields adjacent to coal-powered power plants. Based on initial investigations, they proposed that sulfur dioxide in high concentrations would reduce reproduction in soybeans. They designed an experiment to test this hypothesis (Figure 1.1). In this experiment, 48 soybean plants, just beginning to produce flowers, were divided into two groups, treatment and no treatment. The 24 treated plants were divided into four groups of 6. One group of 6 treated plants was placed in a fumigation chamber and exposed to 0.6 ppm (parts per million) of sulfur dioxide for 4 hours to simulate sulfur dioxide emissions from a power plant. The experiment was repeated on the remaining three treated groups. The no-treatment plants were divided similarly into four groups of 6. Each group in turn was placed in a second fumigation chamber and exposed to filtered air for 4 hours. Following the experiment, all plants were returned to the greenhouse. When the beans matured, the number of bean pods, the number of seeds per pod, and the weight of the pods were determined for each plant.

*Using the soybean example and the questions in Lab Study A, lead a class discussion of dependent variables, independent variables, controlled variables, level of treatment, replication, control treatments, and predictions.*

*As you lead this discussion, do not review all of the text. Students should be prepared to discuss the material and answer questions.*

## Lab Study A. Determining the Variables

Read the description of each category of variable; then identify the variable described in the preceding investigation. The variables in an experiment must be clearly defined and measurable. The investigator will identify

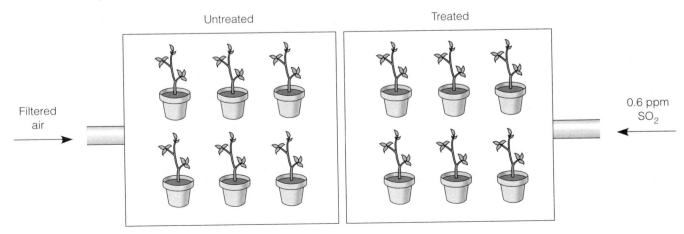

**Figure 1.1.**
**Experimental design for soybean experiment.** The experiment was repeated four times. Soybeans were fumigated for 4 hours.

and define *dependent, independent,* and *controlled variables* for a particular experiment.

## The Dependent Variable

Within the experiment, one variable will be measured or counted or observed in response to the experimental conditions. This variable is the **dependent variable.** For the soybeans, several dependent variables are measured, all of which provide information about reproduction. What are they?

> *number of bean pods, number of seeds per pod, and weight of pods*

## The Independent Variable

The scientist will choose one variable, or experimental condition, to manipulate. This variable is considered the most important variable by which to test the investigator's hypothesis and is called the **independent variable.** What was the independent variable in the investigation of the effect of sulfur dioxide on soybean reproduction?

> *sulfur dioxide*

Can you suggest other variables that the investigator might have changed that would have had an effect on the dependent variables?

> *light, fertilizer, water, variety of soybean, size of plants, temperature*

Although other factors, such as light, temperature, time, and fertilizer, might affect the dependent variables, only one independent variable is usually chosen. Why is it important to have only one independent variable?

> *You need to know which factor is affecting the dependent variable(s).*

*There are complex experimental designs that include more than one independent variable. However, students at this level should limit their experiments to one independent variable.*

Why is it acceptable to have more than one dependent variable?

> *Each of the variables can be a way of measuring the characteristic that is affected—in this case, reproduction.*

## The Controlled Variable

*Students often confuse controlled variables with the control experiment or treatment.*

Consider the variables that you identified as alternative independent variables. Although they are not part of the hypothesis being tested in this investigation, they would have significant effects on the outcome of this experiment. These variables must, therefore, be kept constant during the course of the experiment. They are known as the **controlled variables.** The underlying assumption in experimental design is that the selected independent variable is the one affecting the dependent variable. This is only true if all other variables are controlled. What are the controlled variables in this experiment? What variables other than those you may have already listed can you now suggest?

> *temperature; size of plants; age of plants; variety of soybeans; length of day; amount of fertilizer; amount of light; watering regime; air flow; humidity and temperature in the fumigation chamber*

# Lab Study B. Choosing or Designing the Procedure

The **procedure** is the stepwise method, or sequence of steps, to be performed for the experiment. It should be recorded in a laboratory notebook before initiating the experiment, and any exceptions or modifications should be noted during the experiment. The procedures may be designed from research published in scientific journals, through collaboration with colleagues in the lab or other institutions, or by means of one's own novel and creative ideas. The process of outlining the procedure includes determining control treatment(s), levels of treatments, and numbers of replications.

## Level of Treatment

The value set for the independent variable is called the **level of treatment.** For this experiment, the value was determined based on previous research and preliminary measurements of sulfur dioxide emissions. The scientists may select a range of concentrations from no sulfur dioxide to an extremely high concentration. The levels should be based on knowledge of the system and the biological significance of the treatment level. In some experiments however, independent variables represent categories that do not have a level of treatment (for example, gender). What was the level of treatment in the soybean experiment?

*0.6 ppm/4 hours*

## Replication

Scientific investigations are not valid if the conclusions drawn from them are based on one experiment with one or two individuals. Generally, the same procedure will be repeated several times (**replication**), providing consistent results. Notice that scientists do not expect exactly the same results inasmuch as individuals and their responses will vary. Results from replicated experiments are usually averaged and may be further analyzed using statistical tests. Describe replication in the soybean experiment.

*The plants were divided into four groups of 6 plants each. The entire procedure was replicated four times.*

## Control

The experimental design includes a **control** in which the independent variable is held at an established level or is omitted. The control or control treatment serves as a benchmark that allows the scientist to decide whether the predicted effect is really due to the independent variable. In the case of the soybean experiment, what was the control treatment?

*fumigating with filtered air, in other words, air free of sulfur dioxide*

What is the difference between the control and the controlled variables discussed previously?

*Controlled variables are independent variables that can affect the results of the experiment unless they are kept constant. The control is setting the independent variable under investigation to zero or a standard value.*

# Lab Study C. Making Predictions

The investigator never begins an experiment without a prediction of its outcome. The **prediction** is always based on the particular experiment designed to test a specific hypothesis. Predictions are written in the form of if/then statements: "If the hypothesis is true, then the results of the experiment will be . . ."; for example, "**if** extracts of marigold and rosemary are more effective than DEET in repelling insects, **then** there will be fewer bites on the arm sprayed with the plant extract compared to the arm sprayed with DEET after a 5-minute exposure to mosquitoes." Making a prediction provides a critical analysis of the experimental design. If the predictions are not clear, the procedure can be modified before beginning the experiment. For the soybean experiment, the hypothesis was: "Exposure to sulfur dioxide reduces reproduction." What should the prediction be? State your prediction.

> *If sulfur dioxide reduces reproduction, then the number of pods, the number of seeds per pod, and the weight of the pods per plant should decrease in plants exposed to sulfur dioxide compared with control plants.*

To evaluate the results of the experiment, the investigator always returns to the prediction. If the results match the prediction, then the hypothesis is supported. If the results do not match the prediction, then the hypothesis is falsified. Either way, the scientist has increased knowledge of the process being studied. Many times the falsification of a hypothesis can provide more information than confirmation, since the ideas and data must be critically evaluated in light of new information. In the soybean experiment, the scientist may learn that the prediction is true (sulfur dioxide does reduce reproduction at the concentration tested). As a next step, the scientist may now wish to identify the particular level at which the effect is first demonstrated.

Return to p. 4 and review your hypotheses for the numbered questions. Consider how you might design an experiment to test the first hypothesis. For example, you might count the number of eggs per nest or count the number of chicks successfully raised in a summer. The prediction might be:

> **If** supplemental feeding of birds at backyard feeders increases their reproductive success (*a restatement of the hypothesis*), **then** there will be more eggs per nest for birds with supplemental feeding compared to birds who receive no supplemental feeding (*predicting the results from the experiment*).

Now consider an experiment you might design to test the second hypothesis on p. 4. How will you measure "better verbal skills"?

State a prediction for this hypothesis and experiment. Use the if/then format.

> *If preschool boys in classes that include girls develop better verbal skills than boys in all-male classes, then boys in coed classes will have more words in their vocabulary than boys in all male classes.*

The actual test of the prediction is one of the great moments in research. No matter the results, the scientist is not just following a procedure but truly testing a creative explanation derived from an interesting question.

## Discussion

1. From this exercise, list the components of scientific investigations from asking a question to carrying out an experiment.

    *1. Asking a question*

    *2. Developing hypotheses or explanations based on observations and previous knowledge*

    *3. Designing an experiment*

    *4. Making predictions*

2. From this exercise, list the variables that must be identified in designing an experiment.

    *independent variable, dependent variable, controlled variable*

3. What are the components of an experimental procedure?

    *levels of treatment, control, and method of replication*

EXERCISE 1.3
# Designing an Experiment

*Since this concept will be new to many students, after you have introduced the exercise and the students have read the introduction, ask questions to check their understanding of the differences in allometric and isometric growth.*

## Materials

meter sticks or yard sticks
short metric rulers (approximately 30 cm or 1 foot)
cloth metric measuring tapes or cotton string
calipers (optional)
calculators

## Introduction

The shape of an organism depends in part on the relative growth rates of different body parts during development. The terms **allometry** and **allometric growth** are used when describing the changing relative rates of growth. Specifically, allometry is the phenomenon whereby parts of an organism grow at different rates, or the disproportionate growth of a part in relation to the entire organism. This can be observed in humans as seen in Figure 1.2, an illustration of differential growth rates. Notice in this figure, in which the newborn and adult are scaled to the same height, that the head is larger in proportion to the body length in a newborn than in an adult.

Allometric growth contrasts with **isometric** growth when two parts grow at the same rate. This takes place when proportions between body parts remain constant as the organism grows. You can imagine this type of growth by taking a photo of the baby in Figure 1.2 and simply enlarging it as it grows to an adult. For example, if the arms and legs of a human grow isometrically, then their lengths will be the same relative to the body in a newborn as in an adult.

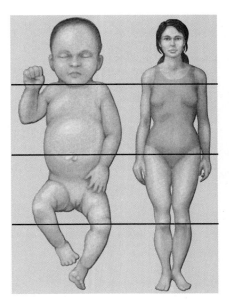

**Figure 1.2.**
**Allometric growth in humans.** In this figure, the body of the newborn and adult have been scaled to the same height.

**Table 1.1**
Average (50th percentile) Body Part Sizes for Newborn (38 Weeks Gestation) North American Humans, Both Sexes

| Body Part | Description of Measurement | Size (cm) (rounded to nearest 0.5) |
|---|---|---|
| Length/height | From sole of foot to top of head | 48 |
| Head circumference | Head at largest diameter | 33 |
| Upper limb (total arm) | From shoulder joint to tip of middle finger | 20 |
| Hand length | From wrist to tip of middle finger | 6 |
| Span | Between fingertips of middle finger with arms outstretched | 51 |
| Lower limb (total leg) | From sole of foot to joint where leg meets hip laterally | 18 |
| Crown-rump length | In seated position, from top of head to surface of a flat-bottom chair or bench | 33 |
| Foot length | Measure from heel to toe standing on a tape measure | 7.5 |

(Data from Hall et al., 2007)

*Hall et al. (2007) provides a compendium of data derived from numerous studies. For further information on methodologies and additional measurements see references in this source.*

In the following exercise you will collect data to compare ratios of height to selected body parts for newborns with the same ratios in students (yourselves). This should provide information to determine if selected body parts grow allometrically or isometrically in humans. In the process you will practice designing experiments, collecting data, and processing that data. The data for newborns are given in Table 1.1. These data are taken from Hall *et al.* (2007), a handbook of physical measurements for health professionals.

In your research teams, examine Table 1.1 and take a few minutes to discuss several *specific* questions that you can ask about growth rates (allometric, isometric) of body parts in relation to height in humans. Choose two body parts and write your questions in the margin of the lab manual. Discuss with your teammates a testable hypothesis for each question. Choose one question and hypothesis to add to a class list recorded by the instructor.

*Divide the class into teams of four students. Ask each student team to report its best question and hypothesis. List these on an overhead or on the board, and have the class decide which questions and hypotheses to investigate.*

 The class decides on two body parts to be compared with total height. For each, the class selects a question and hypothesis and then develops the experimental design for each body part. They then predict the results of the experiments. The same two body parts and height are measured by all teams.

Record the *body parts* and the *corresponding questions* chosen by the class.

*Head circumference: Does the head grow isometrically or allometrically in relation to height?*

*Arm length, (etc): Does the arm grow isometrically or allometrically in relation to height?*

## Hypothesis

Record a hypothesis for each of the questions chosen by the class.

*Possible hypotheses:*

*The circumference of the human head increases allometrically in relation to height.*

*The length of the human arm increases isometrically in relation to height.*

## The Experiment

In this experiment you will compare the ratios of *height to selected body parts* in newborns and students. As a class, design your experiment. Describe how you will measure each selected body part for each team member. Include data collection and analysis. To determine height to body part ratio (H/BP), students may self-report their height. Be sure all teams understand the protocol.

*Choose two body parts and measure the selected body parts for each member of your team. See Table 1.1 for suggested methods.*

*1. Determine the height of each member of your team. (Some students can self-report their height.)*

*2. Divide the height of the individual by the size of the measured body part. This determines the height to body part ratio (H/BP).*

*3. Analyze data in tables and graphs.*

What is the dependent variable in your experiments?

*The ratio of height to body part derived from measurements of height and body parts.*

What is the independent variable?

*The body parts selected.*

Controlled variables:

*Age. All subjects may not have achieved full growth. Sex. Ethnicity.*

Control:

*The measurements for the newborn.*

Level of treatment:

*This particular experiment has no level of treatment.*

*We suggest you encourage students to choose head circumference, which is allometric. Any other body part will make an interesting comparison.*

*Students must decide how they will measure the body parts. This is a valuable activity, as they design a protocol for how all teams make measurements and record data. Do not assist or correct. Keep teams on time for completing the experiment.*

Replication:

*Number of students in class*

## Prediction

Predict the results of each experiment based on your hypotheses (if/then). Write two predictions based on the two body parts chosen.

> *If the human arm grows isometrically in relation to the height of the individual,* then *the ratio of height to arm length will be equivalent in the newborn and adult.*
>
> *If the head grows allometrically in relation to the height of the individual,* then *the ratio of height to head circumference in a newborn will not be equivalent (be greater or less) than the ratio of height to head circumference in an adult.*

## Performing the Experiment

Following the procedures established by your class for each body part, perform the experiment.

## Results

1. List the *first* body part chosen in the title for Table 1.2a.
2. Record results for each member of your team for the first body part in Table 1.2a. (Remember, to convert inches to cm, multiply by 2.54.)
3. Calculate H/BP ratios for each member of your team.
4. Record data for newborns from Table 1.1 for the first body part.
5. Calculate the H/BP ratio for newborns.
6. Follow the same procedure for the second body part chosen, and record all data in Table 1.2b.
7. Record ratios for every student in Table 1.3.
8. Determine the *class average* for H/BP and record in Table 1.3.

*Note: Data tables in Excel format are available to download from the website, masteringbiology.com.*

**Table 1.2a**

Height to *(Head Circumference)* Ratio for Students and Newborns

| Subject | Height (cm) | Body Part Size (cm) | H/BP Ratio |
|---|---|---|---|
| Team member 1 | | | |
| Team member 2 | | | |
| Team member 3 | | | |
| Team member 4 | | | |
| Newborns | *48* | *33* | *1.45* |

**Table 1.2b**
Height to *(Arm Length)* Ratio for Students and Newborns

| Subject | Height (cm) | Body Part Size (cm) | H/BP Ratio |
|---|---|---|---|
| Team member 1 | | | |
| Team member 2 | | | |
| Team member 3 | | | |
| Team member 4 | | | |
| Newborns | *48* | *20* | *2.4* |

**Table 1.3**
Ratios of Height to Two Body Parts for Each Student

| Student # | 1 | 2 | 3 | 4 | 5 | 6 | 7 | 8 | 9 | 10 | 11 | 12 |
|---|---|---|---|---|---|---|---|---|---|---|---|---|
| Body Part # 1 | | | | | | | | | | | | |
| Body Part # 2 | | | | | | | | | | | | |

**Table 1.3, continued**

| Student # | 13 | 14 | 15 | 16 | 17 | 18 | 19 | 20 | 21 | 22 | 23 | 24 | Average Ratio |
|---|---|---|---|---|---|---|---|---|---|---|---|---|---|
| Body Part # 1 | | | | | | | | | | | | | |
| Body Part # 2 | | | | | | | | | | | | | |

## EXERCISE 1.4
# Presenting and Analyzing Results

Once the data are collected, they must be organized and summarized so that scientists can determine if the hypothesis has been supported or falsified. In this exercise, you will design **tables** and graphs; the latter are also called **figures**. Tables and figures have two primary functions. They are used (1) to help you analyze and interpret your results and (2) to enhance the clarity with which you present the work to a reader or viewer.

*If laboratory computers are available, download Table 1.3 in Excel and have students record their results there. Or create an overhead acetate of Table 1.3. Then calculate the class average. This table is based on 24 students in a class. Modify to fit your class size.*

## Lab Study A. Tables

You have collected data from your experiment in the form of a list of numbers that may appear at first glance to have little meaning. Look at your data. How could you organize the data set to make it easier to interpret? You could *average* the data set for each treatment, but even averages can be rather uninformative. Could you use a summary table to convey the data (in this case, averages)?

Table 1.4 is an example of a table using data averages of the number of seeds per pod and number of pods per plant as the dependent variables and exposure to sulfur dioxide as the independent variable. Note that the number of replicates and the units of measurement are provided in the table title.

Tables are used to present results that have a few too many data points. They are also useful for displaying several dependent variables and when the quantitative values rather than the trends are the focus. For example, average number of bean pods, average number of seeds per pod, and average weight of pods per plant for treated and untreated plants could all be presented in one table.

The following guidelines will help you construct a table.

- All values of the same kind should read down the column, not across a row. Include only data that are important in presenting the results and for further discussion.

- Information and results that are not essential (for example: test-tube number, simple calculations, or data with no differences) should be omitted.

- The headings of each column should include units of measurement, if appropriate.

- Tables are numbered consecutively throughout a lab report or scientific paper. For example: Table 1.4 would be the fourth table in your report.

- The **title**, which is located at the top of the table, should be clear and concise, with enough information to allow the table to be understandable apart from the text. Capitalize the first and important words in the title. Do not capitalize articles (a, an, the), short prepositions, and conjunctions. The title does not need a period at the end.

- Refer to each table in the written text. Summarize the data and refer to the table; for example, "The plants treated with sulfur dioxide produced an average of 1.96 seeds per pod (Table 4)." Do not write, "See the results in Table 4."

- If you are using a database program, such as Excel, you should still sketch your table on paper before constructing it on the computer.

**Table 1.4**
Effects of 4-Hour Exposure to 0.6 ppm Sulfur Dioxide on Average Seed and Pod Production in Soybeans (24 Replicates)

| Treatment | Seeds per Pod | Pods per Plant |
|-----------|---------------|----------------|
| Control | 3.26 | 16 |
| SO$_2$ | 1.96 | 13 |

## Application

1. Using the student *average data* from Table 1.3, design a table to present the results for students in your class for H/BP for each body part measured. Include in your table ratios for newborns.

2. Label this Table 1.5. Compose a title for your table. Refer to the guidelines in the previous section of this lab topic for composing titles.

**Table 1.5**

*Average Ratios of Height to Head Circumference and to Arm Length for Newborns and Students*

|  | Head Circumference (cm) | Arm Length (cm) |
|---|---|---|
| Newborns |  |  |
| Students |  |  |

# Lab Study B. Figures

Graphs, diagrams, drawings, and photographs are all called *figures*. The results of an experiment usually are presented graphically, showing the relationships among the independent and dependent variable(s). A graph or figure provides a visual summary of the results. Often, characteristics of the data are not apparent in a table but may become clear in a graph. By looking at a graph, then, you can visualize the effect that the independent variable has on the dependent variable and detect trends in your data. Making a graph may be one of the first steps in analyzing your results.

The presentation of your data in a graph will assist you in interpreting and communicating your results. In the final steps of a scientific investigation, you must be able to construct a logical argument based on your results that either supports or falsifies your starting hypothesis. Your graph should be accurately and clearly constructed, easily interpreted, and well annotated.

The following guidelines will help you to construct such a graph.

* Use graph paper and a ruler to plot the values accurately. If using a database program, you should first sketch your axes and data points before constructing the figure on the computer.

* The independent variable is graphed on the *x* axis (horizontal axis, or abscissa), and the dependent variable, on the *y* axis (vertical axis, or ordinate).

* The numerical range for each axis should be appropriate for the data being plotted. Generally, begin both axes of the graph at zero (the extreme left corner). Then choose your intervals and range to maximize the use of the graph space. Choose intervals that are logically spaced and therefore will allow easy interpretation of the graph, for example, intervals of 5s or 10s. To avoid generating graphs with wasted space, you may signify unused graph space by two perpendicular tic marks between the zero and your lowest number on one or both axes.

* Label the axes to indicate the variable and the units of measurement. Include a legend if colors or shading is used to indicate different aspects of the experiment.

* Choose the type of graph that best presents your data. Line graphs and bar graphs are most frequently used. The choice of graph type depends on the nature of the variable being graphed.

• Compose a title for your figure, and write it below your graph. Figures should be numbered consecutively throughout a lab report or scientific paper. Each figure is given a caption or title that describes its contents, giving enough information to allow the figure to be self-contained. Capitalize only the first word in a figure title and place a period at the end.

## The Line Graph

**Line graphs** show changes in the quantity of the chosen variable and emphasize the rise and fall of the values over their range. Use a line graph to present continuous quantitative data. For example, changes in the dependent variable, soybean weight, measured over time would be depicted best in a line graph.

• Plot data as separate points.

• Whether to connect the dots or draw a best fit curve depends on the type of data and how they were collected. To show trends, draw smooth curves or straight lines to fit the values plotted for any one data set. Connect the points dot to dot when emphasizing meaningful changes in values on the x axis.

• If more than one set of data is presented on a graph, use different colors or symbols and provide a key or legend to indicate which set is which.

• A boxed graph, instead of one with only two sides, makes it easier to see the values on the right side of the graph.

Note the features of a line graph in Figure 1.3, which shows the decline in smoking by high school seniors since 1998.

## The Bar Graph

**Bar graphs** are constructed following the same principles as for line graphs, except that vertical bars, in a series, are drawn down to the horizontal axis. Bar graphs are often used for data that represent separate or discontinuous groups or nonnumerical categories, thus emphasizing the discrete differences between the groups. For example, a bar graph might be used to depict differences in number of seeds per pod for treated and untreated soybeans. Bar graphs are also used when the values on the x axis are numerical but grouped together. These graphs are called histograms.

Note the features of a bar graph in Figure 1.4, which shows the increase of diabetes in older citizens of three racial groups.

You will be asked to design graphs throughout this laboratory manual. Remember, the primary function of the figure is to present your results in the clearest manner to enhance the interpretation and presentation of your data.

## Application

1. Using data from your experiments and the grid provided on the next page, design a *bar graph* that shows the relationship between the dependent and independent variables for both body parts. Discuss with your teammates how to design one figure so that it includes the data for the independent variable for each experiment.

   a. What was the independent variable for each experiment? On which axis would you graph this?

   *the body parts measured; on the X axis*

   b. What was the dependent variable? Write this on the appropriate axis.

   *The ratio of height to body part; write this on the Y axis.*

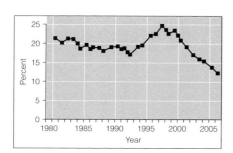

**Figure 1.3.**
**Percent of 12th graders who smoked cigarettes daily.** (*Forum on Child and Family Statistics*, 2009)

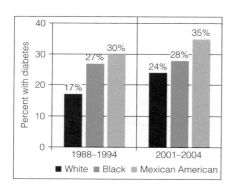

**Figure 1.4.**
**Percent of people 65 or older in three racial groups with diabetes in the U.S.** (*Centers for Disease Control*, 2008)

*If students will construct graphs and figures with a software program, then use this opportunity to provide a handout or instructions. (See Knisely, 2005.)*

2.  Add a *legend* to your figure to distinguish newborn and student ratios.

3.  Draw, label, and compose a *title* for your figure.

4.  Imagine an experiment similar to the one you have performed where it would be appropriate to use a line graph.

> *If measurements were taken of body parts at 5-year intervals, then a line graph would be appropriate and the independent variable would be age. The dependent variable would be the ratio.*

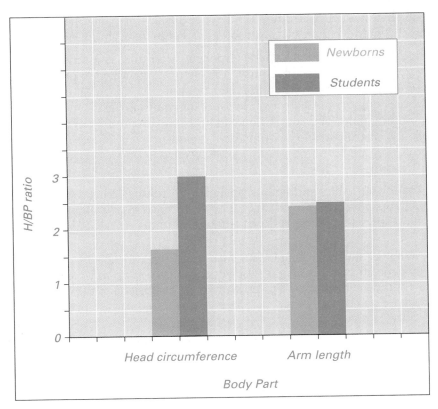

**Figure 1.**

*Comparison of ratios of height to head circumference and to arm length for newborns and students.*

---

## EXERCISE 1.5
# Interpreting and Communicating Results

The last component of a scientific investigation is to interpret the results and discuss their implications in light of the hypothesis and supporting literature. The investigator studies the results, including tables and figures and determines if the hypothesis has been supported or falsified. If the hypothesis has been falsified, the investigator must suggest alternate hypotheses for testing. If the hypothesis has been supported, the investigator suggests additional experiments to strengthen the hypothesis, using the same or alternate methods.

> *Mention that statistical tests can be used to determine if results support or falsify hypotheses. The statistical test chi-square will be used later in the laboratory program.*

Scientists will thoroughly investigate a scientific question, testing hypotheses, collecting data, and analyzing results. In the early stages of a scientific study, scientists review the scientific literature relevant to their topic. They continue to review related published research as they interpret their results and develop conclusions. The final phase of a scientific investigation is the communication of the results to other scientists. Preliminary results may be presented within a laboratory research group and at scientific meetings where the findings can be discussed. Ultimately, the completed project is presented in the form of a scientific paper that is reviewed by scientists within the field and published in a scientific journal. The ideas, procedures, results, analyses, and conclusions of all scientific investigations are critically scrutinized by other scientists. Because of this, science is sometimes described as *self-correcting,* meaning that errors that may occur are usually discovered within the scientific community.

Scientific communication, whether spoken or written, is essential to science. During this laboratory course, you often will be asked to present and interpret your results at the end of the laboratory period. Additionally, you will write components of a scientific paper for many lab topics. In Appendix A at the end of the lab manual, you will find a full description of a scientific paper and instructions for writing each section.

## Application

*Have one or two students present summary statements and figures to be critiqued by the class. Supply acetates of blank grids and pens for students to use in overhead projection.*

1. Using your tables and figures, analyze your results. What relationships are apparent between variables? Look for trends in your figures and tables. Discuss your conclusions with your group.

2. Write a summary statement for your experiments incorporating evidence from your results. Use your results to support or falsify your hypotheses. Be prepared to present your conclusions to the class.

*Critiquing the experiment is one of the most valuable aspects of this investigation and will improve the design of future investigations.*

3. Critique your experiment. What weaknesses do you see in the experiment? Suggest improvements.

   *Students should brainstorm any possible weaknesses in their experiment and suggest improvements.*

| **Weaknesses in Experiment** | **Improvement** |
|---|---|
| 1. *Data only analyzed by one figure* | *Use multiple figures to analyze data* |
| 2. *Did not control for other independent variables: ethnicity, age, nutrition, sex, environment* | *Eliminate other independent variables* |
| 3. *Small sample size* | *Expand the experiment to a larger sample size* |
| 4. *No statistical analysis* | *Perform statistical tests appropriate for ratios* |
| 5. | |

4. Suggest additional and modified hypotheses that might be tested in the future. Briefly describe your next experiment.

5. Refer to Appendix A "Scientific Writing," at the end of your lab manual. Briefly describe the four major parts of a scientific paper. What is the abstract? What information is found in a References Cited section? What sections of a scientific paper always include references?

## Reviewing Your Knowledge

1. Review the major components of an experiment by matching the following terms to the correct definition: *control, controlled variables, level of treatment, dependent variable, replication, procedure, prediction, hypothesis, independent variable.*

   *You may choose to give the remainder of the exercise as a take-home assignment.*

   a. Variables that are kept constant during the experiment (variables not being manipulated)

   *controlled variables*

   b. Tentative explanation for an observation

   *hypothesis*

   c. What the investigator varies in the experiment (for example, time, pH, temperature, concentration)

   *independent variable*

   d. Process used to measure the dependent variable

   *procedure*

   e. Appropriate values to use for the independent variable

   *level of treatment*

   f. Treatment that eliminates the independent variable or sets it at a standard value

   *control*

   g. What the investigator measures, counts, or records; what is being affected in the experiment

   *dependent variable*

   h. Number of times the experiment is repeated

   *replication*

   i. Statement of the expected results of an experiment based on the hypothesis

   *prediction*

2. Identify the dependent and independent variables in the following investigations. (*Circle* the dependent variable and *underline* the independent variable.)

    a. Scientists investigating the effects of increased temperatures on plants in urban environments measured the size of ragweed flowers in Baltimore city lots and rural fields.

       *independent—increased temperatures; dependent—size of ragweed flowers*

    b. Scientists determined the abundance of apple aphids on leaves of 120 apple trees on 5 days in October. The autumn leaf color of the apple trees varied with 40 red trees, 40 yellow trees, and 40 green trees.

       *independent—autumn leaf color of apple trees; dependent—abundance of apple aphids*

    c. Synapse number and level of synaptic proteins in the brains of fruit flies are measured during wakeful periods and periods of sleep.

       *independent—awake or asleep; dependent—number of synapses and level of synaptic proteins*

3. Suggest a control treatment for each of the following experiments.

    a. Geneticists are studying the inheritance of "wolfman syndrome," where the body and face are covered with dark hair. They studied three families in which 16 individuals had the syndrome. They discovered DNA deletions in four genes in each of the 16 persons.

       Control treatment:

       *Determine the DNA deletions in family members without the syndrome.*

    b. Nutrition experts investigate if eating yogurt every day will reduce gum disease.

       Control treatment:

       *Half the people in the experiment do not eat yogurt.*

    c. The number of T-lymphocytes are counted in the blood of vultures that feed on carcasses of livestock treated with antibiotics.

       Control treatment:

       *Vultures that feed on wild prey carcasses with low levels of antibiotics.*

4. In a recent study of 10,000 women (Velicer *et al.*, 2004) scientists reported that women who had breast cancer had a history of heavier antibiotic use than women who did not have breast cancer. What possible explanations for this correlation can you suggest?

    *1. Antibiotics cause breast cancer.*

    *2. Bacterial infections trigger breast cancer.*

    *3. Some women have weaker immune systems and therefore are more likely to have infections or breast cancer, but the two are not related.*

5. What is the essential feature of science that makes it different from other ways of understanding the natural world?

*Science uses the scientific method, utilizing hypotheses that can be tested and falsified by experimentation, observations, or other methods.*

# Applying Your Knowledge

### Interpreting Graphed and Tabular Data

1. The use of DDT for malaria control stopped being funded by the World Health Organization (WHO) in the 1980s. The World Bank required a ban on DDT for developing countries seeking loans. At the same time there has been an increase in resistance by the malarial parasite to the most common antimalarial drugs. Since 2001, WHO has allowed the spraying of DDT in Africa on interior walls to kill mosquitoes. Review the graph in Figure 1.5 and information provided and then answer the following questions.

   What is the independent variable?

   *time, measured in years*

   What is the dependent variable?

   *number of cases of DDT*

   Why was a bar graph selected to present these data? Could the authors have used a line graph?

   *The results are given in number of cases of malaria per year. The authors wanted to show these as discrete data. Yes, since time is a continuous variable, this could also be a line graph.*

   Write a statement summarizing the results. Specifically address trends from 1972–1992; 1995–2000; and 2000–2004.

   *From 1972–1980s DDT spraying effectively reduced the number of cases of malaria to less than 5,000, but in the 1990s the number of cases rose to 10,000. From 1995 when DDT spraying was banned the number of cases of malaria rose rapidly to a high of more than 60,000 cases in 2000. Since 2001 with the resumption of DDT spraying the number of malaria cases has been decreasing toward 10,000.*

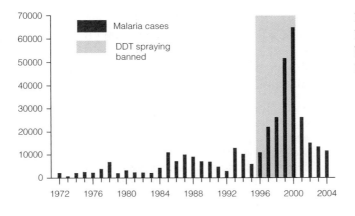

**Figure 1.5.**
**Malaria cases in South Africa before, during, and after banned DDT spraying.** *(After Opar, 2006)*

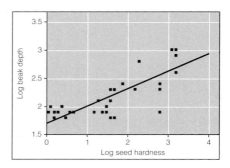

**Figure 1.6.**
**The relationship of beak depth in ground finches and the maximum hardness of seeds they can crack.**
*(After Grant and Grant, 2003)*

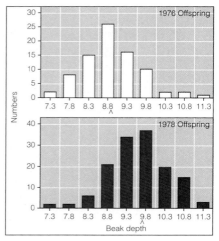

**Figure 1.7.**
**Evolutionary change in beak depth in the population of the medium ground finch on the Galápagos Island of Daphne Major.** The carets indicate the means for the population.
*(After Grant and Grant, 2003)*

2. Rosemary and Peter Grant have studied the Galápagos finches on the Island of Daphne Major for 30 years. They have identified, marked, and measured every finch on the island. They recorded the parents and offspring for generations of finch species. The medium ground finches, *Geospiza fortis*, are medium-sized, seed-eating birds. The Grants determined the relationship between seed size and hardness with beak depth (Figure 1.6). Which finches are able to crack the hardest seeds? State the trend.

*As the size of the beaks increases in a population so does the maximum size and hardness of the seeds they can crack.*

In 1977, the Galápagos Islands experienced a severe drought that resulted in the death of many plants, severe competition for seeds among the finches, and death of many birds. Beak size for the medium ground finch offspring born in 1976 before the drought and those born in 1978, the year following the drought, are shown in Figure 1.7 (Grant and Grant, 2003).

Compare the average beak depth for the offspring from 1976 and 1978.

*The average beak depth was 8.8 mm in 1976 versus 9.8 mm in 1978.*

Approximately how many birds in the population each year had beaks larger than 9.3 mm?

*In 1976 ~ 15 greater than 9.3*

*In 1978 ~ 70 greater than 9.3*

During the drought of 1977, which group of medium ground finches were more successful in finding food and reproducing, those with large beaks or those with small beaks? Explain.

*The finches with large beaks were more successful at surviving and reproducing. Those with small beaks did not reproduce as well. There was a change in the structure of the population, so that in the following generation there were more offspring with large beaks.*

Based on Figures 1.6 and 1.7, what kind of seeds were available in 1977, soft or hard?

*hard*

Using guidelines for graphs on pp. 15–17, how could you improve the axes labels on this figure?

*Units of measurement should be included for beak depth.*

3. Review the guidelines for graphs on pp. 15–17 and critique Figure 1.8. This figure illustrates the changes in risk factors that are important in chronic diseases such as coronary heart disease, lung cancer, diabetes and stroke.

*Students may list some of the following problems: The title should be below the figure and only the first word capitalized. The title is vague and incomplete. The y axis is not clearly labeled. There are no increment marks between the numbers on the x axis. There are no data points, even though the changes in the angle of the line indicate that data points have been omitted. The labeling of the lines is confusing.*

**Figure 1.8.**
**RISK-FACTOR PREVALENCE IN U.S.**

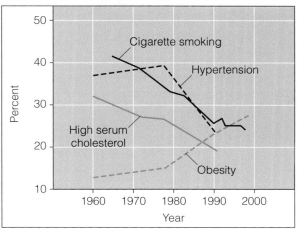

*Source: Centers for Disease Control and Prevention. (After Doyle, 2001)*

## Practicing Experimental Design

1.  Beginning in 2006, scientists have been alarmed by the decline in bat populations in the northeastern U.S. Bat populations in some caves have declined by 50% or more. Bats (dead and living) are found with a white fungus on their faces, ears, and wings, and thus the name "white nose syndrome." Many bats have come out of hibernation early and during daylight; without insects to eat they are emaciated or starve to death. White nose syndrome is spreading into caves in Virginia and West Virginia. The main question is whether the white fungus, *Geomyces* sp., is the primary cause of death. Scientists are investigating other explanations. For example, other pathogens (viruses or bacteria) may be the primary infective agents, and the white fungus may be secondary. Scientists also have found that the digestive systems of affected bats have fewer bacteria necessary for the digestion of insects. Therefore, less energy is available to them during hibernation. This could result in starvation and increased susceptibility to the fungus (Zimmerman, 2009). Using the criteria in Lab Study B, Developing Hypotheses, select the hypothesis you would pursue as a scientist and justify your choice.

    *Students should use the criteria that a hypothesis must be testable and that it must be falsifiable. All three hypotheses are potentially testable and falsifiable. For example: The primary cause of death could be the white fungus, other pathogens, or starvation due to fewer bacteria necessary for digesting insects.*

2.  "Grains of paradise" plants grow in the swampy region inhabited by the western lowland gorilla, make up 80–90% of their diet, and are even utilized in constructing their nests each night. Captive lowland gorillas in zoos are not fed grains of paradise, but rather have a complex diet of processed vitamin-rich food plus fruits and vegetables available in the marketplace. Recently scientists identified potent anti-inflammatory and antimicrobial compounds in grains of paradise that may hold the key to a puzzling question: "Why do western lowland gorillas in zoos have an alarmingly high rate of cardiomyopathy (a type of heart disease)?" Hypothesize about the effect of diet (grains of paradise) on rates of heart disease. Describe a simple preliminary experiment to test your hypothesis, and state a prediction.

## Hypothesis:

*Grains of paradise provide key compounds that prevent or reduce cardiomyopathy in wild gorillas.*

## Experiment:

*Any experiment that compares zoo gorillas with and without grains of paradise in their diet will work. For example: Select 20 zoo dwelling (captive) gorillas and feed them grains of paradise in their diet two meals a day. Select 20 captive gorillas and feed them their normal zoo diet. Compare rates and indicators of heart disease in both groups.*

## Prediction:

*If grains of paradise provide key compounds that prevent or reduce cardiomyopathy in gorillas, then increasing grains of paradise in captive gorilla diets will result in reduced heart disease compared to rates of heart disease in captive gorillas without grains of paradise in their diet.*

3.  The red-cockaded woodpecker (*Picoides borealis*), listed as a federally-endangered species, lives in old growth longleaf pine forests of southern Georgia and other southeastern states (Figure 1.9). Populations of these birds are declining because of loss of suitable habitat. Red-cockaded woodpeckers excavate nesting holes primarily in living longleaf pines with red heart disease, a fungus that affects the tree's heartwood. The ideal forest has short, sparse undergrowth, usually maintained by fire. Establishing new populations of these woodpeckers is limited by their preference to colonize sites with existing nesting holes, rarely moving into new territories, perhaps because of the high energy demands of excavating new cavities, which can take from 10 months to several years. Hoping to find management techniques to help increase populations of these birds, four researchers from North Carolina State University questioned if artificially constructed nesting cavities could be used to

**Figure 1.9.**
**Red-cockaded woodpecker.** A federally endangered species that lives in old growth longleaf pine forests of the southeastern U.S.

increase family groups of birds. *They hypothesized that artificially con-structed cavities in unoccupied habitats or those abandoned because of unsuit-able cavities would increase groups of birds* (Walters *et al.*, 1992).

To test this hypothesis they identified 20 suitable sites that had not been previously occupied and contained trees suitable for cavities. In 10 of these sites they constructed clusters of new cavities. In the remaining 10 they constructed no cavities. In addition to the new sites, they chose 20 abandoned sites with trees suitable for cavities. In 10 of the sites they constructed new cavities, but in the remaining 10 they constructed no cavities.

What prediction might the researchers make based on their hypothesis and experiment?

**If** *new artificially constructed cavities in unoccupied and abandoned habitats increase numbers of groups of birds,* **then** *there will be more groups of birds in the habitats with artificially constructed cav-ities than in similar habitats with no artificial cavities.*

The results of the experiment show, of the 20 sites with constructed artificial cavities, 18 were subsequently occupied (Table 1.6).

In this experiment, what is the independent variable?

*artificial cavity construction*

What is the dependent variable?

*number of cavities occupied*

Identify the controls.

*the sites where no cavities were constructed*

How many replicates were used?

*20—10 unoccupied sites, 10 abandoned sites*

What controlled variables might need to be considered in designing this experiment?

*suitability of habitats: sparse undergrowth, age and species of pine*

**Table 1.6**
Family Groups of Red-cockaded Woodpeckers Colonizing Previously Unoccupied or Abandoned Habitats with and Without Artificially Constructed Cavities

| | Artificial Cavities | | No Artificial Cavities | |
|---|---|---|---|---|
| | Unoccupied | Abandoned | Unoccupied | Abandoned |
| **Total Number of Sites** | 10 | 10 | 10 | 10 |
| **Groups Nesting** | 9 | 9 | 0 | 0 |

Did the results support the hypothesis?

*yes*

Based on these results, what would you recommend for conservation of this species?

*This experiment confirms that artificially constructed cavities can be used to encourage birds to occupy new habitats, offering hope that the continuing decline of the species can be reversed.*

 **Student Media: BioFlix, Activities, Investigations, and Videos**
www.masteringbiology.com (select Study Area)

**Activities**—Ch.1: Graph It! An Introduction to Graphing
**Investigations**—Ch.1: How Does Acid Precipitation Affect Trees?

# References

Barnard, C., F. Gilbert, and P. McGregor. *Asking Questions in Biology,* 3rd ed. Harlow, England: Pearson, 2007.

Center for Disease Control, 2008, "Older Person's Health" http://www.cdc.gov/nchs.fastats/. Accessed July, 2009.

Engelman, R. "Population and Sustainability." *Scientific American Earth 3.0,* 2009, vol. 19(2), pp. 22–29.

Forum on Child and Family Statistics, 2008, "America's Children in Brief: Key National Indicators of Well-Being, 2008" http://www.childstats.gov/. Accessed July, 2009.

Grant, B. R. and P. R. Grant. "What Darwin's Finches Can Teach Us about the Evolutionary Origins of Biodiversity." *BioScience,* 2003, vol. 53, pp. 965–975.

Hall, Jo, J. Allanson, K. Gripp, and A. Slavotinck. *Handbook of Physical Measurements,* 2nd ed. Oxford, England: Oxford University Press, 2006.

Knisely, K. *A Student Handbook for Writing in Biology,* 3rd ed. Sunderland, MA: Sinauer Associates, 2009. (Instructions for making graphs using Excel, Appendix 2.)

Moore, J. *Science as a Way of Knowing.* Cambridge, MA: Harvard University Press, 1993.

Opar, A. "The Return of DDT." *SEED,* 2006, vol. 2(8) p. 20.

Pechenik, J. *A Short Guide to Writing about Biology,* 7th ed. San Francisco, CA: Addison Wesley Longman, 2009.

Reece, J., et al. *Campbell Biology,* 9th ed. San Francisco, CA: Pearson Education, 2011.

Velicer, C. M., S. R. Heckbert, J. W. Lampe, J. D. Potter, C. A. Robertson, and S. H. Taplin. "Antibiotic Use in Relation to the Risk of Breast Cancer." *JAMA,* 2004, vol. 291, pp. 827–835.

Walters, J. R., C. K. Copeyon, and J. H. Carter. "Test of the Ecological Basis of Cooperative Breeding in Red-cockaded Woodpeckers." *The Auk,* 1992, 1009(1), pp. 90–97.

Zimmerman, R. "Biologists Struggle to Solve Bat Deaths." *Science,* 2009, vol. 324, pp. 1134–1135.

# LAB TOPIC 1

# Scientific Investigation
## Teaching Plan for Laboratories

## Main Concepts and Objectives

1. Concept: the scientific process. Students will practice asking questions that can be answered by using scientific investigation, developing testable hypotheses, and producing experimental procedures. (Exercises 1.1 and 1.2)

2. Concept: designing experiments. Students will design an original experiment investigating allometric and isometric growth. (Exercise 1.3)

3. Concept: processing data. Students will practice developing tables to record data and composing line and bar graphs to explain data. (Exercise 1.4)

4. Concept: interpreting graphs. Students will practice using graphs to analyze and interpret data. (Exercise 1.4)

5. Concept: discussing results. Students will discuss the results of their original experiment in light of their hypothesis. (Exercise 1.5)

6. Concept: writing to learn. Students will use writing as a learning tool as they formulate answers to questions and discuss results of experiments. (Exercises 1.1–1.5)

7. Concept: scientific communication. Students will practice interpreting and communicating results. (Exercise 1.5)

8. Specific content. Students will be able to describe or define: *hypothesis, scientific method, dependent variable, independent variable, controlled variable, levels of treatment, replication, control, prediction, table, line graph, bar graph, procedure, allometric growth, isometric growth.*

## Order of the Lab

| | |
|---|---|
| 1. Introduce scientific investigation. | (15 min) |
| 2. Lead class in a discussion of Exercises 1.1 and 1.2. | (30 min) |
| 3. Student teams propose questions and hypotheses. Class designs an experiment. | (30 min) |
| 4. Student teams perform the experiment and collect data. | (30 min) |
| 5. Students graph and analyze data. | (30 min) |
| 6. Students interpret and summarize results. | (25 min) |
| 7. One or two students read summary statements to be critiqued by the class. | (15 min) |
| 8. Assign Reviewing Your Knowledge and Applying Your Knowledge; if time permits, begin discussion of experimental design problems in lab. | (5 min) |

**Total lab time:** 3 hours

**For a 2-hour lab:** Reduce the total time allotted to items 1 and 2 from 45 minutes to 30 minutes. Assign item 6 to be completed by research teams outside the lab. If possible, have student reports (item 7) at the beginning of the next lab.

# Classroom Management

*Students must read Exercises 1.1 and 1.2 before coming to lab and should be prepared to discuss the applications and questions in these exercises.* Use questions in the exercises to lead a class discussion of the items in Exercises 1.1 and 1.2. As you begin Exercise 1.3, have students collaborate for no more than 10 minutes; then lead the class as students determine their hypothesis and design their experiment. Allow them to develop their experimental design and reach a class consensus. Remember, at the end of the exercise they will critique not only their results, but their design as well. One of the advantages of this exercise is the collaboration and activity involved.

If time permits, students should work in teams to answer questions from Practicing Experimental Design in Applying Your Knowledge. If the problems are assigned as homework, encourage students to collaborate on these problems.

# Student Development

This lab will set the tone of the laboratory experience for your class. Students should begin to develop an investigative mentality. They should begin to function as research teams, dividing labor and assigning tasks. They will begin to develop styles that they will use throughout the term. Encourage them to practice recording data in a neat, accurate format. Students may initially be passive in this lab. Encourage an interactive lab experience as students participate in class discussions and work as members of collaborative research teams. As students develop hypotheses and design experiments, encourage and welcome their ideas. Remember, there are many acceptable hypotheses. The principles learned in this lab will be used throughout the year. Students must master the scientific process and the elements of scientific investigation.

# Lab Safety Precautions

No unusual safety hazards in this laboratory.

# Discussion and Summary

Lead the class in discussion throughout the lab. At the end of the lab, one or two students should read summary statements to be critiqued by the class. Students should summarize the conclusions of their experiment in their lab manuals.

# Evaluation

You may require that students finish the activities at the end of the lab at home, to be returned for evaluation. Students can be assigned a component of a scientific paper as part of the writing program (see Appendix A). On a written test, ask students to apply their knowledge to new but similar problems to those in the lab topic. Exam questions can be developed from earlier editions of the lab manual. For a 2-hour lab, ask students to design tables and graphs outside of class, using data from their original experiment. Informally note students' participation and make an effort to include all students in discussions. Success in this lab topic will be reflected in students' ability to follow investigative methods in future labs.

# Lab Topic 2
# Microscopes and Cells

## Laboratory Objectives

After completing this lab topic, you should be able to:
1. Identify the parts of compound and stereoscopic microscopes and be proficient in their correct use in biological studies.
2. Describe procedures used in preparing materials for electron microscopy and compare these with procedures used in light microscopy.
3. Identify cell structures and organelles from electron micrographs and state the functions of each.
4. Describe features of specific cells and determine characteristics shared by all cells studied.
5. Distinguish between eukaryotic and prokaryotic cells.
6. Discuss the evolutionary significance of increasing complexity from unicellular to multicellular organization and provide examples from the lab.

*For a 2-hour lab:*
*Omit the electron microscope and micrographs study (Exercise 2.4) and the unknowns (Exercise 2.5, Lab Study D). Assign Applying Your Knowledge questions as take-home as a substitute for class discussion.*

## Introduction

According to cell theory, the *cell* is the fundamental biological unit, the smallest and simplest biological structure possessing all the characteristics of the living condition. All living organisms are composed of one or more cells, and every activity taking place in a living organism is ultimately related to metabolic activities in cells. Thus, understanding the processes of life necessitates an understanding of the structure and function of the cell.

The earliest known cells found in fossilized sediments 3.5 billion years old (called **prokaryotic** cells) lack nuclei and membrane-bound organelles. Cells with a membrane-bound nucleus and organelles (**eukaryotic** cells) do not appear in the fossil record for another 2 billion years. But the eventual evolution of the eukaryotic cell and its internal compartmentalization led to enormous biological diversity in single cells. The evolution of loose aggregates of cells ultimately to colonies of connected cells provided for specialization, so that groups of cells had specific and different functions. This early division of labor included cells whose primary function was locomotion or reproduction. The evolution of multicellularity appears to have originated more than once in eukaryotes and provided an opportunity for extensive adaptive radiation as organisms specialized and diversified, eventually giving rise to fungi, plants, and animals. This general trend in increasing complexity and specialization seen in the history of life will be illustrated in Lab Topic 2.

Given the fundamental role played by cells in the organization of life, one can readily understand why the study of cells is essential to the study of life.

Cells, however, are below the limit of resolution of the human eye. We cannot study them without using a microscope. The microscope has probably contributed more than any other instrument to the development of biology as a science and continues today to be the principal tool used in medical and biological research. There are four types of microscopes commonly used by biologists. You will learn how to use two of these microscopes, the compound microscope and the stereoscopic microscope, in today's laboratory. Both of these microscopes use visible light as the source of illumination and are called light microscopes. Two other microscopes, the scanning electron microscope and the transmission electron microscope, use electrons as the source of illumination. Electron microscopes are able to view objects much smaller than those seen in a light microscope. Although these microscopes are not used in this laboratory, you will be given the opportunity to learn more about them in Exercise 2.4.

Microscopes are used by biologists in numerous subdisciplines: genetics, molecular biology, neurobiology, cell biology, evolution, and ecology. The knowledge and skills you develop today will be used and enhanced throughout this course and throughout your career in biology. It is important, therefore, that you take the time to master these exercises thoroughly.

# EXERCISE 2.1
# The Compound Light Microscope

## Materials

compound microscope

## Introduction

The microscope is designed to make objects visible that are too difficult or too small to see with the unaided eye. There are many variations of light microscopes, including phase-contrast, darkfield, polarizing, and UV. These differ primarily in the source and manner in which light is passed through the specimen to be viewed.

The microscopes in biology lab are usually compound binocular or monocular light microscopes, some of which may have phase-contrast attachments. **Compound** means that the scopes have a minimum of two magnifying lenses (the ocular and the objective lenses). **Binocular microscopes** have two eyepieces, **monoculars** have only one eyepiece, and **light** refers to the type of illumination used, that is, visible light from a lamp.

Your success in and enjoyment of a large portion of the laboratory work in introductory biology will depend on how proficient you become in the use of the microscope. When used and maintained correctly, these precision instruments are capable of producing images of the highest quality.

Although there are many variations in the features of microscopes, they are all constructed on a similar plan (Figure 2.1). In this exercise you will be introduced to the common variations found in different models of compound microscopes and asked to identify those features found on your microscope.

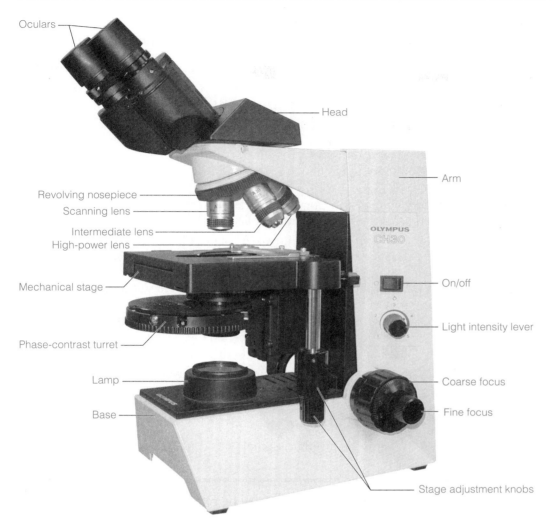

Oculars

Head

Arm

Revolving nosepiece

Scanning lens

Intermediate lens

High-power lens

OLYMPUS
CH30

Mechanical stage

On/off

Light intensity lever

Phase-contrast turret

Lamp

Coarse focus

Base

Fine focus

Stage adjustment knobs

**Figure 2.1a.**
**The compound binocular light microscope.** Locate the parts of your microscope described in Exercise 2.1 and label this photograph. Indicate in the margin of your lab manual any features unique to your microscope.

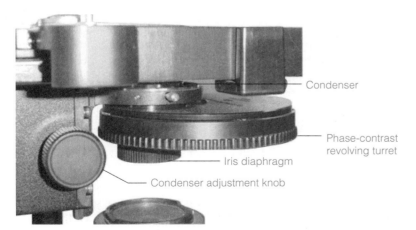

Condenser

Phase-contrast
revolving turret

Iris diaphragm

Condenser adjustment knob

**Figure 2.1b.**
**Enlarged photo of compound light microscope as viewed from under the stage.** This microscope is equipped with phase-contrast optics. Locate the condenser, condenser adjustment knob, phase-contrast revolving turret, and iris diaphragm on your microscope (if present) and label them on the diagram.

**Please treat these microscopes with the greatest care!**

## Procedure

1. Obtain a compound light microscope, following directions from your instructor. To carry the microscope correctly, hold the arm with one hand, and support the base with your other hand. Remove the cover, but do not plug in the microscope.

2. Locate the parts of your microscope, and label Figure 2.1. Refer to the following description of a typical microscope. In the spaces provided, indicate the specific features related to your microscope.

   a. The **head** supports the two sets of magnifying lenses. The **ocular** is the lens in the eyepiece, which typically has a magnification of 10×. If your microscope is binocular, the distance between the eyepieces (**interpupillary distance**) can be adjusted to suit your eyes. Move the eyepieces apart, and look for the scale used to indicate the distance between the eyepieces. Do not adjust the eyepieces at this time. A pointer has been placed in the eyepiece and is used to point to an object in the **field of view,** the circle of light that one sees in the microscope.

   *If you are using binocular microscopes, demonstrate the movement of the oculars, and indicate the scale for interpupillary distance. Students should not adjust the oculars at this time.*

   Is your microscope monocular (one eyepiece) or binocular (two eyepieces)?

   What is the magnification of your ocular(s)?

 **Although the eyepiece may be removable, it should not be removed from the microscope.**

   *If your students' microscopes have an oil immersion objective, you may want to point it out and demonstrate its correct use in the next section. Oil immersion is not required for any of the exercises in this manual.*

   b. **Objectives** are the three lenses on the **revolving nosepiece.** The shortest lens is typically 4× and is called the **scanning lens.** The **intermediate lens** is 10×, and the longest, the **high-power lens,** is 40× (the fourth position on the nosepiece is empty). It is important to clean both the objective and ocular lenses before each use. Dirty lenses will cause a blurring or fogging of the image. Always use lens paper for cleaning! Any other material (including Kimwipes®) may scratch the lenses.

   What is the magnification of each of your objectives? List them in order of increasing magnification.

   c. The **arm** supports the stage and condenser lens. The **condenser lens** is used to focus the light from the **lamp** through the specimen to be viewed. The height of the condenser can be adjusted by an **adjustment knob.** The **iris diaphragm** controls the width of the circle of light and, therefore, the amount of light passing through the specimen.

If your microscope has phase-contrast optics, the condenser may be housed in a **revolving turret.** When the turret is set on 0, the normal optical arrangement is in place. This condition is called **brightfield microscopy.** Other positions of the turret set phase-contrast optics in place. To use phase-contrast, the turret setting must correspond to the magnifying power of the objective being used.

Is your microscope equipped with phase-contrast optics?

The **stage** supports the specimen to be viewed. A mechanical stage can be moved right and left and back and forth by two **stage adjustment knobs.** With a stationary stage, the slide is secured under **stage clips** and moved slightly by hand while viewing the slide. The distance between the stage and the objective can be adjusted with the **coarse** and **fine focus adjustment knobs.**

Does your microscope have a mechanical or stationary stage?

d.  The **base** acts as a stand for the microscope and houses the lamp. In some microscopes, the intensity of the light that passes through the specimen can be adjusted with the **light intensity lever.** Generally, more light is needed when using high magnification than when using low magnification. Describe the light system for your microscope.

---

## EXERCISE 2.2
# Basic Microscope Techniques

---

## Materials

clear ruler
coverslips
prepared slides: letter and crossed thread
lens paper
blank slides
Kimwipes®
dropper bottle with distilled water

## Introduction

In this exercise, you will learn to use the microscope to examine a recognizable object, a slide of the letter *e*. Recall that microscopes vary, so you may have to omit steps that refer to features not available on your microscope. Practice adjusting your microscope to become proficient in locating a specimen, focusing clearly, and adjusting the light for the best contrast.

## Procedure

1.  Clean microscope lenses.

    Each time you use the microscope, you should begin by cleaning the lenses. Using lens paper moistened with a drop of distilled water, wipe the ocular, objective, and condenser lenses. Wipe them again with a piece of dry lens paper.

Use only lens paper on microscope lenses. Do not use
Kimwipes®, tissues, or other papers.

2. Adjust the focus on your microscope.
   a. Plug your microscope into the outlet.
   b. Turn on the light. Adjust the light intensity to mid-range if your
      microscope has that feature.
   c. Rotate the 4× objective into position using the revolving nosepiece
      ring, not the objective itself.
   d. Take the letter slide and wipe it with a Kimwipe® tissue. Each time
      you study a prepared slide, you should first wipe it clean. Place the
      letter slide on the stage, and center it over the stage opening.

 Slides should be placed on and removed from the stage
only when the 4× objective is in place. Removing a slide
when the higher objectives are in position may scratch
the lenses.

   e. Look through the ocular and bring the letter into rough focus by
      slowly focusing upward using the coarse adjustment.
   f. For binocular microscopes, looking through the oculars, move the
      oculars until you see only one image of the letter *e*. In this position,
      the oculars should be aligned with your pupils. In the margin of
      your lab manual, make a note of the **interpupillary distance** on the
      scale between the oculars. Each new lab day, before you begin to use
      the microscope, set this distance.
   g. Raise the condenser to its highest position, and fully close the iris
      diaphragm.
   h. Looking through the ocular, slowly lower the condenser just until
      the graininess disappears. Slowly open the iris diaphragm just until
      the entire field of view is illuminated. This is the correct position for
      both the condenser and the iris diaphragm.
   i. Rotate the 10× objective into position.
   j. Look through the ocular and slowly focus upward with the coarse
      adjustment knob until the image is in rough focus. Sharpen the
      focus using the fine adjustment knob.

 Do not turn the fine adjustment knob more than two
revolutions in either direction. If the image does not come
into focus, return to 10× and refocus using the coarse
adjustment.

k. For binocular microscopes, cover your left eye and use the fine adjustment knob to focus the fixed (right) ocular until the letter *e* is in maximum focus. Now cover the right eye and, using the diopter ring on the left ocular, bring the image into focus. The letter *e* should now be in focus for both of your eyes. Each new lab day, as you begin to study your first slide, repeat this procedure.

l. You can increase or decrease the contrast by adjusting the iris diaphragm opening. Note that the maximum amount of light provides little contrast. Adjust the aperture until the image is sharp.

m. Move the slide slowly to the right. In what direction does the image in the ocular move?

*left*

n. Is the image in the ocular inverted relative to the specimen on the stage?

*yes*

o. Center the specimen in the field of view; then rotate the 40× objective into position while watching from the side. *If it appears that the objective will hit the slide, stop and ask for assistance.*

**Most of the microscopes have parfocal lenses, which means that little refocusing is required when moving from one lens to another. If your scope is *not* parfocal, ask your instructor for assistance.**

p. After the 40× objective is in place, focus using the fine adjustment knob.

**Never focus with the coarse adjustment knob when you are using the high-power objective.**

q. The distance between the specimen and the objective lens is called the **working distance.** Is this distance greater with the 40× or the 10× objective?

*10×*

3. Compute the total magnification of the specimen being viewed. To do so, multiply the magnification of the ocular lens by that of the objective lens.

a. What is the total magnification of the letter as the microscope is now set?

*400×*

    b. What would be the total magnification if the ocular were 20× and the objective were 100× (oil immersion)? This is the magnification achieved by the best light microscopes.

      *2000×*

4. Measure the diameter of the field of view. Once you determine the size of the field of view for any combination of ocular and objective lenses, you can determine the size of any structure within that field.

    a. Rotate the 4× objective into position and remove the letter slide.

    b. Place a clear ruler on the stage, and focus on its edge.

*Students must estimate the distance between mm marks on the ruler. These may not be in sharp focus.*

    c. The distance between two lines on the ruler is 1 mm. What is the diameter (mm) of the field of view?

      *4 mm*

*If your microscopes have a mechanical stage, you can tape a section cut from a clear ruler to a microscope slide. Try this with your microscope first.*

    d. Convert this measurement to micrometers (μm), a more commonly used unit of measurement in microscopy (1 mm = 1,000 μm).

    e. Measure the diameters of the field of view for the 10× and 40× objectives, and enter all three in the spaces below to be used for future reference.

      *4× ≈ 4,000 μm*    *10× ≈ 1,900 μm*    *40× ≈ 800 μm*

    f. What is the relationship between the size of the field of view and magnification?

      *As the magnification increases, the field of view decreases.*

5. Determine spatial relationships. The **depth of field** is the thickness of the specimen that may be seen in focus at one time. Because the depth of focus is very short in the compound microscope, focus up and down to clearly view all planes of a specimen.

*Encourage students to adjust the light so that they have good contrast for fibers in the threads.*

    a. Rotate the 4× objective into position and remove the ruler. Take a slide of crossed threads, wipe it with a Kimwipe®, and place the slide on the stage. Center the slide so that the region where the two threads cross is in the center of the stage opening.

    b. Focus on the region where the threads cross. Are both threads in focus at the same time?

      *yes*

    c. Rotate the 10× objective into position and focus on the cross. Are both threads in focus at the same time?

      *no*

    Does the 4× or the 10× objective have a shorter depth of field?

      *10×*

d. Focus upward (move the stage up) with the coarse adjustment until both threads are just out of focus. Slowly focus down using the fine adjustment. Which thread comes into focus first? Is this thread lying under or over the other thread?

*under*

e. Rotate the 40× objective into position and slowly focus up and down, using the fine adjustment only. Does the 10× or the 40× objective have a shorter depth of field?

*40×*

*Use a mixture of slides in the lab that vary for the color of thread on top.*

6. At the end of your microscope session, use the following procedures to store your microscope.
   a. Rotate the 4× objective into position.
   b. Remove the slide from the stage.
   c. Return the phase-contrast condenser to the 0 setting if you have used phase-contrast.
   d. Set the light intensity to its lowest setting and turn off the power.
   e. Unplug the cord and wrap it around the base of the microscope.
   f. Replace the dust cover.
   g. Return the microscope to the cabinet using two hands; one hand should hold the arm, and the other should support the base.

These steps should be followed every time you store your microscope.

EXERCISE 2.3
# The Stereoscopic Microscope

## Materials

stereoscopic microscope
dissecting needles
living *Elodea*

microscope slides
droppers of water
coverslips

## Introduction

The stereoscopic (dissecting) microscope has relatively low magnification, 7× to 30×, and is used for viewing and manipulating relatively large objects. The binocular feature creates the stereoscopic effect. The stereoscopic microscope is similar to the compound microscope except in the following ways: (1) The depth of field is much greater than with the compound microscope, so objects are seen in three dimensions, and (2) the light source can be directed down onto as well as up through an object, which permits the viewing of objects too thick to transmit light. Light directed down on the object is called **reflected** or **incident light.** Light passing through the object is called **transmitted light.** Refer to the Websites section at the end of this lab topic for a video of stereomicroscope use.

### Procedure

1. Remove your stereoscopic microscope from the cabinet and locate the parts labeled in Figure 2.2. Locate the switches for both incident and transmitted light. In the margin of your lab manual, note any features of your microscope that are not shown in the figure. What is the range of magnification for your microscope?

2. Observe an object of your choice at increasing magnification. Select an object that fits easily on the stage (e.g., ring, coin, fingertip, pen, ruler).

    a. Place the object on the stage and adjust the interpupillary distance (distance between the oculars) by gently pushing or pulling the oculars until you can see the object as a single image.

    b. Change the magnification and note the three-dimensional characteristics of your object.

    c. Adjust the lights, both reflected and transmitted. Which light gives you the best view of your object?

**Figure 2.2.**
**The stereoscopic (dissecting) microscope.** Locate the parts of your microscope by referring to this photograph. Note in the margin any features of your microscope that are not shown in the photograph.

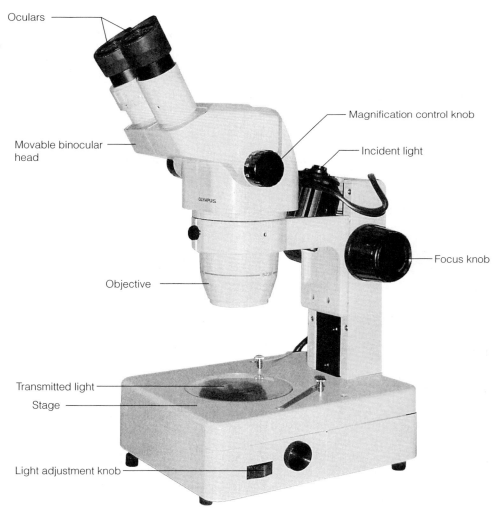

Oculars

Movable binocular head

Objective

Magnification control knob

Incident light

Focus knob

Transmitted light

Stage

Light adjustment knob

3. Prepare a **wet mount** of *Elodea.* Living material is often prepared for observation using a wet mount. (The material is either in water or covered with water prior to adding a coverslip.) You will use this technique to view living material under the dissecting and compound microscopes (Figure 2.3).

   a. Place a drop of water in the center of a clean microscope slide.

   b. Remove a single leaf of *Elodea,* and place it in the drop of water.

   c. Using a dissecting needle, place a coverslip at a 45° angle above the slide with one edge of the coverslip in contact with the edge of the water droplet, as shown.

   d. Lower the coverslip slowly onto the slide, being careful not to trap air bubbles in the droplet. The function of the coverslip is threefold: (1) to flatten the preparation, (2) to keep the preparation from drying out, and (3) to protect the objective lenses. Over long periods of time, the preparation may dry out, at which point water can be added to one edge of the coverslip.

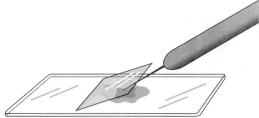

**Figure 2.3.**
**Preparation of a wet mount.** Place a drop of water and your specimen on the slide. Using a dissecting needle, slowly lower a coverslip onto the slide, being careful not to trap air bubbles in the droplet.

 Specimens can be viewed without a coverslip using the stereoscopic microscope, but a coverslip must always be used with the compound microscope.

*Students will view bacteria slides without a coverslip in Lab Topic 13.*

4. Observe the structure of the *Elodea* leaf at increasing magnification.

   a. Place the leaf slide on the stage and adjust the focus. Change the magnification and note the characteristics of the leaf at increased magnification.

   b. Sketch the leaf in the margin of your lab manual and list, in the space below, the structures that are visible at low and high magnification.

   Low:

   High:

   Is it possible to see cells in the leaf using the stereoscopic microscope?

   *yes, cell outlines*

   Organelles?

   *no*

   c. Save your slide for later study. In Exercise 2.5, Lab Study C, you will be asked to compare these observations of *Elodea* with those made while using the compound microscope.

*This exercise can be assigned to complete outside of the laboratory using resources and electron micrographs in the textbook, library reserve, and online.*

EXERCISE 2.4

# The Transmission Electron Microscope

## Materials

demonstration resources for the electron microscope
electron micrographs

## Introduction

*If you are using a video of electron microscopy, have students review this exercise and use the list of terms in question #3 in the procedure as they view the video.*

The transmission electron microscope (TEM) magnifies objects approximately 1,000× larger than a light microscope can (up to 1,000,000×). This difference depends on the **resolving power** of the electron microscope, which allows the viewer to see two objects of comparable size that are close together and still be able to recognize that they are two objects rather than one. Resolving power, in turn, depends on the wavelength of light passed through the specimen: the shorter the wavelength, the greater the resolution. Because electron microscopes use electrons as a source of illumination and electrons have a much shorter wavelength than does visible light, the resolving power of electron microscopes is much greater than that of light microscopes. Both the electron and light microscopes can be equipped with lenses that allow for tremendous magnification, but only the electron microscope has sufficient resolving power to make these lenses useful.

## Procedure

1. Compare the features of the light and electron microscopes (Figure 2.4).
   a. Name three structures found in both microscopes.

      *ocular lens, objectives, condenser*

   b. What is the energy source for the electron microscope?

      *electrons*

      For the compound microscope?

      *light*

   c. Describe how the lenses differ for the two microscopes.

      *The light microscope has glass objectives; the electron microscope has electromagnets.*

2. Using the resources provided by your instructor, review the procedures and materials for preparing a specimen for electron microscopy. Websites that describe electron microscopy are listed at the end of this lab topic

3. Define the following terms on separate paper or in the margin of your lab manual.

*Students may also check websites for definitions of terms.*

   | | |
   |---|---|
   | *fixation* | *staining with heavy metals* |
   | *embedding* | *electromagnetic lenses* |
   | *ultramicrotome* | *fluorescent screen* |
   | *boat on diamond or glass knife* | *vacuum* |
   | *copper grids* | *electron micrographs* |

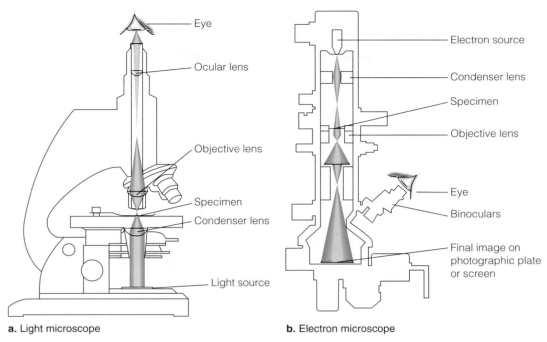

**a.** Light microscope        **b.** Electron microscope

**Figure 2.4.**
**Comparison of light microscope and electron microscope.** The source of illumination is light for the light microscope and electrons for the electron microscope. The image is magnified by glass objectives in light microscopy and by electromagnets in electron microscopy.

4. When an electron microscope is used, cells are usually studied using electron micrographs, photographs taken of the image seen on the fluorescent screen. Observe electron micrographs on demonstration in the laboratory, in your textbook, or on websites listed at the end of this lab topic. For the following list of cellular organelles, *underline* those organelles and structures that you predict may be seen *only* with the electron microscope.

   *plasma membrane, cell wall, nucleus, chloroplast,*

   *mitochondria, vacuole, Golgi apparatus,*

   *peroxisome, lysosome, endoplasmic reticulum,*

   *ribosome, flagella, cilia*

# EXERCISE 2.5
# The Organization of Cells

In this exercise, you will examine the features common to all eukaryotic cells that are indicative of their common ancestry. However, you will observe that all cells are not the same. Some organisms are **unicellular** (single-celled), with all living functions (respiration, digestion, reproduction, and excretion) handled by that one cell. Others form random, temporary **aggregates,** or clusters, of cells. Clusters composed of a consistent and predictable number of cells are called **colonies.** Simple colonies are clusters of cells of similar types with a predictable structure, but the cells have no physiological connections. More complex colonies have cells of different types. In some colonial algae the cells are called *somatic cells* (cells that are not reproductive) and cells that specialize in reproduction. In these colonies, if either type of cell is isolated from the colony, they may be reproductive, dividing and producing new colonies.

Other algae may contain both cell types, somatic and reproductive, but their somatic cells *never* become reproductive, even when isolated, and reproductive cells cannot persist independently, but must be associated with somatic cells to live. These organisms are described as **multicellular** because they have two or more types of cells with specialized structure and function, and these cell types, when isolated, are not capable of prepetuating the species in the wild. In more complex algae, fungi, plants, and animals, specialized cells may be organized into tissues that perform particular functions for the organism. Tissues, in turn, may combine to form organs, and tissues and organs combine to form a coordinated single organism.

In this exercise, you will examine selected unicellular, aggregate, colonial, and multicellular organisms. (See Color Plates 1–7.)

# Lab Study A. Unicellular Organisms

## Materials

| | |
|---|---|
| microscope slides | coverslips |
| culture of *Amoeba* | dissecting needles |
| living termites | insect Ringers |
| forceps | |

## Introduction

Unicellular eukaryotic organisms may be **autotrophic** (photosynthetic) or **heterotrophic** (deriving food from other organisms or their by-products). These diverse organisms, called protists, will be studied in detail in Lab Topic 14.

## Procedure

1. Examine a living *Amoeba* (Figure 2.5) under the compound microscope. Amoebas are aquatic organisms commonly found in ponds. To transfer a specimen to your slide, follow these procedures:

    a. Place the culture dish containing the amoeba under the dissecting microscope, and focus on the bottom of the dish. The amoeba will appear as a whitish, irregularly shaped organism attached to the bottom.

    b. Using a clean pipette (it is important not to interchange pipettes between culture dishes), transfer a drop with several amoebas to your microscope slide. To do this, squeeze the pipette bulb *before* you place the tip under the surface of the water. Disturbing the culture as little as possible, pipette a drop of water with debris from the *bottom* of the culture dish. You may use your stereoscopic microscope to scan the slide to locate amoebas before continuing.

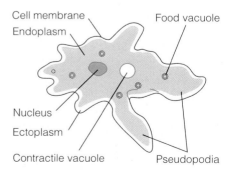

**Figure 2.5.**
**Amoeba.** An amoeba moves using pseudopodia. Observe the living organisms using the compound microscope (Color Plate 1).

c. Cover your preparation with a clean coverslip.

d. Under low power on the compound scope, scan the slide to locate an amoeba. Center the specimen in your field of view; then switch to higher powers.

e. Identify the following structures in the amoeba:

**Cell membrane** is the boundary that separates the organism from its surroundings.

**Ectoplasm** is the thin, transparent layer of cytoplasm directly beneath the cell membrane.

**Endoplasm** is the granular cytoplasm containing the cell organelles.

The **nucleus** is the grayish, football-shaped body that is somewhat granular in appearance. This organelle, which directs the cellular activities, will often be seen moving within the endoplasm.

**Contractile vacuoles** are clear, spherical vesicles of varying sizes that gradually enlarge as they fill with excess water. Once you've located a vacuole, watch it fill and then empty its contents into the surrounding environment. These vacuoles serve an excretory function for the amoeba.

**Food vacuoles** are small, dark, irregularly shaped vesicles within the endoplasm. They contain undigested food particles.

**Pseudopodia** ("false feet") are fingerlike projections of the cytoplasm. They are used for locomotion as well as for trapping and engulfing food in a process called **phagocytosis.**

*Amoebas appear gray and may be near debris. The fast-swimming protozoans are food. Students having trouble locating an amoeba should look at other students' specimens so they have a good "search image." If a microscope/camera/TV setup is available, use a successful student's slide to show an amoeba to all students.*

 **Student Media Videos—Ch. 28: Amoeba; Amoeba Pseudopodia**

2. Examine *Trichonympha* under a compound microscope. You will first have to separate the *Trichonympha* (Figure 2.6) from the termite with which it lives in a symbiotic relationship. *Trichonympha* and other organisms occupy the gut of the termites, where they digest wood particles eaten by the insect. Termites lack the enzymes necessary to digest wood and are dependent on *Trichonympha* to make the nutrients in the wood available to them. *Trichonympha* has become so well adapted to the environment of the termite's gut that it cannot survive outside of it.

To obtain a specimen:

a. Place a couple of drops of **insect Ringers** (a saline solution that is isotonic to the internal environment of insects) on a clean microscope slide.

b. Using forceps or your fingers, transfer a termite into the drop of Ringers.

c. Place the slide under the dissecting microscope.

d. Place the tips of dissecting needles at either end of the termite and pull in opposite directions.

e. Locate the long tube that is the termite's intestine. Remove all the larger parts of the insect from the slide.

f. Using a dissecting needle, mash the intestine to release the *Trichonympha* and other protozoa and bacteria.

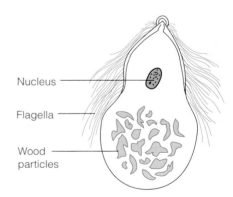

**Figure 2.6.**
*Trichonympha.* A community of microorganisms, including *Trichonympha*, inhabits the intestine of the termite. Following the procedure in Exercise 2.5, Lab Study A, disperse the microorganisms and locate the cellular structures in *Trichonympha* (Color Plate 2).

   g. Cover your preparation with a clean coverslip.

   h. Transfer your slide to the compound microscope and scan the slide under low power. Center several *Trichonympha* in the field of view and switch to higher powers.

---

 Several types of protozoans and bacteria will be present in the termite gut.

---

   i. Locate the following structures under highest power.

     **Flagella** are the long, hairlike structures on the outside of the organism. The function of the flagella is not fully understood. Within the gut of the termite, the organisms live in such high density that movement by flagellar action seems unlikely and perhaps impossible.

     The **nucleus** is a somewhat spherical organelle near the middle of the organism.

     **Wood particles** may be located in the posterior region of the organism.

## Lab Study B. Aggregate and Colonial Organisms

### Materials

microscope slides
dissecting needles
forceps

coverslips
cultures of *Protococcus* and
    *Scenedesmus*

### Introduction

Unlike unicellular organisms, which live independently of each other, colonial organisms are cells that live in groups and are to some degree dependent on one another. The following organisms show an increasing degree of interaction among cells.

*Protococcus cells are small and clumped. Students should disperse the cells well and view only a few green cells.*

### Procedure

1. Examine *Protococcus* under the compound microscope. *Protococcus* (Figure 2.7) is a terrestrial green alga that grows on the north sides of trees and is often referred to as "moss."

   a. To obtain a specimen, use a dissecting needle to brush off a small amount of the green growth on the piece of tree bark provided into a drop of water on a clean microscope slide. Avoid scraping bark onto the slide. Cover the preparation with a clean coverslip.

   b. Observe at highest power that these cells are **aggregates:** The size of the cell groupings is random, and there are no permanent connections between cells. Each cell is surrounded by a cell membrane and an outer **cell wall.**

   c. Observe several small cell groupings and avoid large clumps of cells. Cellular detail may be obscure.

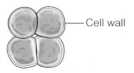

Cell wall

**Figure 2.7.**
***Protococcus.*** *Protococcus* is a terrestrial green alga that forms loose aggregates on the bark of trees (Color Plate 3).

2. Examine living *Scenedesmus* under the compound microscope. *Scenedesmus* (Figure 2.8) is an aquatic green alga that is common in aquaria and polluted water.

   a. To obtain a specimen, place a drop from the culture dish (using a clean pipette) onto a clean microscope slide, and cover it with a clean coverslip.

   b. Observe that the cells of this organism form a **simple colony:** The cells always occur in groups of from four to eight cells, and they are permanently united.

   c. Identify the following structures.

      The **nucleus** is the spherical organelle in the approximate middle of each cell.

      **Vacuoles** are the transparent spheres that tend to occur at either end of the cells.

      **Spines** are the transparent projections that occur on the two end cells.

      **Cell walls** surround each cell.

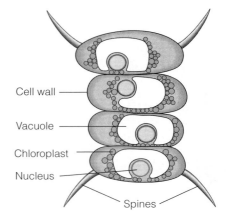

**Figure 2.8.**
***Scenedesmus.*** *Scenedesmus* is an aquatic alga that usually occurs in simple colonies of four cells connected by an outer cell wall (Color Plate 4).

# Lab Study C. Multicellular Organisms

## Materials

microscope slides
dropper bottles of water
toothpicks
coverslips
*Elodea*

methylene blue
finger bowl with disinfectant
broken glass chips
*Volvox* cultures

## Introduction

Review the criteria for characterizing an organism as multicellular in the introduction of Exercise 2.5 on pp. 41–42. In multicellular organisms, there are two or more cell types with specialized structure and function that cannot persist when isolated from other cells in the organism. If these cells are isolated, they are not capable of perpetuating the species. In this lab study, you will examine an example of a green alga, a plant, and an animal to investigate the criteria for multicellularity and observe cells that compose basic tissue types.

## Procedure

### *Volvox*

*Volvox* (Figure 2.9) is an aquatic green alga that is common in aquaria, ponds, and lakes. In older literature this organism was described as colonial and was not considered to be multicellular. Today, however, scientists have concluded that it is more accurate to call *Volvox* multicellular. In this activity you will look for evidence that supports this conclusion.

1. Examine living *Volvox* under the compound microscope. To obtain a specimen, prepare a wet mount as you did for *Scenedesmus* with the following addition: Before placing a drop of the culture on your slide, place several glass chips on the slide. This will keep the coverslip from crushing these spherical organisms.

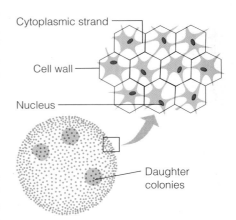

**Figure 2.9.**
***Volvox.*** In this organism, the individual cells are interconnected by cytoplasmic strands to form a sphere. Small clusters of cells, called daughter colonies, are specialized for reproduction (Color Plate 5).

2. Observe that the cells of this organism lie in a transparent matrix forming a large hollow sphere. The approximately 500 to 50,000 (depending on the species) nonreproductive somatic cells are permanently united by cytoplasmic connections. These cells have chloroplasts for photosynthesis and flagella that beat in a coordinated motion to move the colony like a ball. During asexual reproduction, certain cells in the sphere (reproductive cells) enlarge and migrate inward to become daughter colonies.

3. Identify the following structures: **somatic cells** with **cytoplasmic connections** and **flagella**. Depending on the magnification of your microscope, you may be able to distinguish **cell walls** and **nuclei** in the cells. **Daughter colonies** are smaller spheres within the larger colony. These are released when the parent colony disintegrates.

 Student Media Video—Ch. 28: *Volvox* Colony

## Plant Cells

1. The major characteristics of a typical plant cell are readily seen in the leaf cells of *Elodea*, a common aquatic plant (Figure 2.10). Prepare a wet mount and examine one of the youngest (smallest) leaves from a sprig of *Elodea* under the compound microscope.

2. Identify the following structures.

   The **cell wall** is the rigid outer framework surrounding the cell. This structure gives the cell a definite shape and support. It is not found in animal cells.

   **Protoplasm** is the organized contents of the cell, exclusive of the cell wall.

   **Cytoplasm** is the protoplasm of the cell, exclusive of the nucleus.

   The **central vacuole** is a membrane-bound sac within the cytoplasm that is filled with water and dissolved substances. This structure serves to store metabolic wastes and gives the cell support by means of turgor pressure. Animal cells also have vacuoles, but they are not as large and conspicuous as those found in plants.

   **Chloroplasts** are the green, spherical organelles often seen moving within the cytoplasm. These organelles carry the pigment chlorophyll that is involved in photosynthesis. As the microscope light heats up the cells, cytoplasm and chloroplasts may begin to move around the central vacuole in a process called *cytoplasmic streaming,* or *cyclosis.*

   The **nucleus** is the usually spherical, transparent organelle within the cytoplasm. This structure controls cell metabolism and division.

3. What three structures observed in *Elodea* are unique to plants?

   *chloroplast, cell wall, large central vacuole*

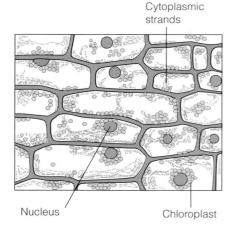

Cytoplasmic strands

Nucleus

Chloroplast

**Figure 2.10.**
*Elodea.* *Elodea* is an aquatic plant commonly grown in freshwater aquaria. The cell structures may be difficult to see because of the three-dimensional cell shape and the presence of a large central vacuole (Color Plate 6).

*The nucleus may be difficult to see because the chloroplasts are abundant in the cytoplasm, obstructing the view of the nucleus. The majority of the cell is filled by the clear vacuole. Strands of grainy cytoplasm may be visible in cells with few chloroplasts. The nucleus will appear as a clear oval, often in a corner of the cell.*

4. Compare your observations of *Elodea* using the compound scope with those made in Exercise 2.3 using the stereoscopic scope. List the structures seen with each:

   Stereoscopic:

   Compound:

---

 Student Media Video—Ch. 6: Cytoplasmic Streaming

---

## *Animal Cells*

1. Animals are multicellular heterotrophic organisms that ingest organic matter. They are composed of cells that can be categorized into four major tissue groups: epithelial, connective, muscle, and nervous tissue. In this lab study, you will examine epithelial cells. Similar to the epidermal cells of plants, **epithelial cells** occur on the outside of animals and serve to protect the animals from water loss, mechanical injury, and foreign invaders. In addition, epithelial cells line interior cavities and ducts in animals. Examine the epithelial cells (Figure 2.11) that form the lining of your inner cheek. To obtain a specimen, follow this procedure:

   a. With a clean toothpick, gently scrape the inside of your cheek several times.

   b. Roll the scraping into a drop of water on a clean microscope slide, add a small drop of methylene blue, and cover with a coverslip. Discard the used toothpick in disinfectant.

   c. Using the compound microscope, view the cells under higher powers.

2. Observe that these cells are extremely flat and so may be folded over on themselves. Attempt to locate several cells that are not badly folded, and study their detail.

3. Identify the following structures.

   The **cell membrane** is the boundary that separates the cell from its surroundings.

   The **nucleus** is the large, circular organelle near the middle of the cell.

   **Cytoplasm** is the granular contents of the cell, exclusive of the nucleus.

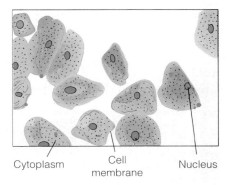

Cytoplasm        Cell membrane        Nucleus

**Figure 2.11.**
**Human epithelial cells.** The epithelial cells that line your cheek are thin, flat cells that you can remove easily from your cheek by scraping it with a toothpick (Color Plate 7).

## Lab Study D. Unknowns

### Materials

microscope slides
coverslips
pond water or culture of unknowns

### Introduction

Use this lab study to see if you have met the objectives of this lab topic. As you carry out this lab study, (1) think carefully about using correct microscopic techniques; (2) distinguish organisms with different cellular organization or configuration (unicellular, colonial, etc.); (3) note how the different organisms are similar yet different; and (4) note cell differences.

*Ask students to bring their own pondwater samples. This is the students' favorite study. Mixed protozoan cultures are available from supply houses for those students who are unsuccessful with their own cultures. See the Preparation Guide.*

All of the cells studied to this point in this lab topic have been examples of **eukaryotic** cells. As you examine drops of pond water as described in the procedure following, you may observe examples of **prokaryotic** cells in colonies or filaments. Eukaryotic cells have a true nucleus containing chromosomes with genetic material separated from the remainder of the cell by a nuclear envelope. All cellular organelles are also bound by membranes. In prokaryotic cells genetic material is not bound by a nuclear envelope, and no membrane-bound organelles are present. Prokaryotic cells will be studied in more detail in Lab Topic 13, Bacteriology.

### Procedure

*Identifying organisms is not important; however, some students will be curious. Provide resources or diagrams of common organisms.*

1. Examine several drops of the culture of pond water that you collected, or examine the unknown culture provided by the instructor.
2. Record in Table 2.1 the characteristics of at least four different organisms.
3. Determine if a well-defined nucleus and organelles are present (eukaryote).

**Table 2.1**
Characteristics of Organisms Found in Pond Water

| Tube Unknown | Means of Locomotion | Cell Wall (+/−) | Chloroplasts (+/−) | Cellular Organization | Eukaryote (yes or no) |
|---|---|---|---|---|---|
| 1 | | | | | |
| 2 | | | | | |
| 3 | | | | | |
| 4 | | | | | |
| 5 | | | | | |

# Reviewing Your Knowledge

1. Describe at least two types of materials or observations that would necessitate the use of the stereoscopic microscope.

   *small insects, small flower structures, surface features of plants and animals*

2. What characteristics do all eukaryotic cells have in common?

   *nucleus, cytoplasm, plasma membrane, mitochondria, etc.*

3. a.   What cellular features differentiate plants from animals?

   *chloroplast, cell wall, large central vacuole*

   b. How are the structures that are unique to plants important to their success?

   *Chloroplast is necessary for photosynthesis; cell wall provides rigid support provided in animals by a skeleton; large vacuole provides for storage and flexible support by turgor pressure.*

4. Return to question #4, pp. 41–42. Based on your observations in today's laboratory, *circle* those organelles that are visible in the light microscope. Compare your observations with your initial predictions.

5. Review the criteria used to distinguish between colonial and multicellular organisms. Why is *Volvox* now considered multicellular?

   *Volvox has two cell types (somatic and reproductive) specialized for different functions. Isolated differentiated somatic cells are not capable of perpetuating the species.*

## Applying Your Knowledge

1. In your own words, describe the evolutionary trend for increasing organismal complexity, using examples from this lab to illustrate your answer.

   *In the history of life, the organization of cells and organisms has become increasingly complex. The first eukaryotic organisms that evolved were single-celled, represented in this lab by the amoeba. Aggregates (loose clusters of unconnected cells) existed later, represented by* Protococcus. *Simple colonies evolved with structural connections (*Scenedesmus*), and eventually more complex colonies and multicellularity evolved. Multicellularity evolved several times, giving rise to plants (*Elodea*), animals (humans), and fungi.*

2. We often imply that multicellular organisms are more advanced (and therefore more successful) than unicellular or colonial organisms. Explain why this is not true, using examples from this lab or elsewhere.

   *Complex organisms should not be thought of as advanced in the sense of better adapted; rather, they originated later in the history of life than simple organisms. Extant organisms, such as humans and amoebas, are obviously adapted to a successful way of life today.*

3. Following is a list of tissues that have specialized functions and demonstrate corresponding specialization of subcellular structure. Match the tissue with the letter of the cell structures and organelles listed to the right that would be abundant in these cells.

| *Tissues* | *Cell Structures and Organelles* |
|---|---|
| • Enzyme (protein)-secreting cells of the pancreas  *c, e, h* | a. plasma membrane |
| • Insect flight muscles  *b* | b. mitochondria |
| • Cells lining the respiratory passages  *f* | c. Golgi apparatus<br>d. chloroplast |
| • White blood cells that engulf and destroy invading bacteria  *i* | e. endoplasmic reticulum<br>f. cilia and flagella |
| • Leaf cells of cacti  *d, g* | g. vacuole<br>h. ribosome<br>i. lysosome<br>j. peroxisomes |

4. One organism found in a termite's gut is *Mixotricha paradoxa*. This strange creature looks like a single-celled swimming ciliate under low magnification. However, the electron microscope reveals that it contains spherical bacteria rather than mitochondria and has on its surface, rather than cilia, hundreds of thousands of spirilla and bacilla bacteria. You are the scientist who first observed this organism. How would you describe this organism—single-celled? aggregate? colony? multicellular? Review definitions of these terms on pp. 41–42. Can the structure of this organism give you any insight into the evolution of eukaryotic cells? (*Hint:* See the discussion of the endosymbiosis hypothesis in your text.)

   *Accept any reasonable speculation about answers to these questions.*

5. *Pleodorina* is an aquatic green alga that is common in ponds, lakes, and roadside ditches. This organism is made up of 32 to 128 cells that are embedded in a gel-like matrix. In mature colonies two types of cells can be distinguished, small somatic cells and larger reproductive cells that divide to form new colonies. Somatic cells carry on photosynthesis, but may become reproductive if isolated from the colony.

   Review the criteria used to determine multicellularity, and decide if *Pleodorina* should be classified as multicellular or colonial.

   *Phycologists consider* Pleodorina *colonial and not multicellular because somatic cells in* Pledorina, *if they are separated from the reproductive cells, can redifferentiate as reproductive cells, divide, and produce small offspring with both somatic and reproductive cells.*

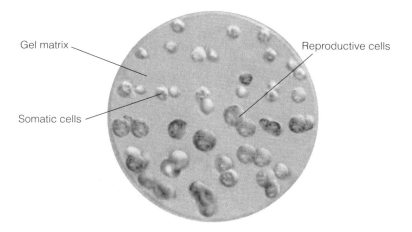

Gel matrix

Reproductive cells

Somatic cells

## Investigative Extensions

1. Survey bodies of water surrounding your campus and assess the environmental conditions. Obtain samples of organisms in ponds, lakes, or rivers from several sites with different environmental conditions. Are some sites more polluted than others? Are there temperature differences? If a fountain is available on campus, compare this site with water in a pond that has no fountain (less water movement). You may find a site near the power-generating facility of your campus.

   Measure and note as many independent variables as possible—temperature of the water, relative levels of pollution, surroundings (near an open space, between buildings, etc.).

List types of organisms observed. Describe the characteristics of organisms. Note differences in density and diversity of organisms in samples taken from different sources.

2. Investigate additional organelles in plants. Plants have specific cellular structures related to their lives as plants. For example, plants use chlorophyll pigments bound in the membranes of **chloroplasts** to absorb wavelengths of light that fuel photosynthesis. The starch synthesized in photosynthesis is later stored in special organelles called **leucoplasts** or **amyloplasts.** Plants are colorful (think of the many functions of color in plants). Some of these colorful pigments are water soluble and stored in vacuoles, but others, including some of the brightly colored carotenoids (reds, yellows, oranges) are membrane bound in organelles called **chromoplasts.** Chloroplasts, chromoplasts, and leucoplasts are three kinds of **plastids,** all of which are similar in structure and typically found in plant cells.

Consider the structure and function of these three plastids. Where in the plant body would you expect to find them—in roots, stems, leaves, flowers, or fruits? Hypothesize which of these plastids might be found in green pepper skin, tomato skin, potato tubers, or other plant material.

To test for chromoplasts and chloroplasts, peel the thin epidermis from the outside of the plant or make very thin sections of plant tissue. These can be placed on a slide with water and a coverslip then viewed under high power. To visualize leucoplasts, try adding a drop of iodine to the slide before covering. Iodine stains starch a dark blue or black.

## Student Media: BioFlix, Activities, Investigations, and Videos
www.masteringbiology.com (select Study Area)

**BioFlix**—Ch. 6: Tour of an Animal Cell; Tour of a Plant Cell

**Activities**—Ch. 6: Build an Animal Cell and a Plant Cell; Build a Chloroplast and a Mitochondrion; Cilia and Flagella; Review: Animal Cell Structure and Function; Review: Plant Cell Structure and Function

**Investigations**—Ch. 6: What Is the Size and Scale of Our World?

**Videos**—Ch. 6: Cytoplasmic Streaming; *Paramecium* Cilia Ch. 28: Volvox Colony; Volvox Flagella; Amoeba; Amoeba Pseudopodia Discovery Video: Ch. 6: Cells

# References

Alberts, B., D. Bray, J. Lewis, M. Raff, K. Roberts, and P. Walter. *Molecular Biology of the Cell,* 4th ed. New York: Garland, 2002.

Becker, W. M., L. J. Kleinsmith, and J. Hardin. *The World of the Cell,* 6th ed. Redwood City, CA: Benjamin Cummings, 2006.

Kirk, D. L. "A Twelve-step Program for Evolving Multicellularity and a Division of Labor." *BioEssays,* 2005, vol. 27, pp.299–310.(The treatment of *Volvox* as multicellular and the inclusion of *Pleodorina* as a colonial organism was based on discussions with D. L. Kirk.)

Margulis, L., and D. Sagan. "The Beast with Five Genomes," *Natural History,* 2001, vol. 110, pp. 38–41.

Reece, J., et al. *Campbell Biology,* 9th ed. San Francisco, CA: Pearson Education, 2011.

## Websites

Cartoon depictions of cellular structures as seen in electron micrographs:
http://www.cellsalive.com

Detailed procedures for fixing, embedding, and staining tissues in electron microscopy:
http://www.bristol.ac.uk/vetpath/cpl/emtechs.htm

A short video showing the use of the stereoscopic microscope:
http://videos.howstuffworks.com/hsw/12618-the-world-of-the-microscope-the-stereomicroscope-video.htm

Structure and function of the electron microscope. Good overview of the electron microscope:
http://www.youtube.com/videos

Search for the structure function of the electron microscope at:
fToTFjwUc5M

Students collect water from a pond and use the compound microscope to view the diversity of organisms in a drop of water:
http://videos.howstuffworks.com/hsw/12622-the-world-of-the-microscope-organisms-in-pond-water-video.htm

Ultrastructure of the cell (electron micrographs); excellent electron micrographs of many cells. Click on the blue boxes to view enlarged portions of different cell types:
http://www.bu.edu/histology/m/t_electr.htm

## LAB TOPIC 2

# Microscopes and Cells
## Teaching Plan for Laboratories

## Main Concepts and Objectives

1. Concept: microscopy. Students will become proficient in the use of the light microscope. Students will recognize the differences in light and electron microscopy. (Exercises 2.1–2.4)

2. Concept: cells as the basic structural and functional unit of life. Students will describe the structure of cells in both plant and animal tissue. They will describe the structure and function of organelles observed with light microscopy and in electron micrographs. (Exercise 2.5)

3. Concept: unity in the diversity of life. Students will describe similarities and differences in cells of different organisms. (Exercise 2.5)

4. Concept: evolutionary history. Students will describe evidence that supports the theory that over time organisms have evolved from single cells to aggregates, to colonial organisms, and, finally, to multicellular organisms. (Exercise 2.5)

5. Specific content. Students will be able to describe or define: parts of compound and dissecting microscopes, comparison of light and electron microscopes, names of all organisms, differences in plant and animal cells, names of all cell structures, functions of organelles, definitions and examples of *unicellular, aggregate, simple* or *complex colony,* and *multicellular.*

## Order of the Lab

1. Introduce major concepts and objectives. You may choose to introduce the concept of classification, including domains and kingdoms, here. (See the Introduction, Lab Topic 13 Bacteriology, in this manual and relevant pages in the text.)

If so, point out that organisms in the domain Bacteria
will be studied in Lab Topic 13 and fungi in Lab Topic 14.          (15 min)

2. Microscope technique.                                           (45 min)

3. Electron micrographs and review of electron microscopy
   (may be observed anytime during lab).                           (20 min)

4. Study organisms and unknown samples.                            (60 min)

5. Discussion, conclusions.                                        (30 min)

6. Remind students about proper microscope storage.                (10 min)

For a 2-hour lab: Omit the electron microscope and micrographs study (Exercise 2.4)
and the unknowns (Exercise 2.5, Lab Study D), although the latter is a highlight of the
lab. The final discussion can be reduced or left to the discussion questions at the end of
the lab topic.

## Classroom Management

Students work independently on all activities. It is very important that students begin to
feel comfortable using the microscope, which can be accomplished only if each person
works independently when viewing slides. However, encourage students to discuss their
observations and conclusions with their lab partners.

## Student Development

Students must organize time and resources effectively. They develop basic microscope
skills that are necessary throughout the remaining labs. They must record and
summarize observations and apply their knowledge using the unknowns and discussion
questions.

## Discussion and Summary

Use discussion questions at the end of the lab topic to encourage students to summarize
laboratory concepts and observations.

## Evaluation

Informally evaluate microscope skills. Review answers to discussion questions. Laboratory material should be covered on the next lab test.

## Lab Topic 3
# Diffusion and Osmosis

## Laboratory Objectives

After completing this lab topic, you should be able to:

1. Describe the mechanism of diffusion at the molecular level.
2. List several factors that influence the rate of diffusion.
3. Describe a selectively permeable membrane, and explain its role in osmosis.
4. Define *hypotonic, hypertonic,* and *isotonic* in terms of relative concentrations of osmotically active substances.
5. Discuss the influence of the cell wall on osmotic behavior in cells.
6. Explain how incubating plant tissues in a series of dilutions of sucrose can give an approximate measurement of osmolarity of tissue cells.
7. Explain why diffusion and osmosis are important to cells.
8. Apply principles of osmotic activity to medical, domestic, and environmental activities.
9. Discuss the scientific process, propose questions and hypotheses, and make predictions based on experiments to test hypotheses.
10. Practice scientific persuasion and communication by constructing and interpreting graphs.

*For a 2-hour lab: Begin the potato osmolarity experiment immediately. Omit student reports. Have students make their graphs as an outside assignment. See Teaching Plan.*

*Instructors and/or students can download data tables in Excel format. Go to www .masteringbiology.com. Look in the Study Area under Lab Media.*

## Introduction

Maintaining the steady state of a cell is achieved only through regulated movement of materials through cytoplasm, across organelle membranes, and across the plasma membrane. This regulated movement facilitates communication within the cell and between cytoplasm and the external environment. The cytoplasm and extracellular environment of the cell are aqueous solutions. They are composed of water, which is the **solvent,** or dissolving agent, and numerous organic and inorganic molecules, which are the **solutes,** or dissolved substances. Organelle membranes and the plasma membrane are **selectively permeable,** allowing water to freely pass through but regulating the movement of solutes.

The cell actively moves some dissolved substances across membranes, expending adenosine triphosphate (ATP) (biological energy) to accomplish the movement. Other substances move passively, without expenditure of ATP from the cell, but only if the cell membrane is permeable to those substances. Water and selected solutes move passively through the cell and cell

*Adopters of previous editions of this laboratory manual will note that we have changed the sequence of lab topics to better reflect the sequence of topics in most general biology texts. In this new sequence, this lab topic is the student's next opportunity to practice experimental design and process. Stress to your students the importance of reading carefully and following procedures in a meticulous, step-by-step fashion. Stress that they collect data carefully and report results accurately in tables. Since students are just beginning to learn how to collect and record data, the instructor should review the design of their tables and graphs.*

membranes by **diffusion**, a physical process in which molecules move from an area where they are in high concentration to one where their concentration is lower. The energy driving diffusion comes only from the intrinsic kinetic energy (energy of motion) in all atoms and molecules. If nothing hinders the movement, a solute will diffuse until it reaches equilibrium.

**Osmosis** is a type of diffusion; in cells it is the diffusion of water through a selectively permeable membrane from a region where it is highly concentrated to a region where its concentration is lower. The difference in concentration of water occurs if there is an unequal distribution of at least one dissolved substance on either side of a membrane and the membrane is impermeable to that substance. For example, if a membrane that is impermeable to sucrose separates a solution of sucrose from distilled water, water will move from the distilled water, where it is in higher concentration, through the membrane into the sucrose solution, where it is in lower concentration.

*We use the terms* hypertonic, hypotonic, *and* isotonic *in the broad definition: comparing one solution to another relative to osmotically active substances (solute concentrations). The strictest definitions of these terms refer only to cells and their environments.*

Three terms, **hypertonic, hypotonic,** and **isotonic,** are used when referring to two solutions separated by a selectively permeable membrane (Figure 3.1). The hypertonic solution (Figure 3.1a) has a greater concentration of solutes that cannot cross the membrance (nonpenetrating solutes) than the solution on the other side of the membrane. It is described, therefore, as having a greater osmolarity (solute concentration expressed as molarity). The hypotonic solution (Figure 3.1b) has a lower concentration of nonpenetrating solutes, or a lower osmolarity, than the solution on the other side of the membrane. When the two solutions are in equilibrium, the concentration of nonpenetrating solutes being equal on both sides of the membrane, the osmolarities are equal and the solutions are said to be isotonic (Figure 3.1c). The *net flow* of water is from the hypotonic to the hypertonic solution. When the solutions are isotonic, there is no net flow of water across the membrane.

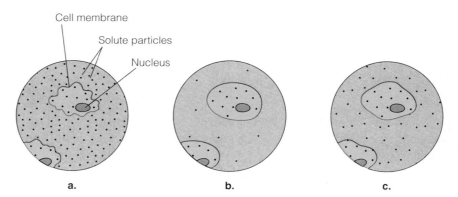

**Figure 3.1.**
**Diagrammatic representation of cells in (a) hypertonic, (b) hypotonic, and (c) isotonic solutions.** The hypertonic solution has a greater concentration of nonpenetrating solutes than the solution on the other side of the membrane, the hypotonic solution has a lower concentration of nonpenetrating solutes than the solution on the other side of the membrane, and the concentration of nonpenetrating solutes is equal on both sides of the membrane in isotonic solutions.

EXERCISE 3.1
# Diffusion of Molecules

In this exercise you will investigate characteristics of molecules that facilitate diffusion, factors that influence diffusion rates, and diffusion of solutes through a selectively permeable membrane.

## Experiment A. Kinetic Energy of Molecules

### Materials

dropper bottle of water
carmine powder
dissecting needle

slide and coverslip
compound microscope

*If you have a microscopy/video system, you could do this as a demonstration.*

### Introduction

Molecules of a liquid or gas are constantly in motion because of the intrinsic kinetic energy in all atoms and molecules. In 1827, Robert Brown, a Scottish botanist, noticed that pollen grains suspended in water on a slide appeared to move by a force that he was unable to explain. In 1905, Albert Einstein, searching for evidence that would prove the existence of atoms and molecules, predicted that the motion observed by Brown must exist, although he did not realize that it had been studied for many years. Only after the kinetic energy of molecules was understood did scientists ask if the motion observed by Brown and predicted by Einstein could be the result of molecular kinetic energy being passed to larger particles. We now know that intrinsic molecular kinetic energy is the driving force of diffusion. In this experiment, you will observe large particles suspended in water in motion similar to that observed by Brown, traditionally called **Brownian movement.** You will relate the motion observed to the forces that bring about diffusion.

*India ink can be substituted for carmine; or if you have pollen available, use this and duplicate Brown's original experiment.*

### Procedure

Work in pairs. One person should set up the microscope while the other person makes a slide as follows:

1. Place a drop of water on the slide.
2. Touch the tip of a dissecting needle to the drop of water and then into the dry carmine.
3. Add the carmine on the needle to the drop of water on the slide, mix, cover with a coverslip, and observe under the compound microscope.
4. Observe on low power and then high power. Focus as much as possible on one particle of carmine.
5. Record your findings in the Results section, and draw conclusions based on your results in the Discussion section.

### Results

Describe the movement of single carmine particles.

1. Is the movement random or directional?
2. Does the movement ever stop?

3. Do smaller particles move more rapidly than larger particles? Other observations?

*The movement appears to be random, and it never stops. The critical student will note that smaller particles seem to move faster than larger ones and that a given particle appears equally likely to move in any direction.*

### Discussion

1. Are you actually observing molecular movement? Explain.

*No. Molecules and atoms are submicroscopic particles. The kinetic energy in the surrounding water molecules is causing them to collide with each other and to bombard the carmine particles, bringing about their movement.*

2. How can molecular movement bring about diffusion?

*The random motion of molecules will result in a net movement of molecules away from a region of high concentration to regions of lower concentration.*

3. Speculate about the importance of diffusion in cell metabolism.

*Nutrients, hormones, and wastes move within and among cells by diffusion.*

## Experiment B. Diffusion of Molecules through a Selectively Permeable Membrane

### Materials

string or rubber band
wax pencil
30% glucose solution
starch solution
I₂KI solution
Benedict's reagent
hot plate

500-mL beaker one-third filled with water
handheld test tube holder
3 standard test tubes
disposable transfer pipettes
2 400-mL beakers to hold dialysis bag
30-cm strip of moist dialysis tubing

*The plastic transfer pipettes that come with living specimens work best.*

### Introduction

Dialysis tubing is a membrane made of regenerated cellulose fibers formed into a flat tube. If two solutions containing dissolved substances of different molecular weights are separated by this membrane, some substances may readily pass through the pores of the membrane, but others may be excluded.

Working in teams of four students, you will investigate the selective permeability of dialysis tubing. You will test the permeability of the tubing to the reducing sugar, glucose (molecular weight 180), starch (a variable-length polymer of glucose), and iodine potassium iodide (I₂KI). You will place a solution of glucose and starch into a dialysis tubing bag and then place this bag into a solution of I₂KI. Sketch and label the design of this experiment in the margin of your lab manual to help you develop your hypotheses.

You will use two tests in your experiment:

1. *I₂KI test for presence of starch.*

   When I₂KI is added to the unknown solution, the solution turns purple or black if starch is present. If no starch is present, the solution remains a pale yellow-amber color.

2. *Benedict's test for reducing sugar.*

   When Benedict's reagent is added to the unknown solution and the solution is heated, the solution turns green, orange, or orange-red if a reducing sugar is present (the color indicates the sugar concentration). If no reducing sugar is present, the solution remains the color of Benedict's reagent (blue).

## Question

Remember that every experiment begins with a question. Review the design of this experiment in the Introduction above. Formulate a question about the permeability of dialysis tubing. The question may be broad, but it must propose an idea that has measurable and controllable elements.

> *Does the molecular weight of a solute impact its movement through a membrane?*

## Hypothesis

Hypothesize about the selective permeability of dialysis tubing to the substances being tested.

> *Molecules will diffuse through pores in the dialysis tubing bag based on their molecular weights. (Accept any testable hypothesis.)*

*As students begin to practice developing questions, hypotheses, and predictions, you may choose to review criteria for "good" questions, hypotheses, and predictions, giving examples. Ask to see students' responses here before they proceed.*

## Prediction

Predict the results of the I₂KI and Benedict's tests based on your hypothesis (if/then).

> *If molecules diffuse through pores in the dialysis tubing based on their molecular weights, then the solution inside the bag will turn purple or black as I₂KI diffuses in and reacts with the starch. Benedict's test will indicate the presence of a reducing sugar inside and outside the bag as glucose diffuses out.*

## Procedure

1. Prepare the dialysis bag with the initial solutions.

   a. Fold over 3 cm at the end of a 25- to 30-cm piece of dialysis tubing that has been soaking in water for a few minutes, pleat the folded end "accordion style," and close the end of the tube with the string or a rubber band, forming a bag. This procedure must secure the end of the bag so that no solution can seep through.

   b. Roll the opposite end of the bag between your fingers until it opens, and add 4 pipettesful of 30% glucose into the bag. Then add 4 pipettesful of starch solution to the glucose in the bag.

   c. Hold the bag closed and mix its contents. Record its color in Table 3.1 in the Results section. Carefully rinse the outside of the bag in tap water.

   d. Add 300 mL of water to a 400- to 500-mL beaker. Add several droppersful of I₂KI solution to the water until it is visibly yellow-amber. Record the color of the H₂O + I₂KI solution in Table 3.1.

*There should be enough glucose and starch in the bag to allow diffusion and still be concentrated enough to react when tested. Leave enough space in the bag so the top of the bag can be draped over the beaker edge.*

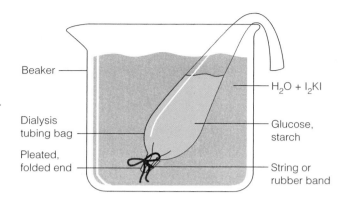

**Figure 3.2.**
**Setup for Exercise 3.1, Experiment B.**
The dialysis tubing bag, securely closed at one end, is placed in the beaker of water and I₂KI. The open end of the bag should drape over the edge of the beaker.

*If the contents of the bag have not begun to turn purple within about 15 minutes, tell students to add more I₂KI to the beaker.*

e. Place the bag in the beaker so that the untied end of the bag hangs over the edge of the beaker (Figure 3.2). *Do not allow the liquid to spill out of the bag!* If the bag is too full, remove some of the liquid and rinse the outside of the bag again. If needed, place a rubber band around the beaker, holding the bag securely in place. If some of the liquid spills into the beaker, dispose of the beaker water, rinse, and fill again.

2. Leave the bag in the beaker for about 30 minutes. (You should go to another lab activity and then return to check your setup periodically.)

3. After 30 minutes, carefully remove the bag and stand it in a dry beaker.

4. Record in Table 3.1 the final color of the solution in the bag and the final color of the solution in the beaker.

5. Perform the Benedict's test for the presence of sugar in the solutions.

   a. Label three clean test tubes: control, bag, and beaker.
   b. Put 2 pipettesful of water in the control tube.
   c. Put 2 pipettesful of the bag solution in the bag tube.
   d. Put 2 pipettesful of the beaker solution in the beaker tube.
   e. Add 1 dropperful of Benedict's reagent to each tube.
   f. Heat the test tubes in a boiling water bath for about 3 minutes.
   g. Record your results in Table 3.1.

6. Review your results in Table 3.1 and draw your conclusions in the Discussion section.

### Results

Complete Table 3.1 as you observe the results of Experiment B.

### Table 3.1
Results of Experiment Investigating the Permeability of Dialysis Tubing to Glucose, I₂KI, and Starch

| Solution Source | Original Contents | Original Color | Final Color | Color after Benedict's Test |
|---|---|---|---|---|
| **Bag** | *30% glucose and starch* | *no color, chalky* | *purple* | *orange-red* |
| **Beaker** | *H₂O + I₂KI* | *pale yellow* | *pale yellow* | *orange-red* |
| **Control** | *H₂O* | *—* | *—* | *blue* |

## Discussion

1. What is the significance of the final colors and the colors after the Benedict's tests? Did the results support your hypothesis? Explain, giving evidence from the results of your tests.

   *The purple color in the bag indicates that I$_2$KI has diffused into the bag and reacted with the starch there. The fact that the beaker remains pale yellow supports the hypothesis that starch cannot diffuse from the bag through the tubing into the beaker. The positive Benedict's test in both bag and beaker supports the hypothesis that the glucose in the bag diffused through the tubing into the beaker.*

2. How can you explain your results?

   *The results can be explained by concluding that the dialysis tubing is permeable to glucose and I$_2$KI but not to starch.*

3. From your results, predict the size of I$_2$KI molecules relative to glucose and starch.

   *I$_2$KI molecules are smaller than starch. We have no evidence to predict if they are larger or smaller than glucose.*

4. What colors would you expect if the experiment started with glucose and I$_2$KI inside the bag and starch in the beaker? Explain.

   *The solution in the beaker would turn purple as I$_2$KI diffused out and reacted with starch. The glucose would diffuse out of the bag, and both solutions would test positive for sugar.*

# EXERCISE 3.2
# Osmotic Activity in Cells

All organisms must maintain an optimum internal osmotic environment. Terrestrial vertebrates must take in and eliminate water using internal regulatory systems to ensure that the environment of tissues and organs remains in osmotic balance. Exchange of waste and nutrients between blood and tissues depends on the maintenance of this condition. Plants and animals living in fresh water must control the osmotic uptake of water into their hypertonic cells.

In this exercise, you will investigate the osmotic behavior of plant and animal cells placed in different molar solutions. What happens to these cells when they are placed in hypotonic or hypertonic solutions? This question will be investigated in the following experiments.

# Experiment A. Osmotic Behavior of Animal Cells

## Materials

On demonstration:
  test tube rack
  3 test tubes with screw caps, each containing one of the three solutions
    of unknown osmolarity
  ox blood
  newspaper or other printed page

For microscopic observations:
  4 clean microscope slides and coverslips
  wax pencil
  dropper bottle of ox blood
  dropper bottles with three solutions of unknown osmolarity

## Introduction

Mature red blood cells (erythrocytes) are little more than packages of hemoglobin bound by a plasma membrane permeable to small molecules, such as oxygen and carbon dioxide, but impermeable to larger molecules, such as proteins, sodium chloride, and sucrose. In mammals these cells even lack nuclei when mature, and as they float in isotonic blood plasma, their shape is flattened and pinched inward into a biconcave disk. Oxygen and carbon dioxide diffuse across the membrane, allowing the cell to carry out its primary function, gas transport, which is enhanced by the increased surface area created by the shape of the cell. Scientists questioning what happens to red blood cells in different molar solutions observed that the cells respond dramatically if they are not in an isotonic environment. When water moves into red blood cells placed in a hypotonic solution, the cells swell and the membranes burst, or undergo **lysis.** When water moves out of red blood cells placed in a hypertonic solution, the cells shrivel and appear bumpy, or **crenate.** In this experiment, you will investigate the behavior of red blood cells when the osmolarity of the environment changes from isotonic to hypertonic or hypotonic. (See Color Plate 8.)

## Hypothesis

Hypothesize about the behavior of red blood cells when they are placed in hypertonic or hypotonic environments.

> *Differences in osmolarity cause water to move into cells placed in hypotonic environments and move out of cells placed in hypertonic environments.*

## Prediction

Predict the results of the experiment based on your hypothesis (if/then).

> *If differences in osmolarity cause water to move into cells placed in hypotonic environments and move out of cells in hypertonic environments, then cells placed in a hypotonic solution will swell and burst, and cells placed in a hypertonic solution will shrink and become crenated.*

## Procedure

1. Observe the three test tubes containing unknown solutions and blood on demonstration. These tubes have been prepared in the following way.

   Test tube 1: 15 mL of unknown solution A

   Test tube 2: 15 mL of unknown solution B

   Test tube 3: 15 mL of unknown solution C

   Your instructor has added 5 drops of ox blood to each test tube.

   Observe the appearance of each test tube. Is it opaque? Is it translucent? Describe your observations in Table 3.2 in the Results section.

2. Be sure each test tube cap is securely tightened, then hold each test tube flat against the printed newspaper article or page of text.

3. Attempt to read the print. Describe in Table 3.2 in the Results section.

   Continue your investigation of osmotic behavior of animal cells by performing microscopic observations of cells in the three unknown solutions.

 **Have your microscope ready, and observe slides immediately after you have prepared them. Do one slide at a time.**

4. Label four clean microscope slides A, B, C, and D.

5. Place a drop of blood on slide D, cover with a coverslip, and observe the shape of red blood cells with no treatment. Record your observations in Table 3.2 in the Results section.

6. Put a drop of solution A on slide A and add a coverslip. Place the slide on the microscope stage and carefully add a small drop of blood to the edge of the coverslip. The blood cells will be drawn under the coverslip by capillary action.

7. As you view through the microscope, carefully watch the cells as they come into contact with solution A; record your observations in Table 3.2.

8. Repeat steps 3 and 4 with solutions B and C.

9. Record your observations in Table 3.2. Draw your conclusions in the Discussion section.

*The instructor should check the ox blood before lab. If the blood is crenated, you will need to explain to the students that the cells are crenated. The cells will still respond to solutions. They will swell and burst in a hypotonic solution, return to normal in an isotonic solution, and become more crenated in a hypertonic solution.*

*If it is available, have students use phase-contrast to view slides. The lysed membranes or "ghosts" will be visible in the hypotonic solution. These are not visible without phase-contrast.*

## Results

1. Record your observations of the demonstration test tubes in Table 3.2.

**Table 3.2**
Appearance of Unknown Solutions A, B, and C

|  | Appearance of the Solution | Can You Read the Print? |
|---|---|---|
| **Test tube 1 (unknown A)** | opaque | no |
| **Test tube 2 (unknown B)** | transparent | yes |
| **Test tube 3 (unknown C)** | opaque | no |

*Possible results; instructor may vary solutions.*

2. Record your microscopic observations of red blood cell behavior in Table 3.3.

**Table 3.3**
Appearance of Red Blood Cells in Test Solutions

*Possible results; instructor may vary solutions.*

| Solution | Appearance/Condition of Cell |
|---|---|
| **D (blood only)** | *biconcave disks* |
| **A** | *cells shrink (crenation)* |
| **B** | *cells swell, burst (lyse), and disappear* |
| **C** | *cells appear as biconcave disks* |

**Discussion**

Explain your results in terms of your hypothesis.

1. Explain the appearance of the three test tubes on demonstration.

   *The test tubes containing hypertonic and isotonic solutions will be opaque as the cells remain intact. Students will not be able to read the print through the solution. The test tube with the hypotonic solution will appear translucent and the solution will be reddish because the cells have lysed, releasing the hemoglobin. Students should be able to read the print through the solution.*

2. Based on the demonstration and your microscopic investigation, which of the three solutions is hypotonic to the red blood cells?

   *B*

*Instructor may vary solutions.*

   Hypertonic?

   *A*

   Isotonic?

   *C*

   Verify your conclusions with the laboratory instructor.

3. What conditions might lead to results other than those expected?

   *Contamination on the slide, in the pipette, or in the containers. Ox blood often arrives with cells already crenated (shrunken) during the preparation and shipping process.*

# Experiment B. Osmotic Behavior in Cells with a Cell Wall

## Materials

On demonstration: 2 compound microscopes labeled A and B
1 slide of *Elodea* in a hypertonic salt solution
1 slide of *Elodea* in distilled water

## Introduction

In their natural environment, cells of freshwater plants and algae are bathed in water containing only dilute concentrations of solvents. The net flow of water is from the surrounding medium into the cells. To understand this process, review the structure of *Elodea* cells from Lab Topic 2.

The presence of a cell wall and a large fluid-filled central vacuole in a plant or algal cell will affect the cell's response to solutions of differing molarities. When a plant cell is placed in a hypertonic solution, water moves out of the cell; the protoplast shrinks and may pull away from the cell wall. This process is called **plasmolysis**, and the cell is described as **plasmolyzed** (Figure 3.3). In a hypotonic solution, as water moves into the cell and ultimately into the cell's central vacuole, the cell's **protoplast** (the plant cell exclusive of the cell wall—the cytoplasm enclosed by plasma membrane) expands. The cell wall, however, restricts the expansion, resulting in **turgor pressure** (pressure of the protoplast on the cell wall owing to uptake of water). A high turgor pressure will prevent further movement of water into the cell. This process is a good example of the interaction between pressure and osmolarity in determining the direction of the net movement of water. The hypertonic condition in the cell draws water into the cell until the membrane-enclosed cytoplasm presses against the cell wall. Turgor pressure begins to force water through the membrane and out of the cell, changing the direction of net flow of water (Figure 3.4).

Scientists call the combined force created by solute concentration and physical pressure **water potential**. For a detailed explanation of water potential, see a discussion of plant transport mechanisms in your text (e.g., Chapter 36 in *Campbell Biology*, 9th ed.). In contrast to an animal cell, the

*Some instructors may choose to introduce the concept of water potential in this experiment and in Exercise 3.3. For a discussion of water potential, consult a text, e.g., Campbell Biology, 9th ed., Chapter 36.*

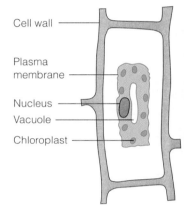

**Figure 3.3.**
**Plant cell placed in a hypertonic solution.** Water leaves the central vacuole and the cytoplasm shrinks, a process called plasmolysis.

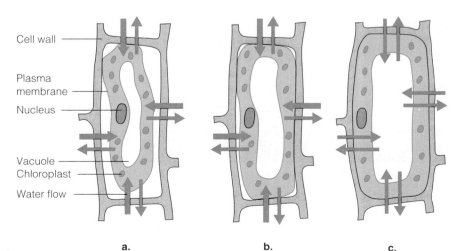

**a.**          **b.**          **c.**

**Figure 3.4.**
**The effect of turgor pressure on the cell wall and the direction of net flow of water in a plant cell.** A plant cell undergoes changes in a hypotonic solution. (a) Low turgor pressure. The net flow of water comes into the cell from the surrounding hypotonic medium. (b) Turgor pressure increases. The protoplast begins to press on the cell wall. (c) Greatest turgor pressure. The tendency to take up water is ultimately restricted by the cell wall, creating a back pressure on the protoplast. Water enters and leaves the cell at the same rate.

*Because this is demo only, students may tend to skip this experiment. Stress that understanding the concept of turgor (cell wall) pressure will be important for interpreting the results of Exercise 3.3.*

ideal state for a plant cell is turgidity. When a plant cell is turgid, it is not isotonic with its surroundings but is hypertonic, having a higher solute concentration than its surroundings. In this state, the plant cell protoplast presses on the cell wall. The pressure of the protoplast on the cell wall is an important force in plant activity. For example, it may cause young cells to "grow" as the elastic cell wall expands.

For this experiment, two slides have been set up on demonstration microscopes. On each slide, *Elodea* has been placed in a different molar solution: One is hypotonic (distilled water) and one is hypertonic (concentrated salt solution).

### Question

Propose a question about the movement of water in *Elodea* leaves placed in different molar solutions.

> *Will water move in or out of cells of* Elodea *leaves placed in different molar solutions?*

### Hypothesis

Hypothesize about the movement of water in cells with a cell wall when they are placed in hypertonic or hypotonic environments.

> *Water moves into plant cells placed in a hypotonic solution. Water moves out of plant cells placed in a hypertonic solution.*

### Prediction

Predict the appearance of *Elodea* cells placed in the two solutions (if/then).

> *If water moves into plant cells placed in a hypotonic solution, then the cells will expand and the protoplast will press on the cell wall. If water moves out of plant cells in a hypertonic solution, then the protoplast will shrink and form a clump in the center of the cell.*

### Procedure

*For a striking demonstration of the effect of a hypertonic solution on plant cells, set beakers labeled A and B beside the corresponding demonstration microscopes. Partially fill each beaker with the respective solution and add a celery stalk or carrot to each an hour or so before lab begins. Ask students to feel the celery or carrot and describe as further evidence to support their conclusions about the* Elodea *solutions. Be sure to have paper towels nearby.*

1. Observe the two demonstration microscopes with *Elodea* in solutions A and B.
2. Record your observations in Table 3.4 in the Results section, and draw your conclusions in the Discussion section.

### Results

Describe the appearance of the *Elodea* cells in Table 3.4.

**Table 3.4**
Appearance of *Elodea* Cells in Unknown Solutions A and B

| Solution | Appearance/Condition of Cells |
|---|---|
| A *(concentrated salt solution)* | *Cell walls retain their shape, but protoplast forms a clump in the center of the cell; plasmolysed* |
| B *(double-distilled water)* | *Protoplast presses on cell wall, and cell walls may bulge slightly; turgid* |

**Discussion**

1. Based on your predictions and observations, which solution is hypertonic?

   *A*

   Hypotonic?

   *B*

2. Which solution has the greatest osmolarity?

   *A*

3. Would you expect pond water to be isotonic, hypertonic, or hypotonic to *Elodea* cells? Explain.

   *Clean pond water is hypotonic. Although one would expect the organism to be in osmotic balance with its environment, the fact is that organisms living in fresh water must constantly regulate water balance by removing fresh water moving in by osmosis. In the case of some protozoa, contractile vacuoles remove excess water. Kidneys take care of this for fish. For plants and algae, the cell wall prevents protoplast expansion, pressure builds, and excessive water is prevented from entering.*

4. Verify your conclusions with your laboratory instructor.

 Student Media Videos—Ch. 7: Turgid *Elodea;* Plasmolysis

# EXERCISE 3.3
# Investigating Osmolarity of Plant Cells

Knowing the solute concentration of cells has both medical and agricultural applications. In plants, scientists know that for normal activities to take place, the amount of water relative to solute concentration in cells must be maintained within a reasonable range. If plant cells have a reduced water content, all vital functions slow down.

In the following experiments, you will estimate the osmolarity (solute concentration) of potato tuber cells using two methods, change in weight and change in volume. You will incubate pieces of potato tuber in sucrose solutions of known molarity. The object is to find the molarity at which weight or volume of the potato tuber tissue does not change, indicating that there has been no net loss or gain of water. This molarity is an indirect measure of the solute concentration of the potato tuber. This measure is indirect because water movement in plant cells is also affected by the presence of cell walls (see Figure 3.4c).

Work in teams of four. Each team will measure either weight change or volume change. Time will be available near the end of the laboratory period for each team to present its results to the class for discussion and conclusions.

*In this exercise, when the cell is in an isotonic solution the pressure component of water potential is essentially zero. Therefore, solute concentration is an estimate of osmolarity.*

# Experiment A. Estimating Osmolarity by Change in Weight

## Materials

1 large potato tuber
7 250-mL beakers (disposable cups may be substituted)
wax marking pencil
forceps
balance that weighs to the nearest 0.01 g
aluminum foil
petri dish

sucrose solutions: 0.1, 0.2, 0.3, 0.4, 0.5, 0.6 molar (*M*)
razor blade
cork borer
deionized (DI) water (0 molar)
paper towels
metric ruler
calculator

## Introduction

In this experiment, you will determine the weight of several potato tuber cylinders and incubate them in a series of sucrose solutions. After the cylinders have incubated, you will weigh them and determine if they have gained or lost weight. This information will enable you to estimate the osmolarity of the potato tuber tissue.

## Question

What question is being investigated in this experiment?

*What is the osmolarity of potato tuber tissue?*

## Hypothesis

Hypothesize about the osmolarity of potato tuber tissue in relation to the sucrose solutions.

*The osmolarity of potato tuber cells is approximately equivalent to that of a 0.3 M sucrose solution. (Accept any testable hypothesis.)*

## Prediction

Predict the results of the experiment based on your hypothesis (if/then).

*If the osmolarity of potato tuber tissue and cells is equivalent to a 0.3 M sucrose solution, then the potato samples should lose weight in 0.4, 0.5, and 0.6 M solutions. They should not change weight in 0.3 M solution and should gain weight in 0, 0.1, and 0.2 M solutions.*

## Procedure

1. Obtain 100 mL of DI water and 100 mL of each of the sucrose solutions. Put each solution in a separate, appropriately labeled 250-mL beaker or paper cup.

Cork borers and razor blades can cut! Use them with extreme care! To use the cork borer, hold the potato in such a way that the borer will not push through the potato into your hand.

2. Use a sharp cork borer to obtain seven cylinders of potato. Push the borer through the length of the potato, twisting it back and forth. When the borer is filled, remove from the potato and push the potato cylinder out of the borer. You must have seven complete, undamaged cylinders at least 5 cm long.

*If you are using large potatoes, have students push the borer through the width. The borer must cut completely through the potato.*

3. Line up the potato cylinders and, using a sharp razor blade, cut all cylinders to a uniform length, about 5 cm, removing the peel from the ends.

4. Place all seven potato samples in a petri dish, and keep them covered to prevent their drying out.

 In subsequent steps, treat each sample individually. Work quickly. To provide consistency, each person should do one task to all cylinders (one person wipe, another weigh, another slice, another record data).

5. Remove a cylinder from the petri dish, and place it between the folds of a paper towel to blot sides and ends.

6. Weigh it to the nearest 0.01 g on the aluminum sheet on the balance. Record the weight in Table 3.5 in the Results section.

7. Immediately cut the cylinder lengthwise into two long halves.

8. Transfer potato pieces to the water beaker.

9. Note what time the potato pieces are placed in the water beaker. Time: _____ .

10. Repeat steps 5 to 8 with each cylinder, placing potato pieces in the appropriate incubating solution from 0.1 to 0.6 M.

 Be sure that the initial weight of the cylinder placed in each test solution is accurately recorded.

11. Incubate 1.5 to 2 hours. (As this takes place, you will be performing other lab activities.)

12. Swirl each beaker every 10 to 15 minutes as the potato pieces incubate.

13. At the end of the incubation period, record the time when the potato pieces are removed. Time: _____ .
    Calculate the approximate incubation time in Table 3.5.

14. Remove the potato pieces from the first sample. Blot the pieces on a paper towel, removing excess solution only.

15. Weigh the potato pieces and record the final weight in Table 3.5.

16. Repeat this procedure until all samples have been weighed in the chronological order in which they were initially placed in the test solutions.

17. Record your data in the Results section, and complete the questions in the Discussion section.

### Results

1. Complete Table 3.5. To calculate percentage change in weight, use this formula:

$$\text{Percentage change in weight} = \frac{\text{weight change}}{\text{initial weight}} \times 100$$

If the sample gained in weight, the value should be positive. If it lost in weight, the value should be negative.

**Table 3.5**
Data for Experiment Estimating Osmolarity by Change in Weight

| Approximate time in solutions: _____ | | | | | | | |
|---|---|---|---|---|---|---|---|
| | **Sucrose Molarity** | | | | | | |
| | **0.0** | **0.1** | **0.2** | **0.3** | **0.4** | **0.5** | **0.6** |
| **Final weight (g)** | | | | | | | |
| **Initial weight (g)** | | | | | | | |
| **Weight change (g)** | | | | | | | |
| **% change in weight** | | | | | | | |

*Be sure that students follow these directions carefully because the 0 is in the middle of the y axis*

*This is the first graph that students will create since they learned about graphs in Lab Topic 1. Check their graph design and figure title before they proceed with the discussion questions.*

2. Plot percentage change in weight as a function of the sucrose molarity in Figure 3.5.

   a. Place a 0 in the middle of the y axis. Choose appropriate scales.

   b. Label the axes of the graph: Determine dependent and independent variables, and place each on the appropriate axis (see Lab Topic 1 for assistance in graphing).

   c. Graph your results. Weight increase (positive values) should be above the zero change line on the "percentage change in weight" axis. Weight decrease should be below the zero change line.

   d. Construct a curve that best fits the data points. Use this curve to estimate the osmolarity of the potato tuber.

   e. Compose an appropriate figure title.

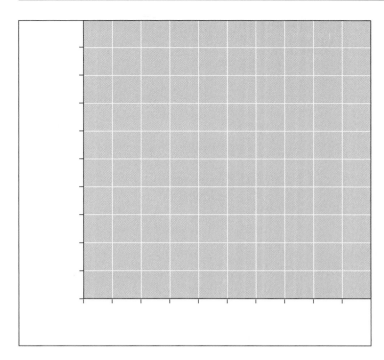

**Figure 3.5.**

_____

_____

_____

## Discussion

1. At what sucrose molarity does the curve cross the zero change line on the graph?

   _The point at which this happens usually falls between 0.2 and 0.3 M. It can vary, depending on the variety and age of the potato. As potatoes age, they lose water and some varieties contain more water than others._

2. Explain how this information can be used to determine the osmolarity of the potato tuber tissue.

   _If the weight does not change at 0.3 M sucrose, then water was neither lost nor gained. At this point the pressure created by the cell wall is negligible. Water potential and osmolarity are essentially equal. This molarity is, therefore, equivalent to the osmolarity (solute concentration) of the potato tuber tissue._

3. In more dilute concentrations of sucrose, the weight of the potato pieces _____ (increases/decreases) after incubation. What forces other than solute concentration will have an impact on the amount of water taken up by the potato pieces (see Figure 3.4c)?

   _When a plant cell is bathed in a hypotonic solution, the cell wall creates (turgor) pressure that has an impact on the amount of water crossing the cell membrane._

4. Estimate the osmolarity of the potato tuber tissue.

# Experiment B. Estimating Osmolarity by Change in Volume

## Materials

1 large potato tuber
vernier caliper
7 250-mL beakers (disposable
   cups may be substituted)
wax marking pencil
forceps
petri dish
razor blade

cork borer (0.5-cm diameter)
sucrose solutions: 0.1, 0.2, 0.3,
   0.4, 0.5, 0.6 *M*
DI water (0 *M*)
metric ruler
paper towels
calculator

## Introduction

*Science educators are urging those in science curriculum development to look for opportunities to integrate science and math. This potato exercise is an excellent opportunity to do this. Students may balk at the integrative aspects of this exercise (processing data, using math), but it is an excellent one pedagogically.*

In this experiment, you will determine the volume of several potato tuber cylinders by measuring the length and diameter of each. You will then incubate them in a series of sucrose solutions. After the cylinders have incubated, you will again measure their length and diameter and determine if they have increased or decreased in size. This information will enable you to estimate the osmolarity (solute concentration) of the potato tuber tissue.

## Question

What question is being investigated in this experiment?

*What is the osmolarity of potato tuber tissue?*

## Hypothesis

Hypothesize about the osmolarity of potato tuber tissue.

*The osmolarity of potato tuber cells is approximately equivalent to that of a 0.3 M sucrose solution. (Accept any testable hypothesis.)*

## Prediction

Predict the results of the experiment based on your hypothesis (if/then).

*If the osmolarity of potato tuber tissue cells is equivalent to that of a 0.3 M solution, then the potato samples should shrink in 0.4, 0.5, and 0.6 M solutions. They should not change size in a 0.3 M solution and should enlarge in 0, 0.1, and 0.2 M solutions.*

## Procedure

*We require that all students in the class learn how to use the vernier caliper.*

1. Practice measuring with the vernier caliper (Figure 3.6a, b).
   a. Identify the following parts of the caliper and add these labels on Figure 3.6a: *stationary arm, movable arm, ruler, vernier scale.* Notice that the numbers on the bottom ruler scale are centimeters; each graduated line is 1 mm.
   b. Choose a small object (a coin will work) and place it between the two arms, adjusting the movable arm until both arms just touch the object.
   c. Note the 0 mark on the vernier scale (Figure 3.6b). The graduated line on the ruler just to the left of the 0 mark is the distance between

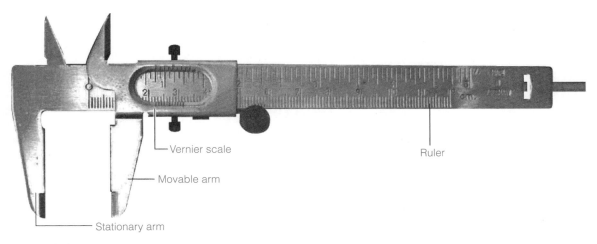

**Figure 3.6a.**
**Vernier caliper.** Identify the stationary arm, movable arm, ruler, and vernier scale.

the caliper arms measured in whole millimeters. In Figure 3.6b, that number is 22 mm. Write that number for your object as the answer in blank (1), below.

d.  Look at the graduated lines between 0 and 10 on the vernier scale. Note the line on the vernier scale that exactly matches with a line on the ruler. That line on the vernier scale is the measurement in tenths of a millimeter, which should be added to the whole-millimeter reading. In Figure 3.6b, that number is 4. Write the measurement in tenths of a millimeter for your object as the answer in blank (2) below.

What is the size of your object?

(1) _____

(2) _____

Total measurement: _____

When you know how to measure using the caliper, proceed to the next step.

2.  Obtain 100 mL of DI water and 100 mL of each of the sucrose solutions. Put each solution in a separate, appropriately labeled 250-mL beaker or paper cup.

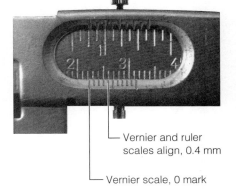

Vernier and ruler scales align, 0.4 mm

Vernier scale, 0 mark

**Figure 3.6b.**
**Enlarged vernier scale.** The correct measurement is 22.4 mm.

 Use cork borers and razor blades with extreme care! To use the cork borer, hold the potato in such a way that the borer will not push through the potato into your hand.

3.  Use a sharp cork borer to obtain seven cylinders of potato. Push the borer through the length of the potato, twisting it back and forth. When the borer is filled, remove it from the potato and push the potato cylinder out of the borer. You must have seven complete, undamaged cylinders at least 5 cm long.

4.  Line up the potato cylinders and, using a sharp razor blade, cut all cylinders to a uniform length, about 5 cm, removing the peel from the ends.

5. Place all seven potato samples in a petri dish, and keep them covered to prevent their drying out.

 In subsequent steps, treat each sample individually. Work quickly. To provide consistency, each person should do one task to all cylinders (one person wipe, another measure, another record data).

6. Remove a cylinder from the petri dish, and place it between the folds of a paper towel to blot sides and ends.
7. Using the caliper, measure the length and diameter of the cylinder to the nearest 0.1 mm, and record these measurements in Table 3.6 in the Results section. To measure, both arms of the caliper should touch but not compress the cylinder.
8. Transfer the cylinder to the 0 M (water) beaker.
9. Note the time the cylinder is placed in the 0 M beaker. Time: _____.
10. Repeat steps 6 to 8 with each cylinder, placing the cylinders in the appropriate incubating solution from 0.1 to 0.6 M.

 Be sure that the initial length and diameter of the cylinder placed in each test solution are accurately recorded.

11. Incubate from 1.5 to 2 hours. (During this time period, you will be performing other lab activities.)
12. Swirl each beaker every 10 to 15 minutes as the cylinders incubate.
13. At the end of the incubation period, record the time each cylinder is removed from a solution. Time: _____.

    Calculate the approximate incubation time in Table 3.6.
14. Remove the cylinders in the chronological order in which they were initially placed in the test solutions.
15. Blot each cylinder as it is removed (sides and ends), and use the vernier caliper to measure the length and diameter to the nearest 0.1 mm.
16. Finish recording your data in the Results section, and answer the questions in the Discussion section.

## Results

1. Complete Table 3.6. To calculate the volume of a cylinder, use this formula:

$$\text{Volume of a cylinder (mm}^3) = \pi(\text{diameter}/2)^2 \times \text{length}$$
$$(\pi = 3.14)$$

To calculate percentage change in volume, use this formula:

$$\text{Percentage change in volume} = \frac{\text{change in volume}}{\text{initial volume}} \times 100$$

If the sample increases in volume, the value will be positive. If it decreases in volume, the value will be negative.

**Table 3.6**

Data for Experiment Estimating Osmolarity by Change in Volume

| Approximate time in solutions: _____ | | | | | | | |
|---|---|---|---|---|---|---|---|
| | **Sucrose Molarity** | | | | | | |
| | **0.0** | **0.1** | **0.2** | **0.3** | **0.4** | **0.5** | **0.6** |
| **Final diameter (mm)** | | | | | | | |
| **Final length (mm)** | | | | | | | |
| **Final volume (mm³)** | | | | | | | |
| **Initial diameter (mm)** | | | | | | | |
| **Initial length (mm)** | | | | | | | |
| **Initial volume (mm³)** | | | | | | | |
| **Change in volume (mm³)** | | | | | | | |
| **% change in volume** | | | | | | | |

2. Plot percentage change in volume as a function of the sucrose molarity in Figure 3.7.

   a. Place a 0 in the middle of the y axis. Choose appropriate scales.

   b. Label the axes of the graph: Determine dependent and independent variables, and place each on the appropriate axis (see Lab Topic 1).

   c. Graph your results. Volume increase should be above the zero change line on the "percentage change in volume" axis. Volume decrease should be below the zero change line.

*Be sure that students follow these directions carefully because the 0 is in the middle of the y axis*

*This is the first graph that students will create since they learned about graphs in Lab Topic 1. Check their graph design and figure title before they proceed with the discussion questions.*

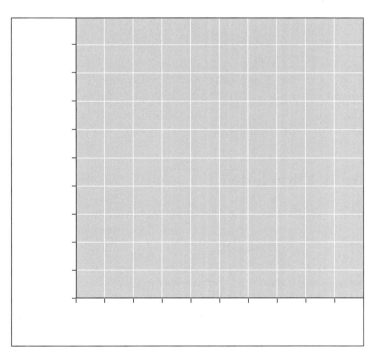

Figure 3.7.

_____

_____

_____

d. Construct a curve that best fits the data points. Use this curve to estimate the osmolarity of the potato tuber.

e. Compose an appropriate figure title.

### Discussion

1. At what sucrose molarity does the curve cross the zero change line on the graph?

   *The point at which this happens usually falls between 0.2 and 0.3 M. It can vary, depending on the variety and age of the potato. Older potatoes lose water and some varieties contain more water than others.*

2. Explain how this information can be used to determine the osmolarity of the potato tuber tissue.

   *If the weight does not change at 0.3 M sucrose, then water was neither lost nor gained. At this point the pressure of the cell wall is negligible. Water potential and osmolarity are essentially equal. This molarity is, therefore, equivalent to the osmolarity of the potato tuber tissue.*

3. In more dilute concentrations of sucrose, the volume of the potato pieces _____ (increases/decreases) after incubation. What forces other than solute concentration will have an impact on the amount of water taken up by the pieces?

   *When a plant cell is bathed in a hypotonic solution, the cell wall creates (turgor) pressure that has an impact on the amount of water crossing the cell membrane.*

4. Estimate the osmolarity of the potato tuber tissue.

## Reviewing Your Knowledge

1. Once you complete this lab topic, you should be able to define and use the following terms. Provide examples if appropriate.

   *selectively permeable, solvent, solute, diffusion, osmosis, nonpenetrating solute, hypotonic, hypertonic, isotonic, turgor pressure, osmolarity, water potential, Brownian movement, lysis, crenate, plasmolysis, plasmolyzed, turgid*

2. Compare the response of plant and animal cells placed in hypertonic, isotonic, and hypotonic solutions.

## Applying Your Knowledge

1. Unlike animals, plants never absorb enough water that their cells burst, but they frequently lose enough water to wilt. Describe plant wilting in terms of turgor pressure. What is the optimum environment for growing plants?

   *As water evaporates outside the cells, water will diffuse out and evaporate, reducing turgor pressure in the cells. This leads to the collapse of cells, and the leaf will fold and drop. Plants grow best in a hypotonic environment. This provides maximum turgor pressure, which gives the plant support and allows optimum water movement.*

2. The pond water samples you observed in Lab Topic 2, Lab Study D probably contained a variety of multicellular, colonial, and single-celled organisms. Some of these may have had cell walls, but others were lacking cell walls. What adaptations for osmoregulation are found in single-celled organisms, such as the *Amoeba,* and multicellular organisms that lack cell walls but live in a hypotonic environment?

*Students will remember the contractile vacuoles of the amoeba and speculate about similar organelles in other single-celled organisms. For multicellular organisms, they may suggest that they have kidneys or other excretory organs.*

3. Shrimp fishing off the coast of Georgia was closed in 2001, due to a drastic reduction in the shrimp population. Harvest of blue crab plummeted in 2002 and 2003 following 5 years of drought. In 2008 and 2009, however, rainfall amounts reached near normal and the shrimp population returned to predrought numbers. Blue crab harvest has more than doubled compared with 2002 totals. Rainfall amounts in upland regions of the state have a significant effect on the salinity of coastal estuaries, regions where open ocean and fresh water from underground aquifers and rivers mix. These estuaries are the "nurseries" for many marine animals. Speculate about possible causes for the decline of shrimp in 2001 and blue crab in 2002 and 2003, and the recovery of these species in 2009.

*Blue crab and shrimp spend much of their life cycle in estuaries. Drought leads to an increase in salinity in estuaries resulting in high mortality in larval stages and drastic reductions in populations of these organisms, two of the most important fisheries on the Georgia coast.*

4. Each year in the U.S., millions of people become ill and thousands die from eating food contaminated with bacteria. Controlling the growth of microorganisms on food is a particular challenge and over the course of many centuries various methods have been developed for controlling microorganisms in food. Some of these include preservation by irradiation, drying, freezing, canning, and salting. Preservation by salting is based on the principle of osmosis. One example of this is the preservation of ham by applying large amounts of salt ("salt-cured") or sugar ("sugar-cured") to the meat within 48 hours after slaughter. Using information you learned in this lab, speculate about how this process works to preserve the ham.

*A salt or salt-sugar mixture is rubbed over the ham. This causes water to move from the higher water concentration in microorganisms on the surface of the meat to the high salt concentration and lower water concentration, dehydrating the microorganisms and killing them.*

## Investigative Extensions

1. Organisms that live in marine environments are described as being *euryhaline* (able to live in waters of a wide range of salinity) or *stenohaline* (unable to withstand wide variation in salinity of the surrounding water). This characteristic often determines the range of habitat for a marine animal. For example, one would predict that a sessile (attached) marine invertebrate growing on a pier in a tidal river where

*See Prep Guide for suggested vendors.*

the salinity changes as the tide floods and ebbs is more euryhaline than an organism living in the open ocean.

Design an experiment to test the range of tolerance of two or more marine invertebrates (for example, barnacles, sea squirts, small crabs, marine mussels, periwinkles) available from biological supply houses. Determine if they may be described as euryhaline or stenohaline organisms.

2. General science teachers have long known that white vinegar (or a 10% acetic acid solution) will dissolve the shell of a chicken egg, leaving intact the membranes surrounding the albumin and yolk.

   a. Design and perform an experiment to allow you to estimate the *osmolarity* of the chicken egg cell. Hint: It will take 24 to 36 hours to completely remove the shell. If the process goes too slowly, move the eggs to a fresh acid solution after about 12 hours.

   b. Design and perform an experiment to answer the question, "Does the *concentration* of the solute have an effect on the *rate of movement* of the solvent (water) across the membrane of a chicken egg?"

   c. Design and perform an experiment to answer the question, "Does temperature have an effect on the *rate* of osmosis?"

## Student Media: BioFlix, Activities, Investigations, and Videos
**www.masteringbiology.com (select Study Area)**

**BioFlix**—Ch. 7: Membrane Transport
**Activities**—Ch. 1: Graph It! An Introduction to Graphing
   Ch. 7: Membrane Structure; Selective Permeability of Membranes; Diffusion; Osmosis and Water Balance in Cells
**Investigations**—Ch. 7: How Do Salt Concentrations Affect Cells?
**Videos**—Ch. 7: Turgid *Elodea;* Plasmolysis

## References

Exercise 4.1, Experiment A, was adapted from D. R. Helms and S. B. Miller, *Principles of Biology: A Laboratory Manual for Biology 110.* Apex, NC: Contemporary Publishing, 1978. Used by permission.

Lang, F., and S. Waldegger. "Regulating Cell Volume." *American Scientist,* 1997, vol. 85, pp. 456–463.

Reece, J., et al. *Campbell Biology,* 9th ed. San Francisco, CA: Pearson Education, 2011.

## Websites

A discussion of osmosis and diffusion in kidney dialysis:
http://www.rockwellmed.com/hemodefn.htm

A discussion of reverse osmosis and its applications:
http://en.wikipedia.org/wiki/Reverse_osmosis

Includes sections entitled "Real-Life Applications" and Osmosis and Medicine:
http://www.answers.com/topic/osmosis

# LAB TOPIC 3

# Diffusion and Osmosis
## Teaching Plan for Laboratories

## Main Concepts and Objectives

1. Concept: diffusion. Students will describe the mechanism of diffusion at the atomic level and list factors that influence the rate of diffusion. (Exercise 3.1)

2. Concept: osmosis. Students will explain that the movement of water in and out of cells through the selectively permeable cell membrane is due to differences in osmolarity of solutions on either side of the membrane. (Exercise 3.1)

3. Concept: pressure in the cell influencing movement of water by osmosis. Students will compare differences in responses of plant and animal cells placed in different molar solutions. They will explain the impact of the cell wall presence in plant and algal cells. (Exercise 3.2)

4. Concept: osmolarity. Students will determine the osmolarity of potato tuber cells and explain the concept. (Exercise 3.3)

5. Concept: scientific method. Students will propose hypotheses, make predictions, collect data, determine dependent and independent variables, plot data points, and come to conclusions based on results. (Exercises 3.1–3.3)

6. Specific content. Students will be able to describe or define: *selectively permeable membrane* (allows movement of water and other selected solutes; some texts use *semipermeable membrane*, which refers to a membrane that allows only water to cross), *solvent, solute, diffusion, osmosis, nonpenetrating solutes* (those solutes that cannot cross a membrane), *hypotonic, hypertonic, isotonic, turgor pressure, osmolarity, Brownian movement, plasmolysis, lysis, crenate.*

## Order of the Lab

1. Begin potato osmolarity experiment (Exercise 3.3). The potato samples should incubate for at least 1.5 to 2 hours. As they incubate, proceed with the introduction and other experiments and observations.    (30 min)

2. Introduce concepts and objectives.    (20 min)

3. Begin dialysis experiment (Exercise 3.1, Experiment B). Setup will take about 10 minutes. The experiment will run concurrently with the later exercises.    (15 min)

4. Observe carmine slide (Exercise 3.1, Experiment A).    (10 min)

5. Observe red blood cells of oxen.    (20 min)

6. Observe demonstration of *Elodea* slides (Exercise 3.2, Experiment B).    (10 min)

7. Collect and process data from potato osmolarity experiment.    (45 min)

8. Students report on results of potato osmolarity experiment.    (20 min)

9. Assignment. Instruct students on writing a Discussion section describing the potato osmolarity experiment.    (10 min)

**For a 2-hour lab:** Begin the potato osmolarity experiment immediately. Omit student reports (step 8). Have students collect their data and then make their graphs as an outside assignment to be handed in at the next lab period. Assign the Discussion section of the writing program (step 9) using a handout.

As students wait for the potato tuber pieces to incubate, they can observe the carmine slide and perform the ox blood experiment. The *Elodea* is a demonstration.

## Classroom Management

Success in these exercises will depend on the organizational and laboratory skills of your students. Stress that the experiments will require that they concentrate, take careful measurements, and keep on task. Students work independently on the ox blood and *Elodea* experiments. Students work in pairs for the carmine/Brownian movement observation. All other activities are in teams of four students.

For the potato osmolarity experiment (Exercise 3.3), divide students into teams of four. Half of the teams will study the change in weight of the potato samples (Experiment A), half the change in volume (Experiment B). Only the students carrying out Experiment B will actually use the vernier caliper; however, all students must learn how to use it.

After both experiments are completed, students should record the data for both experiments. You can facilitate this by having a blank transparency on the overhead projector. A volunteer from each team should fill in the results for his or her team, and other students should copy their results.

For slides of ox blood, students work independently and must work quickly. The cells burst (hemolyse) very rapidly.

## Student Development

Science educators agree that for the maximum pedagogical impact, science curricula must begin to integrate math and the sciences. This lab topic accomplishes this goal perhaps better than any other in this manual. Students are required to use math, physics, and chemistry to understand osmosis and diffusion, important concepts that underlie many biological processes. Some students are uncomfortable with studies such as these, which require them to analyze, integrate, and organize. As an instructor, resist the temptation to simplify labs that require more instructor and student preparation and participation. Teaching students to integrate and synthesize concepts should be an important objective of any science laboratory course and is an important objective of this lab.

Students continue to practice scientific processes, proposing hypotheses, making predictions, testing hypotheses, and recording and processing results. Students will continue to develop laboratory skills, such as learning microscopical technique, using an electronic balance, and using a vernier caliper. Students practice organizational skills, teamwork, and communication of scientific discovery to other scientists. Writing the Discussion section to be turned in for evaluation will give students practice in scientific writing and continue to prepare them to write a complete scientific paper.

## Lab Safety Precautions

1. *Trimming potato cylinders.* Use only single-edged razor blades and warn students to use them with extreme care. Have disposable gloves in the room, and use them if it is necessary to clean up any blood. Use paper towels and a disinfectant such as 10% Clorox to clean up blood. Discard the paper towels and the gloves in a plastic bag, and close the bag with a twist tie. Dispose of this bag as your institution suggests.

2. *Using cork borers.* Warn students that *cork borers* can cause cuts if not used correctly. Students should hold the potato in such a way that the borer will not push through the potato into their hand.

3. *Using hot plates.* Students should avoid contacting the hot surface. Students cannot tell by appearance if the hot plate is hot. (They expect it to turn red as a stove burner does.) A beaker of water on the hot plate will usually steam, providing a visual cue that the hot plate is hot.

## Discussion and Summary

Discuss results of each experiment if time permits. Choose two students to report the results of the potato osmolarity experiment. One student should report the results of Experiment A, and one, of Experiment B. Ask if other teams have conflicting results. Discuss possible sources of error. Be ready to ask questions and to clarify any misconceptions about the concepts or conclusions.

## Evaluation

Ask students to submit a discussion of the results of the potato osmolarity experiment, either Experiment A or Experiment B, to be graded. By the end of the term, students will have submitted Results, Discussion, Materials and Methods, and Introduction sections in scientific paper format.

## Lab Topic 4
# Enzymes

## Laboratory Objectives

After completing this lab topic, you should be able to:

1. Define *enzyme* and describe the activity of enzymes in cells.
2. Differentiate competitive and noncompetitive inhibition.
3. Discuss the effects of varying environmental conditions such as pH and temperature on the rate of enzyme activity.
4. Discuss the effects of varying enzyme and substrate concentrations on the rate of enzyme activity.
5. Discuss the scientific process, propose questions and hypotheses, and make predictions based on hypotheses and experimental design.
6. Practice scientific thinking and communication by constructing and interpreting graphs of enzyme activity.

*For a 2-hour lab: Omit student reports and the instructor's reviews of the scientific method and scientific writing. The assignment could be provided as a handout.*

*The success of this lab topic depends on using clean glassware, fresh enzymes, and careful preparation of all materials and solutions. See the latest* Preparation Guide *for Investigating Biology for prep instructions.*

## Introduction

Living cells perform a multitude of chemical reactions very rapidly because of the participation of enzymes. **Enzymes** are biological **catalysts**, compounds that speed up a chemical reaction without being used up or altered in the reaction. The material with which the catalyst reacts, called the **substrate**, is modified during the reaction to form a new product (see Figure 4.1). But because the enzyme itself emerges from the reaction unchanged and ready to bind with another substrate molecule, a small amount of enzyme can alter a relatively enormous amount of substrate.

The **active site** of an enzyme will bind with the substrate, forming the **enzyme-substrate complex.** It is here that catalysis takes place, and when it is complete, the complex dissociates into enzyme and product or products.

Enzymes are, in part or in whole, proteins and are highly specific in function. Because enzymes lower the energy of activation needed for reactions to take place, they accelerate the rate of reactions. They do not, however, determine the direction in which a reaction will go or its final equilibrium.

Enzyme activity is influenced by many factors. Varying environmental conditions, such as pH or temperature, may change the three-dimensional shape of an enzyme and alter its rate of activity. Specific chemicals may also bind to an enzyme and modify its shape. Chemicals that must bind for the enzyme to be active are called **activators. Cofactors** are nonprotein substances that usually bind to the active site on the enzyme and are essential for the enzyme to work.

*Remind students that enzyme reactions can be either catabolic or anabolic. All of the examples in this lab topic are catabolic reactions.*

*Instructors and/or students can download data tables in Excel format. Go to www .masteringbiology.com. Look in the Study Area under Lab Media.*

**Figure 4.1.**
**Enzyme activity.** A substrate or substrates bind to the active site of the enzyme, forming the enzyme-substrate complex, which then dissociates into enzyme and product(s). The enzyme may catalyze the addition or removal of a molecule or a portion of a molecule from the substrate to produce the product (a), or the enzyme may catalyze the splitting of a substrate into its component subunits (b).

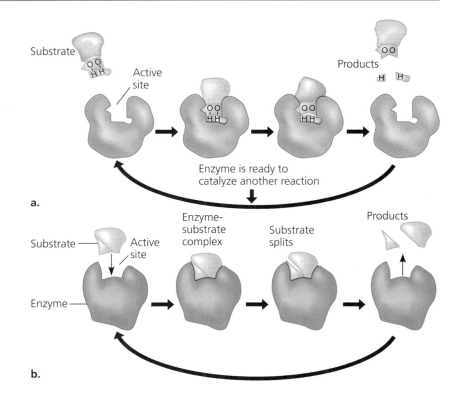

Organic cofactors are called **coenzymes,** but other cofactors may simply be metal ions. Chemicals that shut off enzyme activity are called **inhibitors,** and their action can be classified as **competitive** or **noncompetitive inhibition.**

Review Figure 4.1, illustrating enzyme activity. There are two ways to measure enzyme activity: (1) Determine the rate of disappearance of the substrate, and (2) determine the rate of appearance of the product.

In this laboratory, you will use both methods to investigate the activity of two enzymes, **catechol oxidase** and **amylase.** You will use an inhibitor to influence the activity of catechol oxidase and determine if it is a competitive or noncompetitive inhibitor. Additionally, you will investigate the effect of changing environmental conditions on the rate of amylase activity.

EXERCISE 4.1

# Experimental Method and the Action of Catechol Oxidase

## Materials

| | |
|---|---|
| test-tube rack | pipette filler |
| 3 small test tubes | pipette bulb |
| small Parafilm™ squares | distilled or deionized (DI) water |
| calibrated 5-mL pipette | potato extract |
| 3 calibrated 1-mL pipettes | catechol |
| disposable pasteur pipettes | disposable gloves (optional) |

# Introduction

This exercise will investigate the result of catechol oxidase activity. In the presence of oxygen, catechol oxidase catalyzes the removal of electrons and hydrogens from **catechol,** a phenolic compound found in plant cells. Catechol is converted to benzoquinone, a pigment product. The hydrogens combine with oxygen, forming water (Figure 4.2). The pigment products are responsible for the darkening of fruits and vegetables, such as apples and potatoes, after exposure to air.

In this exercise you will use an extract of potato tuber to test for the presence of catechol oxidase and to establish the appearance of the products when the reaction takes place.

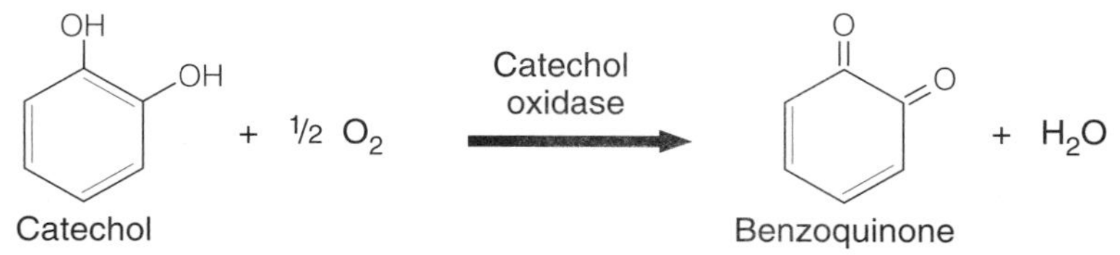

*Additional names for catechol oxidase include tyrosinase and diphenol oxidase. Catechol is also called 1,2 dihydroxybenzene.*

**Figure 4.2.**
**The oxidation of catechol.** In the presence of catechol oxidase, catechol is converted to benzoquinone. Hydrogens removed from catechol combine with oxygen to form water.

# Question

Remember that every experiment begins with a question. Review the information given above about the activity of catechol oxidase. You will be performing an experiment using potato extract.

Formulate a question about catechol oxidase and potato extract. The question may be broad, but it must propose an idea that has measurable and controllable elements.

*In this and the previous lab topics we formally included space and directions for stating the question for each experiment. In subsequent labs, encourage students to pose questions before stating hypotheses. These may be written in the margin of the lab manual.*

> *Catechol oxidase is present in many plants. Is it present in potato extract?*

# Hypothesis

Construct a hypothesis for the presence or absence of catechol oxidase in potato extract. Remember, the hypothesis must be testable. It is possible for you to propose one or more hypotheses, but all must be testable.

*Review criteria for "good" hypotheses and predictions, if needed.*

> *Catechol oxidase is present in potato extract.*

# Prediction

Predict the result of the experiment based on your hypothesis. To test for the presence or absence of catechol oxidase in potato extract, your prediction would be what you expect to observe as the result of this experiment (if/then).

> *If catechol oxidase is present in potato extract, then a pigment will develop in a mixture of potato extract and catechol.*

Catechol is a poison! Avoid contact with all solutions. Do not pipette any solutions by mouth. Wash hands thoroughly after each experiment. If a spill occurs, notify the instructor. If the instructor is unavailable, wear disposable gloves and use dry paper towels to wipe up the spill. Follow dry towels with towels soaked in soap and water. Dispose of all towels in the trash.

## Procedure

1. Using Table 4.1, prepare the three experimental tubes. Note that all tubes should contain the same total amount of solution. Do not cross-contaminate pipettes! After each tube is prepared, use your finger to hold a Parafilm™ square securely over the tube mouth and then rotate the tube to mix the contents thoroughly. Use a fresh square for each tube.

*Demonstrate how to use a calibrated pipette and pipette filler.*

**Table 4.1**
Contents of the Three Experimental Tubes

| Tube | Distilled Water | Catechol | Distilled Water | Potato Extract |
|------|-----------------|----------|-----------------|----------------|
| 1 | 5 mL | 0.5 mL (10 drops) | 0.5 mL (10 drops) | — |
| 2 | 5 mL | 0.5 mL (10 drops) | — | 0.5 mL (10 drops) |
| 3 | 5 mL | — | 0.5 mL (10 drops) | 0.5 mL (10 drops) |

*Less expensive pasteur pipettes may be substituted for calibrated pipettes. In this case, use 0.5 mL = 10 drops.*

2. Explain the experimental design: What is the purpose of each of the three test tubes? Which is the control tube? Is more than one control tube necessary? Explain. Which is the experimental tube? Why is an additional 0.5 mL of distilled water added to tubes 1 and 3, but not tube 2?

   1. *Control, no enzyme*

   2. *Experimental tube*

   3. *Control, no substrate*

3. Observe the reactions in the tubes, and record your observations in the Results section below. Explain your conclusions in the Discussion section.

## Results

Design a simple table to record results (Table 4.2).

## Discussion

Explain your results in terms of your hypothesis.

*The results indicate that catechol oxidase is present in the potato extract. This conclusion is supported by the development of pigment in the presence of catechol.*

**Table 4.2**
Results of Catechol Oxidation Experiment

| Tube | Color (+) or (−) |
|------|------------------|
| 1 | − |
| 2 | + |
| 3 | − |

---

## EXERCISE 4.2
# Inhibiting the Action of Catechol Oxidase

---

### Materials

test-tube rack
3 small test tubes
small Parafilm™ squares
calibrated 5-mL pipette
4 calibrated 1-mL pipettes
disposable pasteur pipettes

pipette bulb
distilled water
potato extract
catechol
phenylthiourea (PTU)
disposable gloves (optional)

### Introduction

This exercise will investigate the inhibition of enzyme activity by specific chemicals called **inhibitors.** The specific inhibitor used will be **phenylthiourea (PTU).** To be active, catechol oxidase requires copper as a cofactor. PTU is known to combine with the copper in catechol oxidase and inhibit its enzymatic activity.

An inhibitor molecule affects an enzyme in one of two ways. **Competitive inhibition** takes place when a molecule that is structurally similar to the substrate for a particular reaction competes for a position at the active site on the enzyme. This ties up the enzyme so that it is not available to the substrate. Competitive inhibition can be reversed if the concentration of substrate is raised to sufficiently high levels while the concentration of the inhibitor is held constant (Figure 4.3).

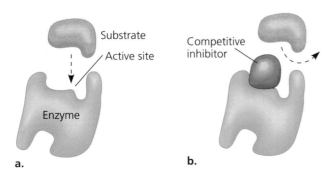

**Figure 4.3.**
**Action of a competitive inhibitor.**
(a) Substrate normally can bind to the active site of an enzyme.
(b) A competitive inhibitor mimics the substrate and competes for the position at the active site on the enzyme.

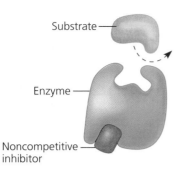

Substrate

Enzyme

Noncompetitive inhibitor

**Figure 4.4.**

**Action of a noncompetitive inhibitor.** The noncompetitive inhibitor binds to the enzyme at a location away from the active site, either blocking access to the active site or changing the conformation of the enzyme, rendering it inactive.

In **noncompetitive inhibition,** the inhibitor binds to a part of the enzyme that is *not* the active site. In so doing, it changes the nature of the enzyme so that its catalytic properties are lost. This can happen in two ways. Either the noncompetitive inhibitor itself physically blocks the access to the active site, or it causes a conformational change in the protein, thus inactivating the active site. In noncompetitive inhibition the inhibitor can become unbound, reversing the inhibition. However, unlike competitive inhibition, adding additional substrate will not reverse the inhibition (Figure 4.4).

In the following experiment, you will determine if PTU is a competitive or noncompetitive inhibitor.

### Question

Pose a question about the activity of PTU.

*What type of inhibitor is PTU, competitive or noncompetitive?*

### Hypothesis

Hypothesize about the nature of inhibition by PTU.

*PTU is a noncompetitive inhibitor. Students may hypothesize about whether it is competitive or noncompetitive. Do not correct them.*

### Prediction

Predict the results of the experiment based on your hypothesis (if/then).

*If PTU is a noncompetitive inhibitor, then increasing the concentration of substrate will not reverse the inhibition and the color of tubes 1 and 2 will be the same. The prediction should match the hypothesis.*

### Procedure

 PTU and catechol are poisons! Avoid contact with solutions. Do not pipette any solutions by mouth. Wash hands thoroughly after the experiment. If a spill occurs, notify the instructor. If the instructor is unavailable, wear disposable gloves and use dry paper towels to wipe up the spill. Follow dry towels with towels soaked in soap and water. Dispose of all towels in the trash.

1. Using Table 4.3, prepare three experimental tubes. Be sure to add solutions in the sequence given in the table (water first, potato extract next, PTU next, etc.). Cover each tube with a fresh Parafilm™ square and mix.

*Warn students about mixing pipettes. Tell them to return pipettes to the appropriate solution or discard them if in doubt. You may choose to label pipettes for reuse.*

*Less expensive pasteur pipettes may be used for 1-mL calibrated pipettes. In this case, use 0.5 mL = 10 drops.*

**Table 4.3**
Contents of the Three Experimental Tubes

| Tube | Distilled Water | Potato Extract | PTU | Distilled Water | Catechol |
|------|-----------------|----------------|--------|-----------------|----------|
| 1 | 5 mL | 0.5 mL | 0.5 mL | 0.5 mL | 0.5 mL |
| 2 | 5 mL | 0.5 mL | 0.5 mL | — | 1 mL |
| 3 | 5 mL | 0.5 mL | — | 1 mL | 0.5 mL |

2. Which test tube is the control?

   *tube 3 (no inhibitor; pigment will develop; positive control)*

3. Why was the concentration of catechol increased in test tube 2?

   *to test whether increased concentration of the substrate will overcome inhibition*

4. Why should the catechol be added to the test tubes last?

   *The inhibitor must be added first; otherwise, the reaction takes place regardless of the presence of the inhibitor.*

5. Record your observations in the Results section, and explain your results in the Discussion section.

## Results

Design a table to record your results (Table 4.4).

**Table 4.4**
Results of Inhibition Experiment

| Tube | Color (+) or (−) |
|:---:|:---:|
| 1 | − |
| 2 | − |
| 3 | + |

*Have students hold the three test tubes together to compare colors. Test tubes 1 and 2 will be noticeably similar and lighter than test tube 3, indicating that the additional substrate in test tube 2 did not overcome the inhibition.*

## Discussion

1. Explain your results in terms of your hypothesis.

   *PTU inhibition is noncompetitive. Increasing substrate concentration does not reduce inhibition.*

2. One member of your team is not convinced that you have adequately tested your hypothesis. How could you expand this experiment to provide additional evidence to strengthen your conclusion?

   *Students will suggest that the increase in catechol concentration in test tube 2 is inadequate to give conclusive results. They may suggest additional treatments with increased substrate.*

EXERCISE 4.3

# Influence of Concentration, pH, and Temperature on the Activity of Amylase

## Introduction

In the following exercise, you will investigate the influence of enzyme concentration, pH, and temperature on the activity of the enzyme **amylase.** Amylase is found in the **saliva** of many animals, including humans, that utilize **starch** as a source of food. Starch, the principal reserve carbohydrate stores of plants, is a polysaccharide composed of a large number of glucose monomers joined together. Amylase is responsible for the preliminary digestion of starch. In short, amylase breaks up the chains of glucose molecules in starch into maltose, a two-glucose-unit compound. Further digestion of this disaccharide requires other enzymes present in pancreatic and intestinal secretions. To help us follow the digestion of starch into maltose by salivary amylase, we will take advantage of the fact that starch, but not maltose, turns a dark purple color when treated with a solution of $I_2KI$ (this solution is normally yellow-amber in color). Draw equations to help you remember these reactions in the margin of your lab manual.

*Starch + $I_2KI$ → purple-black*
*Maltose + $I_2KI$ → yellow-amber*

In the following experiments, the **rate of disappearance** of starch in different amylase concentrations allows a quantitative measurement of reaction rate. Recall that the rate of appearance of the product (in this case, maltose) would give the same information, but the starch test is simpler.

You will be assigned to a team with three or four students. Each team will carry out only one of the experiments. However, each student is responsible for understanding all experiments and results. Be prepared to present your results to the entire class. Your instructor may require you to write a component of a scientific paper. (See Appendix A.)

## Experiment A. The Influence of Enzyme Concentration on the Rate of Starch Digestion

### Materials

| | |
|---|---|
| test-tube rack | 1 calibrated 1-mL pipette |
| 10 standard test tubes | 2 calibrated 5-mL pipettes |
| wax pencil | disposable pasteur pipettes |
| test plate | pipette bulb |
| flask of distilled or DI water | buffer solution (pH = 6.8) |
| beaker of distilled or DI | $I_2KI$ solution |
|    rinse water | 1% starch solution |
| 5-mL graduated cylinder | 1% amylase solution |

### Introduction

In this experiment you will vary the concentration of the enzyme amylase to determine what effect the variation will have on the rate of the reaction. You will make serial dilutions of the amylase resulting in a range of enzyme concentrations. For serial dilutions, you will take an aliquot (sample) of the original enzyme and dilute it with an equal amount of water for a 1 : 1 dilution (50% of the original concentration). You will then take an aliquot of the resulting 1 : 1 solution and add an equal amount of water for a 1 : 3 dilution of the original concentration. You will continue this series of dilutions until you have four different amylase concentrations.

## Question

Pose a question about enzyme concentration and reaction rate.

> *What effect does enzyme concentration have on enzymatic reaction rate?*

## Hypothesis

Hypothesize about the effect of changing enzyme concentration on the rate of reaction.

> *As enzyme concentration increases, the rate of reaction increases. Other hypotheses are acceptable if testable.*

## Prediction

Predict the results of the experiment based on your hypothesis (if/then).

> *If the concentration of the enzyme is increased, then the rate of starch disappearance will increase.*

## Procedure

1. Prepare the amylase dilution (test-tube set 1):
   a. Number five standard test tubes 1 through 5.
   b. Using the 5-mL graduated pipette, add 5 mL distilled water to each test tube.
   c. Make serial dilutions as follows (use the graduated cylinder):

      Tube 1: Add 5 mL amylase and mix by rolling the tube between your hands. (Dilution: 1 : 1; 0.5% amylase)

      Tube 2: Add 5 mL amylase solution from tube 1 and mix. (Dilution: 1 : 3; 0.25% amylase)

      Tube 3: Add 5 mL amylase solution from tube 2 and mix. (Dilution: 1 : 7; 0.125% amylase)

      Tube 4: Add 5 mL amylase solution from tube 3 and mix. (Dilution: 1 : 15; 0.063% amylase)

      Tube 5: Add 5 mL amylase solution from tube 4 and mix. (Dilution: 1 : 31; 0.031% amylase)

      Rinse the graduated cylinder thoroughly.

   *Students may be confused by the dilutions. A 1 : 7 dilution indicates 1 part of the original enzyme in a total volume of 8 parts. Note that concentrations are provided for ease in graphing.*

2. Prepare the experimental test tubes (test-tube set 2):
   a. Number a second set of five standard test tubes 1 through 5.
   b. Beginning with tube 5 of the first set, transfer 2 mL of this dilution into tube 5 of the second set. Use a 5-mL pipette for the transfer. Rinse the pipette in distilled water, and repeat the procedure for tubes 4, 3, 2, and 1, transferring 2 mL of tube 4 (first set) into tube 4 (second set), etc. After these transfers have been carried out, test-tube set 1 will no longer be used.

c. Add 40 drops of pH 6.8 buffer solution to each of the tubes in the second set. Mix by rolling the tubes between your hands. Set these tubes aside.

d. Add 1 or 2 drops of $I_2KI$ to each compartment of four rows of a test plate. You will use a separate row for each concentration of amylase.

*Remind students to begin with tube 5. This has the lowest concentration of enzyme and will take the longest to react. Proceeding so gives students an opportunity to see the starch test in advance of the most rapid reaction.*

e. Using the second set of tubes, proceed with the tests beginning with tube 5.

  (1) Using a clean 1-mL pipette, add 1 mL of the 1% starch solution to tube 5 and mix by rolling the tube between your hands. One team member should immediately record the time. This is time 0.

  (2) Quickly remove 1 drop of the mixture with a disposable pasteur pipette, and add it to a drop of $I_2KI$ in the first compartment on the test plate (time 0).

Remember, when the enzyme and substrate are together, the reaction has begun!

  (3) Sample the reaction mixture at 10-second intervals, *each time using a new compartment of the test plate.* Continue until a blue color is no longer produced and the $I_2KI$ solution remains yellow-amber (indicating the digestion of all the starch). Record the time required for the digestion of the starch in Table 4.5.

  (4) Repeat steps 1 through 4 for the other four concentrations (tubes 4, 3, 2, and 1 of set 2).

3. Finish recording your findings in the Results section, and state your conclusions in the Discussion section.

### Results

1. Complete Table 4.5 as you determine rates of digestion (time of starch disappearance) in different enzyme concentrations.

 **Table 4.5**
Time of Starch Disappearance in Different Concentrations of the Enzyme Amylase

| Tube | % Amylase | Time of Starch Disappearance (in seconds) |
|:---:|:---:|:---:|
| 1 | 0.50 | |
| 2 | 0.25 | |
| 3 | 0.125 | |
| 4 | 0.063 | |
| 5 | 0.031 | |

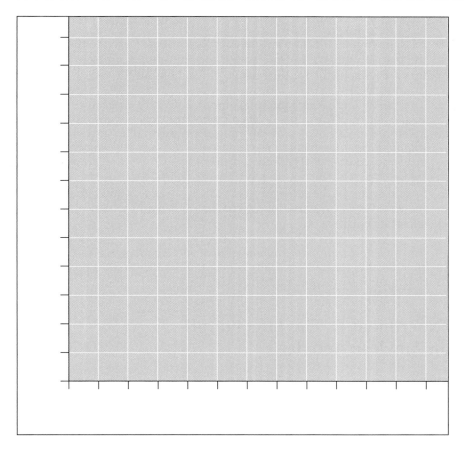

*Students' results should produce a curve with approximately this shape:*

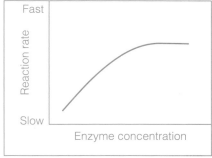

**Figure 4.5.**
**Enzyme reaction rate (time of starch disappearance) for different concentrations of amylase.**

2. Construct a graph (Figure 4.5) to illustrate your results. See Lab Topic 1 for assistance in graph construction.

   a. What is the independent variable? Which is the appropriate axis for this variable?

   *concentration of enzyme or amylase; the x axis (horizontal)*

   b. What is the dependent variable? Which is the appropriate axis for this variable?

   *reaction rate (time of starch disappearance); the y axis (vertical)*

   c. Label the axes of the graph. Using Table 4.5, note the maximum number of seconds in your results, and choose an appropriate scale for the dependent variable. Reaction rate, the dependent variable, was measured as time of starch (product) disappearance. The data must, therefore, be graphed in reverse order because the highest values indicate the slowest reaction rate. You should place "0" at the end of the axis and write "fast" by your 0. Place your highest number near the origin (where the x and y axes cross). Write "slow" near the origin. Choose an appropriate scale for the independent variable (% amylase), and label this axis.

*Since reversing the axis may be confusing to students, the instructor should check students' graphs before they proceed.*

## Discussion

1. Explain your results in terms of your hypothesis. Describe the shape of the reaction rate curve as substrate concentration increases. What factors are responsible for the change in reaction rate?

   *The curve increases rapidly as more substrate molecules are available. Then the curve levels off when the concentration of the substrate results in all enzyme molecules having their active sites engaged, and the maximum rate at which the enzyme can convert substrate to product is reached (saturation).*

2. Speculate about the shape of a curve measuring reaction rate if you had increased the concentration of enzyme, but held the concentration of substrate constant.

   *The reaction rate would increase until all substrate is used and then decline. Regardless of the concentration of enzyme, when the substrate is gone, the reaction stops.*

*Use 7 buffers to show the extremes. Approximate times for digestion:*

*pH 4 = > 420 sec*
 *5 = > 420*
 *6 = 70*
 *7 = 40*
 *8 = 270*
 *9 = > 420*

# Experiment B. The Effect of pH on Amylase Activity

## Materials

| | |
|---|---|
| test-tube rack | 1% amylase solution |
| 6 standard test tubes | $I_2KI$ solution |
| test plate | 1% starch solution |
| wax pencil | beaker of distilled or DI rinse water |
| pipette bulb | 6 buffer solutions |
| 3 5-mL calibrated pipettes |  (pH = 4, 5, 6, 7, 8, 9) |
| disposable pasteur pipettes | pH paper |

## Introduction

The environmental factor pH can influence the three-dimensional shape of an enzyme. Every enzyme has an optimum pH at which it is most active. In this experiment you will determine the optimum pH for the activity of amylase. What was the source of the amylase used in this experiment? (Check the Introduction to this exercise.)

*saliva*

## Question

Pose a question about pH and reaction rate.

*What effect does pH have on enzymatic reaction rate?*

## Hypothesis

*Students may choose to use pH paper to check the pH of saliva.*

Hypothesize about the rate of activity of amylase at various pHs.

*Hopefully, students will conclude that saliva is about pH 7 and will hypothesize that amylase will be most active at this neutral range. A number of hypotheses are acceptable if testable.*

## Prediction

Predict the results of the experiment based on your hypothesis (if/then).

*If amylase is most active at pH 7, then, in a given time, starch will disappear more rapidly at this pH than at lower or higher pHs.*

## Procedure

1. Using a wax pencil, number six standard test tubes 1 through 6. Beginning with tube 1 and pH 4, mark one tube for each pH of buffer (4, 5, 6, 7, 8, 9). After you mark the test tubes, use a 5-mL graduated pipette to add 5 mL of the appropriate buffer to each test tube (5 mL buffer 4 to tube 1, 5 mL buffer 5 to tube 2, etc.). Rinse the pipette with distilled water after dispensing each buffer.

Buffers can burn skin! Avoid contact with all solutions. Do not pipette any solutions by mouth. Wash hands thoroughly after each experiment. If a spill occurs, notify the instructor. If the instructor is unavailable, wear disposable gloves and use dry paper towels to wipe up the spill. Follow dry towels with towels soaked in soap and water. Dispose of all towels in the trash.

2. Using a clean 5-mL graduated pipette, add 1.5 mL amylase solution to each tube and mix by rolling the tubes in your hands.
3. Introduce 1 or 2 drops of $I_2KI$ into the compartments of several rows of the test plate.
4. Using only tube 1, add 2.5 mL of the 1% starch solution with a clean 5-mL pipette. Leave the pipette in the starch solution. Mix by rolling the tube in your hands. One team member should immediately record the time. This is time 0. Start testing immediately (next step).

Remember, when the enzyme and substrate are together, the reaction has begun!

5. Using a disposable pasteur pipette, remove a drop of the reaction mixture from tube 1. Add to a drop of $I_2KI$ on the test plate.
6. Sample the reaction mixture at 10-second intervals, *each time using a new compartment of the test plate.* Continue until a blue color is no longer produced and the $I_2KI$ solution remains yellow-amber (indicating the digestion of all the starch).
7. Record the time required for the digestion of the starch in Table 4.6. If after 7 minutes there is no color change, terminate the experiment with that reaction mixture.
8. Repeat steps 4 through 7 using the other five test tubes. Use separate rows on the test plate for each pH. Rinse the pipette between uses. Record results in Table 4.6.
9. Graph your observations in the Results section, and explain your results in the Discussion section.

### Results

1. Complete Table 4.6 as rates of digestion (time of starch disappearance) in different pHs are determined.

 **Table 4.6**

Time of Starch Disappearance in Different pH Environments for the Enzyme Amylase

| Tube | pH | Time of Starch Disappearance (in minutes) |
|------|-----|------------------------------------------|
| 1 | 4 | |
| 2 | 5 | |
| 3 | 6 | |
| 4 | 7 | |
| 5 | 8 | |
| 6 | 9 | |

2. Construct a graph using Figure 4.6 to illustrate your results. See Lab Topic 1 for assistance in graph construction.

   a. What is the independent variable? Which is the appropriate axis for this variable?

   *pH; the x axis*

*Students' results should produce a curve with approximately this shape:*

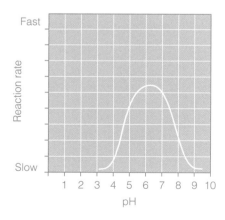

**Figure 4.6.**
**Reaction rate (time of starch disappearance) of amylase in different pH environments.**

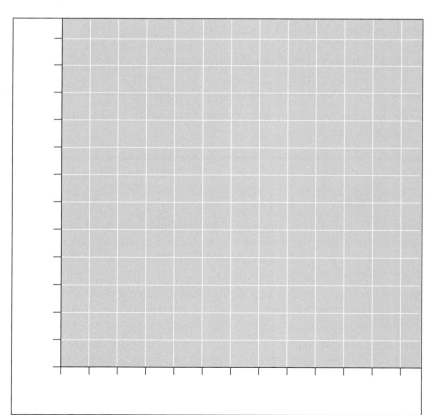

b. What is the dependent variable? Which is the appropriate axis for this variable?

*reaction rate (time of starch disappearance); the y axis*

c. Label the axes of the graph. Using Table 4.6, note the maximum number of minutes in your results, and choose an appropriate scale for the dependent variable. Reaction rate, the dependent variable, was measured as time of starch (product) disappearance. The data must, therefore, be graphed in reverse order because the highest values indicate the slowest reaction rate. You should place "0" at the end of the axis and write "fast" by your 0. Place your highest number near the origin (where the *x* and *y* axes cross). Write "slow" near the origin. Choose an appropriate scale for the independent variable (pH) and label the axis.

*Since reversing the axis may be confusing to students, the instructor should check students' graphs before they proceed.*

## Discussion

Explain your results in terms of your hypothesis. Describe the shape of the reaction rate curve obtained with change in pH. What factors are responsible for the shape of this curve?

*Enzymes have a 3-D structure (conformation) at which they are most active. That structure changes in different pH environments, resulting in a negative effect on enzyme activity. So the curve will peak at the pH where the shape of the enzyme is optimum.*

# Experiment C. The Effect of Temperature on Amylase Activity

## Materials

| | |
|---|---|
| 8 standard test tubes | buffer solution (pH = 6.8) |
| test-tube rack | 1% amylase solution |
| 2 5-mL calibrated pipettes | flask of DI water |
| 2 1-mL calibrated pipettes | On the demonstration table: |
| disposable 7.5-inch pasteur | water bath at 80°C |
|    pipettes | water bath at 37°C |
| pipette bulb | test-tube rack at room |
| wax pencil |    temperature |
| 1% starch solution | beaker of crushed ice for |
| $I_2KI$ solution |    ice bath |

## Introduction

Chemical reactions accelerate as temperature rises, partly because increased temperatures speed up the motion of molecules. This means that substrates collide more frequently with enzyme active sites. Generally, a 10° rise in temperature results in a two- to threefold increase in the rate of a particular reaction. However, at high temperatures, the integrity of proteins can be irreversibly denatured. The activity of enzymes is dependent on the proper tertiary and quaternary structures; the optimum temperature for activity, therefore, may vary, depending on the structure of the enzyme.

What was the source of the amylase used in this experiment? (Check the Introduction to this exercise.)

*saliva*

## Question

Pose a question about temperature and reaction rate.

*What effect do different temperatures have on enzyme reaction rate?*

## Hypothesis

Hypothesize about the rate of activity of amylase at various temperatures.

*An enzyme will react optimally at the temperature in which it is normally found ("body temperature"). If students remember that the human body is about 37°C, they may hypothesize that amylase activity will be optimum at that temperature. A number of hypotheses are acceptable if testable.*

## Prediction

Predict the results of the experiment based on your hypothesis (if/then).

*If amylase activity is optimum at the temperature at which it is normally found (in this case, 37°C), then starch digestion will be most rapid at this temperature. The prediction should indicate the hypothesis and the results of the appropriate test.*

## Procedure

1. Number four standard test tubes 1 through 4.
2. Using the 5-mL calibrated pipette, add 2 mL of the 1% starch solution to each tube.
3. Using a clean 5-mL pipette, add 4 mL DI water to each tube.
4. Add 1 mL of 6.8 buffer to each tube.
5. Place the test tubes as follows:

   Tube 1: 80°C water bath

   Tube 2: 37°C water bath

   Tube 3: test-tube rack (room temperature, or about 22°C)

   Tube 4: beaker of crushed ice (4°C)

6. Number and mark a second set of standard test tubes 1A through 4A. Use the 1-mL calibrated pipette to add 1 mL amylase to each tube, and place as follows (do not mix together the solutions in the two sets of tubes until instructed to do so):

   Tube 1A: 80°C water bath

   Tube 2A: 37°C water bath

   Tube 3A: test-tube rack (room temperature, or about 22°C)

   Tube 4A: beaker of crushed ice (4°C)

7. Let all eight tubes sit in the above environments for 10 minutes. You should have one tube of amylase and one of starch at each temperature.
8. Fill several rows of the test plate with 1 or 2 drops of $I_2KI$ per compartment.

*Reaction rate for 37°C is only slightly greater than 22°C— approximately 2.5 min. and 3 min., respectively. There will be no digestion in 80°C and ice. If you have additional water baths, add a temperature between 37°C and 80°C.*

Remember, when the enzyme and substrate are together, the reaction has begun!

9. Leaving the tubes in the above environments as they are being tested, mix tubes 1 and 1A, record the time (this is time 0), and use a disposable pipette to immediately add 1 or 2 drops of the mixture to a drop of $I_2KI$ on the test plate.

10. Continue adding mixture drops to *new wells* of $I_2KI$ at 30-second intervals until a blue color is no longer produced and the $I_2KI$ solution remains yellow-amber (indicating that all the starch is digested). If within 10 minutes there is no color change, terminate the experiment with that particular reaction mixture. Record your results in Table 4.7.

11. Repeat steps 9 and 10 for the other reaction mixtures (mix tubes 2 and 2A and test, mix 3 and 3A and test, etc.).

12. If time permits, transfer tube 4 (in ice, containing the enzyme and substrate) into the 37°C water bath. After 2 minutes, test the contents of the tube at 10-second intervals. Record these results in the margin of your lab manual and refer to them as you answer Discussion Question 2 below.

13. Finish recording and graphing your observations in the Results section, and explain your results in the Discussion section.

## Results

1. Complete Table 4.7 as rates of digestion (time of starch disappearance) in different temperatures are determined.

 **Table 4.7**
Time of Starch Disappearance in Different Temperatures for the Enzyme Amylase

| Tube | Temp. °C | Time of Starch Disappearance (in minutes) |
|------|----------|-------------------------------------------|
| 1    | 80°      |                                           |
| 2    | 37°      |                                           |
| 3    | 22°      |                                           |
| 4    | 4°       |                                           |

2. Construct a graph using Figure 4.7 on the next page to illustrate your results. See Lab Topic 1 for assistance in graph construction.

a. What is the independent variable? Which is the appropriate axis for this variable?

*temperature; the x axis*

b. What is the dependent variable? Which is the appropriate axis for this variable?

*reaction rate (time of starch disappearance); the y axis*

*Students' results should produce a curve with approximately this shape:*

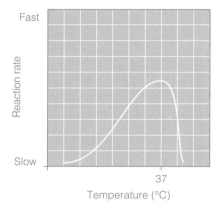

**Figure 4.7.**
**Reaction rate (time of starch disappearance) of amylase in various temperatures.**

*Since reversing the axis may be confusing to students, the instructor should check students' graphs before they proceed.*

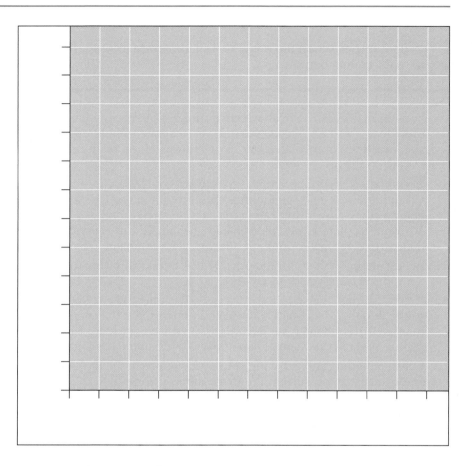

c. Label the axes of the graph. Using Table 4.7, note the maximum number of minutes in your results, and choose an appropriate scale for the dependent variable. Reaction rate, the dependent variable, was measured as the time of starch disappearance. The data must, therefore, be graphed in reverse order because the highest values indicate the slowest reaction rate. You should place "0" at the end of the axis and write "fast" by your 0. Place your highest number near the origin (where the *x* and *y* axes cross). Write "slow" near the origin. Choose an appropriate scale for the independent variable (temperature) and label the axis.

## Discussion

1. Explain your results in terms of your hypothesis. Describe the shape of the reaction rate curve as temperatures increase. What factors are responsible for the shape of this curve?

   *The reaction rate will first increase and then decrease rapidly. As temperature increases substrate molecules collide with the enzyme active sites more frequently up to a point at which the enzyme is denatured. Each enzyme has its optimum temperature.*

2. What do you think would happen to reaction rate in the tube incubated in ice if this tube, with enzyme and substrate already mixed, were placed in the 37°C water bath?

   *Although there will be no enzymatic activity in crushed ice, after a couple of minutes in the 37°C water bath, enzymatic activity will be restored.*

Explain your answer in terms of the effect of various temperatures on enzyme structure and the rate of enzyme activity.

*At freezing temperatures the velocity of the enzymatic reaction is exceedingly slow because substrate molecules are not colliding with active sites. However, the enzyme is not denatured, and when returned to a favorable temperature, the enzymatic reaction will resume. At extremely high temperatures, however, the enzyme is denatured as bonds are disrupted.*

## Reviewing Your Knowledge

1. Define and use the following terms, providing examples if appropriate. *catalyst enzyme, substrate, active site, cofactor, coenzyme, competitive inhibition, noncompetitive inhibition.*

   a. Compare and contrast competitive and noncompetitive inhibition.

   *In competitive inhibition, a molecule (the inhibitor) that is structurally similar to the substrate for a particular reaction competes for a position at the active site on the enzyme. This type of inhibition is completely reversible. In noncompetitive inhibition, the inhibitor binds to a site that is not the active site, blocking access to the active site or changing the configuration of the enzyme, rendering it inactive.*

   b. Why does adding additional substrate overcome competitive but not noncompetitive inhibition?

   *Because the inhibitor and substrate compete for the active site in competitive inhibition, the result of competition depends on how many of each are present. By increasing the concentration of substrate, the inhibitor is outcompeted. Because a noncompetitive inhibitor binds at some site other than the active site, once this binding has taken place, the tertiary structure of the enzyme is disrupted. The substrate can no longer bind to the active site, regardless of the concentration.*

2. Review changes in reaction rate as substrate concentration, pH, and temperature change. Explain the conditions or factors that are responsible for these changes.

## Applying Your Knowledge

1. In the first experiment, the enzyme catechol oxidase was extracted from potato. However, this was not a purified preparation; it contained hundreds of enzymes. What evidence supports the assumption that catechol oxidase was the enzyme studied?

   *Although many enzymes were present, the only substrate provided was catechol. Since enzymes react with a specific substrate, the catechol oxidase was the active enzyme. Also, the color change indicated the formation of the product from this reaction.*

2. Well-preserved mammoths have been found in ice and frozen soil in northern Siberia. Using information about enzyme activity learned in this lab, explain why these animal carcasses have survived all of these years.

   *Activity of bacterial enzymes and enzymes that would destroy tissues of the mammoth has been slowed by the freezing temperatures.*

3. Ethylene glycol, the main ingredient in antifreeze, is an odorless, colorless, sweet-tasting water-soluble chemical that, when ingested, is a toxic poison. The treatment for ethylene glycol poisoning involves emergency procedures that include intravenous doses of ethanol for several hours. To understand how this treatment works, you need to know that ethylene glycol is metabolized by the enzyme alcohol dehydrogenase (ADH) into four products, and these products are the actual toxic agents. If this reaction can be inhibited, then the kidneys will eliminate the ethylene glycol intact before the toxins are produced. Chemists know that ADH has 100x greater affinity for ethanol than for ethylene glycol.

Given your knowledge of enzyme activity and inhibitors, speculate about the mechanism by which ethanol prevents ethylene glycol poisoning.

*Ethanol is a competitive inhibitor for ADH. When infused into the blood, it will outcompete ethylene glycol for the active site of the enzyme. The intact ethylene glycol will then be excreted by the kidneys.*

4. There is an enzyme that catalyzes the production of the pigment responsible for dark fur color in Siamese cats and Himalayan rabbits. This enzyme is *thermolabile,* meaning that it does not function at higher temperatures. Rabbits raised at 5°C are all black. If raised at 20°C, they are white with black paws, ears, and noses; they are all white when raised at 35°C. Which of the following best represents an activity curve for this enzyme?

*(answer = a)*

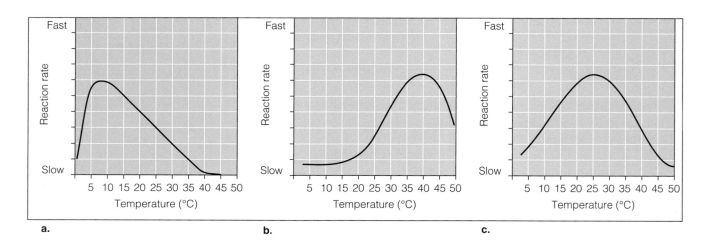

a.                                    b.                                    c.

5. You are investigating a specific compound known to inhibit the activity of an enzyme. You hypothesize that the compound is a competitive inhibitor of the enzyme's activity. To test your hypothesis, you design an experiment where you add increasing amounts of the enzyme's substrate to a solution containing the enzyme, and then you test the reaction rate. Which of the following graphs represents the predicted results of your experiment, based on your hypothesis?

*(answer = a)*

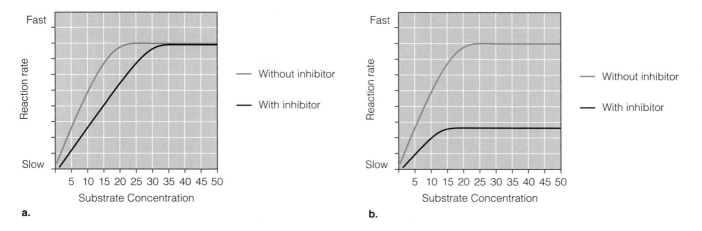

a.

b.

## Investigative Extensions

1. In this lab topic you investigated the activity of the enzyme amylase, a carbohydrate-digesting enzyme found in the digestive tract of many animals. Other digestive enzymes in animal digestive tracts include those that digest proteins, proteases. Although we usually think of protein-digesting enzymes as being found in the digestive tract of animals, it is interesting to note that plants also contain proteases. For example, pineapple fruits and stems contain a protease, **bromelain,** and the fruit of papaya (*Carica papaya*) contains a protease, **papain.** Both of these enzymes are reported to have health benefits, and papain is used in meat tenderizers, for example, Adolph's Meat Tenderizer.™

   Using protocols and materials presented in this lab topic, design an experiment to investigate the activity of one of these enzymes. What is the optimum pH for its activity? At what temperature is it denatured? Does it digest a range of common proteins, for example, gelatin or albumin? Use library resources or the Web to research other questions you might have about proteases in plants; for example, what role do they play in plant metabolism?

2. Cottage cheese and many other cheeses are produced by adding the enzyme rennin to milk, producing curds and whey. Using information you learned in this lab, design an experiment to determine the pH at which rennin is most effective. How will you measure the results? Using sources from the Web or library, find tests that can be used to detect protein, sugars, and fats, and use these tests to determine the chemical makeup of curds and whey.

3. The experiments using catechol oxidase may be quantified by measuring results using a Spectronic 20. Modify the instructions in Exercises 4.1 and 4.2 according to the following instructions.

   a. Rather than using three experimental tubes, use four cuvettes, and label them B (the blank), 1, 2, and 3.

   b. Prepare the solutions in cuvettes as directed for test tubes 1, 2, and 3, omitting catechol until ready to take the reading. The blank will contain 5 mL of distilled water, 20 drops of distilled water, and 10 drops of potato extract only. Do not add catechol to the blank.

   c. Design a table to record your data.

d. Record absorbance of the solutions in the four cuvettes. (For steps (1)–(3) following, see the diagram of the Spectronic 20D+ and details of its operation in Lab Topic 6 Photosynthesis. Steps (4)–(6) are self-explanatory.)

(1) Set the wavelength at 540 nm.

(2) Zero the instrument.

(3) Calibrate the instrument.

(4) Add the catechol to the appropriate cuvettes.

(5) Insert cuvette 1 into the sample holder and record absorbance in your data table. Immediately repeat with cuvettes 2 and 3.

(6) Continue to take readings at 5-minute intervals for 30 minutes or an appropriate time.

 **Student Media: BioFlix, Activities, Investigations, and Videos**
www.masteringbiology.com (select Study Area)

**Activities**—Ch. 1: Graph It! An Introduction to Graphing.
Ch. 8: How Enzymes Work

## References

Klabunde, T., C. Eicken, J. Sacchettini, and B. Krebs. "Crystal Structure of a Plant Catechol Oxidase Containing a Dicopper Center." *Nature Structural Biology,* 1998, vol. 5(12), pp. 1084–1090. Full text available online. (Discusses activity of catechol oxidase and the specific mode of inhibition by PTU.)

Mathews, C. K., K. E. Van Holde, and K. G. Ahern. *Biochemistry.* San Francisco, CA: Benjamin Cummings, 2000.

Nelson, D. and M. Cox. *Lehninger Principles of Biochemistry,* 4th ed. New York: Worth Publishers, 2005.

## Websites

Describes enzyme preparations approved as food additives:
http://www.fda.gov/Food/FoodIngredientsPackaging/ucm084292.htm

Good discussions of enzymes used in cheese production:
http://www.vegsoc.org/info/cheese.html

# LAB TOPIC 4

# Enzymes
## Teaching Plan for Laboratories

## Main Concepts and Objectives

1. Concept: scientific method. Students will be able to describe the typical format used in scientific investigation. (Exercises 4.1–4.3)

2. Concept: enzyme inhibition. Students will differentiate competitive and noncompetitive inhibition. (Exercise 4.2)

3. Concept: activity of enzymes. Students will describe enzyme activity and how it can be modified by varying such factors as enzyme concentration, pH, and temperature. (Exercise 4.3)

4. Concept: scientific writing. Students will practice writing the Results section of a scientific paper.

5. Specific content. Students will be able to describe or define: *enzyme, substrate, catalyst, cofactor, coenzyme, competitive* and *noncompetitive inhibition, dependent variable, independent variable, control.*

## Order of the Lab

1. The instructor introduces enzymes, major concepts, and objectives.                                                           (15 min)

2. The instructor reviews using the scientific method, formulating hypotheses, and making predictions (refer to Lab Topic 1).    (5 min)

3. Perform Exercises 4.1 and 4.2.                                                (30 min)

4. Discuss results of Exercises 4.1 and 4.2.                                     (10 min)

5. Each student team performs one of the three experimentson amylase activity (Exercise 4.3, A, B, or C).                         (75 min)

6. Designated students report and discuss results of the amylase experiments. All experiments should be discussed.               (25 min)

7. The instructor reviews scientific writing, specifically, writing a Results section and making graphs (refer to Lab Topic 1 and Appendix A).                                                (15 min)

8. Describe the assignment due next week. Students will use data from one of the experiments in Exercise 4.3 and write a Results section. You may vary the paper section assigned to deter students from sharing old papers.                              (5 min)

**For a 2-hour lab:** Omit student reports and the instructor's reviews of the scientific method and scientific writing. The assignment could be provided as a handout.

## Classroom Management

Students work in groups of four. All groups carry out Exercises 4.1 and 4.2. Assign one experiment (A, B, or C) from Exercise 4.3 to each student group. Students will choose one person to report their group's results to the class. You will probably have two groups doing each experiment. Have only one report unless groups have conflicting results. Supply overhead acetates and Vis-à-Vis pens for students to use in their presentations. If laboratory computers are available, students may graph their results using Excel or other software.

## Student Development

Students, in addition to learning enzyme concepts, will develop skills in scientific method, organizational and laboratory techniques, and data processing. Those students who give the oral reports will practice communication skills. Students will be asked to give reports in various labs throughout the year. Try to provide the opportunity for each student to present at least once.

## Lab Safety Precautions

Instruct students to:

1. Avoid contact with all solutions. The buffers can burn the skin. Catechol and PTU are both poisons. Disposable gloves are unnecessary while performing experiments if precautions are taken.

2. Wash hands thoroughly after each experiment.

3. Notify the instructor immediately if a spill occurs. The instructor should use disposable gloves, paper towels, and soap and water to wipe up spills. Dispose of all towels in a plastic bag, and place it in the trash. If the instructor is not available, students should clean up spills as directed in the student lab manual.

## Discussion and Summary

The instructor leads students in discussing the results of Exercises 4.1 and 4.2. Students report and discuss the results of Exercise 4.3.

## Evaluation

The instructor should informally note the quality of students' laboratory skills. Students will submit a Results section in the format of a scientific paper to be graded. Test concepts on the next laboratory exam. To reduce opportunities for cheating, vary assignments among lab sections and/or in subsequent years. For example, have one lab section write a Discussion section.

## Lab Topic 5
# Cellular Respiration and Fermentation

 This lab topic gives you another opportunity to practice the scientific process introduced in Lab Topic 1. Before going to lab, review scientific investigation in Lab Topic 1 and carefully read Lab Topic 5. Be prepared to use this information to design an experiment in fermentation or cellular respiration.

## Laboratory Objectives

After completing this lab topic, you should be able to:

1. Describe alcoholic fermentation, naming reactants and products.
2. Describe cellular respiration, naming reactants and products.
3. Explain oxidation-reduction reactions in cellular respiration.
4. Name and describe environmental factors that influence enzymatic activity.
5. Explain spectrophotometry and describe how this process can be used to measure aerobic respiration.
6. Propose hypotheses and make predictions based on them.
7. Design and execute an experiment testing factors that influence fermentation or cellular respiration.
8. Practice scientific thinking and communication by analyzing, interpreting, and presenting results.

*For a 2-hour lab: Perform Experiment A in each exercise and have students design, but not perform, an original experiment. Alternatively, perform Exercise 5.1, Experiment A, and design and perform an independent investigation with yeast.*

*Instructors and/or students can download data tables in Excel format. Go to www.masteringbiology.com. Look in the Study Area under Lab Media.*

## Introduction

You have been investigating cells and their activities in Lab Topics 2–4: cellular structure and evolution (Lab Topic 2), movement across cell membranes (Lab Topic 3), and enzymatic activities (Lab Topic 4). This lab topic and the following one (Photosynthesis, Lab Topic 6) investigate energy transformations in cells. Photosynthesis is the process of transferring the sun's radiant energy to organic molecules, namely, glucose (Figure 5.1). This lab topic investigates **fermentation** and **cellular respiration**, cellular processes that release the energy in glucose to synthesize **adenosine triphosphate** (ATP). The energy in ATP can then be used to perform cellular work. Fermentation is an anaerobic (without oxygen) process; cellular respiration is aerobic (utilizing oxygen). *All living organisms, including bacteria, protists, plants, and animals, produce ATP in fermentation or cellular respiration and then use ATP in their metabolism.*

*If possible in your laboratory program, 1 week before this lab, instruct students to read the lab topic and meet with their investigative team to discuss possible independent investigations before coming to the lab. Ask them to develop preliminary questions and hypotheses. They may choose to do a preliminary literature search using their text and references for this lab topic or electronic sources. Provide students with a list of available materials before lab so that students can submit a request for any additional materials prior to lab.*

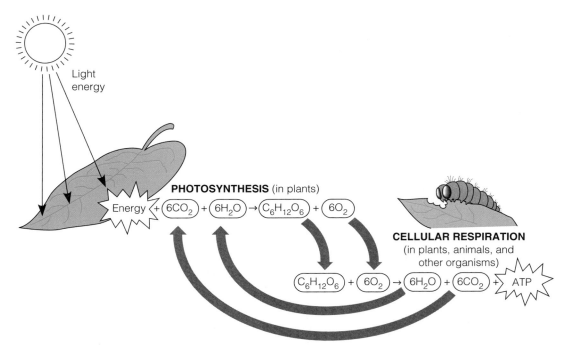

**Figure 5.1.**
**Energy flow through photosynthesis and cellular respiration.** Light
energy from the sun is transformed to chemical energy in photosynthesis.
This chemical energy is used to synthesize glucose and oxygen from carbon
dioxide and water. The energy stored in plant organic molecules—glucose, for
example—can be utilized by plants or by consumers. The energy in organic
molecules is released to form ATP during cellular respiration in plants, animals,
and other organisms.

Fermentation and cellular respiration involve oxidation-reduction reactions (redox reactions). Redox reactions are always defined in terms of electron transfers, oxidation being the *loss* of electrons and reduction the *gain*
of electrons. In cellular respiration, two hydrogen atoms are removed from
glucose (oxidation) and transferred to a coenzyme called nicotinamide
adenine dinucleotide (**NAD$^+$**), reducing this compound to **NADH.** Think
of these two hydrogen atoms as 2 electrons and 2 protons. NAD$^+$ is the
oxidizing agent that is reduced to NADH by the addition of 2 electrons and
1 proton. The other proton (H$^+$) is released into the cell solution. NADH
transfers electrons to the electron transport chain. The transfer of electrons
from one molecule to another releases energy, and this energy can be used
to synthesize ATP.

Cellular respiration is a sequence of three metabolic stages: **glycolysis** in
the cytoplasm, and the **Krebs cycle** and the **electron transport chain** in
mitochondria (Figure 5.2). Fermentation involves glycolysis but does not
involve the Krebs cycle and the electron transport chain, which are inhibited at low oxygen levels. Two common types of fermentation are **alcoholic
fermentation** and **lactic acid fermentation.** Animals, certain fungi, and
some bacteria convert pyruvate produced in glycolysis to lactate. Plants and
some fungi, yeast in particular, convert pyruvate to ethanol and carbon
dioxide. Cellular respiration is much more efficient than fermentation in
producing ATP. Cellular respiration can produce a maximum of 38 ATP
molecules; fermentation produces only 2 ATP molecules.

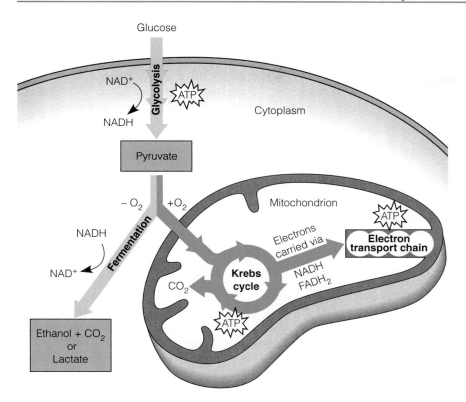

**Figure 5.2.**
**Stages of cellular respiration and fermentation.** Cellular respiration consists of glycolysis, the Krebs cycle, and the electron transport chain. Glycolysis is also a stage in fermentation.

Before you begin today's lab topic, refer to the preceding paragraph and Figure 5.2 as you review major pathways, reactants, and products of fermentation and cellular respiration by answering the following questions.

1. Which processes are anaerobic?

   *glycolysis and fermentation*

2. Which processes are aerobic?

   *Krebs cycle and electron transport chain*

3. Which processes take place in the cytoplasm of the cell?

   *glycolysis and fermentation*

4. Which processes take place in mitochondria?

   *Krebs cycle and electron transport chain*

5. What is the initial reactant in cellular respiration?

   *glucose*

6. What is (are) the product(s) of the anaerobic processes?

   *ethanol, lactate, $CO_2$, $NAD^+$, and 2 ATP*

7. What is (are) the product(s) of the aerobic processes?

   *$CO_2$, water, and 36–38 ATP*

8. Which gives the greater yield of ATP, alcoholic fermentation or cellular respiration?

   *cellular respiration*

In this lab topic you will investigate alcoholic fermentation first and then cellular respiration. Working in teams of two to four students, you will first perform two introductory experiments (Experiment A of each exercise). Experiment B in each exercise provides questions and background to help you propose one or more testable hypotheses based on questions from the experiments or your prior knowledge. Your team will then design and carry out an independent investigation based on your hypotheses, completing your observations and recording your results in this laboratory period. After discussing the results, your team will prepare an oral presentation in which you will persuade the class that your experimental design is sound and that your results support your conclusions. If assigned by the lab instructor, *each of you* independently will submit Results and Discussion sections describing the results of your experiment (see Appendix A).

 First complete Experiment A in each exercise. Then discuss possible questions for investigation with your research team. Be certain you can pose an interesting question from which to develop a testable hypothesis. Design and perform the experiment today. Prepare to report your results in oral and/or written form.

## EXERCISE 5.1
# Alcoholic Fermentation

For centuries, humans have taken advantage of yeast fermentation to produce alcoholic beverages and bread. Consider the products of fermentation and their roles in making these economically and culturally important foods and beverages. Alcoholic fermentation begins with glycolysis, a series of reactions breaking glucose into two molecules of **pyruvate** with a net yield of 2 ATP and 2 NADH molecules. In anaerobic environments, in two steps the pyruvate (a 3-carbon molecule) is converted to ethyl alcohol (ethanol, a 2-carbon molecule) and $CO_2$. In this process the 2 NADH molecules are oxidized, replenishing the $NAD^+$ used in glycolysis (Figure 5.2).

## Experiment A. Alcoholic Fermentation in Yeast

### Materials

4 respirometers:
  test tubes, 1-mL graduated
  pipettes, aquarium tubing,
  flasks, binder clips
pipette pump
3 5-mL graduated pipettes,
  labeled "DI water," "yeast,"
  and "glucose"

3-inch donut-shaped metal weights
yeast solution
glucose solution
DI water
water bath
wax pencil

## Introduction

In this experiment, you will investigate alcoholic fermentation in a yeast (a single-celled fungus), *Saccharomyces cerevisiae,* or "baker's yeast." When oxygen is low, some fungi, including yeast and most plants, switch from cellular respiration to alcoholic fermentation. In bread making, starch in the flour is converted to glucose and fructose, which then serve as the starting compounds for fermentation. The resulting carbon dioxide is trapped in the dough, causing it to rise. Ethanol is also produced in bread making but evaporates during baking. Alcoholic fermentation is also an economically and culturally important process in the production of alcoholic beverages, such as beer and wine. Recently, interest in alternative biofuels has created a new industry in which fermentation is central to the conversion of energy crops (e.g., corn), forestry and municipal waste, and agricultural residues to ethanol.

In this laboratory experiment, the carbon dioxide ($CO_2$) produced, being a gas, bubbles out of the solution and can be used as an indication of the relative rate of fermentation taking place. Figure 5.3 shows the respirometers you will use to collect $CO_2$. The rate of fermentation, a series of enzymatic reactions, can be affected by several factors, for example, concentration of yeast, concentration of glucose, or temperature. In this experiment you will investigate *the effects of yeast concentration.* In your independent study, Exercise 5.3, you may choose to investigate other independent variables.

*Although this is not an oxygen-free environment, yeast in the pipette switch to alcoholic fermentation as oxygen is depleted.*

## Hypothesis

Hypothesize about the effect of different concentrations of yeast on the rate of fermentation.

*Fermentation will be greater in the fermentation tube with more yeast.*

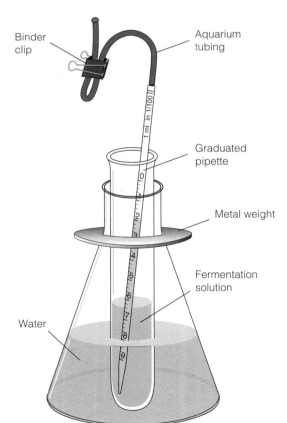

**Figure 5.3.**
**Respirometer used for yeast fermentation.**

Labels: Binder clip · Aquarium tubing · 1 ml in 1/100 · Graduated pipette · Metal weight · Fermentation solution · Water

## Prediction

Predict the results of the experiment based on your hypothesis (if/then).

*If fermentation is greater in the fermentation tube with more yeast, then more gas will be produced in this tube.*

## Procedure

*Check the fermentation rates of your yeast concentrations before lab. If the respiration rate proceeds too slowly or too rapidly, adjust the yeast concentration and/or the temperature of the water bath up or down. Do not exceed 37°C in the water bath.*

1. Obtain four flasks and add enough tap water to keep them from floating in a water bath (fill to about 5 cm from the top of the flask). Label the flasks 1, 2, 3, and 4. To stabilize the flasks, place a 3-inch donut-shaped metal weight over the neck of the flasks.

2. Obtain four test tubes (fermentation tubes) and label them 1, 2, 3, and 4. Add solutions as in Table 5.1 to the appropriate tubes. Rotate each tube to distribute the yeast evenly in the tube. Place tubes in the corresponding numbered flasks.

3. To each tube, add a 1-mL graduated pipette to which a piece of plastic aquarium tubing has been attached.

4. Place the flasks with the test tubes and graduated pipettes in the water bath at 30°C. Allow them to equilibrate for about 5 minutes with the tubing unclamped.

**Table 5.1**
Contents of Fermentation Solutions (volumes in mL)

| Tube | DI Water | Yeast Suspension | Glucose Solution |
|------|----------|------------------|------------------|
| 1 | 4 | 0 | 3 |
| 2 | 6 | 1 | 0 |
| 3 | 3 | 1 | 3 |
| 4 | 1 | 3 | 3 |

*Students will need to practice adjusting the solution to the 0 mark. Encourage teamwork!*

5. Attach the pipette pump to the free end of the tubing on the first pipette. Use the pipette pump to draw the fermentation solution up into the pipette. Fill it past the calibrated portion of the tube, but do not draw the solution into the tubing. Fold the tubing over and clamp it shut with the binder clip so the solution does not run out. Open the clip slightly, and allow the solution to drain down to the 0-mL calibration line (or slightly below). If the level is below the zero mark, open the clamp slightly while another student adjusts the level using the pipette pump. Be patient. This may require a couple attempts! Quickly do the same for the other three pipettes.

6. In Table 5.2, quickly record your initial readings for each pipette in the "Initial reading" row in each "Actual (A)" column. This will be the *initial reading (I)*.

7. Two minutes after the initial readings for each pipette, record the actual readings (A) in mL for each pipette in the "Actual (A)" column. Subtract I from A to determine the total amount of $CO_2$ evolved (A − I). Record this value in the "$CO_2$ Evolved (A − I)" column. *From now on, you will subtract the initial reading from each actual reading to determine the total amount of $CO_2$ evolved.*

8. Continue taking readings every 2 minutes for each of the solutions for 20 minutes. Remember, take the actual reading from the pipette and

subtract the initial reading to get the total amount of $CO_2$ evolved in each test tube.

*Students should calculate the results as they record readings.*

9. Record your results in Table 5.2.

## Results

1. Complete Table 5.2.
2. Using Figure 5.4, construct a graph to illustrate your results. The graph can also be completed in Excel.
   a. What is (are) the independent variable(s)? Which is the appropriate axis for this variable?

   *The independent variable being tested is yeast concentration. Time is also an independent variable. Graph time on the x axis. Your graph will have four lines, one for each tube.*

   b. What is the dependent variable? Which is the appropriate axis for this variable?

   *amount of $CO_2$ evolved; the y axis*

   c. Choose an appropriate scale and label the *x* and *y* axes.

**Table 5.2**
Total $CO_2$ Evolved by Different Concentrations of Yeast. Actual values are the graduated pipette readings. For $CO_2$ evolved values, subtract the initial reading from the actual reading. This is the amount of $CO_2$ accumulated over time.

| Time (min) | Tube 1 Actual (A) | Tube 1 $CO_2$ Evolved (A−I) | Tube 2 Actual (A) | Tube 2 $CO_2$ Evolved (A−I) | Tube 3 Actual (A) | Tube 3 $CO_2$ Evolved (A−I) | Tube 4 Actual (A) | Tube 4 $CO_2$ Evolved (A−I) |
|---|---|---|---|---|---|---|---|---|
| Initial reading (I) | | | | | | | | |
| 2 | | | | | | | | |
| 4 | | | | | | | | |
| 6 | | | | | | | | |
| 8 | | | | | | | | |
| 10 | | | | | | | | |
| 12 | | | | | | | | |
| 14 | | | | | | | | |
| 16 | | | | | | | | |
| 18 | | | | | | | | |
| 20 | | | | | | | | |
| | | | | | | | | |

**Figure 5.4.**

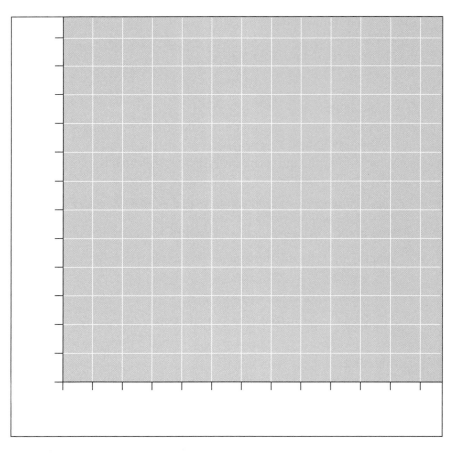

_____

_____

_____

d. Should you use a legend? If so, what would this include?

_The legend should indicate the designation for each test tube (line color, symbol for points, etc.)._

e. Compose a figure title.

**Discussion**

1. Explain the experimental design. What is the purpose of each test tube? Which is (are) the control tube(s)?

   _Test tube 1 is a control (no yeast). Test tube 2 is a control (no glucose). Test tubes 3 and 4 are the experimental tubes. Tube 4 has three times as much yeast as tubes 2 and 3._

2. Which test tube had the highest rate of fermentation? Explain why.

   _Test tube 4 contains both the substrate (glucose) and the greatest amount of yeast (containing the enzymes), so this test tube should have the highest rate of fermentation._

3. Which test tube had the lowest rate of fermentation? Explain why.

   _There should be no fermentation in test tube 1 because there is no yeast. Test tube 2 with yeast, but no substrate might have a small amount of fermentation because of residual respiration in the yeast, but the amount should be negligible._

4. Why were different amounts of water added to each fermentation solution?

   _to maintain the appropriate concentrations of yeast in each tube_

# Experiment B. Student-Designed Investigation of Alcoholic Fermentation

## Materials

all materials from Experiment A
beakers
graduated pipettes of various sizes
different substrates: sucrose, saccharin, Nutrasweet™, Splenda™,
    fructose, starch, glycogen, honey, corn syrup, pyruvate, maltose, stevia,
    agave nectar
different types of yeast: dry active, quick rise, Pasteur champagne (for
    wine making), sourdough yeast (*Candida milleri*)
various fermentation inhibitors: sodium fluoride, ethyl alcohol, Na benzoate
various salt solutions
various pH buffers
spices: ground cinnamon, cloves, caraway, ginger, cardamom, nutmeg,
    mace, thyme, dry mustard, hot peppers, cayenne pepper, celery seed,
    turmeric, oregano, sage
disposable gloves
additional glassware
mortar and pestle

## Introduction

If your team chooses to study alcoholic fermentation for your independent
investigation and report, design a simple experiment to investigate some
factor that affects alcoholic fermentation. Use the available materials or ask
your instructor about the availability of additional materials.

## Procedure

1. Collaborating with your research team, read the following potential
   questions, and choose a question to investigate using this list or an idea
   from your prior knowledge. You may want to check your text and other
   sources for supporting information. You should be able to explain the
   rationale behind your choice of question. For example, if you choose to
   investigate *starch* as a substrate, you should be able to explain that the
   yeast must first digest starch before the glucose can be used in alcoholic
   fermentation and the impact this might have on the experiment.

   a. Would other substrates be as effective as glucose in alcoholic fer-
      mentation? Possible substrates:

      sucrose (table sugar—glucose and fructose disaccharide)
      honey (mainly glucose and fructose)
      corn syrup (fructose and sucrose)
      starch (glucose polymer in plants)
      saccharin, Equal™, Splenda™
      fructose
      maltose
      pyruvate
      stevia
      agave nectar

   b. Would fermentation rates change with different types of yeasts, for
      example, quick rise, Champagne yeast used in wine making and
      *Candida milleri* used in sour dough?

*Do not use brewer's yeast from health food stores. These yeast are dead.*

   c. What environmental conditions are optimum for alcoholic fermentation? What temperature ranges?

      What pH ranges?

   d. What is the maximum amount of ethyl alcohol that can be tolerated by yeast cells?

---

 If you select toxins or fermentation inhibitors for your investigation, ask the instructor about safety procedures. Post safety precautions and follow safety protocol, including wearing gloves and protective eyewear. Notify the instructor of any spills.

---

   e. Sodium fluoride, commonly used to prevent tooth decay, inhibits an enzyme in glycolysis. At what concentration is it most effective?

   f. Would adding $MgSO_4$ enhance glycolysis? $MgSO_4$ provides $Mg^{++}$, a cofactor necessary to activate some enzymes in glycolysis.

   g. Does a high concentration of sucrose inhibit fermentation?

   h. An old German baker's wisdom says, "A pinch of ginger will make your yeast work better." Some spices enhance yeast activity while others inhibit it (Corriher, 1997). Do spices enhance or inhibit fermentation? Are the effects of spice dependent on the concentration, enhancing at low concentrations and inhibiting at high? Try ginger, ground cardamom, caraway, cinnamon, mace, nutmeg, thyme, dry mustard, cayenne pepper, hot peppers, turmeric, oregano, celery seed, sage, or others.

   i. Salt is often used as a food preservative to prevent bacterial and fungal growth (for example, in country ham). But salt is also important to enhance the flavor of bread when added in small amounts. At what concentration does salt begin to inhibit yeast fermentation?

   j. Does the food preservative Na benzoate inhibit cellular respiration?

   k. How do fermentation rates compare for baker's yeast (*Saccharomyces cervisiae*) and sourdough yeast (*Candida milleri*) in different pH environments?

2. Design your experiment, proposing hypotheses, making predictions, and determining procedures as instructed in Exercise 5.3.

---

## EXERCISE 5.2
# Cellular Respiration

---

Most organisms produce ATP using cellular respiration, a process that involves glycolysis, the Krebs cycle, and the electron transport chain. In cellular respiration, many more ATP molecules are produced than were produced in alcoholic fermentation (potentially 38 compared to 2). After the series of reactions in the cytoplasm (glycolysis), pyruvate enters the mitochondria, where enzymes for the Krebs cycle and the electron transport chain are located. The Krebs cycle is a series of eight steps, each catalyzed by a specific enzyme. As one compound is converted to another, $CO_2$ is given off and hydrogen ions and electrons are removed. The electrons and hydrogen ions are passed to $NAD^+$ and another electron carrier, FAD (flavin adenine dinucleotide). NADH and $FADH_2$ carry the electrons to the electron transport chain, where the electrons pass along the chain to the final electron acceptor, oxygen. In the process, ATP molecules and water are produced (Figure 5.2).

## Experiment A. Oxidation-Reduction Reactions in a Mitochondrial Suspension

### Materials

| | |
|---|---|
| mitochondrial suspension | 4 cuvettes or small test tubes |
| succinate | Parafilm® squares |
| buffer | Kimwipes® |
| DPIP solution | spectrophotometer |
| 1-mL graduated pipette | wax pencil |
| pipette pump | |

### Introduction

In this experiment, you will investigate cellular respiration in isolated mitochondria. Your instructor has prepared a mitochondrial suspension from pulverized lima beans. The suspension has been kept on ice to prevent enzyme degradation, and the Krebs cycle will continue in the mitochondria as in intact cells. Sucrose has been added to the mitochondrial suspension to maintain an osmotic balance.

One step in the Krebs cycle is the enzyme-catalyzed conversion of succinate to fumarate in a redox reaction. In intact cells, succinate loses hydrogen ions and electrons to FAD, and, in the process, fumarate is formed (Figure 5.5).

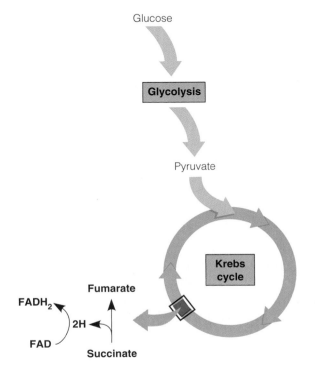

**Figure 5.5.**

**At one point in the Krebs cycle, succinate is converted to fumarate.** Hydrogens from succinate pass to FAD, reducing it to $FADH_2$.

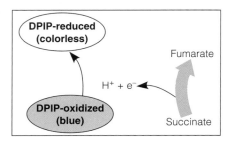

**Figure 5.6.**
**DPIP intercepts the hydrogen ions and electrons as succinate is converted to fumarate.** DPIP changes from blue to colorless.

We will utilize this step in the Krebs cycle to study the rate of cellular respiration under different conditions. To perform this study, we will add a substance called DPIP (di-chlorophenol-indophenol), an electron acceptor that intercepts the hydrogen ions and electrons released from succinate, changing the DPIP from an oxidized to a reduced state. DPIP is *blue* in its oxidized state but changes from blue to *colorless* as it is reduced (Figure 5.6).

We can use this color change to measure the respiration rate. To do this, however, we must have some quantitative means of measuring color change. An instrument called a **spectrophotometer** will allow us to do this. A spectrophotometer measures the amount of light absorbed by a pigment. In the spectrophotometer, a specific wavelength of light (chosen by the operator) passes through the pigment solution being tested—in this case, the blue DPIP. The spectrophotometer then measures the proportion of light *transmitted* or, conversely, *absorbed* by the DPIP and shows a reading on a calibrated scale. As the DPIP changes from blue to clear, it will absorb less light and more light will pass through (be transmitted through) the solution, and the transmittance reading will increase. As aerobic respiration takes place, what should happen to the percent transmittance of light through the DPIP?

*Transmittance should increase as the blue color changes to colorless.*

Our experiment will involve using succinate as the substrate and investigating the effect that *changing the amount of succinate will have on the cellular respiration rate*.

## Hypothesis

Hypothesize about the effect of an increased amount of substrate on the rate of cellular respiration.

*Increased substrate will increase the rate of cellular respiration.*

## Prediction

Predict the results of the experiment based on your hypothesis (if/then).

*If increased substrate increases the rate of cellular respiration, then respiration will be greater in the solution with more succinate and this tube should turn from blue to clear and percent transmittance will increase more rapidly.*

## Procedure

For the analog Spectronic 20 operating procedures, see the Preparation Guide, 7th ed.

1. Prepare the spectrophotometer.

   The instructions that follow are for a Thermo Scientific Spectronic 20D+ (Figure 5.7). Turn on the machine (power switch C) at least 15 minutes before beginning.

   a. Using the wavelength control knob (A), select the wavelength: 600 nm. Your instructor has previously determined that this wavelength is absorbed by DPIP.

   b. Press the mode selection button to select "Transmittance."

c. Zero the instrument by adjusting the control knob (the same as power switch C) so that the meter reads 0% transmittance. There should be no cuvette in the instrument, and the sample holder cover must be closed. Once it is set, do not change this setting.

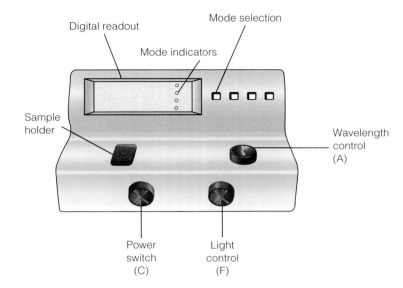

**Figure 5.7.**
**The Thermo Scientific Spectronic 20D+.** A spectrophotometer measures the proportion of light of different wavelengths absorbed and transmitted by a pigment solution. Inside the spectrophotometer, light is separated into its component wavelengths and passed through a sample.

2. Obtain four cuvettes and label them B, 1, 2, and 3. The B will be the blank.

3. Prepare the blank first by measuring 4.6 mL buffer, 0.3 mL mitochondrial suspension, and 0.1 mL succinate into the B cuvette. Cover the cuvette tightly with Parafilm and invert it to mix the reactants thoroughly.

4. Calibrate the spectrophotometer as follows: Wipe cuvette B with a Kimwipe and insert it into the sample holder. Be sure you align the etched mark on the cuvette with the line on the sample holder. Close the cover. Adjust the light control (F) until the meter reads 100% transmittance. Remove cuvette B. You are now ready to prepare the experimental cuvettes. The blank corrects for differences in transmittance due to the mitochondrial solution.

5. Measure the buffer, DPIP, and mitochondrial suspension into cuvettes 1, 2, and 3 as specified in Table 5.3.
   ***Do not add the succinate yet!***

6. Perform the next two steps as *quickly* as possible. First, add the succinate to each cuvette.

**Table 5.3**
Contents of Experimental Tubes (volumes in mL)

| Tube | Buffer | DPIP | Mitochondrial Suspension | Succinate (add last) |
|------|--------|------|--------------------------|----------------------|
| 1 | 4.4 | 0.3 | 0.3 | 0 |
| 2 | 4.3 | 0.3 | 0.3 | 0.1 |
| 3 | 4.2 | 0.3 | 0.3 | 0.2 |

7. Cover the opening to tube 1 with Parafilm, wipe it with a Kimwipe, insert it into the sample holder, and record the percent transmittance in Table 5.4 in the Results section. Repeat this step for tubes 2 and 3.

*If you do not see a color change, add additional succinate to tubes 2 and 3, keeping relative proportions.*

***If the initial reading is higher than 30%, tell your instructor immediately.** You may need to add another drop of DPIP to each tube and repeat step 7. The reading must be low enough (the solution dark enough) to give readings for 20–30 minutes. If the solution is too light (the transmittance is above 30%), the reactions will go to completion too quickly to detect differences in the tubes.*

8. Before each set of readings, insert the blank, cuvette B, into the sample holder. Adjust to 100% transmittance if necessary.

*Have students take readings at 2-minute intervals if the reactions are proceeding rapidly.*

9. Continue to take readings at 5-minute intervals for 20–30 minutes. *Each time, before you take a reading, cover the tube opening with Parafilm and invert it to mix the contents.* Record the results in Table 5.4.

### Results

1. Complete Table 5.4. Compose a title for the table.

 **Table 5.4**

| Tube | Time (min) | | | | | | |
|:---:|:---:|:---:|:---:|:---:|:---:|:---:|:---:|
| | **0** | **5** | **10** | **15** | **20** | **25** | **30** |
| **1** | | | | | | | |
| **2** | | | | | | | |
| **3** | | | | | | | |

2. Using Figure 5.8, construct a graph to illustrate your results. The graph can also be completed in Excel.

   a. What is (are) the independent variable(s)? Which is the appropriate axis for this variable?

   *The independent variable being tested is succinate concentration. Time is also an independent variable. Graph time on the x axis. Your graph will have three lines, one for each tube.*

   b. What is the dependent variable? Which is the appropriate axis for this variable?

   *percent transmittance; the y axis*

   c. Choose an appropriate scale and label the *x* and *y* axes.

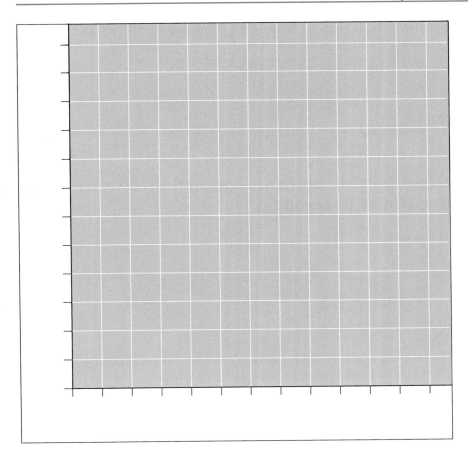

**Figure 5.8.**

_____

_____

_____

   d. Should you use a legend? If so, what would this include?

      *The legend should indicate the designation for each test tube*
      *(line color, symbols for points, etc.).*

   e. Compose a figure title.

## Discussion

1. Explain the experimental design. What is the role of each of the components of the experimental mixtures?

      *lima bean extract—provides the enzymes in the mitochondria*
      *succinate—the substrate*
      *DPIP—the electron acceptor*
      *buffer—maintains the optimum pH for the reactions*

2. Which experimental tube is the control?

      *Test tube 1 is a control (no substrate).*

3. In which experimental tube did transmittance increase more rapidly? Explain.

      *There is twice as much succinate in tube 3 as in tube 2. Transmittance*
      *in tube 3 goes to 100% more rapidly than it does in tube 2. This indicates*
      *that the reaction rate is faster when more substrate is available.*

4. Why should the succinate be added to the reaction tubes last?

      *Because once the succinate has been added, the reactions will begin.*

5. Was your hypothesis falsified or supported by the results? Use your data to support your answer.

6. What are some other independent variables that could be investigated using this technique?

   *effects of different pH; different concentrations of enzymes; different temperature effects, etc.*

## Experiment B. Student-Designed Investigation in Cellular Respiration

### Materials

all materials from Experiment A
additional substrates: glucose, fructose, maltose, artificial sweeteners, starch, glycogen, stevia, agave nectar
inhibitors: rotenone, oligomycin, malonate, antimycin A
spices: cinnamon, capsaicin (hot pepper), turmeric, cloves peaches, cassava root
plants containing coumarins: sweet clover, Cassia cinnamon sticks
extracts of red oak bark with lignins
different pH buffers
ice bath
water bath
disposable gloves
hammer
mortar and pestle

### Introduction

If your team chooses to study cellular respiration for your independent investigation and report, design a simple experiment to investigate some factor that affects cellular respiration. Use the available materials, or ask your instructor about the availability of additional materials.

 If you select toxins or respiratory inhibitors for your investigation, ask the instructor about safety procedures. Post safety precautions and follow safety protocol, including wearing gloves and protective eyewear. Notify the instructor of any spills.

### Procedure

1. Collaborating with your research team, read the following potential questions, and choose a question to investigate using this list or an idea from your prior knowledge. You may want to check your text and other sources for supporting information.

   a. Would other substrates be as effective as succinate in cellular respiration? Possible substrates:

   glucose

   sucrose (table sugar—disaccharide of glucose and fructose)

starch (glucose polymer in plants)

saccharin, Nutrasweet™, Splenda™, or other artificial sweeteners

fructose

agave nectar

stevia

b. What environmental conditions are optimum for cellular respiration?

What temperature ranges?

What pH ranges?

c. What inhibitors of cellular respiration are most effective? Consider the following list:

*Rotenone,* an insecticide, inhibits electron flow in the electron transport chain.

*Oligomycin,* an antibiotic, inhibits ATP synthesis.

*Malonate* blocks the conversion of succinate to malate. How would you determine if this is competitive or noncompetitive inhibition?

*Antimycin* is an antibiotic that inhibits the transfer of electrons to oxygen.

Crushed peach seeds and cassava root contain cyanide, a respiratory inhibitor.

Some spices: tumeric (contains curcumin), cloves, hot peppers (contains capsaicin), cinnamon

Plants with a high concentration of coumarin: sweet clover or Cassia cinnamon sticks

Lignins found in extracts from red oak bark

2. Design your experiment, proposing hypotheses, making predictions, and determining procedures as instructed in Exercise 5.3.

---

## EXERCISE 5.3
# Designing and Performing Your Open-Inquiry Investigation

## Materials

See each Experiment B materials list in Exercises 5.1 and 5.2.

## Introduction

Now that you have completed both introductory investigations, your research team should decide if you will investigate fermentation or cellular respiration. Return to the investigation of your choice and review the suggestions in Experiment B. Use Lab Topic 1 Scientific Investigation as a reference for designing and performing this independent investigation. You will need to think critically and creatively as you ask questions and formulate your hypothesis. As a team, review and modify the procedures, determine any additional required materials, review the techniques, and assign tasks to all members of your research team. Your experiments will be successful if you plan carefully, think critically, perform lab techniques accurately and systematically, and record and report data accurately. The following outline will assist you in designing and performing your original investigation.

## Procedure

1. **Decide on one or more questions to investigate.** Suggested questions are included in Experiment B as a starting point. (Refer to Lab Topic 1, Exercise 1.1, Lab Study A. Asking Questions.)

   **Question:**

2. **Formulate a testable hypothesis.** (Refer to Exercise 1.1, Lab Study B. Developing Hypotheses.)

   **Hypothesis:**

3. **Summarize the essential elements of the experiment.** (Use separate paper.)

4. **Predict the results of your experiment based on your hypothesis.** (Refer to Lab Topic 1, Exercise 1.2, Lab Study C. Making Predictions.)

   **Prediction:** (If/then)

5. **Outline the procedures to be used in the experiment.** (Refer to Lab Topic 1, Exercise 1.2, Lab Study B. Choosing or Designing the Procedure.)

   a. Review and modify the procedures used in Experiment A (see either Fermentation or Cellular Respiration). List each step in your procedure in numerical order.

   b. Create a table with contents of the experimental tubes (model this after either Table 5.1 for fermentation or Table 5.3 for cellular respiration). If computers are available, create or copy your new table into Excel.

   c. Critique your procedure: check for replicates, levels of treatment, controls, time intervals, total volume of experimental tubes, experimental conditions, glassware, and equipment.

   d. If your experiment requires materials other than those provided, ask your laboratory instructor about their availability. If possible, submit requests in advance.

   e. Create a table for data collection, using Table 5.2 (fermentation) or Table 5.4 (cellular respiration) as a model. If computers are available, create or copy your new data table into Excel.

6. **Perform the experiment**, making observations and collecting data for analysis.

 If your experiment involves the use of toxins or respiration inhibitors, use them only in liquid form as provided by the instructor. Wear protective gloves and eyewear. Ask your instructor about proper disposal procedures. If a spill occurs, notify your instructor immediately for proper cleanup.

7. **Record results including observations and data** in your data table. Be thorough when collecting data. Make notes about experimental conditions and observations. Do not rely on your memory for information that you will need when reporting your results.

8. **Prepare your discussion.** Discuss your results in light of your hypothesis.

    a. Review your prediction. Did your results correspond to the prediction you made? If not, explain how your results are different from your predictions, and why this might have occurred.

    b. Review your hypothesis. Review your results (tables and graphs). Do your results support or falsify your hypothesis? Explain your answer, using your data for support.

    c. If you had problems with the procedure or questionable results, explain how they might have influenced your conclusion.

    d. If you had an opportunity to repeat and expand this experiment to make your results more convincing, what would you do?

    e. Summarize the conclusion you have drawn from your results.

9. **Be prepared to report your results to the class.** Prepare to persuade your fellow scientists that your experimental design is sound and that your results support your conclusions.

10. If your instructor requires it, **submit Results and Discussion sections** of a scientific paper (see Appendix A). Keep in mind that although you have performed the experiments as a team, you must turn in a lab report of *your original writing*. Your tables and figures may be similar to those of your team members, but your Results and Discussion sections must be the product of your own literature search and creative thinking.

*Consider having a symposium day for teams to present their results. A poster session also would allow students to communicate results in ways that are similar to scientific meetings.*

## Reviewing Your Knowledge

1. Having completed this lab topic, you should be able to define, describe, and use the following terms: *aerobic, anaerobic, substrate, reactants, products, spectrophotometer, respirometer, NAD$^+$, NADH, FAD, FADH$_2$, ATP.*

2. State the beginning reactants and the end products of glycolysis, alcoholic fermentation, the Krebs cycle, and the electron transport chain. Describe where these processes take place in the cell and the conditions under which they operate (aerobic or anaerobic).

    glycolysis:

    *glucose; pyruvate, NADH, ATP; cytoplasm; aerobic or anaerobic*

    alcoholic fermentation:

    *glucose; $CO_2$ and ethanol; cytoplasm; anaerobic*

    Krebs cycle:

    *pyruvate (pyruvate is converted to acetyl CoA before entering Krebs cycle); $CO_2$, ATP, NADH, FADH$_2$; mitochondria; aerobic*

    electron transport chain:

    *electrons from NADH, FADH$_2$; ATP, $H_2O$; mitochondria; aerobic*

3. Suppose you do another experiment using DPIP to study cellular respiration in isolated mitochondria, and the results using the spectrophotometer

show a final percent transmittance reading of 42% in tube 1 and 78% in tube 2. Both tubes had an initial reading of 30%. In which tube did the greater amount of cellular respiration occur? Explain your answer in terms of the changes that take place in DPIP.

*Tube 2 had a greater increase in percent transmittance, so respiration must have been greater in this tube. The DPIP picked up the hydrogen atoms from Krebs cycle reactions and changed from blue to colorless. The greater the color change, the more cellular respiration has taken place.*

4.  How do you know that the electrons causing the change in color of DPIP are involved in the succinate–fumarate step?

    *Initially, electrons from other steps could be utilized, but these substrates would quickly be used up. We supplied succinate to the mixture, replenishing the substrate.*

## Applying Your Knowledge

1.  Your mother has been making yeast bread all afternoon, and she has just put two loaves in the oven. You open the oven door to see what is baking. Your mother yells, "Don't slam the door!" Why?

    *The bread rises by trapping $CO_2$ in the dough, creating pockets of the gas. If you slam the door, those pockets may collapse, causing the bread to fall.*

2.  Two characteristics of natural wines are that they have a maximum alcohol content of 14% and are "sparkling" wines. Apply your understanding of alcoholic fermentation to explain these characteristics.

    *In natural wines, after the ethyl alcohol concentration reaches 14%, the alcohol kills the yeast cells, causing fermentation to cease. The fizz or sparkle in wines is the accumulation of $CO_2$.*

3.  Increasing concerns about dependence on fossil fuels has generated research and development of renewable fuels, such as ethanol. Ethanol can be produced from grains (for example, corn) or from plant biomass (tree chips and sawdust, composted food waste or paper products). The production of ethanol is an industrial application of fermentation studied in introductory biology laboratories. Using information from your laboratory experiments, propose the components necessary for converting corn to ethanol. (Consider reactants, products, and environmental conditions.) Corn plants store energy as starch, which cannot enter the fermentation pathway directly. Based on your knowledge of the molecular structure of starch, include components necessary to convert starch to simple sugars.

    *Reactants: cornstarch, which requires an enzyme like amylase to convert starch to simple sugars*

    *Microbes: yeast*

    *Products: ethanol and $CO_2$*

    *Environmental conditions: anaerobic environment, optimal temperature and pH*

4.  Skunk cabbage is a plant that is able to generate heat and regulate its body temperature, like a warm-blooded animal. Botanists have suggested that

the ability to produce heat is important in these plants because it provides a warm environment for pollinators. The heat may also help to dissipate the carrion-like scent produced by some skunk cabbage flowers. Clearly, these plants must have a high respiratory rate to produce temperatures as high as 37°C. How could you determine if the temperature is the result of cellular respiration? What features of the plant surface and cell structure might be present if respiration is actively occurring in the flowers?

*One method scientists have used is to measure oxygen consumption in various portions of the flowering stalk and in flowers along the stalk. Oxygen consumption is quite high and the tissues have an increase in mitochondria and a high number of stomates (pores for gas exchange on the surface of the plant).*

 **Student Media: BioFlix, Activities, Investigations, and Videos**
www.masteringbiology.com (select Study Area)

**BioFlix**—Ch. 9: Cellular Respiration
**Activities**—Ch. 9: Overview of Cellular Respiration; Glycolysis; The Citric Acid Cycle; Electron Transport; Fermentation
**Investigations**—Ch. 9: How Is the Rate of Cellular Respiration Measured?

# References

Corriher, S. O. *Cookwise: The Hows and Whys of Successful Cooking.* New York: William Morrow, 1997.

Gänzle, M. G., M. Ehmann, and W. P. Hammes. "Modeling of Growth of *Lactobacillus sanfarciscensis* and *Candida milleri* in Response to Parameters of Sourdough Fermentation." *Appl. Environ. Microbiol.,* 1998, vol. 64, pp. 2616–2623.

Nelson, D. L., and M. M. Cox. *Lehninger, Principles of Biochemistry,* 4th ed. New York: Worth Publishers, 2005.

Reece, J., et al. *Campbell Biology,* 9th ed. San Francisco, CA: Pearson Education, 2011.

Seymour, R. S. "Biophysics and Physiology of Temperature Regulation in Thermogenic Flowers." *Bioscience Reports,* 2001, vol. 21, pp. 223–236.

Shelef, L. A. "Antimicrobial Effects of Spices." *Journal of Food Safety,* 1984, vol. 6, pp. 29–44.

Simpson, B. B., and M. Conner-Ogorzaly. *Economic Botany,* 3rd ed. New York: McGraw-Hill, 2001. Includes information on spices, medicinal plants, toxins, and other possible choices for independent investigations.

Some procedures and many ideas in this lab topic were based on an exercise written by Jean Dickey, published in J. Dickey, *Laboratory Investigations for Biology.* Menlo Park, CA: Addison Wesley Longman, 1995.

The procedure used to assay mitochondrial activity was based on a procedure from "Succinic Acid Dehydrogenase Activity of Plant Mitochondria," in F. Witham, D. Blaydes, and R. Devlin, *Exercises in Plant Physiology.* Boston, MA: Prindle, Weber & Schmidt, 1971.

# Websites

Cornell University Poisonous Plants Page:
http://www.ansci.cornell.edu/plants/index.html

Metabolic poisons:
http://www.ruf.rice.edu/~bioslabs/studies/mitochondria/mitopoisons.html

Oxidative Phosphorylation: Uncouplers and Inhibitors, University of Leeds, School of Biochemistry and Molecular Biology:
http://www.bmb.leeds.ac.uk/illingworth/oxphos/poisons.htm

U.S. Dept. of Energy, Energy Efficiency and Renewable Energy:
http://www.eere.energy.gov/biomass/abcs_biofuels.html

# LAB TOPIC 5

# Cellular Respiration and Fermentation
## Teaching Plan for Laboratories

## Main Concepts and Objectives

1. Concept: alcoholic fermentation. Students will be able to describe the process of alcoholic fermentation, identifying reactants and products. (Exercise 5.1)

2. Concept: cellular respiration. Students will describe the process of cellular respiration, identifying reactants and products, and contrast this process with alcoholic fermentation. (Exercise 5.2)

3. Concept: oxidation-reduction reactions in cellular respiration. Students will describe how DPIP can substitute for fumarate in living cells and receive electrons from succinate. The extent of DPIP color change indicates the level of cellular respiration. (Exercise 5.2)

4. Concept: spectrophotometry. Students will use a Spectronic 20 to measure the transmittance of light through DPIP samples as a measure of the rate of cellular respiration. (Exercise 5.2)

5. Concept: factors influencing fermentation and cellular respiration.* Students will list factors (inhibitors, pH, temperature, salt concentration, and concentrations of enzyme and substrate). (Exercises 5.1–5.3)

6. Concept: the scientific process. Students will design and perform an original experiment asking questions, proposing hypotheses, predicting results of the experiment based on hypotheses, presenting their results in tables and figures, interpreting and discussing the results, and stating conclusions. (Exercise 5.3)

7. Specific content. Students will be able to describe or define: reactants and products of alcoholic fermentation and aerobic respiration; factors affecting enzyme activity; *fumarate, succinate, DPIP, substrate, reactant, $NAD^+$, NADH, FAD, $FADH_2$, spectrophotometry, transmittance, absorption*; scientific process.

*May vary depending on the investigations students choose.

## Order of the Lab

If possible in your laboratory situation, 1 week before this lab, instruct students to read the lab topic and meet with their investigative team to discuss possible independent investigations *before* coming to the lab. You may ask them to submit a materials list prior to the lab. We require our students to bring a draft proposal with materials list to lab.

This exercise is designed to be completed in one 3-hour laboratory period. Being able to complete the entire lab will depend on your students being prepared and organized.

1. Introduce the main concepts and explain the design of the exercise. Give an overview of fermentation and cellular respiration using Figure 5.2. Review enzyme activity if necessary.                    (15 min)

2. Perform Exercise 5.1, Experiment A.                    (45 min)

3. Introduce the use of the spectrophotometer. This instrument will be used in the next lab topic, Photosynthesis, so do not explain changing wave-length at this time.                    (20 min)

4.  Perform Exercise 5.2, Experiment A.                                    (45 min)

5.  Perform Exercise 5.3.                                                   (55 min)

**For a 2-hour lab:** Perform Experiment A in each exercise and have students design, but not perform, an original experiment. Alternatively, perform one Experiment A and design an independent investigation for that study.

## Classroom Management

Students work in teams of two or four. Demonstrate the setup of the respirometer as you introduce its use. Student research teams cooperate in planning their independent investigations. Encourage group discussion and cooperative learning.

## Student Development

In the introductory studies, students will practice asking questions, proposing hypotheses, making predictions, and collecting and interpreting data. As they design and execute their original experiments, they will practice the entire scientific process. If a writing assignment is required, they will practice organizing data in tables and graphs and using results to support conclusions. If oral reports are required, students will practice using persuasion and presenting data.

## Lab Safety Precautions

If toxins or respiration inhibitors will be used in the independent investigations, do not allow students to work with powders. The prep person should make up solutions in the hood using hand, eye, and clothing protection. Contact your campus biosafety personnel about use and disposal. Prepare warnings to be posted in the lab and on each inhibitor. Then warn students verbally. Have students wear disposable gloves and protective eyewear when working with the inhibitors. Be prepared to clean up any spill as directed in the Materials Safety Data Sheets supplied with each inhibitor.

## Discussion and Summary

Students will discuss results and complete questions in the exercises. The instructor may choose to have a representative from each student team report the results of the teams' experiment to the class in lab time or at some future meeting. Students may write a section of a scientific paper.

## Evaluation

You may evaluate the quality of the students' experiments and also grade them on their oral presentations. You may choose to require a written report in scientific format for evaluation. If you are following the plan to integrate scientific writing, as suggested in this manual, this lab topic may provide the opportunity to have students practice writing Results and Discussion sections. Writing a complete scientific paper for this lab might be premature because this lab takes place early in the first term of the course. However, if your students have written individual sections of a paper and have been evaluated on these sections, you may choose to have them write a complete paper.

# Lab Topic 6
# Photosynthesis

## Laboratory Objectives

After completing this lab topic, you should be able to:

1. Describe the roles played by light and pigment in photosynthesis.
2. Name and describe pigments found in photosynthesizing tissues.
3. Explain the separation of pigments by paper chromatography, based on their molecular structure.
4. Demonstrate an understanding of the process of spectrophotometry and the procedure for using the spectrophotometer.

*For a 2-hour lab: Separate and extract pigments before lab. Collect the data for the absorption spectrum and provide students with a copy of the data set. See Teaching Plan.*

*Instructors and/or students can download data tables in Excel format. Go to www .masteringbiology.com. Look in the Study Area under Lab Media.*

## Introduction

Without photosynthesis, there could be no life on Earth as we know it. The Earth is an open system constantly requiring an input of energy to drive the processes of life. All energy entering the biosphere is channeled from the sun into organic molecules via the process of photosynthesis. As the sun's hydrogen is converted to helium, energy in the form of photons is produced. These photons pass to Earth's surface and those with certain wavelengths within the visible light portion of the electromagnetic spectrum are absorbed by pigments in the chloroplasts of plants, initiating the process of photosynthesis.

In photosynthesis, light energy is transformed into chemical energy. That chemical energy is used to synthesize organic compounds (glucose) from $CO_2$, and in the process water is used and $O_2$ is released. Glucose, a primary source of energy for all cells, may be converted to sucrose and transported or stored in the polymer starch. These organic molecules are building blocks for plant growth and development. Animals consume plants and convert the plant molecules into their own organic molecules and energy sources—the ultimate in recycling. Oxygen, also produced by photosynthesis, is necessary for aerobic respiration in the cells of plants, animals, and other organisms (Figure 6.1).

In this laboratory, you will investigate cellular and environmental components utilized in the process of photosynthesis. In several experiments, you will determine photosynthetic activity by testing for the production of starch, using iodine potassium iodide ($I_2KI$), which stains starch purple-black. A change from the yellow-amber color of the iodine solution to a purple-black solution is a positive test for the presence of starch. This is the same test for the presence of starch that you used to study starch digestion by amylase in Lab Topic 4 Enzymes.

*In Lab Topic 4, students used the iodine test to observe the disappearance of the substrate starch. In this lab topic, the iodine test will show the appearance of the product starch.*

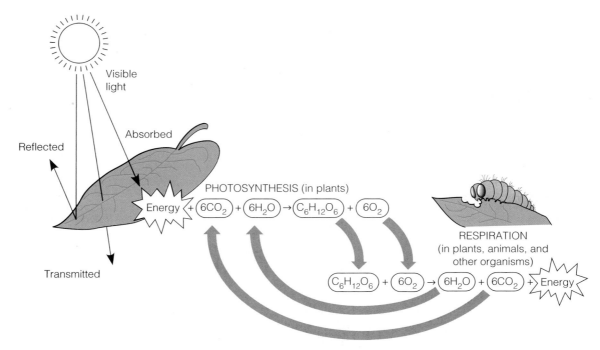

**Figure 6.1.**
**Energy flow through photosynthesis and respiration.** Energy flows from the sun into the biological systems of Earth, and visible light may be reflected, transmitted, or absorbed. Plants absorb light energy and convert it to chemical energy during photosynthesis. In this process, carbon dioxide and water are used to synthesize glucose (and ultimately other organic compounds) and to release oxygen. These organic molecules and the energy stored in them can be utilized by animals and other organisms that consume them. The energy in organic molecules is released to form ATP during cellular respiration in plants, animals, and other organisms.

*This exercise should be done along with Exercise 6.2 so that all leaves are boiled at the same time.*

E X E R C I S E   6 . 1
# The Wavelengths of Light for Photosynthesis

## Materials

black construction paper
green, red, and blue plastic filters
paper clips
forceps
1 geranium plant with at least 4 good leaves per 8 students
1 1,000-mL beaker filled with 300 mL of water
1 400-mL beaker filled with 200 mL of 80% ethyl alcohol
dropper bottle with concentrated $I_2KI$ solution

hot plate
petri dish
squirt bottle of water
scissors

## Introduction

In this exercise, you will investigate the photosynthetic activity of different wavelengths of light. You will determine if products of photosynthesis are present in leaf tissue that has been exposed to different wavelengths of light for several days. Working with other students in groups of eight, you will cover small portions of different leaves of a geranium plant with pieces of black paper and green, red, and blue plastic filters. Each pair of students will be responsible for one of the four treatments. Later you will determine photosynthetic activity by testing for the presence of starch under the paper or filters. Starch is synthesized from glucose molecules and can be easily tested as an indirect product of photosynthesis. Starch will turn dark in the presence of iodine ($I_2KI$).

Plastic filters used in this exercise are designed to reflect and transmit the appropriate wavelengths of light to correspond to the visible light spectrum. For example, green filters reflect and transmit green light. What wavelengths of light will be absorbed by the green filter? Refer to Color Plate 9 at the end of the lab manual, which shows the electromagnetic spectrum. What wavelength of light will be reflected and transmitted by the black paper and each of the colored filters? Note that the same wavelengths of light are reflected and transmitted by the filters.

> *Black reflects and transmits none. Green reflects and transmits green wavelengths, wavelengths around 550 nanometers (nm). Red reflects and transmits red wavelengths (around 700 nm). Blue reflects and transmits blue (around 500 nm).*

*See the Preparation Guide for source of special plastic filters. Do not buy locally. Plastic filters used in this exercise are designed to reflect and transmit the appropriate wavelengths of light to correspond to the visible spectrum. For example, green filters transmit green light to the leaf. The wavelengths of light that would be most efficient in photosynthesis are absorbed by this filter.*

## Hypothesis

Hypothesize about photosynthetic activity in leaf cells covered with different colored filters as described.

> *Photosynthesis will occur in areas where red and blue light are transmitted and reflected by the filter.*

*Students may pose a variety of hypotheses. As long as the hypotheses are testable, do not make corrections.*

## Prediction

Predict the results of the experiment based on your hypothesis (if/then).

> *If photosynthesis will occur in areas where red and blue light are transmitted and reflected by the filter, then the leaf area under red and blue filters will turn dark in the presence of $I_2KI$ and starch, but the leaf area under black and green filters will remain light.*

## Procedure

1. Four to 5 days before the experiment is to be carried out, cut a piece from each color of plastic filter and one from the black construction paper. Each piece should be a rectangle approximately 2.5 cm by 5 cm. Double over the strip and slide the edge of a healthy geranium leaf, still attached to the plant, between the folded edges. Carefully slip a slightly sprung paper clip over the paper, securing the paper to the leaf. The paper should be on both sides of the leaf. Follow this procedure with the other colors and the black construction paper, using a different leaf for each strip. Return the plant with treated leaves to bright light. Your instructor may have already carried out this step for you.

2. On the day of the lab, carry the plant with leaves covered to your desk. You will have to be able to recognize each leaf after the paper is removed and the leaf is boiled. To facilitate this, with your teammates devise a way to distinguish each leaf, and write the distinction in the space provided. Differences in size or shape may distinguish different leaves, but it may be necessary to introduce distinguishing features, such as by

*You can bring the plants into the lab the previous week and have students cover the leaves, or your prep staff can complete this step 4 to 5 days before this lab. For best results, keep the plants in light 24 hrs/day.*

cutting the petioles to different lengths or cutting out small notches in a portion of the leaf not covered by the paper. Record below the distinguishing differences for each treatment.

black paper:

green filter:

red filter:

blue filter:

3. After you have distinguished each leaf, sketch the leaf in the Results section, showing the position of the paper or filter on the leaf.

*Set up baths and turn on hot plates after introducing exercise. Each alcohol bath will accommodate two geranium leaves and two Coleus leaves from the next exercise.*

4. Set up the boiling alcohol bath. Place a 1,000-mL beaker containing 300 mL of water on the hot plate. Carefully place the 400-mL beaker containing 200 mL of 80% ethyl alcohol into the larger beaker of water. Turn on the hot plate and bring the nested beakers to a boil. Adjust the temperature to maintain slow boiling. Do not place the beaker of alcohol directly on the hot plate.

 **Ethyl alcohol is highly flammable! Do not place the beaker of alcohol directly on the hot plate. To bring it to a boil, raise the temperature of the hot plate until the alcohol just boils, and then reduce the temperature to maintain slow boiling. Do not leave boiling alcohol unattended.**

*Plastic filters are expensive and can be reused. The students should organize themselves so that each pair is working with one geranium treatment and one Coleus leaf, and they should do these exercises simultaneously. Turn off the hot plate when finished with both sets of leaves.*

*If you are using dilute iodine solutions, do not cover leaves with water in step 6. Prep notes are for concentrated iodine.*

5. Remove the paper and filters from each leaf; using forceps, carefully drop all the leaves into the boiling alcohol solution to extract the pigments. Save the plastic filters.

6. When the leaves are almost white, use forceps to remove them from the alcohol. Place them in separate petri dishes, rinse with distilled water, and add enough distilled water to each dish to just cover the leaf. Turn off the hot plate if all teams have completed boiling leaves for this exercise (and Exercise 6.2).

7. Add drops of $I_2KI$ solution to the water until a pale amber color is obtained. $I_2KI$ reacts with starch to produce a purple-black color.

8. Wait about 5 minutes and sketch each leaf in the Results section, showing which areas of the leaf tested positive for starch.

### Results

*For optimum results in this experiment, the filters must be properly secured so no light leaks under the filters. The plants also must be treated long enough for previously stored starch to be utilized in cellular respiration.*

1. Sketch and label each leaf before boiling, showing the location of the paper or filter.

2. Sketch and label each leaf after staining to show the location of the stain.

## Discussion

1. Which treatment allowed the greatest photosynthetic activity? (Explain your results in terms of your hypothesis.)

   *The leaves covered with black paper or a green filter tested negative for starch under the paper and filter. Red and blue will have a positive, but greatly reduced, response. Ask students why.*

2. When the red filter is placed on a leaf, what wavelengths of light pass through and reach the leaf cells below? (Check wavelengths in Color Plate 9, which shows the electromagnetic spectrum.)

   *Students' answers will vary.*

   *approximately 620 to 700 nm*

   Green filter?

   *approximately 500 to 580 nm*

   Blue filter?

   *approximately 425 to 500 nm*

3. Was starch present under the black construction paper? Explain this in light of the fact that black absorbs all wavelengths of light.

   *No. The paper absorbed all the light and did not allow any to pass through.*

---

## E X E R C I S E   6 . 2
# Pigments in Photosynthesis

---

## Materials

*Coleus* plant with multicolored leaves
forceps
1 1,000-mL beaker filled with 300 mL of water
1 400-mL beaker filled with 200 mL of 80% ethyl alcohol
dropper bottle with concentrated $I_2KI$ solution
hot plate
squirt bottle of water

## Introduction

A variety of pigments are found in plants, as anyone who visits a botanical garden in spring or a deciduous forest in autumn well knows. A pigment is a substance that absorbs light. If a pigment absorbs all wavelengths of visible light, it appears black. The black construction paper used in Exercise 6.1 is colored with such a pigment. Other pigments absorb some wavelengths and reflect others. Yellow pigments, for example, reflect light wavelengths in the yellow portion of the visible light spectrum, green reflects in the green portion, and so on.

Some colors are produced by only one pigment, but an even greater diversity of colors can be produced by the cumulative effects of different pigments in cells. Green colors in plants are produced by the presence of chlorophylls *a* and *b* located in the chloroplasts. Yellow, orange, and bright red colors are produced by carotenoids, also in chloroplasts. Blues, violets, purples, pinks, and dark reds are usually produced by a group of water-soluble pigments, the anthocyanins, that are located in cell vacuoles and do not contribute to photosynthesis. Additional colors may be produced by mixtures of these pigments in cells.

*This exercise should be done along with Exercise 6.1 so that all leaves are boiled at the same time.*

Working with one other student, you will use the $I_2KI$ test for starch as in Exercise 6.1 to determine which pigment(s) in a *Coleus* leaf support photosynthesis. Before beginning the experiment, examine your *Coleus* leaf and hypothesize about the location of photosynthesis based on the leaf colors. (See Color Plate 10.)

### Hypothesis

Hypothesize about the location of photosynthesis based on the leaf colors.

*Photosynthesis occurs in the presence of green chlorophyll pigment.*

### Prediction

Predict the results of the experiment based on your hypothesis (if/then).

*If photosynthesis occurs in the presence of green chlorophyll pigment, then an $I_2KI$ test will produce a black stain in the green and purple areas but not in the other areas. Most students will hypothesize that an $I_2KI$ test will produce a black stain in the green areas only, which will promote an interesting discussion.*

### Procedure

*Select leaves that are pink, purple, green, and white (or yellow).*

1. Remove a multicolored leaf from a *Coleus* plant that has been in strong light for several hours.
2. In Table 6.1, list the colors of your leaf, predict the pigments present to create that color, and predict the results of the $I_2KI$ starch test in each area of the leaf.
3. Sketch the leaf outline in the Results section, mapping the color distribution before the $I_2KI$ test.
4. Extract the pigments as previously described in Exercise 6.1, and test the leaf for photosynthetic activity using $I_2KI$.

 Ethyl alcohol is highly flammable! Do not place the beaker of alcohol directly on the hot plate. To bring it to a boil, raise the temperature of the hot plate until the alcohol just boils, and then reduce the temperature to maintain slow boiling. Do not leave boiling alcohol unattended.

**Table 6.1**
Predicted and Observed Results for the Presence of Starch in Colored
Regions of the *Coleus* Leaf

| Color | Pigments | Starch Present (predicted) + or − | Starch Present (actual results) + or − |
|---|---|---|---|
| Green | chlorophyll a chlorophyll b | + | + |
| Purple | chlorophylls anthocyanin | + | + |
| Pink | anthocyanin | − | − |
| White | no pigments | − | − |
| Other | | | |

*The* Coleus *you use may not have all of these colors.*

5. Sketch the leaf again in the Results section, outlining the areas showing a positive starch test.

## Results

1. Record the results of the $I_2KI$ test in Table 6.1.
2. Compare the sketches of the *Coleus* leaf before and after the $I_2KI$ test.

   **Before $I_2KI$ Test:**                **After $I_2KI$ Test:**

3. Which pigments supported photosynthesis? Record your results in Table 6.1.

   *chlorophyll in both green and purple regions*

## Discussion

Describe and explain your results based on your hypothesis.

*As hypothesized, photosynthesis took place in chlorophyll-containing regions (green and purple). The pink and yellow areas lack photosynthetic pigments and therefore do not support photosynthesis.*

# EXERCISE 6.3
# Separation and Identification of Plant Pigments by Paper Chromatography

## Materials

| | |
|---|---|
| capillary tube | forceps |
| beakers | scissors |
| extractions of leaf pigments in acetone | acetone |

chromatography paper stapled into a cylinder marked with a pencil line about 1 cm from one end

quart jar with lid, containing solvent of petroleum ether and acetone

*Extracting pigments from grass, spinach, or other greens is an easy process (see* Preparation Guide, *7th ed.). As a demonstration, show students the grass or spinach soaking in acetone. The experiment will be more meaningful if students see that the green color of plants is actually a complex of pigments.*

## Introduction

Your instructor has prepared an extract of chloroplast pigments from fresh green grass or fresh spinach. A blender was used to rupture the cells, and the pigments were then extracted with acetone, an organic solvent. Working with one other student, begin this exercise by separating the pigments extracted using paper chromatography. To do this, you will apply the pigment extract to a cylinder of chromatographic paper. You will then place the cylinder in a jar with the organic solvents petroleum ether and acetone. The solvents will move up the paper and carry the pigments along; the pigments will move at different rates, depending on their different solubilities in the solvents used and the degree of attraction to the paper. The leading edge of the solvent is called the **front.** Discrete pigment bands will be formed from the front back to the point where pigments were added to the paper.

The following information will be helpful to you as you make predictions and interpret results:

1. **Polar molecules** or substances dissolve (or are attracted to) polar molecules.
2. **Nonpolar molecules** are attracted to nonpolar molecules to varying degrees.
3. Chromatography paper (cellulose) is a polar (charged) substance.
4. The solvent, made of petroleum ether and acetone, is relatively nonpolar.
5. The *most nonpolar* substance will dissolve in the nonpolar solvent *first*.
6. The *most polar* substance will be attracted to the polar chromatography paper; therefore, it will move *last*.

Use this information and the molecular structure of major leaf pigments to predict the relative solubilities and separation patterns for the pigments and to identify the pigment bands. Study the molecular structure of the four common plant pigments in Figure 6.2. As you study these diagrams, rank the pigments according to polarity in the space provided. To determine polarity, count the number of polar oxygens present in each molecule.

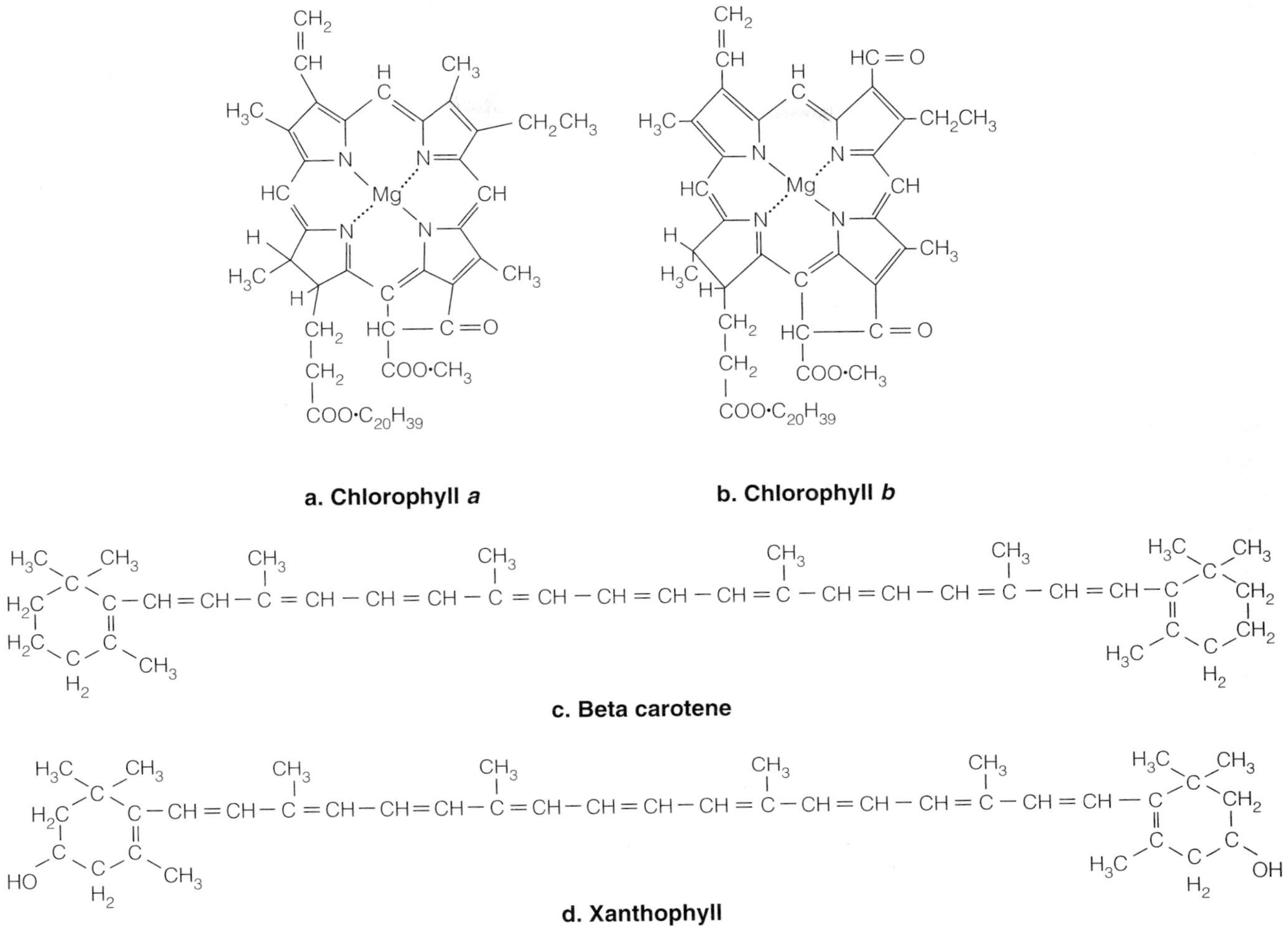

**a. Chlorophyll *a***

**b. Chlorophyll *b***

**c. Beta carotene**

**d. Xanthophyll**

**Figure 6.2.**
**Molecular structure of major leaf pigments.** The molecular structure of
chlorophyll (*a*), chlorophyll (*b*), carotene (*c*), and xanthophyll (*d*). To determine
polarity, count the number of polar oxygens present in each molecule.

Most polar:   *chlorophyll b: six polar groups*
              *chlorophyll a: five polar groups*
              *xanthophyll: two polar groups*
Least polar:  *carotene: no polar groups*

## Hypothesis

State a hypothesis relating polarities and solubilities of pigments.

*Carotene is the least polar pigment and will be the most soluble in
the solvent, followed by xanthophyll and chlorophyll a. Chlorophyll
b is the most polar and will be the least soluble in the solvent.*

## Prediction

Predict the results of the experiment based on your hypothesis (if/then).

*If the hypothesis is true, then carotene will travel the greatest
distance on the paper, moving with the solvent front. Xanthophyll
will be next, followed by chlorophyll a. If chlorophyll b is the least
soluble, then it will travel the least distance.*

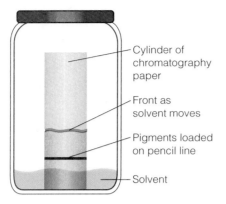

**Figure 6.3.**

**Paper chromatography of photosynthetic pigments.** Add the pigment solution to the paper cylinder along the pencil line. Then carefully place the cylinder into a jar containing a small amount of solvent. Close the lid and watch the pigments separate according to their molecular structures and solubilities.

*While waiting for separation of pigments, organize students into groups for Exercise 6.4 and review procedures.*

## Procedure

1. Using a capillary tube, streak the leaf pigment extract on a pencil line previously drawn 1 cm from the edge of the paper cylinder. Allow the chlorophyll to dry. Repeat this step three or four times, allowing the extract to air-dry each time. You should have a band of green pigments along the pencil line. The darker your band of pigments, the better the results of your experiment will be.

---

 Perform the next step in a hood or in a well-ventilated room. Do not inhale the fumes of the solvent. *NO SPARKS!* Acetone and petroleum ether are extremely flammable. Avoid contact with all solutions. Wash hands with soap and water. If a spill occurs, notify the instructor. If an instructor is not available, do not attempt to clean up. Leave the room.

---

2. Obtain the jar containing the petroleum ether and acetone solvent. Using forceps, carefully lower the loaded paper cylinder into the solvent, and quickly cover the jar tightly with the lid (Figure 6.3). *Avoid inhaling the solvent.* The jar should now contain a saturated atmosphere of the solvent. Allow the chromatography to proceed until the solvent front has reached to within 3 cm of the top of the cylinder.
3. Remove the cylinder from the jar, allow it to dry, and remove the staples.
4. Save your paper with the separated pigments for the next exercise.

## Results

Sketch the chromatography paper. Label the color of the various bands. The front, or leading edge of the paper, should be at the top. The pencil line where pigment was added originally should be at the bottom.

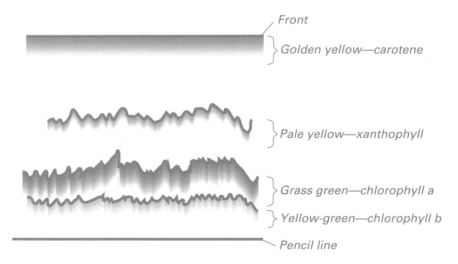

*If there are other dull gray bands, these may be chlorophyll decomposition products. Discard those sections of paper.*

## Discussion

Based on your hypothesis and predictions, identify the various pigment bands. *The entire class should come to a consensus about the identifications.* Label your drawing in the Results section above, indicating the correct identification of the pigment bands.

# EXERCISE 6.4
# Determining the Absorption Spectrum for Leaf Pigments

## Materials

spectrophotometer
Kimwipes®
2 cuvettes
20-mL beakers to elute pigments

1 150-mL beaker to hold cuvettes
acetone
cork stoppers for cuvettes

## Introduction

In Exercise 6.1, you applied colored plastic filters and black paper to leaves to determine which wavelengths of light would support photosynthesis. Review your conclusions from that exercise and from Exercise 6.2 about pigments used in photosynthesis. Which pigments did you conclude support photosynthesis?

*green pigment chlorophyll* a

In Exercise 6.4, you will work in teams of four or five students, carrying your investigation a step further by plotting the absorption spectrum of leaf pigments separated by paper chromatography. The **absorption spectrum** is the absorption pattern for a particular pigment, showing relative absorbance at different wavelengths of light. For example, we know that chlorophyll *a* is a green pigment, and we know that it reflects or transmits green wavelengths of light. We do not know, however, the relative proportions of wavelengths of light absorbed by chlorophyll *a*. This information is of interest because it suggests that those wavelengths showing greatest absorbance are important in photosynthesis.

The absorption spectrum can be determined with an instrument called a **spectrophotometer,** or **colorimeter**. A spectrophotometer measures the proportions of light of different wavelengths (colors) absorbed and transmitted by a pigment solution. It does this by passing a beam of light of a particular wavelength (designated by the operator) through the pigment solution being tested. The spectrophotometer then measures the proportion of light transmitted or, conversely, absorbed by that particular pigment and shows the reading on the calibrated scale.

Before measuring the absorption spectrum of the four pigments separated by paper chromatography, consult the diagram of the electromagnetic spectrum (Color Plate 9), and predict the wavelengths of light at which absorption will be greatest for each pigment. Record your predictions in Table 6.2.

**Table 6.2**
Predicted/Pigments Wavelengths of Greatest Absorption for the Photosynthetic

| Pigment | Wavelengths of Greatest Absorption (predicted) |
|---------|------------------------------------------------|
| **1. Chlorophyll *a*** | *430, 662* |
| **2. Chlorophyll *b*** | *453, 642* |
| **3. Carotene** | *approx. 450–550* |
| **4. Xanthophyll** | *approx. 450–550* |

*Student predictions should approximate these wavelengths.*

## Hypothesis

State a hypothesis that describes the general relationship of each of the pigments to the color of light that it absorbs.

*The pigments will reflect and transmit light of the color seen and absorb all others.*

*Demonstrate reflected and transmitted light by holding up a piece of cellophane.*

## Prediction

Predict the results based on your hypothesis (if/then).

*If the pigments reflect and transmit light of the color seen and absorb all others, then chlorophylls a and b, being green pigments, should absorb red and blue; xanthophyll and carotene, being yellow and orange, will absorb blue and green.*

## Procedure

1. Cut out the pigments you separated by paper chromatography, and distribute the paper strips as follows:

   *Team 1:* carotene

   *Team 2:* xanthophyll

   *Team 3:* chlorophyll *a*

   *Team 4:* chlorophyll *b*

   *Teams 5 and 6:* will determine the absorption spectrum of the total pigment solution

 Perform the next three steps in a hood or in a well-ventilated room. Do not inhale the fumes of the solvent. *NO SPARKS!* Acetone is extremely flammable. Avoid contact with all solutions. If a spill occurs, notify the instructor. Wash hands with soap and water.

2. *Teams 1 to 4.* Dilute the pigments as follows: Cut up the chromatography paper with your assigned pigment into a small (20-mL) beaker. Add 10 mL of acetone to the beaker and swirl. This solution containing a single pigment will be your solution B, to be used to determine

the absorption spectrum for that pigment. Your reference material will be acetone with no pigments, solution A.

3. *Teams 5 and 6.* Add drops of the original chlorophyll extract solution (acetone pigment mixture) to 10 mL of acetone until it looks pale green. This will be your pigment solution for cuvette B. Your reference material will be acetone with no pigment. This will be in cuvette A.

4. Each team should fill two cuvettes two-thirds full, one (B) with the pigment solution, the other (A) with the reference material (acetone only). Wipe both cuvettes with a Kimwipe to remove fingerprints, and handle cuvettes only with Kimwipes as you proceed.

   What is the purpose of the cuvette with reference material only?

   *This cuvette corrects for differences in absorption due to the acetone absorption. Your readings are only for the pigment, not the pigment and the acetone. It is similar to a control.*

5. Measure the absorption spectrum. Record your measurements in Table 6.3. The instructions that follow are for a Thermo Scientific Spectronic 20D+ (Figure 6.4). Turn on the machine (power switch C) for at least 15 minutes before beginning.

   a. Press the mode selection button to select "Transmittance."

   b. *Select the beginning wavelength* using the wavelength control knob (A). Begin measurements at 400 nanometers (nm).

   c. *Zero the instrument* by adjusting the 0 control knob (same as the power switch C) so that the meter reads 0% transmittance. There should be no cuvette in the instrument, and the sample holder cover must be closed.

   d. *Calibrate the instrument.* Insert cuvette A into the sample holder and close the lid. (Be sure to align the etched mark on the cuvette with the line on the sample holder.) Adjust the light control (F) until the meter reads 100% transmittance, or 0 absorption. You are now ready to make your first reading.

   e. *Change the mode.* Push the mode selection button to change the mode to "Absorbance."

*If you have two or three spectrophotometers, then have students work in groups reading and recording samples. Have the team measuring absorption in the total pigment go first because they do not have to elute their pigments. As a spectrophotometer is free, have the next group ready to read their samples. If you only have one spectrophotometer, you may choose to separate the pigments prior to lab and determine the absorption spectrum. These data could be provided to the class. If students are careful and clearly mark their tubes, they could read all four tubes at each wavelength before going on to the next increment. This would be tedious but would take less time than reading each spectrum four times. You might also consider beginning with this exercise, letting students read their samples while completing the other exercises.*

*If your students are using a different type of spectrophotometer, place a copy of the instructions next to each spectrophotometer.*

*For the analog Spectronic 20 operating procedures, see the preparation guide, 7th ed.*

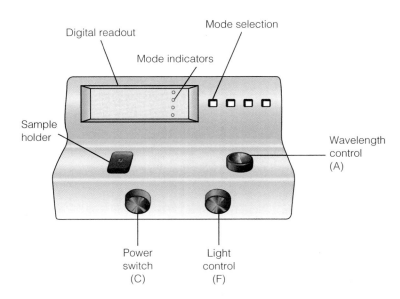

**Figure 6.4.**
**The Thermo Scientific Spectronic 20D+.** A spectrophotometer measures the proportion of light of different wavelengths absorbed and transmitted by a pigment solution. Inside the spectrophotometer, light is separated into its component wavelengths and passed through a sample. The graph of absorption at different wavelengths for a solution is called an *absorption spectrum.*

f. *Begin your readings.* Remove cuvette A and insert cuvette B. (Align the etched mark.) Close the cover. Record the reading on the absorbance scale. Remove cuvette B.

g. *Recalibrate the instrument.* Insert cuvette A into the sample holder, and set the wavelength to 420 nm. Again, calibrate the instrument to 100% transmittance (0 absorption) with cuvette A in place, using the light control (F). *Note: Each time you calibrate to 100% transmittance use the mode selection button to choose "Transmittance" and then switch back to "Absorbance" for your reading.*

h. *Take the second reading.* Remove cuvette A and insert cuvette B. Record absorbance at 420 nm. Remove cuvette B.

i. *Continue your observations,* increasing the wavelength by 20-nm increments until you reach 720 nm. Be sure to recalibrate each time you change the wavelength.

6. Pool data from all teams to complete Table 6.3. The table is available in Excel format.

**Table 6.3**
Absorbance of Photosynthetic Pigments Extracted from Fresh —————————————*

| Wavelength | Chlorophyll *a* | Chlorophyll *b* | Xanthophyll | Carotene | Total Pigment |
|:---:|:---:|:---:|:---:|:---:|:---:|
| 400 | | | | | |
| 420 | | | | | |
| 440 | | | | | |
| 460 | | | | | |
| 480 | | | | | |
| 500 | | | | | |
| 520 | | | | | |
| 540 | | | | | |
| 560 | | | | | |
| 580 | | | | | |
| 600 | | | | | |
| 620 | | | | | |
| 640 | | | | | |
| 660 | | | | | |
| 680 | | | | | |
| 700 | | | | | |
| 720 | | | | | |

*Complete title with name of plant used for extract, for example, beans.

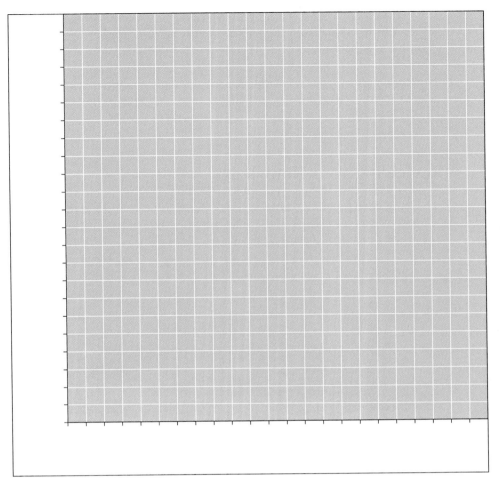

**Figure 6.5.**
**Absorption spectrum for chlorophyll *a*, chlorophyll *b*, carotene, and**
**xanthophyll.** Plot your results from Exercise 6.4. Label all axes, and draw
smooth curves to fit the data. Label the graph for easy identification of pigments.

## Results

1. Using the readings recorded in Table 6.3, plot in Figure 6.5 the absorption spectrum for each pigment.

2. Refer to Lab Topic 1 for information about constructing graphs. Choose appropriate scales for the axes, determine dependent and independent variables, and plot data points. Draw smooth curves to fit the values plotted. Label the graph for easy identification of pigments plotted, or prepare a legend and use colored pencils.

*Project the table and graph using the Excel version or use an overhead transparency of Table 6.3 to pool class results.*

## Discussion

1. List in the margin or on another page the pigments extracted and the optimum wavelength(s) of light for absorption for each pigment.

2. Which pigment is most important in the process of photosynthesis? Support your choice with evidence from your results.

*chlorophyll a, because it absorbs the greatest amount of visible light*

3. Chlorophyll *b* and carotenoids are called *accessory pigments*. Using data from your results, speculate about the roles of these pigments in photosynthesis.

*These pigments increase the spectrum of light available for use by the plant. The energy absorbed by these pigments is passed to chlorophyll a.*

## Reviewing Your Knowledge

1. Using your previous knowledge of photosynthesis and the results from today's exercises, explain the role, origin, or fate of each factor involved in the process of photosynthesis.

2. A pigment solution contains compound A with 4 polar groups and compound B with 2 polar groups. You plan to separate these compounds using paper chromatography with a nonpolar solvent. Predict the location of the two bands relative to the solvent front. Explain your answer.

*Compound B with only 2 polar groups will be closer to the front. The lesser polar molecule (B) will be more soluble in the solvent and move with the solvent front. The more polar (A) will be less soluble, have the greater attraction to the paper, and therefore move more slowly.*

## Applying Your Knowledge

1. Rice is the number one food crop, feeding over 50% of the world's population. Some scientists estimate that at the current rate of population growth, rice farmers will need to produce 50% more rice per hectare by 2050. Researchers are working to increase the photosynthetic efficiency of rice to meet the concern over food shortage. Using your knowledge of photosynthesis, suggest features of the plant and photosynthetic process that could be modified. Think creatively!

*Modifications may include increasing the concentration of chloroplasts and chlorophyll in leaves, maximizing the size of leaves, increasing drought resistance, and genetically modifying the pigment system to include more accessory pigments that expand the absorption spectrum.*

2. In response to shortened day length and cool temperatures in the fall, many trees begin a period of senescence when the breakdown of chlorophyll exceeds chlorophyll production. The leaves of these trees appear to change to yellow and orange. Using your knowledge of photosynthetic pigments, explain the source of these yellow-orange hues.

   *Chlorophyll breaks down, the green color disappears, and the accessory pigments, including the yellow and orange carotenoids, become visible. Other pigments, particularly brilliant reds, are synthesized in the autumn.*

3. Duckweed is a tiny floating aquatic plant that can reproduce so rapidly that it can completely cover the surface of small fish hatchery ponds. One proposal to limit the growth of this invasive plant is to use colored panels to shield the ponds, inhibit photosynthesis, and kill this pest. What color filters would you order for the panels in this project? What wavelengths of light will be absorbed by these filters? Do you think this proposal will be effective in killing the duckweed?

   *You should order green panels, which reflect and transmit green light and absorb all other colors, including red and blue, which are effective in photosynthesis. Green filters transmit approximately 500–580 nm of light. This is an engineering opportunity that would be fun to see!*

4. In a classic experiment in photosynthesis performed in 1883 by the German botanist Thomas Engelmann, he surrounded a filament of algae with oxygen-requiring bacteria. He then exposed the algal strand to the visible-light spectrum along its length. In which wavelengths of light along the algal strand would you expect the bacteria to cluster? Explain.

   *This experiment was the first to demonstrate the action spectrum for photosynthesis. The largest number of bacteria clustered around the algal segments illuminated by red and blue wavelengths of light. In these regions, cells of the photosynthesizing algal filament were releasing the most oxygen, attracting the bacteria.*

5. An **action spectrum** is a graph that illustrates the efficacy of different wavelengths of light in promoting photosynthesis. Engelmann's elegant experiment in question #4 is an example of an action spectrum where the efficacy of various wavelengths of light was measured by the production of oxygen and the clustering of bacteria. How does an **action spectrum** differ from the **absorption spectrum** that you created in Exercise 6.4?

   *An absorption spectrum measures the wavelengths of light absorbed by isolated pigments independent of the role played by these pigments in photosynthesis. An action spectrum shows the efficacy of wavelengths of light in the process of photosynthesis.*

## Investigative Extensions

Design and perform comparative investigations of plant pigment systems. The technique in this lab topic can be used to compare plant pigments for cultivated varieties of a species, plants growing in different environments, different species, or even seasonal differences for a single tree or several species.

Suggestions for comparative studies:

- Plants grown in shade versus sun
- Varieties of hostas or other plants that have blue-green, yellow-green, or bright green leaves
- Variegated plants that have combinations of yellow-green and bright green leaves compared to plants with uniformly green leaves
- Fresh fall leaves from red maples, hickories, or other species
- Fresh leaves from hardwoods collected and tested early in the fall while still green, then at 1-week intervals as they change colors

Note: Remember that anthocyanins are water soluble pigments located in cell vacuoles and they do not contribute to photosynthesis.

## Student Media: BioFlix, Activities, Investigations, and Videos
www.masteringbiology.com (select Study Area)

**BioFlix**—Ch. 10: Photosynthesis
**Activities**—Ch. 10: The Sites of Photosynthesis; Overview of Photosynthesis; Light Energy and Pigments; The Light Reactions; The Calvin Cycle
**Investigations**—Ch. 10: How Is the Rate of Photosynthesis Measured? How Does Paper Chromatography Separate Plant Pigments?

## References

Graham, L. E., J. H. Graham, and L. W. Wilcox. *Plant Biology*, 2nd ed. San Francisco: Benjamin Cummings, 2005.

Mauseth, James. *Botany*, 4th ed. Sudbury, MA: Jones and Bartlett, 2008.

Motten, Alex. "Diversity of Photosynthetic Pigments." *Tested Studies for Laboratory Teaching* (Volume 25), Proceedings of the 25th Workshop/Conference of the Association for Biology Laboratory Education (ABLE), Michael A. O'Donnell, Editor, 2003.

Nelson, D., and M. Cox. *Lehninger Principles of Biochemistry*, 4th ed. New York: Worth Publishers, 2005.

Taiz, L., and E. Zeigler. *Plant Physiology*, 4th ed. Sunderland, MA: Sinauer, 2006.

## Websites

Arizona State University Center for Bioenergy and Photosynthesis:
http://bioenergy.asu.edu/
http://photoscience.la.asu.edu/photosyn/education/learn.html

# LAB TOPIC 6

# Photosynthesis
## Teaching Plan for Laboratories

## Main Concepts and Objectives

1. Concept: the roles of light and pigment in photosynthesis. Students will investigate and be able to discuss the wavelengths of light used in photosynthesis and the relationships among light, wavelengths, pigments, and colors as related to photosynthesis. (Exercises 6.1–6.4)

2. Concept: starch and oxygen as products of photosynthesis. Students will use the $I_2KI$ test for starch and the production of oxygen as indications of photosynthetic activity. Students will discuss the conversion of glucose to starch and the role of light in photosynthesis. (Exercises 6.1 and 6.2)

3. Concept: paper chromatography. Students will describe the process and discuss how it may be used to separate leaf pigments. (Exercise 6.3)

4. Concept: spectrophotometry. Students will describe and use this process to measure the absorption spectrum of leaf pigments. (Exercise 6.4)

5. Specific content: the reactants and products of photosynthesis, details of the processes of paper chromatography and spectrophotometry, *absorption spectrum, pigment.*

## Order of the Lab

This is a very busy laboratory. Students will be working on several experiments simultaneously. Encourage the students to organize their time and teams to complete the exercises and to pool data where needed.

1. Introduce the material. Using the summary equation for photosynthesis, briefly give an overview of the process. Discuss laboratory objectives. Briefly introduce paper chromatography and spectrophotometry.                                          (20 min)

2. At this point students should turn on hot plates. They will be needed for Exercises 6.1 and 6.2. Explain procedures for Exercises 6.1 and 6.2, cautioning students about safety procedures when using hot plates to boil alcohol.                              (10 min)

3. Perform Exercises 6.1 and 6.2.                                            (45 min)

4. Discuss results of Exercises 6.1 and 6.2.                                 (15 min)

5. Separate leaf pigments by paper chromatography.                          (30 min)

6. Measure absorption spectrum of leaf pigments (Exercise 6.4).             (30 min)

7. Graph and discuss results of Exercise 6.3.                               (25 min)

8. Assign Introduction and Bibliography sections for students to complete outside of class.                                              (5 min)

**For a 2-hour lab:** If time and spectrophotometers are limited in the lab, you can separate and extract the pigments before lab begins. Collect the data for the absorption spectrum and then provide students with a copy of the data set. An alternative is to begin the lab with the chromatography experiment (Exercise 6.3). Four pairs of students can be assigned the task of completing the spectrophotometry during the remainder of the lab period while the rest of the class completes the other exercises. Be sure that everyone understands the spectrophotometer and that all team members collect data and discuss the results from all experiments.

## Classroom Management

Students work in teams of eight for Exercise 6.1, in teams of four for Exercise 6.4, and in teams of two for Exercises 6.2 and 6.3. Each student team should record data from all teams in the spectrophotometry exercise.

## Student Development

Students practice using the scientific method, asking questions, developing hypotheses, making predictions, and recording data. Students must practice organizational skills. They learn two new investigative processes, paper chromatography and spectrophotometry.

## Lab Safety Precautions

1. *Hot plates.* Students should avoid contacting the hot surface. Students cannot tell by appearance if a hot plate is hot. (They may expect it to turn red as a stove burner does.) A beaker of water on the hot plate will usually steam, providing a visual cue that the hot plate is hot.

2. *Boiling ethyl alcohol.* Use nested beakers as described in the lab exercises. Place a large beaker of water on the hot plate. Place a smaller beaker of alcohol inside the large beaker. Slowly heat the water in the large beaker until it is just boiling, and watch carefully that it does not overheat. Use forceps to add and remove leaves. Do not place alcohol on the hot plate.

3. *Acetone.* Flammable. Harmful if swallowed or inhaled. Use in adequate ventilation. Causes drying and irritation of skin. Wash skin with soap and water if contact is made. If a spill occurs, eliminate all sources of ignition. Cover the spill with vermiculite or other flammable-solvent absorbent material, scoop it up, and dispose of it (or call your school's safety officer).

4. *Petroleum ether.* Extremely flammable. Harmful if swallowed or inhaled. Keep in tightly closed container. Avoid breathing vapors. Use in hood or with adequate ventilation. Avoid contact with eyes, skin, and clothing. Wash hands thoroughly after handling. If a spill occurs, eliminate all sources of ignition. Evacuate the area. Cover the spill with an activated carbon absorbent, take it up, and place it in a closed container. Transport it outdoors. Ventilate the area and wash the spill site after material pickup is complete. Wear heavy rubber gloves for cleanup. Use a self-contained breathing apparatus.

## Discussion and Summary

Because understanding later exercises depends on results in earlier exercises, each exercise and its results should be discussed when appropriate as the laboratory progresses. Students should present the results of each exercise to be discussed by the entire class.

## Evaluation

Cover concepts on the next laboratory test. If you are following the plan to integrate scientific writing as suggested in this manual, ask your students to submit Introduction and References sections. Vary assignments from year to year.

## Lab Topic 7
# Mitosis and Meiosis

## Laboratory Objectives

After completing this lab topic, you should be able to:

1. Describe the activities of chromosomes, centrioles, and microtubules in the cell cycle, including all phases of mitosis and meiosis.
2. Recognize human chromosomes in leukocytes.
3. Identify the phases of mitosis in root tip and whitefish blastula cells.
4. Describe differences in mitosis and cytokinesis in plant and animal cells.
5. Describe differences in mitosis and meiosis.
6. Explain crossing over, and describe how this can bring about particular arrangements of ascospores in the fungus *Sordaria*.

*For a 2-hour lab: Divide the lab topic into two labs, studying mitosis in the first lab and meiosis in the second lab. See the Teaching Plan for additional suggestions.*

## Introduction

The nuclei in cells of eukaryotic organisms contain chromosomes with clusters of **genes,** discrete units of hereditary information consisting of duplicated deoxyribonucleic acid (DNA). Structural proteins in the chromosomes organize the DNA and participate in DNA folding and condensation. When cells divide, chromosomes and genes are duplicated and passed on to daughter cells. Single-celled organisms divide for reproduction. Multicellular organisms have reproductive cells (eggs or sperm), but they also have somatic (body) cells that divide for growth or replacement.

In somatic cells and single-celled organisms, the nucleus divides by **mitosis** into two daughter nuclei, which have the same number of chromosomes and the same genes as the parent cell. For example, the epidermis or outer layer of skin tissue is continuously being replaced through cell reproduction involving mitosis. All of these new skin cells are genetically identical. Yeast and amoeba are both single-celled organisms that can reproduce asexually through mitotic divisions to form additional organisms—genetically identical clones.

Cancerous cells are characterized by uncontrolled mitotic and cell division, and therefore, the study of mitosis and its regulation is key to developing new cancer treatments. In 2009, three scientists studying chromosomes and the regulation of mitosis were awarded the Nobel Prize in medicine for their discovery of **telomeres.** Telomeres are DNA sequences on the ends of chromosomes that become shorter during every mitotic cycle of somatic cells. Without telomeres protecting the chromosome ends, important genes located at the ends of chromosomes might be lost in mitosis. The loss of telomeres in each division of somatic cells may be one of many regulatory mechanisms

*The time required for these exercises will depend on how much understanding students bring from their lecture class.*

*Excellent audiovisuals on cell reproduction, mitosis, and meiosis are available in the Study Area of masteringbiology.com. See the listings near the end of this lab topic under Student Media: BioFlix, Activities, Investigations, and Videos.*

**Figure 7.1.**
**The cell cycle.** In interphase ($G_1$, S, $G_2$), DNA replication and most of the cell's growth and biochemical activity take place. In the M phase, the nucleus divides in mitosis, and the cytoplasm divides in cytokinesis.

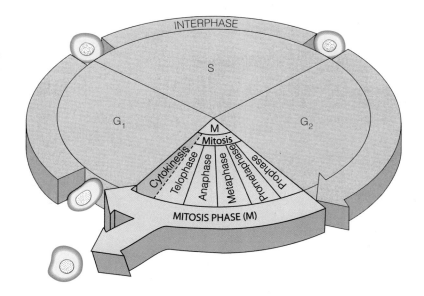

that limit the number of mitotic cycles that cells can undergo. If telomeres fail to shorten, cells may continue to divide, as in cancer cells.

In multicellular organisms, in preparation for sexual reproduction, a type of nuclear division called **meiosis** takes place. In meiosis, certain cells in ovaries or testes (or sporangia in plants) divide twice, but the chromosomes only replicate once. This process results in the four daughter nuclei with new combinations of chromosomes. Eggs or sperm (or spores in plants) are eventually formed. In contrast to mitosis, the process of meiosis contributes to the genetic variation that is important in sexual reproduction. Generally in both mitosis and meiosis, after nuclear division the cytoplasm divides, a process called **cytokinesis.**

Events from the beginning of one cell division to the beginning of the next are collectively called the **cell cycle.** The cell cycle is divided into two major phases: interphase and mitotic phase (M). The M phase represents the division of the nucleus and cytoplasm (Figure 7.1).

E X E R C I S E   7 . 1

# Modeling the Cell Cycle and Mitosis in an Animal Cell

## Materials

60 pop beads of one color
60 pop beads of another color

4 magnetic centromeres
4 centrioles

## Introduction

Scientists use models to represent natural structures and processes that are too small, too large, or too complex to investigate directly. Scientists develop their models from observations and experimental data, usually accumulated from a variety of sources. Building a model can represent the culmination of a body of scientific work, but most models represent a well-developed hypothesis that can then be tested against the natural system and modified.

Linus Pauling's novel and successful technique of building a physical model of hemoglobin was based on available chemical data. This technique was later adopted by Francis Crick and James Watson to elucidate the nature of the hereditary material, DNA. Watson and Crick built a wire model utilizing evidence collected by many scientists. They presented their conclusions about the structure of the DNA helix in the journal *Nature* in April 1953 and were awarded the Nobel Prize for their discovery in 1962.

Today in lab you will work with a partner to build models of cell division: mitosis and meiosis. Using these models will enhance your understanding of the behavior of chromosomes, centrioles, membranes, and microtubules during the cell cycle. After completing your model, you will consider ways in which it is and is not an appropriate model for the cell cycle. You and your partner should discuss activities in each stage of the cell cycle as you build your model. After going through the exercise once together, you will demonstrate the model to each other to reinforce your understanding.

In the model of mitosis that you will build, your cell will be a **diploid** cell ($2n$) with four chromosomes. This means that you will have two homologous pairs of chromosomes. In diploid cells, **homologous chromosomes** are the same length, have the same centromere position, and contain genes for the same characters. One pair will be long chromosomes, the other pair, short chromosomes. (**Haploid** cells have only one of each homologous pair of chromosomes, denoted $n$.)

## Lab Study A. Interphase

During interphase, a cell performs its specific functions: Liver cells produce bile; intestinal cells absorb nutrients; pancreatic cells secrete enzymes; skin cells produce keratin. Interphase consists of three subphases, $G_1$, S, and $G_2$, which begin as a cell division ends. As interphase begins, there is approximately half as much cytoplasm in each cell as there was before division. Each new cell has a nucleus that is surrounded by a **nuclear envelope** and that contains chromosomes in an uncoiled, or decondensed, state. In this uncoiled state, the mass of DNA and protein is called **chromatin**. Located outside the nucleus is the **centrosome**, a granular region that contains a pair of **centrioles** in animal cells. The centrosome is the organizing center for microtubules (Figure 7.2).

*Pipe cleaners or clay can be used to substitute for the pop bead models. Use different colors to represent maternally and paternally derived homologous chromosomes.*

### Procedure

1. Build a homologous pair of single chromosomes using 10 beads of one color for one member of the long pair and 10 beads of the other color for the other member of the pair. Place the magnetic centromere at any position in the chromosome, but note that it must be in the same position on homologous chromosomes. The centromere appears as a constricted region when chromosomes are condensed. Build the short pair with the same two different colors, but use fewer beads. You should have enough beads left over to duplicate each chromosome later.

2. Model **interphase** of the cell cycle:

   a. Pile all the assembled chromosomes in the center of your work area to represent the decondensed chromosomes as a mass of chromatin in $G_1$ (**gap 1**).

   b. Position two centrioles as a pair just outside your nucleus. Have the two members of the centriole pair at right angles to each other. (Recall, however, that most plant cells do not have centrioles.)

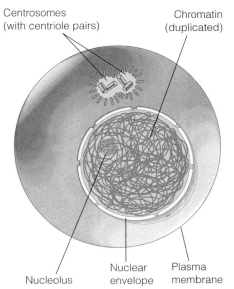

Centrosomes (with centriole pairs)   Chromatin (duplicated)

Nucleolus   Nuclear envelope   Plasma membrane

**Figure 7.2. Interphase.**

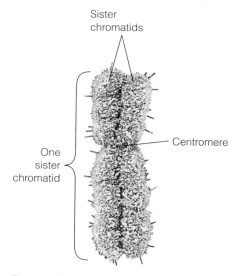

**Figure 7.3.**
**Duplicated chromosome composed of two sister chromatids held together at the centromere; condensed as in prometaphase.**

In the $G_1$ phase, the cytoplasmic mass increases and will continue to do so throughout interphase. Proteins are synthesized, new organelles are formed, and some organelles such as mitochondria and chloroplasts grow and divide in two. Throughout interphase one or more dark, round bodies, called **nucleoli** (singular, **nucleolus**), are visible in the nucleus.

c. Duplicate the centrioles: Add a second pair of centrioles to your model; again, have the two centrioles at right angles to each other. **Centriole duplication** begins in late $G_1$ or early S phase.

d. Duplicate the chromosomes in your model cell to represent DNA replication in the **S (synthesis) phase:** Make a second strand that is identical to the first strand of each chromosome. In duplicating chromosomes, you will use two magnets to form the new centromere. In the living cell, the centromere in a cell is a single unit until it splits in metaphase. In your model, consider the pair of magnets to be the single centromere.

Unique activities taking place during the S phase of the cell cycle are the replication of chromosomal DNA and the synthesis of chromosomal proteins. DNA synthesis continues until chromosomes have been duplicated. Each strand of a duplicated chromosome is called a **sister chromatid.** Sister chromatids are identical to each other and are held together tightly at the centromere (Figure 7.3).

e. Do not disturb the chromosomes to represent $G_2$ **(gap 2).**

During the $G_2$ phase, in addition to continuing cell activities, cells prepare for mitosis. Enzymes and other proteins necessary for cell division are synthesized during this phase.

f. Separate your centriole pairs, moving them toward opposite poles of the nucleus to represent that the $G_2$ phase is coming to an end and mitosis is about to begin.

How many pairs of homologous chromosomes are present in your cell during this stage of the cell cycle?

*two pairs of homologous chromosomes*

## Lab Study B. M Phase (Mitosis and Cytokinesis)

In the M phase, the nucleus and cytoplasm divide. Nuclear division is called *mitosis.* Cytoplasmic division is called *cytokinesis.* Mitosis is divided into five subphases: prophase, prometaphase, metaphase, anaphase, and telophase.

### Procedure

1. To represent **prophase,** leave the chromosomes piled in the center of the work area.

Prophase begins when chromosomes begin to coil and condense. At this time they become visible in the light microscope. Centrioles continue to move to opposite poles of the nucleus, and as they do so, a fibrous, rounded structure tapering toward each end, called a **spindle,** begins to form between them. Nucleoli begin to disappear (Figure 7.4).

What structures make up the fibers of the spindle? (Check text if necessary.)

*microtubules*

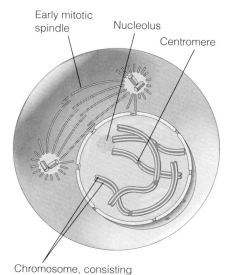

**Figure 7.4.**
**Prophase mitosis.**

2. At **prometaphase,** the centrioles are at the poles of the cell. To represent the actions in prometaphase, move the centromeres of your chromosomes to lie on an imaginary plane (the equator) midway between the two poles established by the centrioles.

During prometaphase chromosomes continue to condense (Figure 7.3). The nuclear envelope breaks down as the spindle continues to form. Some spindle fibers become associated with chromosomes at protein structures called **kinetochores.** Each sister chromatid has a kinetochore at the centromere. These spindle microtubules now extend from the chromosomes to the centrosomes at the poles. The push and pull of spindle fibers on the chromosomes ultimately leads to their movement to the equator. When the centromeres lie on the equator, prometaphase ends and the next phase begins (Figure 7.5).

How many duplicated chromosomes are present in your prometaphase nucleus?

*four*

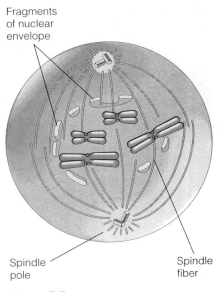

Fragments of nuclear envelope

Spindle pole

Spindle fiber

**Figure 7.5.**
**Prometaphase mitosis.**

 Students often find it confusing to distinguish between chromosome number and chromatid number. To simplify this problem, count the number of centromeres. The number of centromeres represents the number of chromosomes. In your model, the pair of joined magnets represents one centromere.

3. To represent **metaphase,** a relatively static phase, leave the chromosomes with centromeres lying on the equator.

In metaphase, duplicated chromosomes lie on the equator (also called the metaphase plate). The two sister chromatids are held together at the centromere. Metaphase ends as the centromere splits.

Label Figure 7.6 with *chromosome, spindle fibers, centrosome, centrioles, kinetochore, equator.*

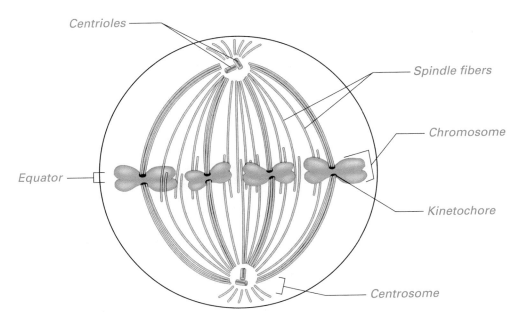

Centrioles

Spindle fibers

Chromosome

Equator

Kinetochore

Centrosome

**Figure 7.6.**
**The mitotic spindle at metaphase.**

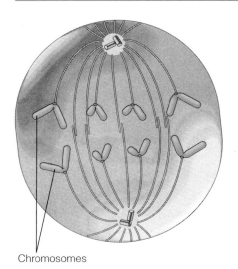

Chromosomes

**Figure 7.7.**
**Anaphase mitosis.**

4. Holding on to the centromeres, pull the magnetic centromeres apart and move them toward opposite poles. This action represents **anaphase.**

   After the centromere splits, sister chromatids separate and begin to move toward opposite poles. Chromatids are now called **chromosomes.** Anaphase ends as the chromosomes reach the poles (Figure 7.7).

   Describe the movement of the chromosome arms as you move the centromeres to the poles.

   *Arms trail passively behind the centromeres moving to the poles.*

   Certain biologists are currently investigating the role played by spindle fibers in chromosome movement toward the poles. Check your text for a discussion of one hypothesis, and briefly summarize it here.

   *The hypothesis that microtubules between centromeres and the poles depolymerize at the centromere (kinetochore) end, rendering microtubules shorter and shorter is discussed in Chapter 12 of Campbell Biology, 9th ed. Somehow the kinetochore remains attached to the microtubules, and chromosomes are pulled to the poles. The mechanism may be related to motor protein activities similar to those bringing about the movement of cilia and flagella.*

5. Pile your chromosomes at the poles to represent **telophase.**

   As chromosomes reach the poles, anaphase ends and telophase begins. The spindle begins to break down. Chromosomes begin to uncoil, and nucleoli reappear. A nuclear envelope forms around each new cluster of chromosomes. Telophase ends when the nuclear envelopes are complete (Figure 7.8).

   How many chromosomes are in each new nucleus?

   *four*

   How many chromosomes were present in the nucleus when the process began?

   *four*

   How would the condition of telomeres have changed from a previous cell cycle? (See the Introduction to this lab topic.)

   *The number of telomeres is reduced in every mitotic cycle of somatic cells.*

6. To represent cytokinesis, leave the two new chromosome masses at the poles.

   The end of telophase marks the end of nuclear division, or mitosis. Sometime during telophase, the division of the cytoplasm, or cytokinesis, results in the formation of two separate cells. In cytokinesis in cells of animals, fungi, and slime molds, a **cleavage furrow** forms at the equator and eventually pinches the parent cell cytoplasm in two (Figure 7.9a). Actin and myosin, the same molecules found in muscle cells, contribute to the formation of the cleavage furrow. In plant cells, membrane-bound vesicles migrate to the center of the equatorial plane and fuse to form the **cell plate.** This eventually extends across the cell,

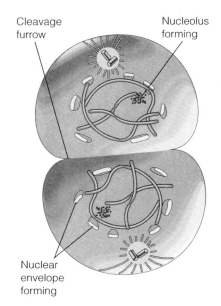

Cleavage furrow

Nucleolus forming

Nuclear envelope forming

**Figure 7.8.**
**Telophase mitosis.**

dividing the cytoplasm in two. Cell wall materials in the vesicles are released into the space between the membranes of the cell plate forming the new cell wall (Figure 7.9b, c).

a.

**Figure 7.9.**
**Cytokinesis in animal and plant cells.** (a) In animal cells, a cleavage furrow forms at the equator and pinches the cytoplasm in two. (b) In plants, a cell plate forms in the center of the cell and grows until it divides the cytoplasm in two. (c) Photomicrograph of cytokinesis in a plant cell.

Membrane-bound vesicles

Double membrane enclosing cell plate

New cell wall material

b.

Vesicles forming cell plate

Nucleus

Wall of parent cell

Nucleus

c.

# EXERCISE 7.2
# Observing Mitosis and Cytokinesis in Plant Cells

## Materials

prepared slide of onion root tip
compound microscope

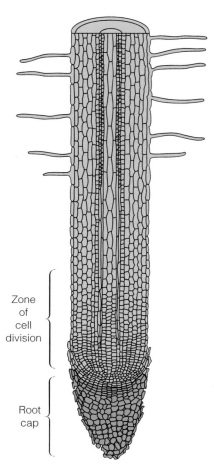

**Figure 7.10.**
**Longitudinal section through a**
**root tip.** Cells are dividing in the
zone of cell division just behind the
root cap.

## Introduction

The behavior of chromosomes during the cell cycle is similar in animal and plant cells. However, differences in cell division do exist. Plant cells have no centrioles, yet they have bundles of microtubules that converge toward the poles at the ends of a spindle. Cell walls in plant cells dictate differences in cytokinesis. In this exercise, you will observe dividing cells in the zone of cell division of a root tip (Color Plate 11).

## Procedure

1. Examine a prepared slide of a longitudinal section through an onion root tip using low power on the compound microscope.
2. Locate the region most likely to have dividing cells, just behind the root cap (Figure 7.10).

   At the tip of the root is a root cap that protects the tender root tip as it grows through the soil. Just behind the root cap is the **zone of cell division.** Notice that rows of cells extend upward from this zone. As cells divide in the zone of cell division, the root tip is pushed farther into the soil. Cells produced by division begin to mature, elongating and differentiating into specialized cells, such as those that conduct water and nutrients throughout the plant.
3. Focus on the zone of cell division. Then switch to the intermediate power, focus, and switch to high power.
4. Survey the zone of cell division and locate stages of the cell cycle: interphase, prophase, prometaphase, metaphase, anaphase, telophase, and cytokinesis.
5. As you find a dividing cell, speculate about its stage of division, read the following descriptions given of each stage to verify that your guess is correct, and, if necessary, confirm your conclusion with the instructor.
6. Draw the cell in the appropriate boxes provided. Label nucleus, nucleolus, chromosome, chromatin, mitotic spindle, and cell plate when appropriate.

## Interphase ($G_1$, S, $G_2$)

Nuclear material is surrounded by a nuclear envelope. Dark-staining bodies, nucleoli, are visible. Chromosomes appear only as dark granules within the nucleus. Collectively, the chromosome mass is called *chromatin*. The chromosomes are not individually distinguishable because they are uncoiled into long, thin strands. Chromosomes are replicated during this phase.

## Prophase

Chromosomes begin to coil and become distinguishable thin, threadlike structures, widely dispersed in the nucleus during prophase. Although there are no centrioles in plant cells, a spindle begins to form. Nucleoli begin to disappear. The nuclear envelope is still intact.

## Prometaphase

By prometaphase, the chromosomes are thick and short. Each chromosome is duplicated consisting of two chromatids held together by a centromere. The nuclear membrane and nucleoli break down in prometaphase. Chromosomes move toward the equator.

## Metaphase

Metaphase begins when the centromeres of the chromosomes lie on the equator of the cell. The arms of the chromatids extend randomly in all directions. A spindle may be apparent. Spindle fibers are attached to kinetochores at the centromere region and extend to the poles of the cell. As metaphase ends and anaphase begins, the centromeres split.

### Anaphase

The splitting of centromeres marks the beginning of anaphase. Each former chromatid is now a new single chromosome. These chromosomes are drawn apart toward opposite poles of the cell. Anaphase ends when the migrating chromosomes reach their respective poles.

### Telophase and Cytokinesis

Chromosomes have now reached the poles. The nuclear envelope re-forms around each compact mass of chromosomes. Nucleoli reappear. Chromosomes begin to uncoil and become indistinct. Cytokinesis is accomplished by the formation of a cell plate that begins in the center of the equatorial plane and grows outward to the cell wall.

---

# EXERCISE 7.3
# Observing Chromosomes, Mitosis, and Cytokinesis in Animal Cells

In this exercise, you will look at the general shape and form of human chromosomes and observe chromosomes and the stages of mitotic division in the whitefish. You will also compare these chromosomes with the plant chromosomes studied in Exercise 7.2. Chromosome structure in animals and plants is basically the same in that both have centromeres and arms. However, plant chromosomes are generally larger than animal chromosomes.

# Lab Study A. Mitosis in Whitefish Blastula Cells

## Materials

prepared slide of sections of whitefish blastulas
compound microscope

## Introduction

The most convenient source of actively dividing cells in animals is the early embryo, where cells are large and divide rapidly with a short interphase. In blastulas (an early embryonic stage), a large percentage of cells will be dividing at any given time. By examining cross sections of whitefish blastulas, you should be able to locate many dividing cells in various stages of mitosis and cytokinesis (Color Plate 12).

## Procedure

1. Examine a prepared slide of whitefish blastula cross sections. Find a blastula section on the lowest power, focus, switch to intermediate power, focus, and switch to high power.

2. As you locate a dividing cell, identify the stage of mitosis. *Be able to recognize all stages of mitosis in these cells.*

3. Identify the following in several cells:

   **nucleus**, **nuclear envelope**, and **nucleolus**

   **chromosomes**

   **mitotic spindle**

   **asters**—an array of microtubules surrounding each centriole pair at the poles of the spindle

   **centrioles**—small dots seen at the poles around which the microtubules of the spindle and asters appear to radiate

   **cleavage furrow**

## Results

1. List several major differences you have observed between mitosis in animal cells and mitosis in plant cells:

   |  | *Animal* | *Plant* |
   |---|---|---|
   | *cytokinesis:* | *cleavage furrow* | *cell plate* |
   | *centrioles:* | *present* | *absent* |
   | *asters:* | *present* | *absent* |

2. Locate, draw, and label in the space provided a blastula cell in metaphase and a cell in telophase/cytokinesis to illustrate these differences.

|  |  |
|---|---|
| | |

Metaphase                                      Telophase/Cytokinesis

Student Media BioFlix and Videos—Animal Mitosis (time-lapse); Sea Urchin Embryonic Development (time-lapse)

## Lab Study B. Human Chromosomes in Dividing Leukocytes

### Materials

slides of human leukocytes (white blood cells) on demonstration with compound microscopes

### Introduction

Cytogeneticists examining dividing cells of humans can frequently detect chromosome abnormalities that lead to severe mental retardation. To examine human chromosomes, leukocytes are isolated from a small sample of the patient's blood and cultured in a medium that inhibits spindle formation during mitosis. As cells begin mitosis, chromosomes condense and become distinct, but in the absence of a spindle they cannot move to the poles in anaphase. You will observe a slide in which many cells have chromosomes condensed as in prometaphase or metaphase, but they are not aligned on a spindle equator (see Color Plate 13).

### Procedure

*If they are available, have on demonstration photographs of karyotypes or chromosome slide preparations showing chromosome abnormalities. These may be obtained from mental health and genetic counselors. See the Websites section of this lab topic for an excellent online source of normal and abnormal karyotypes that can be downloaded for teaching purposes.*

1. Attempt to count the chromosomes in one cell in the field of view. Normally, humans have 46 chromosomes. Persons with trisomy 21 (three copies of chromosome 21), or Down syndrome, have 47 chromosomes. Are the cells on this slide from a person with a normal chromosome number?

2. Notice that each chromosome is duplicated, being made up of two sister chromatids held together by a single centromere. In very high magnifications, bands can be seen on the chromosomes. Abnormalities in banding patterns can also be an indication of severe mental retardation.

## EXERCISE 7.4
# Modeling Meiosis

### Materials

| | |
|---|---|
| 60 pop beads of one color | 4 centrioles |
| 60 pop beads of another color | letters *B*, *D*, *b*, and *d* printed on |
| 8 magnetic centromeres | mailing labels |

# Introduction

Meiosis takes place in all organisms that reproduce sexually. In animals, meiosis occurs in special cells of the gonads; in plants, in special cells of the sporangia. Meiosis consists of *two* nuclear divisions, **meiosis I** and **II**, with an atypical interphase between the divisions during which cells do not grow and synthesis of DNA does not take place. This means that meiosis I and II result in four cells from each parent cell, each containing half the number of chromosomes, one from each homologous pair. Recall that cells with only one of each homologous pair of chromosomes are haploid (*n*) cells. The parent cells, with pairs of homologous chromosomes, are diploid (2*n*). The haploid cells become sperm (in males), eggs (in females), or spores (in plants). One advantage of meiosis in sexually reproducing organisms is that it prevents the chromosome number from doubling with every generation when fertilization occurs.

What would be the consequences in successive generations of offspring if the chromosome number were not reduced during meiosis?

> *Although limited polyploidy (more than two complete chromosome sets) is not uncommon in plants, it is rare in animals. Increasing numbers of chromosomes lead to abnormalities in mitotic cycles, leading to the death of the organism.*

Meiosis involves the very precise movement and sorting of chromosomes (in the case of humans, 23 homologous pairs, 46 chromosomes, or 92 chromatids). This complicated process is not always perfect. Sometimes the homologous pairs do not separate properly or the sister chromatids fail to separate. When this happens the result is that the final nuclei may have either one too many chromosomes or one too few. As you observed in Exercise 7.3 Lab Study B, individuals with Down syndrome have 47 chromosomes as a result of an error in meiosis, in which a gamete from one of the parents had two copies of chromosome 21.

# Lab Study A. Interphase

Working with another student, you will build a model of the nucleus of a cell in interphase before meiosis. Nuclear and chromosome activities are similar to those in mitosis. You and your partner should discuss activities in the nucleus and chromosomes in each stage. Go through the exercise once together, and then demonstrate the model to each other to reinforce your understanding. Compare activities in meiosis with those in mitosis as you build your model.

## Procedure

1. Build the premeiotic interphase nucleus much as you did the mitotic interphase nucleus. Have two pairs of chromosomes (2*n* = 4) of distinctly different sizes and different centromere positions. Have one member of each pair of homologues be one color, the other, a different color.

2. To represent $G_1$ (gap 1), pile your four chromosomes in the center of your work area. The chromosomes are decondensed.

    Cell activities in $G_1$ are similar to those activities in $G_1$ of the interphase before mitosis.

    In $G_1$, are chromosomes single or duplicated?

*single*

3. Duplicate the chromosomes to represent DNA duplication in the S (synthesis) phase. Recall that in living cells, the centromeres remain single, but in your model you must use two magnets. What color should the sister chromatids be for each pair?

*the same color as the original chromatid, to represent exact duplication*

4. Duplicate the centriole pair.
5. Leave the chromosomes piled in the center of the work area to represent $G_2$ (gap 2).

   As in mitosis, in $G_2$ the cell prepares for meiosis by synthesizing proteins and enzymes necessary for nuclear division.

## Lab Study B. Meiosis I

Meiosis consists of two consecutive nuclear divisions, called *meiosis I* and *meiosis II.* As the first division begins, the chromosomes coil and condense, as in mitosis. Meiosis I is radically different from mitosis, however, and the differences immediately become apparent. In your modeling, as you detect the differences, make notes in the margin of your lab manual.

### Procedure

1. Meiosis I begins with the chromosomes piled in the center of your work area.

   As chromosomes begin to coil and condense, prophase I begins. Each chromosome is duplicated, made up of two sister chromatids. Two pairs of centrioles are located outside the nucleus.

2. Separate the two centriole pairs and move them to opposite poles of the nucleus.

   The nuclear envelope breaks down and the spindle begins to form as in mitosis.

3. Move each homologous chromosome to pair with its partner. You should have four strands together.

   Early in prophase I, each chromosome finds its homologue and pairs in a tight association. The process of pairing is called **synapsis.** Because the chromosomes are duplicated, this means that each paired duplicated chromosome complex is made of four strands. This complex is called a **tetrad.**

   How many tetrad complexes do you have in your cell, which is $2n = 4$?

   *two*

4. Represent the phenomenon of **crossing over** by detaching and exchanging identical segments of any two nonsister chromatids in a tetrad.

   Crossing over takes place between nonsister chromatids in the tetrad. In this process, a segment from one chromatid will break and exchange with the exact same segment on a nonsister chromatid in the tetrad. The crossover site forms a **chiasma** (plural, **chiasmata**).

5. Return the exchanged segments of chromosome to their original chromosomes before performing the crossing-over activity in the next step.

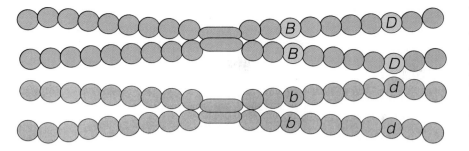

**Figure 7.11.**
**Arrangement of alleles *B*, *b*, *D*, and *d* on chromosome models.** One duplicated homologous chromosome has *B* alleles and *D* alleles on each chromatid. The other has *b* and *d* alleles on each chromatid.

Genes (traits) are often expressed in different forms. For example, when the gene for seed color is expressed in pea plants, the seed may be green or yellow. Alternative forms of genes are called **alleles.** Green and yellow are alleles of the seed-color gene. It is significant that crossing over produces new allelic combinations among genes along a chromatid. To see how new allelic combinations are produced, proceed to step 6.

6. Using the letters printed on mailing labels, label one bead (gene locus) on each chromatid of one chromosome *B* for brown hair color. Label the beads in the same position on the two chromatids of the other member of the homologous pair *b* for blond hair color.

   The *B* and *b* represent alleles, or alternate forms of the gene for hair color.

   On the chromatids with the *B* allele, label another gene *D* for dark eye color. On the other member of the homologous pair of chromosomes, label the same gene *d* for pale eyes. In other words, one chromosome will have *BD*, the other chromosome, *bd* (Figure 7.11).

7. Have a single crossover take place involving only two of the four chromatids between the loci for hair color and eye color. Remember, the crossover must take place between nonsister chromatids.

   What combinations of alleles are now present on the chromatids?

   *BD, bD, Bd, bd*

8. Confirm your results with your laboratory instructor.

*Print letters B, b, D, d on mailing labels and then cut them out. Have them available in small petri dishes.*

---

 **If you are having difficulty envisioning the activities of chromosomes in prophase I and understanding their significance, discuss these events with your lab partner and, if needed, ask questions of your lab instructor before proceeding to the next stage of meiosis I.**

---

9. Move your tetrads to the equator, midway between the two poles.

   Late in prophase I, tetrads move to the equator.

10. To represent metaphase I, leave the tetrads lying at the equator.

    During this phase, tetrads lie on the equatorial plane. *Centromeres do not split as they do in mitosis.*

11. To represent anaphase I, separate each duplicated chromosome from its homologue, and move one homologue toward each pole. In our model, the two magnets in sister chromatids represent one centromere holding together the two sister chromatids of the chromosome.

How does the structure of chromosomes in anaphase I differ from that in anaphase in mitosis?

*In anaphase I, chromosomes are duplicated. In anaphase of mitosis, they are single.*

12. To represent telophase I, place the chromosomes at the poles. You should have one long and one short chromosome at each pole, representing a homologue from each pair.

Two nuclei now form, followed by cytokinesis. How many chromosomes are in each nucleus?

*two*

*One potential drawback of the pop bead model is that in some cases one centromere must be represented by two magnets. Remind students that in duplicated chromosomes, two magnets represent one centromere. Students working with the model seldom have a problem with this.*

 **The number of chromosomes is equal to the number of centromeres. In this model, two magnets represent one centromere in duplicated chromosomes.**

Would you describe the new nuclei as being diploid (2*n*) or haploid (*n*)?

*Haploid; the number of chromosomes is halved.*

How has crossing over changed the combination of alleles in the new nuclei?

*In the new nuclei, B may be on the same chromosome with d, and b on the same chromosome with D.*

Are both chromosomes of the same color in the same nucleus? Compare your results with others.

*Students should note that a nucleus may contain recombinations of two chromosomes of the same color, or two chromosomes of different colors.*

13. To represent meiotic interphase, leave the chromosomes in the two piles formed at the end of meiosis I.

The interphase between meiosis I and meiosis II is usually short. There is little cell growth and no synthesis of DNA. All the machinery for a second nuclear division is synthesized, however.

14. Duplicate the centriole pairs.

## Lab Study C. Meiosis II

The events that take place in meiosis II are similar to the events of mitosis. Meiosis I results in two nuclei with half the number of chromosomes as the parent cell, but the chromosomes are duplicated (made of two chromatids), just as they are at the beginning of mitosis. The events in meiosis II must change duplicated chromosomes into single chromosomes. As meiosis II begins, two new spindles begin to form, establishing the axes for the dispersal of chromosomes to each new nucleus.

## Procedure

1. To represent prophase II, separate the centrioles and set up the axes of the two new spindles. Pile the chromosomes in the center of each spindle.

   The events that take place in each of the nuclei in prophase II are similar to those of a mitosis prophase. In each new cell the centrioles move to the poles, nucleoli break down, the nuclear envelope breaks down, and a new spindle forms. The new spindle forms at a right angle to the axis of the spindle in meiosis I.

2. Align the chromosomes at the equator of their respective spindles.

   As the chromosomes reach the equator, prophase II ends and metaphase II begins.

3. Leave the chromosomes on the equator to represent metaphase II.

4. Pull the two magnets of each duplicated chromosome apart.

   As metaphase II ends, the centromeres finally split and anaphase II begins.

5. Separate sister chromatids (now chromosomes) and move them to opposite poles.

   In anaphase II, single chromosomes move to the poles.

6. Pile the chromosomes at the poles.

   As telophase II begins, chromosomes arrive at the poles. Spindles break down. Nucleoli reappear. Nuclear envelopes form around each bunch of chromosomes as the chromosomes uncoil. Cytokinesis follows meiosis II.

   a. What is the total number of nuclei and cells now present?

      *four*

   b. How many chromosomes are in each?

      *two*

   c. How many cells were present when the entire process began?

      *one*

   d. How many chromosomes were present per cell when the entire process began?

      *four*

   e. How many of the cells formed by the meiotic division just modeled are genetically identical? (Assume that alternate forms of genes exist on homologues.)

      *none*

f. Explain your results in terms of independent assortment and crossing over. (Refer to your textbook.)

*Crossing over can result in new allelic combinations not present in the parent cells. Mendel's law of independent assortment states that each homologous pair will separate independently of all other pairs. As a result, the final nuclei could have any combination of colors, but will have one long and one short chromosome.*

### Results

Summarize the major differences between mitosis and meiosis in Table 7.1.

**Table 7.1**
Comparing Nuclear and Chromosomal Activities in Mitosis and Meiosis

| | **Mitosis** | **Meiosis** |
|---|---|---|
| **Synapsis** | *no* | *yes* |
| **Crossing over** | *no* | *yes* |
| **When centromeres split** | *as anaphase begins* | *as anaphase II begins* |
| **Chromosome structure and movement during anaphase** | *single chromosomes move to poles* | *I duplicated chromosomes move to poles*<br>*II single chromosomes move to poles* |
| **No. of divisions** | *1* | *2* |
| **No. of cells resulting** | *2* | *4* |
| **No. of chromosomes in daughter cells** | *same as in parent cell* | *1/2 number in parent cell* |
| **Genetic similarity of daughter cells to parent cells** | *same as parent cell* | *new gene combinations not in parent cell* |

EXERCISE 7.5

# Meiosis in *Sordaria fimicola*: A Study of Crossing Over

### Materials

petri dish containing mycelia resulting from a cross between *Sordaria* with black and tan spores
slides and coverslips
dropper bottles of water
matches

wire bacterial transfer loop
alcohol lamp

## Introduction

In the study of meiosis, you demonstrated that genetic recombination may occur as a result of the exchange of genetic material between homologous chromosomes in the process of crossing over. Crossing over occurs during prophase I, when homologous chromosomes synapse. While they are joined in this complex, nonsister chromatids may break at corresponding points and exchange parts. A point at which they appear temporarily joined as a result of this exchange is called a **chiasma** (Figure 7.12).

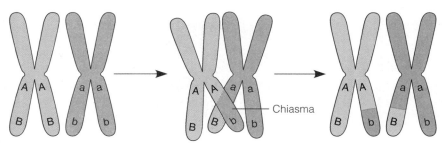

**Figure 7.12.**
**Crossing over.** Chromatid arms break and rejoin with a nonsister member of the tetrad, forming a chiasma between nonsister chromatids. This process results in the exchange of genetic material.

*Sordaria fimicola* is a fungus that spends most of its life as a haploid **mycelium,** a mass of cells arranged in filaments. When conditions are favorable, cells of filaments from two different mating types fuse (see Figures 7.13a and b); ultimately, the nuclei fuse (Figure 7.13c) and $2n$ zygotes are produced, each inside a structure called an **ascus** (plural, **asci**) (Figure 7.13d). Asci are protected within a **perithecium.** Each $2n$ zygote undergoes meiosis, and the resulting cells (ascospores) remain aligned, the position of an ascospore within the ascus depending on the orientation of separating chromosomes on the equatorial plane of meiosis I. After meiosis, each resulting ascospore divides once by mitosis (Figure 7.13e), resulting in eight ascospores per ascus (Figure 7.13f). This unique sequence of events means that it is easy to detect the occurrence of crossing over involving chromatids carrying alleles that encode for color of spores and mycelia.

If two mating types of *Sordaria,* one with black spores and the other with tan spores, are grown on the same petri dish (Color Plate 14), mycelia from the two may grow together, and certain cells may fuse. Nuclei from two fused cells then fuse, and the resulting zygote contains one chromosome carrying the allele for black spores and another carrying the allele for tan spores. After meiosis takes place, one mitosis follows, and the result is eight ascospores in one ascus: four black spores and four tan spores. If no crossing over has taken place, the arrangement of spores will appear as in Figure 7.14.

**Figure 7.13.**
**Abbreviated diagram of the life cycle of *Sordaria fimicola*.**
(a) Cells from filaments of two different mating types fuse. (b) One cell with two nuclei is formed. (c) The two nuclei fuse, forming a 2*n* zygote. (d) The zygote nucleus begins meiosis, and an ascus begins to form in a perithecium. (e) Meiosis continues, followed by mitosis. (f) The mature ascus contains eight ascospores. (g) Micrograph of crushed perithecium with asci containing ascospores.

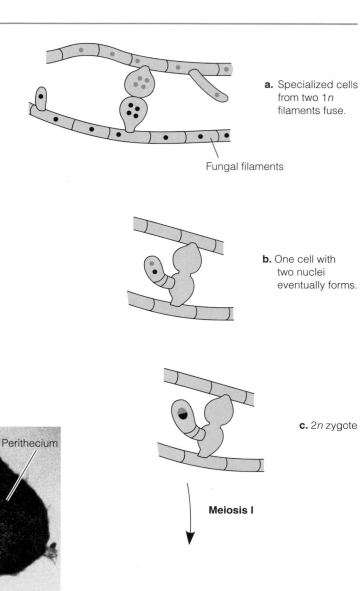

**a.** Specialized cells from two 1*n* filaments fuse.

Fungal filaments

**b.** One cell with two nuclei eventually forms.

**c.** 2*n* zygote

**Meiosis I**

**d.** Two 1*n* nuclei

**Meiosis II**

**e.** Four 1*n* nuclei

Young ascus

**Mitosis**

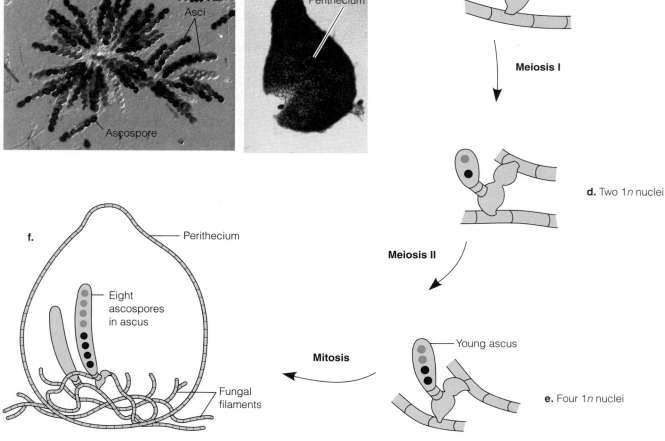

**g.**

Asci

Ascospore

Perithecium

**f.**

Perithecium

Eight ascospores in ascus

Fungal filaments

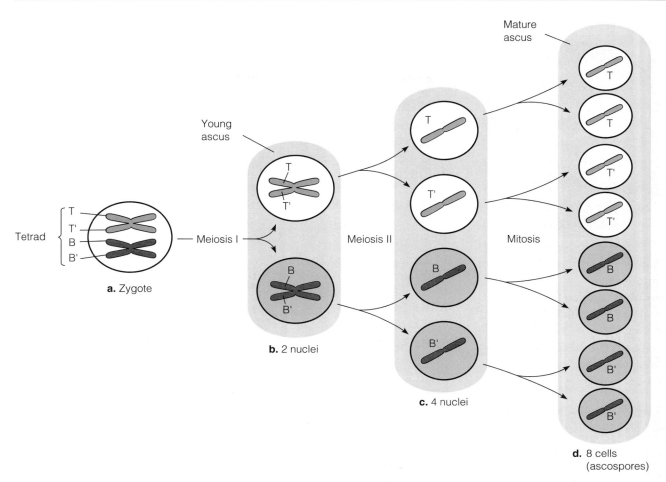

**Figure 7.14.**
**Arrangement of spores in asci resulting from a cross between fungi with black spores and fungi with tan spores when no crossing over takes place.** (a) In the zygote nucleus, the light homologous chromosome has chromatids labeled T and T′. Each chromatid has identical tan alleles for spore color. The dark homologous chromosome (chromatids labeled B and B′) has black alleles. (b) During meiosis I, the two homologous chromosomes separate into two different nuclei retained in one developing ascus. (c) Meiosis II produces four nuclei, two containing a chromosome with the tan allele and two containing a chromosome with the black allele, still within the one ascus. (d) Now each nucleus divides by mitosis, followed by cytokinesis, resulting in eight cells, called ascospores. The ascus now contains eight ascospores. Four of the spores have the tan allele in their nuclei and appear tan. Four ascospores have the black allele and appear black.

If crossing over does take place, the arrangement of spores will differ. In the spaces provided, using Figure 7.14 as a reference, draw diagrams that illustrate the *predicted* arrangement of spores in the ascus when crossing over takes place between the following chromatids and the alleles for color are exchanged: (a) T and B, (b) T and B′, (c) T′ and B, and (d) T′ and B′.

In lab today, you will observe living cultures of crosses between black and tan *Sordaria*. You will look for asci with spores arranged as in your predictions.

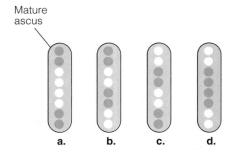

## Procedure

1. Place a drop of water on a clean slide, and carry it and a coverslip to the demonstration table.
2. Light the alcohol lamp and flame a transfer loop.

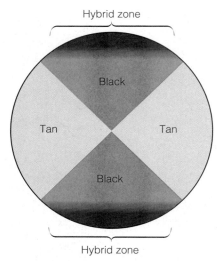

Hybrid zone

Black

Tan          Tan

Black

Hybrid zone

**Figure 7.15.**

**Most likely location for perithecia containing asci with hybrid spores.** Following the procedure, collect dark, round perithecia from zones of hybridization along the petri dish perimeter as indicated (Color Plate 14).

*A scalpel, dissecting needle, fine-tipped forceps, or a toothpick can be substituted for the transfer loop. A sterile technique is necessary only if you plan to use the cultures for several days. One culture is usually enough for several labs. Just be sure that it does not dry out. If using a culture on consecutive days, refrigerate it between labs to prevent premature dispersal of spores.*

3. Open the lid of the *Sordaria* culture slightly, and use the loop or other instrument to remove several perithecia from the region near the edge of the dish where the two strains have grown together (Figure 7.15).

4. Place the perithecia in the drop of water on your slide, and cover it with the coverslip.

5. Return to your work area.

6. Using the eraser end of a pencil, tap lightly on the coverslip to break open and flatten out the perithecia.

7. Systematically scan back and forth across the slide using the intermediate power of the compound microscope. When you locate clusters of asci, focus, switch to high power, count the asci, and determine if crossing over has taken place. Record your numbers in Table 7.2.

### Results

In Table 7.2, record the numbers of asci with (a) spores all of one color (indicating that the zygote was formed by fusion of cells of the same strain), (b) black and tan spores with no crossover, and (c) black and tan spores with a crossover.

**Table 7.2**
Microscopic Observations of Crossing Over

| Asci Types | Number of Asci in Each Category |
|---|---|
| **Spores all one color** | |
| **Crossover absent** | |
| **Crossover present** | |

### Discussion

1. What percentage of asci observed resulted from the fusion of cells from different strains?

2. What percentage of those asci resulting from the fusion of different strains demonstrates crossovers?

## Reviewing Your Knowledge

1. Define the following terms and use each in a meaningful sentence. Give examples when appropriate.

   *mitosis, meiosis, cytokinesis, chromosome, chromatin, centromere, centriole, centrosome, kinetochore, spindle, aster, homologous chromosome, synaptonemal complex, synapsis, tetrad, chiasma, sister chromatid, nucleolus, cell plate, cleavage furrow, diploid, haploid, crossing over, mycelium, perithecium, ascus*

2. Describe the activity of chromosomes in each stage of mitosis.

3. In the photomicrograph of dividing root cells at right, identify interphase and the following phases of mitosis: prophase, metaphase, anaphase, telophase, and cytokinesis.

4. Describe the activity of chromosomes in each stage of meiosis I and meiosis II.

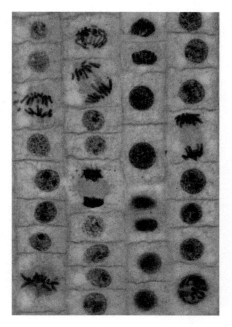

5. Observe the drawing of several phases of meiosis below.

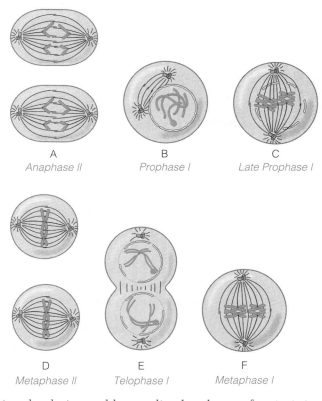

| A | B | C |
|---|---|---|
| *Anaphase II* | *Prophase I* | *Late Prophase I* |

| D | E | F |
|---|---|---|
| *Metaphase II* | *Telophase I* | *Metaphase I* |

a. Using the designated letters, list the phases of meiosis in sequence.

   *B, C, F, E, D, A*

b. Label each stage (include I or II).

c. At what stage would crossing over occur?

   *C*

d. What is the diploid number for this organism?

   *4*

6. Mitosis is important as organisms, both animals and plants, increase in size and grow new tissues and organs. Unlike animals, plants continue to grow throughout their life. Where would you expect mitosis to be most common in the body of a mature plant?

*Plants grow from the tips of their stems and roots, in tissues called meristematic tissues. In these regions mitosis takes place rapidly, producing new cells that then expand to bring about growth. In addition, most long-lived plants (such as trees) also grow in girth. Layers of cells (called cambium) divide by mitosis to bring about this growth.*

7. What role would mitosis play in the body of an adult animal?

*In an adult, mitosis takes place in tissues that replace aging cells with new cells, for example, skin, blood, hair follicles, and bone.*

8. What advantage does the process of crossing over bring to reproduction?

*It produces new combinations of alleles, leading to genetic variation in offspring. Genetic variation is important to a population as the raw material on which natural selection works.*

9. Why would the method of cytokinesis in animal cells not work in plant cells?

*Because of the presence of the cell wall, the plant cell cannot pinch in two by a constriction furrow. The plant cell cytoplasm must divide by a process that allows membranes to traverse the cell from cell wall to cell wall and then allows cell wall materials to be deposited between the membranes.*

## Applying Your Knowledge

1. Explain why models are important to scientific study of biological systems. Provide two examples of models other than those described in the exercises.

*Students may check their texts for ideas. Some suggestions: biological membranes, muscle contraction, molecules, atoms, protein synthesis, water transport in plants.*

2. You have probably heard that the liver of an adult human can "regenerate itself." How is the process of regeneration related to mitosis? Why is it possible to have a living donor for a transplanted liver?

*Liver cells, even in an adult, can be stimulated to divide by external clues such as an injury. Portions of an adult liver can be surgically removed and transplanted into a recipient. The donor's liver will regenerate and the transplanted portion will grow into a functioning liver in the recipient.*

3. Identical twins Jan and Fran were very close sisters. So, when Jan died suddenly, Fran moved in to help take care of Jan's daughter (her niece), Millie. Some time later Fran married her brother-in-law and became Millie's stepmother. When Fran announced that she was pregnant, poor Millie became confused and curious. "So," Millie asked, "who is this baby? Will she be my twin? Will she be my sister, my stepsister, my cousin?" Can you answer her questions? What is the genetic relationship between Millie and the baby? What processes are involved in the formation of gametes and how do they affect genetic variation?

*Jan and Fran are identical twins and therefore have the same genes. Millie and the new baby have the same father. All persons receive half their genes from their father and half from their mother. Therefore, Millie and the baby will be sisters genetically. However, they will not be twins. Through independent assortment and crossing over during meiosis, chromosomes are shuffled so that no two gametes are the same. Secondly, at fertilization, two unique gametes will combine, forming a zygote with new gene combinations, thus promoting genetic variation.*

4. Two natural plant products, vinblastine from the rosy periwinkle and paclitaxel (taxol) from the Pacific yew, have been used successfully in the treatment of a wide range of cancers. These chemicals work by interfering with mitosis but by different methods. Vinblastine inhibits mitosis by preventing the assembly of the spindle. Paclitaxel promotes microtubule synthesis and binds to the microtubles preventing the depolymerization (disassembly) of the spindle during mitosis. Based on your knowledge of mitosis, at what phase do you expect cell division to be interrupted by each of these cancer-fighting compounds? Explain your answer based on the activities occurring at the mitotic stage.

*Vinblastine must inhibit cell division during prometaphase when the chromosomes attach to the spindle. Without the spindle, the chromosomes would not be able to align at the equator of the cell in metaphase. Paclitaxel (taxol) would inhibit cell division during anaphase when the microtubules depolymerize, moving the chromosome closer to the centrosomes located at the poles of the cell. Without microtubule disassembly, the chromosomes cannot move to the poles and mitosis cannot continue.*

5. Lonesome George is the last Galápagos tortoise of his species, which once inhabited the island of Pinta but is now considered one of the rarest reptiles on Earth. After years of searching for a mate for George on the islands and in zoos, geneticists recently found a male tortoise on a neighboring island whose genetic information indicated that one of his parents had to be a Pinta tortoise. What evidence do you think scientists utilized to conclude that the newly discovered male tortoise was the offspring of a Pinta tortoise? Relate your answer to your understanding of meiosis and sexual reproduction.

*Half of the chromosomes (and therefore DNA) of the male tortoise must be the same as George's chromosomes. In meiosis the chromosomes are reduced in number by one half. Each nucleus that results from meiosis contains one member of each pair of chromosomes (haploid). Two haploid gametes (egg and sperm) combine at fertilization to restore the diploid number.*

The San Diego Zoo Conservation and Research for Endangered Species (CRES) program maintains a "Frozen Zoo," which contains cell lines for endangered animal species. Geneticists eventually may be able to clone Lonesome George from frozen stem cells if a female tortoise cannot be found. Explain the difference between the two approaches to preserving a rare species: finding a mate and cloning. Your answer should reflect the outcomes of meiosis and mitosis and their contribution to genetic variation.

*In finding a rare mate, scientists hope to maintain the genetic information from two genetically different individuals. Through the process of meiosis these two individuals would produce a large number of genetically unique gametes that would then potentially combine to form countless genetically unique individuals. Meiosis, through crossing over and independent assortment, promotes genetic variation. Fertilization would further contribute to genetically distinct tortoises. Mitosis is the process involved in cloning, and all cells produced as a result of mitosis will be genetically identical. Scientists have the potential to produce additional Georges, but they will all still be lonesome.*

6. The sheep "Dolly" was the first mammal cloned by transplanting a nucleus from a cell in an adult sheep into an egg of a donor sheep. Dolly appeared to grow normally for several years, but at age 6, she developed

complications from several physical conditions normally seen only in older sheep. There are many hypotheses about causes for Dolly's deterioration. Propose one hypothesis about the role of telomeres in dividing cells as she aged.

*The telomeres of chromosomes in the cloned egg were in the adult condition, with many telomeres having been lost in mitosis. This removed their protective function, leading to the early loss of important genes on her chromosomes, a condition found in elderly mammals.*

## Investigative Extension/Case Study

*Have students work in teams of two or three students and assign one of the three patients to each group. Alternatively, students can be assigned all three patients to investigate.*

1. **Case Study:** Three patients have been referred to the Center for Genetic Counseling to be evaluated for possible genetic disorders. The patients have prepared a short history and provided a blood sample for chromosome analysis, beginning with the microscopic observation of the chromosomes (similar to Exercise 7.3, Lab Study B). Digital images of the chromosomes are available and may be used to develop a karyotype showing the chromosomes aligned in pairs based on the location of centromeres, overall size, and banding patterns from various staining techniques.

   With your laboratory partner(s) enter the University of Arizona website for the Biology Project "Karyotyping Activity" (http://www.biology .arizona.edu/human_bio/activities/karyotyping/karyotyping.html). Select one or more of the three patients (ask your instructor), then read the patient history, complete the karyotype, make a diagnosis, and research the genetic syndrome. Your analysis of the patient should include a comparison with karyotypes for normal men and women. Resources are available to assist you with your genetic counseling, including Internet resources.

   At the conclusion of your investigation submit a final report that will be used in counseling the patient. The report should include a copy of the labeled karyotype, an analysis of the findings, a description of the genetic syndrome, and an explanation of the underlying cellular processes. Provide citations and helpful references with your final report.

*Instructions for utilizing living materials in the study of mitosis and meiosis in plants are provided in the Preparation Guide.*

2. **Investigation:** Plants can be used to investigate similarities and differences among species for chromosome numbers and morphology during mitosis (root tips) and meiosis (anthers). Spiderwort (*Tradescantia* spp.) and society garlic (*Tulbaghia violacea*) are readily available cultivated species that are well-suited for these studies. See your instructor for instructions on preparing slides of living plant material.

## Student Media: BioFlix, Activities, Investigations, and Videos
www.masteringbiology.com (select Study Area)

**BioFlix**—Ch. 12: Mitosis
   Ch. 13: Meiosis
**Activities**—Ch. 12: Roles of Cell Division; The Cell Cycle; Mitosis and Cytokinesis Animation; Causes of Cancer
   Ch. 13: Asexual and Sexual Life Cycles; Meiosis Animation; Origins of Genetic Variation
   Ch. 15: Mistakes in Meiosis
**Investigations**—Ch. 12: How Much Time Do Cells Spend in Each Phase of Mitosis?
   Ch. 13: How Can the Frequency of Crossing Over Be Estimated?
**Videos**—Ch. 12: Animal Mitosis; Sea Urchin Embryonic Development

# References

Becker, W. M., L. Kleinsmith, J. Hardin and G. Bertoni. *The World of the Cell,* 7th ed. Redwood City, CA: Benjamin Cummings, 2003 to 2009.

Bold, H. C., C. J. Alexopoulos, and T. Delevoryas. *Morphology of Plants and Fungi.* New York: Harper & Row, 1980.

Costello, M. and K. Kellmel. "Medical Attributes of *Taxus brevifolia*—The Pacific Yew" [online] available at http://wilkesl.wilkes.edu/~kklemow/Taxus.html, 2003.

Holden, C. "A Long-lost Relative," *Science,* 2007, vol. 316, p. 669.

Olive, L. S. "Genetics of *Sordaria fimicola*. I. Ascospore color mutants." *American Journal of Botany,* 1956, vol. 43, p. 97.

Snyder, Lucy. "Pharmacology of Vinblastine, Vincristine Vindesine and Vinorelbine." *Cyberbotanica.* [online] available at http://biotech.icmb.utexas.edu/botany/vvv.html, 2004.

# Websites

Additional investigations of cell reproduction at University of Arizona Biology Project:
http://www.biology.arizona.edu/

Animation of meiosis and independent assortment: Google youtube.com and search for meiosis animation

Current research on mitosis and videos of plant and animal cells dividing:
http://www.bio.unc.edu/faculty/salmon/lab/mitosis

Description of the cell cycle and mitosis, relating these processes to cancer:
http://cancerquest.emory.edu

Normal and abnormal karyotypes may be downloaded for teaching purposes:
http://worms.zoology.wisc.edu/zooweb/Phelps/karyotype.html

Photos of live cancer cells dividing in real time:
http://www.cellsalive.com/cam.htm

See this website for animations of the cell cycle, mitosis, and meiosis:
http://www.cellsalive.com

# LAB TOPIC 7

# Mitosis and Meiosis
## Teaching Plan for Laboratories

## Main Concepts and Objectives

1. Concept: modeling. Students will describe and give examples of scientists' use of models to increase understanding of natural processes. (Exercises 7.1 and 7.4)

2. Concept: the cell cycle. Students will describe cellular activities that take place in each stage of a typical cell life cycle. (Exercise 7.1)

3. Concept: mitosis and cytokinesis. Students will describe the activities of chromosomes, centrioles, membranes, and microtubules during the division of the cell nucleus and subsequent division of cytoplasm. They will be able to identify stages of mitosis in plant and animal cells and be able to compare differences in mitosis and cytokinesis in plant and animal cells. (Exercises 7.1–7.3)

4. Concept: meiosis. Students will describe the activities of chromosomes in each stage of meiosis I and II. They will compare chromosomal behavior in meiosis and mitosis. (Exercise 7.4)

5. Concept: crossing over. Students will describe the activities in chromosomes as crossing over takes place and observe the results of crossing over in the powdery mildew *Sordaria fimicola* ascospore arrangements. (Exercise 7.5)

6. Specific content. Students will be able to describe the stages of the cell cycle, mitosis, and meiosis and be able to define: *chromatin, chromosome, centromere, centrosome, kinetochore, spindle, aster, homologous chromosome, synaptonemal complex, synapsis, tetrad, chiasma, sister chromatid, nucleolus, cell plate, cleavage furrow, diploid, haploid, crossing over, mycelium, perithecium, and ascus.*

| | | | |
|---|---|---|---|
| Centromere and arm | The centromere is not always located in the middle | It can be displaced toward one end | or directly on the end |
| | Arms / Centromere | Arms / Centromere | Arm / Centromere |
| Single chromosome | | | Centromere |
| Duplicated chromosome | During duplication chromosomes form two identical sister chromatids joined at the centromere Chromatid 1 / Chromatid 2 / Centromere / Centromere | | |
| Homologous chromosomes | Referred to as homologues, one is derived from each parent. Length and centromere position are identical. The same genes, but not necessarily the same alleles, are present on the two homologous chromosomes. They have no physical association with each other except in prophase I of meiosis. Homologues / Homologues / Homologues | | |
| Tetrad | In prophase I, the duplicated homologous pairs synapse, forming a four-stranded complex known as a tetrad. Tetrad / Tetrads | | |

# Order of the Lab

1. Using appropriate diagrams, review the cell cycle and introduce new terminology.                         (15 min)
2. View video of mitosis, if available.                         (10 min)
3. Students model mitosis, demonstrating the process to each other.                         (30 min)
4. Students identify stages of mitosis in onion root tip and whitefish blastula slides. Students observe slides of human chromosomes in dividing leukocytes.                         (30 min)
5. Review significant events in meiosis.                         (10 min)
6. Students model meiosis, demonstrating the process to each other.                         (30 min)
7. Introduce meiosis and crossing over in *Sordaria*.                         (10 min)

8. Students make slides and interpret results
   of spore configurations in *Sordaria*.                           (40 min)

9. Class discussion of differences in mitosis and meiosis.          (5 min)

**For a 2-hour lab:** Divide the lab topic into two labs, studying mitosis in the first lab and meiosis in the second lab. The extra time will allow more class discussion. Having more time is particularly useful if the topics have not been adequately covered in lecture. With a total of 4 hours, you can have students prepare root tip squashes from onion root tips or roots of germinating *Vicia faba* seeds rather than use prepared slides. If materials and procedures are available, students can make anther squashes to study meiosis in *Tradescantia* (spiderwort) or *Tulbaghia violacea* flowers. *Tradescantia* is available from Carolina Biological Supply, and *Tulbaghia* is commercially available for flower gardens in many regions of the United States. See the *Preparation Guide* for procedures for root tip and anther squashes.

## Classroom Management

It is critical that the appropriate stage be set for the model building in this laboratory; otherwise, students will spend their study time playing with the pop beads. Emphasize how important it is that students understand mitosis and meiosis before they attempt to understand genetics and many other concepts in biology. Point out that, at first glance, these processes may appear simple, but they can easily become confusing. By approaching the learning of these processes from several perspectives—class lecture, concept review in lab introduction, visual presentation in video or film loop, modeling and peer teaching, and identifying stages in slides—students should complete this lab with a clear understanding of mitosis and meiosis.

Students work in pairs on modeling. Have them work through the model once together, discussing the activities of chromosomes, centrioles, microtubules, and other structures at each stage. They should then practice "peer teaching," demonstrating the process to each other.

As students finish building each model, have the class collectively answer the questions in the manual relating to the process just completed. Take care that the same students do not answer all the questions. Call on students who are less likely to participate.

When students make and observe slides, they should work independently. Stages of mitosis need not be drawn consecutively, but as students make the correct identification, they should make the drawing in the box provided.

## Student Development

Students improve skills in observation and interpretation. They develop communication skills as they practice peer teaching. They develop writing skills as they answer the questions in the exercise.

## Discussion and Summary

Encourage your students to discuss differences in mitosis in animal and plant cells. They should also discuss differences between mitosis and meiosis. They should answer the questions at the end of the exercises.

## Evaluation

Students can complete the Case Study outside of the laboratory and submit a report.

You may choose to require that your students answer the questions at the end of the exercise to be handed in. Test major concepts and identification of stages of mitosis on a laboratory exam.

## Lab Topic 8
# Mendelian Genetics I: Fast Plants

## Laboratory Objectives

After completing this lab topic, you should be able to:

1. Describe the mode of inheritance of three traits in *Brassica rapa* (Wisconsin Fast Plants™).
2. Use the Mendelian model to test for patterns of inheritance.
3. Describe and use the chi-square test to compare observed and expected results from genetic crosses.
4. Describe at least one type of non-Mendelian inheritance.
5. Define terminology used in the study of genetics.

*For a 2-hour lab:*
*The lab introduction and seed planting activity can be completed in 2 hours. See Teaching Plan for suggestions for completing the lab project.*

*Instructors and/or students can download data tables in Excel format. Go to www .masteringbiology.com. Look in the Study Area under Lab Media.*

## Introduction

Gregor Mendel's study of inheritance in garden peas provided evidence that traits were inherited as "particles" rather than by blending, which was the generally accepted model of that time. Although the Mendelian model does not explain every type of inheritance, it remains the cornerstone of genetics because it provides a testable model from which a scientist can generate hypotheses with predictable results.

In this lab topic, you will investigate the inheritance of traits in a small plant, *Brassica rapa,* using the Mendelian model to develop your hypotheses. You will actually follow many of Mendel's procedures as you work with a **monohybrid** (single trait) **cross** and a **dihybrid** (two traits) **cross**. Each cross begins with true-breeding parents, each of which is **homozygous** (has two identical alleles for a given trait). If parents with different traits (for example, yellow and green pea pods) are cross-pollinated, under the Mendelian model the offspring will be **heterozygous** (receiving one allele from each parent). However, the offspring will express only one allele, which Mendel described as **dominant**; that is, only one allele needs to be present to be expressed. The masked trait is **recessive**; two alleles must be present for a recessive trait to be expressed. The parents in this cross are referred to as the **P generation**, and the hybrid offspring as the $F_1$ (first filial or hybrid) **generation**. Two $F_1$ hybrids are then crossed to produce the next group of offspring, or $F_2$ **generation**. The predicted Mendelian results for the $F_2$ generation from a monohybrid cross would be one-fourth of the offspring with the recessive trait and three-fourths with the dominant trait (Figure 8.1). To explain these results, Mendel formulated the **law of segregation**, which states that allele pairs separate during the formation of gametes (during meiosis), with the paired condition being restored during fertilization to form the zygote.

*This lab topic must be started early in the term to allow enough time for students to complete the crosses and write a final report. Refer to the Preparation Guide and Teaching Plan for suggested timetables.*

*In the Study Area of masteringbiology .com, see Student Media for suggested activities, investigations, and videos.*

**Figure 8.1.**

**Mendel's law of segregation.** Mendel proposed that allele pairs separate during the formation of gametes and that the paired condition is restored during fertilization. The expected results for the F₂ generation of a monohybrid cross would be 75% dominant (green pea pods) and 25% recessive (yellow pea pods).

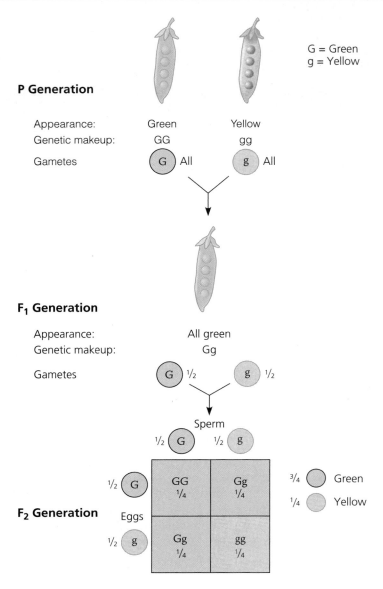

The results of Mendel's dihybrid crosses indicated that traits did segregate and that pairs of alleles assorted themselves independently of other pairs of alleles during meiosis. This is Mendel's **law of independent assortment,** which integrates the processes studied in Lab Topic 7, mitosis and meiosis, with the inheritance of alleles that is carried on chromosomes (Figure 8.2).

Part of the intrigue of genetics comes from the non-Mendelian results of experiments, which require modified hypotheses and novel procedures. Imagine the excitement in Thomas Hunt Morgan's lab when crosses involving white-eyed fruit flies suggested that the inheritance of eye color might somehow be related to the sex of the fly! (This is an example of what is called **sex linkage.**) Mendel could not have investigated sex linkage in peas because peas, like most plants, do not have sex chromosomes but produce both sexes in one flower.

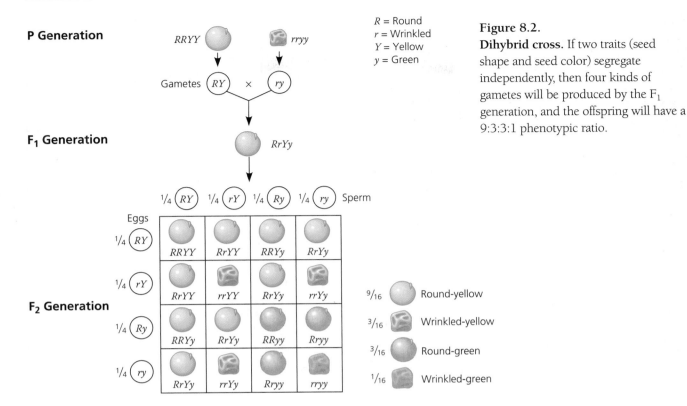

**P Generation**    RRYY    rryy

Gametes (RY) × (ry)

$R$ = Round
$r$ = Wrinkled
$Y$ = Yellow
$y$ = Green

**F₁ Generation**    RrYy

¼ (RY)    ¼ (rY)    ¼ (Ry)    ¼ (ry)    Sperm

Eggs
¼ (RY)

**F₂ Generation**

¼ (rY)

¼ (Ry)

¼ (ry)

| | | | |
|---|---|---|---|
| RRYY | RrYY | RRYy | RrYy |
| RrYY | rrYY | RrYy | rrYy |
| RRYy | RrYy | RRyy | Rryy |
| RrYy | rrYy | Rryy | rryy |

⁹⁄₁₆    Round-yellow

³⁄₁₆    Wrinkled-yellow

³⁄₁₆    Round-green

¹⁄₁₆    Wrinkled-green

**Figure 8.2.**
**Dihybrid cross.** If two traits (seed shape and seed color) segregate independently, then four kinds of gametes will be produced by the F₁ generation, and the offspring will have a 9:3:3:1 phenotypic ratio.

**Cytoplasmic inheritance** is another example of non-Mendelian inheritance. Not all genetic information is determined by nuclear genes. Both the mitochondria and chloroplasts carry genetic information, which is passed to the next generation in the cytoplasm rather than in the nucleus. These genes are inherited from the maternal parent because the cytoplasm of the zygote comes entirely from the egg. The sperm contributes only nuclear material. Each person carries the mitochondrial DNA of his or her mother, and, like-wise, mitochondrial and chloroplast DNA in plants is maternally inherited.

Genetics has a unique vocabulary that you must master if you are to understand and communicate the concepts. You began developing this vocabulary when you studied mitosis and meiosis. Before you begin today's lab, review defini-tions of *trait, gene, allele,* and *chromosome* from Lab Topic 7; use your textbook to define any terms you do not recognize in Table 8.1 on the next page.

In this laboratory, you will investigate inheritance of three traits in Wisconsin Fast Plants. These plants, developed by Dr. Paul Williams and a team of sci-entists in the Department of Plant Pathology, University of Wisconsin, Madison, are strains of *Brassica rapa* (RCBr, rapid-cycling brassicas) that complete their entire breeding cycle from seed to seed in 35 days (see Color Plate 15 and Figure 8.3). Because of their rapid breeding cycle, plants in the Brassica, or mustard, family are emerging as model plants for teaching and research, much as *Drosophila*, the fruit fly, has historically been the animal of choice for research in eukaryotic genetics. The Wisconsin scientists have developed several strains of plants with strongly contrasting traits. In this lab topic, you will investigate the inheritance of three of these traits: pres-ence or absence of anthocyanin, yielding green or purple plants; color of plant—yellow-green or green; and presence or absence of variegated (green and white) leaves.

**Table 8.1**
Genetic Terminology

| Term | Definition |
|---|---|
| Genotype | |
| Phenotype | |
| Wild-type trait | |
| Mutant trait | |
| $F_1$ | |
| $F_2$ | |
| Dominant trait | |
| Recessive trait | |
| Hybrid | |
| Homozygous | |
| Heterozygous | |

 Student Media Videos—Ch. 30: Flowering Plant Life Cycle

The Crucifer Genetics Cooperative for *Brassica* has developed guidelines for designating symbols to represent particular genotypes of *Brassica* that are similar to symbols used in other model systems, such as *Drosophila* (see Lab Topic 9, Mendelian Genetics II: *Drosophila*). Students who wish to perform extended investigations using *Brassica* should refer to the *Wisconsin Fast Plants Manual* (Carolina Biological Supply, 1989) for guidelines for allelic and genotypic symbols. However, for the purposes of this lab, you should use the common symbols of Mendelian inheritance, in which the dominant allele is designated by the capitalized first letter of the trait name, and the recessive allele by the lowercase letter.

**Figure 8.3.** (☞)
**Life cycle of *Brassica rapa*, Fast Plant.** The entire life cycle of this Fast Plant is completed from seed to seed in 35 days. Note the stages of development and the day in the life cycle.

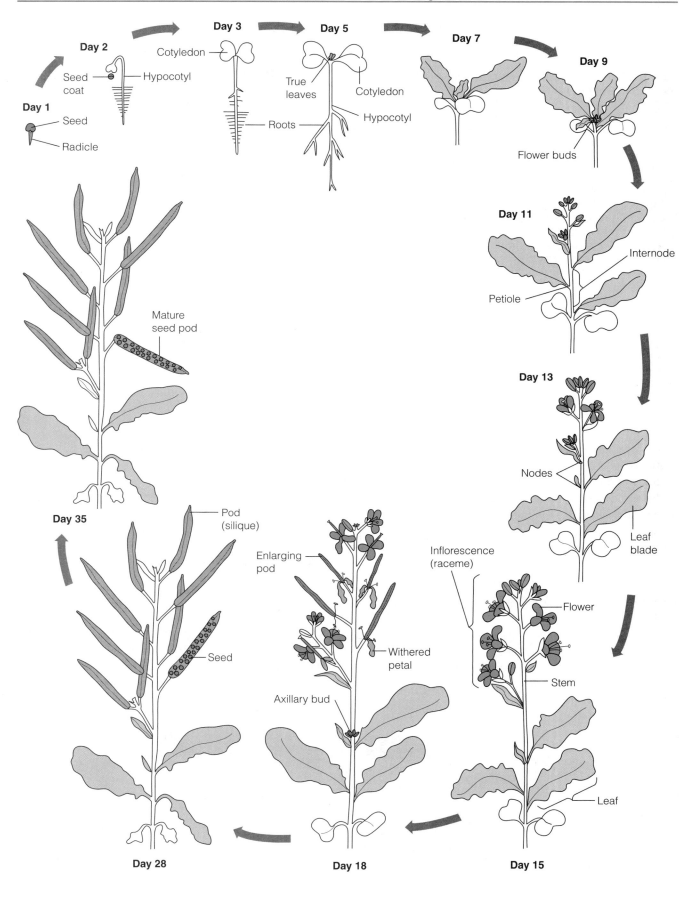

*Materials needed for this lab topic are conveniently available from Carolina Biological Supply as part of its commitment to provide Fast Plants to educators. However, the materials for this lab can also be supplied from inexpensive and readily available resources. For example, the light system can be purchased or constructed from a builder's light fixture and PVC pipe; 35-mm film canisters can substitute for the Styrofoam™ quads. See the Preparation Guide for suggested alternatives to purchasing all materials. See also Wisconsin Fast Plants website: http://www .fastplants.org/intro.php*

## Materials for All Exercises

seeds of designated phenotypes
seed-collecting pan
small envelopes
wicks
label tape
Styrofoam "quads" used to germinate seeds (see Figure 8.4)
fluorescent light bank (see Figure 8.5)

water reservoir for petri dish (plastic dish of appropriate size or cut-off base of 2-L bottle)
water and dropper
potting mix
fertilizer pellets
watering tray
petri dish with filter paper

# Overview of Exercises

The crosses for each exercise are briefly outlined below, along with the general procedures that will be used in all exercises. These include planting seeds, pollinating flowers, and harvesting and germinating seeds. Specific instructions are provided within each exercise.

## Exercises 8.1 and 8.2

You will begin with $F_1$ hybrid seeds resulting from a cross previously made between homozygous wild-type plants and homozygous mutant plants. You will germinate these $F_1$ seeds, cross-pollinate the flowers, allow the flowers to mature and produce $F_2$ seeds, collect the mature $F_2$ seeds, germinate the $F_2$ seeds, and record phenotypic ratios in this generation.

**Figure 8.4.**
**Planting and thinning Fast Plants.**
(a) Place absorbent wicks in the quad, with the tips extending from holes in the bottom. (b) Place potting soil, fertilizer pellets, and more soil in each cell. Make a small depression, and plant two or three seeds in each cell; then cover with soil. (c) Water seeds with a dropper until water drips from the wicks. Place on watering tray. (d) After recording seedling phenotypes, remove all but one large healthy seedling in each cell.

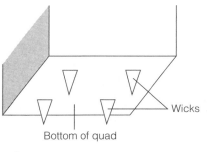

**a.** Quad and wicks

**b.** Planting

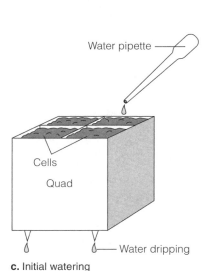

**c.** Initial watering

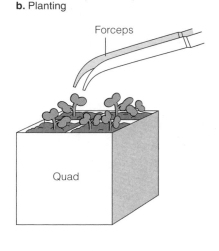

**d.** Thinning

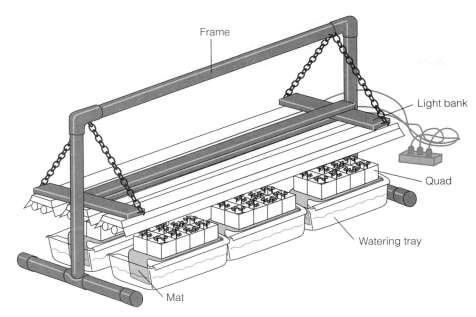

Frame

Light bank

Quad

Watering tray

Mat

**Figure 8.5.**
**Growing system for Fast Plants.**
Fast Plants require high light to
complete their life cycle in 35 days.
The light bank should be adjusted
to remain 2–3 inches above plants.
The light should be on
24 hours a day.

### Exercise 8.3

You will grow green plants and variegated plants, cross these parent plants,
and then grow the offspring from this cross and record the phenotypes.

You will begin planning for this lab and germinating all seeds (unless otherwise
noted) *6 to 8 weeks* before the laboratory period designated for making final observations. Before coming to lab on this preliminary date, plan your schedule of
activities carefully. Notice that several activities will take place outside of sched-
uled lab time. You are personally responsible for the success of this laboratory,
and you will be responsible for maintaining plants and carrying out activities.

## General Procedures for All Exercises

Complete and detailed procedures are provided with Exercises 8.1–8.3,
starting on the next page; procedures common to all exercises are described
for planting seeds, pollinating flowers, and harvesting and germinating
seeds of the next generation.

### Planting Seeds

Plant the appropriate seeds in quads (Figure 8.4).

1. Add one wick to each cell in a quad; wicks should extend from the base.
2. Add potting soil to each cell until half full.
3. Add three fertilizer pellets to each cell.
4. Add more soil and press to make a depression.
5. Add two or three seeds to each cell and barely cover with potting mix.
6. Using a dropper, water each cell until water drips from the wick.
7. Place quad on watering tray under fluorescent light bank (Figure 8.5).
   The lights should be maintained 2–3 inches above the growing plants
   and should be on 24 hours a day. Be sure that the wicks in the quad
   make good contact with the mat on the watering tray. Check the watering system regularly throughout the experiment.
8. Use tape to label your quad with your name, the date, and the plant type(s).

*Fast Plants only follow the
shortened life cycle when they
are correctly grown under the
recommended bank of fluorescent
lights. Do not substitute growing
in a sunny window or greenhouse.
The height of the lights above the
plants should be constant; therefore,
lights will have to be raised as the
plants grow. See the* Preparation
Guide *for details.*

9. After 4 or 5 days, record phenotypes of plants in the appropriate table and thin plants, leaving only the most vigorous single plant in each cell (Figure 8.4d).

## Pollinating Flowers

After two or three flowers open on most plants (approximately day 14), pollinate as follows (Figure 8.6):

*Fast Plants are self-incompatible so students must make crosses among plants. Buds are removed to prevent indiscriminate pollinations and to stop flowering and promote fruit and seed formation.*

1. Using a "bee-stick" (Williams, 1980), made by gluing a dry honeybee thorax to the top of a toothpick, transfer pollen from one plant to another. (You can also use a small, soft paintbrush.)
2. Save the stick by inserting it into one cell of the quad. Use it to pollinate again 2 and 4 days later.
3. After the third pollination, pinch off all unopened buds.
4. Remove and discard all new buds and shoots for the next 2 weeks.

## Harvesting Seeds

Seeds are ready to harvest approximately 21 days after pollination.

1. Remove quad with plants from the watering tray and dry for 5 days.
2. Remove dry seed pods and roll them between your hands over a collecting pan to free the seeds from the pod.
3. Store seeds in an envelope labeled with your name, the date, and the seed type.

## Germinating Seeds

Seeds in Exercises 8.2 and 8.3 will be germinated in quads by planting as described above for initial seeds (p. 187). Seedlings can be scored as soon as the first leaves are produced, and the phenotype of interest can be observed.

Seeds of offspring in Exercise 8.1 will be germinated in petri dishes as described below (Figure 8.7).

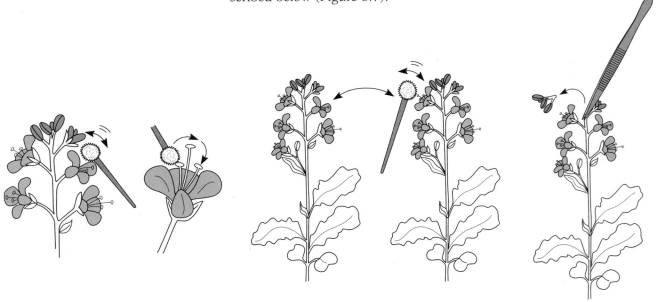

**a.** Pollen transfer to bee-stick        **b.** Cross-pollination        **c.** Removal of unopened buds

**Figure 8.6.**
**Pollination of Fast Plants.** (a, b) Using a bee-stick, transfer pollen from one plant to another. Plants are self-incompatible and must be cross-pollinated. (c) After pollinating on 3 days, remove all remaining flowers and buds.

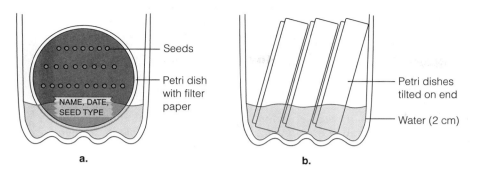

**Figure 8.7.**
**Germination of seeds in petri dishes.** In Exercise 8.1, seeds are germinated on moist filter paper in petri dishes.

1. Moisten a piece of filter paper in a petri dish, labeling the paper with your name, the date, and the seed type. Pour off excess water.
2. Place 25 of the harvested seeds in neat rows in the upper two-thirds of the filter paper.
3. Place the petri dish tilted on end in a water reservoir. Add about 2 cm water.
4. Place the dish and reservoir under the light bank.
5. In approximately 48–96 hours, observe seedlings and record phenotypes in the appropriate table.
6. Tabulate results, and perform a chi-square test on class results.

# EXERCISE 8.1
# Inheritance of Anthocyanin Gene

## Materials

12 $F_1$ seeds resulting from a cross previously made between a homozygous plant that has anthocyanin present and a true-breeding plant that has no anthocyanin
1 quad and related planting supplies (refer to Materials for All Exercises section)
2 mature plants with parental phenotypes

## Introduction

Wild-type *Brassica rapa* has a red-purple pigment present, called *anthocyanin*. Plants with even a faint red or purple coloration in the stems or leaf stalks are considered to show the phenotype for anthocyanin. Mutant plants have no anthocyanin and are completely green. In this exercise, you will investigate the inheritance of this trait. Formulate a hypothesis for this trait that describes a mode of inheritance based on the Mendelian model. Will these traits conform to Mendel's laws? Is one trait dominant or recessive? Refer to Figure 8.6 and your text if necessary.

*We have chosen a mutation that lacks anthocyanin for both the monohybrid cross (Exercise 8.1) and the dihybrid cross (Exercise 8.2). We think it is important to have students recognize and observe the phenotype before adding a second trait. However, if you would like to change mutants from one term to the next, $F_1$ seeds are available from Carolina Biological Supply for a dwarf mutant, rosette. See the Preparation Guide.*

## Hypothesis

Hypothesize about the inheritance of these traits.

*These traits follow Mendelian laws, and the presence of purple pigment is dominant.*

### Prediction

Predict the results based on your hypothesis (if/then). Outline the cross(es) for your predictions in the margin of your lab manual, and provide expected results for each in the form of a Punnett square if appropriate. Refer to Figures 8.1 and 8.2 and your textbook if needed.

*If purple is inherited as a dominant Mendelian trait, then the F$_2$ results will show a phenotypic ratio of three purple plants for every green plant.*

### Procedure

 Refer to General Procedures (pp. 187–189) and appropriate figures. All days are approximate. Monitor plants for initiation of flowering and seed set.

1. Refer to the checklist of activities in Table 8.2, and record dates, procedures, and any observations as you follow the procedure for this exercise. Initial the final column when you have completed each step.
2. *Day 1.* Plant F$_1$ hybrid seeds resulting from crosses previously made between homozygous purple plants and homozygous green plants. See Planting Seeds, pp. 187–188.

**Table 8.2**
Checklist of Activities for Exercise 8.1. Record the date and initial the final column indicating that you have completed each step.

| Approximate Day | Date | Activity | Initials |
|---|---|---|---|
| Day 1 | | Plant F$_1$ hybrid seeds. See Planting Seeds, pp. 187–188. Regularly check water. | |
| Day 4 or 5 | | Observe seedlings and record numbers of each phenotype in Table 8.3. Thin plants to one per cell. | |
| Days 14, 16, 18 | | Pollinate on 3 days; pollinate at least 6–8 flowers. See Pollinating Flowers, p. 188. | |
| Days 20 to 39 | | Remove buds and shoots. | |
| Day 39 | | Harvest F$_2$ seeds and germinate in petri dishes. See Harvesting Seeds, p. 188, and Germinating Seeds, pp. 188–189. | |
| Day 42 | | Count and record numbers of each phenotype in Table 8.4. | |

3. *Day 4 or 5.* Observe germinating seedlings and compare phenotypes with parental plants on demonstration. Record numbers of each phenotype in Table 8.3 in the Results section and on the class data sheet. Thin plants to one per cell. Remember to keep plants watered.

4. *Days 14, 16, 18.* Cross the $F_1$ plants. Pollinate plants three times, with 1 day between pollinations. Pollinate at least six to eight flowers. See Pollinating Flowers, p. 188.

5. *Days 20 to 39.* Remove new buds and shoots. Keep plants watered.

6. *Day 39.* (It should be a minimum of 21 days after the last pollination.) Harvest the $F_2$ seeds and germinate them in the petri dish as directed in the General Procedures section, Harvesting Seeds, p. 188, and Germinating Seeds, pp. 188–189.

7. *Day 42 (or next lab period).* Count the numbers of each phenotype, and record these data in Table 8.4 in the Results section.

*The instructor should monitor plants daily because temperature and light conditions may affect flowering times. You may need to adjust the laboratory schedule accordingly. If the flowers open before the next lab period, you may choose to do the first series of pollinations for the students.*

 You and your lab partner are responsible for the care and maintenance of your plants. Remember to check the reservoir for water, and be sure that the wicks make good contact with the watering tray, ensuring moisture flow to quads. A scientist has to remember the organism!

## Results

1. Record the phenotypes of the $F_1$ seeds in Table 8.3 and the phenotypes of the $F_2$ offspring in Table 8.4. Compile results from all teams.

2. Note any changes in procedure or problems with the experiment in the margin of your lab manual or on the schedule of activities (Table 8.2).

 **Table 8.3**
Phenotypes of $F_1$ *B. rapa* Seedlings from Crosses between Plants with Anthocyanin (Purple) and Without Anthocyanin (Green)

*Enlarge this table and Tables 8.4, 8.7, 8.8, and 8.11, and make transparencies for students to fill in and copy.*

*The results for the $F_1$ generation in Table 8.3 should be all purple. Have students post their results before continuing the crosses. This is an important check on their ability to score the phenotypes.*

| Number of Plants | | | | | |
|---|---|---|---|---|---|
| Team No. | Purple | Green | Team No. | Purple | Green |
| 1 | | | 7 | | |
| 2 | | | 8 | | |
| 3 | | | 9 | | |
| 4 | | | 10 | | |
| 5 | | | 11 | | |
| 6 | | | 12 | | |
| Class total | Purple: | | | Green: | |

**Table 8.4**
Phenotypes of F$_2$ Seedlings

| Number of Plants | | | | | |
|---|---|---|---|---|---|
| Team No. | Purple | Green | Team No. | Purple | Green |
| 1 | | | 7 | | |
| 2 | | | 8 | | |
| 3 | | | 9 | | |
| 4 | | | 10 | | |
| 5 | | | 11 | | |
| 6 | | | 12 | | |
| Class total | Purple: | | | Green: | |

*The results for the F$_2$ generation in Table 8.4 should be approximately a 3:1 ratio of purple to green. Students should post their results before discarding their offspring to allow for rescoring plants if needed. Amazingly, students will sometimes score their plants to match their hypotheses—right or wrong.*

3. Compare your observed and expected results by calculating the chi-square test in Table 8.5. Refer to your hypothesis and predictions and use the class data from Table 8.4. Refer to Appendix B for an explanation of the chi-square test.

## Discussion

1. What do the results suggest is the inheritance pattern of this trait? Which is the dominant trait, anthocyanin present (purple) or anthocyanin absent (green)? Assign symbols and write genotypes and phenotypes of all F$_1$ and F$_2$ offspring.

   *Anthocyanin present (purple) is a dominant trait, and the absence of anthocyanin (green) is recessive. These traits follow Mendel's laws, and the results are those predicted for a monohybrid cross.*

**Table 8.5**
Chi-Square Calculations

| | Anthocyanin Present (Purple) | Anthocyanin Absent (Green) |
|---|---|---|
| Observed value ($o$)` | | |
| Expected value ($e$) | | |
| Deviation ($o - e$) or $d$ | | |
| Deviation$^2$ ($d^2$) | | |
| $d^2e$ | | |
| Chi-square, $\chi^2 = \Sigma d^2/e$ | | |

What evidence supports your conclusion?

*The $F_1$ offspring should all be purple; 25% of the $F_2$ offspring should be green (no purple seen) and 75% should be purple, as predicted.*

2. Did the results support your hypothesis? Explain, describing the results of the chi-square test.

   *The observed numbers are consistent with those expected under Mendel's laws. The class data should be close to the predicted ratios, and the chi-square value should be small, indicating that there is no significant difference between the expected and observed results.*

3. Explain the significance of the *p* value (see Appendix B).

   *The p value selected by most biologists is 0.05, indicating that there is less than a 5% chance of error of stating that there is a difference when, in fact, there is no significant difference.*

4. If the results do not support expected Mendelian ratios, what are other possible explanations for these results? What other patterns of inheritance could explain the results?

   *Possible sources of error would include problems with scoring plants or carefully following procedures.*

5. What additional experiments can you suggest to further test your original or modified hypothesis? Outline your crosses.

---

EXERCISE 8.2

# Inheritance of Plant Color: Green, Yellow-Green, and Purple (Anthocyanin Present)

## Materials

12 $F_1$ seeds resulting from a cross previously made between a true-breeding, homozygous normal bright green parent (no purple) and a true-breeding homozygous yellow-green with purple parent
1 quad and related planting supplies (see Materials for All Exercises section)
2 mature plants with the paternal phenotypes

## Introduction

The wild-type plant of *Brassica rapa* is green with purple; anthocyanin is present in the stems and/or leaf stalks. However, variants exist that are bright green (no purple), yellow-green with purple, and yellow-green (no purple) (Color Plate 15). In this exercise, you will investigate the inheritance of these phenotypes. You will determine if one or two genes are involved. You will determine if the inheritance pattern suggests simple dominance and recessiveness, intermediate inheritance, or other modes of inheritance. Formulate a hypothesis for these traits that describes a mode of inheritance based on the Mendelian model. Will these traits conform to Mendel's laws? Is one trait dominant or recessive?

## Hypothesis

Hypothesize about the inheritance of these phenotypes.

> *There are two Mendelian genes. Purple and bright green are dominant, whereas the absence of purple and yellow-green color is recessive. Accept any testable hypothesis. Remember that the data (not the instructors) should modify hypotheses.*

## Prediction

Predict the results of the experiment based on your hypothesis. Outline the cross(es) for your predictions in the margin of your lab manual. Provide expected offspring for each cross in a Punnett square if appropriate. Refer to Figures 8.1 and 8.2 and your text if necessary.

> *If this is an example of a dihybrid cross, then the $F_1$ should all be green with purple and the $F_2$ should have a 9:3:3:1 phenotypic ratio. Students should sketch a Punnett square.*

## Procedure

 Refer to General Procedures (pp. 187–189) and appropriate figures. All days are approximate. Monitor plants for initiation of flowering and seed set.

*Temperature and light conditions may affect flowering times. The instructor may need to adjust the laboratory schedule accordingly.*

1. Refer to the checklist of activities in Table 8.6, and record dates, procedures, and your observations as you follow the procedure for this exercise. Initial the final column when you have completed each step.
2. *Day 1.* Plant $F_1$ hybrid seeds resulting from previously made crosses between homozygous bright green (no purple) and homozygous yellow-green (with purple) plants. See Planting Seeds, pp. 187–188.
3. *Day 4 or 5.* Observe germinating seedlings and compare phenotypes with the adult parental plants on demonstration (Color Plate 15). Record the numbers of each phenotype in Table 8.7 and on the class data sheet. Modify the table to accommodate your results. Write in the headings for the phenotypes and draw in columns. Thin plants to one per cell. Remember to keep plants watered.
4. *Days 14, 16, 18.* Cross the $F_1$ plants. Pollinate plants three times, with 1 day between pollinations. Pollinate at least six to eight flowers. See Pollinating Flowers, p. 188.
5. *Days 20 to 39.* Remove and discard new buds and shoots. Keep plants watered.

**Table 8.6**
Checklist of Activities for Exercise 8.2. Record the date and initial the final column indicating that you have completed each step.

| Approximate Day | Date | Activity | Initials |
|---|---|---|---|
| Day 1 | | Plant $F_1$ hybrid seeds. See Planting Seeds, pp. 187–188. Regularly check water. | |
| Day 4 or 5 | | Observe seedlings and record numbers of each phenotype in Table 8.7. Thin plants to one per cell. | |
| Days 14, 16, 18 | | Pollinate on 3 days; pollinate at least 6–8 flowers. See Pollinating Flowers, p. 188. | |
| Days 20 to 39 | | Remove buds and shoots. | |
| Day 39 | | Harvest $F_2$ seeds and germinate in quads. See Harvesting Seeds, p. 188, and Planting Seeds, pp. 187–188. | |
| Days 47 to 52 | | Count and record numbers of each phenotype in Table 8.8. | |

6. *Day 39.* (It should be a minimum of 21 days after the last pollination.) Harvest the $F_2$ seeds and begin the next generation. Germinate seeds in a quad. See Harvesting Seeds, p. 188, and Planting Seeds, pp. 187–188.

7. *Days 47 to 52.* (Plants should be large enough to score phenotypes.) Count the numbers of each phenotype, and record these data in Table 8.8. Modify the table to accommodate your results. Write in the headings for the phenotypes and draw in columns.

*On day 39 you may want to germinate seeds of the parental phenotypes. Even though students have seen the phenotypes before beginning the project, they often will have difficulty distinguishing the leaf color (green, yellow-green) and the presence of anthocyanin (purple, not purple) in the four possible combinations.*

 You and your lab partner are responsible for the care and maintenance of your plants. Remember to check the reservoir for water, and be sure that the wicks make good contact with the watering tray, ensuring moisture flow to quads. The organism must survive to reproduce!

## Results

1. Record the $F_1$ offspring in Table 8.7 and the $F_2$ offspring in Table 8.8. Modify these tables to accommodate observed phenotypes. Compile results from all teams.

2. Note any modifications of procedures or problems in the experiment in the margin of your lab manual or on your schedule of activities (see Table 8.6).

3. Compare your observed and expected results by calculating the chi-square test in Table 8.9. Refer to your hypothesis and predictions and use the class data from Table 8.8. Refer to Appendix B for an explanation of chi-square analysis.

*The results for the F₁ generation in Table 8.7 should be all bright green with purple. Students should set up the top of the table with the four combinations of phenotypes (bright green with purple; bright green without purple; yellow-green with purple; yellow-green without purple). Have students post their results before continuing the project to be certain they are scoring the phenotypes properly.*

**Table 8.7**

Phenotypes of F₁ *B. rapa* Seedlings from a Cross between a Homozygous Normal Green Parent and a Yellow-Green with Purple Parent. Write in headings for the phenotypes and draw in columns.

| Team No. | Phenotypes | | | |
|---|---|---|---|---|
| | *Bright Green Purple* | *Bright Green No Purple* | *Yellow-Green Purple* | *Yellow-Green No Purple* |
| 1 | | | | |
| 2 | | | | |
| 3 | | | | |
| 4 | | | | |
| 5 | | | | |
| 6 | | | | |
| 7 | | | | |
| 8 | | | | |
| 9 | | | | |
| 10 | | | | |
| 11 | | | | |
| 12 | | | | |
| **Class total** | | | | |

*The results for the F₂ generation in Table 8.8 should be approximately a 9:3:3:1 ratio for the four phenotypes. Have students post their results before discarding their plants. Scoring is sometimes difficult, and, of course, each team's individual results will deviate from the expected due to the small sample size. Students will sometimes score their plants to match their hypotheses—right or wrong.*

### Discussion

1. What do the results suggest is the inheritance pattern of these traits? Assign symbols and write genotypes and phenotypes of all F₁ and F₂ offspring.

   *This is an example of a dihybrid cross. Anthocyanin purple and no purple are alleles of one gene. Purple is a dominant trait, and the absence of anthocyanin is recessive. Green and yellow-green are alleles of a second gene, and green is dominant. These traits follow Mendel's laws, and the results are those predicted for a dihybrid cross. Students should include genotypes and phenotypes.*

   What evidence supports your conclusion?

   *The F₁ offspring should all be purple and bright green, and the F₂ offspring should be 9 bright green, purple; 3 bright green, no purple; 3 yellow-green, purple; 1 yellow-green, no purple.*

**Table 8.8**
Phenotypes of $F_2$ Seedlings. Write in headings for the phenotypes and draw in columns.

| | Phenotypes | | | |
|---|---|---|---|---|
| **Team No.** | *Bright Green Purple* | *Bright Green No Purple* | *Yellow-Green Purple* | *Yellow-Green No Purple* |
| 1 | | | | |
| 2 | | | | |
| 3 | | | | |
| 4 | | | | |
| 5 | | | | |
| 6 | | | | |
| 7 | | | | |
| 8 | | | | |
| 9 | | | | |
| 10 | | | | |
| 11 | | | | |
| 12 | | | | |
| **Class total** | | | | |

2. Did the results support your hypothesis? Explain, describing the results of the chi-square test.

   *The observed numbers are consistent with those expected under Mendel's laws. The class data should be close to the predicted ratios, and the chi-square value should be small, indicating that there is no significant difference between the expected and observed results.*

3. Write a statement explaining the significance of the $p$ value (see Appendix B).

   *The p value selected by most biologists is 0.05, indicating that there is less than a 5% chance of error of stating that there is a difference when, in fact, there is no significant difference.*

4. If the results do not support expected Mendelian ratios, what are other possible explanations for these results? What other patterns of inheritance could explain the results?

**Table 8.9**

Chi-Square Calculations. Write in headings for the phenotypes and draw in columns.

| | Phenotypes | | | |
|---|---|---|---|---|
| | *Bright Green Purple* | *Bright Green No Purple* | *Yellow-Green Purple* | *Yellow-Green No Purple* |
| **Observed value ($o$)** | | | | |
| **Expected value ($e$)** | | | | |
| **Deviation ($o - e$) or $d$** | | | | |
| **Deviation$^2$ ($d^2$)** | | | | |
| $d^2/e$ | | | | |
| **Chi-square, $x^2 = \Sigma d^2/e$** | | | | |

5. What additional experiments can you suggest to further test your original or modified hypothesis? Outline your crosses.

EXERCISE 8.3

# Inheritance of Variegated and Nonvariegated Leaf Color

## Materials

6 seeds from homozygous green plants
6 seeds from homozygous variegated plants
2 quads and related planting supplies (see Materials for All Exercises section)
5-in. by 8-in. index cards
stakes (small wooden applicator sticks)

## Introduction

In one strain of *Brassica rapa* plants, wild-type plants are entirely green (nonvariegated). However, other plants have white or yellow blotches or streaks on the green leaves and stems (variegated). In this exercise, you will investigate the inheritance of this trait for leaf color. Which trait is dominant, variegated or nonvariegated? Does the inheritance of this trait follow predicted Mendelian ratios? Formulate a hypothesis for these traits that describes a mode of inheritance based on the Mendelian model.

## Hypothesis

Hypothesize about the inheritance of this trait.

*The traits follow Mendelian inheritance; green leaves are dominant, and variegated leaves are recessive. This is the most likely student hypothesis. Any testable hypothesis will work.*

*Students will make their hypotheses at the time they plant their seeds. It is unlikely that they will have any idea that this is an example of cytoplasmic inheritance. The excitement for the students comes when their data do not match their hypotheses. This will be one of the best experiences of investigating biology.*

## Prediction

Predict the results of the experiment based on your hypothesis. Outline the cross(es) for your predictions in the margin of your lab manual. Provide expected offspring for each cross in a Punnett square if appropriate. Refer to Figures 8.1 and 8.2 and your text if necessary.

*If green leaves are a dominant Mendelian trait, then the results of the cross should be all green heterozygous offspring.*

## Procedure

 Refer to the General Procedures section (pp. 187–188) and relevant figures. All days are approximate. Monitor plants for initiation of flowering and seed set.

1. Refer to the checklist of activities in Table 8.10 and record dates, procedures, and your observations as you follow the procedure for this exercise. Initial the final column when you complete each step.
2. About 3 or 4 days *before* day 1 in the previous exercises, plant three seeds from variegated plants in two cells of the quad. Leave the other two cells empty. (Your instructor may have already done this.) Variegated plants grow more slowly than totally green plants and must be planted earlier. See Planting Seeds, pp. 187–188.
3. *Day 1.* Plant three seeds from green plants in each of the two empty cells of your quad. See Planting Seeds, pp. 187–188.
4. *Day 4 or 5.* Thin the green seedlings to one per cell. Thin variegated seedlings, leaving one plant that is definitely variegated in each cell.
5. *Days 10 to 14.* To prevent "like" plants from pollinating each other, place a 5-in. by 8-in. card supported by two stakes across the quad, separating green from green and variegated from variegated plants. One variegated and one green plant should be on each side of the card. Be sure flowers do not extend above the card. If they do, substitute a larger card. Keep plants watered.
6. *Days 16, 18, 20.* Using a bee-stick, transfer pollen from a flower on a white or variegated stem to flowers on the totally green plant. Using a different bee-stick, carry out the **reciprocal pollination;** that is, transfer pollen from a flower of a green plant to a flower of a white or variegated plant. Pollinate plants three times, with three consecutive pollinations, with 1 day between pollinations. See Pollinating Flowers, p. 188.

*Depending on your students and prep time, you may have students plant the variegated seeds on their lab day and come back to plant the green seeds, or you can plant the variegated seeds 3 or 4 days before lab and let students plant the seeds from green plants on lab day. If both types of seeds are planted at the same time, the flowering will be out of phase for pollinating.*

*Students should choose plants that strongly express the variegated trait; otherwise, the variegation may not extend into the reproductive structures.*

**Table 8.10**

Checklist of Activities for Exercise 8.3. Record the date and initial the final column indicating that you have completed each step.

| Approximate Day | Date | Activity | Initials |
|---|---|---|---|
| Before day 1 | | Plant seeds from variegated parents in two cells. See Planting Seeds, pp. 187–188. | |
| Day 1 | | Plant seeds from green parents in remaining two cells. See Planting Seeds, pp. 187–188. | |
| Day 4 or 5 | | Thin seedlings to one per cell. Leave clearly variegated seedlings in two cells. | |
| Days 10 to 14 | | Place card in quad so each side has a green and a variegated plant. Prevent pollination between like plants. | |
| Days 16, 18, 20 | | Cross-pollinate green and variegated flowers. Use a different bee-stick for reciprocal pollinations. Pollinate on 3 days. See Pollinating Flowers, p. 188. | |
| Days 20 to 39 | | Remove buds and shoots. | |
| Day 39 | | Harvest F$_1$ seeds and germinate in quads. See Harvesting Seeds, p. 188, and Planting Seeds, pp. 187–188. | |
| Days 47 to 52 | | Count and record numbers of each phenotype in Table 8.11. | |

*Temperature and light conditions may affect flowering times. The instructor may need to adjust the laboratory schedule accordingly.*

7. *Days 20 to 39.* Remove and discard new buds and shoots. Keep plants watered.

8. *Day 39.* (It should be a minimum of 21 days after the last pollination. This may be performed during the next lab period.) Harvest seeds from green plants. Next, *keeping seeds separate,* harvest seeds from variegated plants. See Harvesting Seeds, p. 188.

 Collect and store seeds in separate envelopes labeled with the date and seed type. Throw away any seeds if you are unsure of the parents.

*Remind students to collect and store seeds separately from green parents and variegated parents.*

9. Begin the second generation. Germinate seeds in a quad. Plant five seeds from the green plants in each of two cells and five seeds from the variegated plants in each of the other two cells. Label quads to indicate the parents: "green" or "variegated." See Planting Seeds, pp. 187–188.

10. *Days 47 to 52.* Record phenotypes of plants from green parents and phenotypes of plants from variegated parents in Table 8.11. Modify the table to accommodate the phenotypes for each cross.

You and your lab partner are responsible for the care and maintenance of your plants. Remember to check the reservoir for water, and be sure that the wicks make good contact with the watering tray, ensuring moisture flow to quads.

## Results

1. Record the offspring phenotypes in Table 8.11. Record your results separately for the reciprocal crosses: (A) seeds from green plants pollinated with pollen from variegated plants and (B) seeds from variegated plants pollinated with pollen from green plants. Modify the table to accommodate the observed phenotypes for each cross. Compile results from all teams.

2. Note any modifications of procedures or problems in the experiment in the margin of your lab manual or on your schedule of activities (see Table 8.10).

*The variegated trait is inherited in the cytoplasm; therefore, the offspring from maternal parents with the green phenotype can produce only green offspring. The variegated maternal parents have combinations of cells, some of which are green (have chloroplasts) and some of which are white (lack chloroplasts). The offspring from these plants can be either green or variegated.*

 **Table 8.11**

Phenotypes of Germinated Seeds Taken from Green Plants Pollinated with Variegated Pollen (A) and Variegated Plants Pollinated with Green Pollen (B). Write in headings for the phenotypes and draw in columns.

| Team No. | Phenotypes A | | Phenotypes B | |
|---|---|---|---|---|
| | *Green* | *Variegated* | *Green* | *Variegated* |
| 1 | | 0 | | |
| 2 | | 0 | | |
| 3 | | 0 | | |
| 4 | | 0 | | |
| 5 | | 0 | | |
| 6 | | 0 | | |
| 7 | | 0 | | |
| 8 | | 0 | | |
| 9 | | 0 | | |
| 10 | | 0 | | |
| 11 | | 0 | | |
| 12 | | 0 | | |
| Class total | | 0 | | |

3. Compare your observed and expected results for cross **A** and cross **B**. Refer to your hypothesis and predictions and use the class data from Table 8.11.

## Discussion

1. Did the results support your hypothesis? Explain, comparing your predictions and results.

   *The results do not support the common student hypothesis of Mendelian inheritance. If students hypothesized Mendelian inheritance, then they predicted that all offspring would be either green or variegated, depending on which trait they predicted to be dominant. The observed results for the reciprocal crosses should have differed. All offspring from the (A) crosses should have been green, and many of the offspring from the (B) crosses should have been variegated. Usually, not all offspring will be variegated, depending on the amount of variegation in the flower pollinated. The observed results differ from expected.*

2. If the results do not support expected Mendelian ratios, what are other possible patterns of inheritance that could explain the results?

   *These traits appear to be inherited from the seed parent. This is an example of cytoplasmic inheritance of chloroplast genes.*

3. Why would it be inappropriate to use the chi-square test for these data?

   *Because 100% of the offspring were expected to be green, one category (variegated) would not exist for the expected (resulting in a zero value).*

4. What additional experiments can you suggest to further test your original or modified hypothesis? Outline your crosses.`

## Reviewing Your Knowledge

1. Review all terminology used in this laboratory. Define and provide examples of all terms indicated in bold type.

2. The fruit fly, *Drosophila,* is an extremely useful organism in the study of genetics today. *Brassica rapa,* a Fast Plant, is also becoming a popular organism for study. Based on your knowledge and experience, what are some important characteristics of fruit flies and Fast Plants that make them ideal for genetic experiments?

   *short life cycle; easy to manipulate for crosses; small organisms, requiring little space for large numbers of individuals*

3. In general terms, explain why a scientist would use the chi-square test as part of his or her data analysis. What important information does it provide?

   *Chi-square analysis tests the hypothesis that the observed and expected results are not significantly different from each other. Although some deviation from the expected results might be expected owing to chance alone, the question arises as to how much deviation is acceptable. The chi-square test provides a way of determining how much deviation is acceptable with a probability of 0.05.*

For additional genetics problems see www.masteringbiology.com or websites at the end of the lab topic.

4. Cyndi grows thousands of *B. rapa* plants from seed and carefully watches for any unusual plants that might be a new mutation. She has

been watching a small seedling whose leaves continue to develop, but the stem is not elongating. The result is a cluster or rosette of leaves sitting just above the soil surface. After 15 days the plant is beginning to produce flowers but is still a rosette. Cyndi is encouraged and hypothesizes that rosette is a recessive mutation. Describe the first cross she should make and the predicted results.

*She should cross rosette with a normal* B. rapa. *The predicted results are all offspring will be normal.*

Since *B. rapa* is an annual plant, the one rosette individual dies. Cyndi would like to continue her study of this mutation, but all of the offspring from her original cross were normal. There are no more plants with the rosette mutation. What can Cyndi do to continue her investigation of rosette? What would be the results of these "next steps"?

*All the offspring from her original cross are normal height, and their genotype should be heterozygous (Rr). Cyndi can cross two of these heterozygous plants, and the predicted results are 75% normal height and 25% rosette (a 3:1 ratio). This assumes that rosette is a recessive trait.*

What results would you expect if Cyndi crosses her new generation of rosette plants with a known heterozygous plant of normal height?

*50% normal height, 50% rosette*

## Applying Your Knowledge

1.  If two traits, dwarfism and stem color, were both on the same pair of chromosomes, would you expect an $F_2$ generation of 9:3:3:1 for a dihybrid cross? Use Mendel's laws and your understanding of meiosis to explain your answer. Refer to Figure 8.2, if needed.

    *We would not expect a 9:3:3:1 ratio if the genes were linked (occurred on the same chromosome). The gametes could not randomly produce all possible combinations. Alleles of different genes would be inherited together.`*

2.  A budding plant breeder discovers a rare mutant of *Brassica* that is almost entirely white. The plastids in the leaves have a mutation that results in white plastids. If the breeder uses the pollen from this mutant to pollinate a plant with normal green leaves, what kind of leaves would you expect to see in the offspring? Explain.

    *The offspring would be green because the maternal parent would contribute all of the plastids in the cytoplasm.*

    If the normal green parent is used as the pollen parent, would you expect the results to be the same? Explain.

    *The results would be different. In the reverse cross, the offspring would be white or die since the maternal parent would contribute only the mutant—white plastids.*

3.  Stefanie is a young farmer growing new and heritage pumpkins. In her rows of pumpkins she finds a dark orange pumpkin with warts and a green pumpkin with smooth skin. She knows that orange skin color is dominant in pumpkins, but she cannot find any information on the inheritance of warty skin in pumpkins. She decides to cross the pumpkins to determine if warty skin is a dominant or recessive trait. State a hypothesis for the inheritance of warty skin in pumpkins.

### Hypothesis

*Warty skin is a dominant trait. (Students can hypothesize either dominant or recessive.) Example assumes trait is dominant.*

Then based on your hypothesis, predict the results of the $F_1$ and $F_2$ dihybrid crosses between the *orange warty* pumpkin and the *green smooth* pumpkin. (Assume these pumpkins are homozygous for both traits.)

*Skin color: orange (Or)*        *Skin surface: warty (Wt)*

*green (or)*                      *smooth (wt)*

*Parental cross*                 *orange warty × green smooth*

*Genotypes*                      *OrOr WtWt      oror wtwt*

  *$F_1$ offspring—orange warty (Oror Wtwt)*

*$F_2$ cross—Oror Wtwt × OrorWtwt*

### Punnett Square of F2 Offspring

|  | Or Wt | Or wt | or Wt | or wt |
|---|---|---|---|---|
| **Or Wt** | OrOr WtWt | OrOr Wtwt | Oror WtWt | Oror Wtwt |
| **Or wt** | OrOr Wtwt | OrOr wtwt | Oror Wtwt | Oror wtwt |
| **or Wt** | Oror WtWt | Oror Wtwt | oror WtWt | oror WtWt |
| **or wt** | Oror Wtwt | Oror wtwt | oror Wtwt | oror wtwt |

*Phenotypic ratios expected if warty is dominant.*

*9/16 Orange warty*

*3/16 Orange smooth*

*3/16 Green warty*

*1/16 Green smooth*

The results of Stefanie's crosses were not revealed until the following summer. All of the pumpkins grown from her crosses were orange warty pumpkins. Do these results support your hypothesis?

*Yes, if students hypothesized that warty was dominant. No, if students hypothesized that warty was recessive. The trait is dominant.*

4. Molecular geneticists interested in the evolutionary history of the human race have concentrated their research on samples of DNA from women representing all races and continents. Why might the DNA of women—and not men—be of interest?

*Researchers at the University of California at Berkeley and Emory University are studying mitochondrial DNA, which has a rapid mutation rate and is inherited only from the mother's egg.*

## Investigative Extensions

1. Wisconsin Fast Plants, like the fruit fly and Mendel's peas, offer a model system for investigating inheritance. Other interesting traits in *B. rapa* are available for extended study. Select from the following list of traits.

   • *Rosette* is a mutation that results in a small plant in which the leaves remain at soil level, because the stem does not elongate.

- *Petite/Astro* plants were developed to meet the "5-10-15" NASA specifications of 5 seed pods, no taller than 10 cm, and flowers by 15 days, so that they could be studied in space!

- *Tall* plants have an elongated internode (stem length between leaves) and the average plant height at 15 days is 28 cm.

- In addition there are plants that have been developed to investigate two traits in dihybrid crosses: *purple stem, hairy* (the standard trait) and *non-purple stem, hairless* (two mutations). The presence of anthocyanins produces the purple stem color, and the hairless trait is for the absence of hairs on leaves and petioles.

Follow the procedures and planning schedules provided in the General Procedures for All Exercises and in Exercises 8.1, 8.2, and 8.3.

2. Genetic variation and its relationship to the environment can be investigated using the *hairy/hairless* trait. You may choose to begin your investigation by simply growing the seeds and scoring the traits to determine the range and distribution of genetic variation in your sample of plants. These plants can also be used to investigate selection and genetic variation through the selective breeding for either hairy or hairless plants and recording your results over several generations. Working with these interesting organisms will raise additional questions and hypotheses as you carefully observe, measure, and analyze the parents and offspring. Seed stocks, additional suggestions, and resources are available at www.fastplants.org and at Carolina Biological Supply (http://www.carolina.com/fastplants/default.asp).

3. What did Thomas Hunt Morgan do when his student Calvin Bridges found a mutant fly with white eyes in his *Drosophila* cultures? How would *you* determine the mode of inheritance for unknown traits? Using the BioQUEST simulation program "Genetics Construction Kit" (GCK), you will be faced with the same dilemma. Unlike most genetic problems, for this simulation program there is no answer key! You will have to do the crosses, analyze the data, and make a logical and convincing argument based on your evidence to convince your colleagues and instructor that you have uncovered the genetics of these traits. In GCK the computer generates field populations of "flies" with designated phenotypes. As the geneticists, your team must consider the possible modes of inheritance (the hypotheses), then design and complete crosses (testing your hypothesis, and analyzing the results). The simulation program can be used to investigate dominant, recessive, and incomplete dominant traits, sex-linked traits and linked autosomal traits, and also multiple alleles. The instructor can set the range of traits and the level of difficulty ("Genetics Construction Kit," BioQUEST Library Online).

*The modes of inheritance, suggestions for investigations, and details of procedures are available at the Carolina Biological Supply website. The selection investigations with hairy/hairless are particularly interesting since the presence/absence of hairs is a single gene mutation, but the degree of hairiness is a polygenetic trait. For students who have done crosses with single genes, this offers an opportunity to ask very different questions, with new techniques and the opportunity to bridge genetics, ecology, and evolutionary biology. See the Fast Plants Resources at http://www.carolina.com/category/teacher+resources/wisconsin+fast+plants.do*

*The BioQUEST Library VI is available at numerous vendors, and the entire library plus new projects is available online. Check the BioQUEST home page at www.bioquest.org for details on a downloadable version.*

# Student Media: BioFlix, Activities, Investigations, and Videos
**www.masteringbiology.com (select Study Area)**

**Activities**—Ch. 14: Monohybrid Cross; Dihybrid Cross; Gregor's Garden
   Ch. 30: Angiosperm Life Cycle
**Investigations**—Biology Labs On-Line: FlyLab: Performing Monohybrid, Dihybrid, and Trihybrid Crosses
**Videos**—Ch. 30: Time Lapse Flowering Plant Life Cycle; Flower Blooming Time Lapse; Bee Pollinating

## References

Heitz, J. and C. Giffen. *Practicing Biology: A Student Workbook,* 4th ed. San Francisco, CA: Pearson Benjamin Cummings, 2011. Chapters 14 and 15 provide a guide to solving genetics problems and practice problems.

Jungck, J., and J. Calley. "Genetics Construction Kit," *BioQUEST Library Online.* http://www.bioquest.org/BQLibrary/library_result.php (software available to download)

Klug, W. S., M. R. Cummings, C. Spencer, and M. A. Palladino. *Essentials of Genetics,* 7th ed. San Francisco, CA: Benjamin Cummings, 2010.

Williams, Paul H. "Bee-Sticks, an Aid in Pollinating Cruciferae." 1980, *HortScience,* vol. 115 (6), pp. 802–803.

*Wisconsin Fast Plants Manual.* Burlington, NC: Carolina Biological Supply, 1989. Procedures for exercises adapted by permission of Carolina Biological Supply, Burlington, NC 27215, and Wisconsin Alumni Research Foundation, Madison, WI 53707. © 1989 Wisconsin Alumni Research Foundation.

## Websites

BioQUEST home:
www.bioquest.org/

Carolina Biological Supply Co. information on Fast Plants seed stock descriptions:
http://www.carolina.com/category/teacher+resources/wisconsin+fast+plants/ wisconsin+fast+plant+seed+descriptions+profiles.do

Genetic problems and tutorials:
http://www.biology.arizona.edu/

How Fast Are Wisconsin Fast Plants? Youtube video:
http://www.youtube.com/watch?v=73S63xV_o34

Human genes and diseases:
http://www.ncbi.nlm.nih.gov/books/bv.fcgi

Wisconsin Fast Plants:
http://www.fastplants.org/

Wisconsin Fast Plants Life Cycle Time Lapse. Youtube video:
http://www.youtube.com/watch?v=JumEfAbjBjk

# LAB TOPIC 8

## MENDELIAN GENETICS I: FAST PLANTS

# Teaching Plan for Laboratories

This laboratory requires careful advance planning. We recommend that you develop a calendar detailing dates for planting demonstration plants and all student work. See the *Preparation Guide* for detailed growing instructions.

## Main Concepts and Objectives

1. Concept: genetic terminology. Students will acquire and use the language of genetics. (Introduction; Exercises 8.1–8.3)

2. Concept: Mendelian inheritance of single genes. Students will make monohybrid crosses and determine dominant and recessive traits by analyzing results. (Exercise 8.1)

3. Concept: Mendelian inheritance of two genes. Students will make dihybrid crosses and determine the dominant and recessive traits for each of two unlinked genes. (Exercise 8.2)

4. Concept: non-Mendelian inheritance. Students will make reciprocal crosses between variegated and nonvariegated individuals and compare the results with those expected with Mendelian inheritance. They will explain their results in terms of cytoplasmic inheritance. (Exercise 8.3)

5. Concept: chi-square test. Students will analyze their results of crosses using the chi-square test. (Exercises 8.1, 8.2, and 8.3)

6. Concept: scientific method. Students will propose hypotheses and predictions, collect data, and compare observed and expected results. (Exercises 8.1, 8.2, and 8.3)

7. Concept: science not quick and easy. Students will be responsible for the care and maintenance of their organisms and will continue their research over 6–8 weeks. (Exercises 8.1, 8.2, and 8.3)

8. Concept: scientific writing. Students will continue to develop scientific writing skills. (Exercises 8.1, 8.2, and 8.3)

9. Specific content. Students will be able to describe and define: *homozygous, heterozygous, dominant, recessive, monohybrid and dihybrid crosses, $F_1$ and $F_2$ generations, law of segregation, law of independent assortment, cytoplasmic inheritance, Fast Plants, chi-square.*

## Order of the Lab

This laboratory requires 6–8 weeks to complete and therefore should be considered a continuing research project for students. The actual laboratory time can be organized in several ways. A 1- to 2-hour laboratory period is required to introduce the plants, the traits that are to be investigated, and the procedures, plus setting up the experiments. Students are responsible for the care of their plants, completing the crosses, and collecting the data. This can be done outside of lab or by allocating a few minutes at the beginning or end of lab periods for students to work with their plants. Finally, students will need an introduction to chi-square analysis at some point if they have not used this statistical test before. The outline below is for the initiation of the lab topic.

1. Introduce concepts and objectives. Review genetics terminology. (5 min)

2. Describe Fast Plants, their life cycle, and the genetic traits to be investigated. (10 min)

3. Organize students into teams of two to four students, and record the names of members of each team. (10 min)

4. Review each genetic cross and have students work on developing hypotheses and predictions. (30 min)

5. Demonstrate procedures for planting quads with Fast Plants. Remind students to label everything with their name, date, and seed type, and record information in schedules of activities. (15 min)

6. Plant seeds in quads. For Exercise 8.3, the variegated seeds should have been planted in advance; if the students are planting, they should return in 3 or 4 days to plant green seeds. (30 min)

7. Review responsibilities of students and location of materials. (5 min)

Over the next 6 weeks, remind students to care for their plants and to follow the schedule of activities. At some point prior to week 6, students should be introduced to the chi-square test. A final laboratory period can be used for oral reports from teams. Depending on your lab schedule, students can write a scientific paper and no additional lab time is required.

**For a 2-hour lab:** The initial introduction to the lab and the planting of seeds can be accomplished in 2 hours. See suggestions above for completing work and presenting results.

## Classroom Management

The success of this exercise depends on the ability of students to sustain a continuing project, keeping their plants alive and following the procedures and activities that span many weeks. Students must be responsible for keeping plants watered and for organizing and labeling seeds following the first set of crosses. Ask students about their project during the ensuing weeks to ensure their constant care for the plants. Because some students will initially fail to check their plants to be sure they are surviving, it is important to check on the plants regularly, particularly before a weekend. A student assistant might be assigned this task. For large classes, the space required for this lab may seem large, but students will need to use only 1 square foot per team. Students should not take plants back to their rooms, since the plants require high light to reproduce.

## Student Development

Laboratory exercises are designed to be completed in a 2- or 3-hour period. Yet we know that science is a continuing process, which seldom matches an introductory lab syllabus. The actual care and maintenance of organisms and the preparation of materials are accomplished by a hidden army that supports the work of each lab period. Students need to be responsible for the work of science, including following a procedure over several weeks for caring for their plants and scheduling their activities, which may depend on how their plants grow. Students expect all their results to match their expectations. They will work very hard to make their results from the last cross (cytoplasmic inheritance) match the Mendelian model. Some will attribute their results to errors in their procedure. Without explaining their results, encourage them to think about the data and possible explanations. Students are now faced with a beautiful hypothesis ruined by an ugly fact. Students continue to practice the scientific method, proposing hypotheses, making predictions, testing hypotheses, and recording and analyzing data.

## Discussion and Summary

Students can discuss results and complete questions in the exercises. Students can present results in a final lab summary period. Students can write a scientific paper. Students will be collaborating and therefore should discuss results with team members.

## Evaluation

Students can be tested on the material, especially terminology and expected results. Students can be graded on oral or written presentations. If you are following the plan for integrating scientific writing in the laboratory program, as suggested in this manual, students should compose a title page and a Materials and Methods section for one exercise in this lab topic. If students have completed lab topics in which they have written all the individual parts of a paper, consider having them write a complete paper using the final exercise in the lab.

## Lab Topic 9
# Mendelian Genetics II: *Drosophila*

## Laboratory Objectives

After completing this lab topic, you should be able to:

1. Discuss why *Drosophila* is one of the most important organisms used in eukaryotic genetics.
2. Explain how a biochemical assay can be used as an indication of biochemical phenotypes.
3. Describe the inheritance pattern of the gene for aldehyde oxidase.
4. Name genes using the convention recommended in *Drosophila* genetics.
5. Determine parental genotypes by investigating offspring.
6. Use the chi-square test to evaluate experimental results.
7. Describe gene mapping.

*For a 2-hour lab:* Omit Exercise 9.4. Or, rather than omitting the important concept of gene linkage, you may choose to perform this lab topic in two 2-hour lab sessions. See the Teaching Plan.

*Instructors and/or students can download data tables in Excel format. Go to www .masteringbiology.com. Look in the Study Area under Lab Media.*

## Introduction

The exercises in this lab topic exemplify some of the classic areas of genetic research that led to an understanding of the principles governing the inheritance of specific traits. Initial experiments were concerned with the transmission of hereditary factors from generation to generation and led to the discovery of Mendel's laws, which define the pattern of inheritance of individual genes. Later experiments identified chromosomes as the physical structures wherein the units of heredity reside and provided firm cytological evidence for the theorems of Mendelian genetics. More recent investigations have addressed the biochemical and molecular basis of gene expression.

One of the classic tools of genetic research is the fruit fly, *Drosophila melanogaster*. This organism has played an important role in the development of our knowledge of heredity since the famous scientist, Thomas Hunt Morgan, began using fruit flies for genetics studies in 1907. One characteristic of *Drosophila melanogaster* that makes this organism easy to study is its very low chromosome number. The haploid (*n*) number of chromosomes is 4, and the chromosomes are designated X(1), 2, 3, and 4 (Figure 9.1). The 2, 3, and 4 chromosomes are the same in both sexes and are referred to as **autosomes** to distinguish them from the X and Y **sex chromosomes.** *Drosophila* females are characterized by two X chromosomes whereas *Drosophila* males have an X and a Y chromosome. Chromosome 4 and the Y chromosome contain so few genes that, for all practical purposes, they can be ignored. Thus, almost the entire genetic content of the *Drosophila* genome resides on only three chromosomes: X, 2, and 3.

*We are assuming that students performing this laboratory will have been introduced to dominant/recessive alleles, sex linkage, and Mendel's laws of inheritance. If this is not true for your students, briefly introduce these concepts in your introductory remarks.*

**Figure 9.1.**
**Metaphase chromosomes from a dividing cell in *Drosophila melanogaster*.** The haploid number of chromosomes is 4. Females have two X chromosomes whereas males have an X and a Y chromosome.

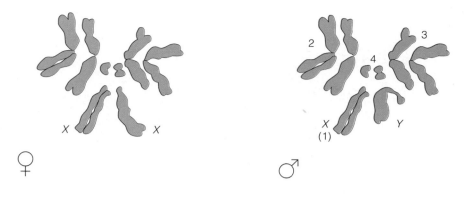

Another characteristic of *Drosophila* that makes it an excellent genetic research tool is its short generation time. At 25°C, a *Drosophila* culture will produce a new generation in 10 days: 1 day in the egg (embryo) stage, 5 days in the larval stage, and 4 days in the pupal stage (Figure 9.2).

You will use *Drosophila melanogaster* in each of the following exercises. You will be asked to investigate the inheritance of a gene called *aldox*, and you will determine the position of this gene on its chromosome; that is, you will map the gene.

---

EXERCISE 9.1

# Establishing the Enzyme Reaction Controls

---

## Materials

stereoscopic microscope
vials 1a and 1b
ether dropper bottles or FlyNap
re-etherizer
2 spot assay plates
large and small white index cards

toothpicks
pestle
Kimwipes®
assay mixture dropper bottles
water bottle

## Introduction

The trait to be studied in each exercise of this lab topic is the presence or absence of the enzyme **aldehyde oxidase (AO),** which catalyzes the oxidation of a number of aldehydes, including acetaldehyde and benzaldehyde. AO activity is controlled by one gene, the **aldox** gene. Although *Drosophila* flies possess AO activity, its physiological importance to the organism is not well understood. Mutant strains that exhibit no AO activity are available, and their viability and fertility are normal. This latter observation indicates that AO activity is not a vital enzyme activity for a fly that is reared in a laboratory setting.

To test for AO activity, you will use an **enzyme spot test,** or **spot assay.** This test works on the following principle: In the presence of AO, the substrate benzaldehyde, when mixed with the color indicator nitroblue tetrazolium

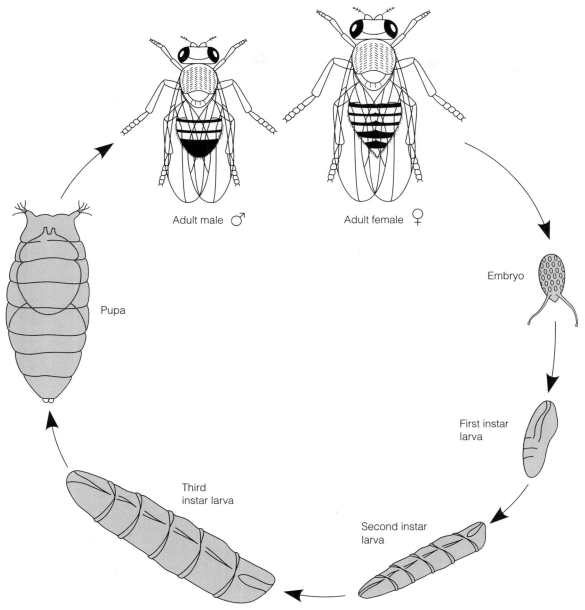

**Figure 9.2.**
**Developmental stages of *Drosophila melanogaster*.** An embryo hatches to a
larva, which undergoes two molts and then pupates. The pupae develop into
adult winged flies.

(NBT)–phenazine methosulfate (PMS), will oxidize to form benzoic
acid and a blue color. The blue color indicates that the enzyme is present
and active.

This reaction can be diagrammed as follows:

| **Substrate** | **Enzyme AO** | **Product** |
|---|---|---|
| benzaldehyde + NBT + PMS | $\longrightarrow$ | benzoic acid + blue products |
| (assay mixture) | | |

Without AO, the reaction will not proceed, and no blue color will be produced.

This first exercise will demonstrate the positive and negative enzyme reactions as seen in the spot assay. Vial 1a contains flies that have the enzyme present. What will be the results if flies from this vial are homogenized in the assay mixture? Will they demonstrate AO activity? (Remember that AO activity produces a blue color with the assay.)

Vial 1b contains flies that do not have the enzyme present. What will be the results if these flies are homogenized in the assay mixture? Will they demonstrate AO activity?

### Hypothesis

Hypothesize about AO activity in flies from vial 1a and flies from vial 1b.

> *Vial 1a flies have the enzyme and will demonstrate AO activity. The flies in vial 1b do not have the enzyme and will not demonstrate AO activity.*

### Prediction

Predict the results of the experiment (test) based on your hypothesis (if/then).

> *If vial 1a flies demonstrate AO activity, then a blue color will develop when these flies are homogenized with the assay mixture (positive reaction). Flies in vial 1b, lacking the enzyme, will not demonstrate AO activity and will not produce a color change (negative reaction).*

### Procedure

*FlyNap can be substituted for ether. Follow directions given on the bottle label.*

*Since students are working in pairs, one student should anesthetize the flies in vial 1a, the other, those in vial 1b. Students then share results. It is important that both students view flies to learn how to determine their sex.*

1. Anesthetize the flies in vials 1a and 1b as follows:

   a. From your ether dropper bottle, place 2 or 3 drops of ether on the cotton plug of your vial. Be sure to recap the ether bottle tightly.

 Remember that ethyl ether fumes are explosive. Use in a well-ventilated room. No flames or sparks! If a spill occurs, call an instructor.

   b. Invert your vial so that the adult flies will fall asleep on the cotton plug rather than on the culture medium.

   c. When flies have become immobilized on the cotton plug, remove them for examination.

   d. Using the stereoscopic microscope, examine the flies on a white card using a toothpick to turn them.

   e. A fly "re-etherizer" petri dish with a gauze pad taped in the lid is provided in case the adults begin to awaken before phenotype classification is concluded. Place the lid/gauze pad saturated with ether over the flies when needed.

2. From each vial, identify two or three females and two or three males, and return the rest to their appropriate vial, keeping the vial on its side until the flies wake up. (Vial 1a will be used again in later experiments.) Use the following criteria to distinguish adult males and females (Figure 9.3).

a. *Size*. The female is generally larger than the male.

b. *Shape of abdomen*. The female abdomen is larger and more pointed than the male abdomen.

c. *Abdominal pigmentation*. In dorsal view, the alternating dark and light bands on the entire rear portion of the female abdomen are visible; the last few segments of the male are uniformly pigmented.

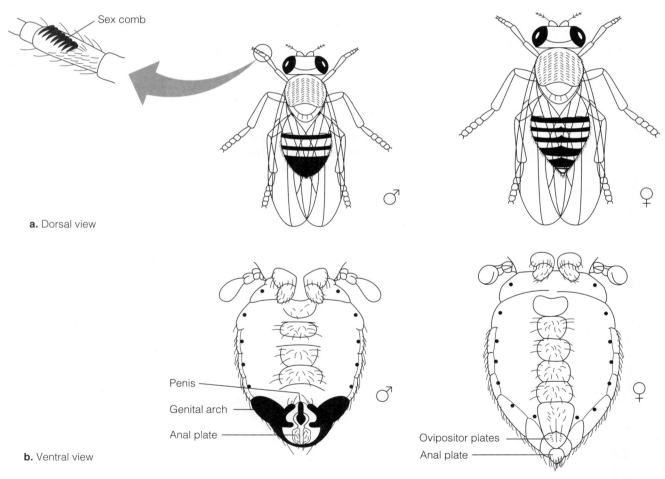

**Figure 9.3.**
**Characteristics of male and female *Drosophila melanogaster*.** The female is larger and has alternating dark and light dorsal bands on the abdomen. The dorsal abdomen of the male is uniformly pigmented. Conspicuous sex combs are visible on each foreleg of the male. In ventral view, the male's darkly pigmented genital arch and penis are visible.

    d. *Sex comb*. On males, there is a tiny brushlike tuft of hairs on the basal tarsal segment of each foreleg. This is the most reliable characteristic for sexing males accurately.

    e. *External genitalia*. On the ventral portion of the abdomen, the female has anal plates and lightly pigmented ovipositor plates. The male has anal plates and a darkly pigmented genital arch and penis.

3. Keeping track of the sexes, place the flies from vial 1a in one row of a spot assay plate, one fly per well.

4. Place the flies from vial 1b in a different row.

 The next two steps need to be done quickly, since the assay mixture is light sensitive. Cover the spot plate with an index card if there is any delay between steps.

5. Add 1 drop of assay mixture to each well.

 The assay mixture contains carcinogens. Do not allow it to contact skin. Wash hands thoroughly after performing the tests. Disposable gloves may be provided for your use. Notify an instructor if a spill occurs. If an instructor is unavailable, wipe up the spill wearing disposable gloves and using dry paper towels. Follow dry towels with towels soaked in soap and water. Dispose of all towels and gloves in a plastic bag in the trash.

6. Homogenize the flies with a pestle. Wipe off the pestle after each fly to avoid contamination. Why is it necessary to homogenize the flies?

    *Homogenizing releases the enzymes.*

7. Place the spot plate in a place away from light, such as a desk drawer.

8. After 5 minutes, check the reactions, and record your results in Table 9.1 in the Results section.

9. Thoroughly rinse your spot plate, and shake off the excess water.

*A positive test produces a blue color. One reviewer suggested that the positive result looks like a "squished blueberry." A negative test remains the color of the assay mixture, pale yellow.*

### Results

Complete Table 9.1 as the results of the assay tests are determined.

**Table 9.1**
AO Activity in Male and Female Flies from Vials 1a and 1b

| Vial 1a (AO Present) | | | Vial 1b (AO Absent) | | |
|---|---|---|---|---|---|
| Fly No. | Sex (F, M) | AO Activity (+/−) | Fly No. | Sex (F, M) | AO Activity (+/−) |
| 1 | | | 1 | | |
| 2 | | | 2 | | |
| 3 | | | 3 | | |
| 4 | | | 4 | | |
| 5 | | | 5 | | |
| 6 | | | 6 | | |

**Discussion**

1. Do your results match your predictions?

   *The flies from vial 1a should all show a positive test and turn blue. The flies from vial 1b should show a negative test and remain yellow, the color of the assay mixture.*

2. Does the sex of the fly appear to have an impact on the results of the assay test?

   *There should be no differences between results of males and females. Both sexes from vial 1a should test positive. Both sexes from vial 1b should test negative.*

3. Which two characteristics are most useful to your group in determining the sex of flies?

   *Sex combs and external genitalia are usually the most reliable. Newly emerging flies are not fully pigmented.*

EXERCISE 9.2

# Determining the Pattern of Inheritance of the Aldox Gene

**Materials**

vial 1a
vial 2
remaining materials from Exercise 9.1

## Introduction

This exercise initiates our study of the pattern of inheritance of the aldox gene, which determines the activity of aldehyde oxidase (AO).

---

 The gene is called the *aldox gene;* the enzyme produced by this gene is *aldehyde oxidase* and is abbreviated AO.

---

In this exercise, you will determine which **allele** (form of the gene) is dominant: enzyme-present or enzyme-absent. In addition, you will determine if the gene is **autosomal** or **sex-linked.** A sex-linked gene is located on one sex chromosome but not on the other. An autosomal gene is located on any chromosome *except* a sex chromosome. Finally, you will use this information to name the gene according to conventional naming procedures used by *Drosophila* geneticists.

The organisms you will use are in vial 2, which contains the offspring (called $F_1$ *progeny*) from a mating between a female that *lacks the enzyme* and a male that *has the enzyme*. For now, name the enzyme-present allele *AO-present* and the enzyme-absent allele *AO-absent*. The female is from a stock of flies that, when inbred, *consistently* lacks the enzyme from generation to generation. We say that this stock **breeds true** for AO-absent. The male is from a stock that *consistently* has the enzyme from generation to generation; that is, it breeds true for AO-present. This stock, which has the enzyme, is genetically like flies most commonly found in nature and is called the **wild type.**

It is necessary to understand this background information to be able to predict the outcome of this cross. To help you in your predictions, answer the following questions.

1. What is the genotype of the female (maternal parent—lacks the enzyme) in our cross?

   *AO-absent AO-absent*

2. What would be the genotype of the male (paternal parent—has the enzyme) in our cross if the gene is *not* sex-linked?

   *AO-present AO-present*

3. What would be the genotype of the male if the gene is sex-linked?

   *AO-present Y*

4. What would be the genotypes of the $F_1$ progeny if the gene is *not* sex-linked?

   *All offspring would be AO-present AO-absent.*

5. What would be the genotype if the gene *is* sex-linked?

   *females: AO-present AO-absent; males: AO-absent Y*

## Hypothesis

Hypothesize about the inheritance of this gene. Is it sex-linked? According to your hypothesis, what will be the genotypes of the parents?

> *The inheritance of this trait is not sex-linked, and the genotypes of the parents are AO-absent AO-absent (female parent) and AO-present AO-present (male parent). Other hypotheses are acceptable if testable.*

## Prediction

Predict the offspring from the hypothesized parents. (*Hint:* Set up a Punnett square in the margin.)

> *See the Teaching Plan.*

## Procedure

1.  Anesthetize the flies in vial 2.
2.  Count out 10 males and 10 females. Keep track of the sexes of flies as you place them in the spot plate. Record the sex of each fly in Table 9.2.

---

 It is easier to keep accurate records if you put all males in one row of the spot plate and all females in another row. Do not write on the spot plates!

---

3.  Perform the spot assay as described in Exercise 9.1 with the following addition: From vial 1a, select a single fly and assay it with this and all following experiments. This will provide you with a *positive control* to compare with your unknown assays.

---

 The assay mixture is light sensitive. Cover the spot plate with an index card between steps.

---

4.  When you see that the positive control has turned blue, indicating that the assay is working, record the results in Table 9.2.
5.  Rinse out the spot plate carefully.

## Results

Record the sex of each fly and the results of the spot assay tests in Table 9.2.

**Table 9.2**
Data Sheet for Exercise 9.2: Results of Spot Assay Tests on $F_1$ Progeny from a Female That Lacks AO Activity and a Male That Has AO Activity

| Fly No. | Sex (F, M) | AO Activity (+/−) | Fly No. | Sex (F, M) | AO Activity (+/−) |
|---------|-----------|-------------------|---------|-----------|-------------------|
| 1 | | | 11 | | |
| 2 | | | 12 | | |
| 3 | | | 13 | | |
| 4 | | | 14 | | |
| 5 | | | 15 | | |
| 6 | | | 16 | | |
| 7 | | | 17 | | |
| 8 | | | 18 | | |
| 9 | | | 19 | | |
| 10 | | | 20 | | |
| Positive control: | | | | | |

## Discussion

1. What allele appears to be dominant?

   *AO-present*

2. What evidence of your results in Table 9.2 supports your answer?

   *A blue color develops with each fly, indicating that all flies have tested positive. Thus, AO-present must be dominant over AO-absent.*

3. Determine the conventional way to name this gene. The following information will assist you. For any gene with two alleles, there exists the wild type and the mutation. Recall that the wild type is the allele most commonly found in nature. In this case, the wild type is AO-present. AO-absent is considered a **mutation,** a change in the DNA of the gene. In naming genes and alleles, give the gene an appropriate name using a word or letter derived from the phenotype affected by the gene. (In this case, the gene is named *aldox* because the enzyme that it produces catalyzes the oxidation of aldehydes.) If the mutation is dominant, capitalize the first letter in the name. If the mutation is recessive, do not capitalize. Based on the results of your experiment (see question 1), should the aldox gene be written beginning with a capital or a lowercase letter?

   *Since the mutation (AO-absent) is recessive, write the name beginning with a lowercase letter,* aldox.

4. Correctly write the names of the wild-type and mutant alleles. By convention, wild-type alleles are designated by a superscript + after the name.

   *Wild type is* aldox$^+$*, the mutant is* aldox.

5. Did you hypothesize that the gene was sex-linked (on a sex chromosome) or autosomal (not on a sex chromosome)?

   *Students generally will propose an autosomal gene.*

6. Describe how your results either support or falsify your hypothesis.

   *Had the gene been sex-linked, the dominant allele would have been expressed in all of the females but none of the males. Since all offspring expressed the dominant allele, the gene must not be sex-linked. See the Teaching Plan.*

7. Write a statement describing your conclusions from this experiment.

   *The mutant allele, AO-absent (aldox), is recessive, and the wild-type allele, AO-present (aldox$^+$), is dominant. This trait is autosomal, not sex-linked.*

---

EXERCISE 9.3

# Determining Parental Genotypes Using Evidence from Progeny

## Materials

vial 1a
vial 3
remaining materials from Exercise 9.1

## Introduction

Once the pattern of inheritance has been determined, this information can be used to predict genotypes and phenotypes of individuals by observing parents, siblings, and offspring. Vial 3 contains the first-generation ($F_1$) progeny from a mating between unknown parents. The objective of this exercise is to determine the genotypes of the parents. At this stage in the investigation, you have no data from observations on which to base a hypothesis. The following procedure will allow you to collect preliminary data, then, using this data, to propose and test a hypothesis about the genotypes of the parents.

## Procedure, Preliminary Observations

1. Anesthetize the flies in vial 3.
2. Count out 24 flies and place them in individual spot plate wells.
3. Perform the spot assay, including a positive control, a fly from vial 1a.
4. Record the results on *two* data sheets, Table 9.3, and the data sheet for the total class (provided by the instructor).

*You may download the data sheet for the whole class (Class Data Sheet, Exercise 9.3) in Excel format from www.masteringbiology.com. Look in the Study Area under Lab Media. Project as students fill in their data. Alternatively, this data sheet may be found in Lab Topic 9 of the* Preparation Guide *that accompanies this laboratory manual. Make an acetate of this sheet, and project.*

## Results, Preliminary Observations

1. Record results of assay tests on progeny of the unknown parents in Table 9.3.

**Table 9.3**
Data Sheet for Exercise 9.3: Results of Assay Tests on Progeny of Unknown Parents

| Fly No. | AO Activity (+/−) | Fly No. | AO Activity (+/−) |
|---------|-------------------|---------|-------------------|
| 1 | | 13 | |
| 2 | | 14 | |
| 3 | | 15 | |
| 4 | | 16 | |
| 5 | | 17 | |
| 6 | | 18 | |
| 7 | | 19 | |
| 8 | | 20 | |
| 9 | | 21 | |
| 10 | | 22 | |
| 11 | | 23 | |
| 12 | | 24 | |
| Positive control: | | | |

2. Total the number of offspring in each phenotype category.

|  | **Your Totals** | **Class Totals** |
|--|-----------------|------------------|
| AO-present | _____ | _____ |
| AO-absent | _____ | _____ |

3. Review observations made in previous experiments.
   a. Is the trait sex-linked?

   *no (from Exercise 9.2)*

   b. Which allele is dominant?

   *From Exercise 9.2, we know that AO-present is dominant over AO-absent.*

## Hypothesis

Using all observations, hypothesize the genotypes of the parent flies, making sure to name the alleles correctly.

*The parents are heterozygous: aldox⁺ aldox. (Students may have other hypotheses, especially if the data from this last exercise do not indicate an obvious ratio.)*

## Prediction

Predict the results of the experiment (if/then).

*If the parents are heterozygous, then the ratio of phenotypes in the progeny will be 3:1, with 3 AO-present (the dominant trait) to 1 AO-absent (the recessive trait).*

## Procedure—Testing Your Hypothesis

Using the total class data and Table 9.4, refer to Appendix B and perform the chi-square test to determine if the results of the exercise support or falsify your hypothesis. Calculate the expected values for each trait based on the total number of flies counted in your class. (For a discussion of the chi-square test, see Appendix B.)

## Results

Complete the chi-square calculations in Table 9.4.

**Table 9.4**
Chi-Square Calculations to Evaluate Results of Exercise 9.3
(Observed value represents total class data.)

| | AO Activity (+) | AO Activity (−) |
|---|---|---|
| Observed value (*o*) | | |
| Expected value (*e*) | | |
| Deviation (*o* − *e*) or *d* | | |
| Deviation$^2$ (*d*$^2$) | | |
| *d*$^2$/*e* | | |
| Chi-square $\chi^2 = \Sigma d^2/e$ | | |
| Degrees of freedom (*df*) | | |
| Probability (*p*) (see a $\chi^2$ table) | | |

*See the Teaching Plan for hypothetical results. If this is the first time students have used the chi-square calculation, you will need to explain it here, using valuable lab time. If possible, discuss the chi-square concept in lecture or in a study session.*

## Discussion

1. Do the class results support or falsify your hypothesis?

   *The observed results should be consistent with a 3:1 ratio. The chi-square value should be small (less than 3.84; p = 0.05), indicating that any deviation from the expected is due to chance alone.*

2. Does this experiment support or contradict your conclusions concerning the pattern of inheritance derived from Experiment 9.2?

   *These results representing the F$_2$ generation provide support for the inheritance of the mutation (AO-absent) as an autosomal recessive trait.*

---

EXERCISE 9.4

# Mapping Genes

---

## Materials

vial 1a
vial 4
remaining materials from Exercise 9.1

## Introduction

*Students must have been presented information in lecture about dihybrid inheritance, a testcross, and linkage before they can understand this experiment. Previous information about mapping is optional; students usually understand this concept after performing this experiment.*

In this exercise, you will investigate the inheritance of two genes in *Drosophila*: the aldox gene and a gene that influences eye color named **sepia.** In the wild-type fly (sepia$^+$), eye color is red. In the mutant fly, eye color is dark brown. The wild-type allele is dominant over the mutant. The flies you will be studying are the offspring from a cross in which the parents differ in these two genes: a **dihybrid cross.** You will ask if these two genes are transmitted from parent to offspring *linked* together or if each gene is inherited independently of the other. If genes are transmitted from parent to offspring linked together, this means that they are located on the same chromosome. Then if the location of one gene is known, the pattern of inheritance can provide evidence that will allow you to determine the location of the second gene. With this information, you can construct a map showing gene locations.

Vial 4 contains the F$_1$ progeny from a mating between parents having the following genotypes:

*The line below sepia$^+$ aldox$^+$ indicates that these two alleles are linked and that the sepia aldox alleles are linked.*

| Parent 1 | | Parent 2 |
|:---:|:---:|:---:|
| $\dfrac{sepia^+\,aldox^+}{sepia\;aldox}$ | $\times$ | $\dfrac{sepia\;aldox}{sepia\;aldox}$ |

where sepia$^+$ represents the dominant, wild-type eye color allele that produces red eye color and sepia represents the mutant, recessive allele that produces a dark-brown eye color.

## Hypothesis

Hypothesize about the inheritance of these two genes. Are they inherited linked together or independently of each other?

> *The two genes are inherited independently of each other. (This is the most probable student hypothesis; accept any testable hypothesis.)*

## Prediction

Predict the ratios of phenotypic classes of offspring resulting from the mating described above.

> *If the two genes are inherited independently of each other, then a cross between an individual that is heterozygous for the two genes and a homozygous recessive individual will yield offspring in a 1:1:1:1 ratio.*

*In this cross, a homozygous recessive individual is crossed with an individual that is heterozygous for both genes.*

Use a Punnett square to illustrate your prediction.

|  | *sepia aldox* |
|---|---|
| *sepia⁺ aldox⁺* | *sepia⁺ aldox⁺* <br> *sepia aldox* |
| *sepia⁺ aldox* | *sepia⁺ aldox* <br> *sepia aldox* |
| *sepia aldox⁺* | *sepia aldox⁺* <br> *sepia aldox* |
| *sepia aldox* | *sepia aldox* <br> *sepia aldox* |

*This would be the typical student prediction for traits inherited independently. Some students might predict linkage. Do not correct either group. Encourage different hypotheses. The results are the test!*

¼ *red eyes, AO-present*
¼ *red eyes, AO-absent*
¼ *sepia eyes, AO-present*
¼ *sepia eyes, AO-absent*

## Procedure

1. Anesthetize the flies in vial 4.
2. Count out 50 flies, and classify them on the basis of eye color (red or sepia).
3. Keeping the eye colors separate, perform the spot assay, including a positive control from vial 1a.
4. Record the results in Table 9.5.
5. Rinse out the spot plate.

# Results

*You may download the data sheet for the whole class (Class Data Sheet, Exercise 9.4) in Excel format from www.masteringbiology.com. Look in the Study Area under Lab Media. Project as students fill in their data. Alternatively, this data sheet may be found in Lab Topic 9 of the* Preparation Guide *that accompanies this laboratory manual. Make an acetate of this sheet, and project.*

*Typical class totals for each phenotype:*

*Red eyes, AO-present = 202*
*Red eyes, AO-absent = 91*
*Sepia eyes, AO-present = 103*
*Sepia eyes, AO-absent = 204*

1. Record the results of the eye color classification and the spot test for each fly in Table 9.5. Total your results below. Add your results to the total class data sheet provided by the instructor. Total the class data and record below.

|                          | Your Totals | Class Totals |
|--------------------------|-------------|--------------|
| **red eyes, AO-present**   | _____  | _____   |
| **red eyes, AO-absent**    | _____  | _____   |
| **sepia eyes, AO-present** | _____  | _____   |
| **sepia eyes, AO-absent**  | _____  | _____   |

**Table 9.5**
Data Sheet for Exercise 9.4: Mapping Genes, Recording Eye Color and AO Activity for 50 Flies

| Fly No. | Eye Color | AO (+/−) | Fly No. | Eye Color | AO (+/−) |
|---------|-----------|----------|---------|-----------|----------|
| 1 |  |  | 26 |  |  |
| 2 |  |  | 27 |  |  |
| 3 |  |  | 28 |  |  |
| 4 |  |  | 29 |  |  |
| 5 |  |  | 30 |  |  |
| 6 |  |  | 31 |  |  |
| 7 |  |  | 32 |  |  |
| 8 |  |  | 33 |  |  |
| 9 |  |  | 34 |  |  |
| 10 |  |  | 35 |  |  |
| 11 |  |  | 36 |  |  |
| 12 |  |  | 37 |  |  |
| 13 |  |  | 38 |  |  |
| 14 |  |  | 39 |  |  |
| 15 |  |  | 40 |  |  |
| 16 |  |  | 41 |  |  |
| 17 |  |  | 42 |  |  |
| 18 |  |  | 43 |  |  |
| 19 |  |  | 44 |  |  |
| 20 |  |  | 45 |  |  |
| 21 |  |  | 46 |  |  |
| 22 |  |  | 47 |  |  |
| 23 |  |  | 48 |  |  |
| 24 |  |  | 49 |  |  |
| 25 |  |  | 50 |  |  |

2. What phenotypic classes were observed in the total class data, and in what approximate ratio?

   *The results will not be 1:1:1:1 as expected; there should be significantly more red eyes, AO-present, and more sepia eyes, AO-absent. See the Teaching Plan.*

3. On separate paper, using class totals, perform the chi-square test to determine if the results support or falsify your hypothesis (see Appendix B).

   *The chi-square test may be a homework assignment.*

## Discussion

1. Do the data support your predicted results?

   *yes, if predicted linkage*
   *no, if predicted independent assortment*

2. If the results differ from what was expected, can you suggest an explanation for these differences?

   *could be due to incorrect scoring or could be that traits are linked on the same chromosome*

3. Suppose that the aldox gene and the sepia gene are located on the same chromosome. This means that when meiosis takes place, the two genes will not assort independently but will be linked together, moving into the same gamete *unless crossing over has taken place* (refer to Lab Topic 7 to review independent assortment in meiosis). Only if crossing over takes place will **recombinant classes** of phenotypes be observed. A **recombinant chromosome** is one emerging from meiosis with a combination of alleles not present on the chromosomes entering meiosis. What are the recombinant classes of phenotypes for this cross?

   *sepia eyes, enzyme present*
   *red eyes, enzyme absent*

4. The distance between two genes is related to the frequency of recombinants produced by crossing over during meiosis. The closer two genes are, the fewer recombinants. Geneticists use an arbitrary measure, or map unit, to represent the distance between two genes. A 1% recombinant frequency equals one map unit. Calculate the relationship between map units and recombinants as follows:

   *Geneticists have determined that these two genes are exactly 31.2 map units apart. This means that about 31% of the offspring should be recombinant classes. Students always find recombinant classes, but they usually find fewer than 31%. Do not correct them. The objective of this exercise is to learn how to map chromosomes. Determining the exact location of aldox is secondary.*

   $$\text{Map units} = \frac{\text{number of recombinants}}{\text{total}} \times 100$$

5. In your experiment, did recombinant classes exist? If they did, how frequent were they?

   *About 31% of the offspring should be recombinants.*

6. Do your data suggest that aldox and sepia are on the same chromosome, and if so, how far is the aldox locus from the sepia locus?

   *yes, about 31 map units*

7. What would be the exact map position of the aldox locus if the sepia locus is 26.0? Are you sure?

*The actual map position for aldox is exactly 57.2. Students' results should approximate this.*

*Theoretically, the aldox gene could be on either side of the sepia gene. However, since it is 31.2 map units away, and sepia is at 26, it can only be at 57.2.*

## Reviewing Your Knowledge

*Earlier editions of this laboratory manual included an additional exercise where students investigated the developmental profile of the aldox gene. Suggestions for performing this exercise have now been moved to item 1 in Investigative Extensions. For the complete instructions for performing this investigation, see Lab Topic 9 of the Preparation Guide that accompanies this laboratory manual.*

*Throughout this lab manual, check older editions for additional questions that may be used for practice or on tests.*

1. List the most obvious characteristics used to determine the sex of a fruit fly.

   *males: sex comb; dark dorsal abdomen; dark genital arch on ventral tip of abdomen*

   *females: alternating light and dark bands on dorsal abdomen; larger body than male*

2. A fruit fly geneticist discovered a genetic mutation that resulted in pupae and young flies with dark pigment granules in the nuclei and cytoplasm of their fat cells. After studying the inheritance of the mutation, the geneticist named the gene **Frd** (for **Freckled**). What does this name tell you about the inheritance of this gene?

   *The mutant (freckled) is dominant to wild type.*

   In another strain of fruit flies, geneticists discovered a mutation producing flies with no gut muscles. They named the gene controlling this phenotype **jeb** (for **jelly belly**). What does this tell about the inheritance of this gene?

   *The mutant (no gut muscles) is recessive to wild type (with gut muscles).*

3. Is the genetic map or **linkage map** produced by determining recombination frequencies an actual physical map of the chromosome? Refer to your textbook.

   *No. Map units do not have absolute sizes. A linkage map tells the relative positions of genes on a chromosome but not the actual physical location.*

## Applying Your Knowledge

1. In *Drosophila* there is a gene where a mutant named ebony produces flies with a shiny black body color, darker than wild type. A gene for eye color has a scarlet mutant with eyes that are much brighter red than wild type.

   A cross between wild type flies that breed true and ebony body, scarlet-eyed flies produces an $F_1$ generation with the wild type phenotype for both traits. Predict the ratio of offspring phenotypes in the next generation of a cross between these $F_1$ flies and ebony, scarlet-eyed flies. Assume the genes are not linked.

*If the genes are not linked, then the ratio of offspring will be 1:1:1:1*

   *1 wild-type body, wild-type eye color*

   *1 wild-type body, scarlet eye color*

   *1 ebony body, wild-type eye color*

   *1 ebony body, scarlet eye color*

The results of the cross between the $F_1$ and ebony body, scarlet-eyed flies are as follows:

   288 wild-type body, wild-type eye color

   102 wild-type body, scarlet eye color

   112 ebony body, wild-type eye color

   300 ebony body, scarlet eye color

Are the genes linked? If so, how many map units apart are they located? Show your work.

$$map\ units = \frac{number\ of\ recombinants}{total} \times 100 = \frac{214}{800} \times 100$$

*The genes are linked 26.7 map units apart.*

If the ebony gene is located at map unit position 70.7 on chromosome 3, where is the gene for scarlet eye color located? (Give all possible positions.)

*Position 44 is the actual map position on chromosome 3, but students should say that 97.4 is also possible.*

2. Explain why we cannot use a testcross to detect linkage between two genes located on the same chromosome 50 map units apart.

   *If the two genes were 50 map units apart, the percentage of recombination for these genes would be 50%. These two linked genes would behave exactly as two genes on different chromosomes because there would be a crossover event 50% of the time. This would produce a 1:1:1:1 ratio of phenotypes, the same as expected for unlinked genes.*

# Investigative Extensions

1. Students may investigate the developmental profile of the aldox gene. All genes are not active in all cells, nor are genes in a cell lineage active at all times in the developmental cycle of the organism. Cells constantly turn genes on and off, bringing about the patterns of development and differentiation in the organism. There are four distinct developmental stages in the fruit fly: the embryo, larva, pupa, and adult (Figure 9.2).

   Using procedures and spot assay tests described in Exercise 9.1, design an experiment that will investigate the expression of the aldox gene during the developmental cycle of the fruit fly.

*Procedures for this investigation may be found in Lab Topic 9 of the Preparation Guide that accompanies this laboratory manual.*

*The BioQUEST Library VI is available at numerous vendors, and the entire library plus new projects is available online. Check the BioQUEST home page at www.bioquest.org for the later information.*

2. "Genetics Construction Kit" is a computer-based, problem-solving program based on fly genetics. Students perform investigative studies in classical genetics, including linkage. This program provides sets of organisms with unknown patterns of inheritance. Students may propose hypotheses, make predictions, and then cross unknown organisms and analyze crosses to discover these inheritance patterns.

## Student Media: BioFlix, Activities, Investigations, and Videos
www.masteringbiology.com (select Study Area)

**Activities**—Ch. 15: Linked Genes and Crossing Over; Sex-Linked Genes
**Investigations**—Ch. 15: What Can Fruit Flies Reveal About Inheritance? Biology Labs On-Line: FlyLab

## References

This lab topic was first published as J. G. Morgan and V. Finnerty, "Inheritance of Aldehyde Oxidase in *Drosophila melanogaster*" in *Tested Studies for Laboratory Teaching* (Volume 12), Proceedings of the 12th Workshop/Conference of the Association for Biology Laboratory Education (ABLE), Corey A. Goldman, Editor. Used by permission.

Heitz, J. and C. Giffen. *Practicing Biology: A Student Workbook*, 4th ed. San Francisco, CA: Benjamin Cummings, 2010.

Jungck, J. and J. Calley, "Genetics Construction Kit," *BioQUEST Library Online*. http://www.bioquest.org/BQLibrary/library_result.php (software available to download).

Klug, W.S., M. R. Cummings, C. Spencer, and M. A. Palladino. *Essentials of Genetics*, 7th ed. San Francisco, CA: Benjamin Cummings, 2010.

## Websites

A directory of Internet resources for research on *Drosophila*. The WWW Virtual Library: *Drosophila*: http://www.ceolas.org/fly

Interactive fruit fly genetics lab, requires registration: http://biologylab.awlonline.com. Select Flylab.

## LAB TOPIC 9

# Mendelian Genetics II: *Drosophila*
## Teaching Plan for Laboratories

## Main Concepts and Objectives

1. Concept: *Drosophila* as a genetic tool. Students should be able to discuss why *Drosophila* is one of the most important organisms used in eukaryotic genetics. (Exercises 9.1–9.4)
2. Concept: assays frequently used in genetic research as an indication of biochemical phenotypes. Students will gain experience using biochemical assays. (Exercises 9.1–9.4)
3. Concept: naming genes. Students will learn the convention used in *Drosophila* genetics when naming genes. (Exercise 9.2)
4. Concept: Mendelian patterns of inheritance. Students will determine the phenotypes of parents by observing offspring. (Exercise 9.3)

5. Concept: statistical analysis of data. Students will use the chi-square test to evaluate how well their observed data fit the expected ratio of phenotypes in a given genetic cross. (Exercise 9.3)

6. Concept: gene mapping. Students will describe how data collected from a cross involving linked genes can be used to map the location of those genes. (Exercise 9.4)

7. Specific content. Students will be able to describe or define: sexes of fruit flies, fly handling, assays, chi-square test, naming genes, *Punnett square, mapping, crossing over, pupae, larvae*, genetics vocabulary.

## Order of the Lab

| | |
|---|---|
| 1. Introduce major concepts and objectives. | (10 min) |
| 2. Discuss procedures and precautions. | (10 min) |
| 3. Students perform Exercise 9.1. | (20 min) |
| 4. Class discussion of results. | (5 min) |
| 5. Introduce Exercise 9.2 (determining inheritance patterns of aldox). Discuss the conventional way of naming mutations in *Drosophila*. | (15 min) |
| 6. Students perform Exercise 9.2. | (25 min) |
| 7. Introduce Exercise 9.3 (unknown parents). | (5 min) |
| 8. Students perform Exercise 9.3. | (30 min) |
| 9. Discuss results and conclusions. | (10 min) |
| 10. Introduce Exercise 9.4 (linked genes). | (10 min) |
| 11. Students perform Exercise 9.4. | (30 min) |
| 12. Discuss results; discuss mapping of genes. | (10 min) |

The times required for each exercise are based on optimum conditions: Students arrive prepared for lab, knowledgeable about concepts of Mendelian genetics (monohybrid inheritance, dihybrid inheritance, and linkage) from lecture and having previously read the exercises.

**For a 2-hour lab:** Omit steps 10 to 12 (Exercise 9.4). If you do this, the prep will be much easier, but students will not learn how to map genes. If your laboratory schedule will permit, you may wish to perform this lab topic in two 2-hour sessions and spend more time with analysis of data and discussion of the significance of results. If you choose this latter organization, adjust the fly prep schedules in the *Preparation Guide* to ensure that the dihybrid test flies will mature on the appropriate days.

## Classroom Management

Students do all experiments in pairs. Collect data from the entire class for Exercises 9.3 and 9.4. Have all students perform exercises simultaneously to facilitate class discussions at the end of each exercise.

## Student Development

After completing this exercise, students will understand some important concepts in genetics and will be better prepared for an upper-level genetics course. They will have improved skills in data processing, laboratory techniques working with *Drosophila*, testing the Mendelian model, statistical analysis, and organization and collaborative investigation.

## Lab Safety Precautions

### Ethyl Ether

Instruct students to:

1. Use with adequate ventilation. No sparks!
2. Avoid inhaling fumes. Some students may develop a headache after exposure to ethyl ether. FlyNap can be substituted, but this substance also has an unpleasant smell.

### Assay Mixture

The assay mixture is a carcinogenic poison. Although the small amounts used in this lab do not pose a health hazard, you should use disposable gloves to avoid contact with the solution. You may also choose to have gloves available for your students; however, it is unlikely that students will come into contact with the liquid, since they dispense the drops from small amber dropper bottles. Nevertheless, warn students to:

1. Wash hands thoroughly after each exercise.
2. Avoid eating, drinking, or smoking in lab or in the building.
3. Inform the instructor in the event of a spill. The instructor should wear gloves and wash the area thoroughly using soap, water, and disposable towels. The soiled towels should be placed in a plastic bag and discarded in the trash.
4. Thoroughly wash the spot plates.

## Discussion and Summary

Students report results and discuss their significance at the end of each exercise.

## Evaluation

Specific content can be evaluated on a laboratory test. Students can be tested on all terms, characteristics of flies, the chi-square test and its significance, mapping, monohybrid and dihybrid crosses, working with unknowns, and application of these results to similar types of problems. You can evaluate other objectives by assigning take-home problems from the course text or problems of your design.

## Notes on the Exercises

The vocabulary is important. The gene is commonly called *aldox*. The enzyme is aldehyde oxidase and is abbreviated AO. Assigning symbols is discussed below.

### Exercise 9.1

Establish the controls; demonstrate how to distinguish male and female flies; and demonstrate differences in positive and negative spot test results. You may choose to make a transparency showing male and female fly characteristics to use in your introduction. A transparency that reviews spot assay procedures and safety precautions is useful. Leave this transparency on the overhead projector as the students perform the exercises.

Results of Exercise 9.1. The flies in vial 1a have the enzyme AO, and all flies should turn blue when homogenized in the assay mixture. Vial 1b flies should show no color change because the enzyme is not present in these organisms. *Summary*: vial 1a—AO-present; vial 1b—AO-absent.

## Exercise 9.2

*Objective 1.* Determine the pattern of inheritance of the aldox gene. Students should be able to answer two questions: (1) Which allele is the dominant allele, AO-present or AO-absent? (2) Is the gene sex-linked?

*Objective 2.* Students learn the conventional symbols used to designate alleles in *Drosophila* genetics.

**The cross.** The maternal parent is homozygous and does not show AO activity. She has two copies of the AO-absent allele: AO-absent AO-absent. (Do not place a slash mark [/] between the alleles because this mark in this usage indicates linked genes.) The paternal parent is homozygous and shows AO activity. If the gene is not sex-linked, he has two copies of the allele: AO-present AO-present. If the gene is sex-linked, he has one copy of the allele and a Y chromosome: AO-present Y.

When naming genes and alleles, *Drosophila* geneticists use the following convention: Give the gene an appropriate name, depending on its function. In this case, the gene is aldox because the enzyme controls the oxidation of several aldehydes. If the allele that is the mutation is dominant, write the name of the gene beginning with a capital letter: Aldox. If the mutation is recessive, write the name beginning with a lowercase letter: aldox. Since the AO-present allele is most frequently seen in natural populations, and AO-absent is the mutant allele, designate AO-present with a$^+$. Therefore, after this experiment, the students will know that the mutation (AO-absent) is recessive and that the names should be written aldox$^+$ for the wild-type allele and aldox for the mutant allele.

To answer question 2 (Is the gene sex-linked?): If the gene is autosomal, all the offspring of this cross have the enzyme activity, as can be seen in Punnett square (a). If the gene is sex-linked, all females have the enzyme activity and all males lack the enzyme activity, as can be seen in Punnett square (b).

|       | aldox$^+$ | aldox$^+$ |       | aldox$^+$ | Y |
|-------|-----------|-----------|-------|-----------|---|
| aldox | aldox$^+$ aldox | aldox$^+$ aldox | aldox | aldox$^+$ aldox | aldox Y |
| aldox | aldox$^+$ aldox | aldox$^+$ aldox | aldox | aldox$^+$ aldox | aldox Y |

a.                                                                  b.

**Results.** (a) is correct. The gene is not sex-linked.

## Exercise 9.3

The offspring in vial 3 are from a cross between two parents of unknown phenotypes and genotypes. Exercise 9.3 possibilities:

1.  The following crosses will produce all offspring with enzyme activity:

     aldox$^+$ aldox$^+$ × aldox$^+$ aldox$^+$
     aldox$^+$ aldox$^+$ × aldox$^+$ aldox
     aldox$^+$ aldox$^+$ × aldox aldox

2.  The following cross will produce a 3:1 ratio, 3 with the enzyme activity and 1 without the enzyme activity:

     aldox$^+$ aldox × aldox$^+$ aldox

3.  The following cross will produce a 1:1 ratio:

$$\text{aldox}^+ \text{aldox} \times \text{aldox aldox}$$

**Results of Exercise 9.3.** The results will show a ratio of offspring ($F_1$) approximately 3:1. The students will then conclude that both parents must have been heterozygous, as demonstrated in this Punnett square:

|  | aldox$^+$ | aldox |
|---|---|---|
| aldox$^+$ | aldox$^+$ aldox$^+$ | aldox$^+$ aldox |
| aldox | aldox$^+$ aldox | aldox aldox |

3:1 phenotypic ratio

The students should calculate the chi-square using data from the entire class.

## Hypothetical Results and Chi-Square Calculations

Chi-square is a statistical tool that can be used to test if the results obtained in an experiment are consistent with the hypothesis. The formula for the calculation is

$$\chi^2 = \Sigma \, \frac{(\text{observed} - \text{expected})^2}{\text{expected}}$$

where $\Sigma$ stands for summation and applies to all classes in the experiment. For our hypothetical experiment, assume 10 groups of students collecting data, 25 flies each, for a total of 250 flies counted.

| | AO-present: 190 | AO-absent: 60 |
|---|---|---|
| Observed: | | |
| Expected: | $0.75 \times 250 = 187.5$ | $0.25 \times 250 = 62.5$ |
| Deviation: (d)  (observed − expected) | 2.5 | 2.5 |
| $d^2$: | 6.25 | 6.25 |
| $d^2$/expected | 6.25/187.5 = 0.0333 | 6.25/62.5 = 0.1 |
| $\chi^2$ | 0.0333 + 0.1 = 0.1333 | |

## Degrees of Freedom (df)

df = $n - 1$, where $n$ = number of categories. Here we have two categories, AO-present and AO-absent. Therefore, the degrees of freedom = 1.

Now go to a chi-square distribution table and read across 1 df. Note that 0.133 falls between 0.06 and 0.15. Therefore, the probability is between 0.8 and 0.7 (80% and 70%) that the deviation is due to chance alone.

Generally in biological research, statisticians agree that if the probability is greater than 0.05 (5%), the results have a satisfactory fit to the expected. If the probability is less than 0.05, then the deviation from the expected is too great, something other than chance is causing the deviation, and the hypothesis is not valid.

## Exercise 9.4

This exercise investigates gene mapping. The objective is to determine the location of the aldox gene by determining the map distance between it and a gene at a known location on the same chromosome. We know that the sepia (brown eye color) gene is located at map unit 26.0 on chromosome 3. Aldox is on the same chromosome. The testcross (heterozygote crossed with a homozygous recessive) is as illustrated:

$$
P: \quad \frac{\text{sepia}^+\ \text{aldox}^+}{\text{sepia aldox}} \times \frac{\text{sepia aldox}}{\text{sepia aldox}}
$$

$P_1$  　　　　　$P_2$

sepia$^+$ = wild type (red eyes)

sepia = mutant (brown eyes)

$P_1$ gametes:　sepia$^+$ aldox$^+$

　　　　　　　sepia aldox

　　　　　　　sepia$^+$ aldox

　　　　　　　sepia aldox$^+$

The last two gametes are recombinant gametes.

$P_2$ gametes: Only sepia aldox is possible.

How recombinants occur (crossover between $s^+$ and $a^+$ loci):

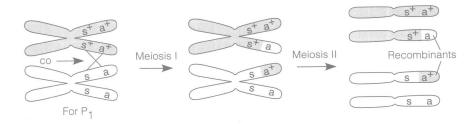

For $P_2$, even if a crossover takes place, all gametes will be:

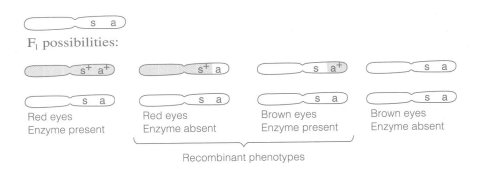

$F_1$ possibilities:

Red eyes　　　　　　Red eyes　　　　　　Brown eyes　　　　　Brown eyes
Enzyme present　　　Enzyme absent　　　 Enzyme present　　　Enzyme absent

Recombinant phenotypes

$$
\text{Frequency of recombinant classess} = \frac{\text{number of recombinant flies}}{\text{total number of flies}}
$$

$$
\text{Map units} = \text{frequency of recombinant classes} \times 100
$$

The map position of the aldox locus will be 26.0 + the map units of aldox from sepia. At this point, we do not know the definite position of the aldox locus, but we do know it relative to sepia. Students should determine that aldox is 31 map units away from sepia. Theoretically, it could be on either side of sepia (26 + 31 or 26 − 31). However, the locus must be a positive number, so the map position of aldox is about 57.

# Lab Topic 10
# Molecular Biology

## Laboratory Objectives

After completing this lab topic, you should be able to:

1. Describe the function of restriction enzymes.
2. Discuss the basic principles of electrophoresis in general and for DNA specifically.
3. Construct a tentative map of DNA molecules based on restriction fragments.
4. Explain the use of enzymes to map DNA molecules, and discuss the importance of mapping.
5. Use gel results to estimate DNA fragment sizes.
6. Discuss the universality of the genetic code.
7. Describe ways in which the technology of molecular biology is being used in industry, medicine, criminal justice, agriculture, and basic research.

*For a 2-hour lab: Exercise 10.1 can be completed in a 2-hour lab period if some materials (for example, the digests) are prepared in advance by the instructor.*

*The restriction enzyme digestion should incubate for 1 hour. To begin the lab, have students practice pipetting, prepare the digestions, and pour the gel (Procedure, pp. 241–243, steps 1–3). Once the digestions are incubating and the gels are poured, introduce the Lab Topic and applications and review Preliminary Questions, pp. 240–241. See Teaching Plan.*

## Introduction

"Golden rice" with β carotene, herbicide-resistant corn, insect-resistant cotton, zebrafish that glow in the dark, bacteria that produce large quantities of human growth hormone, rice plants containing milk proteins that can help treat diarrhea—all of these examples represent organisms that have had their DNA modified through **recombinant DNA technology,** a process of introducing specific genes from one organism into another (Figure 10.1). This process creates a **transgenic organism,** or an organism with a gene from another species. This technology was made possible by the discovery in the 1970s of **restriction enzymes** in bacteria. These enzymes function to cut DNA molecules at specific sites and have become one of the basic tools of molecular biology. Once fragments of DNA have been isolated, the process of **polymerase chain reaction,** or **PCR,** enables scientists to make many copies (molecular clones) of the isolated DNA. As can be viewed in many popular television programs, the criminal justice system has come to rely heavily on DNA technology, as **DNA fingerprinting** is used to identify criminals in cases of rape and murder.

*In the Study Area of masteringbiology.com, see Student Media for suggested activities, investigations, and videos.*

Essentially, the age of molecular genetics was launched in 1953, when James Watson and Francis Crick revealed the structure of the DNA double helix, and in the following decade the essence of the genetic code was solidly established. Since techniques for making recombinant DNA were developed, this field has revolutionized genetics and molecular biology. Scientists could now analyze the nucleotide sequence of many genes, and in June 2000, the announcement was made that the major portions of the **human genome** had

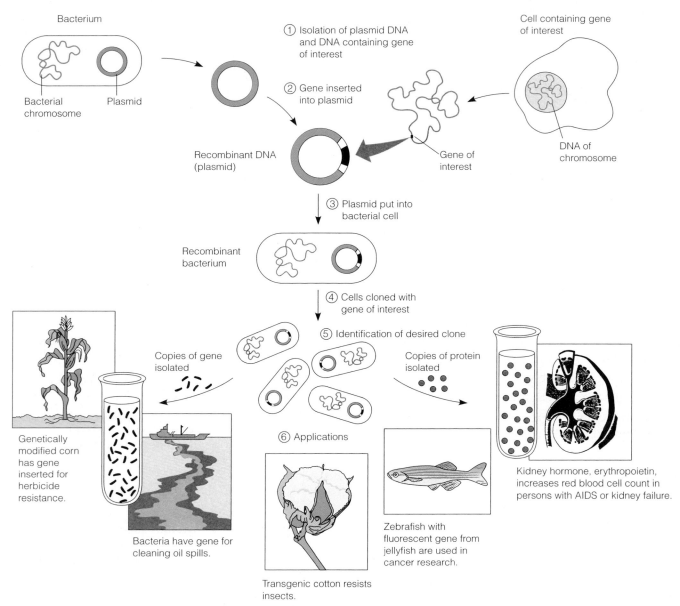

**Figure 10.1.**

**Overview of recombinant DNA technology. 1.** The genetic engineer isolates plasmid DNA from bacteria and purifies DNA containing the gene of interest from another cell. **2.** A piece of DNA containing the gene is inserted into the plasmid, producing recombinant DNA. **3.** The plasmid is put back into a bacterial cell. **4.** This genetically engineered bacterium is then grown in culture. The bacterial culture now contains many copies of the gene (one per cell in this example). **5.** The cell clone carrying the gene of interest is isolated. **6.** The bottom of the figure illustrates a few of the current applications of genetically engineered bacteria.

been mapped. By 2003, all 3 billion nucleotide base pairs of the human genome had been sequenced. More than 2,000 scientists had been involved in this monumental accomplishment. More recently the National Institutes of Health officially launched the first stages of the Cancer Genome Atlas, an effort to catalog genomic changes in cancer cells in humans. In 2008, scientists reported the successful use of gene therapy (inserting a functional gene into a patient with a defective gene) to treat a form of congenital blindness (LCA).

As the human genome was being sequenced, laboratories all over the world began sequencing genomes of other animals and plants, with more than

5,000 species being sequenced to date. In 2009, the National Center for Biotechnology Information reported that complete genomes had been sequenced for over 1,000 prokaryotes. Many of these organisms are important in biological research, for example *Drosophila*, *E. coli*, *C. elegans* (a roundworm), *Zea mays* (corn), and the flowering plant, *Arabadopsis*. Exciting questions related to evolution are being investigated using DNA sequencing. The chimpanzee genome has been sequenced, and recent work on sequencing genes in Neanderthal will allow comparisons of human, chimp, and Neanderthal DNA.

Understanding the universality of the genetic code has allowed information gleaned from the study of organisms such as yeasts and flies to be applied to human beings. You will learn in later lab topics that molecular biology has been used successfully to explore the evolutionary relationships among many living organisms of Earth.

In the following exercise you will employ some of the basic molecular techniques used in laboratories worldwide to study everything from the virus that causes AIDS to DNA extracted from mummies. Following the procedures outlined in Exercise 10.1, you will use restriction enzymes to cut bacterial DNA and propose a tentative map of the location of genetic material. Then you will be asked to answer a series of questions that will lead you through a process of analyzing your results and constructing a restriction map.

Student Media Video—Ch. 20: Biotechnology Lab; Discovery Videos: Cloning; DNA Forensics

EXERCISE 10.1

# Mapping DNA Using Restriction Enzymes and Electrophoresis

## Materials

for practice pipetting:
  petri dish with sample 1% agar
    gel with wells
  empty microtubes
  practice pipette solution
gel electrophoresis apparatus:
  gel plates
  comb to make wells
  chamber for electrophoresis
  chamber cover
power supply with electrodes
heat-resistant gloves
250-mL Erlenmeyer flask
100-mL graduated cylinders
microwave or hot plate

deionized water
agarose
pUC 19 (plasmid) DNA
ice
restriction enzymes:
  *Ava* II, *Pvu* II
restriction buffers
molecular weight markers ($\lambda$ DNA)
gel loading dye (bromophenol
  blue)
plastic wrap
15 1.5-mL microtubes (various
  colors)
thermometer
metric rulers

*If funds are limited, consider using a few micropipettors for practice and demonstration, but have most measurements done with microcapillary pipettes. A microcentrifuge is helpful to get a small volume to the bottom of a tube, but it can also be accomplished by gently tapping the bottom of the tube. If students are not diluting their own TBE, then put out 1X TBE. Agarose can be prepared in advance, or students can prepare their own solutions as suggested in the Procedures section. See note on staining; ethidium bromide (a carcinogen) can be used in place of methylene blue.*

*TAE buffer may also be used. See Preparation Guide.*

TBE (Tris base; boric acid; ethylenediaminetetraacetic acid, or EDTA; NaOH), 10× to be diluted in lab to 1×
microcentrifuge (helpful, not necessary)
micropipettors and tips or microcapillary pipettes and plunger (various sizes from 1 to 100 μL)
37°C incubator or water bath
55°C water bath for agarose
portable freezer box or ice chest
Polaroid camera with Polaroid 667 film

| methylene blue stain: | alternative stain: ethidium bromide |
| --- | --- |
| 0.025% methylene blue | ethidium bromide |
| staining trays | UV-protective goggles |
| light box | UV light box |
| deionized water | disposable gloves |

## Introduction

The first step in any refined DNA analysis, such as DNA sequencing or expressing a gene in another organism, is to construct a map of the molecule. Scientists use naturally occurring enzymes to cut large DNA molecules into smaller pieces. These fragments are sorted and separated by size using a technique called *electrophoresis*. The results are then used to reconstruct the DNA molecule. This initial process of DNA analysis is called **mapping.**

Molecular biologists use some of nature's own tools to do this mapping. In this experiment, you will use restriction enzymes called **restriction endonucleases** to help manipulate DNA molecules. Restriction enzymes recognize a specific DNA sequence wherever it occurs in a DNA molecule (Figure 10.2) and cut the DNA at or near that site; thus, the name. *Restriction* refers to cutting, *endo* to inside a molecule (as opposed to *exo*, which refers to the ends of a molecule), and *nuclease* to the digestion of a nucleic acid such as DNA. Each restriction enzyme is named for the species of bacteria from which it is isolated. For example, *Eco*RI was discovered in *Escherichia coli*, strain R, Roman numeral "I."

*Other palindromes:*

*radar*
*star rats*
*damn mad*
*Was it a rat I saw?*
*Poor Dan is in a droop.*

Cutting (also called **restricting,** or **digesting**) requires energy in the form of adenosine triphosphate (ATP) and involves a physical cleaving of chemical bonds. The specific recognition sites where the cuts occur are often **palindromic**; that is, the sequences of the complementary strands read the same backward and forward (Figure 10.2). (The phrase *race car* is an example of a palindrome. Can you think of others?) In nature, bacteria use restriction enzymes to recognize and metabolize foreign DNA. This constitutes a primitive immune system, since it recognizes and rids the organism of an invader.

In this exercise, you will use restriction endonucleases in conjunction with gel electrophoresis to map the 2,686-**base-pair** (bp) pUC 19 **plasmid.** A plasmid is a relatively small extrachromosomal and circular molecule of DNA found in bacteria and yeasts. pUC 19, a plasmid found in *E. coli*, is one of the most significant cloning tools used in molecular biology labs. Because of its size and DNA sequence, it is an excellent system to study.

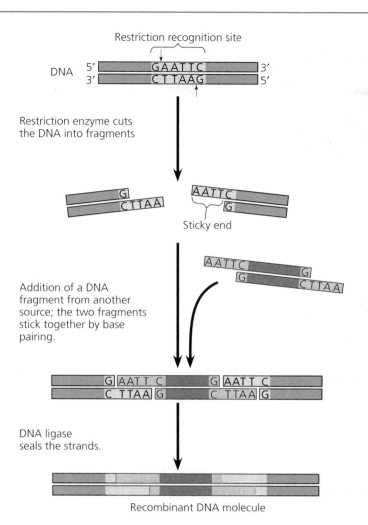

**Figure 10.2.**

**Using a restriction enzyme and DNA ligase to make recombinant DNA.** The restriction enzyme (*Eco*RI) recognizes a six-base-pair sequence and makes staggered cuts in the sugar-phosphate backbone within this sequence. Notice that the recognition sequence along one DNA strand is the exact reverse of the sequence along the complementary strand (that is, they are palindromic). Complementary ends will stick to each other by hydrogen bonding, rejoining fragments in their original combinations or in new recombinant combinations. The enzyme DNA ligase can then catalyze the formation of bonds joining the fragment ends. If the fragments are from two different sources, the result is recombinant DNA.

Once you have cut pUC 19 into discrete fragments, you will need a method of detecting the digested products. Agarose gel electrophoresis is commonly used to separate these fragments (Figure 10.3). The DNA fragments are placed in the gel, and an electric current runs through the matrix of the gel-like agarose. The fragments will move through the gel at different rates, depending on their charge and size. How does the charge of a DNA molecule vary with its size? How is this different from what you see with proteins?

*The charge-to-mass ratio remains the same for all fragments of DNA. There is one negative charge associated with the phosphate group of each nucleotide. Proteins, however, vary in charge-to-mass ratio to a greater degree because their final charge is dependent on the amino acid sequence.*

(For another application of electrophoresis, refer to Lab Topic 12, Population Genetics II: Determining Genetic Variation.) After running the DNA samples in a gel, the gel is stained with methylene blue and viewed on a light box, allowing the visualization of discrete DNA bands on the gel. Alternatively, the gel is stained with ethidium bromide, which binds the DNA and fluoresces under ultraviolet (UV) light.

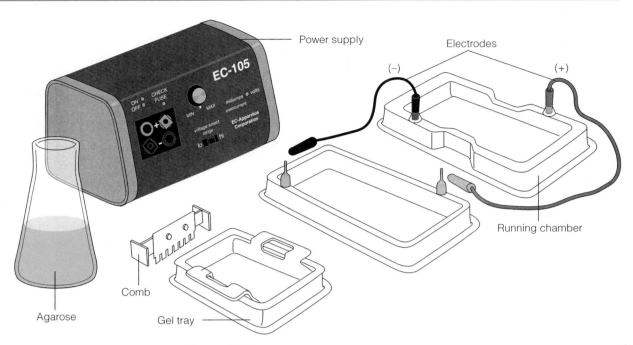

**Figure 10.3.**
**Electrophoresis apparatus.** The apparatus should include the gel tray, comb, agarose, running chamber with electrodes, and power supply.

*Students may not be able to answer these questions at first, but they should have enough information to think through them. They need to begin to think about what they will be doing before beginning the exercise. Their hypotheses and predictions will be based on these questions.*

## Preliminary Questions

Based on your knowledge of the structure and function of DNA and the technique of gel electrophoresis, answer the following questions before beginning your investigation.

1. DNA will move through the gel along the electric current, its direction depending in part on the molecule's charge. In which direction, positive to negative or vice versa, will DNA move in the electric field? Why?

   *Molecules move toward the opposite charge. DNA is negatively charged (from the phosphate groups) and will move from the negative electrode (black) toward the positive electrode (red) at the pH of the gel and solution.*

2. Given that all DNA molecules have similar charge-to-mass ratios, what property do you think is most important in determining migration within the gel? Why?

   *Size is the most important property. Similar charge-to-mass ratios might lead one to the guess that all molecules would migrate similarly because larger molecules have a correspondingly greater negative charge. However, we know that this guess would be wrong, because larger fragments of DNA migrate more slowly. The reason for slower migration of larger fragments is the increased length of the molecule. The DNA molecule is believed to weave through the gel matrix in a manner similar to the movement of a snake. The longer the DNA molecule, the more likely it is to get caught up on the gel matrix. The DNA molecule must then unravel itself, and this requires time.*

3. To map the plasmid, you will need to know the size of the fragments on the gel. Can you suggest a way to determine the size of these fragments?

   *Fragment bands must be matched to a known standard. Molecular weight markers of different DNA fragments of known sizes are run in the gel simultaneously with samples prepared by students.*

4. Given that the distance a fragment of DNA moves into the gel is inversely proportional to the log of the size of the fragment in base pairs, what will a graph of fragments 0.5, 1, 5, 10, and 20 kilobases look like? (The size of DNA fragments is measured in nucleotide bases, and 1,000 bp equals 1 kilobase, Kb.) What do the axes represent?

   *If "fragment size" is on the y axis and "cm traveled" is on the x axis, then the graph will look like this:*

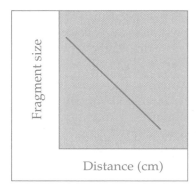

## Hypothesis

Based on your understanding of restriction enzymes and electrophoresis, state a hypothesis concerning the migration of DNA fragments.

> *DNA fragments will move along the gel, according to their size and charge.*

## Prediction

Predict the banding patterns you will observe in the gel based on your hypothesis.

> *If DNA fragments move according to their size and charge, then a series of bands will be seen at different distances along the gel. The smallest fragments will travel the greatest distance on the gel.*

## Procedure

### 1. Practice pipetting and loading a gel

If you have not used a micropipettor or microcapillary pipette before, practice the technique. You may want to have your instructor check your technique because errors as small as one-half microliter can result in inappropriate digests. Although a microliter (μL) seems very small, it can be an amazing amount in molecular biology. For example, 1 μL might contain the amount of DNA in all the genes in a human cell!

*Immediately begin the lab session with Procedure steps 1 and 2.*

Micropipettors are designed to measure small volumes of liquid and each pipettor can be adjusted for a range of volumes, for example, from 2 to 20 μL. *Never* use a pipettor without first attaching the plastic disposable tip.

a. Examine the micropipettor. What is the maximum volume for the pipettor (should be numbered and color coded)? Locate the window that

*Prepare sample gels so students may practice pipetting and loading gels. Add 1% agar (not agarose) to 100-mm × 15-mm petri dishes in which one or two combs have been positioned. When the gel is hardened, squirt water on the plate and remove the combs. Have students practice with "used" pipette tips. Have these in labeled containers on the lab bench.*

*Taking time to practice pipetting can prevent errors and wasting of expensive materials.*
*Once students have practiced loading the gel, remove the practice mixture to ensure that it is not used later in error.*

displays the volume currently set for the pipettor. Set the micropipettor to 20 µL by rotating the dial until the correct volume appears in the display window. Sketch this window in the margin of your lab manual and show the setting to your instructor before continuing.

b. Locate the disposable tips that correspond to your pipettor. Attach the pipette tip and check that it is securely in place.

c. Using the practice solution of water, glycerin, and blue dye, practice drawing the liquid into the pipette tip. *Before placing the tip into the liquid,* depress the pipette plunger with your thumb to the *first stop* to eject any air.

d. Place the tip into the practice solution and *slowly* release the plunger. Remove the tip from the liquid.

e. Locate an empty microtube from your materials. Insert the pipette tip into the microtube so that the tip is close to the bottom of the tube. Touch the tip to the side of the tube. Slowly press the plunger down to the first stop and then continue to *press all the way down to the second stop* to release all the liquid from the tip. Remove the pipettor from the microtube. *Do not release the plunger yet!*

f. When the tip is completely out of the tube, *slowly* release the plunger to the starting position. *Never release the plunger inside the tube* or you could withdraw an unknown amount of solution back into the tip.

g. Discard the tip, using the release button on the pipettor. Use a new tip each time you use the pipettor, even if you are pipetting the same liquids. Allow other members of your team to practice pipetting before continuing to the next step.

h. Reset the pipettor to 2 µL. Sketch the window for this volume in the margin of your lab manual. Before pipetting any liquid, depress the plunger to locate the first and second stops. Note the short distance to the first stop for this extremely small volume. Note that there is very little distance between stop 1 and stop 2.

i. Repeat steps c–g above, being careful to depress the plunger only to stop 1 before withdrawing the liquid. Observe the volume of liquid in the pipette tip before releasing the liquid into the microtube.

j. Practice loading a gel. Your instructor has prepared practice gels in a petri dish. The gel should be covered with water. **Reset the pipette volume to 20 µL.** (Compare the volume window to your original sketch.) Attach a new tip to the pipettor. Following steps c–g above, draw 20 µL of the blue practice mixture into the pipette tip and hold the micropipettor with two hands—one hand to deliver the sample and the other to stabilize the end. Be sure that the sample is all the way down in the tip of the pipette and that there is no air between the sample and the tip. Carefully place the tip just inside the well but not piercing the side or bottom of the well. Slowly release the liquid into the well. All students should practice this procedure several times.

In all cases, a fresh, unused pipette tip or microcapillary pipette should be used every time a pipetting task is performed. This is true even if you are pipetting repeatedly from the same solution, especially in the case of enzymes, which are easily contaminated.

## 2. Prepare the digestions

a. Your instructor has already prepared three color-coded microtubes containing deionized (DI) water and an appropriate buffer. Obtain these three microtubes and label them **A** (for *Ava* II), **P** (for *Pvu* II), and **AP** (for both). Be sure that your labels correspond to the color key provided by the instructor. Write that color in the appropriate cell in the first column of Table 10.1. When prompted, you will add the amounts of each substance indicated in Table 10.1 to these microtubes. Read and understand steps b–d before you prepare your microtubes. *Do not begin to prepare your microtubes until step e.*

b. The buffers added by your instructor are endonuclease buffers for the respective restriction enzymes. They contain a Tris buffer to maintain the pH and salts such as NaCl to maintain optimal ionic strength and $MgCl_2$ required by the enzymes for catalytic activity.

c. The concentration of pUC 19 DNA solution in Table 10.1 has been prepared by your instructor to contain 1 µg/2 µL of DNA (or 0.5 µg/µL). This DNA will be cut into smaller pieces by the restriction enzymes.

d. You will use restriction enzymes *Ava* II (found in *Anabena variabilis*) and *Pvu* II (found in *Proteus vulgaris*). Enzymes are assigned units based on the amount required to digest 1 mg of DNA in 1 hour. Thus, 5 µg of DNA requires 5 units of enzyme to completely digest the DNA in 1 hour. Given that we will be digesting 1 µg of DNA, at the appropriate time, you will add 1 µL of the appropriate enzyme to each microtube.

 The restriction endonucleases must be added last to the mixture. Adding the enzyme early will result in immediate and inappropriate digestion of the DNA!

*Color-code these microtubes so each tube is the same throughout the class.*

*We chose to use enzymes Ava II and Pvu II because each produces only two bands. Other enzymes may be substituted. Sin I is another enzyme that cuts the DNA at the same sequence as Ava II. See suggestions in the Preparation Guide.*

*Use the respective buffers supplied with the enzymes for the single digests. Use only the Ava II buffer with the double digest.*

## Table 10.1

Contents of the Three Tubes. Write the appropriate color for each tube, A, P, and AP. These tubes will be incubated in a 37°C water bath for at least 30 minutes. Some items in the tube have already been added by the instructor. Note that each tube should have a total of 30 µL of solution.

| Tube Label/Color | Instructor Adds | Student Adds before Incubating Tube |
|---|---|---|
| A/ _____ | 24 µL DI water<br>3 µL *Ava* II buffer | 2 µL pUC 19 DNA<br>1 µL *Ava* II enzyme |
| P/ _____ | 24 µL DI water<br>3 µL *Pvu* II buffer | 2 µL pUC 19 DNA<br>1 µL *Pvu* II enzyme |
| AP/ _____ | 23 µL DI water<br>3 µL *Ava* II* buffer | 2 µL pUC 19 DNA<br>1 µL *Ava* II enzyme<br>1 µL *Pvu* II enzyme |

*Use the *Ava* II buffer only.

*Use colored microtubes for all of these preparations. Keep the microtubes in the freezer until just before lab. Transport the microtubes to the lab in ice or in portable freezer boxes. For larger classes we suggest that you prepare the three color-coded microtubes by adding the deionized water and the respective buffers to each before lab. One set of these three tubes may then be placed at each student team's work area before lab. At the appropriate time, the students carry these tubes to the instructor's desk to add the DNA and the enzymes from stock solutions there. Be sure that students add the DNA first and then the restriction enzyme last, just before they place the tubes in the water bath. For smaller classes, have students add all components to each tube.*

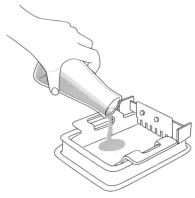

**a.** Pour gel with comb in position

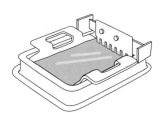

**b.** Gel covers half of comb teeth

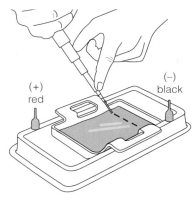

**c.** Load samples

**Figure 10.4.**

**Preparing, loading, and running the gel.** (a) Pouring the gel into the gel tray. Note position of comb. (b) The gel is poured so that it covers approximately half the height of the comb teeth. The comb will be removed before the samples are loaded. (c) Loading samples and dye using a micropipettor. (d) Gel running with electrodes attached. Note that the dye bands are moving through the gel.

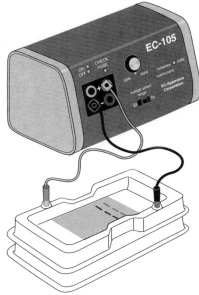

**d.** Gel running—note loading dye moving on gels

*If the concentration of pUC 19 DNA is 0.5 µg/µL, then use 2 µL. However, if the concentration is different, calculate the appropriate volume for 1 µg pUC 19 and adjust the volume of dH$_2$O accordingly.*

*If the concentration of the enzyme requires the students to use less than 1 µL of solution to add the appropriate units of restriction enzyme, you may have them add 1 µL. The excess enzyme will not affect the results, but the digestion will be faster. Alternatively, you can dilute the enzyme before providing it to the students. Dilute with the respective 1× buffer.*

*Once you understand each component added to the microtubes, proceed with step e.*

e. Add 2 µL of pUC 19 DNA to each tube. Tap the bottom of the tube on the lab bench to move the DNA to the bottom of the tube. Return the stock supply of DNA to the cold box.

f. Add the appropriate enzyme(s) to each tube. Tap the bottom of the tube on the lab bench. Be sure all 20 µL of the reaction mix is together at the bottom of the tube. If it is not, gently tap the bottom of the tube on the lab bench until all the liquid accumulates at the bottom. Return stock supplies of enzymes to the cold box.

g. Label the top of the microtube with a symbol to designate your team. Place the tubes at 37°C in a water bath or incubator for at least 45 minutes but preferably 1 hour.

### 3. Prepare and pour the agarose gel

While the DNA digestions are incubating, your instructor will demonstrate how to make a gel, or you will make your own gel. Refer to Figure 10.4 as you make and pour your gel. Your instructor may have already performed steps a–c. If so, skip to step d.

a. Put 0.8 g of agarose in a 250-mL Erlenmeyer flask and add 100 mL of 1× Tris-borate-EDTA, or TBE (if the TBE is 10×, be sure to dilute 10-fold with deionized water).

b. Cover the flask loosely with plastic wrap, and place it in a microwave oven for 2–3 minutes or on a hot plate until the solution boils. In either case, swirl the solution intermittently while wearing a protective glove. Continue until the solution is clear, indicating that all of the agarose is in solution.

 Watch closely to ensure that the solution does not boil over!

c. Allow the gel solution to cool to about 50°C. This process can be expedited by carefully swirling the flask under cool water from the sink. Be careful not to let the gel solidify in the flask.

d. Set the comb over the gel plate so that the teeth rest just above the plate (Figure 10.4a).

e. Slowly pour the gel onto the plate until the solution covers about one-half the height of the comb's teeth (Figures 10.4a, b). Avoid creating bubbles. If any form, quickly and carefully pop them with a sharp object, such as a micropipette tip. (Why would bubbles be a problem?) If the gel is too thin, it will fall apart; if it is too thick, it will take too long to run.

f. Allow the gel to solidify. It will become opaque.

g. When the gel has hardened, squirt some water around the comb and then pull the comb teeth from the gel slowly, carefully, and evenly. If you pull too fast, the suction will break out the bottom of the wells created by the comb.

### 4. Load the gel and perform electrophoresis of samples

a. Place your gel in the running chamber with the wells nearest the negative (black) electrode, and completely cover the gel with 1×TBE. Orient the running chamber so you can see in the wells. *Do not move the running chamber from this point on.*

b. Before you load the gel, you must assemble three additional samples. Obtain these tubes from the instructor. One sample (labeled 1) is for loading dye only. A second sample (labeled 2) contains **molecular weight markers,** and the third (labeled 6) is for uncut pUC 19 DNA. Your instructor will give you the key for tube colors. Write the color for each of these tubes below. These tubes contain the following, already added by your instructor:

1 (color _____): 20 µL of DI water

2 (color _____): 2 µL of λ DNA, 18 µL of DI water (the molecular weight markers)*

6 (color _____): 18 µL of DI water

*The molecular weight markers (tube 2) are used to determine the size of a DNA fragment. These markers, purchased commercially, are of known sizes with which you can compare your results. You will analyze these markers simultaneously in the same gel with your fragments of unknown size.

c. Assemble the six microtubes with respective solutions that you will load on your gel and arrange in this order in a microtube rack:
   (1) Tube 1 from step b (DI water)
   (2) Tube 2 from step b (molecular weight markers)
   (3) Digestion sample A (*Ava* II digest)
   (4) Digestion sample P (*Pvu* II digest)
   (5) Digestion sample AP (*Ava* II and *Pvu* II digest)
   (6) Tube 6 from step b (DI water)

d. Add 2 µL of loading dye to each of the six tubes. Tap the end of each tube on the table to mix.

*You can prepare the gels in advance, but there is value in having students make their own so they understand the materials and techniques involved. We make a stock supply of 0.8% agarose and then divide it into individual flasks with the appropriate amount for one gel, one flask for each student team. We keep these flasks in a 55°C water bath until the students pour their own gels.*

*For instructors who wish to stain with ethidium bromide, we recommend the stain cards. See Alternative Staining Procedure on p. 247–249.*

*See the Preparation Guide for information about preparing the molecular weight markers.*

*Use the same pUC 19 DNA as previously used.*

e. Add 2 µL of pUC 19 DNA that has not been digested (uncut DNA) to tube 6. Tap the tube on the table. What is the purpose of the uncut or undigested DNA in tube 6?

*This serves as a control to indicate the band formed by undigested DNA.*

*To reduce the number of gels that are run, choose a gel comb with 14 wells, then have two groups load onto the same gel. Leave two lanes open in the middle of the gel.*

f. Carefully load each of the six samples into their corresponding wells in your gel, as shown in Figure 10.5. Tube 1 will be in well 1, tube 2 in well 2, the *Ava* II digest in well 3, the *Pvu* II digest in well 4, the *Ava* II/*Pvu* II digest in well 5, and the uncut DNA in well 6. *Set your micropipettor to 20 µL.* Load 20 µL of each sample.

Hold the micropipettor with two hands, one hand to deliver the sample and the other to stabilize the end. Be sure that the sample is all the way down in the tip of the pipette and that no air is between the sample and the tip. Place the tip of the pipette below the surface of the buffer and just into the well. Be careful not to puncture the well bottom. Slowly release the sample into the well. The density of the dye will help it sink into the well (Figure 10.4c).

g. Carefully attach positive (red) and negative (black) electrodes to the corresponding terminals (red into red, black into black) on the power supply and on the gel box (Figure 10.4d).

*Have students write the appropriate microtube color above each well.*

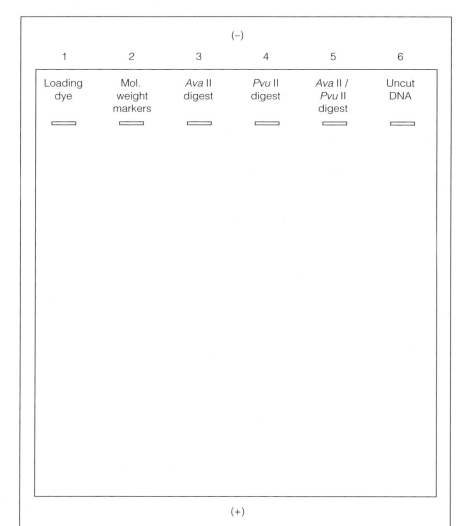

**Figure 10.5.**
**Wells for loading the gel.** Well 1 is used only to practice loading dye. Refer to this figure to ensure that the correct sample is loaded into the appropriate well. Write the appropriate microtube color above each well.

h. Turn the power on to about 100 volts, and make sure that small bubbles arise from the electrodes in the gel buffer, verifying current flow. Check the power source to ensure it is set to volts. Check the loading dye to make sure that the samples are running in the correct direction, toward the positive electrode (see Figure 10.4d).

*We have had success running the gels at 100 volts for 20 minutes followed by 20 minutes at 120 volts. Try different voltages before lab. Use times and voltages that do not melt the gels.*

**Turn off the power to the gel before making any adjustments to the electrophoresis setup.**

i. Run the gel until the loading dye (bromophenol blue) moves down the gel approximately 6–8 cm (this will take 45–60 minutes). Watch the gel carefully. After 10 minutes, first turn off the power; then check to make sure that the gel is not hot. Accidentally making the gel or gel buffers from the wrong concentration of TBE can result in gel and buffer overheating. If this occurs, your experiment will fail.

## 5. Practice mapping

While the gel is running, work the practice mapping problem (Exercise 10.2) at the end of the lab topic.

## 6. Visualize the DNA in the gel

The dye you see on the gel does not stain the DNA but is added so that the movement of the current can be verified. The DNA must be stained by another means, by adding either methylene blue or ethidium bromide. To visualize the DNA, proceed as follows.

*If you are using ethidium bromide stain, you must use a camera to photograph gels so that students can measure on the photograph the distance traveled by their fragments, thus minimizing their exposure to UV light.*

### Methylene Blue Staining Procedure

1. Turn off the power. Remove the gel tray from the running chamber.
2. Slide the gel out of the tray into a 0.025% solution of methylene blue in a staining tray for 30 minutes.
3. Transfer the gel to a destaining tray and destain for several hours or overnight in enough water to just cover the gel. DNA bands will become visible when the gel is viewed over a visible light box.
4. Photograph the gel using a Polaroid camera.
5. If you do not photograph the gel, measure the distance traveled for each band on the gel, sketch the bands in Figure 10.6, and record your measurements next to the corresponding bands.
6. Ask your instructor for directions for disposal of your gel.

*Given the hazards of using ethidium bromide, we highly recommend using methylene blue. See the Preparation Guide for suggestions. We routinely use the methylene blue stain with great success. Check your gels after a couple of hours. Bands can be seen in less time than overnight. You may not need a light box.*

### Alternative Staining Procedure—Ethidium Bromide

**Ethidium bromide is carcinogenic and should be handled with gloves that are thrown away immediately after use. Ethidium bromide should be used only in a confined area, and all lab equipment that comes in contact with the ethidium should be treated with bleach. Once ethidium bromide is added to the gel, you must wear gloves and avoid handling the gel.**

*Ethidium bromide should be handled with great care. Avoid contact with the stain. Everyone in the lab should wear gloves at all times. Used gloves should be discarded immediately following contact with the chemical. Wipe down area with 10% bleach. To dispose of ethidium bromide cards and gels, check with your institution's chemical safety officer.*

1. Turn off the power. **Wear disposable gloves!**
2. Remove the gel tray from the running chamber. Slide the gel out of the tray onto a piece of plastic wrap. Using a disposable pipette, moisten the gel with a small amount of buffer.
3. Place the ethidium bromide staining paper on the gel with the unprinted surface down. Firmly smooth the entire surface of the paper onto the gel. Repeat the smoothing step several times. Place a small beaker on top of the gel to assure good contact between the stain and the gel.
4. After 2 minutes, remove the stain card. Discard the card in the receptacle provided by your instructor. Carefully lift the plastic wrap to move the gel to the UV light source in a darkroom. Wear protective goggles! Observe the DNA bands with the darkroom lights off.

UV-safe goggles must be worn at all times when viewing gels on the UV box.

**Figure 10.6.**

**Results of gel electrophoresis.**

Sketch the bands seen on your gel. Record next to each band the distance traveled, as measured on the photograph, and the size of each fragment, determined from the molecular weight marker standards.

*The fragments generated upon digestion with Pvu II will be 322 and 2,364 base pairs in length (0.322 and 2.364 Kb), while Ava II will generate two fragments of 222 and 2,464 base pairs. Although it may be more difficult to determine the exact length of the larger fragments for either of these two digestions, the students can use the values they get for the smaller fragments and subtract them from the total length.*

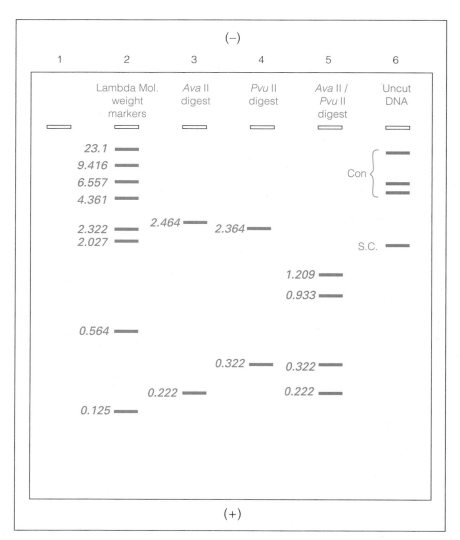

*Lengths given in Kb*

*S.C. = Supercoiled DNA*

*Con = Concatamers, or DNA rings with more than one copy of the DNA found in a single pUC plasmid*

5. Photograph the gel using a camera with shields for UV photography. This will minimize exposure to the UV light and provide a permanent record of your gel.

6. Ask your instructor for directions for proper disposal of your gel. This is particularly important for gels stained with ethidium bromide.

7. Using the photograph, measure the distance traveled for each band on the gel, sketch the bands in Figure 10.6, and record your measurements next to the corresponding bands.

## Results

1. What are the "controls" in the gel you ran?

   *For this experiment, for every DNA sample cut, the uncut DNA must be run alongside the cut DNA to interpret whether the cut DNA actually is cut. If the same DNA were cut with two enzymes simultaneously, for controls the DNA should also be cut with each enzyme individually to ensure that both enzymes worked.*

2. In the wells where uncut DNA is run, you will see more than one band. Hypothesize: What types of DNA molecules might be represented in the different bands? How could you test your ideas? (*Hint:* Remember that the agarose gel is a matrix and consider if your uncut pUC 19 is linear or circular.)

   *Some uncut DNA may be supercoiled, some nicked circular, and some linear. All these molecules move differently through a gel matrix. Only the linear bands can be sized relative to the DNA size standard, since that DNA is all linear. However, a comparison of uncut DNA and the same molecule cut with restriction enzymes allows one to see if and how much of the cut DNA was digested.*

   *When uncut plasmid migrates in the gel, the resulting numerous bands can be confusing. There are three major types of uncut plasmid: supercoiled, nicked circular, and linear. The migration of these three types varies slightly, depending on the particular gel buffer used. Inside bacteria, the plasmid is maintained in a supercoiled form. Of the three uncut forms, this tight form moves easiest, and thus fastest, through the gel matrix. The slower-moving linear form results from a complete break in the DNA during preparation. Another type of plasmid seen in uncut plasmid preparations is called nicked circular and results when one of the two DNA strands is broken, resulting in an open circle, as the supercoil relaxes. Nicked circles can also form several circles linked together, which will appear to be another slower-running type in the gel.*

3. On the photo of your gel, measure the distance traveled for each band on the gel. Sketch the bands in Figure 10.6. Write the distance beside each band. (If you did not photograph the gel, you will already have made these measurements and sketches.)

4. Obtain the known sizes of the DNA molecular weight marker fragments from your instructor. Write these sizes (in bp or Kb) beside the molecular weight marker bands in Figure 10.6. In Figure 10.7, graph the distance traveled by the marker fragments (bands) on your gel (in cm on the *x* axis) versus the size of the corresponding fragments (in bp or Kb on the *y* axis). Draw a best-fit line through these points on the graph.

5. Using the graph you constructed in Figure 10.7, determine the size of every other band on the gel. To do this, on the line drawn in Figure 10.7, locate the distance traveled by each restriction fragment (measured and recorded in step 3). Follow this point over to the *y* axis to determine the number of base pairs in the restriction fragment. Record your results next to the corresponding band in Figure 10.6.

Use the following Discussion section as a guide to analyzing your data. If you have not already worked the practice problem, Exercise 10.2, do it now, before continuing to the Discussion section. Your final analysis should include a complete restriction map of the pUC 19 plasmid.

**Figure 10.7.**

**Relationship of distance traveled (cm) and the size of the restriction fragment (bp or Kb) for known DNA molecular weight markers.** Note the distances traveled for each marker, and choose and add an appropriate scale for the *x* axis. Obtain the known marker sizes from your instructor, and graph them with the corresponding distances from your gel. Note this is graphed on semi-log paper.

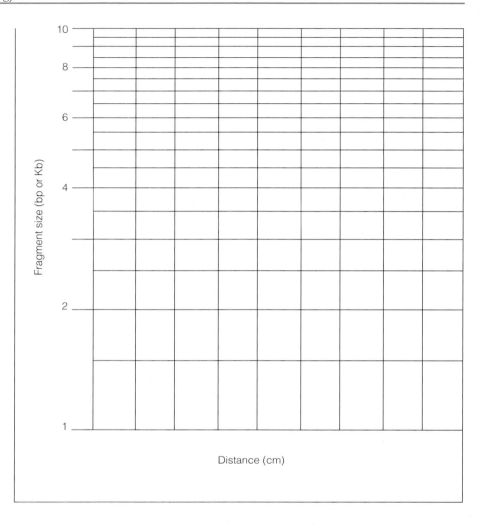

Distance (cm)

**Discussion**

Do not begin the discussion questions until you understand the practice problem. Then refer to your results as you answer the following questions about the map of pUC 19.

1. How many Kb is pUC 19?

   *2.7 Kb, or 2,686 bp—this was mentioned on page 238.*

2. How many DNA fragments were produced by the enzyme digestions? By *Ava* II?

   *2*

   By *Pvu* II?

   *2*

   By the double digest with *Ava* II and *Pvu* II?

   *4*

3. Note that both the small fragments generated in the single digest are still present in the double digest. What does this mean?

   *Pvu II did not cut within the small fragment of Ava II.*

4. Draw the two restriction fragments for *Ava* II in the space provided. Label the two ends "*Ava* II." Indicate the size of the fragments that were produced in the double digest, and place the *Pvu* II sites on the large *Ava* II fragment. Refer to the practice problem for assistance.

   *The solution for the results differs slightly from what is presented in the practice problem. The restriction sites do not alternate around the plasmid but are located on opposite sides of the DNA.*

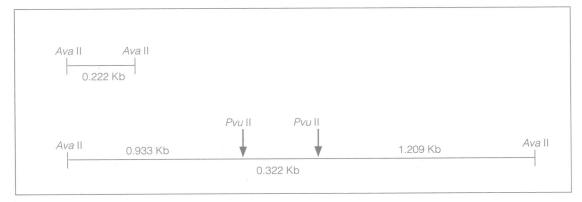

   *The students, by looking at the products of the double digest, should realize that the large fragment generated by the Ava II digest is further digested into three pieces by Pvu II in the double digest. Since, as discussed earlier, linear pieces will produce (n + 1) fragments for n restriction sites, we can assume there are two Pvu II sites in the large Ava II fragment. Moreover, we can see from the Pvu II digest that the sites are located 322 bp apart. We can, therefore, position our sites as shown in the previous diagram and put the 0.933 Kb and 1.209 Kb fragments on either side.*

5. Join the fragments you have drawn above to re-create the original plasmid.

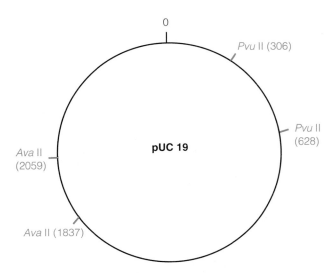

   *The students should generate the above plasmid just by joining the ends of the two fragments shown in question 4.*

6. Now that you have mapped some DNA fragments, what could you do to further characterize the DNA you have isolated? Refer back to the introduction to this lab topic.

*Further map the DNA; make RNA and protein from the DNA in vitro; sequence the DNA.*

---

## EXERCISE 10.2
# Practice Problem for Mapping DNA

*While the gels are running, lead the discussion of the practice mapping problem. Students will need to work through the logic of what the bands/fragments represent and how they can use the information to place the puzzle pieces back together.*

Mapping and sequencing genomes of organisms has led to a new field of genetics called **genomics.** One approach to determining the gene sequence is mapping the DNA fragments produced by restriction enzymes. These restriction sites can serve as markers to align the DNA. An example of a DNA map is shown in Figure 10.8 for a small plasmid similar to pUC 19, LITMUS 28i. The numerous restriction sites are indicated by the abbreviations around the perimeter of the circle. Can you find a restriction site for *Pvu* II? The location of each enzyme's restriction site was determined using a process similar to the one you will be using today. Fortunately, you are working with only two enzymes!

The construction of a DNA map is really a logic puzzle. You must compare results for the single digests with those of the double digests (those cut with two enzymes) to correctly orient the sites on the DNA molecule. As with anything else, the process becomes easier with practice. The experience you will gain as you solve the following problem will help prepare you to analyze your exercise results.

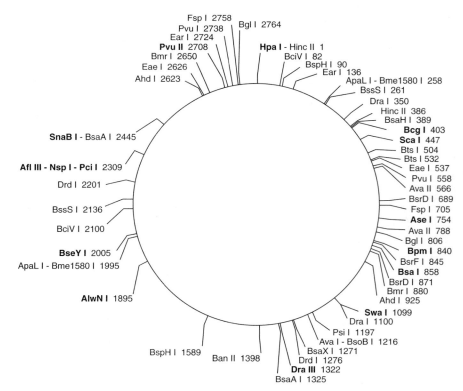

**Figure 10.8.**
**Restriction enzyme map of LITMUS 28i, a small *E. coli* plasmid.**

You are investigating the pathogenic organism *Bacillus anthrax*, which has recently been isolated from infected individuals. It is known that this organism contains a plasmid, and because these genetic elements often contain antibiotic resistance genes, you decide to begin your investigation with the plasmid. Initial studies indicate that the DNA molecule is 4,000 bp (4 Kb) long.

## Procedure

1. Inspect the gel diagrammed in Figure 10.9. The restriction pattern shown in the gel diagram was produced by digestion of the bacterial plasmid with two restriction enzymes, *A* and *B*.

2. Determine the number of fragments produced by each enzyme, and determine the number of fragments produced by the double digest.

   *We can see that each enzyme generates two fragments when used alone. The double digest produces four.*

   How does the number of fragments correlate with the number of restriction sites in the DNA molecule? Remember that plasmids are circular.

   *For circular pieces of DNA, the number of restriction fragments is equal to the number of restriction sites in that DNA. With linear DNA, the number of restriction fragments is equal to the number of restriction sites plus one.*

3. Note that the size of the fragments has to be determined by comparison to the molecular weight markers. In this problem, the size of the fragments has already been determined and is recorded next to each band.

4. Make a restriction map of the isolated plasmid by determining the relative positions of the restriction sites. The best approach is to realize that the double digest is really an extension of the single digests. For example, the double digest is analogous to taking the bands you see in the *A* restriction digest and digesting these bands with the *B* enzyme. The

*Students have difficulty visualizing circular DNA cut into fragments, as well as the alignment steps. Consider using a small child's round plastic bracelet or other object to demonstrate these concepts.*

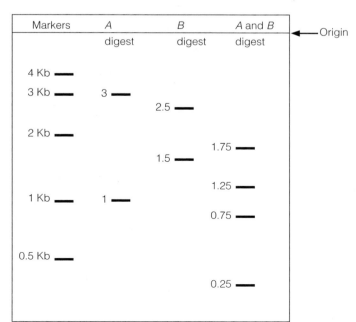

**Figure 10.9.**
**Restriction pattern generated by restriction digest of the *Bacillus anthrax* plasmid.** Note the position of the origin and the molecular weight markers.

smaller bands you see in the double digest should add up to the bands seen in the *A* digest alone.

a. Determine which bands of the double digest come together to form the small *A* restriction fragment.

 *The small A restriction fragment is 1 Kb in length. The two double digest fragments that must make up this 1 Kb fragment are the 0.25 Kb and 0.75 Kb fragments.*

b. Determine which bands of the double digest make up the large *A* fragment.

 *The 1.75 Kb and 1.25 Kb fragments form the large A fragment that is 3 Kb long.*

c. Draw the two *A* restriction fragments in the space provided, and indicate where the *B* restriction sites fall on the fragments. Indicate the size of the fragments that are produced in the double digest.

*Have students work through this problem. Use the notation shown, which is standard for restriction analysis.*

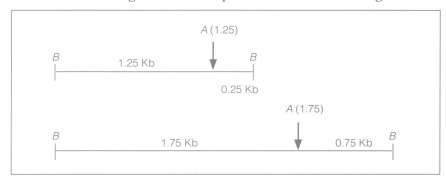

d. Do the same analysis on the *B* fragments. Compare the double digest with the single *B* digest.

e. Draw the *B* restriction fragments in the space provided, and indicate where the *A* restriction sites fall on the fragments. Indicate the size of the fragments that are produced in the double digest.

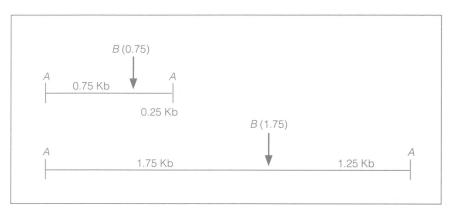

f. Align the fragments from the preceding analysis. You will notice from your drawings that there are double digest fragments of the same size. These fragments are the key to solving our puzzle because they represent overlap. If we align the large *B* fragment and the large *A* fragment, we find that we can match up restriction sites (Figure 10.10).

g. Continue to align fragments. As shown in Figure 10.10, the alignment is used to draw out one longer fragment. By continuing to

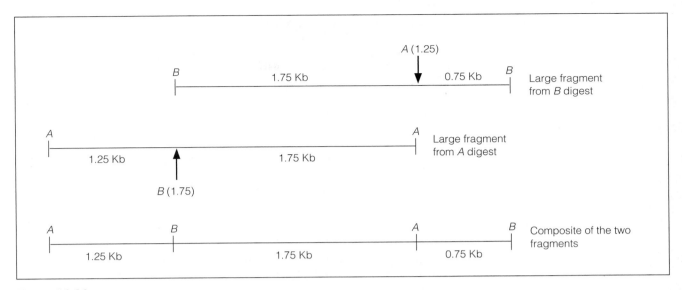

**Figure 10.10.**
**Alignment of the large fragments from single digests.** The large fragments from both single digests can be aligned so that one longer fragment is drawn.

align fragments, it is possible to re-create the entire restriction map. Which fragments would you try to line up next?

*fragment containing either the 1.25 Kb fragment or the 0.75 Kb fragment*

Since we are working with a circular DNA molecule, would you expect to eventually repeat yourself?

*Yes, you would. At that point you would know that you had finished putting together the restriction map. Ends would be joined, and a circular restriction map would result.*

h. Continue to align fragments until you have finished your complete map.

i. Draw the map in the space provided. *Hint:* Arbitrarily place one of the A sites at position 0 (the very top of your circle) and use that as your reference site.

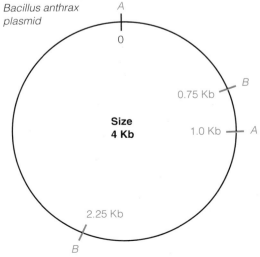

After completing the practice problem, resume your work to complete Exercise 10.1.

# Reviewing Your Knowledge

1. a. State the role of each of the following in molecular biology and provide an example if appropriate: *restriction enzyme, plasmid DNA, molecular weight markers,* and *micropipettors.*

   b. State the role of each of the following in gel electrophoresis: *gel buffer, comb, power pack, agarose, methylene blue* (or alternatively, *ethidium bromide*), *loading dye.*

2. The treatment of diabetes was revolutionized when transgenic bacteria were utilized to produce human insulin. What are transgenic organisms? What feature of DNA is responsible for a human gene in bacteria effectively producing the insulin protein?

   *Transgenic organisms have a gene or segment of DNA from a different species inserted into their genome. Because the genetic code is universal, bacteria and humans produce the same amino acid sequence from the same corresponding nucleotide sequence. For example, in both bacteria and humans, the mRNA code, UUU, produces the same amino acid, phenylalanine.*

3. Hypothesize about why restriction enzyme digestions are performed at 37°C.

   *This is the optimal performance temperature for most restriction enzymes, as most function in nature at around this temperature.*

4. Restriction endonucleases are enzymes. What type of macromolecule are endonucleases? Think of a general way that these enzymes might go about the business of cutting DNA.

   *Endonucleases are proteins. This is an open-ended question designed to help students begin thinking about proteins concretely that is, about physically doing something in the cell or in the test tube. These enzymes physically bind to the DNA at their recognition sequence, then, in a very specific chemical reaction, cut the DNA.*

5. BamHI and EcoRI are two restriction enzymes used to digest a plasmid DNA in an experiment similar to the one completed in the laboratory. The results of the gel electrophoresis are shown in the diagram below. The number next to each band represents the size of the DNA fragment.

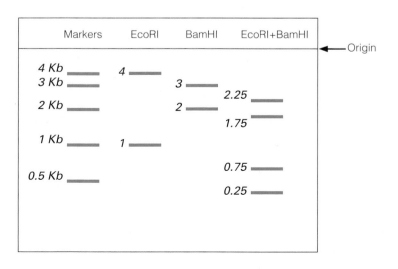

Using this information, answer the following questions.

a. What is the total length (in kilobases, Kb) of the plasmid DNA used in this experiment?

*5 Kb*

b. How many BamHI restriction sites are present in this DNA?

*2*

How many EcoRI restriction sites are present?

*2*

c. How many fragments are produced by the double digest with BamHI and EcoRI?

*4*

d. Using this information construct the map for the plasmid DNA showing the location of all the restriction sites. Follow the procedure in the practice mapping problem, Exercise 10.2.

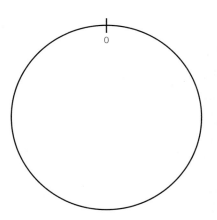

# Applying Your Knowledge

1. Scientists are developing genetically engineered foods that are insect-resistant, are more nutritious than traditional foods, and can grow in poorer soils. Many scientists believe that genetically modified (GM) foods will become essential as human populations continue to grow.

   The development of GM foods is not without controversy, however. Using resources from the Web, the library, and your text, choose one of these topics and prepare an essay discussing the controversy involving GM foods or scientists' concerns for potentially harmful effects of GM foods.

   a. Intellectual property rights: Monsanto Canada Inc. v. Schmeiser
   b. Environmental concerns—populations and biodiversity:
      - Will GM plants that are toxic to insects have a negative impact on other insects such as monarch butterflies?
      - Will the decline in insect "pests" have an effect on other animals, such as birds?
      - What could be the impact if crops that are herbicide-tolerant are sprayed to the extent that no wild plants are able to survive in the area?
      - Genes have been inserted into chromosomes of a fish species to make the animal grow larger. Would this give these fish a reproductive advantage over natural populations if these fish are more successful at attracting mates? What if the offspring of the genetically modified fish are less able to survive in natural environments?
   c. Human Health:

      Are safety assessment standards of the United States FDA adequate to protect against potential human health problems that might arise from eating GM foods?

2. In 2005, efforts began to catalog genomic changes of all human cancers. Efforts such as this may eventually revolutionize cancer treatment. Gene therapy has already been used successfully in other medical applications,

*Students may pursue these or other ethical issues using a variety of resources from publications in science and/or ethics.*

for example in severe combined immunodeficiency (SCID). But ethical questions arise. Should gene therapy be restricted to somatic cells, or should individuals be able to alter their germ-cell line (those cells that will be passed to their offspring)? What about the expense involved with these treatments? What effect will this have on health-care systems? Will only the wealthy be able to afford treatments? What effect will making gene treatments available to everyone have on our already out-of-hand health-care costs?

3. More and more, in cases of rape or murder, forensic scientists are using DNA fingerprinting as evidence for the guilt or innocence of a suspect. In this process, a few selected portions of DNA from small amounts of blood or semen from the crime scene can be analyzed and compared with DNA from a suspect. If gel electrophoresis shows that two samples match, the probability that the two samples are *not* from the same person can be anywhere from one chance in 100,000 to one in 1 billion, depending on how the test was performed. And yet, even with compelling DNA evidence to the contrary, suspects are sometimes found "not guilty." What problems with this type of evidence can arise to create doubt about the guilt of the person?

*The jury may be convinced that human error, insufficient statistical data, or simply someone tampering with the evidence jeopardized the tests. Sloppy lab technique, inexperienced researchers, and sloppy police procedures—all could be used to create doubt.*

4. The map for pUC 19 is available online at New England Biolabs, www.neb.com; in the search window type "pUC 19 map." Compare your map using *Pvu* II and *Ava* II with that published on the Web.

Locate the restriction sites for *Pvu* I and *Acl* I. How many restriction sites are present for each enzyme?

*2 for each*

How many fragments would be produced from digestions of pUC 19 using these two enzymes?

*Each enzyme would cut the plasmid DNA into 2 fragments. A double digest with both enzymes would produce 4 fragments.*

## Investigative Extensions

Students may extend this lab topic to produce recombinant DNA by isolating a DNA fragment and inserting it into a plasmid DNA that is used to transform bacteria. See Morgan and Carter, *Investigating Biology*, 1st ed., 1993, Exercise 9.2, "Recombinant DNA."

 **Student Media: BioFlix, Activities, Investigations, and Videos**
**www.masteringbiology.com (select Study Area)**

**BioFlix**—Ch. 16: DNA Replication
**Activities**—Ch. 20: Applications of DNA Technology; Restriction Enzymes; Cloning a Gene in Bacteria; Gel Electrophoresis of DNA; Analyzing DNA Fragments Using Gel Electrophoresis; DNA Fingerprinting; Making Decisions About DNA Technology Golden Rice
**Investigations**—Ch. 20: How Can Gel Electrophoresis Be Used to Analyze DNA? LabBench: Molecular Biology
**Videos**—Ch. 20: Biotechnology Lab; Discovery Videos: Cloning; DNA Forensics

# References

We acknowledge the contributions of Dr. Arri Eisen and Dr. Alex Escobar, Biology Department, Emory University in writing early versions of this lab topic.

Bloom, M. V., G. A. Freyer, and D. A. Micklos. *Laboratory DNA Science.* Menlo Park, CA: Benjamin Cummings, 1996.

Briggs, A. et al. "Targeted Retrieval and Analysis of Five Neanderthal mtDNA Genomes." *Science,* 2009, vol. 325, pp. 318–321.

Collins, F., and A. Barker. "Mapping the Cancer Genome." *Scientific American,* 2007, vol. 296, (3), pp. 50–57.

Feuillet, C. and K. Eversole. "Solving the Maze." *Science,* 2009, vol. 326, pp. 1071–1072.

Phillips, T. "Genetically Modified Organisms. (GMOs): Transgenic Crops and Recombinant DNA Technology." *Nature Education,* 2008, vol. 1(1), published online at http://www.nature.com/scitable/topicpage/Genetically-Modified-Organisms-GMOs-Transgenic-Crops-and-732.

Pray, L. "Recombinant DNA Technology and Transgenic Animals." *Nature Education,* 2008, vol. 1(1), published online at http://www.nature.com/scitable/topicpage/Recombinant-DNA-Technology-and-Transgenic-Animals-34513.

Raney, T. and P. Pingali. "Sowing a Gene Revolution." Scientific American, 2007, vol. 297, pp. 104–111.

Russell, P. J. *iGenetics: A Molecular Approach*, 3rd ed. San Francisco, CA: Benjamin Cummings, 2009.

# Websites

The Cancer Genome Atlas (TCGA), National Cancer Institute; studying cancer through biotechnology, including sequencing genomes:
http://cancergenome.nih.gov/index.asp

Issues in Biotechnology, American Institutes of Biological Sciences, provides resources, links, and interviews:
http://www.actionbioscience.org/biotech/

National Eye Institute, NIH, provides information on the biotechnology and success of gene therapy for Leber Congenital Amaurosis (LCA), a form of congenital blindness:
http://www.nei.nih.gov/lca/

A Science Primer: A Basic Introduction to the Science Underlying NCBI Resources:
http://www.ncbi.nlm.nih.gov/About/primer/genetics_molecular.html

# LAB TOPIC 10

# Molecular Biology
## Teaching Plan for Laboratories

While some instructors may be hesitant to try molecular biological investigations, the investment in effort and initial funds is extremely worthwhile for several reasons. The students are thoroughly intrigued by the idea of visualizing DNA and using the techniques of biotechnology. The experiment is very straightforward, involving simple techniques. The concepts involved range from the specific structure and function of DNA to the larger theme of the unity and diversity of life. Finally, after the initial financial investment is made, annual costs are minimal.

## Main Concepts and Objectives

1. Concept: use of restriction enzymes to map DNA. Students will use restriction enzymes to digest DNA. They will be able to discuss the mechanisms of this process. They will analyze their results and construct a tentative map of the DNA. (Exercises 10.1 and 10.2)

2.  Concept: electrophoresis as a technique for separating molecules. Students will use electrophoresis to separate DNA fragments resulting from restriction enzyme digestion. Students will be able to discuss the separation of molecules based on specific properties of those molecules, whether they are DNA, RNA, or protein. (Exercise 10.1)

3.  Concept: use of naturally occurring proteins (restriction enzymes) to investigate and explore DNA. Students will investigate DNA using enzymes that have specific functions and that can be harnessed for use in the molecular biology laboratory. (Exercise 10.1)

4.  Concept: universality of the genetic code and conservation of the functions of proteins. Students will be able to describe the use of viral and bacterial genomes and enzymes in recombinant DNA technology as examples of molecular evidence in support of the theory of evolution. (Exercises 10.1 and 10.2)

5.  Specific content. Students will be able to describe or define: *restriction enzymes (endonucleases), digestion, gel electrophoresis, recombinant DNA, mapping, plasmid, pUC, palindromic, DNA molecular weight markers.*

## Order of the Lab

1.  Begin Exercise 10.1; set up digestions. Briefly explain procedures for preparing digestions. If students have not used micropipettes before, spend 10 minutes practicing (p. 241–242).               (30 min)

2.  Prepare and pour the gels (p. 244–245).                                      (10 min)

3.  As the digestions incubate and the gels solidify, introduce the topic and applications. Review Preliminary Questions on p. 240–241               (20 min)

4.  After about 1 hour for digestions, load gel.                                 (15 min)

5.  Run gel and do practice problem, Exercise 10.2.                              (60 min)

6.  Stain gel.                                                                    (30 min)

7.  Construct restriction map.                                                    (15 min)

During the downtime after the gels have been poured and the digestions are finishing, discuss the mechanisms of electrophoresis and other questions the students might have. During the electrophoresis, work the practice mapping problem.

In lecture or recitation sections, students should review concepts and terminology prior to the lab and participate in follow-up discussions of their results.

**For a 2-hour lab:** Exercise 10.1 can be completed in a 2-hour lab period if some materials (for example, the digests and the gels) are prepared in advance by the instructor. Have students complete Exercise 10.2 outside of class.

## Classroom Management

Students work in teams of four for Exercise 10.1 as they digest the DNA and separate the fragments using gel electrophoresis. If some groups' digestions do not work, or if there are other problems, groups that have successful digestions can share their good samples and gels for graphing analysis.

For Exercise 10.2, the practice mapping problem, have students work independently or in pairs, discussing the solution of the problem.

With the methylene blue stain, destaining the gels takes place overnight. Have students return to lab the next day to see their gels and make a photograph. The team should designate one person to make the photograph. Then students can determine the sizes of the bands on their gel as a take-home assignment or perhaps in a recitation section. If you have labs on successive days, students can view the previous day's gels and complete the lab in one 3-hour session. Instructors may prepare gels in advance for the first lab section.

# Student Development

The students enter a new realm of research: the very small molecule. They learn to work with and analyze molecules. They learn to grasp the importance of care and accuracy when studying molecules. They are asked to use critical thinking skills as they apply their knowledge of the structure and function of DNA to answer questions and develop hypotheses and solve mapping problems.

# Lab Safety Precautions

Methylene blue provides good results but requires more time. The alternative staining method, ethidium bromide and ultraviolet light, presents safety concerns in this lab. Instruct students to take appropriate precautions if you use ethidium bromide.

1. Ethidium bromide is a carcinogen. Wear rubber gloves throughout the lab. Anything that comes in contact with ethidium needs to be cleaned thoroughly. Follow the procedure provided in the marginal notes and the *Preparation Guide*.

2. Ultraviolet light is a mutagen. UV light boxes can be purchased with a protective see-through cover. Regardless, everyone in the room with the light box should be wearing UV-protective goggles. Not all safety goggles are UV-safe.

3. When working with hot plates, avoid hot surfaces. It is not obvious if the plates are hot. Large preparations of several gels worth of agarose can be heated in a microwave to avoid hot plate problems. Handle flask with hot gloves and watch for boiling over of the agarose solution.

4. Electrophoretic voltage (100 volts) is dangerous. Electrophoresis apparatus with plastic covers can be purchased for optimal safety. *No adjustments should be performed on the gel or gel apparatus while the power is on.*

# Discussion and Summary

Each step in this lab topic builds on the previous one; therefore, at each step, before continuing, review the preceding steps.

# Evaluation

Students should be evaluated on lab reports describing their findings and on how well they relate these findings to the important concepts in the lab. Grasp of concept can be evaluated on lab tests. Students can be asked to solve additional mapping problems.

## Lab Topic 11

# Population Genetics I: The Hardy-Weinberg Theorem

## Laboratory Objectives

After completing this lab topic, you should be able to:

1. Explain Hardy-Weinberg equilibrium in terms of allelic and genotypic frequencies and relate these to the expression $(p + q)^2 = p^2 + 2pq + q^2 = 1$.

2. Describe the conditions necessary to maintain Hardy-Weinberg equilibrium.

3. Use the bead model to demonstrate conditions for evolution.

4. Test hypotheses concerning the effects of evolutionary change (migration, mutation, genetic drift by either bottleneck or founder effect, and natural selection) using a computer model.

*For a 2-hour lab: Assign fewer modeling scenarios per student team. Omit student oral reports during the laboratory period. Omit computer simulations. See the Teaching Plan.*

*Instructors and/or students can download data tables in Excel format. Go to www.masteringbiology.com. Look in the Study Area under Lab Media.*

## Introduction

Charles Darwin's unique contribution to biology was not that he "discovered evolution" but, rather, that he proposed a mechanism for evolutionary change—**natural selection,** the differential survival and reproduction of individuals in a population. In *On the Origin of Species*, published in 1859, Darwin described natural selection and provided abundant and convincing evidence in support of **evolution,** the change in the genetic structure of populations over time. Evolution was accepted as a theory with great explanatory power supported by a large and diverse body of evidence. However, at the turn of the century, geneticists and naturalists still disagreed about the role of natural selection and the importance of small variations in natural populations. How could these variations provide a selective advantage that would result in evolutionary change? It was not until evolution and genetics became reconciled with the advent of population genetics that natural selection became widely accepted.

*In the Study Area of masteringbiology.com, see Student Media for suggested activities, investigations, and videos.*

*Remind students that natural selection occurs at the level of the individual. Removing "less fit" individuals changes gene frequencies in the population (evolution).*

Ayala (1982) defines evolution as "changes in the genetic constitution of populations." A **population** is defined as a group of organisms of the same species that occur in the same area and interbreed or share a common **gene pool,** all the alleles at all gene loci of all individuals in the population. The population is considered the basic unit of evolution. The small scale changes in the genetic structure of populations from generation to generation is called **microevolution.** *Populations evolve, not individuals.* Can you explain this statement in terms of the process of natural selection?

In 1908, English mathematician G. H. Hardy and German physician W. Weinberg independently developed models of population genetics that showed that the process of heredity by itself did not affect the genetic structure of a population. The **Hardy-Weinberg theorem** states that the frequency of alleles in the population will remain the same from generation to generation. Furthermore, the equilibrium genotypic frequencies will be established after one generation of random mating. This theorem is valid only if certain conditions are met:

1. The population is very large.
2. Matings are random.
3. There are no net changes in the gene pool due to mutation; that is, mutation from *A* to *a* must be equal to mutation from *a* to *A*.
4. There is no migration of individuals into and out of the population.
5. There is no selection; all genotypes are equal in reproductive success.

It is estimated, for example, that at the beginning of the 19th century in Great Britain, more than 95% of the peppered moths were light colored, while less than 5% were dark. Only one or two of the dark forms were seen in collections before mid-century. Under Hardy-Weinberg equilibrium, these proportions would be maintained in each generation for large, random-breeding populations with no change in the mutation rate and migration rate, as long as the environment was relatively stable. The process of heredity would not change the frequency of the two forms of the moth. Later in this laboratory, you will investigate what happened to these moths as the environment changed following the Industrial Revolution.

Basically, the Hardy-Weinberg theorem provides a baseline model in which gene frequencies do not change and *evolution does not occur*. By testing the fundamental hypothesis of the Hardy-Weinberg theorem, evolutionists have investigated the roles of mutation, migration, population size, non-random mating, and natural selection in effecting evolutionary change in natural populations. Although some populations maintain genetic equilibrium, the exceptions are intriguing to scientists.

## Use of the Hardy-Weinberg Theorem

The Hardy-Weinberg theorem provides a mathematical formula for calculating the frequencies of alleles and genotypes in populations. If we begin with a population with two alleles at a single gene locus—a dominant allele, *A*, and a recessive allele, *a*—then the frequency of the dominant allele is $p$, and the frequency of the recessive allele is $q$. Therefore, $p + q = 1$. If the frequency of one allele, $p$, is known for a population, the frequency of the other allele, $q$, can be determined by using the formula $q = 1 - p$.

During sexual reproduction, the frequency of each type of gamete produced is equal to the frequency of the alleles in the population. If the gametes combine at random, then the probability of *AA* in the next generation is $p^2$, and the probability of *aa* is $q^2$. The heterozygote can be obtained two ways, with either parent providing a dominant allele, so the probability would be $2pq$.

These genotypic frequencies can be obtained by multiplying $p + q$ by $p + q$. The general equation then becomes

$$(p + q)^2 = p^2 + 2pq + q^2 = 1$$

*Students may recognize this expression as a binomial expansion.*

To summarize:

$$p^2 = \text{frequency of } AA$$
$$2pq = \text{frequency of } Aa$$
$$q^2 = \text{frequency of } aa$$

Follow the steps in this example.

1. If alternate alleles of a gene, *A* and *a*, occur at equal frequencies, *p* and *q*, then during sexual reproduction, 0.5 of all gametes will carry *A* and 0.5 will carry *a*.
2. Then $p = q = 0.5$.
3. Once allelic frequencies are known for a population, the genotypic makeup of the next generation can be predicted from the general equation. In this case,

$$(0.5A + 0.5a)^2 = 0.25AA + 0.5Aa + 0.25aa = 1$$
$$(p + q)^2 = p^2 + 2pq + q^2 = 1$$

   This represents the results of random mating as shown in Figure 11.1.
4. The genotypic frequencies in the population are specifically

$$p^2 = \text{frequency of } AA = 0.25$$
$$2pq = \text{frequency of } Aa = 0.50$$
$$q^2 = \text{frequency of } aa = 0.25$$

5. The allelic frequencies remain $p = q = 0.5$.

| | | **Eggs** | |
|---|---|---|---|
| | | *A* egg $p = 0.5$ | *a* egg $q = 0.5$ |
| **Sperm** | *A* sperm $p = 0.5$ | *AA* $p^2 = 0.25$ | *Aa* $pq = 0.25$ |
| | *a* sperm $q = 0.5$ | *aA* $pq = 0.25$ | *aa* $q^2 = 0.25$ |

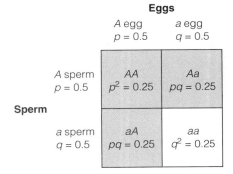

**Figure 11.1.**
**Random mating in a population at Hardy-Weinberg equilibrium.** The combination of alleles in randomly mating gametes maintains the allelic and genotypic frequency generation after generation. The gene pool of the population remains constant, and the populations do not evolve.

In actual populations the frequencies of alleles are not usually equal. For example, in a large random mating population of jimsonweed 4% of the population might be white (a recessive trait), and the frequency of the white allele could be calculated as the square root of 0.04.

1. White individuals $= q^2 = 0.04$ (genotypic frequency); therefore, $q = \sqrt{0.04} = 0.2$ (allelic frequency).
2. Since $p + q = 1$, the frequency of *p* is $(1 - q)$, or 0.8. So 4% of the population are white, and 20% of the alleles in the gene pool are for white flowers and the other 80% are for purple flowers. (Note that you could not determine the frequency of *A* by taking the square root of the frequency of all individuals with purple flowers because you cannot distinguish the heterozygote and the homozygote for this trait.)
3. The genotypic frequencies of the next generation now can be predicted from the general Hardy-Weinberg theorem. First determine the results of random mating by completing Figure 11.2 (refer to Figure 11.1).
4. What will be the genotypic frequencies from generation to generation, provided that alleles *p* and *q* remain in genetic equilibrium?

   *AA = 0.64*
   *Aa = 0.32*
   *aa = 0.04*

| | | **Eggs** | |
|---|---|---|---|
| | | *A egg* $p = 0.8$ | *a egg* $q = 0.2$ |
| **Sperm** | *A sperm* $p = 0.8$ | *AA* $p^2 = 0.64$ | *Aa* $pq = 0.16$ |
| | *a sperm* $q = 0.2$ | *aA* $pq = 0.16$ | *aa* $q^2 = 0.04$ |

**Figure 11.2.**
**Random mating for a population at Hardy-Weinberg equilibrium.** Complete the mating combinations for purple and white flowers.

The genetic equilibrium will continue indefinitely if the conditions of the Hardy-Weinberg theorem are met. How often in nature do you think these conditions are met? Although natural populations may seldom meet all the conditions, Hardy-Weinberg equilibrium serves as a valuable model from which we can predict genetic changes in populations as a result of natural selection or other factors. This allows us to understand quantitatively and in genetic language how evolution operates at the population level.

**Note:** Using the square root to obtain the allelic frequency (step 1 above) is only valid if the population meets the conditions for a population in Hardy-Weinberg equilibrium. Later in the exercise you will use a counting method to calculate allelic frequencies.

---

### EXERCISE 11.1
# Testing Hardy-Weinberg Equilibrium Using a Bead Model

## Materials

plastic or paper bag containing 100 beads of two colors

## Introduction

*The example in Table 11.1 assumes starting frequencies of 0.6 and 0.4. You may choose any starting frequencies.*

Working in pairs, you will test Hardy-Weinberg equilibrium by simulating a population using colored beads. The bag of beads represents the gene pool for the population. Each bead should be regarded as a single gamete, the two colors representing different alleles of a single gene. Each bag should contain 100 beads of the two colors in the proportions specified by the instructor. Record in the spaces provided below the color of the beads and the initial frequencies for your gene pool.

$A =$ _____ color _____ allelic frequency

$a =$ _____ color _____ allelic frequency

1. How many diploid individuals are represented in this population?

   *50*

2. What would be the color of the beads for a homozygous dominant individual?
3. What would be the color of the beads for a homozygous recessive individual?
4. What would be the color of the beads for a heterozygous individual?

## Hypothesis

State the Hardy-Weinberg theorem in the space provided. This will be your hypothesis.

*In a large, randomly mating population with no mutation, migration, or selection, the allelic and genotypic frequencies should remain at equilibrium.*

## Predictions

Predict the genotypic frequencies of the population in future generations (if/then).

*If the population is at Hardy-Weinberg equilibrium, then the frequencies of the beads should not change.*

## Procedure

1. Without looking, randomly remove two beads from the bag. These two beads represent one diploid individual in the next generation. Record in the margin of your lab manual the diploid genotype (*AA*, *Aa*, or *aa*) of the individual formed from these two gametes.

2. *Return the beads to the bag* and shake the bag to reinstate the gene pool. By replacing the beads each time, the size of the gene pool remains constant, and the probability of selecting any allele should remain equal to its frequency. This procedure is called **sampling with replacement.**

3. Repeat steps 1 and 2 (select two beads, record the genotype of the new individual, and return the beads to the bag) until you have recorded the genotypes for 50 individuals who will form the next generation of the population.

## Results

1. *Before calculating the results of your experiment*, determine the *expected* frequencies of genotypes and alleles for the population. To do this, use the original allelic frequencies for the population provided by the instructor. (Recall that the frequency of $A = p$, and the frequency of $a = q$.) Calculate the expected genotypic frequencies using the Hardy-Weinberg equation $p^2 + 2pq + q^2 = 1$. The number of individuals expected for each genotype can be calculated by multiplying 50 (total population size) by the expected frequencies. Record these results in Table 11.1.

*Remind students to first calculate the expected frequencies in Table 11.1.*

**Table 11.1**

*Expected* Genotypic and Allelic Frequencies for the Next Generation Produced by the Bead Model

| Parent Populations | | New Populations | | | | |
|---|---|---|---|---|---|---|
| Allelic Frequency | | Genotypic Number (and Frequency) | | | Allelic Frequency | |
| *A* | *a* | *AA* | *Aa* | *aa* | *A* | *a* |
| 0.6 | 0.4 | 18 (0.36) | 24 (0.48) | 8 (0.16) | 0.6 | 0.4 |

2. Next, *using the results of your experiment*, calculate the *observed* frequencies in the new population created as you removed beads from the bag. Record the number of diploid individuals for each genotype in Table 11.2, and calculate the frequencies for the three genotypes (*AA, Aa, aa*). Add the

*Examples of typical results. Do not expect these numbers to fit Hardy-Weinberg equilibrium exactly. This is the experiment. Although in our example the square root of AA ($\sqrt{0.44} = 0.66$) is approximately the same as the frequency of A (0.66), this may not be the case in the students' experiments.*

numbers of each allele, and calculate the allelic frequencies for *A* and *a*. These values are the observed frequencies in the new population. Genotypic frequencies and allelic frequencies should each equal 1.

**Table 11.2**

*Observed* Genotypic and Allelic Frequencies for the Next Generation Produced by the Bead Model

| Parent Populations | | New Populations | | | | |
|---|---|---|---|---|---|---|
| Allelic Frequency | | Genotypic Number (and Frequency) | | | Allelic Frequency | |
| *A* | *a* | *AA* | *Aa* | *aa* | *A* | *a* |
| *0.6* | *0.4* | *22 (0.44)* | *22 (0.44)* | *6 (0.12)* | *0.66* | *0.34* |

3. To compare your observed results with those expected, you can use the statistical test, chi-square. Table 11.3 will assist in the calculation of the chi-square test. *Note: To calculate chi-square you must use the actual number of individuals for each genotype, not the frequencies.* See Appendix B for an explanation of this statistical test.

*If you are following the sequence of lab topics in this manual, your students will already be familiar with the chi-square test from Lab Topic 8. If this is their first time using this test, be prepared to help students understand its significance. Refer to Appendix B as you give a brief explanation.*

**Table 11.3**

Chi-Square of Results from the Bead Model

|  | *AA* | *Aa* | *aa* |
|---|---|---|---|
| **Observed value (*o*)** | 22 | 22 | 6 |
| **Expected value (*e*)** | 18 | 24 | 8 |
| **Deviation (*o + e*) = *d*** | 4 | 2 | 2 |
| *d²* | 16 | 4 | 4 |
| *d²/e* | 0.89 | 0.16 | 0.5 |
| **Chi-square ($\chi^2$) = $\Sigma d^2/e$** | | 1.55 | |

Degrees of freedom = 2

Level of significance, $p < 0.05$

4. Is your calculated $\chi^2$ value greater or smaller than the given $\chi^2$ value (Appendix B, Table B.2) for the degrees of freedom and *p* value for this problem?

*smaller*

## Discussion

1.  What proportion of the population was homozygous dominant?

    Homozygous recessive?

    Heterozygous?

2.  Were your observed results consistent with the expected results based on your statistical analysis? If not, can you suggest an explanation?

    *Yes. The data support the hypothesis. Your observed distribution is not significantly different from that expected under the Hardy-Weinberg theorem. Any minor differences can be attributed to chance or sampling error.*

3.  Compare your results with those of other students. How variable are the results for each team?

4.  Do your results match your predictions for a population at Hardy-Weinberg equilibrium?

    What would you expect to happen to the frequencies if you continued this simulation for 25 generations?

    *nothing*

    Is this population evolving?

    *no*

    Explain your response.

    *There are no changes in genotypic frequencies after one generation of random mating. The population is at genetic equilibrium.*

5.  Consider each of the conditions for the Hardy-Weinberg model. Does this model meet each of those conditions?

    *Yes. Mating (selection of beads) is random; the population is large (100 beads); and there is no migration, no selection (all reproduce and survive), and no mutation.*

# EXERCISE 11.2
# Simulation of Evolutionary Change Using the Bead Model

Under the conditions specified by the Hardy-Weinberg model (random mating in a large population, no mutation, no migration, and no selection), the genetic frequencies should not change, and evolution should not occur. In this exercise, the class will modify each of the conditions and determine the effect on genetic frequencies in subsequent generations. You will simulate the evolutionary changes that occur when these conditions are not met.

*Stress that students will be simulating evolution as each scenario changes one of the conditions for Hardy-Weinberg equilibrium.*

Working in teams of two or three students, you will simulate *two of the* experimental scenarios presented and, using the bead model, determine the changes in genetic frequencies over several generations. The scenarios include the migration of individuals between two populations, also called **gene flow**; the effects of small population size, called **genetic drift**; and examples of **natural selection**. The effects of mutation take longer to simulate with the bead model, so you will use computer simulation to consider these in Exercise 11.3. *All teams will begin by simulating the effect of genetic drift, specifically, the bottleneck effect* (Experiment A.1). *For the second simulation, you can choose to investigate migration, one of two examples of natural selection, or the founder effect— another example of genetic drift.*

*If you are omitting Exercise 11.3 on computer simulations, have students simulate three scenarios.*

The procedure for investigating each of the conditions will follow the general procedures described as follows. Before beginning one of the simulation experiments, be sure you understand the procedures to be used.

## Procedure

### 1. Sampling with Replacement

Unless otherwise instructed, the gene pool size will be 100 beads. Each new generation will be formed by randomly choosing 50 diploid individuals represented by pairs of beads. After removing each pair of beads (representing the genotype of one individual) from the bag, replace the pair before removing the next set, *sampling with replacement*. Continue your simulations for several generations. For example, if the starting population has 50 beads each of *A* and *a* (allelic frequency of 0.5), then in the next generation you might produce the following results:

Number of individuals: 14*AA*, 24*Aa*, 12*aa*

Number of alleles (beads): 28*A* + 24*A*, 24*a* + 24*a*

Total number of alleles: 100

$$\text{Frequency of } A: \ 28 + 24 = \frac{52}{100} = 0.52$$

$$\text{Frequency of } a: \ 24 + 24 = \frac{48}{100} = 0.48$$

In this example, the frequency should continue to approximate 0.5 for *A* and *a*.

## 2. Reestablishing a Population with New Allelic Frequencies

In some cases, the number of individuals will decrease as a result of the simulation. In those cases, return the population to 100 but reestablish the population with new allelic frequencies. For example, if you eliminate by selection all homozygous recessive (*aa*) individuals in your simulation, then the resulting frequencies would be:

Number of individuals: 14*AA*, 24*Aa*, 0*aa*

Number of alleles (beads): 28*A* + 24*A*, 24*a*

Total number of alleles: 76

$$\text{Frequency of A: } 28 + 24 = \frac{52}{76} = 0.68$$

$$\text{Frequency of a: } 24 = \frac{24}{76} = 0.32$$

To reestablish a population of 100, then, the number of beads should reflect these new frequencies. *Adjust the number of beads* so that *A* is now 68/100 and *a* is 32/100. Then continue the next round of the simulation.

 If this information is not clear to you, ask for assistance before beginning your simulations.

# Experiment A. Simulation of Genetic Drift

## Materials

plastic or paper bag containing 100 beads, 50 each of two colors
additional beads as needed

## Introduction

Genetic drift is the change in allelic frequencies in small populations as a result of chance alone. In a small population, combinations of gametes may not be random, owing to sampling error. (If you toss a coin 500 times, you expect about a 50:50 ratio of heads to tails; but if you toss the coin only 10 times, the ratio may deviate greatly in a small sample owing to chance alone.) **Genetic fixation,** the loss of all but one possible allele at a gene locus in a population, is a common result of genetic drift in small natural populations. Genetic drift is a significant evolutionary force in situations known as the **bottleneck effect** and the **founder effect.** *All teams will investigate the bottleneck effect. You may choose the founder effect (p. 275) for your second simulation.*

## 1. Bottleneck effect

A **bottleneck** occurs when a population undergoes a drastic reduction in size as a result of chance events (not differential selection), such as a volcanic eruption or hurricane. For example, northern elephant seals experienced a bottleneck when they were hunted almost to extinction with only 20 surviving animals in the 1890s. In Figure 11.3, the beads pass through a bottleneck, which results in an unpredictable combination of beads that pass to the other side. These beads would constitute the beginning of the next generation.

**Figure 11.3.**

**The bottleneck effect.** The gene pool can drift by chance when the population is drastically reduced by factors that act unselectively. Bad luck, not bad genes! The resulting population will have unpredictable combinations of genes. What has happened to the amount of variation?

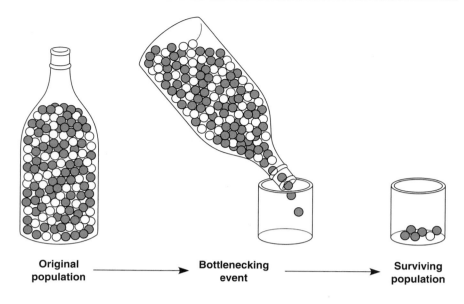

| Original population | → | Bottlenecking event | → | Surviving population |

## Hypothesis

As your hypothesis, either propose a hypothesis that addresses the bottleneck effect specifically or state the Hardy-Weinberg theorem.

*In a large, randomly mating population with no mutation, migration, or selection, the allelic and genotypic frequencies should remain at equilibrium.*

## Prediction

Either predict equilibrium values as a result of Hardy-Weinberg or predict the type of change that you expect to occur in a small population (if/then).

*If the population is at Hardy-Weinberg equilibrium, then the frequencies of the beads should not change.*

*or*

*If a population at Hardy-Weinberg equilibrium is reduced to a very small number, then the genetic frequencies should deviate from equilibrium and may eventually go to fixation for one allele. Most student predictions will not be this specific.*

## Procedure

1. To investigate the bottleneck effect, establish a starting population containing 50 individuals (how many beads?) with a frequency of 0.5 for each allele (Generation 0).

2. *Without replacement*, randomly select 5 individuals (10% of the population), two alleles at a time. This represents a drastic reduction in population size. On a separate sheet of paper, record the genotypes and the number of $A$ and $a$ alleles for the new population.

3. Count the numbers of each genotype and the numbers of each allele. Using these numbers, determine the genotypic frequencies for $AA$, $Aa$, and $aa$ and the new allelic frequencies for $A$ ($p$) and $a$ ($q$) for the surviving 5 individuals. These are your *observed* frequencies. Enter these frequencies in Table 11.4, Generation 1.

*Students will use Table 11.4 to visualize the pattern of changes in allelic and genotypic frequencies in each successive generation.*

**Table 11.4**

Changes in Allelic and Genotypic Frequencies for Simulations of the Bottleneck Effect, an Example of Genetic Drift. First, record frequencies based on the observed numbers in your experiment. Then, using the observed allelic frequencies, calculate the expected genotypic frequencies.

| Generation | Genotypic Frequency Observed | | | Allelic Frequency Observed | | Genotypic Frequency Expected | | |
|---|---|---|---|---|---|---|---|---|
| | *AA* | *Aa* | *aa* | *A (p)* | *a (q)* | $p^2$ | *2pq* | $q^2$ |
| 0 | — | — | — | 0.5 | 0.5 | 0.25 | 0.50 | 0.25 |
| 1 | | | | | | | | |
| 2 | | | | | | | | |
| 3 | | | | | | | | |
| 4 | | | | | | | | |
| 5 | | | | | | | | |
| 6 | | | | | | | | |
| 7 | | | | | | | | |
| 8 | | | | | | | | |
| 9 | | | | | | | | |
| 10 | | | | | | | | |

4. Using the new observed allelic frequencies, calculate the *expected* genotypic frequencies ($p^2$, *2pq*, $q^2$). Record these frequencies in Table 11.4, Generation 1.

5. Reestablish the population to 50 individuals using the new allelic frequencies (refer to the example in the Procedure on p. 271). Repeat steps 2, 3, and 4. Record your results in the appropriate generation in Table 11.4.

6. Reestablish the gene pool with new frequencies after each generation until one of the alleles becomes fixed in the population for several generations.

7. Summarize your results in the Discussion section.

## Results

1. How many generations did you simulate?

2. Using the graph paper at the end of the lab topic, sketch a graph of the change in *p* and *q* over time. You should have two lines, one for each allele.

3. Did one allele go to fixation in that time period? Which allele?

   Remember, genetic fixation occurs when the gene pool is composed of only one allele. The others have been eliminated. Did the other allele ever appear to be going to fixation?

   *One allele should become fixed and the other eliminated. Sometimes one allele and then the other may appear to be headed toward fixation. This should be variable.*

*The expected genotypic frequencies in Table 11.4 will vary as the allelic frequencies change under conditions of genetic drift. Students will see interesting changes in expected genotypic frequencies in the additional simulations.*

*If time permits, have every lab group graph their results on an overhead acetate.*

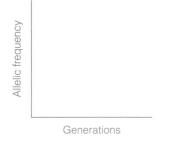

4. Did any of the expected genotypic frequencies go to fixation? If none did, why not?

   *Genotypic frequencies go to fixation only when one allele goes to fixation. Students should see the wide range of variation in genotypes over the course of the experiment.*

5. Compare results with other teams. Did the same allele go to fixation for all teams? If not, how many became fixed for *A* and how many for *a*?

   *The class results represent replicates for simulations of genetic drift. Each allele has a 50% probability of going to fixation. The probability of fixation is equal to the starting allelic frequencies. The results may suggest this conclusion.*

## Discussion

1. Compare the pattern of change for *p* and *q*. Is there a consistent trend or do the changes suggest chance events? Look at the graphs of other teams before deciding.

   *There should not be a consistent pattern.*

2. Explain your observations of genetic fixation for the replicate simulations completed by the class. What would you expect if you simulated the bottleneck effect 100 times?

   *You would expect a close approximation of 50% fixation for each of the alleles, since the starting allelic frequencies were 0.5.*

3. How might your results have differed if you had started with different allelic frequencies, for example, *p* = 0.2 and *q* = 0.8?

   *q would be more likely to go to fixation, since it is already represented by 80% of the alleles. Again, the probability of fixation is equal to the starting frequency.*

4. Since only chance events—that is, the effect of small population size—are responsible for the change in gene frequencies, would you say that evolution has occurred? Explain.

   *Yes. Evolution is the change in gene frequencies over time. Natural selection is important but not the only evolutionary factor. Chance events can have a profound effect on the evolution of populations.*

5. Since the 1890s when northern elephant seals were hunted almost to extinction, the populations have increased to 30,000 animals. In a 1974 study of genetic variation, scientists found that all elephant seals are fixed for the same allele at 24 loci. Explain.

   *The population experienced a bottleneck, and the extreme reduction in the gene pool resulted in some alleles becoming fixed in the population as a result of genetic drift.*

On completion of this simulation, choose one or two of the remaining scenarios to investigate. All scenarios should be completed by at least one team in the laboratory.

## 2. *Founder effect*

When a small group of individuals becomes separated from the larger parent population, the allelic frequencies in this small gene pool may be different from those of the original population as a result of chance alone. This occurs when a group of migrants becomes established in a new area not currently inhabited by the species—for instance, the colonization of an island—and is therefore referred to as the **founder effect.** For example, in 1975, 20 desert bighorn sheep colonized Tiburon Island; by 1999, the population had increased to 650 sheep. A comparison with the other populations of the species in Arizona showed that the island populations had fewer alleles per gene and overall lower genetic variation due to the small size of the colonizing population and inbreeding.

### Hypothesis

As your hypothesis, either propose a hypothesis that addresses founder effect specifically or state the Hardy-Weinberg theorem.

> *In a large, randomly mating population with no mutation, migration, or selection, the allelic and genotypic frequencies should remain at equilibrium.*

### Prediction

Either predict equilibrium values as a result of Hardy-Weinberg or predict the type of change that you expect to occur in a small population (if/then).

> *If the population is at Hardy-Weinberg equilibrium, then the frequencies of the beads should not change.*
>
> *or*
>
> *If a population at Hardy-Weinberg equilibrium is reduced to a very small number, then the genetic frequencies should deviate from equilibrium and may eventually go to fixation for one allele. Most student predictions will not be this specific.*

### Procedure

1. To investigate the founder effect, establish a starting population with 50 individuals with starting allelic frequencies of your choice. Record the frequencies you have selected for Generation 0 in Table 11.5 in the Results section at the end of this exercise.

2. *Without replacement*, randomly select 5 individuals, two alleles at a time, to establish a new founder population. On a separate sheet of paper, record the genotypes and the number of *A* and *a* alleles for the new population.

3. Calculate the new frequencies for *A* ($p$) and *a* ($q$) and the genotypic frequencies for *AA* ($p^2$), *Aa* ($2pq$), and *aa* ($q^2$) for the *founder population*, and record this information as Generation 1 in Table 11.5.

4. Reestablish the population to 50 diploid individuals using the new allelic frequencies. (Refer to the example in the Procedure on p. 271.)

*In the simulation of the bottleneck effect, the population is reduced in size by 90% in each generation. In the simulation of founder effect, the population is only initially reduced and then returns to the starting population size. Each additional generation is sampled with replacement.*

5. Follow the founder population through several generations in the new population. From this point forward, each new generation will be produced by *sampling 50 individuals with replacement*. After each generation, reestablish the new population based on the new allelic frequencies. Continue until you have sufficient evidence to discuss your results with the class.

6. Summarize your results in the Discussion section at the end of this exercise. You will want to compare the founder population with the original population and compare equilibrium frequencies if appropriate. Each group will present its results to the class at the end of the laboratory.

## Experiment B. Simulation of Migration: Gene Flow

### Materials

2 plastic or paper bags, each containing 100 beads of two colors
additional beads as needed

### Introduction

The migration of individuals between populations results in **gene flow.** In a natural population, gene flow can be the result of the immigration and emigration of individuals or gametes (for example, pollen movement). The rate and direction of migration and the starting allelic frequencies for the two populations can affect the rate of genetic change. In the following experiment, the migration rate is equal in the two populations, and the starting allelic frequencies differ for the two. Work in teams of four students.

### Hypothesis

As your hypothesis, either propose a hypothesis that addresses migration specifically or state the Hardy-Weinberg theorem.

> *In a large, randomly mating population with no mutation, migration, or selection, the allelic and genotypic frequencies should remain at equilibrium.*

### Prediction

Either predict equilibrium values as a result of Hardy-Weinberg or predict the type of change that you expect to observe as a result of migration (if/then).

> *If migration occurs between two populations that otherwise meet the conditions for Hardy-Weinberg equilibrium, then the allelic frequencies of the two populations should reach a new equilibrium intermediate between the starting frequencies.*
> *or*
> *If the population is at Hardy-Weinberg equilibrium, then the frequencies of the beads should not change.*

### Procedure

1. To investigate the effects of gene flow on population genetics, establish two populations of 100 beads each. Choose different starting allelic frequencies for *A* and *a* for each population. For example, in Figure 11.4,

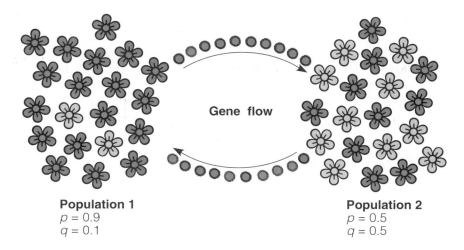

**Population 1**
$p = 0.9$
$q = 0.1$

**Population 2**
$p = 0.5$
$q = 0.5$

Gene flow

**Figure 11.4.**
**Migration.** Migration rates are constant between two populations that differ for starting allelic frequencies.

population 1 might start with 90*A* and 10*a* while population 2 might have 50 of each allele initially. For comparison, you might choose one population with the same starting frequencies as your population in Exercise 11.1. Record the allelic frequencies for each starting population, Generation 0, in Table 11.5 in the Results section at the end of this exercise.

*Four students are a team. One pair records results for one population; the other pair, the second population.*

2. Select 10 individuals (how many alleles?) from each population, and allow them to migrate to the other population by exchanging beads. Do not sample with replacement in this step.

3. Select 50 individuals from each of these new populations, *sampling with replacement*. On a separate sheet of paper, record the genotypes and the number of *A* and *a* alleles for population 1 and population 2.

4. Calculate the new allelic and genotypic frequencies in the two populations following migration. Record your results in Table 11.5 in the Results section. Reestablish each bag based on the new allelic frequencies.

5. Repeat this procedure (steps 2 to 4) for several generations. In doing this, you are allowing migration to take place with each generation. Continue until you have a sufficient number to allow you to discuss your results in class successfully.

6. Summarize your results in the Discussion section at the end of this exercise. Each group will present its results to the class at the end of the laboratory.

## Experiment C. Simulation of Natural Selection

### Materials

plastic or paper bag containing 100 beads of two colors additional beads as needed

### Introduction

**Natural selection,** the differential survival and reproduction of individuals, was first proposed by Darwin as the mechanism for evolution. Although other factors have since been found to be involved in evolution, selection is still considered an important mechanism. Natural selection is based on the observation that individuals with certain heritable traits are more likely to survive and reproduce than those lacking these advantageous traits. Therefore, the proportion of offspring with advantageous traits will increase in the next generation. The genotypic frequencies will change in the population. Whether traits are advantageous in a population depends on the environment and the selective agents (which can include physical and biological factors). Choose one of the following evolutionary scenarios to model natural selection in population genetics.

### 1. Industrial melanism

The peppered moth, *Biston betularia*, is a speckled moth that sometimes rests on tree trunks during the day, where it avoids predation by blending with the bark of trees (an example of cryptic coloration). At the beginning of the 19th century, moth collectors in Great Britain collected almost entirely light forms of this moth (light with dark speckles) and only occasionally recorded rare dark forms (Figure 11.5). With the advent of the Industrial Revolution and increased pollution the tree bark darkened, lichens died, and the frequency of the dark moth increased dramatically in industrialized areas. However, in rural areas, the light moth continued to occur in high frequencies. (This is an example of the relative nature of selective advantage, depending on the environment.)

Color is controlled by a single gene with two allelic forms, dark and light. Pigment production is dominant ($A$), and the lack of pigment is recessive ($a$). The light moth would be $aa$, but the dark form could be either $AA$ or $Aa$.

### Hypothesis

As your hypothesis, either propose a hypothesis that addresses natural selection occurring in polluted environments specifically or state the Hardy-Weinberg theorem.

*In a large, randomly mating population with no mutation, migration, or selection, the allelic and genotypic frequencies should remain at equilibrium.*

a.                                          b.

**Figure 11.5.**
**Two color forms of the peppered moth.** The dark and light forms of the moth are present in both photographs. Lichens are absent in (a) but present in (b). In which situation would the dark moth have a selective advantage?

## Prediction

Either predict equilibrium values as a result of the Hardy-Weinberg theorem or predict the type of change that you expect to observe as a result of natural selection in the polluted environment (if/then).

> *If the population is at Hardy-Weinberg equilibrium, then the frequencies of the beads should not change.*
>
> *or*
>
> *If natural selection occurs in a population that otherwise meets the conditions for Hardy-Weinberg equilibrium, then the allelic and genotypic frequencies will change over time. The allelic frequencies of A and the genotypic frequencies of AA and Aa should increase in a polluted environment. Student predictions may or may not include changes in frequencies.*

## Procedure

1. To investigate the effect of natural selection on the frequency of light and dark moths, establish a population of 50 individuals with allelic frequencies of light (*a*) = 0.9 and dark (*A*) = 0.1. Record the frequencies of the starting population, Generation 0, in Table 11.5 in the Results section at the end of this exercise.

2. Determine the genotypes and phenotypes of the population by selecting 50 individuals, two alleles at a time, by sampling with replacement (see Procedure, p. 270). On a separate sheet of paper, record the genotypes and phenotypes of each individual.

   *This step requires that students determine phenotypes, an important step because selection acts on the phenotype for this trait.*

   Now, assume that pollution has become a significant factor and that in this new population 50% of the light moths but only 10% of the dark moths are eaten. How many light moths must you eliminate from your starting population of 50 individuals? How many dark moths? Can selection distinguish dark moths with *AA* and *Aa* genotypes? How will you decide which dark moths to remove? Remove the appropriate number of individuals of each phenotype.

   *Selection cannot distinguish between the genotypes for this trait. Students can choose to deal with this problem in several different ways; for example, they can remove every other dark moth.*

3. Calculate new allelic frequencies for the remaining population. Reestablish the population with these new allelic frequencies for the 100 alleles. (Refer to the Procedure on p. 271.) Record the new frequencies in Table 11.5 in the Results section.

4. Continue the selection procedure, recording the frequencies, for several generations until you have sufficient evidence to discuss your results with the class.

5. Summarize your results in the Discussion section at the end of this exercise. Each group will present its results to the class at the end of the laboratory.

### 2. Sickle-cell disease

Sickle-cell disease is caused by a mutant allele ($Hb^-$), which, in the homozygous condition, in the past was often fatal to people at quite young ages. The mutation causes the formation of abnormal, sickle-shaped red blood cells that clog vessels, cause organ damage, and are inefficient transporters of oxygen (Figure 11.6). Individuals who are heterozygous ($Hb^+Hb^-$) have the sickle-cell trait (a mild form of the disease), which is not fatal. Scientists were surprised to find a high frequency of the $Hb^-$ allele in populations in Africa until they determined that heterozygous individuals

*Choose your best students for this challenging scenario.*

have a *selective advantage* in resisting malaria. Although the homozygous condition may be lethal, the heterozygotes are under both a positive and a negative selection force. In malarial countries in tropical Africa, the heterozygotes are at an advantage compared to either homozygote.

**Figure 11.6.**
**Effects of the sickle-cell allele.**
(a) Normal red blood cells are disk-shaped. (b) The jagged shape of sickled cells causes them to pile up and clog small blood vessels. (c) The results include damage to a large number of organs. In the homozygous condition, sickle-cell disease can be fatal.

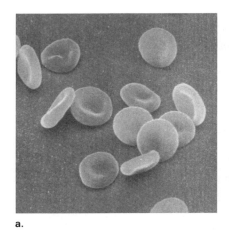

a.

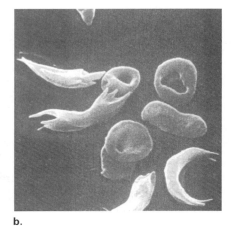

b.

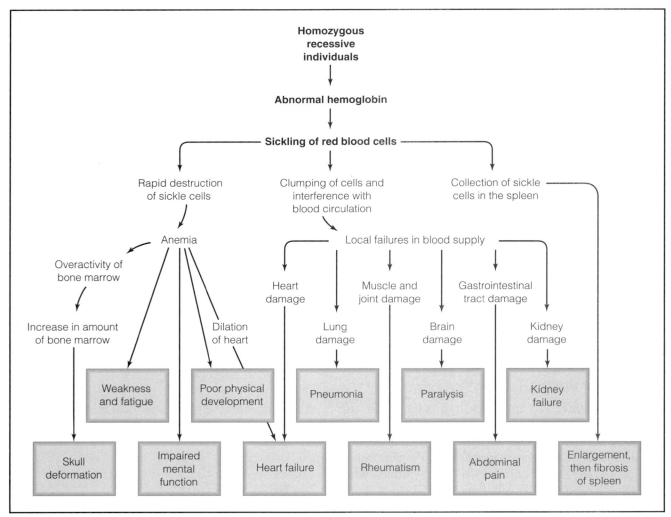

c.

## Hypothesis

As your hypothesis, either propose a hypothesis that addresses natural selection in tropical Africa specifically or state the Hardy-Weinberg theorem.

*In a large, randomly mating population with no mutation, migration, or selection, the allelic and genotypic frequencies should remain at equilibrium.*

## Prediction

Either predict equilibrium values as a result of the Hardy-Weinberg theorem or predict the type of change that you expect to observe as a result of natural selection in tropical Africa (if/then).

*If the population is at Hardy-Weinberg equilibrium, then the frequencies of the beads should not change.*

*or*

*If natural selection occurs in a population that otherwise meets the conditions for Hardy-Weinberg equilibrium, then the allelic and genotypic frequencies will change over time. The allelic frequency of $Hb^+$ and the genotypic frequencies of $Hb^+Hb^+$ and $Hb^+Hb^-$ should increase. Student predictions may or may not include changes in frequencies.*

## Procedure

1. To investigate the role of selection under conditions of heterozygote advantage, establish a population of 50 individuals with allelic frequencies of 0.5 for both alleles. Record the frequencies of the starting population, Generation 0, in Table 11.5 in the Results section at the end of this exercise.

   For our model, assume that the selection force on each genotype is

   > $Hb^+Hb^+$: 30% die of malaria
   > $Hb^+Hb^-$: 10% die of malaria
   > $Hb^-Hb^-$: 100% die of sickle-cell disease

2. Determine the genotypes and phenotypes of the population by selecting 50 individuals, two alleles at a time, by *sampling with replacement* (see Procedure, p. 270). On a separate sheet of paper, record the 50 individuals. Select against each genotype by eliminating individuals according to the percent mortality shown above. How many $Hb^+Hb^+$ must be removed? How many $Hb^+Hb^-$? How many $Hb^-Hb^-$? Remove the appropriate number of beads for each color.

3. Calculate new allelic and genotypic frequencies based on the survivors, and reestablish the population using these new frequencies. (Refer to the procedure on p. 271 if necessary.) Record your new frequencies in Table 11.5 in the Results section.

4. Continue the selection procedure, recording frequencies, for several generations, until you have sufficient evidence for your discussion in class.

5. Summarize your results in the Discussion section. Each group will present its results to the class at the end of the laboratory.

# Results and Discussion for Selected Simulation

### Results

Each team will record the results of its chosen experiment in Table 11.5. Modify the table to match the information you need to record for your simulation. Use the margins to expand the table or make additional notes.

### Discussion

In preparation for presenting your results to the class, answer the following questions about your chosen simulation. Choose one team member to make the presentation, which should be organized in the format of a scientific paper.

1. Which of the conditions that are necessary for Hardy-Weinberg equilibrium were met?

**Table 11.5**
Changes in Allelic and Genotypic Frequencies for Simulations Selected in Exercise 11.2. Calculate expected frequencies based on the actual observed numbers in your experiment.

| Experiment _____ Simulation of _____ | | | | |
|---|---|---|---|---|
| | **Allelic Frequency** | | **Genotypic Frequency** | | |
| **Generation** | $p$ | $q$ | $p^2$ | $2pq$ | $q^2$ |
| 0 | | | | | |
| 1 | | | | | |
| 2 | | | | | |
| 3 | | | | | |
| 4 | | | | | |
| 5 | | | | | |
| 6 | | | | | |
| 7 | | | | | |
| 8 | | | | | |
| 9 | | | | | |
| 10 | | | | | |
| | | | | | |
| | | | | | |
| | | | | | |

2. Which condition was changed?

3. Briefly describe the scenario that your team simulated.

4. What were your predicted results?

5. How many generations did you simulate?

*The expected genotypic frequencies in Table 11.5 will vary as the allelic frequencies change under conditions of genetic drift and migration. However, in simulations of selection, the genotypic frequency of the heterozygote will increase for the sickle-cell trait. By comparison, in the simulation of industrial melanism, students will note the continued appearance of the heterozygote and the recessive homozygote in spite of strong selection against the recessive trait.*

6. Sketch a graph of the change in $p$ and $q$ over time. You should have two lines, one for each frequency.

7. Describe the changes in allelic frequencies $p$ and $q$ over time. Did your results match your predictions? Explain.

8. Describe the changes in the genotypic frequencies.

9. Compare your final allelic and genotypic frequencies with those of the starting population.

10. If you can, formulate a general summary statement or conclusion.

11. Would you expect your results to be the same if you had chosen different starting allelic frequencies? Explain.

12. Critique your experimental design and outline your next simulation.

Be prepared to take notes, sketch graphs, and ask questions during the student presentations. You are responsible for understanding and answering questions for all conditions simulated in the laboratory.

---

## EXERCISE 11.3
# Open-Inquiry Computer Simulation of Evolution

*See the References section for software suggestions. We have used EVOLVE from BioQUEST, and an online version is available at the BioQUEST website (www.bioquest.org).*

*If you are not using computer simulations, then have students complete three or more scenarios.*

*Consider having all students do one simulation on the computer using mutation, then assigning a second problem, which they can choose. Either assign problems or approve computer problems in advance to ensure that a variety of conditions are tested. Students often will be too ambitious, use too many variables, or otherwise complicate the simulations.*

Simulations involving only a few generations are fairly easy with the bead model, but this model is too cumbersome to obtain information about long-term changes or to combine two or more factors. Computer simulations will allow you to model 50 or 100 generations quickly or to do a series of simulations changing one factor and comparing the results. As with any scientific investigation, design your experiment to test a hypothesis and make predictions before you begin.

Several computer simulation programs are available; the instructor will demonstrate the software. From the following suggestions or your own ideas, pursue one of the factors in more detail or several in combination. Consult with your instructor before beginning your simulations. Be prepared to present your results in the next lab period in either oral or written form. Follow the procedures provided with the simulation program.

### Mutation

Evolution can occur only when there is variation in a population, and the ultimate source for that variation is mutation. The mutation rate at most gene loci is actually very small ($1 \times 10^{-5}$ per gamete per generation), and the forward and backward rates of mutation are seldom at equilibrium. Evolutionary change as a result of mutation alone would occur very slowly, thus requiring many generations of simulation. Changes in this condition are best considered using a computer simulation that easily handles the lengthy process. *Do not simulate this condition using the bead model.*

## Other Suggestions for Computer Simulations

1.  The migration between natural populations seldom occurs at the same rate in both directions. Devise a model to simulate different migration rates between two populations. Compare your results with the bead model.

2.  How does the probability of genetic fixation differ for populations that have different starting allelic frequencies? Simulate genetic drift using a variety of allelic frequencies.

3.  How large must a population be to avoid genetic drift?

4.  Devise a model to simulate selection for recessive and lethal traits. Then, using the same starting allelic frequencies, select against the dominant lethals. Compare rate of change of allelic frequency as a result of lethality for dominant and recessive traits.

5.  Mutation alone has only small effects on evolution over long periods of time. Using realistic rates of mutation, compare the evolutionary change for mutation alone with simulations that combine mutation and natural selection.

6.  Combine factors (usually two at a time). For example, combine mutation and natural selection, using realistic estimates of mutation rates. Combine genetic drift and migration. Compare your results to simulations of each condition alone.

# Reviewing Your Knowledge

1.  Define and provide examples of the following terms: *evolution, population, gene pool, gene flow, genetic drift, bottleneck effect, founder effect, natural selection, genetic fixation, genotypic frequency, allelic frequency*, and *model*.

2.  State the five conditions necessary for Hardy-Weinberg equilibrium.

    *a large, randomly mating population, with no mutation, migration, or selection*

3.  Describe how gene flow, genetic drift, and natural selection bring about evolution.

    *In each of these situations, the frequencies of alleles in the original population are changed after the event. In gene flow, individuals move in or out of the population. In genetic drift, an event randomly and drastically reduces the numbers of individuals, changing allelic frequencies. In natural selection, alleles are eliminated because of the survivability of the genotype, again changing allelic frequencies.*

4.  In a population of 100 rock pink plants, 84 individuals have red flowers while 16 have white flowers. Assume that white petals are inherited as a recessive trait ($a$) and red petals as a dominant trait ($A$). What are the frequencies $p$ and $q$? In the next generation, what will be the equilibrium genotypic frequencies?

    *$p = 0.6$, $q = 0.4$, $AA = 0.36$, $Aa = 0.48$, $aa = 0.16$*

5. Migration occurs at a constant rate between two populations of field mice. In one population, 65% of the population are white; in the other population, only 15% are white. What would you expect to happen to the allelic frequencies of these two populations over time?

   *The frequencies should change gradually over time until the two populations reach a new equilibrium that is the same for both populations.*

6. Explain the difference between evolution and natural selection. Use an example to illustrate your answer.

   *Evolution is the change in the gene pool of a population over time. One mechanism responsible for evolution is natural selection, which is the differential survival and reproduction of individuals with advantageous traits. The result in the next generation is a change in the gene pool. An example could be the peppered moth.*

# Applying Your Knowledge

1. Pingelap is a tiny island in the Pacific Ocean that is part of the Federated States of Micronesia. In 1774, a typhoon devastated the island and of the 500 people who inhabited the island only 20 survived. One male survivor was heterozygous for a rare type of congenital colorblindness. After four generations a number of children were born with complete colorblindness and hypersensitivity to light due to defective cones (achromatopsia). The islands remained isolated for over 150 years. Now there are about 3,000 Pingelap descendants, including those on a nearby island and a small town on a third island. The frequency of individuals who are homozygous recessive and have the disease is 5% and another 30% are heterozygous carriers. This compares to the average of 1% carriers for the human population at large. Explain the conditions that are responsible for these differences in genetic frequencies for the Pingelap population and the general human population.

   *The Pingelap population experienced a bottleneck event with the chance survival of a male with the rare allele. Due to the small population size, isolation, and inbreeding (nonrandom mating), the frequency of the allele increased in the population.*

   Three daughters of the male survivor on Pingelap, who were carriers for colorblindness, moved to the nearby island of Mokil. This new island was even smaller and had fewer people. How would this event change the frequency of the colorblindness in the Mokil population? Is this an example of founder effect or bottleneck? Explain.

   *This is an example of founder effect and could result in the increased frequency of colorblindness due to the small population size, the resulting inbreeding, and the frequency of the alleles in the founders. Remember with genetic drift the alleles can become fixed or lost due to chance events.*

   A number of Pingelap citizens were transplanted to a small town on yet a third larger island, the capital of this island group. What factors do you think affected the frequency of colorblindness on this island? Would you expect the frequency of the colorblind allele to be the same on this island as it is on Pingelap? Explain.

*This is an example of migration. The Pingelap colonists are not isolated and would contribute their genes to a larger gene pool. The frequency of the colorblind allele would probably increase, and would be maintained in the population primarily in the heterozygotes or carriers. Inbreeding (nonrandom mating) would not be a factor.*

2. Cystic fibrosis (CF) is caused by a genetic mutation resulting in defective proteins in secretory cells, mainly in the epithelial lining of the respiratory tract. The one in every 2,000 Caucasian babies who has the disease is homozygous for the recessive mutant. Although medical treatment is becoming more effective, in the past most children with CF died before their teens. About 20 Caucasians in 2,000 are carriers of the trait, having one mutant and one normal allele, but they do not develop the disease. According to rules of population genetics, the frequency of the homozygous recessive genotype should be rarer than it is. What is one possible explanation for the unusually high frequency of this allele in Caucasian populations?

*Scientists hypothesize that this situation is similar to the sickle-cell disease scenario and that carriers of the trait have a selective advantage. There is evidence that carriers of the CF trait are protected from bronchial asthma. There is also evidence that carriers are protected from some of the adverse effects of typhoid fever. The defective CF protein reduces the attachment of the bacteria to the intestinal mucosa, thus providing resistance to typhoid fever.*

3. In the figure below, the frequency of dark forms of the peppered moth is shown for populations in northwestern England. Describe the changes in dark moth frequency from 1970 to 2000.

*In the early 1970s, the frequency remained about 90% but then began to decline slowly, then more rapidly, until the frequency was under 10%. It appears that the dark moth could become extinct.*

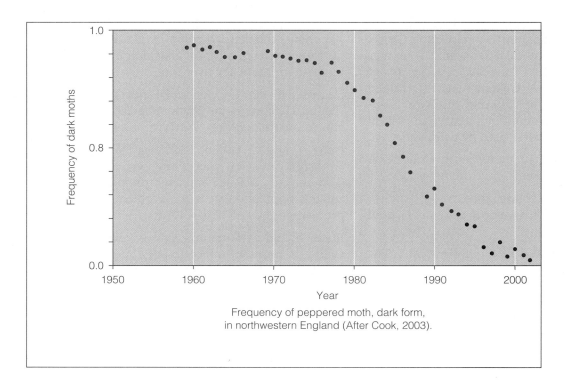

Frequency of peppered moth, dark form,
in northwestern England (After Cook, 2003).

Do these populations appear to be in Hardy-Weinberg equilibrium? Explain.

*If the populations were in Hardy-Weinberg equilibrium, then the allelic frequency would not change. Instead the frequency of the dark form (BB) is decreasing, therefore the frequency of B is decreasing. The populations are not in Hardy-Weinberg equilibrium.*

Using these results for the peppered moth, describe the roles of genetic variation and natural selection in evolutionary change.

*Evolutionary change cannot occur unless genetic variation exists in the population. Evolution depends on change in the genetic structure of the population from generation to generation (the frequency of dark moths is declining, the frequency of light moths is increasing), otherwise selection could simply eliminate the entire population of dark moths. Natural selection acts on the genetic variation. Natural selection is the differential survival and reproduction of individuals with advantageous traits.*

## Student Media: BioFlix, Activities, Investigations, and Videos
**www.masteringbiology.com (select Study Area)**

**BioFlix**—Ch. 23: Mechanisms of Evolution (Animation)
**Activities**—Ch. 23: Causes of Evolutionary Change
**Investigations**—Ch. 22: How Do Environmental Changes Affect a Population? What Are the Patterns of Antibiotic Resistance?
  Ch. 23: How Can Frequency of Alleles Be Calculated? Biology Labs OnLine: Population Genetics Lab: Background, Assignments 1, 2, 4, 5, 6, 7; Evolution Lab
**Videos**—Ch. 22: Discovery Videos: Charles Darwin

## References

Some parts of this lab topic were modified from J. C. Glase, 1993, "A Laboratory on Population Genetics and Evolution: A Physical Model and Computer Simulation," pages 29–41, in *Tested Studies for Laboratory Teaching*, Volume 7/8 (C. A. Goldman and P. L. Hauta, Editors). Proceedings of the 7th and 8th Workshop/Conference of the Association for Biology Laboratory Education (ABLE), 187 pages. Used by permission.

Ayala, F. J. *Population and Evolutionary Genetics: A Primer*. Menlo Park, CA: Benjamin Cummings, 1982.

Cook, L. M. "The Rise and Fall of the *Carbonaria* Form of the Peppered Moth." *The Quarterly Review of Biology*, 2003, vol. 78(4), pp. 399–417.

Futuyama, D. *Evolution*, 2nd ed. Sunderland, MA: Sinauer Associates, 2009.

Halliburton, R. *Introduction to Population Genetics*. San Francisco, CA: Benjamin Cummings, 2004.

Hardy, G. H. "Mendelian Proportions in a Mixed Population." *Science* (July 10, 1908) pp. 49–50. In this interesting early application of the "Hardy-Weinberg Theorem," this letter written by Hardy to the editor of *Science* refutes a report that dominant characters increase in frequency over time.

Hedrick, P. W., G. Gutierrez-Espeleta, and R. Lee. "Founder Effect in an Island Population of Bighorn Sheep." *Molecular Ecology*, 2001, vol. 10(4), pp. 851–857.

Heitz, J. and C. Giffen. *Practicing Biology: A Student Workbook*, 4th ed. San Francisco, CA: Benjamin Cummings, 2010. Chapter 23 provides a guide to solving population genetics problems and practice problems.

Price, F. and V. Vaughn. "Evolve," *BioQuest Library Online*. http://www.bioquest.org/BQLibrary/library_result.php (software available to download).

Sundin, O. H., J. Yang, Y. Li, D. Zhu, J. Hurd, T. Mitchell, E. Silva, and I. Maumenee. "Genetic Basis of Total Colourblindness Among the Pingelapese Islanders." *Nature Genetics*, 2000, vol. 25, pp. 289–293.

## Websites

Issues in Evolution, American Institutes of Biological Sciences, provides resources, interviews, and issues: http://www.actionbioscience.org/evolution/

Library of resources at PBS Evolution website: http://www.pbs.org/wgbh/evolution/library/index.html

Malaria movie showing sickle cell and malaria. Importance of mutation: http://www.pbs.org/wgbh/evolution/library/01/2/l_012_02.html

Resources for evolution and population genetics. Search for population genetics: http://www.pbs.org/wgbh/evolution/library/index.html

Simulation software and resources for population biology: http://www.cbs.umn.edu/populus/ http://dorakmt.tripod.com/evolution/popgen.html

Video of Endler's work on guppies: http://www.pbs.org/wgbh/evolution/sex/guppy/index.html

Video of toxic newt and garter snake—evolutionary arms race: http://www.pbs.org/wgbh/evolution/library/01/3/l_013_07.html

## LAB TOPIC 11

# Population Genetics I: The Hardy-Weinberg Theorem
## Teaching Plan for Laboratories

## Main Concepts and Objectives

1. Concept: allelic and genotypic frequencies. Students will understand and calculate allelic and genotypic frequencies. (Exercise 11.1)

2. Concept: Hardy-Weinberg equilibrium. Students use a bead model to test the Hardy-Weinberg theorem and its conditions. They will determine the relationship between genotypic and allelic frequencies and the general equation $(p + q)^2 = p^2 + 2pq + q^2 = 1$ when all conditions are met. (Exercise 11.1)

3. Concept: conditions for the Hardy-Weinberg theorem. Students use the bead model to determine the effects of changing the conditions specified in the Hardy-Weinberg theorem. They simulate evolutionary scenarios of gene flow (migration), small population size (genetic drift), and natural selection by determining changes in genotypic and allelic frequencies for several generations. (Exercise 11.2)

4. Concept: evolution in populations. Students continue to investigate the changes in gene frequencies (evolution) in populations in which the conditions for the Hardy-Weinberg theorem are not met. Students pursue additional investigations using computer simulations. (Exercise 11.2)

5. Concept: models. Students use the bead model. They also use the computer to simulate evolutionary change that would be difficult, if not impossible, to observe otherwise. (Exercises 11.1–11.3)

6. Concept: scientific communication. Students summarize and present results, critiquing and discussing their simulations. (Exercises 11.2 and 11.3)

7. Specific content. Students will be able to define or describe: *population, gene pool, evolution, allelic frequency, genotypic frequency, gene flow, migration, genetic fixation, genetic drift (founder effect and bottleneck), natural selection.*

## Order of the Lab

1. Introduce the Hardy-Weinberg theorem.                                        (15 min)
2. Model Hardy-Weinberg equilibrium.                                            (30 min)
3. Model genetic drift-bottleneck effect.                                       (20 min)
4. Model additional selected evolutionary scenarios
   and prepare reports.                                                         (45 min)
5. Report results from simulations.                                             (30 min)
6. Introduce computer model and make assignments.                              (15–30 min)

**For a 2-hour lab:** Omit steps 5 and 6, and have students report their results in a short paper or in recitation sections. Alternatively, not all teams would simulate bottleneck effect, but teams would choose any (but only one) scenario to investigate. Students would not do computer simulations. Students would still have time to report results.

## Classroom Management

Students work in pairs on the Hardy-Weinberg model. They may continue to work in pairs or groups of three or four students for the scenarios. We prefer to have all groups be pairs except for the migration model, which requires two populations, so we have four students work on this model. Students should all simulate bottleneck effect and then choose one (or two if computer models will not be used) of the other scenarios to model. Be sure that someone is working on each scenario. Students should work in pairs on the computer model. Introduce the computer simulations in lab and demonstrate the software. Students can write out their hypotheses or simply indicate the problem they plan to pursue on the computer. They should complete their actual simulations outside of class. The instructor may choose to have students report their results in the next lab period or write a short report using scientific writing skills.

This lab can be completed without the computer models. Then students should complete three or more scenarios, depending on the time. The computer simulations are strongly recommended once students understand the Hardy-Weinberg theorem and its conditions. Most students will prefer to generate their own questions, which arise during the course of the lab. Students should be reminded that they are responsible for understanding every scenario.

## Student Development

Students begin to understand the relationship between genetics and evolution. They generate new questions, based on simple models of complex scenarios. Students further develop their ability to model biological phenomena and use computer simulations to investigate complex questions in population genetics. Students enhance their quantitative skills and gain experience analyzing and discussing results.

## Discussion and Summary

Encourage and facilitate discussion during the presentation of groups' results. Suggest questions and problems for computer simulations, which encourage creative thinking.

## Evaluation

Ask students to present results and lead a discussion of the evolutionary scenarios. Students may prepare a written or oral report on computer simulations. Students should be able to calculate genotypic and allelic frequencies. Students should be able to interpret results from similar simulations and models and devise models to investigate problems in population genetics. Students should be able to pose hypotheses and predict results for similar scenarios.

# Population Genetics II: Determining Genetic Variation

## Laboratory Objectives

After completing this lab topic, you should be able to:

1. Use gels to determine the extent of variation in the gene pool of a natural population.
2. Calculate genotypic and allelic frequencies and expected and observed heterozygosity.
3. Use the statistical tool chi-square to analyze results.
4. Describe how enzymes in a tissue extract can be separated by gel electrophoresis.
5. Discuss the relationship between genetic variation and evolution.

*For a 2-hour lab:* Omit Exercise 12.2.

*Instructors and/or students can download data tables in Excel format. Go to www .masteringbiology.com. Look in the Study Area under Lab Media.*

*Note that this entire lab topic may be performed with students using only the information in the lab manual and their personal calculators. However, if starch gel electrophoresis is available, you might demonstrate this research procedure commonly used in population genetics research.*

## Introduction

Rosemary and Peter Grant's 30-year study of ground finches on the island of Daphne in the Galápagos Islands has shown that a variation in bill sizes of ground finches exists and that there is a correlation between feeding behavior and bill size. Birds with larger bills preferentially feed on larger, harder seeds; birds with smaller bills feed on smaller seeds. In dry years when only large seeds are available, natural selection favors ground finches with large bills (Lab Topic 1). What would happen to the ground finch population in a prolonged drought if there were no genetic variation in bill size and all finches had small bills? Without the presence of genetic variation, natural selection might eliminate all birds with small bills, and the population of ground finches could become extinct. Heritable variation is a requirement for evolutionary change in a population in response to changes in the environment.

*Note that answers to most problems in this lab topic are given in Appendix D.*

Variation in populations is easily observed in almost any species; just look around your laboratory and campus. The variation in human populations seems almost limitless. Some variation in populations is the result of environmental differences, while other variation is genetically determined. Although height in humans has a genetic component, evidence suggests that the increase in the average height in humans over the past 50–100 years is primarily due to improvements in diet. Evolutionary biologists are interested in documenting the amount of genetic variation in populations, but separating environmental effects from genetic effects poses a difficult problem. Our earliest estimates of genetic variation came from artificial selection in plant and animal breeding. However, it is not feasible to conduct genetic

*In the Study Area of masteringbiology.com, see Student Media for suggested activities, investigations, and videos.*

crosses and analyses for every species of interest. The problem also is compounded by our inability to determine the genotype of individuals with a dominant phenotype; they could be heterozygous or homozygous.

In the 1960s, molecular biologists developed the technique of gel electrophoresis, which allowed scientists to detect differences in proteins based on their size and charge. These protein differences were the result of changes in the amino acid sequences of the proteins, which in turn were the result of **mutations,** or changes in the DNA sequence. Determining variation in proteins is simply one step from detecting changes in the genes themselves, and gel electrophoresis is much simpler and cheaper than DNA sequencing (determining the nucleotide sequences). In addition, proteins are basically inherited as codominant traits, allowing scientists to distinguish the heterozygous phenotype from the homozygous phenotype on gels. Using gel electrophoresis, population geneticists and evolutionary biologists have been able to survey populations of plants and animals, including humans, to determine the genetic variation present at a large number of genes.

The extent of genetic variation in populations is much greater than expected. One way to estimate genetic variation is to determine the proportion of the population that is heterozygous (having different alleles at a gene locus). On average, invertebrates are heterozygous at 11% of their genes; estimates for vertebrates are 4%, and plants, 16% (Hamrick, Linhart, and Mitton, 1979). Humans are heterozygous at approximately 7% of their genes. Assuming that humans have 100,000 genes in their chromosomes, you would have, on average, two different alleles at 7,000 of your genes. In contrast, the cheetah is heterozygous at only a small percentage (0.07%) of its genes. Understanding genetic variation and the factors that affect variation is essential to the management of this endangered species, which has little potential for evolutionary change because of its enormously reduced genetic variation.

As you begin your study of genetic variation, review the concepts and terminology for population genetics. You will be expected to use and understand the Hardy-Weinberg theorem as you investigate genetic variation in a natural population. Before continuing in this lab topic, complete Table 12.1, providing definitions and examples. Refer to your textbook and Lab Topic 11, Population Genetics I: The Hardy-Weinberg Theorem.

# EXERCISE 12.1
# Interpreting Banding Patterns in a Gel

In this exercise, you will investigate polymorphism in a population of rock pink plants (*Talinum mengesii*) (Color Plate 16), a succulent commonly found growing in shallow soils on rock outcrops in the southeast. You will determine the frequency of (1) different genotypes (genotypic frequency), (2) different alleles (allelic frequency), and (3) heterozygotes (heterozygosity) in the population (Lab Topic 11). These frequencies will allow you to estimate the extent of genetic variation in the rock pink population.

To perform this investigation, you will study diagrammatic reproductions of starch gels that have been stained for selected enzymes found in these plants. (Although starch gels were used for this exercise, acrylamide, cellulose acetate,

**Table 12.1**
Population Genetics Terminology

| |
| --- |
| **Gene locus** |
| **Allele** |
| **Allelic frequency** |
| **Genotypic frequency** |
| **Heterozygote** |
| **Hardy-Weinberg theorem** |
| **Natural selection** |
| **Genetic drift** |
| **Migration** |
| **Mutation** |
| **Random mating** |

and agarose gels can also be used.) Starch gels are made by boiling purified potato starch in a buffer solution with known pH and ionic strength and then allowing it to solidify in a tray. Paper wicks are soaked in a crude extract made by homogenizing individual flower buds from the various rock pink plants, each wick being soaked in a different plant bud homogenate. The wicks are then inserted in a slot cut along the width of the gel; this slot is referred to as the **origin** (Figure 12.1a). The gel is linked to two buffer tanks by contact wicks and connected to an electric current, creating a charged field through which the proteins migrate based on their charge and size (Figure 12.1b and c). Once the enzymes are sufficiently separated to allow the investigator to detect differences in the distances migrated by the enzymes, the gels are removed from the electric field and sliced into three or more thin slabs. Each gel contains all the enzymes present in the original tissue spread throughout the gel, but the investigator stains for only one enzyme in each gel or slice of gel. Since enzymes have specific substrates, each gel can be treated with the substrate for the enzyme of interest plus a salt that reacts with the product of the enzyme reaction to produce a colored band. The result will be one or more colored bands for each individual

**Figure 12.1.**

**Preparation of starch gels and techniques of gel electrophoresis.** (a) Wicks containing the enzymes are placed in a slot cut in the starch gel. This slot is called the *origin*. (b) The gel is linked to two buffer tanks by contact wicks and connected to an electric current. (c) Enzymes migrate to characteristic positions in the gel.

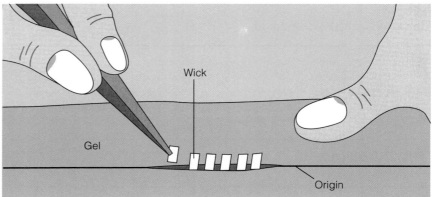

**a.**

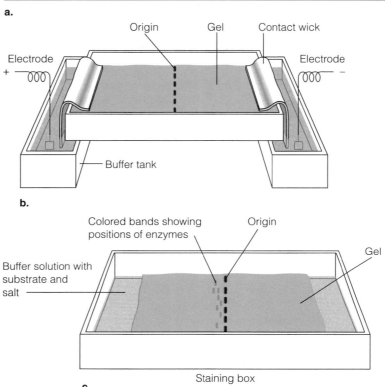

**b.**

**c.**

surveyed in a population. The number of bands and the distance migrated can be used to determine the genotype of individual organisms.

Remember that each protein in the gel migrates in a direction and at a rate that depend on two factors: net electrical charge and molecular size. For example, an enzyme with a greater negative charge will move more quickly toward the anode (the positive pole) than an enzyme with a lesser negative charge. Similarly, a smaller molecule will move more rapidly than a larger molecule. The number and type of amino acids that make up an enzyme determine its size and charge, and the differences in number and type of amino acids depend on mutations in genes, producing alternate forms, or alleles. The enzymes you will study in the rock pink population all migrate toward the anode.

The exercise will be carried out using reproductions of photographs of three slabs sliced from a gel. One slab was stained for alcohol dehydrogenase (ADH), another for phosphoglucoisomerase (PGI), and another for phosphoglucomutase (PGM) (Figure 12.2). This gel contained enzymes from 25 individual plants.

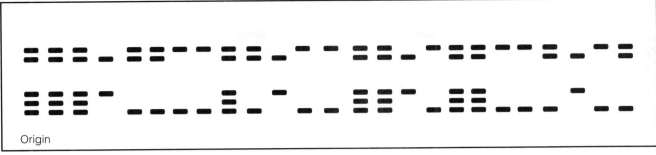

**a. ADH**

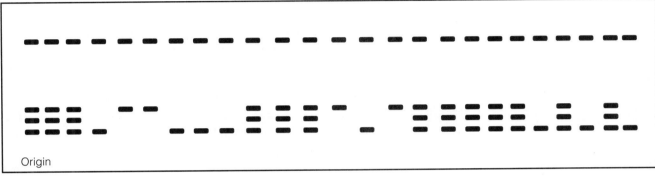

**b. PGI**

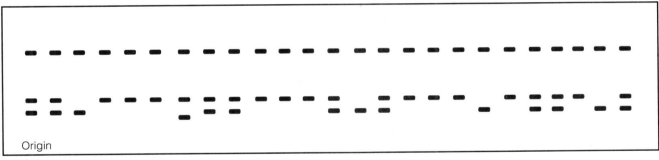

**c. PGM**

**Figure 12.2.**
**The gels to be interpreted in Exercise 12.1.** Each gel has been stained
to reveal a different enzyme: (a) alcohol dehydrogenase (ADH),
(b) phosphoglucoisomerase (PGI), and (c) phosphoglucomutase (PGM).

# Lab Study A. Terminology Associated with Enzymes and Gel Electrophoresis

In this lab study, you will become familiar with the terminology used to describe enzymes and the bands that are visible after separating and staining
the enzymes.

## Materials

No materials are needed for this lab study.

*In this lab, each distinct zone of activity is assumed to be a separate isozyme. Sometimes, however, isozymes and allozymes overlap in their migration, and they can only be distinguished by genetic crosses and analysis. The genetic basis for activity zones in rock pink plants has been established.*

## Procedure

1. Observe the enzyme banding patterns in the three gels in Figure 12.2.

2. Identify two **zones of activity.** A zone of activity is a row of colored bands in the gel, representing the products of one gene. The number of zones of activity represents the number of isozymes for the enzyme present in the gel. **Isozymes** are enzymes with the same function specified by two different genes, so two zones of activity would indicate two different genes coding for the same enzyme.

   a. How many zones of activity are present for each enzyme?

      *two*

   b. Label the zones of activity on each gel in Figure 12.2. Where two or more isozymes are present, the enzyme that migrates the greater distance from the origin is considered to be in the first zone of activity, the lesser distance, in the second zone of activity. To understand the naming of genes, refer to the top of Figure 12.2, stained for ADH. The gene that codes for the faster-migrating isozyme (has moved the greater distance from the origin) is called ADH-1; the gene for the slower-migrating isozyme is ADH-2. If additional isozymes were present, they would be numbered sequentially.

   c. Name each gene by its enzyme and the distance that the isozyme has traveled (isozyme 1 has traveled the greater distance).

   d. Record the number of zones of activity for each enzyme in Table 12.2 in the Results section.

3. Determine whether each gene is monomorphic or polymorphic and identify allozymes. If all individuals in one activity zone have enzymes that migrate the same distance from the origin, the gene is **monomorphic** (for example, PGI-1). All the bands in this activity zone are at the exact same level in the gel. This means that there are no alternate forms of this gene, at least in this population. If the number of bands for individuals of a population varies and there are two or more enzyme bands within an activity zone, the gene is **polymorphic** (for example, PGI-2), and the variant enzymes are called **allozymes** because they are enzymes coded by alleles of a single gene locus.

   a. Designate which genes are monomorphic and which are polymorphic in Table 12.2.

   b. Label allozymes on at least one gel in Figure 12.2.

4. Determine for your study gels (Figure 12.2) whether each enzyme is monomeric, dimeric, or trimeric. Record your answers in Table 12.2 in the Results section. Use Figure 12.3 and the following discussion to help you make your determinations.

When proteins are synthesized from DNA, one polypeptide is made at a time. For some enzymes, this polypeptide is all that is necessary to have a functioning enzyme, and the enzyme is described as being **monomeric.**

Other enzymes may need two or more polypeptides to bind together before the enzyme is active. An enzyme that needs two polypeptides in order to become active is called a **dimeric enzyme,** or **dimer.** If an enzyme requires three polypeptides in order to function, it is called a **trimeric enzyme,** or **trimer;** four polypeptides, a **tetrameric enzyme,** or **tetramer;** and so on.

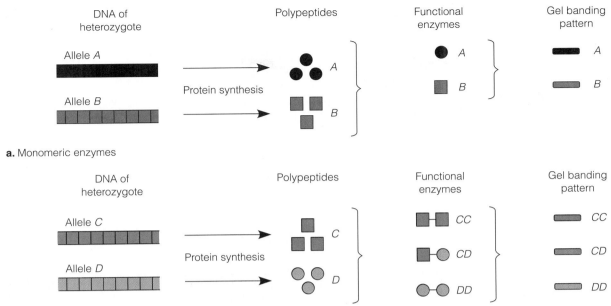

**Figure 12.3.**

**The interpretation of banding patterns for a heterozygote when the enzyme is monomeric or dimeric.**
(a) For monomeric enzymes, two different polypeptides are produced in a heterozygote, but either is a functioning enzyme. Therefore, only two bands will form in the heterozygote: *A* and *B*. (b) For dimeric enzymes, the enzyme can function only if two polypeptides are linked. In a heterozygous individual, the alleles produce two different polypeptides that may link in three configurations. Two like polypeptides can link (*C* to *C* or *D* to *D*), or the two unlike polypeptides can link (*C* to *D*). Each of these three linkages will migrate at different rates on a gel.

Since our staining procedure requires that the enzymes be functioning in order to be seen, we see only enzymes in their active form (whether that form is monomer, dimer, or some other form). Therefore, if some of the individuals on the gel have two different alleles coding for the polypeptides (that is, if the individuals are heterozygous), you can determine if the enzyme is dimeric, trimeric, or tetrameric.

## Distinguishing Monomeric and Dimeric Enzymes

*If an enzyme is monomeric*, in the homozygous individual only one band will appear on the gel, but in the heterozygote two bands will appear on the gel, indicating functional polypeptides, one for each of the codominant alleles (Figure 12.3a).

*If the enzyme is dimeric*, it requires two polypeptides bound together to function. An individual that is homozygous for the gene will produce only one kind of polypeptide, which can only bind to the same kind of polypeptide to give one kind of dimer. This will produce only one band on the gel. Heterozygous individuals, however, produce two polypeptides that can bind together in three different ways. Consider a dimeric enzyme with two different polypeptides, one coded for by allele *C*, the other by allele *D*. If *C* polypeptides run faster than *D* polypeptides on the gel (because of charge), then *CC* dimers will run faster

*Students frequently mistakenly interpret the three bands in an individual heterozygous for a dimeric enzyme as representing a fast, a slow, and a very slow allele. As students perform this exercise, remind them that one diploid individual can have only two alleles at a gene locus and could not have fast, slow, and very slow alleles. Have them review Figure 12.3 if they are confused.*

than *CD* dimers, which will run faster than *DD* dimers. This gives a three-band pattern (Figure 12.3b) and is seen in the dimeric enzyme ADH-2.

In summary, homozygotes for both monomeric and dimeric enzymes produce only one band. Heterozygotes for dimeric enzymes produce three bands. Heterozygotes for monomeric enzymes produce two bands. The third band seen in dimeric enzymes (the hybrid band) will be absent, since each polypeptide is functional and hybrid molecules do not form.

### Results

Record your observations and conclusions in Table 12.2.

### Table 12.2

Observation of Banding Patterns in Figure 12.2. Indicate your answer for each gene locus if there is more than one zone of activity per enzyme

| Enzyme | Number of Zones of Activity (Genes) | Monomorphic or Polymorphic | Monomeric or Dimeric |
|---|---|---|---|
| **ADH** (Fig 12.2a) | 2 | *1: polymorphic* <br> *2: polymorphic* | *1: monomeric* <br> *2: dimeric* |
| **PGI** (Fig 12.2b) | 2 | *1: monomorphic* <br> *2: polymorphic* | *1: cannot determine* <br> *2: dimeric* |
| **PGM** (Fig 12.2c) | 2 | *1: monomorphic* <br> *2: polymorphic* | *1: cannot determine* <br> *2: monomeric* |

### Discussion

1. What evidence do you have that the zones of activity for PGI represent isozymes rather than allozymes? (Note that the only way to be certain that bands are allozymes of the same gene is to perform genetic crosses and analyses.)

   *Differences in charge and size between isozymes can be much greater than between allozymes. Therefore, in general, the differences between migration distances for isozymes are greater than for allozymes, resulting in closer bands in the gel for the latter.*

2. Do any of the genes have a rare allele, that is, an allele that appears only occasionally?

   *PGM-2 has a rare allele.*

3. You see three bands in an individual in the ADH-2 zone of activity. Explain why you conclude that these three bands indicate that this

enzyme is dimeric, that the individual is heterozygous, and that these bands do *not* represent fast, slow, and very slow alleles.

*These bands could not represent three alleles because a diploid organism can have only two alleles. Three alleles can exist in a population but not in an individual.*

## Lab Study B. Scoring the Gel

### Materials

calculator

### Introduction

Scoring a gel involves examining the banding pattern for each individual represented in the gel and determining the genotype and phenotype of that individual. In this lab study, you will summarize these data for all polymorphic genes represented by the enzymes in the gels (see Figure 12.2).

### Procedure

1. Within each zone of activity, identify allozymes according to their rate of migration from the origin ($f$ = fast, or the enzyme band farthest from the origin; $s$ = slow; $vs$ = very slow). Label Figure 12.2 with your results.
2. Determine the genotype and phenotype of each polymorphic individual in the population. Note that at the protein level of expression, alleles are codominant and therefore genotype and phenotype correspond. A heterozygote is immediately known because both alleles are expressed in the form of two bands if the enzyme is monomeric or in the form of three bands if dimeric.

### Results

Summarize and record your results in Table 12.3.

 **Table 12.3**
Number of Individuals of Each Genotype for Each Polymorphic Gene
($f$ = fast allele, $s$ = slow allele, $vs$ = very slow allele)

| Polymorphic Gene Locus | Phenotype: | $f$ | $s$ | $f/s$ | $vs$ | $f/vs$ | $s/vs$ | Total Number of Individuals |
|---|---|---|---|---|---|---|---|---|
| | Genotype: | $f/f$ | $s/s$ | $f/s$ | $vs/vs$ | $f/vs$ | $v/vs$ | |
| ADH-1 | | 8 | 4 | 13 | 0 | 0 | 0 | 25 |
| ADH-2 | | 4 | 13 | 8 | 0 | 0 | 0 | 25 |
| PGI-1 | | 4 | 8 | 13 | 0 | 0 | 0 | 25 |
| PGM-2 | | 11 | 4 | 9 | 0 | 1 | 0 | 25 |

## Discussion

How many bands would be present in a heterozygote if the enzyme is a tetramer? Draw the heterozygote for a tetramer.

*fivebands* *AAAA*
*AAAB*
*AABB*
*ABBB*
*BBBB*

# Lab Study C. Determining Genotypic and Allelic Frequencies

In this lab study, you will use the data from Table 12.3 to determine the frequencies of genotypes (genotypic frequencies) and frequencies of alleles (allelic frequencies) in the rock pink population.

## Materials

calculator

## Procedure

1. Determine genotypic frequencies for ADH-1, ADH-2, PGI-2, and PGM-2. To do this, divide the number of individuals of each genotype by the total number of individuals.

2. Determine allelic frequencies for the same enzymes. Count the number of times the allele appears in the population and divide by the total number of alleles in the population. Record your results in Table 12.4.

   An individual with only a single slow band has two "slow" alleles while a heterozygous fast/slow individual has one "fast" allele and one "slow" allele. The total number of alleles in the sample is twice the number of individuals because each individual carries two alleles.

## Results

Record your conclusions in Table 12.4.

## Discussion

Explain how this population of the diploid rock pink can have three alleles present.

*Although each individual will have only two alleles present, the population can have multiple alleles for any gene locus.*

**Table 12.4**

Genotypic and Allelic Frequencies for ADH-1, ADH-2, PGI-2, and PGM-2

| Polymorphic Gene Locus | Genotypic Frequency | | | | | | Allelic Frequency | | |
|---|---|---|---|---|---|---|---|---|---|
| | *f/f* | *s/s* | *f/s* | *vs/vs* | *f/vs* | *s/vs* | *f* | *s* | *vs* |
| **ADH-1** | (8/25) 0.32 | (4/25) 0.16 | (13/25) 0.52 | — | — | — | (29/50) 0.58 | (21/50) 0.42 | 0 |
| **ADH-2** | 0.16 | 0.52 | 0.32 | — | — | — | 0.32 | 0.68 | 0 |
| **PGI-2** | 0.16 | 0.32 | 0.52 | — | — | — | 0.42 | 0.58 | 0 |
| **PGM-2** | 0.44 | 0.16 | 0.36 | — | 0.04 | — | 0.64 | 0.34 | 0.02 |

# Lab Study D. Determining Heterozygosity

## Materials

calculator

## Introduction

One way to measure the extent of genetic variation in a population is to determine the percentage of genes that are polymorphic—that is, having more than one allele at a gene locus. This estimate of genetic variation can be problematic because a gene with two alleles is counted in the same way as a gene with five alleles. Also, a rare allele (appearing only occasionally in the population) will be included in the same way as an allele that occurs in 50% of the population. Another measure, heterozygosity, appears to provide a better estimate of genetic variation. **Heterozygosity** (*H*) is the average frequency of heterozygous individuals per gene locus.

**To calculate *observed* heterozygosity,** first determine the frequency of heterozygotes at each gene locus (divide the number of heterozygotes by the total number of individuals in the population). Then total the frequency of heterozygous individuals for all gene loci and divide by the total number of loci. (This final average is calculated by including both monomorphic and polymorphic loci in your total number.)

*We will use only a small number of gene loci for determining heterozygosity. A larger sample of the genome, more than ten genes, would be preferable.*

**Table 12.5**
Sample Calculation of Heterozygosity in a Mouse Population

| Gene Locus | Number of Heterozygotes | Frequency of Observed Heterozygotes |
|:---:|:---:|:---:|
| A-1 | 9 | 0.45 (9/20) |
| A-2 | 2 | 0.10 |
| B-1 | 8 | 0.04 |
| C-1 | 0 | 0 |
| C-2 | 4 | 0.20 |
| | | $\Sigma$ 1.15 |

For example, in a population sample of 20 mice, you might study five gene loci and get the results shown in Table 12.5. Then the average heterozygosity for this population would be

$$H = \frac{1.15}{5} = 0.23$$

The *observed heterozygosity* in this population of mice is 0.23.

## Procedure

1. Using data from Table 12.3, record the number of heterozygotes for ADH-1, ADH-2, PGI-2, and PGM-2 in Table 12.6.
2. Divide the number of heterozygotes for each gene locus by the number of individuals in the population. This is the frequency of heterozygotes for that gene locus. Record this information in Table 12.6.
3. Determine the population average heterozygosity. Divide the sum of heterozygosity for all loci by the total number of loci (including both polymorphic and monomorphic gene loci).

## Results

Record all results in Table 12.6.

**Table 12.6**
Heterozygosity for Each Locus and Average Heterozygosity
for Population

| Polymorphic Gene Locus | Number of Individuals | | Frequency of Observed Heterozygotes |
| | Heterozygotes | Total | |
| --- | --- | --- | --- |
| ADH-1 | 13 | 25 | 0.52 |
| ADH-2 | 8 | 25 | 0.32 |
| PGI-2 | 13 | 25 | 0.52 |
| PGM-2 | 10 | 25 | 0.40 |
| Population average heterozygosity: 1.76/6 = 0.29 | | | |

*Remind students that the sum of heterozygosity for all loci is divided by the total number of loci. In this case there are six, four polymorphic and two monomorphic loci.*

### Discussion

List all the heterozygous genotypes for the PGM-2 locus. Did you record the correct number in Table 12.6?

*f/s and f/vs; students count the fast/slow heterozygote but often fail to count the fast/very slow heterozygote.*

# Lab Study E. Comparing Observed and Expected Heterozygosity

In natural populations, the amount of genetic variation might be affected by the reproductive behavior of the organisms. For example, populations that are highly inbred, from either mating between close relatives in animals or self-pollination in plants, might have a lower heterozygosity than expected for a population with the same allelic frequencies with random mating. (Return to Lab Topic 11 to review the guidelines for a population at Hardy-Weinberg equilibrium.)

You can determine the *expected* heterozygosity (expected in a large, randomly mating population) from the allelic frequencies.

To calculate expected heterozygosity for a gene locus with three alleles, the frequency of the alleles would be $f1$, $f2$, and $f3$. The frequency of the homozygotes (from Hardy-Weinberg) would be $(f1)^2$, $(f2)^2$, and $(f3)^2$. Therefore, the expected frequency of heterozygotes would be

$$H_{exp} = 1 - [(f1)^2 + (f2)^2 + (f3)^2 + \cdots]$$

For the mouse example, if two alleles, *a* and *b*, exist for the A-1 gene locus, and the frequency for each of these alleles is known, then the expected heterozygosity for the A-1 gene locus can be calculated. Suppose that the frequency of *a* is 0.6 and the frequency of *b* is 0.4. Then the expected heterozygosity would be

$$H_{exp} = 1 - [(0.6)^2 + (0.4)^2] = 0.48$$

To calculate the expected *number of heterozygotes*, simply multiply the total number of individuals in the population by the expected heterozygosity. For gene locus A-1 in the mouse example, the expected number of heterozygotes in the sample of 20 individuals would be $(20 \times 0.48) = 10$ (rounded off). Using chi-square analysis, you can then determine whether the observed numbers of heterozygotes are significantly different from those expected in a large, randomly mating population.

In this lab study, you will calculate the expected number of heterozygotes for the enzymes in the gels in Figure 12.2 and then use chi-square analysis to determine whether observed and expected numbers of heterozygotes are significantly different in the rock pink population.

## Materials

calculator

## Procedure

1. Calculate the expected heterozygosity for ADH-1, ADH-2, PGI-2, and PGM-2. Record your results below. (Refer to Table 12.4 for allelic frequencies.)

|  | Expected Heterozygosity |
|---|---|
| **ADH-1** | *0.487* |
| **ADH-2** | *0.435* |
| **PGI-2** | *0.487* |
| **PGM-2** | *0.474* |

2. Calculate the expected number of heterozygotes for ADH-1, ADH-2, PGI-2, and PGM-2, rounding to the nearest whole number. Record these data in Table 12.7.
3. Using data from Table 12.6, record the observed number of heterozygotes for these gene loci in Table 12.7.
4. Record the number of homozygous individuals observed for these genes in Table 12.7.
5. Calculate the expected number of homozygous individuals. To do this, subtract the number of expected heterozygotes from the total number of individuals in the population.
6. On a separate paper, use the chi-square test to determine whether the observed and expected numbers are significantly different. (Refer to Appendix B to review the chi-square test.) Record your conclusions in Table 12.7.

## Results

Record data and significance of the chi-square test results in Table 12.7.

**Table 12.7**
Observed Heterozygotes and Homozygotes for ADH-1, ADH-2, PGI-2, and PGM-2 and Significance of Chi-Square Test

*See worksheet in Appendix D, Calculations and Answers to Problems for Lab Topic 12, for Table 12.7.*

| Polymorphic Gene Locus | Number of Heterozygotes | | Number of Homozygotes | | Significant Difference | |
|---|---|---|---|---|---|---|
| | Observed | Expected | Observed | Expected | Yes | No |
| ADH-1 | 13 | 12 | f/f and s/s 12 | 13 | | X |
| ADH-2 | 8 | 11 | 17 | 14 | | X |
| PGI-2 | 13 | 12 | 12 | 13 | | X |
| PGM-2 | 10 | 12 | 15 | 13 | | X |

## Discussion

1. What factors, other than nonrandom mating, could account for a significant difference between observed and expected numbers of heterozygotes in living populations?

   *selection for heterozygotes; migration; genetic drift; mutation*

2. Why do electrophoretic estimates of heterozygosity underestimate genetic variation at the protein level?

   *Genetic variation causes changes in proteins. Most changes in proteins are amino acid substitutions, which do not change the size or charge of the protein. Small changes, then, would not be detected by electrophoresis because electrophoresis depends on detectable size and charge changes.*

EXERCISE 12.2

# Interpreting Results of Gels of a Hypothetical Population

In this exercise, you will apply your knowledge and understanding of genetic variation and gel electrophoresis to a hypothetical problem.

## Problem

Another species of *Talinum*, called *rock red*, has been found on sandstone outcrops in south Georgia. This rare plant is a succulent that inhabits the shallow soils found on rock outcrops. These rock outcroppings are isolated islands of rock found within a sea of forests, fields, and suburban developments. Although the rock red is insect-pollinated, the position of the reproductive structures in the flowers of some populations promotes self-pollination when the flower closes.

In this exercise, you will evaluate gels for the enzymes aconitase (ACN), glyceraldehyde phosphate dehydrogenase (GDH), and PGI. You will pose hypotheses and make predictions, interpret the gels, score the isozymes and allozymes, determine the genotypic and allelic frequencies, and calculate heterozygosity.

### Hypothesis

For this newly discovered species, state a hypothesis concerning the amount of genetic variation present in the population based on the species characteristics (as previously stated) and the factors that affect genetic variation.

*The rock red population appears to be isolated, small, and perhaps inbred; therefore, there should be little genetic variation.*

### Prediction

Predict whether the observed heterozygosity will differ significantly from the expected heterozygosity.

*If these are isolated, inbred populations with little genetic variation, then the observed heterozygosity should be lower than the expected.*

### Procedure

1. Interpret the bands and score the gel in Figure 12.4. Determine isozymes, allozymes, and monomorphic and polymorphic loci, and whether the enzymes are monomeric, dimeric, and so on. (If you need assistance, refer to Exercise 12.1.)

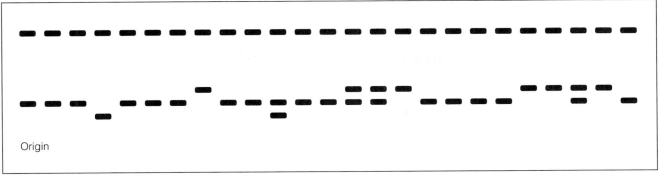

**a. ACN**

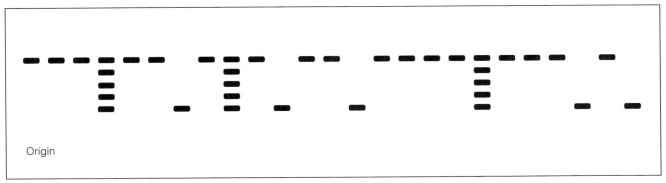

**b. GDH**

**c. PGI**

**Figure 12.4.**
**Gels representing the hypothetical population.** (a) Gel stained for aconitase (ACN). (b) Gel stained for glyceraldehyde phosphate dehydrogenase (GDH). (c) Gel stained for phosphoglucoisomerase (PGI).

2. In the Results section, design appropriate tables to record your data. (Use Tables 12.3 and 12.4 as examples.) You should have one row for each gene locus (include monomorphic and polymorphic genes) and one column for each genotype.

3. Count and record the number of individuals of each genotype for the gene products stained in the gels.

4. Be prepared to present your results in a class discussion.

## Results

1. Determine genotypic and allelic frequencies for the genes, recording your data in tables that you have designed.

**Table IA 12.1**

**Number of Individuals of Each Genotype for Each Polymorphic Gene**
(*f* = fast allele, *s* = slow allele, *vs* = very slow allele)

| Polymorphic Gene Locus | Phenotype: | *f* | *s* | *f/s* | *vs* | *f/vs* | *s/vs* | Total Number of Individuals |
| | Genotype: | *f/f* | *s/s* | *f/s* | *vs/vs* | *f/vs* | *s/vs* | |
|---|---|---|---|---|---|---|---|---|
| ACN-2 | | 5 | 15 | 3 | 1 | 0 | 1 | 25 |
| GDH-1* | | 17 | 5 | 3 | 0 | 0 | 0 | 25 |
| PGI-1 | | 10 | 13 | 2 | 0 | 0 | 0 | 25 |

*\*GDH-1 is a tetramer.*

**Table IA 12.2**

**Genotypic and Allelic Frequencies for ACN-2, GDH-1, and PGI-1**

| Polymorphic Gene Locus | Genotypic Frequency | | | | | | Allelic Frequency | | |
| | *f/f* | *s/s* | *f/s* | *vs/vs* | *f/vs* | *s/vs* | *f* | *s* | *vs* |
|---|---|---|---|---|---|---|---|---|---|
| ACN-2 | 0.2 | 0.6 | 0.12 | 0.04 | — | 0.04 | 0.26 | 0.68 | 0.06 |
| GDH-1 | 0.68 | 0.2 | 0.12 | — | — | — | 0.74 | 0.26 | 0 |
| PGI-1 | 0.40 | 0.52 | 0.08 | — | — | — | 0.44 | 0.56 | 0 |

2. Calculate the frequency of heterozygotes for each locus and the average heterozygosity for the population. Record your data in an appropriate table.

**Table IA 12.3**
Heterozygosity for Each Locus and Average Heterozygosity for Population

| Polymorphic Gene Locus | Number of Individuals | | Frequency of Heterozygotes |
| --- | --- | --- | --- |
| | Heterozygotes | Total | |
| ACN-2 | 4 | 25 | 0.16 |
| GDH-1 | 3 | 25 | 0.12 |
| PGI-1 | 2 | 25 | 0.08 |
| Population average heterozygosity: 0.36/5 = 0.07 | | | |

3. Calculate the expected heterozygosity for each gene locus. Record your data in the margin of your lab manual. Use the formula

$$H_{exp} = 1 - [(f1)^2 + (f2)^2 + \cdots]$$

*ACN-2 = 0.466; GDH-1 = 0.385; PGI-1 = 0.493*

4. Calculate the expected number of heterozygotes for each gene locus. Multiply heterozygosity by the number of individuals in the population. Record your data in an appropriate table.

**Table IA 12.4**
**Observed Heterozygotes and Homozygotes for ACN-2, GDH-1, and PGI-1 and Significance of Chi-Square Test**

| Polymorphic Gene Locus | Number of Heterozygotes | | Number of Homozygotes | | Significant Difference | |
| --- | --- | --- | --- | --- | --- | --- |
| | Observed | Expected | Observed | Expected | Yes | No |
| ACN-2 | 4 | 12 | 21 | 13 | X | |
| GDH-1 | 3 | 10 | 22 | 15 | X | |
| PGI-1 | 2 | 12 | 23 | 13 | X | |

5. On separate paper, use the chi-square test to determine if observed and expected numbers are significantly different.

**Discussion**

1. What was the observed average heterozygosity for this population?

   *See Table IA 12.3.*

2. Discuss the significance of the chi-square test results.

   *All observed heterozygosity is significantly less than the expected.*

3. Was your hypothesis supported? Describe the factors that may affect the genetic variation in this population.

   *Population size is unknown; appears to be self-pollinating and in-bred; migration is probably restricted owing to the isolated nature of the rock outcrops; selection unknown.*

4. What is the average heterozygosity for this population?

   *7%—low for plants in general*

5. Based on your results, what is the evolutionary potential for this species?

   *The species has a low evolutionary potential because of the limited genetic variation present in this population. However, this is only one population. There could be others with more variation.*

# Reviewing Your Knowledge

*See Appendix D for answers to the problems in the Reviewing Your Knowledge and Applying Your Knowledge sections.*

1. Define the following terms, giving examples when appropriate: *heterozygosity, isozymes, allozymes, monomorphic, polymorphic, monomeric, dimeric, trimeric.*

2. Using a separate paper, practice calculating genotypic frequencies, allelic frequencies, and heterozygosity by solving the following problems (from Ayala, 1982).

   a. Three genotypes were observed at the PGM-1 locus in a human population. In a sample of 1,110 individuals, the three genotypes occurred in the following numbers (1 and 2 represent two different alleles):

   | Genotypes: | 1/1 | 1/2 | 2/2 |
   |---|---|---|---|
   | Numbers: | 634 | 391 | 85 |

   Calculate the genotypic and allelic frequencies.

b. Two human serum haptoglobins are determined by two alleles at a single locus. In a sample of 219 Egyptians, the three genotypes occurred in the following numbers (1 and 3 represent the two alleles):

Genotypes:   1/1     1/3    3/3
Numbers:     9     135    75

What are the frequencies of the two alleles?

*1 = 0.35; 3 = 0.65*

c. Calculate the expected frequency of heterozygotes, assuming random mating, from the data given in problems 2a and 2b. Use the chi-square test to determine whether the observed and expected numbers of heterozygous individuals are significantly different.

3. Observe the two gels below (from Hebert and Beaton, 1989). Gel A is stained for the enzyme leucine aminopeptidase (LAP); gel B, for the enzyme phosphoglucose isomerase (PGI). For each gel:

   *For answers to a–e for LAP, see the Teaching Plan, p. 316.*

   a. Identify and label zones of activity.
   b. Indicate isozymes and allozymes.
   c. Tell if the gene is monomorphic or polymorphic.
   d. Determine if the enzyme is monomeric, dimeric, or trimeric.
   e. Calculate allelic and genotypic frequencies for all genes.

**Gel A.** Leucine aminopeptidase (LAP)
(After Hebert and Beaton, 1989)

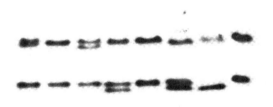

**Gel B.** Phosphoglucoisomerase (PGI)

## Applying Your Knowledge

1. Table 12.8 (from Ayala, 1982) gives the number of individuals in each of the MN blood groups in samples from various human populations. Using a separate paper, calculate the genotypic and allelic frequencies, as well as the expected number of heterozygous individuals, for each population. Test whether the observed and expected numbers of heterozygotes agree.

**Table 12.8**
Number of Individuals in Each of the MN Blood Groups

| Population | M | MN | N | Total |
|---|---|---|---|---|
| Belgians | 896 | 1,559 | 645 | 3,100 |
| English | 121 | 200 | 101 | 422 |
| Egyptians | 140 | 245 | 117 | 502 |
| Ainu | 90 | 253 | 161 | 504 |
| Fijians | 22 | 89 | 89 | 200 |
| Papuans | 14 | 48 | 138 | 200 |

2. Table 12.9 (from Cleland et al., 1996) gives the results of starch gel electrophoresis of fruit fly enzymes from four localities in Arizona.

   Complete Table 12.10 describing populations of *D. mojavensis* studied:

   a. Using the heterozygosity data, determine the number of heterozygotes and homozygotes in the population.

   b. Calculate the number of homozygous *ss* or *ff* individuals in the population. Remember that each individual in the population has two alleles and that heterozygotes will have one *f* and one *s*. (*Hint*: This population is *not* in Hardy-Weinberg equilibrium.)

**Table 12.9**
Frequencies of the Slow (*s*), Fast (*f*), and Very Fast (*vf*) Alleles of Alcohol Dehydrogenase (ADH) in Populations of Adult *Drosophila mojavensis*

| Locality | Number of Individuals | *s* | *f* | *vf* | *H* |
|---|---|---|---|---|---|
| Agua Caliente | 124 | 0.448 | 0.552 | 0.0 | 0.363 |
| Grand Canyon | 41 | 0.659 | 0.341 | 0.0 | 0.195 |
| Guaymas | 49 | 0.755 | 0.163 | 0.082 | 0.326 |
| Mulege | 50 | 0.130 | 0.870 | 0.0 | 0.060 |

*H* = heterozygosity determined by direct count

**Table 12.10**
Numbers of Individuals in Populations for Alcohol Dehydrogenase Alleles

| Locality | Number of Individuals | | | | Number of *vf* Alleles in Population |
| | Heterozygotes | Homozygotes | ss | *ff* | |
| --- | --- | --- | --- | --- | --- |
| **Agua Caliente** | 45 | 79 | 33 | 46 | 0 |
| **Grand Canyon** | 8 | 33 | 23 | 10 | 0 |
| **Guaymas** | 16 | 33 | * | * | 8 |
| **Mulege** | 3 | 47 | 5 | 42 | 0 |

*Can't be determined

How would you describe the *vf* allele?

 *a rare allele*

Can you determine if *vf* alleles are in heterozygous or homozygous individuals?

 *no, not based on these data*

## Investigative Extensions

If you have the equipment or resources in your department (gel trays, power packs, etc.) consider designing a study of populations on your campus or in your local habitats. Allozymes can be used to investigate questions in population ecology, evolution, and conservation biology. Consider the following suggestions for investigations:

- Compare populations of different sizes and/or extent of isolation, exploring the effects of inbreeding and genetic drift on genetic variation.

- Compare closely related species that differ in reproductive biology; for example, outcrossing versus self-pollinating plants.

- Compare genetic variation in populations of a species that inhabit different environments or habitats; for example, white oaks in wetlands, uplands, and protected ravines or pine bark beetles in dry forests versus forests near streams or wetlands.

- Compare genetic variation in a taxonomic group or closely related species; for example, several oak or pine species.

- Compare genetic variation in populations of a native species versus a closely related invasive species.

- Compare the genetic variation in cultivated plants from a nursery versus plants of the same species that are in their native habitat; for example, purple coneflowers.

- Compare populations growing in an extreme environment (temperature, pH, light, or exposed to air, water, or soil pollution) versus populations with moderate environmental conditions.

For additional ideas, procedures, and recipes for stains and gels see the References section.

**Student Media: BioFlix, Activities, Investigations, and Videos**
www.masteringbiology.com (select Study Area)

**BioFlix**—Ch. 23: Mechanisms of Evolution (Animation)
**Activities**—Ch. 23: Causes of Evolutionary Change
**Investigations**—Ch. 22: How Do Environmental Changes Affect a Population? What Are the Patterns of Antibiotic Resistance?
   Ch. 23: How Can Frequency of Alleles Be Calculated? Biology Labs OnLine: Population Genetics Lab; Evolution Lab
**Videos**—Ch. 22: Discovery Videos: Charles Darwin

# References

Ayala, F. J. *Population and Evolutionary Genetics: A Primer*. Menlo Park, CA: Benjamin Cummings, 1982.

Bader, J. M. "Measuring Genetic Variability in Natural Populations by Allozyme Electrophoresis," in *Tested Studies for Laboratory Teaching*, Procedings of the 19th Workshop/Conference of the Association for Biology Laboratory Education (ABLE) 1997, 20, pp. 25–42.

Cleland, S., G. D. Hocutt, C. M. Breitmeyer, T. A. Markow, and E. Pfeiler. "Alcohol Dehydrogenase Polymorphism in Barrel Cactus Populations of *Drosophila mojavensis*." *Genetica*, 1996, vol. 98, pp. 115–117.

Futuyma, D. J. *Evolution*, 2nd ed. Sunderland, MA: Sinauer Associates, 2009.

Grant, Peter. "Natural Selection and Darwin's Finches." *Scientific American*, October, 1991, pp. 82–87.

Hamrick, J. L., Y. B. Linhart, and J. B. Mitton. "Relationships between Life History Characteristics and Electrophoretically Detectable Genetic Variation in Plants." *Annual Review of Ecology and Systematics*, 1979, pp. 173–200.

Hebert, D. N., and M. J. Beaton. *Methodologies for Allozyme Analysis Using Cellulose Acetate Electrophoresis*. Helena Laboratories, 1989.

Heitz, J. and C. Giffen. *Practicing Biology: A Student Workbook*, 4th ed. San Francisco, CA: Benjamin Cummings, 2010. Chapter 23 provides a guide to solving population genetics problems and practice problems.

May, B. "Starch Gel Electrophoresis of Allozymes," in *Molecular Genetic Analysis of Populations—A Practical Approach*, A. R. Hoelzel, editor. 1992, pp. 1–27.

Peroni, P. A. and D. E. McCauley. "Demystifying Hardy-Weinberg: Using Cellulose Acetate Electrophoresis of the *Lap* Locus to Study Population Genetics in White Campion (*Silene latifolia*)," in *Tested Studies for Laboratory Teaching*, Proceedings of the 20th Workshop/Conference of the Association for Biology Laboratory Education (ABLE) 1998, 20, pp. 101–122.

Soltis, D., C. H. Haufler, D. C. Darrow, and G. J. Gastony. "Starch Gel Electrophoresis of Ferns: A Compilation of Grinding Buffers, Gel and Electrode Buffers, and Staining Schedules." *American Fern Journal*, 1983, 73, pp. 9–27.

Soltis, D. E., and P. S. Soltis, eds. *Isozymes in Plant Biology. Advances in Plant Sciences Series*, vol. 4. Portland, OR: Dioscorides Press, 1989.

# Websites

Futuyma, Douglas. *Natural Selection: How Evolution Works*. Interview and resources on evolution, genetic variation, and natural selection:
http://www.actionbioscience.org/evolution/futuyma.html

Issues in Evolution, American Institutes of Biological Sciences, provides resources, interviews, and issues:
http://www.actionbioscience.org/evolution/

Library of resources at PBS Evolution website:
http://www.pbs.org/wgbh/evolution/library/index.html

University of California, Berkley, Museum of Paleontology. *Understanding Evolution*:
http://evolution.berkeley.edu/evosite/evohome.html

# LAB TOPIC 12

# Population Genetics II: Determining Genetic Variation
## Teaching Plan for Laboratories

 This lab topic does *not* require that you perform electrophoresis. The results from actual gels are provided for students to study and interpret. You may choose to demonstrate the procedure if you have the equipment and chemicals. In this lab topic the Hardy-Weinberg theorem is applied to a natural population, and additional evolutionary concepts are explored.

## Main Concepts and Objectives

1. Concept: gel electrophoresis. Students will describe how enzymes in a tissue extract can be separated by gel electrophoresis and then specifically stained, identified, and analyzed. (Exercise 12.1)

2. Concept: genetic variation in populations. Students will determine the extent of variation in the gene pool of a natural population. They will relate genetic variation to evolution. (Exercises 12.1 and 12.2)

3. Concept: genetic variation as estimated by gel electrophoresis. Students will use results of gel electrophoresis to determine allelic and genotypic frequencies in a population. They will calculate heterozygosity for the population. (Exercises 12.1 and 12.2)

4. Concept: statistical analysis of results. Students will compare observed and expected heterozygosity using the chi-square test. (Exercises 12.1 and 12.2)

5. Specific content. Students will be able to describe or define: *heterozygosity, isozymes, allozymes, monomorphic, polymorphic, monomeric, dimeric, trimeric, chi-square.*

## Order of the Lab

1. Introduction to concepts, gel electrophoresis, and how to score gels. Explain terminology. (30 min)

2. Complete Exercise 12.1, Lab Studies A to D. (60 min)

3. Analyze results using the chi-square test (Exercise 12.1, Lab Study E). (30 min)

4. Complete Exercise 12.2. (45 min)

5. Discuss results of Exercise 12.2. (15 min)

6. Assign Reviewing Your Knowledge and Applying Your Knowledge for homework.

For a 2-hour lab: Omit Exercise 12.2.

# Classroom Management

Students work in pairs, discussing questions and benefiting from peer teaching. Move among the groups, asking questions and testing your students' understanding of terminology and theory as they score gels and perform calculations.

# Student Development

Students develop skills in problem solving and statistical analysis. They apply knowledge of enzyme structure and molecular biology to gel electrophoresis and population genetics. Students use a research tool to understand a basic biological concept.

# Discussion and Summary

Lead a discussion of procedure and results. Ask students to present the results of Exercise 12.2. Students should work all problems at the end of the exercise outside of class.

# Evaluation

Test students on an exam. Ask them to interpret and score gels, calculate allelic and genotypic frequencies, calculate heterozygosity (expected and observed), and use and interpret the chi-square test. Students should be able to solve problems similar to those in the lab topic on a laboratory exam.

Answers to Question 3a–e in Reviewing Your Knowledge for LAP.

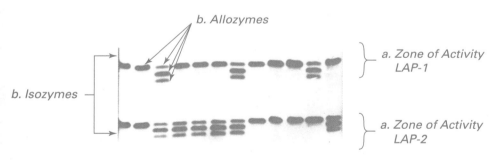

*Gel A. Leucine aminopeptidase (LAP)*

c. Both genes, LAP-1 and LAP-2, are polymorphic.

d. Both isozymes are dimeric.

e. Genotypic frequencies for LAP-1: ff = 0.75; fs = 0.25
   Allelic frequencies for LAP-1: f = 0.875; s = 0.125

   Genotypic frequencies for LAP-2: ff = 0.5; fs = 0.5
   Allelic frequencies for LAP-2: f = 0.75; s = 0.25

# Lab Topic 13
# **Bacteriology**

## Laboratory Objectives

After completing this lab topic, you should be able to:

1. Describe bacterial structure: colony morphology, cell shape, growth patterns.
2. Describe the results of Gram staining and discuss the implications to cell wall chemistry.
3. Describe a scenario for succession of bacterial and fungal communities in aging milk, relating this to changes in environmental conditions such as pH and nutrient availability.
4. Practice aseptic techniques producing bacterial streaks, smears, and lawns.
5. Describe the ecology and control of bacteria, applying these concepts to life situations.

*For a 2-hour lab:* Omit Exercise 13.2, bacterial succession in milk, and possibly the study of bacterial colony characteristics, Exercise 13.1, Lab Study A. See the Teaching Plan.

## Introduction

Humans have named and categorized organisms for hundreds—perhaps even thousands—of years. Taxonomy is an important branch of biology that deals with naming and classifying organisms into distinct groups or categories. Much of the work of early taxonomists included recording characteristics of organisms and grouping them based on appearance, habitat, or perhaps medicinal value. As scientists began to understand the processes of genetics and evolution by natural selection, they realized the value of classifying organisms based on phylogeny, or evolutionary history. Information about phylogeny was obtained from studies of development or homologous features—common features resulting from common genes. In recent years, scientists have begun using biochemical evidence—studies of nucleic acids and proteins—to investigate relationships among organisms, leading to revisions in the taxonomic scheme.

Systematists continue to grapple with the complex challenge of organizing the diversity of life into categories. In the 1960s, Robert Whittaker of Cornell University proposed that scientists use a five-kingdom scheme for classifying living organisms. His scheme was based largely on cell type: prokaryotic or eukaryotic, and nutrition type: autotrophic or heterotrophic. The five-kingdom system places all prokaryotic organisms in the kingdom Monera, and eukaryotic organisms are in kingdoms Protista, Fungi, Plantae, and Animalia. In the late 1970s, the microbiologist Carl Woese and his colleagues at the University of Illinois were studying DNA and ribosomal RNA sequences in prokaryotes when they discovered a group of organisms that were dramatically different from other prokaryotes. Because of their vast differences, Woese proposed a three-domain system of classification that has become widely accepted. In the three-domain system, the three domains—Bacteria, Archaea, and Eukarya—are a taxonomic level higher and include the kingdoms, historically the broadest taxonomic category. In the three-domain system, prokaryotes are placed in one of the two domains Bacteria or Archaea. This classification renders the kingdom Monera obsolete since its members are in two domains. All eukaryotes (organisms that have cells containing true nuclei) are categorized in the domain Eukarya. Three of the kingdoms of the five-kingdom scheme now placed in Eukarya—Fungi, Plantae, and Animalia—are multicellular organisms. Members of the former kingdom Protista are now in the domain Eukarya, but researchers continue to debate the number of kingdoms of protists.

Although many argue that biological categories are subjective in nature and the criteria for designating categories of living organisms have been modified by scientists historically, it is nonetheless true that scientists have set definite criteria or guidelines that form the basis of taxonomic classification. Most organisms may be placed into these designated categories.

In this lab topic, you will study organisms commonly called **bacteria.** In the three-domain system, the common bacteria are classified in the **domain Bacteria.**

Bacteria are small, relatively simple, **prokaryotic,** single-celled organisms. **Prokaryotes,** from the Greek for "prenucleus," have existed on Earth longer and are more widely distributed than any other organismal group. They are found in almost every imaginable habitat: air, soil, and water, in extreme temperatures and harsh chemical environments. They can be photosynthetic, using light, or chemosynthetic, using inorganic chemicals as the source of energy, but most are heterotrophic, absorbing nutrients from the surrounding environment.

Most bacteria have a cell wall, a complex layer outside the cell membrane. The most common component found in the cell wall of organisms in the domain Bacteria is peptidoglycan, a complex protein-carbohydrate polymer. There are no membrane-bound organelles in bacteria and the genetic material is not bound by a nuclear envelope. Bacteria do not have chromosomes as described in Lab Topic 7; their genetic material is a single circular molecule of DNA. In addition, bacteria may have smaller rings of DNA called **plasmids** (see page 238), consisting of only a few genes. They reproduce by a process called **binary fission,** in which the cell duplicates its components and divides into two cells. These cells usually become independent, but they may remain attached in linear chains or grapelike clusters. In favorable environments, individual bacterial cells rapidly proliferate, forming colonies consisting of millions of cells.

Differences in colony morphology and the shape of individual bacterial cells are important distinguishing characteristics of bacterial species. In Exercise 13.1, working independently, you will observe and describe the morphology of colonies and individual cells of several bacterial species. You will examine and describe characteristics of bacteria growing in plaque on your teeth. You and your lab partner will compare results of all lab studies.

# EXERCISE 13.1
# Investigating Characteristics of Bacteria

Because of the small size and similarity of cell structure in bacteria, techniques used to identify bacteria are different from those used to identify macroscopic organisms. Staining reactions and properties of growth, nutrition, and physiology are usually used to make final identification of species. The structure and arrangement of cells and the morphology of colonies contribute preliminary information that can help us determine the appropriate test necessary to make final identification. In this exercise, you will use the tools at hand, microscopes and unaided visual observations, to learn some characteristics of bacterial cells and colonies.

When you are working with bacteria, it is very important to practice certain **aseptic techniques** to make sure that the cultures being studied are not contaminated by organisms from the environment and that organisms are not released into the environment.

1. Wipe the lab bench with disinfectant before and after the lab activities.

2. Wash your hands before and after performing an experiment.

3. Using the alcohol lamp or Bunsen burner, flame all nonflammable instruments used to manipulate bacteria or fungi before and after use.

4. Place swabs and toothpicks in the disposal container immediately after use. *Never place one of these items on the lab bench after use!*

5. Wear a lab coat, a lab apron, or a clean old shirt over your clothes to lessen chances of staining or contamination accidents.

The bacteria used in these exercises are not pathogenic (disease-producing); nevertheless, use appropriate aseptic techniques and work with care! If a spill occurs, notify the instructor. If no instructor is available, wear disposable gloves, and wipe up the spill with paper towels. Follow this by washing the affected area with soap and water and a disinfectant. Dispose of the gloves and soiled towels in the autoclavable plastic bag provided.

# Lab Study A. Colony Morphology

## Materials

disinfectant
stereoscopic microscope
metric ruler
agar plate cultures with bacterial colonies

## Introduction

A **bacterial colony** grows from a single bacterium and is composed of millions of cells. Each colony has a characteristic size, shape, consistency, texture, and color (colony morphology), all of which may be useful in preliminary species identification. Bacteriologists use specific terms to describe colony characteristics. Figure 13.1 illustrates some of the terminology that may be used to describe colony morphology. Use the figure to

**Figure 13.1.**
**Terminology used in describing bacterial colonies.** (a) Common shapes, (b) margins, and (c) surface characteristics are illustrated.

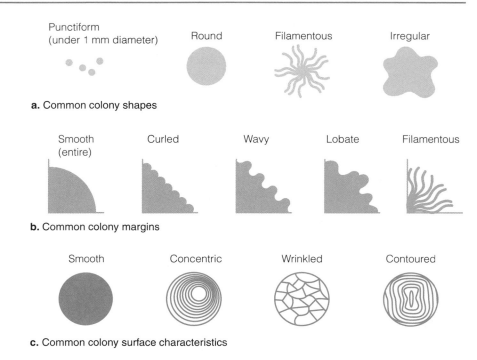

a. Common colony shapes

b. Common colony margins

c. Common colony surface characteristics

*Provide at least six different species, six plates for every four students. Other species may be substituted. See the Preparation Guide for suggestions of bacteria. Seal the plates closed with Parafilm® strips. Student results will vary. Do not expect "correct" answers. The objective is to note distinguishing variations. Do not require students to learn terms used to describe bacteria. If ocular micrometers are available, have students measure colony sizes.*

become familiar with this terminology and describe the bacterial species provided. Occasionally, one or more **fungal colonies** will contaminate the bacterial plates. Fungi may be distinguished from bacteria by the *fuzzy* appearance of the colony (Figure 13.2). The body of a fungus is a mass of filaments called **hyphae** in a network called a **mycelium.** Learn to distinguish fungi from bacteria.

## Procedure

1. Wipe the work area with disinfectant and wash your hands.
2. Set up your stereoscopic microscope.
3. Obtain one of the bacterial plates provided. Leaving the plate closed (unless otherwise instructed), place it on the stage of the microscope.
4. Examine a typical individual, separate colony. Measure the size and note the pigmentation (none or color) of the colony, and record this information in Table 13.1, in the Results section.

**Figure 13.2.**
**(a) Bacteria and (b) fungi growing on nutrient agar plates.** The body of most fungi consists of filaments called *hyphae* in a network called a *mycelium.* The hyphae give fungal colonies a fuzzy appearance. (See Color Plates 17 and 18.)

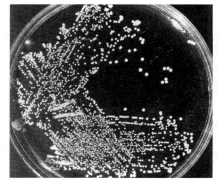

a.

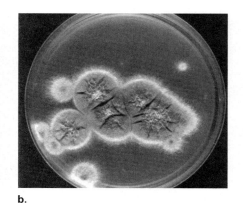

b.

5. Using the diagrams in Figure 13.1, select appropriate terms that describe the colony.

6. Record your observations in Table 13.1.

7. Sketch one colony in the margin of your lab manual, illustrating the characteristics observed.

8. Repeat steps 2 to 6 with two additional species. Your lab partner should examine three different species.

9. Observe Color Plate 17. Describe the shape, margin, surface, and pigmentation (color) of this bacterial species.

## Results

1. Complete Table 13.1 using terms from Figure 13.1 to describe the three bacterial cultures you observed.

2. Compare your observations with those of your lab partner.

## Discussion

1. What are the most common colony shapes, colony margins, and colony surface characteristics found in the species observed by you and your lab partner?

2. Based on your observations, comment on the reliability of colony morphology in the identification of a given bacterial species.

   *Students may conclude that these criteria are not very reliable because they may find it difficult to be certain which features apply. Trained professionals and students, after practice, may be able to make initial identification based on colony morphology. However, incontrovertible identification usually depends on results of additional tests.*

## Table 13.1
Characteristics of Bacterial Colonies

| Name of Bacteria | Size | Shape | Margin | Surface | Pigmentation (none or specific color) |
|---|---|---|---|---|---|
| 1. | | | | | |
| 2. | | | | | |
| 3. | | | | | |

# Lab Study B. Morphology of Individual Cells

## Materials

compound microscope
prepared slides of bacillus,
    coccus, and spirillum bacteria
blank slide
clean toothpick
clothespin

dropper bottle of deionized (DI)
    water
dropper bottle of crystal violet stain
squirt bottle of DI water
alcohol lamp or Bunsen burner
staining pan

## Introduction

Microscopic examination of bacterial cells reveals that most bacteria can be classified according to three basic shapes: **bacilli** (rods), **cocci** (spheres), and **spirilla** (spirals, or corkscrews). In many species, cells tend to adhere to each other and form clusters or chains of cells. In some environments different species may associate in a complex polysaccharide matrix creating a **community**, or assemblage of species of bacteria that adheres to a surface. These communities, called **biofilms**, are found in moist environments where nutrients are plentiful. Examples of environments that support biofilms are soils, water pipes, medical devices such as tubes used in kidney dialysis, and your mouth! In this lab study, you will examine prepared slides of bacteria that illustrate the three basic cell shapes, and then you will examine and describe bacteria growing in your mouth.

## Procedure

*To save time, set this up as a demonstration or show the three types of bacteria using a video-microscopy system.*

1. To become familiar with the basic shapes of bacterial cells, using the compound microscope, examine prepared slides of the three types of bacteria, and make a sketch of each shape in the space provided.

2. Your mouth is a type of ecosystem with at least 600 species of bacteria living in this warm, moist environment. Protein and carbohydrate materials from food particles accumulate at the gum line in your mouth and create an ideal environment for bacteria to grow. This mixture of materials and bacteria is a biofilm called **plaque.** To investigate the forms of bacteria found on your teeth, prepare a stained slide of plaque.

   a. Set out a clean slide.

   b. Place a drop of water on the slide. This must air-dry, so make the drop of water *small*.

   c. Using a fresh toothpick, scrape your teeth near the gum line and mix the scraping in the drop of water.

   d. Spread this plaque-water mixture into a thin film and allow it to air-dry.

   e. When the smear is dry, hold the slide with a clothespin and pass it quickly over the flame of an alcohol lamp or Bunsen burner several times at a 45° angle. This should warm the slide but not cook the bacteria. Briefly touch the warm slide to the back of your hand. If it is too hot to touch, you are allowing it to get too warm.

Keep long hair and loose clothing away from the flame. Extinguish the flame immediately after use.

f. Place the slide on the support of a staining pan or tray and apply three or four drops of crystal violet stain to the smear (Figure 13.3).

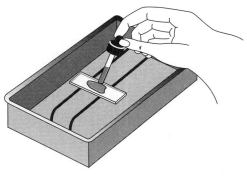

**Figure 13.3.**
Apply several drops of crystal violet stain to the slide supported in a staining pan or tray.

 Crystal violet will permanently stain your clothes, and it may last several days on your hands as well. Work carefully!

g. Leave the stain on the smear for 1 minute.

h. Wash the stain off with a gentle stream of water from a squirt bottle so that the stain goes into the staining pan (Figure 13.4).

i. Blot the stained slide gently with a paper towel. Do not rub hard or you will remove the bacteria.

3. Examine the bacteria growing in the plaque on your teeth and determine bacterial forms. Use the highest magnification on your compound microscope. If you have an oil immersion lens, after focusing on the high-dry power, without changing the focus knobs, rotate the high-dry objective to the side, add a drop of immersion oil directly to the bacterial smear, and carefully rotate the oil immersion objective into place. Focus with the fine adjustment only. After observing the slide, rotate the oil immersion objective away from the slide and wipe the objective carefully, using lens paper to remove all traces of oil.

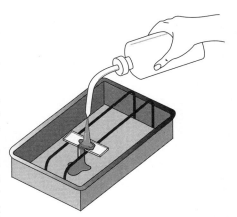

**Figure 13.4.**
**Gently rinse the stain into the staining pan.**

## Results

1. Record the individual cell shapes of bacteria present in plaque.

   *Cocci and bacilli are most common. Species present within a few hours after brushing include Streptococcus mutans (the leading cause of dental caries), S. salivarius, S. sanguis, and lactobacilli. Species in older plaque include Corynebacterium species (a bacillus), Actinomyces species (a filamentous form), and spirochetes. Students may also see yeast and large epithelial cells.*

2. What shapes are absent?

   *Spirilli are generally less common.*

3. Estimate the relative abundance of each shape.

   *Usually the cocci will be in greater abundance (75%) compared to bacilli (25% or less). Proportions will vary, depending on personal oral hygiene. Older plaque has more bacilli and spirilli. We once had a student with good oral hygiene who had an abundance of spirilli. He was told by his dental hygienist that this might be an indication of susceptibility to gum disease.*

## Discussion

1. Discuss with your lab partner information you have learned from your dentist or health class about the relationship among plaque, dental caries (cavities), and gum disease.

   *(You may choose to have students answer this question using online or library references.) In short, scientists have established unequivocally that bacteria cause dental caries. Bacteria convert carbohydrate to lactate, creating an acidic environment that decalcifies the tooth surface. Gum (periodontal) disease results when plaque grows under the gum rim (gingiva). Calcium is deposited in the plaque, forming tartar. The number of bacteria increases, the percentage of actinomycetes increases, and the gums become inflamed. Gums begin to bleed, they recede, and pockets form under the gums. The bone surrounding the teeth is resorbed, and the teeth loosen.*

2. Suggest an explanation for differences in the proportion of each type of bacteria in the bacterial community of plaque.

   *Diet, time since last dental visit, even differences in genetics will bring about differences. People taking antibiotics for other bacterial infections usually have only bacilli in their mouth.*

# Lab Study C. Identifying Bacteria by the Gram Stain Procedure

## Materials

compound microscope
blank slides
alcohol lamp or Bunsen burner
clean toothpicks
staining pan
cultures of *Micrococcus,*
   *Bacillus, Serratia,* and *E. coli*

dropper bottles of Gram
   iodine, crystal violet,
   safranin, DI water, 95% ethyl
   alcohol/acetone mixture
squirt bottle of DI water

## Introduction

*Bacterial colonies older than 24 hours often become gram-variable; use organisms from younger cultures for this study.*

**Gram stain** is commonly used to assist in bacterial identification. This stain, first developed in 1884, separates bacteria into groups, depending on their reaction to this stain. Bacteria react by testing either **gram-positive, gram-negative,** or **gram-variable,** with the first two groups being the most common. Although the exact mechanisms are not completely understood, scientists know that the response of cells to the stain is due to differences in the complexity and chemistry of the bacterial cell wall. Recall that bacterial cell walls contain a complex polymer, **peptidoglycan.** The cell walls of gram-negative bacteria contain less peptidoglycan than gram-positive bacteria. In addition, cell walls of gram-negative bacteria are more complex, containing various polysaccharides, proteins, and lipids not found in gram-positive bacteria. Gram-staining properties play an important role in bacterial classification.

Gram stain relies on the use of three stains: crystal violet (purple), Gram iodine, and safranin (pink/red). *Gram-positive* bacteria (with the thicker peptidoglycan layer) retain the crystal violet/iodine stain and appear blue/purple. *Gram-negative* bacteria lose the blue/purple stain but retain the safranin and appear pink/red (Color Plate 19).

In summary:

| Gram-Negative Bacteria | Gram-Positive Bacteria |
| --- | --- |
| more complex cell wall | simple cell wall |
| thin peptidoglycan cell wall layer | thick peptidoglycan cell wall layer |
| outer lipopolysaccharide wall layer | no outer lipopolysaccharide wall layer |
| retain safranin | retain crystal violet/iodine |
| appear pink/red | appear blue/purple |

In this lab study, you will prepare and stain slides of two different bacterial species. One member of the lab team should stain *Micrococcus* and *Bacillus*. The other member should stain *Serratia* and *E. coli*.

### Procedure

1. Using a *clean* slide, prepare smears as directed for the plaque slide (Lab Study B, steps 2a to 2e), substituting the bacterial species for the plaque. If you are using a liquid bacterial culture, do *not* add water to your slide (step 2b). Label the slide with your initials and the name of the bacterial species being investigated.

2. Support the slide on the staining tray and cover the smear with three or four drops of crystal violet. Wait 1 minute.

3. Rinse the stain gently into the staining pan with water from the squirt bottle.

4. Cover the smear with Gram iodine for 1 minute, setting the stain.

5. Rinse it again with water.

6. Destain (remove the stain) by dropping the 95% alcohol/acetone mixture down the slanted slide one drop at a time. At first a lot of violet color will rinse away. Continue adding drops until only a faint violet color is seen in the alcohol rinse. Do not overdo this step (Figure 13.5). You should be able to see some color in the smear on the slide. If not, you have destained too much. The alcohol/acetone removes the crystal violet stain from the gram-negative bacteria. The gram-positive bacteria will not be destained.

7. Using the water squirt bottle, rinse immediately to prevent further destaining.

*You can substitute broth cultures for agar cultures. If you choose to do this, have students apply two to three loopsful of bacteria directly on the slide and spread it to make a thin film. Dry and heat the smear and proceed to stain it as directed.*

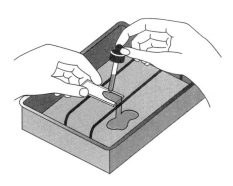

**Figure 13.5.**
**Destain by dropping 95% ethyl alcohol/acetone down the slanted slide until only a faint violet color is seen in the solution.**

*For a discussion of the chemistry of the Gram stain, see Pommerville (2007, pp. 90–92).*

8. Cover the smear with safranin for 30 to 60 seconds. This will stain the destained gram-negative bacteria a pink/red color. The gram-positive bacteria will be unaffected by the safranin (Figure 13.6).

9. Briefly rinse the smear with water as above. Blot it lightly with a paper towel and let it dry at room temperature.

10. Examine each slide using the highest magnification on your microscope.

 If you use oil immersion, remove all traces of oil from the objective after observing the slide.

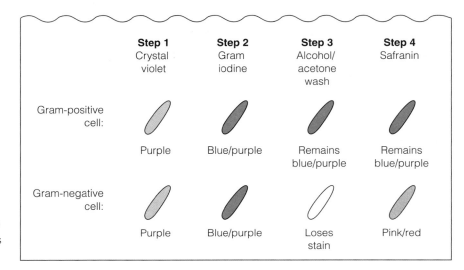

**Figure 13.6.**
**The Gram stain.** Crystal violet and Gram iodine stain all cells blue/purple. Alcohol/acetone destains gram-negative cells. Safranin stains gram-negative cells pink/red.

## Results

Record your observations of the results of the Gram stain in Table 13.2.

### Table 13.2
Bacteria Observed and Results of Gram Stain

*See the* Preparation Guide *for suggestions and tips about bacterial species.*

| Name of Bacteria | Results of Gram Stain |
|---|---|
| 1. *Micrococcus* <br> *Bacillus* | *positive* |
| 2. *Serratia* <br> *E. coli* | *negative* |

## Discussion

1. Which of the bacteria observed are probably more closely related taxonomically?

   *Micrococcus and Bacillus, being gram-positive, are probably more closely related. Serratia and E. coli, both being gram-negative, may be more closely related.*

2. What factors can modify the expected results of this staining procedure?

*Sloppy technique, age of cells (organisms over 24 hours old are often gram-variable, probably because the cell wall changes as cells age). Help students reason through this question because they have no access to the specific information.*

# EXERCISE 13.2
# Ecological Succession of Bacteria in Milk

## Materials

pH paper
flasks of plain and chocolate whole milk aged 1, 4, and 8 days
TGY agar plates of each of the milk types
supplies from Exercise 13.1 for Gram stains

## Introduction

Milk is a highly nutritious food containing carbohydrates (lactose, or milk sugar), proteins (casein, or curd), and lipids (butterfat). This high level of nutrition makes milk an excellent medium for the growth of bacteria. Pasteurizing milk does not sterilize it (sterilizing kills *all* bacteria) but merely destroys pathogenic bacteria, leaving many bacteria that will multiply very slowly at refrigerated temperatures; but at room temperature, these bacteria will begin to grow and bring about milk spoilage. Biologists have discovered that as milk ages, changing conditions in the milk bring about a predictable, orderly succession of microorganism communities (associations of species).

**Community succession** is a phenomenon observed in the organizational hierarchy of all living organisms, from bacterial communities in milk to animal and plant communities in a maturing deciduous forest. In each example, as one community grows, it modifies the environment, and a different community develops as a result.

In this laboratory exercise, you will work in pairs and observe successional patterns in two types of milk, plain whole milk and milk with sucrose and chocolate added. You will record changes in the environmental conditions of the two types of milk as they age. Consider the following information about milk bacteria and their environments as you continue this exercise.

1. *Lactobacillus* (gram-positive rod) and *Streptococcus* (gram-positive coccus) survive pasteurization.

2. *Lactobacillus* and *Streptococcus* ferment lactose to lactate and acetic acid.

3. An acidic environment causes casein to solidify, or curd.

4. Two bacteria commonly found in soil and water, *Pseudomonas* and *Achromobacter* (both gram-negative rods), digest butterfats and give milk a putrid smell.

5. Yeasts and molds (both fungi) grow well in acidic environments.

### Scenario

Propose a scenario (the hypothesis) for bacterial succession in each type of milk.

> *After reading the information given, students might propose that succession in the cultures will begin with a greater proportion of gram-positive coccus bacteria that survive pasteurization with some gram-positive rods. As the milk ages and the pH becomes more acidic, the proportion of gram-negative rods will increase and the gram-positive coccus will decrease. In later stages of succession, yeasts and molds, previously absent, will begin to grow. The additional sucrose in chocolate milk may accelerate the succession process and the growth of yeast. Accept any reasonable scenario.*

On each lab bench are four flasks of plain whole milk and four flasks of chocolate milk. One flask of each has been kept under refrigeration. One flask of each has been at room temperature for 24 hours, one for 4 days, and one for 8 days. On each bench there are also TGY (tryptone, glucose, yeast) agar plate cultures of each of the types of milk.

One team of two students should work with plain milk, another with chocolate milk. Teams will then exchange observations and results.

### Procedure

1. Using the pH paper provided, take the pH of each flask. Record your results in Table 13.3, in the Results section.
2. Record the odor (sour, putrid), color, and consistency (coagulation slight, moderate, chunky) for the milk in each flask.
3. Using the TGY agar plates, observe and describe bacterial/fungal colonies in each age and type of milk. Use the vocabulary you developed while doing Exercise 13.1.
4. Prepare Gram stains of each different bacterial type on each plate using the staining instructions in Exercise 13.1, Lab Study C.
5. Record the results of the Gram stains in Table 13.3, in the Results section.

### Results

Complete Table 13.3, in the Results section, describing the characteristics of each milk culture and the bacteria present in each.

**Table 13.3**
Physical Features and Bacterial/Fungal Communities
of Aged Plain and Chocolate Milk

| Age/Type Milk | Environmental Characteristics (pH, Consistency, Odor, Color) | Organisms Present (Bacteria: Gram +/−, Shapes; Yeasts or Fungi) |
|---|---|---|
| Refrigerated plain | pH 7; no coagulation; no odor | cocci dominant; since bacteria are dilute, few, if any, grow on plates |
| 24-hr plain | pH 5; no coagulation; no odor | cocci dominant; rods present |
| 4-day plain | pH 4; slight coagulation; strong sour odor | cocci dominant; rods present; occasional yeasts |
| 8-day plain | pH 4; heavy coagulation separating solid chunks from liquid; strong odor | cocci and rods present; yeasts more common; filamentous fungi appear |
| Refrigerated chocolate | pH 7; no coagulation; no odor | cocci dominant; since bacteria are dilute, few, if any, grow on plates |
| 24-hr chocolate | pH 5; coagulation beginning; no odor | cocci dominant; rods present |
| 4-day chocolate | pH 4; increased coagulation; strong sour odor | cocci continue to dominate, but rods are proportionally increasing; yeasts very common |
| 8-day chocolate | pH 4; heavy coagulation separating solid chunks; strong odor | filamentous fungi and yeasts dominate; rods more common; cocci declining |

*Results will vary depending on the variety of milk. Let students' observations stand. Do not correct them.*

*Make an overhead acetate of this blank table from the student manual. Have a representative from each group fill in results so that all students have a completed table.*

*For a discussion of changes in bacterial communities as milk spoils, see Pommerville (2007, pp. 768–769).*

## Discussion

1. Describe the changing sequence of organisms and corresponding environmental changes during succession in plain milk. Do the results of your investigation match your hypothesis?

   *As milk ages, the predominant coccus forms feed on milk sugar (lactose), converting it to lactate and lowering the pH. A low pH is unfavorable to the coccus species and favors the bacillus species. Bacillus species multiply as the pH continues to fall, creating an environment that favors the growth of filamentous fungi and yeasts.*

2. Describe the changing sequence of organisms and corresponding environmental changes during succession in chocolate milk. Do the results of your investigation match your hypothesis?

   *The pH rapidly becomes more acidic, which favors the growth of rods, yeasts, and filamentous fungi. Pseudomonas and Achromobacter must be increasing as the putrid smell increases.*

3. Compare succession in plain and chocolate milk. Propose reasons for differences.

   *Additional sugar in chocolate milk encourages more rapid bacterial growth than in plain milk and a more drastic pH change early in the experiment. This seems to favor the rapid growth and ultimate dominance of fungi and the putrid-smelling bacteria.*

4. Propose an experiment to test the environmental factors and/or organisms changing in your proposed scenario for milk succession.

   *Pure cultures of bacteria or fungi could be grown under controlled pH conditions or on media that lack certain nutrients, such as sugar (lactose).*

---

E X E R C I S E   1 3 . 3

# Bacteria in the Environment

---

Bacteria thrive in a wide variety of natural environments. They are present in all environments where higher forms of life exist, and many flourish in extreme environments where no higher life forms exist—boiling hot springs, extremely salty bodies of water, and waters with extreme pH.

In the following experiments, you will sample different environments, testing for the presence of bacteria and fungi. In Experiment A, pairs of students will investigate one of six different environments. Each pair will report results to the entire class. In Experiment B, your team will investigate an environment of your choice.

# Experiment A. Investigating Specific Environments

## Materials

sterile agar plates
wax pencil
2 cotton-tipped sterile swabs
bacterial inoculating loop
alcohol lamp or Bunsen burner
piece of raw chicken in a petri dish
soil samples

samples of pond water
plant leaves or other plant parts
hand soap
Parafilm strips
discard receptacle
forceps

*Have trays labeled for each lab section to store students' cultures.*

## Introduction

The instructor will assign team numbers to each pair of students. Each pair (team) of students will sample bacteria and fungi from one of six environments: food supply, soil, air, pond water, a plant structure, and hands. Read the instructions for all investigations. Think about the following questions, and before you begin your investigation, hypothesize about the relative growth of bacteria and fungi in the different environments.

*In a class of 24, each environment will be tested by two teams.*

Where in the environment would bacteria be more common, and where would fungi be more common? Would any of these environments be free of bacteria or fungi?

 **Seal all plates with Parafilm after preparation! Discard all used swabs in the designated receptacle!**

## Hypothesis

Hypothesize about the presence of bacteria and fungi in the different environments.

*All environments will have some bacteria and fungi present. Bacteria will be more common than fungi on food and hands, and fungi will be more common in soil and water. (Accept any testable hypothesis.)*

## Prediction

Predict the results of the experiment based on your hypothesis (if/then).

*If bacteria and fungi are present in all environments, then after incubation, all plates will show colonies of each. There will be more bacteria on the food and hands cultures and more fungi on the soil and water cultures.*

## Procedure

### Team 1

1. Holding the lid in place, invert an agar plate and label the bottom "chicken."

2. Open the dish containing the piece of chicken, and swab the chicken surface using a sterile cotton swab.

*Use pieces of fresh chicken. Place the pieces in the appropriate number of petri dishes and refrigerate them until needed for the experiment. Return the dishes to the refrigerator after the experiment.*

 **Avoid touching the chicken. Use the swab. Always wash your hands thoroughly after touching raw chicken, owing to the potential presence of *Salmonella*, bacteria that cause diarrhea.**

3. Isolate bacteria by the **streak plate** method.

   a. Carefully lift the lid of the agar plate to 45° and lightly streak the swab back and forth across the top quarter of the agar (Figure 13.7a). Close the lid and *discard the swab in the receptacle provided*. Minimize exposing the agar plate to the air.

   b. Flame the bacterial inoculating loop using the alcohol lamp or Bunsen burner. Allow the loop to cool; starting at one end of the swab streak, lightly streak the microorganism in the pattern shown in Figure 13.7b. Do not gouge the medium.

   c. Reflame the loop and continue to streak as shown in Figure 13.7c and described in the figure legend.

   By the end of the last streak, the bacteria should be separated and reduced in density so that only isolated bacteria remain. These should grow into isolated, characteristic colonies.

4. Write the initials of your team members, the lab room, and the date on the petri dish.

5. Seal the dish with Parafilm and place it in the area indicated by the instructor.

6. Incubate the culture 1 week and observe results during the next laboratory period.

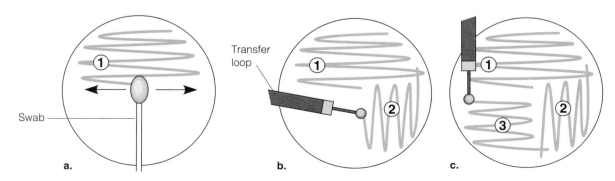

**Figure 13.7.**
**Isolating bacterial colonies using the streak technique.** (a) Streak the swab over the top quarter of the agar plate, region 1. (b) Using the newly flamed loop, pick up organisms from region 1 and streak them into region 2. (c) Reflame the loop and pick up organisms from region 2 and streak them into region 3.

The following week, to avoid exposure to potentially pathogenic bacteria, do not open the petri dish containing the chicken bacteria. Wash hands after handling cultures.

## Team 2

1. Holding the lid in place, invert an agar plate and label the bottom "soil."
2. Using a cotton swab, pick up a small amount of soil from the sample.
3. Prepare a streak culture by following step 3 in the procedure for Team 1.
4. Write the initials of your team members, the lab room, and the date on the petri dish.
5. Seal the dish with Parafilm and place it in the area indicated by the instructor.
6. Incubate the culture 1 week and observe results during the next laboratory period.

*For interest, have one team apply compost.*

## Team 3

1. Holding the lid in place, invert an agar plate and label the bottom "air."
2. Collect a sample of bacteria by leaving the agar plate exposed (lid removed) to the air in some interesting area of the room for 10 to 15 minutes. Possible areas might be near a heat duct or an animal storage bin.
3. If additional agar plates are available, you may choose to sample several sites.
4. Write the initials of your team members, the lab room, and the date on each petri dish.
5. Seal the dish with Parafilm and place it in the area indicated by the instructor.
6. Incubate the culture(s) 1 week and observe results during the next laboratory period.

## Team 4

1. Holding the lid in place, invert an agar plate and label the bottom "pond water."
2. Using a sterile cotton swab, take a sample from the pond water.
3. Prepare a streak culture by following step 3 in the procedure for Team 1.
4. Write the initials of your team members, the lab room, and the date on the petri dish.
5. Seal the dish with Parafilm and place it in the area indicated by the instructor.
6. Incubate the culture 1 week and observe results during the next laboratory period.

## Team 5

1. Obtain a sample from a plant. This might be a leaf, a part of a flower, or another plant structure.

2. Holding the lid in place, invert an agar plate and label the bottom "plant _____ " naming the part you are testing. If you are testing a leaf, indicate if it is the top or bottom of the leaf.

3. Open the plate and place the plant part flat on the agar surface. You may need to use forceps. Be sure there is good contact. Remove and discard the plant part. Close the petri dish.

4. Write the initials of your team members, the lab room, and the date on the petri dish.

5. Seal the dish with Parafilm and place it in the area indicated by the instructor.

6. Incubate the culture 1 week and observe results during the next laboratory period.

## Team 6

1. Draw a line across the center of the bottom of an agar plate. Write "unwashed" on the dish bottom on one side of the line and "washed" on the other side of the line.

2. Select one person who has not recently washed his or her hands to be the test subject. The subject should open the petri dish and *lightly* press three fingers on the agar surface in the half of the dish marked "unwashed." Do not break the agar. Close the petri dish.

3. The subject should wash his or her hands for 1 minute and repeat the procedure, touching the agar with the same three fingers on the side of the dish marked "washed."

4. Write the initials of your team members, the lab room, and the date on the petri dish.

5. Seal the dish with Parafilm and place it in the area indicated by the instructor.

6. Incubate the culture 1 week and observe results during the next laboratory period.

*You may vary this experiment by comparing common soap with antibacterial soap or hand sanitizer.*

## Results

Include results from the entire class.

1. During the following laboratory period, observe your agar cultures of bacteria and fungi from the environment and record your observations in Table 13.4.

2. Place your agar culture on the demonstration table and make a label of the environment being investigated. All students should observe every culture.

3. Observe the agar plates prepared by your classmates. Record observations in Table 13.4.

*Place large labels on a demonstration table and have students place their cultures in the appropriate area.*

**Table 13.4**
Abundance and Types of Colonies Associated with Food
(Raw Chicken), Soil, Air, Water, Plant Structures, and Hands

| Environment | Colony Type(s) and Abundance |
|---|---|
| Chicken | |
| Soil | |
| Air | |
| Pond water | |
| Plant structure | |
| Hands before washing | |
| Hands after washing | |

## Discussion

1. How did the plates differ in the number and diversity of bacterial and fungal colonies?

   *Chicken gives an extensive bacterial growth but has less diversity. The results of cultures from soil, air, and water will vary, with soil usually giving the greatest diversity. What happens with the air culture will depend on your particular facilities. If students place an entire leaf on the agar surface, they may see the leaf silhouette in bacteria!*

2. Did your predictions match your observations? Describe any discrepancies.

   *Students might predict that washing would reduce numbers of bacterial colonies. However, the results may show that after washing, new bacterial species are present and the new species are more common than the original. Washing removes outer cell layers and exposes bacterial species that live deep in underlying dead cells. Point out that washing removes pathogenic bacteria picked up from surfaces. Standard infection control is washing for 1 full minute. Nurses and surgeons scrub and use brushes for 5 minutes before surgery.*

3. What factors might be responsible for your results?

   *Such things as source of substances tested, ventilation systems in room; for stream water, results will depend on degree of pollution.*

4. Based on the results of your experiments, suggest health guidelines for workers in the food industry, as well as for schoolchildren or others who might be concerned with sanitary conditions.

*Students should conclude that the presence of bacteria and fungi in most environmental areas, including human hands, compels us to wash hands before eating and before and after handling food being prepared for eating. Food preparation areas should be washed frequently.*

## Experiment B. Investigating the Environment of Your Choice

### Materials

agar plates                      Parafilm strips
sterile swab                     discard receptacle
capped test tube of sterile water

### Introduction

In the previous experiment, you tested specific environments for the presence of bacteria and fungi. In this lab study, you will study an environment of your choice. If extra agar plates are available, you may choose to investigate bacteria in an environment before and after some treatment, such as bacteria on the water fountain before and after cleaning.

 **Seal all plates with Parafilm after preparation!**

### Hypothesis

Hypothesize about the growth of bacteria in an environment of your choice.

### Prediction

Predict the results of the experiment based on your hypothesis.

## Procedure

1. Decide what environment you will investigate. It might be some environment in the lab room or somewhere in the biology building. Carry the sterile cotton swab and agar plate to the environment, and use the swab to collect the sample. If you are collecting from a dry surface, you should first dip the cotton swab in the sterile water and then swab the surface. If you apply any treatment to the surface, describe the treatment in the margin of your lab manual.

 **Do not do throat or ear swabs! Pathogenic bacteria may be present.**

2. Open the agar plate and lightly streak the swab back and forth across the agar. Discard the swab in the receptacle provided.
3. Label the bottom of the agar plate to indicate the environment tested. Record the environment tested in the Results section.
4. Write the initials of your team members, the lab room, and the date on the petri dish.
5. Seal the dish with Parafilm and place it in the area indicated by the instructor.
6. Incubate the culture 1 week and then observe and describe results during the next laboratory period.

## Results

1. What environment did you investigate? Indicate any treatment you applied.

2. Characterize the bacterial and fungal colonies from your experiment.

## Discussion

1. Do your results match your predictions for the presence of bacteria and fungi in this environment?
2. What factors might be responsible for your results?

# EXERCISE 13.4
# Controlling the Growth of Bacteria

Bacteria are found almost everywhere on Earth, and most species are directly or indirectly beneficial to other organisms. Bacteria are necessary to maintain optimum environments in animal and plant bodies and in environmental systems. However, even beneficial species, if they are reproducing at an uncontrolled rate, are potentially harmful or destructive to their environment. In addition, several species of bacteria and fungi are known to be pathogenic, that is, to cause disease in animals and plants. Their growth must be controlled. Agents have been developed that control bacterial and fungal growth. In this exercise, you will investigate the efficacy of three of these growth-controlling agents: antibiotics, antiseptics, and disinfectants.

## Lab Study A. Using Antibiotics to Control Bacterial Growth

### Materials

*If you have bacteria spreaders, they can be used to produce lawns in this exercise. (One kind commercially available is called Bacti-Spreader.)*

agar plate
metric ruler
sterile swab
wax pencil
Parafilm strips

broth cultures of *Micrococcus*,
    *Bacillus*, *Serratia*, and *E. coli*
antibiotic dispenser with
    antibiotic disks

### Introduction

An **antibiotic** is a chemical produced by a bacterium or fungus that has the potential to control the growth of another bacterium or fungus. Many antibiotics are selective, however, having their inhibiting effect on only certain species of bacteria or fungi. In this lab study, you will apply an assortment of antibiotics to a lawn culture of a bacterial species. Working in pairs, you will determine which antibiotics are able to control the growth of the bacteria. Each pair of students in a group of eight should culture a different bacterium. All four species should be cultured.

A lawn of bacteria is like a lawn of grass—a uniform, even layer of organisms covering an entire surface. Prepare the lawn of bacteria carefully. The success of this experiment will largely depend on the quality of your lawn.

### Hypothesis

Hypothesize about the effect of different antibiotics on the growth of bacteria.

*All antibiotics control the growth of all bacterial species.*

## Prediction

Predict the results of the experiment based on your hypothesis.

*If all antibiotics control the growth of all bacterial species, then the zone of inhibition will be the same around each type of antibiotic disk on all bacterial lawns, regardless of species.*

## Procedure

1. Label the bottom of an agar plate with your initials, the lab room, the date, and a word to indicate the experiment (such as "antibiotic").

2. Prepare a bacterial lawn.

   a. Insert a sterile swab into the bacterial culture in liquid nutrient broth.

   b. Allow the swab to drip for a moment before taking it out of the culture tube, but do *not* squeeze out the tip. The swab should be soaked but not dripping.

   c. Carefully lift the lid of the agar plate to about 45° and swab the *entire* surface of the agar, taking care to swab the bacteria to the edges of the dish (Figure 13.8a).

   d. Rotate the plate 45° and swab the agar again at right angles to the first swab (Figure 13.8b). Close the lid.

3. Carry the agar plate swabbed with bacteria to the demonstration table.

4. Remove the plate lid, place the antibiotic disk dispenser over the plate, and dispense the disks (Figure 13.9). (Each disk has been saturated with a particular antibiotic. The symbol on the disk indicates the antibiotic name. Your instructor will provide a key to the symbols.)

5. Replace the lid, seal the plate with Parafilm, and place the plate in the area indicated by the instructor. Incubate the dishes at 37°C for 24–48 hours and then refrigerate them.

6. Next week, examine cultures to determine bacterial sensitivity to antibiotics. Measure the diameter of the **zone of inhibition** (area around disk where bacteria growth has been inhibited) for each antibiotic (Color Plate 20).

7. Record the measurement for your bacterial species and each antibiotic in Table 13.5. If the antibiotic had no effect on bacterial growth, record the size of the zone as 0.

## Results

1. Using your results and the results from other teams, complete Table 13.5. Record the sizes of the zone of inhibition for all species of bacteria and all antibiotics.

2. Use the following arbitrary criteria to rank relative bacterial sensitivity to antibiotics:

   NS = not sensitive = no zone of inhibition

   S = sensitive = zone size above 0 but less than 1 cm

   VS = very sensitive = zone size greater than 1 cm

   Write the designation in each blank in the table.

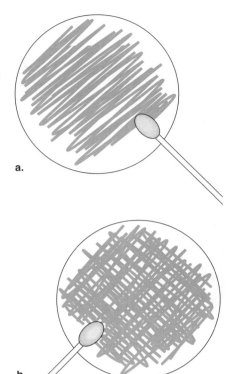

**a.**

**b.**

**Figure 13.8.**
**Preparation of a bacterial lawn.**
(a) Apply the bacteria evenly over the entire agar surface. (b) Rotate the plate and swab at right angles to the first application.

**Figure 13.9.**
**A typical antibiotic dispenser.**

**Table 13.5**

Results of Antibiotic Sensitivity Tests (Size of inhibition zone for each antibioic is given in centimeters.)

| Bacteria (Name, Gram + or −) | Antibiotic | | | | | | |
|---|---|---|---|---|---|---|---|
| | | | | | | | |
| 1. | | | | | | | |
| 2. | | | | | | | |
| 3. | | | | | | | |
| 4. | | | | | | | |

**Discussion**

1. Did your results support your hypothesis? Was the zone of inhibition the same for all bacteria?

2. Were any bacteria very sensitive (greater than 1 cm) to all antibiotics? If so, which bacteria?

3. Based on your results, which antibiotic would you prescribe for each microorganism?

   *Depending on the antibiotics and bacteria chosen, neomycin, kanamycin, and erythromycin are usually more effective than penicillin.*

4. Were the results different for gram-positive and gram-negative bacteria?

   *This will vary depending on which antibiotics and bacteria you use.*

5. Can you think of alternative explanations for the differential effectiveness of some antibiotics?

   *Students should discuss the resistance of bacteria to widely used antibiotics as well as the different modes of action by antibiotics.*

# Lab Study B. Using Antiseptics and Disinfectants to Control Bacterial Growth

## Materials

agar plate
forceps
metric ruler
sterile swab
wax pencil
Parafilm strips

broth cultures of *Micrococcus,
    Bacillus, Serratia,* and *E. coli*
paper disks soaking in disinfectants
paper disks soaking in antiseptics
paper disks soaking in sterile water

## Introduction

Other agents besides antibiotics are often used to control bacterial growth. Those used to control bacteria on living tissues such as skin are called **antiseptics.** Those used on inanimate objects are called **disinfectants.** Antiseptics and disinfectants do not kill all bacteria, as would occur in sterilization, but they reduce the *number* of bacteria on surfaces.

## Hypothesis

Hypothesize about the effect of antiseptics and disinfectants on the growth of bacteria.

*Disinfectants are more effective than antiseptics in controlling bacterial growth.*

## Prediction

Predict the results of the experiment based on your hypothesis.

*If disinfectants are more effective than antiseptics in controlling bacterial growth, then zones of inhibition around disinfectant disks will be larger than those around antiseptic disks.*

## Procedure

1. Holding the lid in place, invert a sterile agar plate and label the bottom with your initials, the lab room, and the date. Draw four circles on the bottom. Number the circles.

2. Using the same bacterial culture as you used in Lab Study A, prepare a lawn culture as instructed in Lab Study A.

3. Carry the closed agar plate swabbed with bacteria to the demonstration table.

4. Open the agar plate; using forceps soaking in alcohol, pick up a disk soaked in one of the antiseptics or disinfectants, shake off the excess liquid, and place the disk on the agar above one of the circles. Repeat this procedure with two more antiseptics and/or disinfectants. Place a disk soaked in sterile water above the fourth circle to serve as a control.

5. Record the name of the agent placed above each numbered circle in Table 13.6. (Example: 1 = Lysol, 2 = Listerine, and so on.) Seal the plate with Parafilm.

*Check the plates after 24–48 hours and refrigerate them if there is adequate growth.*

6. Place the agar plate in the area indicated by the instructor. Incubate the agar plates at 37°C for 24–48 hours and then refrigerate them.

7. Next week, examine the cultures to determine the bacterial sensitivity to disinfectants and antiseptics. Measure the diameter of the zone of inhibition for each agent.

8. Record the measurement for your bacterial species and each inhibiting agent in Table 13.6. If the agent had no effect on bacterial growth, record the size of the zone as 0.

**Table 13.6**
Results of Sensitivity Tests of Antiseptics and Disinfectants (Size of inhibition zones given in centimeters.)

| Bacteria | Antiseptic/Disinfectant/Control | | | |
|---|---|---|---|---|
| | 1. | 2. | 3. | 4. Control |
| 1. | | | | |
| 2. | | | | |
| 3. | | | | |
| 4. | | | | |

**Results**

1. Using your results and the results from other teams, complete Table 13.6. Record sizes of the zone of inhibition for all species of bacteria and all antiseptics and disinfectants.

2. Use the following arbitrary criteria to rank relative bacterial sensitivity to antiseptics and disinfectants:

   NS = not sensitive = no zone of inhibition

   S = sensitive = zone size above 0 but less than 1 cm

   VS = very sensitive = zone size greater than 1 cm

   Write the designation in each blank in the table.

**Discussion**

1. Did your results support your hypothesis? Explain.

   *Yes. Zones of inhibition are larger around disinfectant disks than around antiseptic disks.*

2. Based on your results, which disinfectant is most effective in controlling the growth of bacteria?

   *Household bleach works best.*

3. Which antiseptic is most effective?

   *Rubbing alcohol usually works best.*

4. In which situations is it appropriate to use a disinfectant?

   *Use a disinfectant to control bacterial growth on inanimate objects.*

   An antiseptic?

   *Use an antiseptic to control bacterial growth on living tissues—humans or other animals.*

## Reviewing Your Knowledge

1. Once you have completed this lab topic, you should be able to define and use the following terms, providing examples if appropriate: *sterilize, pasteurize, nutrient broth* and *agar, coccus, bacillus, spirillum, biofilm, antibiotic, antibiotic resistance, antiseptic, disinfectant, peptidoglycan, aseptic technique.*

2. Compare the techniques used to prepare a lawn culture and a streak culture.

3. What is the significance of the clear zones around some antibiotic disks used in Exercise 13.4, Lab Study A?

   *The particular bacteria being tested did not grow in the clear zone indicating that the antibiotic diffused into the agar and inhibited the bacterial growth in that zone.*

4. A liquid that has been sterilized may be considered pasteurized, but one that has been pasteurized may not be considered sterilized. Why not?

   *Sterilization kills all microorganisms. Pasteurization kills only pathogenic bacteria.*

## Applying Your Knowledge

1. Would you expect the community of bacteria in plaque sampled 1 week *after* you have your teeth cleaned to differ from the community of bacteria found 1 week *before* you have your teeth cleaned? Explain. In your answer, consider the results of the milk succession experiment.

   *Initial bacterial communities will produce an environment that will favor the invasion of new species (such as changing pH or creating pockets for anaerobic bacteria to grow), as students observed in the milk experiment.*

2. When you were younger and were sick with a respiratory infection, it is likely that your parents used a portable home humidifier in your room. Room humidifiers are good environments for the growth of biofilms. Using information you learned in Exercise 13.4, what would you suggest as an effective way to clean a humidifier?

   *Students should suggest the disinfectant that was most effective in inhibiting bacterial growth. This was most likely household bleach (sodium hypochlorite).*

3. Bacterial species that are harmful, as well as others that are beneficial, are found living in the human body. To slow the rate of developing antibiotic resistance in bacteria, physicians are being encouraged to use "narrow spectrum" antibiotics—those that target only a few bacterial types. How can the information learned by antibiotic sensitivity testing be used by physicians who must choose antibiotics that inhibit the growth of bacteria causing disease but that do not interfere with beneficial bacteria?

   *Samples of the bacterium could be plated (for example, throat swabs) and antibiotic disks applied. The physician should choose an antibiotic that specifically inhibits the infectious bacterium but does not inhibit beneficial bacteria. Students will see that penicillin, for example, does not inhibit the growth of E. coli, a bacterium that is a symbiotic resident of the human intestine, but is effective in controlling the growth of other bacteria.*

4. Search the Web for information about milk seen in boxes on grocery store shelves. How is this milk prepared? How would you expect bacterial succession in milk prepared in this fashion to differ from succession in milk as investigated in Exercise 13.2?

   *Boxes of milk seen on grocery store shelves are prepared by ultrahigh temperatures (UHT), 285°F (about 140°C) for 1–2 seconds. At this temperature, the milk is sterilized rather than pasteurized.*

5. Death rates due to infectious diseases declined steadily in the United States throughout most of the 20th century. However, since the 1980s, infectious disease-related deaths in the U.S. are increasing significantly. Speculate about possible factors that may be contributing to this increase.

   *Some possible suggestions: Resistance to antibiotics is evolving in many strains of bacteria. Increased population density and international travel make disease transmission much easier. Sexual behavior has changed. Economic development and changes in land use—for example, the rise in Lyme disease may be related to a larger number of homes and recreational areas near forests.*

6. At one time bacterial pneumonia, tuberculosis, streptococcal sore throat, and gonorrhea could easily be controlled with one dose of antibiotic. Now these diseases are among the most difficult to treat. How could you explain this?

   *Use this question to call attention to the serious problem of some bacteria that are developing antibiotic resistance and to encourage students to use antibiotics as directed by their physician. See the CDC web reference at the end of this lab topic.*

7. In 1929, Alexander Fleming, a Scottish physician, discovered the first antibiotic when he noticed that colonies of certain staphylococcus bacteria growing in culture plates appeared to die when the plates became contaminated with the fungus *Penicillium*. Fleming concluded that a substance diffusing from the fungus into the growth medium was causing the bacteria to lyse (break down), and he called this substance penicillin.

   In its natural environment, what would be the adaptive advantage of a fungus producing and secreting a bacterial inhibitor?

   *If the fungus secretes a substance that inhibits the growth of bacteria in its environment, this will reduce competition for nutrients.*

# Investigative Extensions

Topics related to bacteriology are popular studies because they apply to many areas of ecology and human health. Diseases caused by waterborne bacteria cause almost 2 million deaths annually worldwide. In recent years, illnesses caused by contamination of foods such as ground beef, leafy greens, prepackaged cookie dough, and shellfish have been reported, leading to a heightened awareness of potential food contamination and changes in agricultural and food-processing practices.

Recent outbreaks of new flu strains in the U.S. have made the public more aware of the spread of bacteria and viruses by contact with infected people. In May 2009, the CDC issued hand-washing recommendations to reduce disease transmission. This agency recommended providing more hand-washing facilities in public settings and the use of alcohol-based hand sanitizer dispensers if soap and water are not available.

These ideas represent a few of the current popular issues that relate to bacteriology. The opportunity to perform an investigative extension will allow you to ask any number of interesting questions relating to these topics and to design an experiment to investigate one or more of these questions.

## Investigating Soil Bacteria

The study of bacteria is important not only to health professionals, but also to ecologists. Microbial ecologists know the importance of having bacterial diversity in soils in all ecosystems. Certain bacteria in soil serve as decomposers that consume simple carbon compounds releasing inorganic materials useful to other organisms in the soil food web. Other bacteria—for example nitrogen-fixing bacteria, form mutualistic relationships with plants, benefiting both bacteria and plant. Some species of bacteria are important in nitrogen cycling and degradation of pollution, and some soil bacteria are pathogenic and cause plant diseases.

Using ideas and protocols in this laboratory topic, design an experiment to investigate soil microbial diversity. First survey your campus or community and propose possible questions about soil bacteria in your area. For example, you might compare samples of soil taken from the athletic field that is regularly treated with pesticides, from a natural area, a flower garden, a wetland, an organic garden, etc.

Before you begin your experiment, review safety notes and aseptic techniques (Exercise 13.1). Wear gloves when handling plates and bacteria and wash hands frequently. Remember to keep careful records of procedures and results for each component of your experiment.

After preparing and incubating plates (room temperature is usually adequate) record the numbers of fungal and bacterial colonies on each plate and the morphology—shape and color—of bacterial colonies. Then, using techniques for streaking plates, prepare plates of individual colonies. After the plates have incubated, use laboratory protocols for determining cell shapes and Gram stain characteristics. Look for motility in freshly prepared slides.

With your partners, decide how you will present your results. Design tables and graphs to collect and present data. If digital and/or microscope photographic equipment is available, photograph your plates and the bacteria on your slides. Photographs may be used in a written or oral presentation of your experiment.

*If digital cameras are available, allow student teams to check them out to use to document sampling environments, and cultures. With adapters these may be used to document microscopic observations.*

### Investigating Bacteria and Human Health

Using techniques provided in this lab topic and following aseptic techniques and protocols (Exercise 13.1), design an experiment to investigate one of the following:

1. Investigate the efficacy of hand washing by varying the type of soap (liquid, bar, antibacterial, deodorant) or the manner of washing (scrubbing time, use of a brush). (Exercise 13.3)

2. Investigate the efficacy of waterless hand sanitizers in killing bacteria after various activities—for example, using the rest room, touching raw chicken, shaking hands, or cracking a raw egg. Wear disposable gloves when performing this experiment. (Exercise 13.3)

*See Pommerville (2007, p. 714) for a discussion of research relating to some of these natural products.*

3. The chicken industry and the FDA have recently been criticized for having low health and safety standards. Pursue this topic by a survey of brands or handling techniques. Health officials now recommend that all eggs be cooked before eating to avoid *Salmonella*. Determine the extent of contamination in store-bought eggs and in eggs from local sources. (Exercise 13.3)

4. Design an experiment to test bacterial succession in plaque. (Exercise 13.2)

5. Onions, garlic, green tea, cinnamon, honey, wasabi, and grapefruit seeds have all been suggested as having medicinal and antiseptic properties. Design an experiment to test this. (Exercise 13.4)

## Student Media: BioFlix, Activities, Investigations, and Videos
**www.masteringbiology.com (select Study Area)**

**Activities**—Ch. 26: Classification Schemes
   Ch. 27: Prokaryotic Cell Structure and Function; Classification of Prokaryotes
**Videos**—Ch. 27: Discovery Videos: Antibiotics; Bacteria; Tasty Bacteria

# References

Dill, B. and H. Merilles. "Microbial Ecology of the Oral Cavity," in *Tested Studies for Laboratory Teaching (Volume 10). Proceedings of the 10th Workshop/Conference of the Association for Biology Laboratory Education (ABLE)*, Corey Goldman, editor, 1989.

Doolittle, W. F. "Uprooting the Tree of Life." *Scientific American*, 2000, vol. 282, pp. 90–95.

Gillen, A. L. and R. P. Williams. "Pasteurized Milk as an Ecological System for Bacteria." *The American Biology Teacher*, 1988, vol. 50, pp. 279–282.

Hughes, J. M. "Emerging Infectious Diseases: A CDC Perspective." *Emerging Infectious Diseases*, 2001, vol. 7, pp. 494–496.

Levy, S. B. "The Challenge of Antibiotic Resistance." *Scientific American*, 1998, vol. 278, pp. 46–53.

Nester, E. W., C. E. Roberts, N. N. Pearsall, and B. J. McCarthy. *Microbiology*, 2nd ed. Dubuque, IA: WCB/McGraw-Hill, 1998. Good discussion of ecological succession of microbes.

Pommerville, J. C. *Alcamo's Fundamentals of Microbiology*, 8th ed. Boston, MA: Jones and Bartlett Publishers, 2007. Good discussion of the role of bacteria in dental disease.

Reece, J. et al., *Campbell Biology*, 9th ed. San Francisco, CA: Pearson Education, 2011.

## Websites

Articles about bacteria:
http://people.ku.edu/~jbrown/bugs.html

For information on a variety of topics relating to bacteria, visit this site maintained by the American Society for Microbiology:
http://www.microbeworld.org

Information about the Domains Archaea and Bacteria:
http://www.ucmp.berkeley.edu/archaea/archaea.html
http://www.ucmp.berkeley.edu/bacteria/bacteria.html

Information about responsible antibiotic use:
http://www.cdc.gov/getsmart/antibiotic-use/fast-facts
.html

# LAB TOPIC 13

# Bacteriology
## Teaching Plan for Laboratories

## Main Concepts and Objectives

1. Concept: bacteria differing in colony morphology and growth patterns. Students will describe colony morphology of various bacterial species. (Exercise 13.1)

2. Concept: bacteria differing in cell shape. Students will recognize the individual cell shape of various species. (Exercise 13.1)

3. Concept: bacteria identified by chemical and physiological differences. Students will perform and explain the Gram stain. (Exercise 13.1)

4. Concept: microbial succession. Students will describe succession of bacterial and fungal communities in aging milk, relating this to change in environmental conditions such as pH and nutrient availability. (Exercise 13.2)

5. Concept: environmental distribution of bacteria. Students will learn to use correct aseptic techniques to make lawns, streaks, and smears of bacteria from different environments. (Exercise 13.3)

6. Concept: control of bacterial growth. Students will investigate substances that control bacterial growth. (Exercise 13.4)

7. Specific content. Student will be able to describe or define: *prokaryotic, eukaryotic, bacterial lawn culture, bacterial streak culture, bacterial shapes (coccus, bacillus, spirillum), biofilm, antibiotic, antiseptic, disinfectant, microbial ecological succession, peptidoglycan, sterilize, pasteurize, disinfect antibiotic resistance.*

## Order of the Lab

1. Review lab safety and aseptic techniques.                    (10 min)
2. Introduce concepts and objectives.                           (15 min)
3. Provide instructions for techniques.                         (15 min)
4. Students perform exercises.                                   (2 hr)

Allow about 30 minutes at the beginning of lab the following week to observe and record results.

**For a 2-hour lab:** Probably the easiest exercise to omit without disturbing the integrity of the topic is the study of bacterial succession in milk (Exercise 13.2). Although this is a very interesting activity and ties together several concepts in biology, it requires additional discussion and explanation. If you are concerned that the lab is still too long for 2 hours, omit the study of bacteria colony characteristics (Exercise 13.1, Lab Study A). Colony characteristics can be observed as students record results during the second week.

This lab topic could easily be adapted to two 2-hour labs. Postpone the milk bacteria observations until the second week and carry out this exercise after the students observe the results of the first week's experiments.

## Classroom Management

Students work individually or in pairs. Directions are given in the Introduction of each lab study. Occasionally, students are asked to share observations among several pairs.

For the milk succession study, there should be 2 of each of the 6 different TGY agar cultures per laboratory (a total of 12 plates): two 8-day plain and chocolate, two 4-day plain and chocolate, and two 24-hour plain and chocolate. As students study the bacterial succession in milk, have two students work with one plate. The pair should first observe the plate and determine if there are bacteria only, bacteria and fungi, or fungi only. They should then make slides of the bacteria. If several different species of bacteria appear to be present, students should make slides of different bacteria until all species have been studied. If only one species of bacteria appears to be present, both students should make a slide of the one species present. Have students share and discuss results, completing Table 13.3 in the exercise. As students make slides of the bacteria, they should determine characteristics of the bacteria (mainly shape) and exchange data with others in the class until they have enough information to propose a scenario for community succession in each of the two types of milk. They should relate types of bacteria and fungi to environmental conditions (mainly pH) and the smell and texture of the milk. Successional changes occur as the pH and availability of certain foods change. Texts suggest that the sequence is from streptococci to lactobacilli to yeasts and molds to *Bacillus*. We have found that there is a great deal of variation, depending, for example, on the brand of milk used. The point is not that there is a correct answer, but that changes in environment favor new species.

## Student Development

Students are introduced to the aseptic techniques used in microbiology. They improve observational skills and data-collecting skills. They practice hypothesis testing, making predictions, and applying information to new problems. They are asked to synthesize and integrate evidence.

## Lab Safety Precautions

Instruct students to take appropriate precautions, as described.

1. *Bacterial cultures and swabs.* Wear disposable gloves, and clean up spills with a disinfectant. Use disposable towels. Place contaminated towels and gloves in an autoclavable bag, autoclave it, and dispose of it in the trash. Discard all used swabs, toothpicks, and so on in the appropriate disposal containers to be autoclaved.

2. *Alcohol lamp or Bunsen burner.* Keep long hair and loose clothing away from the flame. Extinguish the flame immediately after use.

3. *Clorox, Lysol, strong disinfectants.* These chemicals can irritate skin, and Clorox can bleach clothing. To clean up spills, wear disposable gloves and clean up with paper towels, using lots of water. Rinse exposed skin thoroughly. These used towels can be discarded in the regular trash cans.

4. *Stains.* Stains are permanent on clothing. You may want to wear lab coats or lab aprons, if available, or wear old, clean shirts over your clothes. This will reduce the possibility of staining clothes or contaminating clothing with bacterial cultures.

5. *Uncooked chicken.* Care should be taken with bacteria living on chicken owing to the potential presence of Salmonella. The instructor will use only fresh chicken and will place individual pieces (wings) in a petri dish and leave the dishes in the refrigerator until just before this experiment. After students prepare the culture, return the dishes to the refrigerator. Wash your hands after inoculating the agar plate, and do not open the plate the second week after the bacteria have grown.

6. *Agar plates.* Students should seal all agar plates after inoculating them. Do not open plates the second week.

7. *Used cultures.* Place all agar plates in autoclavable bags, autoclave, and then discard in the trash. Autoclave all broth cultures and discard them in the sink with lots of water.

## Discussion and Summary

Discussion will take place over two lab periods. The first week, discuss colony shapes, bacterial shapes, results of the Gram stain, and observations on bacterial succession in milk. You may choose to require that students turn in their scenario of bacterial succession in milk. At the beginning of the next lab period, have students observe and discuss the results of the environmental cultures and set up demonstrations of successful results with appropriate labels.

## Evaluation

You can test students on concepts and terminology in a lab test. They can submit their milk bacteria succession scenarios for evaluation. Informally evaluate the quality of their aseptic techniques and the results of their experiments.

## Lab Topic 14
# Protists and Fungi

 This lab topic gives you another opportunity to practice the scientific process introduced in Lab Topic 1. Before going to lab, review scientific investigation in Lab Topic 1 and carefully read Lab Topic 14. Be prepared to use this information to design an experiment with protists or fungi.

*If possible in your laboratory situation, 1 week before this lab, instruct students to read the lab topic and meet with their investigative team to discuss possible independent investigations before coming to the lab. You may ask them to develop their preliminary question and hypothesis. Give students a list of available materials. Additional supplies and materials must be requested prior to lab.*

## Laboratory Objectives

After completing this lab topic, you should be able to:

1. Discuss the diversity of protists and fungi, and the current interest in their phylogenetic relationships.
2. Describe the diversity of protists, explaining the means of obtaining nutrition and method of locomotion for each group.
3. Identify representative organisms in several major protistan clades.
4. Discuss the ecological role and economic importance of protists.
5. Describe the characteristics and representative organisms of the green algae and their relationship to land plants.
6. Describe the phyla of the kingdom Fungi, recognizing and identifying representative organisms in each.
7. Describe differences in reproduction in fungal phyla.
8. Discuss the ecological role and economic importance of fungi.
9. Design and perform an independent investigation of a protist or an organism in the kingdom Fungi.

*For a 2-hour lab: Utilize two consecutive lab periods with protists the first week and fungi the second week. See Teaching Plan for additional suggestions.*

*In the Study Area of masteringbiology.com, see Student Media for suggested activities, investigations, and videos.*

## Introduction

Unicellular eukaryotic organisms originated over 2 billion years ago, and today they are found in every habitable region of Earth. The enormous diversity of organisms, their numerous adaptations, and their cellular complexity reflect the long evolutionary history of eukaryotes. For almost 30 years, scientists placed these diverse groups of unicellular organisms into the kingdom Protista. The Protista usually included all organisms not placed in the other eukaryotic kingdoms of Plants, Animals, and Fungi.

This catchall kingdom included not only the unicellular eukaryotes, but also their multicellular relatives, like the giant kelps and seaweeds. However, scientists now agree that the designation kingdom Protista should be abandoned and these eukaryotic organisms that are neither fungi, plants, nor animals, be placed in the domain Eukarya. (Recall from Lab Topic 13 that prokaryotes are placed in domains Bacteria or Archaea.) In this lab topic we will refer to this diverse group as protists meaning a general term, not a taxonomic category.

The most familiar protists, commonly called algae and protozoans, have been well studied since the earliest development of the microscope. Therefore, one might assume that the taxonomic relationships among these groups are well understood. However, their phylogeny (evolutionary history) has been difficult to determine from comparisons of cell structure and function, nutrition, and reproduction. Recent molecular and biochemical research, particularly the ability to sequence ribosomal and transfer RNA genes, has provided strong new evidence for reconstructing the relationships of the protists.

Most recently scientists have suggested that studies of protists using **clades** can be meaningful for indicating evolutionary relationships. A clade is a group of species, all of which are descended from one ancestral species, representing one phylogenetic group. Many characteristics, including molecular and biochemical evidence, are used when organizing clades. As more information from a variety of sources becomes available, major groupings or clades will surely be modified. These investigations into the nature of eukaryotic diversity demonstrate the process of scientific inquiry. New technologies, new ideas, and novel experiments are used to test hypotheses, and the resulting evidence must be consistent with the existing body of knowledge and classification scheme. The results lead to modification of our hypotheses and further research. No matter how many groups or clades are proposed, remember that this is a reflection of the evolution of eukaryotes over the rich history of the earth. It is not surprising that the diversity of life does not easily fit into our constructed categories.

In this lab topic, we will study diverse examples of protists. These protists represent some of the most common clades. In addition to evolutionary relationships, you will give particular attention to nutrition, locomotion, and cellular complexity of each example.

If you complete all of the lab topics in this laboratory manual, you will have studied examples of all the major groups of organisms with the exception of those in domain Archaea. Bacteria are investigated in Lab Topic 13, and several additional protists are introduced in Lab Topic 2. Fungi, one of the kingdoms of the five-kingdom scheme, are studied in this lab topic, and you will investigate plant evolution and animal evolution in subsequent lab topics.

At the end of this lab topic, you will be asked to design a simple experiment to further your investigation of the behavior, ecology, or physiology of one of the organisms studied. As you proceed through the exercises, ask questions about your observations and consider an experiment that you might design to answer one of your questions.

EXERCISE 14.1
# The Protists

In this exercise you will study examples of seven major groups (clades) of protists. (See Table 14.1 below.)

**Table 14.1**
Groups of Protists Investigated in This Exercise

| Group | Lab Study | Examples |
|---|---|---|
| Euglenozoans | A | *Trypanosoma levisi* |
| Alveolates | B | Paramecia Dinoflagellates |
| Stramenopiles | C | Diatoms Brown algae |
| Rhizarians | D | Foraminiferans Radiolarians |
| Amoebozoans | E | *Amoeba* *Physarum* |
| Rhodophyta | F | Red algae |
| Chlorophyta and Charophyta | G | Green algae: *Spirogyra*, *Ulva*, *Chara* |

Protists may be **autotrophic** (photosynthetic) or **heterotrophic** (depending on other organisms for food). Autotrophic organisms are able to convert the sun's energy to organic compounds. The amount of energy stored by autotrophs is called **primary production.** Traditionally, autotrophic protists are called **algae** and heterotrophic protists are called **protozoa**—protists that ingest their food by **phagocytosis** (the uptake of large particles or whole organisms by the pinching inward of the plasma membrane). Some protozoa, euglenoids for example, are **mixotrophic,** capable of photosynthesis and ingestion. As you investigate the diversity of protists and their evolutionary relationships in this exercise, ask questions about the nutritive mode of each. Note morphological characteristics of examples studied. Ask which characteristics are found in organisms in the same clade and those shared with organisms in other clades or groups. Many of these characteristics are examples of evolutionary convergence. Ask questions about the ecology of the organisms. What means of locomotion do they possess, if any? What role do they play in an ecosystem? Do they have any economic value? Where do they live? (Protists live in a diversity of habitats, but most are aquatic. A great variety of protists may be found in **plankton,** the community of organisms found floating in the ocean or in bodies of freshwater.)

## Lab Study A. Euglenozoans—Example: *Trypanosoma levisi*

### Materials

compound microscope
prepared slides of *Trypanosoma levisi*

### Introduction

Organisms in the clade Euglenozoa are grouped together based on the ultrastructure (structure that can be seen only with an electron microscope) of their **flagella** and their mitochondria. Included in this group are some heterotrophs, some autotrophs, and some parasitic species. The many diverse single-celled and colonial flagellates have been a particular challenge to taxonomists. Under the old two-kingdom system of classification, the heterotrophic flagellates were classified as animals, and the autotrophic flagellates (with chloroplasts) were classified as plants. However, euglenozoans include members of each type. The common flagellated, mixotrophic *Euglena* belongs in this clade.

The organism that you will investigate in this exercise, *Trypanosoma levisi*, moves using flagella supported by microtubules. Organisms in the genus *Trypanosoma* are parasites that alternate between a vertebrate and an invertebrate host. *Trypanosoma levisi* lives in the blood of rats and is transmitted by fleas. Its flagellum originates near the posterior end but passes to the front end as a marginal thread of a long undulating membrane. Another organism in this same genus, *T. gambiense*, causes African sleeping sickness in humans. Its invertebrate host is the tsetse fly.

If you did not observe several other examples of flagellates in Lab Topic 2 when you studied the organisms living in a termite's gut, turn to that section of the laboratory manual (p. 43) and, following the procedure, observe these organisms. You may see *Trichonympha* (Color Plate 2) and other flagellates, including *Pyrsonympha* with four to eight flagella, *Trichomonas*, and *Calonympha* with numerous flagella originating from the anterior end of the cell.

*If time permits, you might repeat the exercise with termites in Lab Topic 2, even if you performed it previously, because there are several interesting flagellates present in the termite gut that students may not have observed the first time. Alternatively, you might isolate flagellates from one termite and have these on demonstration microscopes or use a microscopy projection system.*

### Procedure

1. Obtain a prepared slide of *Trypanosoma levisi* (Figure 14.1) and observe it using low, intermediate, and high powers in the compound microscope.
2. Locate the organisms among the blood cells of the parasite's host.
3. Identify the **flagellum,** the **undulating membrane,** and the **nucleus** in several organisms.

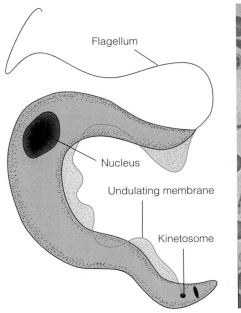

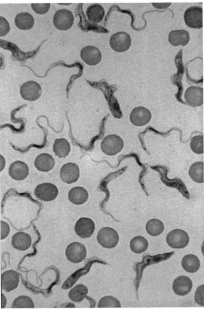

**Figure 14.1.**
*Trypanosoma*, **a euglenozoan,** is a flagellated parasite that lives in the blood of its mammalian host. The flagellum originates near the posterior end, but passes along an undulating membrane to the anterior end.

## Results

1. In the margin of your lab manual, draw several representative examples of *T. levisi* and several blood cells to show relative cell sizes.

2. Turn to Table 14.5 (p. 383) and list the characteristics, ecological roles, and economic importance of *T. levisi*.

# Lab Study B. Alveolates—Examples: Paramecia and Dinoflagellates

## Materials

compound microscope
slides and coverslips
cultures of living *Paramecium caudata*
Protoslo or other quieting agent
solution of yeast stained with Congo red
cultures of *Paramecium caudata* that have been fed yeast stained with
    Congo red (optional)
dropper bottle of 1% acetic acid
transfer pipettes
living cultures or prepared slides of dinoflagellates

## Introduction

Alveolates are single-celled organisms; some are heterotrophic, others autotrophic. The common characteristic of all alveolates is the presence of membrane-bound saclike structures (**alveoli**) packed into a continuous layer just inside and supporting the cell membrane. New groupings of protistans into clades place ciliates and dinoflagellates in the Alveolates.

**Figure 14.2.**
***Paramecium.*** (a) Complete the drawing of a *Paramecium*, labeling organelles and structures. (b) An enlarged view of cilia and the region of alveoli just under the cell membrane.

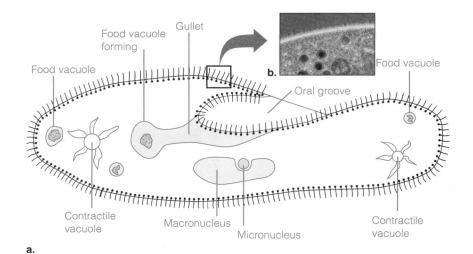

Food vacuole forming

Gullet

Food vacuole

b.

Food vacuole

Oral groove

Contractile vacuole

Macronucleus

Micronucleus

Contractile vacuole

a.

## *Paramecium caudatum*

The first example you will investigate in this lab study is *Paramecium caudatum*, a heterotrophic organism that moves about using cilia (short projections from the cell surface). Cilia are generally shorter and more numerous than flagella. Internally both structures are similar in their microtubular arrangement.

## Procedure

1. Using the compound microscope, examine a living *Paramecium* (Figure 14.2). Place a drop of water from the bottom of the culture on a clean microscope slide. Add a *small* drop of Protoslo or some other quieting solution to the water drop, then add the coverslip.

2. Observe paramecia on the compound microscope using low, then intermediate powers.

3. Describe the movement of a single paramecium. Does movement appear to be directional or is it random? Does the organism reverse direction only when it encounters an object, or does it appear to reverse direction even with no obstruction?

4. Locate a large, slowly moving organism, switch to high power and identify the following organelles:

   **Oral groove:** depression in the side of the cell that runs obliquely back to the mouth that opens into a **gullet.**

   **Food vacuole:** forms at the end of the gullet. Food vacuoles may appear as dark vesicles throughout the cell.

   **Macronucleus:** large, grayish body in the center of the cell. The macronucleus has many copies of the genome and controls most cellular activities, including asexual reproduction.

   **Micronucleus:** often difficult to see in living organisms, this small round body may be lying close to the macronucleus. Micronuclei are involved in sexual reproduction. Many species of paramecia have more than one micronucleus.

   **Contractile vacuole:** used for water balance, two of these form, one at each end of the cell. Each contractile vacuole is made up of a ring of radiating tubules and a central spherical vacuole. Your organism may be under osmotic stress because of the Protoslo, and the contractile vacuoles may be filling and collapsing as they expel water from the cell.

5. Observe feeding in a paramecium. Add a drop of yeast stained with Congo red to the edge of the coverslip and watch as it diffuses around the paramecium. Study the movement of food particles from the oral groove to the gullet to the formation of a food vacuole that will subsequently move through the cell as the food is digested in the vacuole. You may be able to observe the discharge of undigested food from the food vacuole at a specific site on the cell surface.

6. Observe the discharge of **trichocysts,** structures that lie just under the outer surface of the paramecium. When irritated by a chemical or attacked by a predator, the paramecium discharges these long thin threads that may serve as a defense mechanism, as an anchoring device, or to capture prey. Make a new slide of paramecia. Add a drop of 1% acetic acid to the edge of the coverslip and carefully watch a paramecium. Describe the appearance of trichocysts in this species.

*You may choose to have students make slides of paramecia that you have fed yeast stained with Congo red 1 to 2 hours before lab.*

*Students may need to make a fresh paramecium slide for this activity.*

## Results

1. Complete the drawing of a paramecium (Figure 14.2), labeling all the organelles and structures shown in bold in the text.

2. Turn to Table 14.5 (p. 383) and list the characteristics, ecological roles, and economic importance of paramecia.

---

Student Media Videos—Ch. 28: *Paramecium* Vacuole; *Paramecium* Cilia

---

## Dinoflagellates

Swirl your hand through tropical ocean waters at night and you may notice a burst of tiny lights. Visit a warm, stagnant inlet and you might notice that the water appears reddish and dead fish are floating on the surface. Both of these phenomena may be due to activities of dinoflagellates—single-celled organisms that are generally photosynthetic. Some dinoflagellates are able to bioluminesce, or produce light. They sometimes can *bloom* (reproduce very rapidly) and cause the water to appear red from pigments in their bodies. If the organisms in this "red tide" are a species of dinoflagellate that releases toxins, fish and other marine animals can be poisoned. Red tides in the Chesapeake Bay are thought to be caused by *Pfiesteria*, a dinoflagellate that produces deadly toxins resulting in invertebrate and fish kills, and that also may be implicated in human illness and death. Dinoflagellates have a cellulose cell wall often in the form of an armor of numerous plates with two perpendicular grooves, each containing a flagellum. Most of these organisms are autotrophic and play an important role in **primary production** in oceans—photosynthesis that ultimately provides food for all marine organisms.

Dinoflagellates have traditionally been considered algae, but they are now thought to share a common ancestor with ciliates, as evidenced by the presence of alveoli.

## Procedure

1. Obtain a prepared slide or make a wet mount of dinoflagellates (Figure 14.3).

2. Focus the slide on low power and attempt to locate the cells. You may have to switch to intermediate power to see them.

*If you are using living dinoflagellates, have students add a drop of Protoslo to their preparations.*

**Figure 14.3.**
**Dinoflagellates.** The cell wall is made of cellulose plates with two perpendicular grooves, each containing a flagellum.

 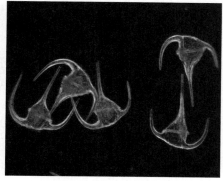

3. Switch to high power.
4. Identify the perpendicular **grooves** and the **cellulose plates** making up the cell wall. Are the plates in your species elongated into spines? **Flagella** may be visible in living specimens.

### Results

1. Draw several examples of cell shapes in the margin of your lab manual. Note differences between the species on your slide and those in Figure 14.3.
2. Turn to Table 14.5 (p. 383) and list the characteristics, ecological roles, and economic importance of dinoflagellates.

 Student Media Video—Ch. 28: Dinoflagellate

## Lab Study C. Stramenopiles—Examples: Diatoms and Brown Algae

### Materials

compound microscope
slides and coverslips
living cultures of diatoms
transfer pipettes
prepared slides of diatomaceous earth (demonstration only)
demonstration materials of brown algae

### Introduction

The clade Stramenopila includes water molds (phylum Oomycetes), diatoms (phylum Bacillariophyta), golden algae (phylum Chrysophyta), and brown algae (phylum Phaeophyta). These organisms are grouped in this clade based on the structure of their flagella (when present). The flagellum has many hairlike lateral projections.

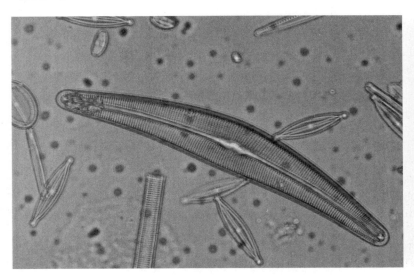
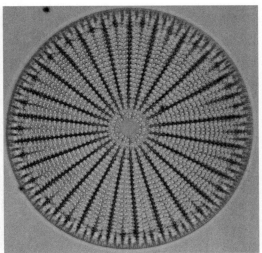

**Figure 14.4.**
**Diatoms are important autotrophs found in plankton.** Many different species
and forms exist. All have cell walls made of silica.

In this lab study you will investigate two examples: diatoms and brown
algae. Both are autotrophic organisms that play an important role in primary production in oceans.

## Diatoms (Bacillariophyta)

Diatoms are important autotrophic organisms in plankton. In fact, they are
the most important photosynthesizers in cold marine waters. They can be
unicellular, or they can aggregate into chains or starlike groups. Protoplasts
of these organisms are enclosed by a cell wall made of silica that persists after
the death of the cell. These cell wall deposits are mined as **diatomaceous
earth** and have numerous economic uses (for example, in swimming pool
filters and as an abrasive in toothpaste and silver polish). Perhaps the greatest value of diatoms, however, is the carbohydrate and oxygen they produce
that can be utilized by other organisms. Ecologists are concerned about the
effects of acid rain and changing climatic conditions on populations of
diatoms and their rate of primary productivity.

Diatom cells are either elongated, boat-shaped, bilaterally symmetrical
**pennate** forms or radially symmetrical **centric** forms (Color Plate 21). The
cell wall consists of two valves, one fitting inside the other, in the manner
of the lid and bottom of a petri dish.

## Procedure

1. Prepare a wet mount of diatoms (Figure 14.4) from marine plankton
   samples or other living cultures.
2. Observe the organisms on low, intermediate, and high powers.
3. Describe the form of the diatoms in your sample. Are they centric,
   pennate, or both?

4. If you are studying living cells, you may be able to detect locomotion. The method of movement is uncertain, but it is thought that contractile fibers just inside the cell membrane produce waves of motion on the cytoplasmic surface that extends through a groove in the cell wall. What is the body form of motile diatoms?

*Some pennate forms demonstrate locomotion.*

5. Observe a single centric form on high power and note the intricate geometric pattern of the cell wall. Can you detect the two valves?

6. Look for chloroplasts in living forms.

7. Observe diatomaceous earth on demonstration and identify pennate and centric forms.

### Results

1. Sketch several different shapes of diatoms in the margin of your lab manual.

2. Turn to Table 14.5 (p. 383) and list the characteristics, ecological roles, and economic importance of diatoms.

 **Student Media Videos—Ch. 28: Diatoms Moving; Various Diatoms**

### Brown Algae (Phaeophyta)

Some of the largest algae, the **kelps**, are brown algae. The Sargasso Sea is named after the large, free-floating brown algae *Sargassum*. These algae appear brown because of the presence of the brown pigment **fucoxanthin** in addition to chlorophyll *a*. Brown algae are perhaps best known for their commercial value. Have you ever wondered why commercial ice cream is smoother in texture than homemade ice cream? Extracts of **algin**, a polysaccharide in the cell wall of some brown algae, are used commercially as thickening or emulsifying agents in paint, toothpaste, ice cream, pudding, and in many other commercial food products. *Laminaria*, known as *kombu* in Japan, is added to soups, used to brew a beverage, and covered with icing as a dessert.

### Procedure

*We suggest that you have on demonstration some preserved specimen or herbarium mounts of* Fucus *or* Laminaria *and* Sargassum.

Observe examples of brown algae on demonstration (Figure 14.5).

### Results

1. In Table 14.2, list the names and distinguishing characteristics of each brown algal species on demonstration. Compare the examples with those illustrated in Figure 14.5 and Color Plate 22.

2. Turn to Table 14.5 (p. 383) and list the key characteristics, ecological roles, and economic importance of brown algae.

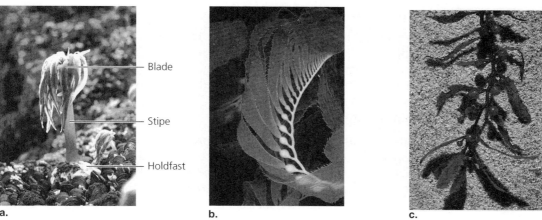

**Figure 14.5.**
**Examples of multicellular brown algae (phylum Phaeophyta).** The body of a brown alga consists of broad blades, a stemlike stipe, and a holdfast for attachment. These body parts are found in the kelps (a) Sea palm (*Postelsia*) and (b) *Nereocystis*. Rounded air bladders for flotation are seen in (c) *Sargassum* and other species of brown algae (See Color Plate 22).

**Table 14.2**
Representative Brown Algae

| Name | Body Form (single-celled, filamentous, colonial, leaflike; broad or linear blades) | Characteristics (pigments, reproductive structures, structures for attachment and flotation) |
|------|------|------|
|  |  |  |
|  |  |  |
|  |  |  |
|  |  |  |

# Lab Study D. Rhizarians—Examples: Foraminiferans and Radiolarians

## Materials

compound microscope
prepared slides of foraminiferans
prepared slides of radiolarian skeletons (demonstration only)

**Figure 14.6.**
**Forams** are heterotrophic organ-isms that move using threadlike pseudopodia. Their shell-like tests are made of calcium carbonate.

*The best way to teach this lab is to have living organisms available. If you are fortunate enough to live near the ocean, you will easily find radiolarians in plankton and forams in plankton and in benthic samples. You may also find radiolarians in plankton from freshwater ponds. The study of plankton tows and bottom samples will be much more exciting than the study of prepared slides, although the slides ensure that students see all organisms. If plankton is available, we suggest that you begin with prepared slides, then have students study the plankton. Add a few drops of formaldehyde to plankton samples if you must keep them longer than a couple of hours.*

*If samples of Indiana limestone are available, foram fossils can be seen with a stereoscopic microscope.*

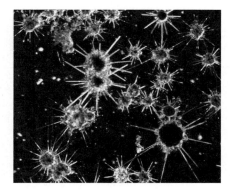

**Figure 14.7.**
**Radiolarians** are supported by a skeleton of silicon dioxide. They use threadlike pseudopodia to obtain food (Color Plate 23).

## Introduction

Rhizarians include foraminiferans and radiolarians, closely related groups composed of ameboid, heterotrophic organisms with **threadlike pseudopodia** or cellular extensions used in feeding and, in some species, locomotion. You will study examples of foraminiferans and radiolarians.

## Foraminiferans

**Foraminiferans,** commonly called **forams,** are another example of organisms that move and feed using pseudopodia. Forams are marine planktonic (freely floating) or benthic (bottom dwelling) organisms that secrete a calcium carbonate shell-like *test* (a hard outer covering) made up of chambers. In many species, the test consists of chambers secreted in a spiral pattern, and the organism resembles a microscopic snail. Although most forams are microscopic, some species, called *living sands*, may grow to the size of several centimeters, an astounding size for a single-celled protist. Threadlike pseudopodia extend through special pores in the calcium carbonate test. The test can persist after the organism dies, becoming part of marine sand. Remains of tests can form vast limestone deposits.

## Procedure

1. Obtain a prepared slide of representative forams (Figure 14.6).
2. Observe the organisms first on the lowest power of the compound microscope and then on intermediate and high powers.
3. Note the arrangement and attempt to count the number of chambers in the test. In most species, the number of chambers indicates the relative age of the organisms, with older organisms having more chambers. Which are more abundant on your slide, older or younger organisms?

   Chambers can be arranged in a single row, in multiple rows, or wound into a spiral. Protozoologists determine the foram species based on the appearance of the test. Are different species present?

## Results

1. Sketch several different forams in the margin of your lab manual. Note differences in the organisms on your slide and those depicted in Figure 14.6.
2. Turn to Table 14.5 (p. 383) and list the characteristics, ecological roles, and economic importance of forams.

## Radiolarians

The **radiolarians** studied here are common in marine plankton. They secrete skeletons of silicon dioxide that can, as with the forams, collect in vast deposits on the ocean floor. Their threadlike pseudopodia, called **axopodia,** extend outward through pores in the skeleton in all directions from the central spherical cell body.

## Procedure

1. Observe slides of radiolarians on demonstration (Figure 14.7).
2. Observe the size and shape of the skeletons and compare your observations with Figure 14.7 and Color Plate 23.

## Results

1. Sketch several different radiolarians skeletons in the margin of your lab manual, noting any differences between the organisms on demonstration and those in the figure and color plate.

2. Turn to Table 14.5 (p. 383) and list the characteristics, ecological roles, and economic importance of radiolarians.

## Lab Study E. Amoebozoans—Examples: *Amoeba*, Slime Molds

### Materials

cultures of *Amoeba proteus* (if amoeba were not studied in Lab Topic 2)
slides and coverslips (for amoeba)
stereoscopic microscopes
*Physarum* growing on agar plates

### Introduction

Amebozoans have pseudopodia as seen in foraminiferia and radiolarians, but the structure is different. Rather than threadlike pseudopodia as seen in these organisms, amebozoans' pseudopodia are *lobe-shaped*. Based on their ameboid characteristics, their phagocytic mode of obtaining nutrition, and molecular systematics, both amoeba and slime molds are included in the clade **Amoebozoa.**

### Amoeba

In Lab Topic 2, you studied ***Amoeba proteus,*** a protozoan species of organisms that move using lobed-shaped pseudopodia (Color Plate 1). Amoeba have no fixed body shape and they are naked; that is, they do not have a shell. Different species may be found in a variety of habitats, including freshwater and marine habitats. Recall that pseudopodia are cellular extensions. As the pseudopod extends, endoplasm flows into the extension. By extending several pseudopodia in sequence and flowing into first one and then the next, the amoeba proceeds along in an irregular, slow fashion. Pseudopodia are also used to capture and ingest food. When a suitable food particle such as a bacterium, another protist, or a piece of detritus (fragmented remains of dead organisms) contacts an amoeba, a pseudopod will flow completely around the particle and take it into the cell by phagocytosis.

If you did not observe *Amoeba proteus* or some other naked amoeba in Lab Topic 2, turn to that section of the laboratory manual (see p. 42) and, following the procedure, observe these organisms.

*Also see amoeba video at website, www.microscopy.fsu.edu/ moviegallery/pondscum/protozoa/ amoeba/*

 **Student Media Videos—Ch. 28: Amoeba; Amoeba Pseudopodia**

## Slime Molds (Mycetozoans)

William Crowder, in a classic *National Geographic* article (April 1926) describes his search for strange creatures in a swamp on the north shore of Long Island. This is his description of his findings: "Behold! Seldom ever before had such a gorgeous sight startled my unexpectant gaze. Spreading out over the bark [of a dead tree] was a rich red coverlet . . . consisting of thousands of small, closely crowded, funguslike growths. . . . A colony of these tiny organisms extended in an irregular patch . . . covering an area nearly a yard in length and slightly less in breadth. . . . Each unit, although actually less than a quarter of an inch in height, resembled . . . a small mushroom, though more marvelous than any I have ever seen."

The creatures described by Crowder are heterotrophic organisms called **slime molds.** They have been called plants, fungi, animals, fungus animals, protozoa, Protoctista, Protista, Mycetozoa, and probably many more names. Classifying slime molds as fungi (as in previous classification schemes) causes difficulties because whereas slime molds are phagocytic like protozoa, fungi are never phagocytic but obtain their nutrition by absorption. Characteristics other than feeding mode, including cellular ultrastructure, cell wall chemistry, and other molecular studies, indicate that slime molds fit better with the ameboid protists than with the fungi. These studies suggest that slime molds descended from unicellular amoeba-like organisms.

There are two types of slime molds, plasmodial slime molds and cellular slime molds. In this lab study, you will observe the plasmodial slime mold *Physarum*. The vegetative stage is called a **plasmodium,** and it consists of a multinucleate mass of protoplasm totally devoid of cell walls. This mass feeds on bacteria as it creeps along the surface of moist logs or dead leaves. When conditions are right, it is converted into one or more reproductive structures, called **fruiting bodies,** that produce spores (Color Plate 25). You may choose to investigate slime molds further in Exercise 14.3.

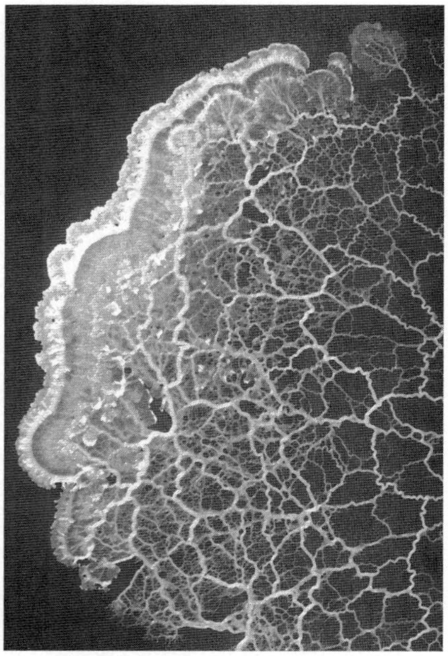

### Procedure

1. Obtain a petri dish containing *Physarum* and return to your lab bench to study the organism. Keep the dish closed.

2. With the aid of your stereoscopic microscope, examine the plasmodium (Figure 14.8, Color Plate 24). Describe characteristics such as color, size, and shape. Look for a system of branching veins. Do you see any movement? Speculate about the source of the movement. Is the movement unidirectional or bidirectional—that is, flows first in one direction and then in the other? Your instructor may have placed oat flakes or another food source on the agar. How does the appearance of the plasmodium change as it contacts a food source?

*The plasmodia do not have a definite size or shape. Physarum plasmodia are often yellow. A network of veins is visible, and protoplasmic streaming may be seen in the veins. The streaming can be very rapid in one direction for a minute or so, then it slows, stops, and flows in the opposite direction. The plasmodium pools around the oat flakes.*

**Figure 14.8.**
**Slime mold.** Slime molds are protists that share some characteristics with both protozoa and fungi. The vegetative stage of a plasmodial slime mold includes an amoeboid phase consisting of a multinucleate mass known as a plasmodium (Color Plate 24).

3. Examine the entire culture for evidence of forming or mature fruiting bodies. Are the fruiting bodies stalked or are they sessile, that is, without a stalk? If a stalk is present, describe it.

*Physarum cinereum has sessile fruiting bodies; P. polycephalum has gyrose, stalked fruiting bodies.*

## Results

1. Sketch the plasmodium and fruiting bodies in the margin of your lab manual. Label structures where appropriate.
2. Turn to Table 14.5 (p. 383) and list the characteristics, ecological roles, and economic importance of slime molds.

**Student Media Videos—Ch. 28: Plasmodial Slime Mold Streaming; Plasmodial Slime Mold**

## Lab Study F: Red Algae (Rhodophyta)

### Materials

examples of red algae on demonstration

### Introduction

The simplest red algae are single celled, but most species have a macroscopic, multicellular body form. The red algae, unlike all the other algae, do not have flagella at any stage in their life cycle. Some scientists suggest that the red algae represent a monophyletic (having a single origin) group and should be placed in their own kingdom. Red algae are autotrophic, containing chlorophyll *a* and the accessory pigments **phycocyanin** and **phycoerythrin** that often mask the chlorophyll, making the algae appear red. These pigments absorb green and blue wavelengths of light that penetrate deep into ocean waters. Many red algae also appear green or black or even blue, depending on the depth at which they are growing. Because of this, color is not always a good characteristic to use when determining the classification of algae. Recall that in Lab Topic 13 you grew bacteria and fungi on plates of agar. This substance, **agar,** is a polysaccharide extracted from the cell wall of red algae. Another extract of red algae cell walls, **carrageenan,** is used to give the texture of thickness and richness to foods such as dairy drinks and soups. In Asia and elsewhere, the red algae *Porphyra* (known as *nori*) are used as seaweed wrappers for sushi. The cultivation and production of *Porphyra* constitute a billion-dollar industry.

*We suggest that you have on demonstration preserved specimen or herbarium mounts of Agardhiella, coralline algae, Polysiphonia, or Porphyra. We do not require that students memorize names of the demonstration materials, or even characteristics of the examples, because many of the latter are variable and the most consistent distinguishing characteristics are related to microscopic and biochemical features and details of life cycles. The value of this activity is that students appreciate the diversity within and among the groups and the difficulty faced by those attempting to classify organisms. The structure of the algae may give evidence of the ecology and distribution of groups.*

### Procedure

Observe the examples of red algae that are on demonstration (Figure 14.9).

### Results

1. In Table 14.3, list the names and distinguishing characteristics of the red algae on demonstration. Compare the demonstration examples with those illustrated in Figure 14.9 and Color Plate 26.
2. Turn to Table 14.5 (p. 383) and list the key characteristics, ecological roles, and economic importance of red algae.

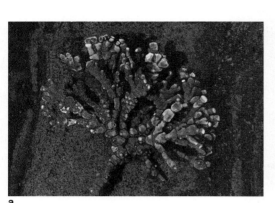

a.
b.
c.

**Figure 14.9.**
**Examples of multicellular red algae (phylum Rhodophyta).** (a) Some red algae
have deposits of carbonates of calcium and magnesium in their cell walls and
are important components of coral reefs. (b) Most red algae have delicate, finely
dissected blades. (c) *Porphyra* (or *nori*) is used to make sushi. (See Color Plate 26.)

**Table 14.3**
Representative Red Algae

| Name | Body Form (single-celled, filamentous, colonial, leaflike) | Characteristics (reproductive structures, structures for attachment or flotation, pigments) |
|---|---|---|
| | | |
| | | |
| | | |
| | | |

## Lab Study G. Green Algae (Chlorophyta and Charophyta)—The Protist-Plant Connection

### Materials

cultures or prepared slides of *Spirogyra* sp.
preserved *Ulva lactuca*
preserved *Chara* sp.

### Introduction

The green algae are divided into two main groups, chlorophytes and charophytes. These groups include the largest numbers of algal species and perhaps
the most important from both ecological and evolutionary perspectives. You

can find examples of unicellular, motile and nonmotile, colonial, filamentous, and multicellular species that inhabit both freshwater and marine environments. All green algae share some characteristics with land plants (for example, the storage of the starch amylase and the presence of chlorophylls *a* and *b*), but it is the charophytes that are considered to be most closely related to land plants because of several unique similarities. The method of cell wall formation during cell division, organization of proteins that synthesize cellulose, and the structure of sperm, when present, are all similarities in charophytes and land plants. Results of recent work in DNA sequencing confirm the close relationship between charophytes and land plants, and have led some scientists to propose that charophytes be included in the Plant kingdom.

Green algae play an important ecological role as primary producers, providing food for animals in aquatic ecosystems; and organisms in one genus, *Ulva*, commonly called sea lettuce, are eaten by humans in salads and cooked in soups. Because of the rapid reproductive rate of some green algae, particularly those in the genus *Chlorella*, some scientists have proposed the use of these green algae for biofuels and for human and livestock food. *Chlorella* has been shown to be a good source of proteins, fats, and carbohydrates, and small chlorella "farms" have been established in various countries. The discovery of a "*Chlorella* growth factor" that is reported to stimulate growth and wound repair in animals has led to the formation of a multibillion-dollar segment of the health food industry.

In this exercise you will view several body forms of green algae on demonstration: single-celled, filamentous, colonial, and multicellular. Finally, you will observe the multicellular, branched green algae *Chara* (the stonewort), believed to be most similar to the green algae that gave rise to land plants over 475 million years ago.

If you completed Lab Topic 2 Microscopes and Cells, you may remember observing aggregates of single-celled algae, *Protococcus*, and the spherical green algae *Volvox*. Your instructor may ask you to review your notes and drawings from Lab Topic 2. In this lab study you will observe the filamentous alga *Spirogyra* and the multicellular algae *Ulva* (chlorophytes) and *Chara* (a charophyte).

## Procedure

1. Using your compound microscope, observe living materials or prepared slides of the filamentous alga *Spirogyra* (Figure 14.10a). This organism is common in small, freshwater ponds. The most obvious structure in the cells of the filament is a long chloroplast. Can you determine how the alga got its name? Describe the appearance of the chloroplast.

   Can you see a nucleus in each cell of the filament?

2. Observe the preserved specimen of *Ulva* sp., commonly called sea lettuce (Figure 14.10b and Color Plate 27). This multicellular alga is commonly found on rocks or docks in marine and brackish water.

   a. Describe the appearance and body form of *Ulva*.

      *The body is a bright green, very thin, broad flat blade.*

   b. Are structures present that would serve to attach *Ulva* to its substrate (dock or rock)? If so, describe them.

      *A small multicellular holdfast that attaches the blade to the substrate may be present.*

   c. Compare your specimen of *Ulva* with that shown in the figure and color plate.

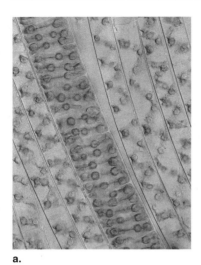
a.

b.

c.

**Figure 14.10.**
**Examples of multicellular green algae.** (a) A filamentous green alga, *Spirogyra*. (b) Some green algae are multicellular as in *Ulva*, sea lettuce (Color Plate 27). (c) A multicellular, branched green alga, *Chara* (Color Plate 28).

*Living* Chara *may be available from local lakes.*

3. Examine the preserved specimen of the multicellular green alga *Chara* (Figure 14.10c). This alga grows in muddy or sandy bottoms of clear lakes or ponds. Its body form is so complex that it is often mistaken for a plant, but careful study of its structure and reproduction confirms its classification as a green alga.

   Note the cylindrical branches attached to nodes. Compare your specimen to Figure 14.10c and Color Plate 28. Sketch the appearance of your specimen in the margin of your lab manual.

## Results

1. In Table 14.4, list the names and distinguishing characteristics of each green algal species studied. Compare these examples with those illustrated in Figure 14.10 and Color Plates 27 and 28.
2. Turn to Table 14.5 (p. 383) and list the key characteristics, ecological roles, and economic importance of green algae.

**Table 14.4**
Representative Green Algae

| Name | Body Form (single-celled, filamentous, colonial, leaflike) | Characteristics (pigments, specialized structures, flagella, structures for attachment) |
|---|---|---|
| *Spirogyra* | | |
| *Ulva* | | |
| *Chara* | | |

## Discussion

1. Describe the mechanism for feeding in amoeboid, flagellated, and ciliated protozoans.

   *Amoeboid protozoa feed by phagocytosis. Flagellated protozoa may simply take up organic carbon or nitrogen compounds by simple diffusion or active transport, or some species use phagocytosis in special regions of the cell. Ciliated protozoa often have an elaborate oral groove and cell "mouth" through which food passes by phagocytosis. Cilia move the food down the groove to the mouth region.*

2. How do you think amoeboid organisms with skeletons, such as the radiolarians, move food to their cell bodies?

   *The pseudopodia that protrude through skeletal pores engulf the food by phagocytosis. Food vacuoles can be detected moving along the long, slender pseudopodia.*

3. Compare the appearance and rate of locomotion in amoeboid, flagellated, and ciliated organisms observed in this exercise.

   *Amoeboid movement is slow and irregular, with many pseudopodia being extended in sequence. Flagellates move rapidly by beating or undulating the flagella, often giving the appearance of irregular movement. Ciliates move rapidly with a gliding motion, the body rotating as it swims.*

4. Describe mechanisms for defense in the organisms studied.

   *Mechanisms for defense include the trichocysts that may play a defensive role in ciliates. In some cases, protozoa retreat when threatened. Some (forams, radiolarians) secrete protective shells.*

5. Compare dinoflagellates and diatoms. What important ecological role is shared by these two groups?

   *primary production*

6. What is one characteristic that you could observe under the microscope to distinguish diatoms and dinoflagellates?

   *two grooves with flagella for dinoflagellates; etched cell walls for diatoms*

7. Slime molds were once placed in the kingdom Fungi. What characteristics suggest that these organisms are protistan?

   *phagocytic mode of nutrition, cellular ultrastructure*

8. What important ecological role is shared by the macroscopic algae (green, red, and brown)?

   *primary production*

9. Based on your observations in the laboratory, what two characteristics might you use to distinguish brown and red algae?

*pigmentation and body form*

10. What characteristics of green algae have led scientists to conclude that this group includes the ancestors of land plants, most likely the charophytes?

*Shared characteristics: storage of amylase, presence of chlorophylls a and b. Plants share with the charophytes similarities in cell wall formation during cell division, similarities in cellulose synthesis. Molecular evidence from ribosomal and transfer RNA gene studies support this conclusion.*

## EXERCISE 14.2
# The Kingdom Fungi

### Introduction

The kingdom Fungi includes a diverse group of organisms that play important economic and ecological roles. These organisms are unicellular (yeasts) or multicellular, heterotrophic organisms that obtain their nutrients by absorption, digesting their food outside their bodies and absorbing the digestion products into their cells. They often have complex life cycles with alternating sexual and asexual (vegetative) reproduction. They may produce spores either asexually by mitosis or sexually by meiosis.

Fungi are beneficial to humans in many ways. The fungus *Penicillium* is used to produce antibiotics. Yeast, a single-celled fungus, is used in the production of wine, beer, and leavened bread. Fungi are also a source of food in many cultures, with truffles being the most expensive. Black truffles are dark, edible subterranean fungi that sell for $350–$500 per pound. The dark green areas seen in Roquefort cheese are patches of the mold *Penicillium roquefortii* growing in cow's milk.

In ecosystems, fungi share with bacteria the essential role of decomposition, returning to the ecosystem the matter trapped in dead organisms. One extremely important ecological role played by fungi is their mutualistic association with roots of most plants, forming "mycorrhizae." Mycorrhizal fungi increase the plant's ability to capture water and provide the plant with minerals and essential elements. In turn, the plants supply the fungi with nutrients such as carbohydrates. This association greatly enhances plant growth, and may have played a role in plant colonization of land.

Although many fungi are beneficial, others play destructive roles in nature. Some species parasitize animals and plants. Athlete's foot and ringworm are diseases commonly known to humans. Histoplasmosis is a respiratory disease in humans caused by a fungus found in soil and in bat and bird droppings. Wheat rust, potato late blight, and sudden oak death (a potentially devastating disease discovered in the United States in 1995) are plant diseases caused by fungi. The ergot fungus that parasitizes rye causes convulsive ergotism in humans who eat bread made with infested grains. The bizarre behavior of young women who were later convicted of witchcraft in Salem Village, Massachusetts, in 1692 has been attributed to convulsive ergotism.

The destructive nature of fungi was dramatically demonstrated following Hurricane Katrina in August 2005. Weeks after the hurricane, many homes, hospitals, and businesses remained flooded, promoting heavy growth of fungi. Molds and mildew of varied species covered walls, ceilings, and building contents, and spores filled air-conditioning and ventilation systems. Many of these structures and their contents had to be destroyed as adequate cleanup was impossible. By November, many residents in New Orleans were suffering from the "Katrina cough," respiratory problems associated with infection by fungus spores. Cleanup from the devastating effects of these organisms totaled millions of dollars.

In this exercise, you will learn about the structure of typical fungi and the characteristics of three important phyla of fungi: Zygomycota, Ascomycota, and Basidiomycota. You will see examples of lichens that are associations between fungi and algae. As you observe these examples, consider interesting questions that might be asked about fungi diversity or ecology. You can choose one of these questions to design a simple experiment in Exercise 14.3.

## Lab Study A. Zygote Fungi—Zygomycota

### Materials

compound microscope
stereoscopic microscope
cultures of *Rhizopus stolonifer*
    with sporangia
cultures of *Pilobolus crystallinus*
    on demonstration

forceps, ethyl alcohol, alcohol lamp
slides and coverslips
dropper bottles of water

### Introduction

One common organism in the phylum Zygomycota is probably growing in your refrigerator right now. The common bread mold, *Rhizopus stolonifer*, grows on many foods as well as bread. In this lab study, you will observe the structure of this species to see many general fungi characteristics. Fungi are made up of threadlike individual filaments, called **hyphae,** which are organized into the body of the fungus, called the **mycelium.** Masses of mycelia from a single spore can grow to be huge, with some reported to cover many acres. This filamentous mass secretes enzymes into the substrate and digests food that will then be absorbed into its cells. Cells of fungi have cell walls made of **chitin** combined with other complex carbohydrates, including cellulose. You may recall that chitin is the main component of insect exoskeletons.

### *Rhizopus stolonifer*

*Rhizopus* reproduces both sexually and asexually. In the zygomycetes (fungi in the phylum Zygomycota), cells of the hyphae are haploid. Hyphae grow over a substrate, for example, a slice of bread, giving the bread a fuzzy appearance. In asexual reproduction, certain hyphae grow upright and develop **sporangia,** round structures, on their tips. Haploid spores develop in the sporangia following mitosis, and when they are mature, they are dispersed through the air. If they fall on a suitable medium, they will absorb water and germinate, growing a new mycelium.

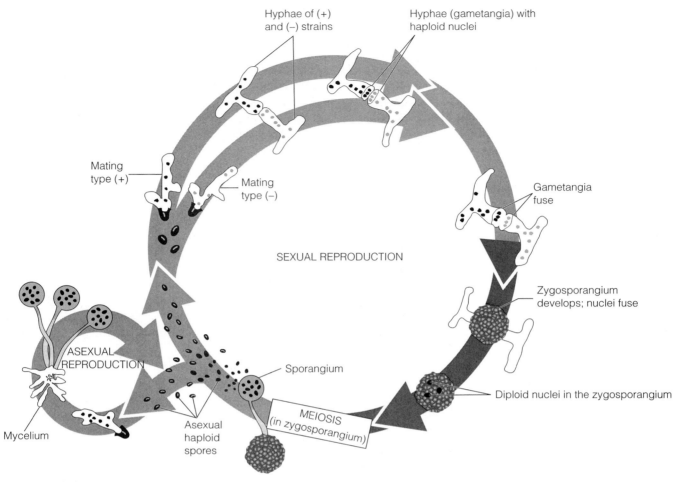

**Figure 14.11.**
***Rhizopus stolonifer.*** *Rhizopus* reproduces both sexually by zygosporangia and asexually by sporangia producing asexual spores. In sexual reproduction, (+) and (−) mating types fuse and a zygosporangium with diploid nuclei ultimately results.

*Rhizopus* also reproduces sexually when compatible mating types designated as (+) and (−) grow side by side. In this case, (+) and (−) hyphae produce extensions called **gametangia** that fuse forming **zygosporangia.** Within the zygosporangia, haploid nuclei fuse (*karyogamy*) producing diploid nuclei. The diploid nuclei then undergo meiosis. Following meiosis, haploid spores are produced in sporangia borne on filaments that emerge from the zygosporangia (Figure 14.11).

### *Pilobolus crystallinus*

*Pilobolus crystallinus* (also called the *fungus gun*, or *shotgun fungus*) is another member of the phylum Zygomycota. This fungus is called a **coprophilous** fungus because it grows on dung. It displays many unusual behaviors, one of which is that it is positively phototropic. Perhaps you can investigate this behavior in Exercise 14.3. Bold et al. (1980) describe asexual reproduction in *Pilobolus*. This species has sporangia as does *Rhizopus*, but rather than similarly dispersing single spores, in *Pilobolus* the sporangium is forcibly discharged as a unit; the dispersion is tied to moisture and diurnal cycles. In nature, in the early evening the sporangia form; shortly after midnight, a swelling appears below the sporangium. Late the following morning, turgor pressure causes the swelling to explode, propelling the sporangium as far as 2 meters. The sticky sporangium will adhere to grass leaves and subsequently may be eaten by an animal—horse, cow, or rabbit. The intact sporangia pass through the animal's digestive tract and are excreted, and the spores germinate in the fresh dung. See the Websites section at the end of this lab topic for an exciting video of *Pilobolus* sporangium discharge.

In this lab study you will investigate *Rhizopus* and observe *Pilobolus* on demonstration.

## Procedure

1. Obtain a culture of *Rhizopus* and carry it to your lab station.
2. Examine it using the stereoscopic microscope.
3. Identify the **mycelia, hyphae,** and **sporangia.**
4. Review the life cycle of *Rhizopus* (Figure 14.11). Locate the structures in this figure that are visible in your culture. Circle the structures involved in asexual reproduction.
5. Using forceps and aseptic technique, remove a small portion of the mycelium with several sporangia and make a wet mount.
6. Examine the hyphae and sporangia using the compound microscope. Are spores visible? How have the spores been produced?

   *by simple mitotic division of the haploid nuclei of the hyphal cells (asexual reproduction)*

   How do the spores compare with the hyphal cells genetically?

   *Genetically, the spores and the hyphal cells are identical.*

   How would spores produced by sexual reproduction differ from spores produced asexually?

   *There will be new genetic combinations in these spores because they have formed as a result of the fusion of haploid nuclei followed by meiosis.*

7. Observe the cultures of *Pilobolus* (Figure 14.12) growing on rabbit dung agar that are on demonstration.
8. Identify the **sporangia, mycelia,** and **hyphae.** What color are the sporangia and spores?

   *black*

## Results

1. Review the life cycle of *Rhizopus* and the structures observed in the living culture and compare with Figure 14.11.
2. Review the structures observed in *Pilobolus* and compare with Figure 14.12.

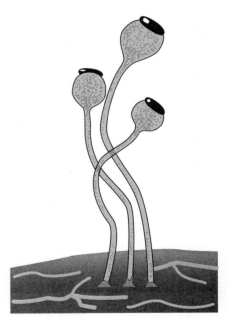

**Figure 14.12.**
*Pilobolus crystallinus.*

## Discussion

1. The body form of most fungi, including *Rhizopus*, is a mycelium composed of filamentous hyphae. Using your observations as a basis for your thinking, state why this body form is well adapted to the fungus mode of nutrition.

   *Hyphae grow among cell structures throughout the food. This allows for close contact over a large surface area for digestion and absorption.*

2. Refer back to the description of *Pilobolus*. Speculate about the adaptive advantage of having a system to propel sporangia, as seen in *Pilobolus*.

   *Since the fungus is growing in dung, were it not for the propulsion system, most newly formed spores would just fall back into the old dung. Propelling sporangia away from the old dung to adhere to grass ensures that the spores will be dispersed in fresh dung.*

# Lab Study B. Sac Fungi—Ascomycota

## Materials

*Use fresh demonstration materials when possible. Check farmers' markets. Make identifying cards to place beside each demonstration with the name of the organism, its phylum, and the section of the lab study that refers to the organism.*

compound microscope
stereoscopic microscope
dried or preserved *Peziza* specimen
prepared slide of *Peziza* ascocarp

cultures of *Penicillium* on demonstration
Roquefort cheese on demonstration
preserved or fresh morels
plastic mounts of ergot in rye or wheat

## Introduction

Fungi in the phylum Ascomycota are called *sac fungi*, or ascopore-producing fungi. This division includes edible fungi, morels, and truffles, but it also includes several deadly plant and animal parasites. For example, chestnut blight and Dutch elm disease have devastated native populations of chestnut and American elm trees. The fungi causing these diseases were introduced into the United States from Asia and Europe. Another sac fungus, *Aspergillus flavus*, is a contaminant of stored nuts and grain. This fungus produces a toxin and the most potent known natural carcinogen. You have already examined one example of the phylum Ascomycota in Lab Topic 7 when you studied meiosis and crossing over in *Sordaria fimicola*.

When present, sexual reproduction in the ascomycota fungi produces either four or eight haploid **ascospores** after meiosis in an **ascus.** Recall that spores in *Sordaria* form after meiosis within asci (Figure 7.13). Asci form within a structure called an **ascocarp.** In *Sordaria* the ascocarp, called a *perithecium*, is a closed, spherical structure that develops a pore at the top for spore dispersal. In some species of sac fungi, the asci are borne on open cup-shaped ascocarps called *apothecia* (sing., *apothecium*). In asexual reproduction, spores are produced, but rather than being enclosed within a sporangium as in zygote fungi, the spores, called **conidia,** are produced on the surface of special reproductive hyphae.

Other features of sac fungi also vary. For example, yeasts are ascomycetes, yet they are single-celled organisms. Yeasts most frequently reproduce asexually by **budding,** a process in which small cells form by pinching off the parent cell. When they reproduce sexually, however, they produce asci, each of which produces four or eight spores. Note that *Penicillium* and some other fungi previously placed in the phylum Deuteromycota are now classified as ascomycetes based on evidence from recent molecular studies.

In this lab study, you will examine a slide of the sac fungi *Peziza* and will observe demonstrations of additional examples of Ascomycota (see Color Plate 29), including *Penicillium* growing in lab cultures and in Roquefort cheese.

*Previous users of this lab manual will note that we have deleted Lab Study D Imperfect Fungi, and moved* Penicillium *into the study of ascomycetes. With application of DNA sequencing to determine the taxonomic status of the imperfect fungi, most authorities now place* Penicillium sp. *in this phylum.*

## Procedure

1. Obtain a dried or preserved specimen of *Peziza* (Figure 14.13a). Notice the open, cup-shaped apothecium, the **ascocarp,** that bears asci within the cup (not visible with the naked eye). Fungi with ascocarps shaped in this fashion are called **cup fungi.** The cup may be supported by a stalk.

2. Examine a prepared slide of *Peziza* using low and intermediate magnifications on the compound microscope. This slide is a section through the ascocarp. Identify **asci.** How many spores are present per ascus? Are they diploid or haploid?

*eight haploid spores*

3. Complete the sketch of the ascocarp section that follows, labeling **asci, spores, hyphae,** and **mycelium.**

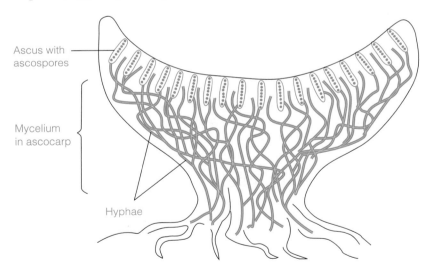

Ascus with ascospores

Mycelium in ascocarp

Hyphae

*Help students visualize the section relative to the Peziza ascocarps on demonstration. Some sections will not show the stalk.*

4. Observe the *Penicillium* on demonstration. You may have observed something similar growing on oranges or other foods in your refrigerator.

5. Describe the texture and the color of the mycelium.

6. Observe the preserved **morels** that are on demonstration (Figure 14.13b). These fungi resemble mushrooms, but the "cap" is convoluted. Asci are located inside the ridges.

7. Observe demonstrations of the mature inflorescence of wheat or rye grass infected with the ascomycete *Claviceps purpurea,* the **ergot** fungus. The large black structures seen among the grains are the ergot.

**Figure 14.13.**
**Examples of sac fungi, phylum Ascomycota.** (a) *Peziza* has a cup-shaped ascocarp with asci within the cup. (b) Morels are cup fungi that resemble mushrooms.

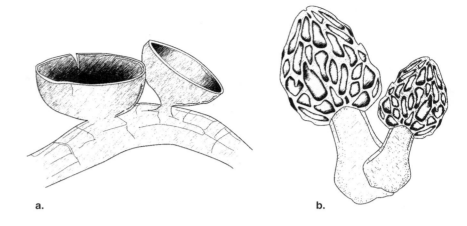

a.                                     b.

## Results

1. Review the structures observed in *Peziza*, morels, and ergot. Modify Figures 14.13a and 14.13b to reflect features of your examples not included in these figures.
2. Sketch ergot examples in the margin of your lab manual.
3. Sketch your observations of *Penicillium* in the margin of your lab manual. Note any features that may be important in distinguishing this organism.

## Discussion

1. What characteristics are common to all sac fungi that reproduce sexually?

   *Spores form by meiosis in asci, giving rise to four or eight spores. Asci are born in structures called ascocarps.*

2. Compare the appearance of *Penicillium* with that of *Rhizopus*.

   *The mycelium of Penicillium is a green, dense, wavy mat of hyphae. The colony is much more dense than that of Rhizopus and the hyphae are more compacted.*

# Lab Study C. Club Fungi—Basidiomycota

## Materials

compound microscope
stereoscopic microscope
fresh, ripe mushroom basidiocarps
prepared slides of *Coprinus* pileus sections

## Introduction

The Basidiomycota phylum (club fungi, or basidiospore-producing fungi) includes the fungi that cause the plant diseases wheat rust and corn smut as well as the more familiar puffballs, shelf fungi, and edible and nonedible mushrooms (the latter often called *toadstools*). A mushroom is actually a reproductive structure called a basidiocarp that produces spores by meiosis. Although most mushrooms are relatively small when mature, one basidiocarp in the Royal Botanic Gardens, Kew, England, in 1996 measured 146 cm (57 inches wide) and weighed 284 kg (625 pounds)! Basidiocarps grow upward from an underground mycelial mass. When they form around the rim of the mass, a "fairy ring" of mushrooms appears. In asexual reproduction, conidia form by mitosis. In this lab study, you will study mushrooms and learn some features of their life cycle.

## Procedure

1. Obtain a fresh mushroom, a **basidiocarp,** and identify its parts: The stalk is the **stipe;** the cap is the **pileus.** Look under the cap and identify **gills.** Spores form on the surface of the gills. Examine the gills with the stereoscopic microscope. Do you see spores? Children often make spore prints in scouts or in elementary school by placing a ripe mushroom pileus with the gill side down on a piece of white paper for several hours, allowing the spores to drop to the paper. Scientists use similar spore prints to accurately identify mushrooms.

2. Label the parts of the mushrooms in Figure 14.14a.

3. Obtain a prepared slide of a section through the pileus of *Coprinus* or another mushroom. Observe it on the compound microscope using low and then intermediate powers. Is your slide a cross section or a longitudinal section through the pileus? Make a sketch in the lab manual

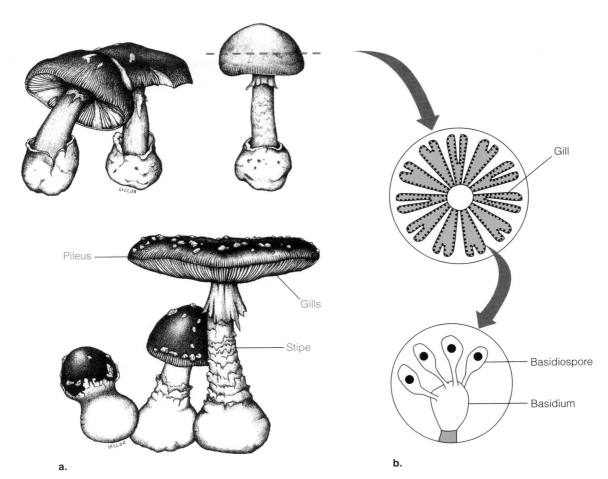

**Figure 14.14.**
**Club fungi, phylum Basidiomycota.** (a) Mushrooms, or basidiocarps, each consisting of a cap, the pileus; and a stalk, the stipe. (b) A section through the gills on a whole basidiocarp reveals basidia and basidiospores.

*Help students visualize the section relative to the basidiocarps.*

margin indicating the plane of your section through the basidiocarp. Compare your section with the fresh mushroom you have just studied and with Figure 14.14b.

4. Using the prepared slide, observe the surface of several gills using high power. Spores are produced at the tips of small club-shaped structures called **basidia.** Locate a basidium and focus carefully on its end. Here you may see four knoblike protuberances. Each protuberance has a haploid nucleus that formed following meiosis, and each becomes a **basidiospore.** When the spores are mature, they are discharged from the basidium and are dispersed by the wind.

### Results

Review the structures observed and label Figure 14.14a. Modify the figure to include features observed in your materials that differ from the figure.

### Discussion

State the characteristics shared by all Basidiomycota.

*They produce basidiospores on the surface of club-shaped basidia within a basidiocarp.*

 Student Media Videos—Discovery Videos: Fungi; Leafcutter Ants

## Lab Study D. Lichens

### Materials

examples of foliose, crustose, and fruticose lichens on demonstration

### Introduction

Lichens are symbiotic associations between fungi and usually algae or photosynthetic bacteria forming a body that can be consistently recognized. The fungal component is usually a sac fungus or a club fungus. The lichen body, called a **thallus,** varies in shape and colors, depending on the species of the components. Reproductive structures can be bright red or pink or green. Photosynthesis in the algae provides nutrients for the fungus, and the fungus provides a moist environment for the algae or bacterium. Because lichens can survive extremely harsh environments, they are often the first organisms to colonize a newly exposed environment such as volcanic flow or rock outcrops, and they play a role in soil formation.

### Procedure

Observe the demonstrations of different lichen types: those with a leafy thallus (**foliose**), a crustlike thallus (**crustose**), or a branching, cylindrical thallus (**fruticose**) (Figure 14.15). Look for cup-shaped or clublike reproductive structures produced by the fungal component of the lichen.

**Figure 14.15.**
**Lichen types.** Lichens may have
(a) a leafy thallus (foliose),
(b) a crustlike thallus (crustose), or
(c) a cylindrical thallus (fruticose).

## Results

1. Sketch the lichens on demonstration in the margins of your lab manual.
2. Label any visible reproductive structures, and, if possible, indicate if the fungal component is a sac fungus or a club fungus.
3. Identify and label each according to lichen type.

## Discussion

Imagine that you are the first scientist to observe a lichen microscopically. What observations would lead you to conclude that the lichen is composed of a fungus and an alga?

*presence of simple photosynthetic cells and filamentous hyphae*

# EXERCISE 14.3
# Designing and Performing an Open-Inquiry Investigation

## Introduction

In this exercise, you will choose one of the organisms observed in this lab topic and design a simple experiment answering a question about its behavior, growth patterns, or interactions with other species.

Use Lab Topic 1 as a reference for designing and performing a scientific investigation. Be ready to assign tasks to members of your lab team. Be sure that everyone understands the techniques that will be used. Your experiment will be successful only if you plan carefully, cooperate with your team members, perform lab techniques accurately and systematically, and record and report data accurately.

## Materials

protozoa and algae cultures

cultures of slime molds *Physarum, Didymium, Dictyostelium*

cultures of *Pilobolus crystallinus, Rhizopus, Penicillium*

sterile agar plates to grow each species

sterile agar with oat flakes

sterile agar with sugar

sterile agar with albumin

sterile agar with pH 6, 7, or 8

aluminum foil

various breads from the health food store—wheat, rye, corn, potato, rice

bread with preservatives

sterilized dung from various animals

mycorrhizae inoculate

## Procedure

1. **Choose a question from this list to investigate or choose a question from your own observations.** *Write your question in the margin of your lab manual.*

   a. Will varying the molarity of the culture medium change the rate of contractile vacuole formation in paramecia?

   b. Do plasmodia of the same species of slime mold unite when growing on the same agar plate? How about different species of slime mold?

   c. Do slime mold plasmodia demonstrate chemotaxis (response to chemical stimuli such as food molecules) or phototaxis (response to light)?

   d. What happens to slime molds if grown in different temperatures?

   e. Do the same fungi grow on different varieties of bread?

   f. How effective are preservatives in preventing fungal growth on foods?

   g. Is *Pilobolus* phototaxic? What about other fungi?

   h. Does succession take place in dung cultures of fungi? Refer to the milk bacteria succession study in Lab Topic 13 and design a similar experiment to investigate this phenomenon in fungi growing on dung.

    i. Is there a difference in the growth of plants growing with and without mycorrhizae?

    j. Can the growth of fungi be altered by supplying different nutrients (e.g., sugar or albumin) in agar culture?

2. **Formulate a testable hypothesis.** (Refer to Exercise 1.1, Lab Study B, Developing Hypotheses.)

   **Hypothesis:**

3. **Summarize the essential elements of the experiment.** (Use separate paper.)

4. **Predict the results of your experiment based on your hypothesis.** (Refer to Lab Topic 1, Exercise 1.2, Lab Study C, Making Predictions.)

   **Prediction:** (If/then)

5. **Outline the procedures used in the experiment.** (Refer to Exercise 1.2, Lab Study B, Choosing or Designing a Procedure.)

   a. On a separate sheet of paper, list in numerical order each exact step of your procedure.

   b. Remember to include the number of replicates (usually a minimum of five), levels of treatment, appropriate time intervals, and controls for each procedure.

   c. If you have an idea for an experiment that requires materials other than those provided, ask your laboratory instructor about availability. If possible, additional supplies will be provided.

   d. When carrying out an experiment, remember to quantify your measurements when possible.

6. **Perform the experiment**, making observations and collecting data for analysis.

7. **Record results, including observations and data** on a separate sheet of paper. Design tables and graphs, at least one of each. Be thorough when collecting data. Do not just write down numbers, but record what they mean as well. Do not rely on your memory for information that you will need when reporting your results.

8. **Prepare your discussion.** Discuss your results in light of your hypothesis.

   a. Review your hypothesis. Review your results (tables and graphs). Do your results support or falsify your hypothesis? Explain your answer, using data for support.

   b. Review your prediction. Did your results correspond to the prediction you made? If not, explain how your results are different from your predictions, and why this might have occurred.

   c. If you had problems with the procedure or questionable results, explain how they might have influenced your conclusion.

   d. If you had an opportunity to repeat and expand this experiment to make your results more convincing, what would you do?

   e. Summarize the conclusion you have drawn from your results.

9. **Be prepared to report your results to the class.** Prepare to persuade your fellow scientists that your experimental design is sound and that your results support your conclusions.

10. If your instructor requires it, **submit a written laboratory report** in the form of a scientific paper (see Appendix A). Keep in mind that although you have performed the experiments as a team, you must turn in a lab report of *your original writing*. Your tables and figures may be similar to those of your team members, but your paper must be the product of your own literature search and creative thinking.

## Reviewing Your Knowledge

1. Complete Table 14.5 comparing characteristics of all protists investigated in Exercise 14.1.

2. Complete Table 14.6 comparing characteristics of fungi (Exercise 14.2).

3. Compare spore formation in sac fungi and club fungi.

*In sac fungi four or eight spores form after meiosis, sometimes followed by mitosis, in structures called asci that are located in an ascocarp that can be a closed, spherical structure (**Sordaria**) or an open cup (**Peziza**). In club fungi, basidia are located on the gills of basidiocarps (mushrooms). After meiosis, four knoblike protuberances form on the basidia and a haploid nucleus moves into each knob, eventually forming a basidiospore in each knob.*

4. Using observations of pigments present, body form, and distinguishing characteristics of the three groups of macroscopic green, brown, and red algae, speculate about where they might be most commonly found in ocean waters.

*Observed characteristics might lead students to draw conclusions about the ecology of the algae, as follows.*

*Green algae: lack strong attachment structures, have thin photosynthetic blades with little protection from desiccation; chlorophylls are the main pigments. They are probably more common near the surface of ocean waters, either floating or attached to something that floats. They probably cannot withstand being alternately exposed and flooded with tidal cycles unless protected by other algae.*

*Red algae: have pigments that absorb light wavelengths that penetrate deeper; the blades are usually finely dissected; they often have structures for attachment. They might be found attached in deep waters, associated with coral reefs, for example. In fact, some red algae deposit calcium carbonate in their cell walls and play a role in building coral reefs.*

*Brown algae: strong attachment structures, long thick blades, flotation structures, mucilaginous secretions to prevent desiccation. All these features are adaptations to living attached along shores where wave action is strong and the algal body is alternately flooded and exposed to air.*

**Table 14.5**
Comparison of Protists Studied in Exercise 14.1

| Group (Clade) | Example(s) | Characteristics | Ecological Role | Economic Importance |
|---|---|---|---|---|
| Euglenozoans | *Trypanosoma levisi* | Microscopic; flagellated, parasitic, heterotrophic | Parasitic | Lives in blood of rats; transmitted by fleas; human diseases |
| Alveolates | Paramecia | Microscopic; single-celled; heterotrophic, alveoli under cell membrane | Food for microscopic consumers in plankton and some animals | Only in role as food for larger organisms |
| | Dinoflagellates | Microscopic; single-celled; autotrophic; cell wall in plates; two perpendicular grooves with flagella | Primary producer; food for consumers in plankton; can be toxic to fish | Secrete toxins that kill fish in red tides |
| Stramenopiles | Diatoms | Microscopic; autotrophic; cell wall of silica; pinnate or centric forms | Primary producer; food for consumers in plankton | Diatomaceous earth has commercial uses |
| | Brown algae | Macroscopic; autotrophic; pigments fucoxanthin and chlorophyll a present | Primary producer | Extract algin used in paint, toothpaste, ice cream, and other food products |
| Rhizarians | Foraminiferans | Microscopic; ameboid; heterotrophic; threadlike pseudopodia; secrete calcium carbonate tests | Food for larger organisms in plankton | Form limestone deposits in ocean floor |
| | Radiolarians | Microscopic; ameboid; heterotrophic; thread like pseudopodia; secrete silicon dioxide skeleton | Food for larger organisms in plankton | Form silicon deposits in ocean floor |
| Amoebozoans | Amoeba | Microscopic; ameboid; lobe-shaped pseudopodia; heterotrophic | Food for larger organisms | Some may be parasitic |
| | *Physarum* | Ameboid plasmodium; heterotrophic | Feeds on bacteria | Sold in biological supply houses |
| Rhodophyta | Red algae | Macroscopic; multicellular body; chlorophyll a, phycocyanin, phycoerythrin pigments | Marine; primary producer | Source of agar and carrageenan; used in food preparation (e.g., sushi) |
| Chlorophyta and Charophyta | Green algae: *Spirogyra, Ulva, Chara* | Microscopic and macroscopic; store amylase; chlorophyll a and b pigments | Primary producer; ancestor to land plants | Food for humans and livestock and proposed medicinal applications |

**Table 14.6**
Comparison of Fungi by Major Features

| Phylum | Example(s) | Sexual Reproductive Structures | Asexual Reproductive Structures |
|---|---|---|---|
| **Zygomycota (Zygote Fungi)** | Rhizopus, Pilobolus | *zygosporangia, zygospores* | *sporangia/spores* |
| **Ascomycota (Sac Fungi)** | Sordaria, Peziza, Penicillium | *morels, ascocarp, asci, ascospores* | *conidia* |
| **Basidiomycota (Club Fungi)** | Coprinus | *basidiocarp, basidia, basidiospores* | *conidia* |

# Applying Your Knowledge

1. In 1950, the living world was classified simply into two kingdoms: plants and animals. More recently, scientists developed the five-kingdom system of classification: plants, animals, monerans, protists, and fungi. In 2000, there was a general consensus among scientists that three domains with more than five kingdoms was a better system for classifying the diversity of life on Earth. However, there is still no consensus on the number of kingdoms or the clustering of organisms that best represents their evolutionary relationships. Using the protists studied in this lab topic, explain why the classification of this diverse group in particular is problematic. How is solving the problem of organizing protistan diversity a model for understanding the process of science?

   *At first, organisms were placed in either the plant or animal kingdom based on the presence or absence of chloroplasts. Observing the general structure, nutrition, and locomotion of many apparently simple organisms, scientists recognized that they did not fit easily into the eukaryotic kingdoms. Therefore, the kingdom Protista was created to include those eukaryotes that were not plants, animals, or fungi. As scientists investigated the cellular structure and function of protists, as well as the molecular evidence provided by RNA and DNA analysis, the relationships of some protists became clearer. However, in some cases the evidence remains inconclusive, thus the tentative groupings for dinoflagellates, ciliates, and even red algae. Science is constantly testing hypotheses (in this case, the classification scheme), providing new evidence from experiments and observations that either falsifies or supports the existing hypothesis (classification). Science is tentative and self-correcting. As new evidence is reported from innovative approaches and techniques, our understanding develops and the classification is modified. This can be both frustrating and exciting for scientists and students.*

*Many websites on protistan systematics are out of date. Your best source for information on this subject may be the latest edition of your text and its website, for example www.masteringbiology.com*

2. Scientists are concerned that the depletion of the ozone layer will result in a reduction of populations of marine algae such as diatoms and dinoflagellates. Recall the ecological role of these organisms and comment on the validity of this concern.

   *Because diatoms and dinoflagellates are important primary producers in oceans, they are the basis of all marine food webs. Environmental changes that have a negative impact on these organisms could ultimately be destructive to most marine animals, leading to a severe reduction in species diversity and in human food supplies.*

3. Imagine an ecosystem with no fungi. How would it be modified?

   *Students should recall the role of decomposition played by fungi. Dead plant and animal matter would remain, body piled upon body. Piles of dead leaves would obscure trees in the forest. Nutrients would run out because they would be tied up in dead organic matter. Without mycorrhizae, plant growth would be reduced.*

4. Speculate about a possible evolutionary advantage to the *fungus* for the following:

   a. *Penicillium* makes and secretes an antibiotic.

      *prevents its food source from being digested by bacteria*

   b. *Ergot* fungus (parasitizes rye grain) produces a chemical that is toxic to animals.

      *discourages animals from eating the grain, the ergot food source*

 ## Student Media: BioFlix, Activities, Investigations, and Videos
**www.masteringbiology.com (select Study Area)**

**Activities**—Ch. 26: Classification Schemes; Ch. 28: Tentative Phylogeny of Eukaryotes; Ch. 31: Fungal Reproduction and Nutrition; Fungal Life Cycles
**Investigations**—Ch. 28: What Kinds of Protists Do Various Habitats Support? Ch. 31: How Does the Fungus *Pilobolus* Succeed as a Decomposer?
**Videos**—Ch. 28: *Paramecium* Vacuole; *Paramecium* Cilia; Dinoflagellate; Diatoms Moving; Various Diatoms; Amoeba; Amoeba Pseudopodia; Plasmodial Slime Mold Streaming; Plasmodial Slime Mold; Discovery Videos: Ch. 31: Fungi; Leafcutter Ants

## References

Ahmadjian, V. "Lichens Are More Important Than You Think." *BioScience*, 1995, vol. 45, p. 124.

Alexopoulos, C., C. Mims, and M. Blackwell. *Introductory Mycology*, 4th ed. New York: John Wiley and Sons, Inc., 1996.

Anderson, R. "What to Do with Protists?" *Australian Systematic Botany*, 1998, vol. 11, p. 185.

Bold, H., C. J. Alexopoulos, and T. Delevoryas. *Morphology of Plants and Fungi*. New York: Harper & Row, 1980, p. 654.

Crowder, W. "Marvels of Mycetozoa." *National Geographic Magazine*, 1926, vol. 49, pp. 421–443.

Doolittle, W. F. "Uprooting the Tree of Life." *Scientific American*, 2000, vol. 282, pp. 90–95.

Litten, W. "The Most Poisonous Mushrooms." *Scientific American*, 1975, vol. 232.

Milius, S. "A Partnership Apart . . . Scientists Dissect and Redefine . . . Lichen Mutualism." *Science News*, 2009, vol. 176(10), pp. 17–20.

Reece, J., et al. *Campbell Biology*, 9th ed. San Francisco, CA: Pearson Education, 2011.

## Websites

Beautiful video of growing fungi from "The Private Life of Plants":
http://www.youtube.com/watch?v=puDKLFcCyI+NR=1&feature=fvwp

A detailed discussion of algae in the phylum Chlorophyta: http://www.scribd.com/doc/2409467/ Chlorophyta

Lichens:
www.lichen.com

Links to pictures of red, brown, and green algae:
https://www.sonoma.edu/users/c/cannon/MarineAlgae.html

Mutualism of fungi and plants:
http://discovermagazine.com/2004/oct/it-takes-a-fungus

Mycological Resources on the Internet:
http://mycology.cornell.edu

Photos and identification of 1,250 species of lichens taken for the book *Lichens of North America* by Irwin Brodo, 2001:
http://www.sharnoffphotos.com/lichens/lichens_home_index.html

Protist Image Data. Excellent page links:
http://megasun.bch.umontreal.ca/protists/protists.html

Seaweeds:
http://www.botany.uwc.ac.za/Envfacts/seaweeds/

See an amoeba video and find interesting information on amoebas:
http://www.microscopy.fsu.edu/moviegallery/pondscum/protozoa/amoeba/

The Tree of Life web project—a collaborative effort of biologists from around the world to present information on diversity and phylogeny of organisms:
http://tolweb.org/tree/

Video of *Pilobolus* spore discharge:
http:www.youtube .com/watch?v=TrKJAojmB1Y

# LAB TOPIC 14

# Protists and Fungi
## Teaching Plan for Laboratories

If you used previous editions of *Investigating Biology*, you will note that this lab topic has been reorganized to reflect recent systems of classification based on molecular data. All organisms studied in the previous edition are included in this edition. We have used the latest edition of *Campbell Biology* as our standard. If you choose to organize your lab as in previous editions, refer to the table at the end of this teaching plan.

## Main Concepts and Objectives

1. Concept: classifying organisms. Classification schemes are assigned by scientists, and the consensus can shift, leading to changes in criteria for taxonomic groups, or clades.

2. Concept: the diversity of protists. Organisms that are single-celled, filamentous, and colonial are in this group. They may be heterotrophic or autotrophic. Multicellular algae and slime molds are considered protists. (Exercise 14.1)

3. Concept: the evolutionary relationship between Chlorophyta and plants. Some systematists suggest that green algae are best placed in the Plant kingdom. (Exercise 14.1, Lab Study G)

4. Concept: the diversity of organisms in the kingdom Fungi. Phyla are designated based on characteristics of reproductive cycles. (Exercise 14.2)

5. Concept: scientific investigation. Students can design and perform a simple investigation of a protistan or fungal organism. (Exercise 14.3)

6. Specific Content. Students will be able to describe or define: names, characteristics, and examples of protists; names and functions of all cell structures in protistan examples; names, characteristics, and examples of phyla in the kingdom Fungi; names of reproductive structures.

## Order of the Lab

1. Introduce concepts and objectives. Describe classroom management.   (20 min)
2. Observe examples of protists: living organisms, prepared slides, demonstrations.   (70 min)
3. Observe examples of fungi.   (70 min)

4. Give instructions about designing the independent
   investigation.                                          (5 min)

5. Students plan their investigations.                     (15 min)

**For 2-hour labs:** Perform activities for two consecutive laboratory periods with protists the first week and fungi the second week. If the lab must be completed in one 2-hour period, omit selected demonstrations and the independent investigation.

## Classroom Management

Each student should have a slide set. Students work individually on prepared slides and in pairs for all other work. Encourage students to share their observations of the demonstrations and slide preparations with their lab partners so they can discuss answers to questions. Students should view demonstrations in staggered fashion to avoid standing and waiting. Question students on phylum characteristics and evidence for environmental adaptations as they work through the lab. This lab should not be a passive experience for the students. Encourage them to think. Ask more questions than you answer. Keep students moving through the material by giving them some time estimates.

The exercise calls for stereoscopic microscopes and compound microscopes. Students should have both close at hand so that they can switch between them. Have the demonstrations set up in order with distinguishing cards. To add interest, use materials from your own region.

Students should work in groups of two or four for their independent investigations. If the laboratory facility permits, students should set up their experiments and return to the lab regularly to collect data. It may be possible for students to take their experiments with them and make observations at home. At the beginning of the next lab, allow students to discuss their results with their lab team and, if possible, report to the class.

## Student Development

Students will improve their skills in microscopy and making observations. If students perform Exercise 14.3, they will continue to practice asking questions, proposing hypotheses, designing and performing experiments, and discussing results. They learn that even observation-based activities can lead to interesting questions. They develop writing skills as they answer the questions in the exercise.

## Discussion and Summary

Students complete tables that require them to summarize their observations. They will discuss with teammates observations made in their independent investigations.

## Evaluation

Students can be evaluated on a written exam with a practical component as part of which they must identify the organisms seen in lab. Instructors must determine the extent to which they will require students to memorize the classification component of the laboratory. If your schedule permits, have students report on the results of their experiments. This could be in the form of an oral or written assignment.

## Alternative Lab Topic Organization

If you choose to organize the protists investigated in this lab topic as in previous editions (for example, the 4th edition) based on nutritive mode and means of locomotion in protozoa, the following table provides an alternative organization for the lab topic.

| Categories of Organisms | Means of Locomotion (If Any) | Examples | Exercise and Lab Study in this Edition (8e) | Clade or Group |
|---|---|---|---|---|
| Heterotrophic Protists (Protozoa) | Ameboid: pseudopodia | Amoeba | 14.1, Lab Study E | Amoebozoans |
| | Ameboid: pseudopodia | Foraminiferans | 14.1, Lab Study D | Forams |
| | Ameboid: pseudopodia | Radiolarians | 14.1, Lab Study D | Radiolarians |
| | Flagella | Trypanosoma | 14.1, Lab Study A | Euglenozoans |
| | Cilia | Paramecia | 14.1, Lab Study B | Alveolates |
| Autotrophic Protists (Algae) | Flagella | Dinoflagellates | 14.1, Lab Study B | Alveolates |
| | Contractile fibers in some | Diatoms | 14.1, Lab Study C | Stramenopiles |
| | Flagellated gametes | Brown algae | 14.1, Lab Study C | Stramenopiles |
| | No flagellated cells | Red algae | 14.1, Lab Study F | Rhodophyta |
| | Some flagellated species; flagellated gametes | Green algae | 14.1, Lab Study G | Chlorophyta Charophyta |
| Funguslike Protists | Ameboid: pseudopodia | Slime mold: Physarum | 14.1, Lab Study E | Amoebozoans |
| Fungi | | Zygote fungi: Rhizopus and Pilobolus | 14.2, Lab Study A | Fungi |
| | | Sac fungi: Peziza and Penicillium | 14.2, Lab Study B | Fungi |
| | | Club fungi: Coprinus and mushrooms | 14.2, Lab Study C | Fungi |
| Lichens | | Foliose, crustose, and fruticose lichens | 14.2, Lab Study D | |

## Lab Topic 15

# Plant Diversity I: Nonvascular Plants (Bryophytes) and Seedless Vascular Plants

## Laboratory Objectives

After completing this lab topic, you should be able to:

1. Describe the distinguishing characteristics of nonvascular plants and seedless vascular plants.

2. Discuss the ancestral and derived features of nonvascular plants and seedless vascular plants relative to their adaptations to the land environment.

3. Recognize and identify representative members of each phylum of nonvascular plants and seedless vascular plants.

4. Describe the general life cycle and alternation of generations in the nonvascular plants and the seedless vascular plants, and discuss the differences between the life cycles of the two groups of plants using examples.

5. Identify fossil members and their extant counterparts in the seedless vascular plants.

6. Describe homospory and heterospory, including the differences in spores and gametophytes.

7. Discuss the ecological role and economic importance of these groups of plants.

*For a 2-hour lab: Omit the demonstration material for liverworts (Exercise 15.1, Lab Study B) and the phyla of seedless vascular plants (Exercise 15.2, Lab Studies A and B). See Teaching Plan for additional suggestions.*

## Introduction

In the history of life on Earth, one of the most revolutionary events was the colonization of land, first by plants, then by animals. Evidence from comparisons of extant land plants and phyla of algae suggests that the first land plants shared a common ancestor with green algae, particularly the charophytes. These first colonists are thought to be most similar to the living, branched, multicellular green alga *Chara* (studied in Lab Topic 14 Protists and Fungi). Once these simple ancestral plants arrived on land over 475 million years ago, they faced new and extreme challenges in their physical environment. Only individuals that were able to survive the variations in temperature, moisture, gravitational forces, and substrate would thrive. Out of this enormous selective regime would come new and different adaptations and new and different life forms: the land plants.

Land plants generally have complex, multicellular plant bodies that are specialized for a variety of functions. Land plants in the Kingdom **Plantae** produce embryos and have evolved specialized structures for protection of the vulnerable stages of sexual reproduction. The plant body is often covered

*If not using Lab Topic 14, Protists and Fungi, consider providing a demonstration of Chara with the information on pp. 367–368.*

*In the Study Area of masteringbiology .com, see Student Media for suggested activities, investigations, and videos.*

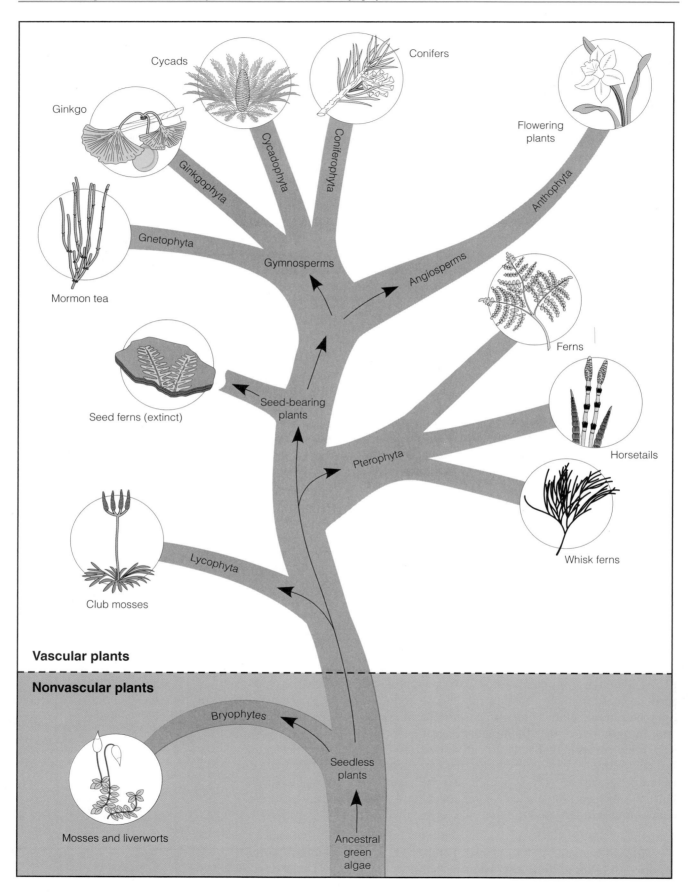

Cycads

Ginkgo

Conifers

Flowering plants

Cycadophyta

Coniferophyta

Ginkgophyta

Anthophyta

Gnetophyta

Gymnosperms

Angiosperms

Mormon tea

Ferns

Seed ferns (extinct)

Seed-bearing plants

Horsetails

Pterophyta

Lycophyta

Whisk ferns

Club mosses

**Vascular plants**

**Nonvascular plants**

Bryophytes

Seedless plants

Mosses and liverworts

Ancestral green algae

**Table 15.1**
Classification of Land Plants

| Classification | Common Name | Illustration |
|---|---|---|
| **Nonvascular Plants (Bryophytes)** | | |
| Phylum Bryophyta | Mosses | |
| Phylum Hepatophyta | Liverworts | |
| Phylum Anthocerophyta | Hornworts | |
| **Vascular Plants** | | |
| **Seedless Plants** | | |
| Phylum Lycophyta | Club mosses | |
| Phylum Pterophyta | Ferns, horsetails, whisk ferns | |
| **Seed Plants** | | |
| Gymnosperms | | |
| Phylum Coniferophyta | Conifers | |
| Phylum Cycadophyta | Cycads | |
| Phylum Ginkgophyta | Ginkgo | |
| Phylum Gnetophyta | Mormon tea | |
| Angiosperms | | |
| Phylum Anthophyta | Flowering plants | |

with a waxy cuticle that prevents desiccation. However, the waxy covering also prevents gas exchange, a problem solved by the presence of openings called **stomata** (sing., **stoma**). Some land plants have developed vascular tissue for efficient movement of materials throughout these complex bodies, which are no longer bathed in water. As described in the following section, the reproductive cycles and reproductive structures of these plants are also adapted to the land environment.

In the two plant diversity labs, you will be investigating the diversity of land plants (Table 15.1 and Figure 15.1), some of which will be familiar to you (flowering plants, pine trees, and ferns) and some of which you may never have seen before (whisk ferns, horsetails, and liverworts). You will study the nonvascular plants and seedless vascular plants in this lab topic, Plant Diversity I, and seed plants in Lab Topic 16 Plant Diversity II.

**(☜) Figure 15.1.**
**Evolution of land plants.** The nonvascular plants and vascular plants probably evolved from ancestral green algae over 475 million years ago. Seedless vascular plants dominated Earth 300 million years ago, and representatives of two phyla have survived until the present. Seed plants replaced the seedless plants as the dominant land plants, and today flowering plants are the most diverse and successful group in an amazing variety of habitats. The representatives studied in Plant Diversity I and II are indicated.

*To maintain your perspective in the face of all this diversity—and to reinforce the major themes of these labs—bear in mind the following questions.*

1. What are the special adaptations of these plants to the land environment?
2. How are specialized plant structures related to functions in the land environment?
3. What are the major trends in the plant kingdom as plant life evolved over the past 500 million years?
4. In particular, how has the fundamental reproductive cycle of alternation of generations been modified in successive groups of plants?

## Plant Life Cycles

All land plants have a common sexual reproductive life cycle called **alternation of generations,** in which plants alternate between a haploid **gametophyte** generation and a diploid **sporophyte** generation (Figure 15.2). In living land plants, these two generations differ in their morphology, but they are still the same species. In all land plants except the bryophytes (mosses and liverworts), the diploid sporophyte generation is the dominant (more conspicuous) generation.

The essential features in the alternation of generations life cycle beginning with the sporophyte are:

- The diploid sporophyte undergoes meiosis to produce haploid **spores** in a protective, nonreproductive jacket of cells called the **sporangium.**

- Dividing by mitosis the spores germinate to produce the haploid gametophyte.

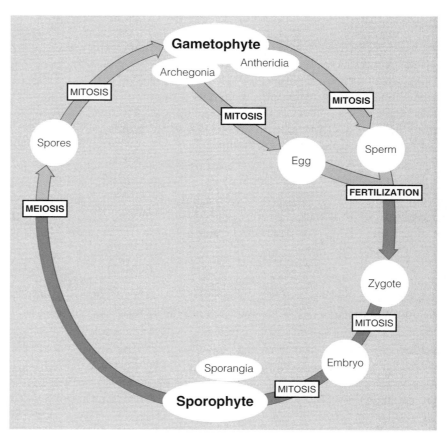

**Figure 15.2.**
**Alternation of generations.** In this life cycle, a diploid sporophyte plant alternates with a haploid gametophyte plant. Note that haploid spores are produced on the sporophyte by meiosis, and haploid gametes are produced in the gametophyte by mitosis. *Using a colored pencil, indicate the structures that are haploid, and with another color, note the structures that are diploid.*

- The gametophyte produces **gametes** inside a jacket of nonreproductive cells forming **gametangia** (sing., **gametangium**).

- **Eggs** are produced by mitosis in **archegonia** (sing., **archegonium**), and **sperm** are produced in **antheridia** (sing., **antheridium**).

- The gametes fuse (**fertilization**) usually by entrance of the sperm into the archegonium, forming a diploid **zygote**, the first stage of the next diploid sporophyte generation.

Note that both gametes and spores are haploid in this life cycle. Unlike the animal life cycle, *the plant life cycle produces gametes by mitosis; spores are produced by meiosis.* The difference between these two cells is that gametes fuse with other gametes to form the zygote and restore the diploid number, while haploid spores germinate to form a new haploid gametophyte plant.

Review the generalized diagram of this life cycle in Figure 15.2. *Using colored pencils, distinguish the structures that are diploid and those that are haploid.* As you become familiar with variations of this life cycle through specific examples, you will want to continue referring to this general model for review.

*Major trends in the evolution of this life cycle include:*

*Throughout the lab topic, students are asked to use colored pencils to indicate the gametophyte and sporophyte generations. Encourage this activity, which reinforces their understanding of life cycles.*

the increased importance of the sporophyte as the photosynthetic and persistent plant that dominates the life cycle; the reduction and protection of the gametophyte within the body of the sporophyte; and the evolution of seeds and then flowers.

## Nonvascular Plants (Bryophytes) and Seedless Vascular Plants

In this lab topic, terrestrial plants will be used to illustrate how life has undergone dramatic changes during the past 500 million years. Not long after the transition to land, plants diverged into at least two separate lineages. One gave rise to the bryophytes, a group of nonvascular plants, including the mosses, and the other to the vascular plants (see Figure 15.1). Nonvascular bryophytes first appear in the fossil record dating over 420 million years ago and remain unchanged, whereas the vascular plants have undergone enormous diversification. As you review the evolution of land plants, refer to the geological time chart for an overview of the history of life on Earth (Figure 15.3).

## EXERCISE 15.1
# Nonvascular Plants (Bryophytes)

The nonvascular plants are composed of three phyla of related plants that share some key characteristics and include mosses (Bryophyta) and liverworts (Hepatophyta). The third phylum, hornworts (Anthocerophyta), will not be seen in lab. (See again Figure 15.1 and Table 15.1.) The term *bryophytes* does not refer to a taxonomic category; rather, bryophytes are an ancient group of nonvascular plants that share a common ancestor, appear to have evolved into several different groups independently, and did not give rise to any other living groups of plants. They are small plants generally lacking vascular tissue (specialized cells for the transport of material),

*All three phyla are sometimes placed in the phylum Bryophyta. If you use this system, have your students modify the headings in Lab Study B and in Table 15.1.*

| Years Ago (millions) | Era Period Epoch | | Life on Earth |
|---|---|---|---|
| | **CENOZOIC** Quaternary | | |
| | | Recent Pleistocene | • Origin of agriculture and artificial selection; *H. sapiens* |
| — 1.8 | Tertiary | | |
| | | Pliocene | • Large carnivores; hominoid apes |
| — 5 | | Miocene | • Forests dwindle; grassland spreads |
| — 23 | | | |
| | | Oligocene | • Anthropoid apes |
| — 35 | | Eocene | • Diversification of mammals and flowering plants |
| — 57 | | | |
| | | Paleocene | • Specialized flowers; sophisticated pollinators and seed distributors |
| — 65 | **MESOZOIC** Cretaceous | | • Flowering plants established and diversified; many modern families present; extinction of many dinosaurs |
| — 145 | Jurassic | | • Origin of birds; reptiles dominant; cycads and ferns abundant; first modern conifers and immediate ancestors of flowering plants |
| — 208 | Triassic | | • First dinosaurs and mammals; forests of gymnosperms and ferns; cycads |
| — 245 | **PALEOZOIC** Permian | | • Diversification of gymnosperms; origin of reptiles; amphibians dominant |
| — 290 | Carboniferous | | • First treelike plants; giant woody lycopods and sphenopsids form extensive forests in swampy areas; evolution of early seeds (seed ferns) and first stages of leaves |
| — 363 | Devonian | | • Diversification of vascular plants; sharks and fishes dominant in the oceans |
| — 409 | Silurian | | • First vascular plants |
| — 439 | Ordovician | | • Diversification of algae and plants colonize land |
| — 510 | Cambrian | | • Diversification of major animal phyla |
| — 570 | **PRECAMBRIAN** Precambrian | | • Origin of bacteria, archaea, and eukaryotes |
| | Earth is about 4.6 billion years old | | |

although water-conducting tubes appear to be present in some mosses. (However, these tubes may be unrelated to the vascular tissue in vascular plants.) The life cycle for the bryophytes differs from all other land plants because the gametophyte is the dominant and conspicuous plant. Because bryophytes are nonvascular, they are restricted to moist habitats for their reproductive cycle and have never attained the size and importance of other groups of plants. The gametophyte plants remain close to the ground, enabling the motile sperm to swim from the antheridium to the archegonium and fertilize the egg. They have a cuticle but lack stomata on the surface of the gametophyte **thallus** (plant body), which is not organized into roots, stems, and leaves. Stomata are present on the sporophyte in some mosses and hornworts.

*Stomata may be present on the sporophyte in some mosses, but they are never found on the gametophyte, the photosynthetic stage of the life cycle.*

Bryophytes are not important economically, with the exception of sphagnum moss, which in its harvested and dried form is known as *peat moss*. Peat moss is absorbent, has an antibacterial agent, and was reportedly once used as bandages and diapers. Today peat moss is used in the horticultural industry, and dried peat is burned as fuel in some parts of the world. Peat lands cover more than 1% of the Earth's surface and store 400 billion metric tons of organic carbon. Harvesting and burning peat releases $CO_2$ to the atmosphere, thus contributing to changes in the global carbon cycle.

## Lab Study A. Bryophyta: Mosses

### Materials

living examples of mosses
prepared slides of *Mnium* archegonia and antheridia
colored pencils

*Since mosses are common in most areas, you might ask students to bring in their own samples.*

### Introduction

The mosses are the most common group of nonvascular plants, occurring primarily in moist environments but also found in dry habitats that are periodically wet (Color Plate 30). Refer to Figure 15.4 as you investigate the moss life cycle, which is representative of the bryophytes.

### Procedure

1. Examine living colonies of mosses on demonstration. Usually you will find the two generations, gametophyte and sporophyte, growing together.
2. Identify the leafy **gametophytes** and the dependent **sporophytes,** which appear as elongated structures growing above them. Tug gently at the sporophyte and notice that it is attached to the gametophyte. Recall that the sporophyte develops and matures while attached to the gametophyte and receives its moisture and nutrients from the gametophyte.

(☜)Figure 15.3.
**Geological time chart.** The history of life can be organized into time periods that reflect changes in the physical and biological environment. Refer to this table as you review the evolution of land plants in Plant Diversity I and II.

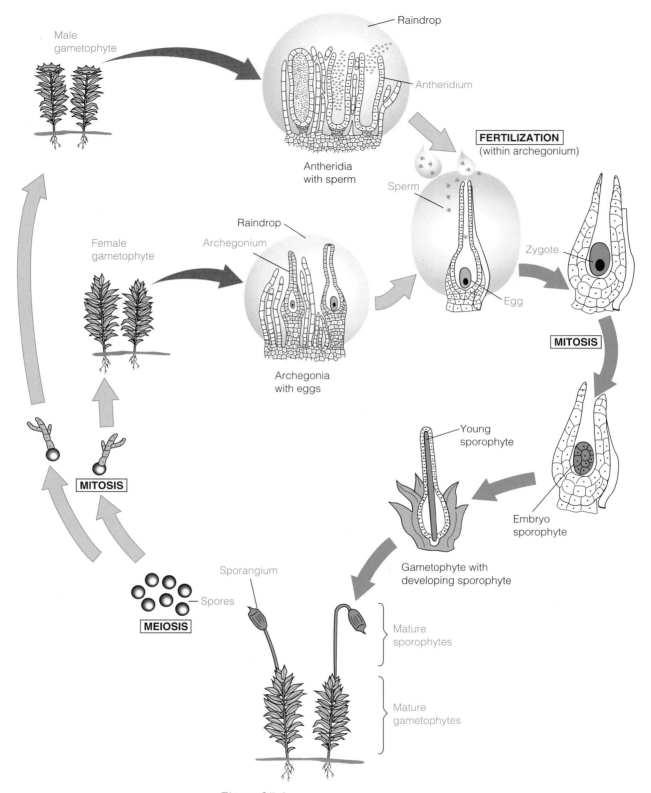

**Figure 15.4.**

**Moss life cycle.** The leafy moss plant is the gametophyte, and the sporophyte is dependent on it, deriving its water and nutrients from the body of the gametophyte. Review this variation of alternation of generations and label the structures described in Lab Study A. *Using colored pencils, highlight the haploid and diploid structures in different colors. Circle the processes of mitosis and meiosis.*

3. The gametes are produced by the gametophyte in **gametangia** *by mitosis*. Gametangia protect the gametes but are not readily visible without a microscope. Observe under the microscope's low-power lens prepared slides containing long sections of heads of the unisex moss *Mnium*, which contain the gametangia. One slide has been selected to show the **antheridia** (male); the other is a rosette of **archegonia** (female). Sperm-forming tissue will be visible inside the antheridia. On the archegonial slide, look for an archegonium. The moss archegonium has a very long neck and rounded base. It will be difficult to find an entire archegonium in any one section. Search for a single-celled **egg** in the base of the archegonium.

4. Refer to Figure 15.4 as you follow the steps of fertilization through formation of the gametophyte in the next generation. The sperm swim through a film of water to the archegonium and swim down the neck to the egg, where fertilization takes place. The diploid zygote divides by mitosis and develops into an embryonic sporophyte within the archegonium. As the sporophyte matures, it grows out of the gametophyte but remains attached, deriving water and nutrients from the gametophyte body. **Spores** develop *by meiosis* in the **sporangium** at the end of the sporophyte. The haploid spores are discharged from the sporangium and in a favorable environment develop into new haploid gametophytes.

## Results

1. Review the structures and processes observed and then label the moss life cycle diagram in Figure 15.4.

2. Using colored pencils, indicate if structures are haploid or diploid and circle the processes of mitosis and meiosis.

*Encourage students to work out the answers by studying the general life cycle. They need to make connections if they are going to see the trends.*

## Discussion

Refer to Plant Life Cycles in the Introduction and Figure 15.2, the generalized diagram of the plant life cycle.

1. Are the spores produced by the moss sporophyte formed by meiosis or mitosis? Are they haploid or diploid?

   *meiosis; haploid*

2. Do the spores belong to the gametophyte or sporophyte generation?

   *gametophyte*

3. Are the gametes haploid or diploid? Are they produced by meiosis or mitosis?

   *haploid; mitosis*

4. Is the dominant generation for the mosses the gametophyte or the sporophyte?

   *gametophyte*

5. Can you suggest any ecological role for mosses?

   *primary producers; few important in primary succession*

6. What feature of the life cycle differs for bryophytes compared with all other land plants?

*The gametophyte is the dominant plant in the life cycle of bryophytes.*

# Lab Study B. Hepatophyta: Liverworts

## Materials

living liverworts

## Introduction

Liverworts are so named because their bodies are flattened and lobed (Color Plate 31). Early herbalists thought that these plants were beneficial in the treatment of liver disorders. Although less common than mosses, liverworts can be found along streams on moist rocks, but because of their small size, you must look closely to locate them.

## Procedure

Examine examples of liverworts on demonstration. Liverworts have a flat **thallus** (plant body). Note the **rhizoids**, rootlike extensions on the lower surface, that primarily anchor plants. Observe the **pores** on the surface of the leaflike thallus. These openings function in gas exchange; however, they are always open since they lack guard cells. On the upper surface of the thallus you should see circular cups called **gemmae cups**, which contain flat disks of green tissue called **gemmae.** The gemmae are washed out of the cups when it rains, and they grow into new, genetically identical liverworts.

## Results

Sketch the overall structure of the liverwort in the margin of your laboratory manual. Label structures where appropriate.

## Discussion

1. Is the plant you observed the gametophyte or sporophyte?

   *gametophyte*

2. Are the gemmae responsible for asexual or sexual reproduction? Explain.

   *Asexual: There is no genetic variation; new plants will be clones of the original.*

3. Why are these plants, like most bryophytes, restricted to moist habitats, and why are they always small?

   *Liverworts, like most bryophytes, have no vascular tissue for transporting water and nutrients. Plants must be small because transport occurs by diffusion and cytoplasmic streaming. Liverworts must live in a moist environment to ensure that the motile sperm can swim to the egg. Although pores are present, the openings are not regulated by guard cells.*

4.  In this lab topic, as in Plant Diversity II (Lab Topic 16) and Plant Anatomy (Lab Topic 20), you are asked to complete tables that summarize features advantageous to the adaptation of plant groups to the land environment. You may be asked to compare these **derived** features with others that have changed little (**ancestral**) in the evolution of land plants. For example, for nonvascular plants, motile sperm might be considered an ancestral feature, while the cuticle would be considered derived.

    Complete Table 15.2, relating the features of nonvascular plants to their success in the land environment. Refer to the lab topic introduction for assistance.

*Students tend to think of ancestral as poorly adapted and derived as successful. Discourage such use of these terms.*

**Table 15.2**
Ancestral and Derived Features of Nonvascular Plants as They Relate to Adaptation to Land

| Ancestral Features | Derived Features |
|---|---|
| *motile sperm that require water* <br> *lack vascular tissue* <br> *no roots, stems, or leaves* <br> *gametophyte dominant* <br> *pores of liverworts* <br> *sporopyte is physically dependent on gametophyte* | *cuticle (in most)* <br> *gametangia and sporangia* <br> *rhizoids* <br> *stomata on sporophyte of mosses* |

---

### EXERCISE 15.2
# Seedless Vascular Plants

---

Seedless, terrestrial plants are analogous to the first terrestrial vertebrate animals, the amphibians, in their dependence on water for external fertilization and development of the unprotected, free-living embryo. Both groups were important in the Paleozoic era but have undergone a steady decline in importance since that time. Seedless plants were well suited for life in the vast swampy areas that covered large areas of the Earth in the Carboniferous period but were not suited for the drier areas of the Earth at that time or for later climatic changes that caused the vast swamps to decline and disappear. The fossilized remains of these swamp forests are the coal deposits of today (Figures 15.3 and 15.7).

Although living representatives of the seedless vascular plants have survived for millions of years, their limited adaptations to the land environment have restricted their range. All seedless vascular plants have vascular tissue, which is specialized for conducting water, nutrients, and photosynthetic products. Their life cycle is a variation of alternation of generations, in which the sporophyte is the dominant plant; the gametophyte is usually independent of the sporophyte. These plants generally have well-developed leaves and roots, stomata, and structural support tissue. However, since they still retain the ancestral feature of motile sperm that require water for fertilization, the gametophyte is small and restricted to moist habitats.

*The phylum Pterophyta includes ferns, horsetails, and whisk ferns. The latter two groups are placed in separate phyla by some botanists—Sphenophyta and Psilophyta, respectively.*

Economically, the only important members of this group are the ferns, a significant horticultural resource.

The phyla included in the seedless vascular plants are Lycophyta and Pterophyta (see again Table 15.1 and Figure 15.1).

The living examples of lycophytes are small club mosses, spike mosses, and quillworts. (Though named "mosses," these plants have vascular tissue and therefore are not true mosses.) The pterophytes include ferns, horsetails, and whisk ferns, which are remarkably different in overall appearance. Current evidence from molecular biology indicates that these diverse plants share a common ancestor and should all be included in the phylum Pterophyta. This evidence also suggests that pterophytes are more closely related to seed plants than they are to lycophytes.

## Lab Study A. Lycophyta: Club Mosses

### Materials

living *Selaginella* and *Lycopodium*
preserved *Selaginella* with microsporangia and megasporangia
prepared slide of *Selaginella* strobilus, l.s.

### Introduction

*Club mosses are available from biological suppliers, and most are easily maintained in a terrarium.*

Living members of Lycophyta are usually found in moist habitats, including bogs and streamsides (Color Plates 32 and 33). However, one species of *Selaginella*, the resurrection plant, inhabits deserts. It remains dormant throughout periods of low rainfall, but then comes to life—resurrects—when it rains. During the Carboniferous period, lycophytes were not inconspicuous parts of the flora but rather formed the forest canopy; they were the ecological equivalent of today's oaks, hickories, and pines (Figure 15.7).

Nonvascular plants and most seedless vascular plants produce one type of spore (**homospory**), which gives rise to the gametophyte by mitosis. One advanced feature occasionally seen in seedless vascular plants is the production of two kinds of spores (**heterospory**). Large spores called **megaspores** divide by mitosis to produce the female gametophyte. The numerous small spores, **microspores,** produce the male gametophytes by mitosis. Heterospory and separate male and female gametophytes, as seen in *Selaginella*, are unusual in seedless vascular plants, but characteristic of seed-producing vascular plants.

### Procedure

1. Examine living club mosses, *Selaginella* and *Lycopodium*. Are they dichotomously branched? (The branches would split in two, appearing to form a Y.) Locate sporangia, which may be present either clustered at the end of the leafy stem tips, forming **strobili,** or **cones,** or dispersed along the leafy stems. Note that these plants have small leaves, or bracts, along the stem.

2. Examine preserved strobili of *Selaginella*. Observe the round sporangia clustered in sporophylls (leaflike structures) at the tip of the stem (Figure 15.5a). These sporangia contain either four megaspores or numerous microspores. Can you observe any differences in the sporangia or spores?

*Depending on the species of* Selaginella, *the microspores may appear red and the megaspores cream colored. Size differences may also be apparent, depending on the stage of maturity.*

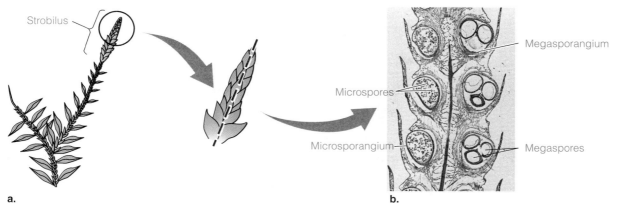

**Figure 15.5.**
*Selaginella.* (a) The leafy plant is the sporophyte. The sporangia are clustered at the tips in strobili. (b) Photomicrograph of a longitudinal section through the strobilus of *Selaginella.*

3. Observe the prepared slide of a long section through the strobilus of *Selaginella.* Begin your observations at low power. Are both microspores and megaspores visible on this slide?

   *yes*

   How can you distinguish these spores?

   *The microspores are small and numerous, while the megaspores are large and usually only two are visible.*

4. Identify the **strobilus, microsporangium, microspores, megasporangium**, and **megaspores** and label Figure 15.5.

## Results

1. Sketch the overall structure of the club mosses in the margin of your lab manual. Label structures where appropriate.
2. Review Figure 15.5 of *Selaginella.* Using a colored pencil, highlight the structures that are haploid and part of the gametophyte generation.

## Discussion

1. Are these leafy plants part of the sporophyte or the gametophyte generation? Do you have any evidence to support your answer?

   *Sporophyte. If sporangia are present, then this must be the sporophyte. The sporophyte is the dominant generation in all seedless vascular plants.*

2. What features would you look for to determine if this were a seedless vascular plant?

   *Presence of vascular tissue; dominant plant is diploid and therefore the sporophyte, which produces haploid spores.*

3. Are microspores and megaspores produced by mitosis or meiosis? (Review the life cycle in Figure 15.2.)

*meiosis*

4. Will megaspores divide to form the female gametophyte or the sporophyte?

*female gametophyte*

---

 Having trouble with life cycles? Return to the Introduction and review the generalized life cycle in Figure 15.2. Reread the introduction to the study of seedless vascular plants. The key to success is to determine where meiosis occurs and to remember the ploidal level for the gametophyte and the sporophyte.

---

## Lab Study B. Pterophyta: Ferns, Horsetails, and Whisk Ferns

### Materials

living and/or preserved horsetails (*Equisetum*)
living and/or preserved whisk ferns (*Psilotum*)
living ferns

### Introduction

*Horsetails are available from biological suppliers but may also be available locally.*

If a time machine could take us back 400 million years to the Silurian period, we would find that vertebrate animals were confined to the seas, and early vascular plants had begun to diversify on land (Figure 15.3). By the Carboniferous period, ferns, horsetails, and whisk ferns grew alongside the lycophytes. Until recently, these three groups of seedless vascular plants were placed in separate phyla: Pterophyta (ferns), Sphenophyta (horsetails), and Psilophyta (whisk ferns). Strong evidence from molecular biology now reveals a close relationship among these three groups, supporting a common ancestor for the group and their placement in one phylum, Pterophyta.

Psilophytes (**whisk ferns**) are diminutive, dichotomously branched (repeated Y branches), photosynthetic stems that reproduce sexually by aerial spores. Today, whisk ferns can be found in some areas of Florida and in the tropics (Color Plate 34). Sphenophytes (**horsetails**) have green jointed stems with occasional clusters of leaves or branches. Their cell walls contain silica that give the stem a rough texture. These plants were used by pioneers to scrub dishes—thus their name, scouring rushes. In cooler regions of North America, horsetails grow as weeds along roadsides (Color Plate 35). **Ferns** are the most successful group of seedless vascular plants, occupying habitats from the desert to tropical rain forests. Most ferns are small plants that lack woody tissue (Color Plate 36). An exception is the tree ferns found in tropical regions. Many cultivated ferns are available for home gardeners.

In this lab study you will investigate the diversity of pterophytes, including whisk ferns, horsetails, and a variety of ferns. The plants on demonstration are sporophytes, the dominant generation in seedless vascular plants. You will investigate the life cycle of a fern in Lab Study C, Fern Life Cycle.

## Procedure

1. Examine a living **whisk fern** (*Psilotum nudum*) on demonstration. This is one of only two extant genera of psilophytes.
2. Observe the spherical structures on the stem. If possible, cut one open and determine the function of these structures. Note the dichotomous branching, typical of the earliest land plants.
3. Examine the **horsetails** (*Equisetum* sp.) on demonstration. Note the ribs and ridges in the stem. Also examine the nodes or joints along the stem where branches and leaves may occur in some species. Locate the **strobili** in the living or preserved specimens on demonstration. These are clusters of **sporangia**, which produce **spores.**
4. Examine the living **ferns** on demonstration. Note the deeply dissected leaves, which arise from an underground stem called a **rhizome**, which functions like a root to anchor the plant. Roots arise from the rhizome. Observe the dark spots, or **sori** (sing, **sorus**), which are clusters of sporangia, on the underside of some leaves, called **sporophylls**. (See Color Plate 37.)

## Results

1. Sketch the overall structure of the whisk fern, horsetail, and fern in the margin of your lab manual. Label structures where appropriate.
2. Are there any leaves on the whisk fern? On the horsetails?
3. Are sporangia present on the whisk fern? On the horsetails? On the ferns?

## Discussion

1. Are the spores in the sporangia produced by mitosis or meiosis?

    *meiosis*

2. Are the sporangia haploid or diploid? Think about which generation produces them.

    *diploid, produced by the diploid sporophyte*

3. Once dispersed, will these spores produce the gametophyte or sporophyte generation?

    *gametophyte*

# Lab Study C. Fern Life Cycle

## Materials

living ferns
living fern gametophytes
 with archegonia and
 antheridia
living fern gametophytes
with young sporophytes
 attached
stereoscopic microscope

compound microscope
prepared slide of fern
 gametophytes with
 archegonia, c.s.
colored pencils
Protoslo®
glycerol in dropping bottle

*Fern gametophytes are fun to grow, but must be started early in the term to be ready for this lab. Students may grow their own*

*An animation of the fern life cycle is available at www.masteringbiology .com.*

## Introduction

In the previous Lab Study you examined the features of the fern sporophyte. In this lab study you will examine the fern life cycle in more detail, beginning with the diploid sporophyte.

## Procedure

1. Examine the sporophyte leaf with sori (sporophyll) at your lab bench (Color Plate 37). Make a wet mount of a sorus, using a drop of glycerol, and do not add a cover slip. Examine the sporangia using a dissecting microscope. You will find the stalked **sporangia** in various stages of development. Find a sporangium still filled with **spores** and observe carefully for a few minutes, watching for movement. The sporangia will open and fling the spores into the glycerol.

*Students sometimes ask about selffertilization in fern gametophytes. An archegonium may produce a hormone that suppresses the formation of antheridia on that gametophyte and induces antheridia production on other gametophytes.*

2. Refer to Figure 15.6 as you observe the events and important structures in the life cycle of the fern. The haploid spores of ferns fall to the ground and grow into heart-shaped **gametophyte** plants. All seedless terrestrial plants depend on an external source of water for a sperm to swim to an egg to effect fertilization and for growth of the resulting sporophyte plant. The gametangia, which bear male and female gametes, are borne on the underside of the gametophyte. Egg cells are produced by mitosis in urnlike structures called **archegonia,** and sperm cells are produced by mitosis in globular structures called **antheridia.** Archegonia are usually found around the notch of the heart-shaped gametophyte, while antheridia occur over most of the undersurface.

*If living gametophytes are not available, use prepared slides or preserved materials.*

3. To study whole gametophytes, make a slide of living gametophytes. View them using the stereoscopic microscope or the scanning lens on the compound microscope. Note their shape and color and the presence of **rhizoids,** rootlike multicellular structures. Locate archegonia and antheridia. Which surface will you need to examine? Sketch in the margin of your lab manual any details not included in Figure 15.6.

*You may choose to make a wet mount of fern sperm with Protoslo and use video projection to demonstrate to the class.*

4. If you have seen antheridia on a gametophyte, remove the slide from the microscope. Gently but firmly press on the coverslip with a pencil eraser. View using the compound microscope first on intermediate and then on high power. Look for motile **sperm** swimming with a spiral motion. Each sperm has two flagella. Add a drop of Protoslo to slow down movement of sperm.

5. Observe the cross section of a fern gametophyte with archegonia. Each archegonium encloses an **egg,** which may be visible on your slide.

6. Make a wet mount of a fern gametophyte with a **young sporophyte** attached. Look for a young **leaf** and **root** on each sporophyte.

7. Share slides of living gametophytes with archegonia, antheridia and sperm, and sporophytes until everyone has observed each structure.

## Results

1. Review the structures and processes observed, and then label the stages of fern sexual reproduction outlined in Figure 15.6.

2. Using colored pencils, circle those parts of the life cycle that are sporophytic (diploid). Use another color to encircle the gametophytic (haploid) stages of the life cycle. Highlight the processes of meiosis and mitosis.

## Discussion

Refer to Figure 15.2, the generalized diagram of the plant life cycle, and Figure 15.6, a representation of the fern life cycle.

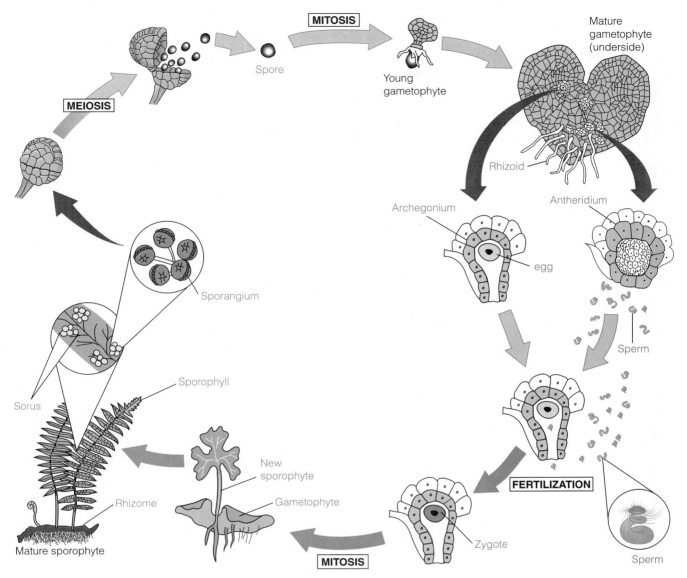

**Figure 15.6.**
**Fern life cycle.** The familiar leafy fern plant is the sporophyte, which alternates with a small, heart-shaped gametophyte. Review this life cycle, a variation of alternation of generations, and label the structures and processes described in Lab Study C. *Using colored pencils, highlight the haploid and diploid structures in different colors.*

1. Are the spores produced by the fern sporophyte formed by meiosis or mitosis?

   *meiosis*

2. Do the spores belong to the gametophyte or the sporophyte generation?

   *gametophyte*

3. Are the gametes produced by mitosis or meiosis?

   *mitosis*

4. Are the archegonia and antheridia haploid or diploid? Think about which generation produces them.

*haploid*

5. Is the dominant generation for the fern the gametophyte or the sporophyte?

*sporophyte*

6. Can you suggest any ecological role for ferns?

*primary producers*

## Lab Study D. Fossils of Seedless Vascular Plants

### Materials

fossils of extinct lycophytes (*Lepidodendron, Sigillaria*)
fossils of extinct sphenophytes (*Calamites*)
fossils of extinct ferns

*Fossil specimens can be obtained from geological suppliers listed in the* Preparation Guide. *Geology departments at universities may supply some as well.*

### Introduction

If we went back in time 300 million years to the Carboniferous period, we would encounter a wide variety of vertebrate amphibians moving about vast swamps dominated by spore-bearing forest trees. Imagine a forest of horsetails and lycophytes the size of trees, amphibians as large as alligators, and enormous dragonflies and roaches! Seedless plants were at their peak during this period and were so prolific that their carbonized remains form the bulk of Earth's coal deposits. Among the most spectacular components of the coal-swamp forest were 100-foot-tall lycophyte trees belonging to the fossil genera *Lepidodendron* and *Sigillaria*, tree ferns, and 60-foot-tall horsetails assigned to the fossil genus *Calamites* (Figures 15.3 and 15.7). In 2004, geologists and paleobotanists recognized a 1,000-hectare fossilized forest from 300 million years ago in Illinois. The ancient forest, which lies above a coal seam, was discovered when tree trunks of lycyophytes and sphenophytes were discovered in the roof of the coal mine. The entire forest had been preserved when an ancient earthquake collapsed the site into a lake bottom. This rare preserved forest has provided a look into the complex relationships within the Carboniferous swamp forest that once dominated the earth.

### Procedure

Examine flattened fossil stems of *Lepidodendron, Sigillaria, Calamites*, and fossil fern foliage, all of which were recovered from coal mine tailings. Compare these with their living relatives, the lycophytes (club mosses), sphenophytes (horsetails), and ferns, which today are diminutive plants found in restricted habitats.

**Figure 15.7.**
**Seedless vascular plants of the Carboniferous period.** (a) Reconstruction of a swamp forest dominated by lycophytes (b) *Lepidodendron* and (c) *Sigillaria*. (d) *Calamites* was a relative of horsetails. (*No. Geo. 7500c, Field Museum of Natural History, Chicago*)

## Results

1. For each phylum of seedless vascular plants, describe those characteristics that are similar for both living specimens and fossils. For example, do you observe dichotomous branching and similar shape and form of leaves, stems, or sporangia? Refer to the living specimens or your sketches.

   Lycophytes:


   Sphenophytes:


   Ferns:


2. Sketch below the overall structure of the fossils. How would you recognize these fossils at a later date? Label structures where appropriate.

## Discussion

The lycophytes, sphenophytes, and ferns were once the giants of the plant kingdom and dominated the landscape. Explain why they are presently restricted to certain habitats and are relatively small in stature.

*These plants all have a small gametophyte, which grows on or in the ground and requires the moisture of rain or dew for the motile sperm to swim freely to the egg. The climate is much drier than it once was, so these plants are confined to moist environments, at least periodically. Their vascular tissue (not studied in this exercise) is less efficient than that of seed plants, which also limits their size.*

## Reviewing Your Knowledge

1. Complete Table 15.3, indicating the ancestral and derived features of seedless vascular plants relative to success in land environments. Recall that in this context the term *ancestral* means a shared trait, while the term *derived* indicates an adaptation to land. For example, traits shared with the nonvascular plants (such as sperm requiring water for fertilization) are ancestral, while the presence of vascular tissue is derived.

**Table 15.3**
Ancestral and Derived Features of Seedless Vascular Plants as They Relate to Adaptation to Land

| Ancestral Features | Derived Features |
|---|---|
| *sperm require water for fertilization*<br>*reduced gametophytes, but usually independent*<br>*homospory* | *vascular tissue*<br>*independent sporophytes*<br>*complex stomata*<br>*strobili*<br>*cuticle (seen in some bryophytes)*<br>*heterospory* |

2. For each of the listed features, describe its contribution, if any, to the success of land plants.

gametangium

*protects the production of gametes from desiccation*

cuticle

*prevents desiccation of the plant body*

rhizoid

*anchors the liverwort to the ground; anchors fern gametophyte* motile sperm

*a liability in the land environment; requires water for fertilization*

vascular tissue

*efficient transport of water, nutrients, and photosynthetic products*

gemma

*production of new plants by asexual reproduction*

3. Complete Table 15.4. Identify the function of the structures listed. Indicate whether they are part of the gametophyte or sporophyte generation, and provide an example of a plant that has this structure.

4. What is the major difference between the alternation of generations in the life cycles of nonvascular plants and seedless vascular plants?

*The gametophyte is the dominant and conspicuous plant for nonvascular plants, and the sporophyte is the dominant and conspicuous plant for all other land plants, including seedless vascular plants.*

**Table 15.4**

Structures and Functions of the Nonvascular Plants and Seedless Vascular Plants

| Structure | Function | Sporophyte/ Gametophyte | Example |
|---|---|---|---|
| Antheridium | *produce and protect sperm* | *gametophyte* | *fern* |
| Archegonium | *produce and protect egg* | *gametophyte* | *fern* |
| Spore | *genetic variation, dispersal* | *gametophyte* | *moss* |
| Gamete | *fuse to restore diploid number* | *gametophyte* | *moss* |
| Rhizome | *anchor plant* | *sporophyte* | *fern* |
| Gemma | *asexual reproduction* | *gametophyte* | *liverwort* |
| Sporangium | *protect spores, site of meiosis* | *sporophyte* | *club moss* |
| Strobilus | *enhance spore dispersal* | *sporophyte* | *horsetail* |
| Sorus | *enhance spore dispersal* | *sporophyte* | *fern* |

# Applying Your Knowledge

1. The fossil record provides little information about ancient mosses. Do you think that nonvascular plants could ever have been large tree-sized plants? Provide evidence from your investigations to support your answer.

   *It is unlikely that nonvascular plants were ever large in size. Bryophytes, in general, do not have vascular tissue. Although some mosses have water-conducting tubes, these tubes do not appear to be vascular tissue. Transport of water and nutrients occurs by diffusion and cytoplasmic streaming, which would not allow material to move great distances against gravity.*

2. On a walk through a botanical garden, you notice a small leafy plant that is growing along the edge of a small stream in a shady nook. You hypothesize that this plant is a lycophyte. What information can you gather to test your hypothesis?

   *A lycophyte would produce spores in sporangia, which are part of the plant, a characteristic that might suggest a moss. In addition, however, a lycophyte would be diploid, and vascular tissue would be visible in microscopic sections of the stem. Ultimately, you could follow the life cycle of the plant and determine the sporophyte and gametophyte generations.*

3. Fern antheridia release sperm that then swim toward archegonia in a watery film. The archegonia release a fluid containing chemicals that attract the sperm. This is an example of chemotaxis, the movement of cells or organisms in response to a chemical. What is the significance of chemotaxis to fern (and moss) reproduction?

   *Chemotaxis assures that the numerous sperm that are produced are directed toward the archegonia rather than swimming randomly, thus increasing the probability of fertilization. Depending on the timing of sperm release and the release of the chemical attractant, chemotaxis may also promote cross-fertilization and genetic variation.*

4. Scientists investigating the evolutionary history (phylogeny) of land plants represent their hypotheses as phylogenetic trees or branching diagrams. Groups of plants that share a common ancestor are called clades, which are represented as lines connected at a branching point (the common ancestor). See the example below.

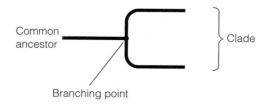

Which two of the phylogenetic trees below best represents the evolutionary history of land plants? Explain your choice of tree using your results from this laboratory topic.

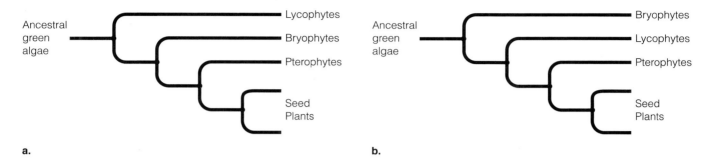

**a.**                                                                                              **b.**

*b.*
*The lycophytes and pterophytes are in the same clade and share a common ancestor that has vascular tissue. Bryophytes would not be in this clade since they do not have vascular tissue.*

5. Heterospory occasionally occurs in lycophytes and ferns, and in all seed plants. Botanists are convinced that heterospory must have originated more than once in the evolution of plants. Can you suggest one or more advantages that heterospory might provide to plants?

   *Heterospory results in a different allocation of resources in the female and male gametophyte, since only a few large megaspores are produced in Selaginella, but many smaller microspores. The megaspore that will produce the egg and support the development of the embryo has more resources. This is similar to the allocation of nutrients and resources in animals.*

   *Heterospory may promote genetic variation since the spores that produce the male and female gametophyte originate from two different meiotic phyla and are dispersed away from the parent plant. There may be more opportunities for cross-fertilization.*

## Investigative Extensions

*C-Ferns*, *Ceratopteris*, are excellent experimental organisms for investigating the alternation of generations life cycle, comparisons of seedless vascular plants with seed plants, and the physiology of ferns. These small ferns have a short life cycle of 12 days from spore to spore, and they can be grown successfully in small laboratory spaces. Amazingly, the motile sperm, easily visible under the microscope, can swim for 2 hours in buffer! The following are questions that you might investigate.

1. Fern archegonia secrete phermones to attract swimming sperm for fertilization, an example of chemotaxis (see question 3 in Applying Your Knowledge). Are sperm attracted to other compounds as well, including organic acids that might be present in the fluid secreted by the archegonia? Are there other compounds that might also be attractants? What are the common characteristics of the compounds that are attractants, for example, type of compound or chemical structure?

2. *C-Ferns* produce two types of gametophytes, hermaphrodites (both archegonia and antheridia present) and males (antheridia only). What factors affect the proportion of hermaphrodite and male gametophytes in the population, for example, temperature, light, or population density?

3. Are sperm from male gametophytes more or less attracted by pheromones compared with sperm from the hermaphrodites? Do sperm from both types of gametophytes respond to the same concentration of pheromones or other organic attractants?

Resources, laboratory procedures and preparation instructions are provided at the *C-Fern* website http://cfern.bio.utk.edu/index.html. Growing materials are available at Carolina Biological Supply, including *C-Fern* Chemotaxis Kit and *C-Fern* Culture Kit.

 **Student Media: BioFlix, Activities, Investigations, and Videos**
www.masteringbiology.com (select Study Area)

**Activities**—Ch. 29: Highlights of Plant Phylogeny; Moss Life Cycle; Fern Life Cycle

**Investigations**—Ch. 26: How Is Phylogeny Discovered by Comparing Proteins?
Ch. 29: What Are the Different Stages of a Fern Life Cycle?

# References

Berg, L. R. *Introductory Botany: Plants, People and the Environment*, 2nd ed. Belmont, CA: Thomson Brooks/Cole, 2007.

*C-Fern Web Manual*, 2009. Can by downloaded at http://www.c-fern.org/

Hickock, L. G. and T. R. Warne. *C-Fern Manual*. Burlington, NC: Carolina Biological Supply, 2000.

Kendrick, P. and P. Davis. *Fossil Plants*. Washington, DC: Smithsonian, 2004.

Mauseth, J. D. *Botany: An Introduction to Plant Biology*, 4th ed. Sudbury, MA: Jones and Bartlett Publishers, 2008.

Ranker, T. A. and C. H. Haufler, editors. *Biology and Evolution of Ferns and Lycophytes*. Cambridge: Cambridge University Press, 2008.

Raven, P. H., R. F. Evert, and S. E. Eichhorn. *Biology of Plants*, 7th ed. New York: W. H. Freeman Publishers, 2004.

# Websites

C-Fern website with information, suggestions, and support:
http://www.c-fern.org/

Fern basics (click on Learn More about Ferns), images and current research:
http://amerfernsoc.org/

Illinois State Geological Survey, "A 300 Million Year Old Pennsylvanian Age Mire Forest." Includes photos and reconstruction illustrations:
http://www.isgs.uiuc.edu/research/coal/fossil-forest/fossil-forest.shtml

An introduction to land plants, including morphology, evolution, and fossils. Images and resources provided:
http://www.ucmp.berkeley.edu/plants/plantae.html

Links to images of plants:
http://botit.botany.wisc.edu/images/130/

The Tree of Life Web Project is a collaborative project of biologists worldwide. Information is provided on all major groups of living organisms including land plants:
http://tolweb.org/Embryophytes

# LAB TOPIC 15

# Plant Diversity I: Nonvascular Plants (Bryophytes) and Seedless Vascular Plants

## Teaching Plan for Laboratories

## Main Concepts and Objectives

1. Concept: All land plants have a similar life cycle (alternation of generations). Students will be able to describe alternation of generations, including structures and processes, and describe in detail the life cycle of a moss and a fern. (Introduction, Exercises 15.1, 15.2)

2. Concept: Variations in reproductive cycles and plant structures of nonvascular and seedless vascular plants represent adaptations to the land environment. Students will identify characteristics of major plant phyla and representative organisms. (Exercises 15.1, 15.2)

3. Concept: Variations in reproductive cycles can be interpreted as evolutionary trends in the history of land plants. Students will compare ancestral and derived features of these plant groups, relating these features to terrestrial life. Students will begin to see trends that will continue in the next lab topic, Plant Diversity II. Trends observed include the following: (Exercises 15.1, 15.2)

   a. The sporophyte becomes the dominant and persistent plant, while gametophytes range from independent, conspicuous plants to reduced, inconspicuous, and dependent plants.

   b. The gametes and spores are increasingly protected within the gametophyte and sporophyte. This is especially true for the egg.

   c. The transfer of sperm requires water in nonvascular and seedless vascular plants but is independent of water in seed plants, which produce pollen.

   d. Nonvascular plants and most seedless vascular plants exhibit homospory. Seed plants and a few lycophytes are heterosporous.

   e. Structural adaptations to terrestrial life are enhanced from nonvascular plants to vascular seed plants.

4. Concept: The evolutionary history of land plants can be understood by a comparative study of living and fossil members of these taxa. Students will compare fossils to living examples of seedless vascular plants. (Exercise 15.2, Lab Study D)

5. Specific content. Students will be able to describe or define: *gametophyte*, *gametangia* (*archegonia* and *antheridia*), *sporophyte*, *spore*, *gametes*, *zygote*, *strobilus*, *homospory*, *heterospory*, *microspore*, *megaspore*.

## Order of the Lab

1. Introduce the objectives of the lab.                                           (5 min)
2. Describe the general life cycle and the alternation of generations.            (15 min)
3. Explain trends to look for throughout this lab and the next.                   (10 min)
4. Preview the lab exercises, indicating the location of materials
   and demonstrations.                                                            (5 min)
5. Review the moss and fern life cycle when students reach the
   appropriate section.                                                           (15 min)
6. Students complete the lab.                                                     (90 min)
7. Ask students to provide evidence to support trends and major
   concepts for this lab.                                                         (10 min)
8. Introduce the lab for next week. Remind students to
   complete Table 16.1 before coming to lab.                                      (5 min)

**For a 2-hour lab:** Omit the demonstration material for liverworts (Exercise 15.1, Lab Study B) and the phyla of seedless vascular plants (Exercise 15.2, Lab Studies A and B) except ferns.

## Classroom Management

Each student should have a slide set. Students work individually on prepared slides and in pairs for all other work. Encourage students to share their observations and slide preparations with one another. Encourage students to view demonstrations in staggered fashion to avoid standing and waiting. Question students about the general life cycle and evidence for trends and adaptations to the land environment as they work through the material. This lab should not be a passive experience for the students. Ask more questions than you answer. Keep students moving through the material by giving them some time estimates.

## Student Development

Students will lose sight of the trends if they try to memorize terms and structures. Stress the structure/function theme, along with adaptation to the land environment. Students should provide evidence to support trends. They should develop observational skills, an ability to view slides and investigate how structures relate to plant life cycles. Most of the material in this lab will be new to students. Encourage them to ask questions.

## Discussion and Summary

Review life cycles and summarize trends, asking students to provide examples. Preview next week's lab by discussing seed plants.

## Evaluation

Lab test should include all organisms, slides, and fossils. Students should be prepared for conceptual and practical questions. This lab provides evidence to support and explain major themes in the course. Encourage students to work with the life cycles and to answer questions at the end of the laboratory.

## Lab Topic 16
# Plant Diversity II: Seed Plants

 Before lab, read the following material on gymnosperms and angiosperms and complete Table 16.1 by listing (and comparing) the traits of each.

## Laboratory Objectives

After completing this lab topic, you should be able to:

1. Identify examples of the phyla of seed plants.
2. Describe the life cycle of a gymnosperm (pine tree) and an angiosperm.
3. Describe features of flowers that ensure pollination by insects, birds, bats, and wind.
4. Describe factors influencing pollen germination.
5. Identify types of fruits, recognize examples, and describe dispersal mechanisms.
6. Relate the structures of seed plants to their functions in the land environment.
7. Compare the significant features of life cycles for various land plants and state their evolutionary importance.
8. Summarize major trends in the evolution of land plants and provide evidence from your laboratory investigations.

*For a 2-hour lab: Set up a demonstration of pollination and fruit keys. Omit the demonstration of gymnosperm phyla (Exercise 16.1, Lab Study A) and omit the pollen germination experiment (Exercise 16.2, Lab Study C). See Teaching Plan for other suggestions.*

*In the Study Area of masteringbiology.com, see Student Media for suggested activities, investigations, and videos.*

## Gymnosperms

For over 500 million years, plants have been adapting to the rigors of the land environment. The nonvascular bryophytes with their small and simple bodies survived in moist habitats, habitats moist at least for part of their life cycle. During the cool Carboniferous period, vascular seedless plants dominated the landscape of the swamp forests that covered much of the earth. Although these plants were more complex and better adapted to the challenges of the land environment, they still were dependent on water for sperm to swim to the egg. During the Mesozoic era, 150 million years ago, Earth became warmer and drier and the swamp forests declined, presenting another challenge to terrestrial plants and animals. Earth at that time was a world dominated by reptilian vertebrates, including the flying, running, and climbing dinosaurs. The landscape was dominated by a great variety of seed-bearing plants called **gymnosperms** (literally, "naked seeds"), which in the Carboniferous period had been restricted to dry sites. During the Mesozoic, a number of distinct gymnosperm groups diversified, and a few of the spore-bearing plants survived (see Figure 15.1 and Table 15.1 in Lab Topic 15). As you review the evolution of land plants, refer to the geological time chart in Figure 15.3 for an overview of the history of life on Earth.

Vertebrate animals became fully terrestrial during the Mesozoic with the emergence of reptiles, which were free from a dependence on water for sexual reproduction and development. The development of the amniotic egg along with an internal method of fertilization made this major transition possible. The amniotic egg carries its own water supply and nutrients, permitting early embryonic development to occur on dry land, a great distance from external water. In an analogous manner, the gymnosperms became free from dependence on water through the development of a process of internal fertilization via the **pollen grain** and development of a **seed**, which contains a dormant embryo with a protective cover and special nutrient tissue.

Several features of the gymnosperms have been responsible for their success. They have reduced (smaller-sized) gametophytes; the male gametophyte is a multinucleated pollen grain, and the female gametophyte is small and retained within the sporangium in the ovule of the sporophyte generation. The pollen grain is desiccation resistant and adapted for wind pollination, removing the necessity for fertilization in a watery medium. The pollen tube conveys the sperm nucleus to an egg cell, and the embryonic sporophyte develops within the gametophyte tissues, which are protected by the previous sporophyte generation. The resulting seed is not only protected from environmental extremes, but also is packed with nutritive materials and can be dispersed away from the parent plant. In addition, gymnosperms have advanced vascular tissues: xylem for transporting water and nutrients and phloem for transporting photosynthetic products. The xylem cells are called *tracheids* and are more efficient for transport than those of the seedless vascular plants.

## Angiosperms

A visit to Earth 60 million years ago, during the late Cretaceous period, would reveal a great diversity of mammals and birds and a landscape dominated by **flowering plants,** or **angiosperms** (phylum **Anthophyta**). Ultimately, these plants would diversify and become the most numerous, widespread, and important plants on Earth. Angiosperms now occupy well over 90% of the vegetated surface of Earth and contribute virtually 100% of our agricultural food plants.

The evolution of the **flower** resulted in enormous advances in the efficient transfer and reception of pollen. Whereas gymnosperms are all wind-pollinated, producing enormous amounts of pollen that reach the appropriate species by chance, the process of flower pollination is mediated by specific agents—insects, birds, and bats—in addition to water and wind.

Pollination agents such as the insect are attracted to the flower with its re-wards of nectar and pollen. Animal movements provide precise placement of pollen on the receptive portion of the female structures, increasing the probability of fertilization. The process also enhances the opportunity for cross-fertilization among distant plants and therefore the possibility of in-creased genetic variation.

Angiosperm reproduction follows the trend for reduction in the size of the gametophyte. The pollen grain is the male gametophyte, and the eight-nucleated **embryo sac** is all that remains of the female gametophyte. This generation continues to be protected and dependent on the adult sporophyte plant. The female gametophyte provides nutrients for the developing sporophyte em-bryo through a unique triploid **endosperm** tissue. Another unique feature of angiosperms is the **fruit.** The seeds of the angiosperm develop within the flower ovary, which matures into the fruit. This structure provides protection and enhances dispersal of the young sporophyte into new habitats.

In addition to advances in reproductive biology, the angiosperms evolved other advantageous traits. All gymnosperms are trees or shrubs, with a large investment in woody, persistent tissue; and their life cycles are long (5 or more years before they begin to reproduce and 2 to 3 years to produce a seed). Flowering plants, on the other hand, can be woody, but many are herbaceous, with soft tissues that survive up to a few years. It is possible for angiosperms to go from seed to seed in less than 1 year. As you perform the exercises in this lab, think about the significance of this fact in terms of the evolution of this group. How might generation length affect the rate of evo-lution? Angiosperms also have superior conducting tissues. Xylem tissue is composed of *tracheids* (as in gymnosperms), but also contains large-diameter, open-ended *vessels.* The phloem cells, called *sieve-tube members,* provide more efficient transport of the products of photosynthesis. The cell structure and organization of plants will be investigated in Lab Topic 20 Plant Anatomy.

Review the characteristics of gymnosperms and angiosperms described in this introduction, and *summarize in Table 16.1 the advantages of these groups relative to their success on land.* You should be able to list several characteris-tics for each. At the end of the lab, you will be asked to modify and com-plete the table, based on your investigations.

You will want to return to this table after the laboratory to be sure that the table is complete and that you are familiar with all these important features.

---

EXERCISE 16.1

# Gymnosperms

The term *gymnosperms* refers to a diverse group of seed plants that do not produce flowers. Although they share many characteristics, including the production of pollen, they represent four distinct groups, or phyla. In this exercise, you will observe members of these phyla and investigate the life cycle of a pine, one of the most common gymnosperms.

**Table 16.1**
Traits for Gymnosperms and Angiosperms
Relative to Their Success on Land

| | **Adaptation to the Land Environment** |
|---|---|
| **Gymnosperms** | *Pollen grain; internal fertilization* <br><br> *Reduced female gametophyte* <br><br> *Ovule; seed* <br><br> *Advanced vascular tissue—tracheids* |
| **Angiosperms** | *Pollen grain; internal fertilization* <br><br> *More reduced female gametophyte = embryo sac* <br><br> *Ovule; seed; carpel; fruit* <br><br> *Double fertilization; endosperm* <br><br> *Flower with one or more carpels* <br><br> *Advanced vascular tissue—tracheids, vessels* <br><br> *Herbaceous and woody habit* <br><br> *Annuals and perennials* |

# Lab Study A. Phyla of Gymnosperms

## Materials

living or pressed examples of conifers, ginkgos, cycads, and Mormon tea

## Introduction

*Examples of these plants are available from biological suppliers or garden centers, depending on your location. You may have to use some pressed plants. See the Preparation Guide for suggestions.*

Gymnosperms are composed of several phyla (see Figure 15.1 and Table 15.1 in Lab Topic 15). The largest and best known phylum is **Coniferophyta,** which includes pines and other cone-bearing trees and shrubs (Color Plate 38). Cycads (**Cycadophyta**), which have a palmlike appearance, are found primarily in tropical regions scattered around the world (Color Plate 39). Ginkgos (**Ginkgophyta**), with their flat fan-shaped leaves, are native to Asia and are prized as urban trees. An extract of Ginkgo is used as an herbal medicine purported to improve memory (Color Plates 40 and 41). **Gnetophyta** is composed of three distinct and unusual groups of plants: gnetums, which are primarily vines of Asia, Africa, and South America; *Welwitschia,* a rare desert plant with two leathery leaves; and Mormon tea (*Ephedra*), desert shrubs of North and Central America (Color Plate 42). Compounds from *Ephedra*, ephedrines, used in diet aids and decongestants, have raised serious concerns due to side effects including cardiac arrest.

## Procedure

1. Observe demonstration examples of all phyla of gymnosperms and be able to recognize their representatives. Note any significant ecological and economic role for these plants.
2. In the margin of your manual, sketch the overall structure of the plants. Label structures where appropriate.

## Results

1. Record your observations in Table 16.2.
2. Are there any reproductive structures present for these plants? If so, make notes in the margin of your lab manual.

**Table 16.2**
Phyla of Gymnosperms

| Phyla | Examples | Characteristics/Comments |
|---|---|---|
| **Coniferophyta** | *Pines, cedars* | *Cones, needle leaves* |
| **Ginkgophyta** | *Maidenhair tree* | *Broad fan-shaped leaves; seeds borne singly on stalks* |
| **Cycadophyta** | *Sago palm, cycads* | *Palmlike leaves; short, woody stems* |
| **Gnetophyta** | *Mormon tea* | *Stem, no leaves; desert plant* |

## Discussion

1. What are the key characteristics shared by all gymnosperms?

   *produce pollen and seeds but no flowers or fruits*

2. What is the ecological role of conifers in forest systems?

   *Primary producers; in some regions they are forest dominants; in others they are important in secondary succession. Students are unlikely to know about succession, but they may have observed conifers in abandoned fields.*

3. What economically important products are provided by conifers?

*Resins, turpentine, wood and paper products, disposable diapers. Students usually have some idea of pine products.*

4. What economically important products are provided by other gymnosperms?

*Ephedrine from Ephedra is used as a decongestant. Ginkgo extract is often taken to improve memory.*

## Lab Study B. Pine Life Cycle

### Materials

living or preserved pine branch, male and female cones (1, 2, and 3 years old)

fresh or dried pine pollen or prepared slide of pine pollen

coverslips

prepared slides of male and female pine cones

colored pencils

slides

 Review the pine life cycle (Figure 16.1) before you begin. Follow along as you complete the exercise.

### Introduction

All gymnosperms are **wind-pollinated** trees or shrubs, most bearing unisexual, male, and female reproductive structures on different parts of the same plant. Gymnosperms are **heterosporous**, producing two kinds of spores: male **microspores**, which develop into **pollen**, and female **megaspores**. The megaspore develops into the female gametophyte, which is not free-living as with ferns but retained within the **megasporangium** and nourished by the sporophyte parent plant. The female gametophye develops archegonia, each containing an egg. Numerous pollen grains (the male gametophytes) are produced in each **microsporangium**, and when they are mature they are released into the air and conveyed by wind currents to the female cone. **Pollen tubes** grow through the tissue of the megasporangium, and the **sperm nucleus** is released to fertilize the egg. After fertilization, development results in the formation of an **embryo**. A **seed** is a dormant embryo embedded in nutrient tissue of the female gametophyte and surrounded by the hardened sporangium wall, or **seed coat**.

 Having trouble with life cycles? Return to Lab Topic 15 (Plant Diversity I) and review the generalized life cycle (Figure 15.2). The key to success is to determine where meiosis occurs and to remember the ploidal level for the gametophyte and sporophyte.

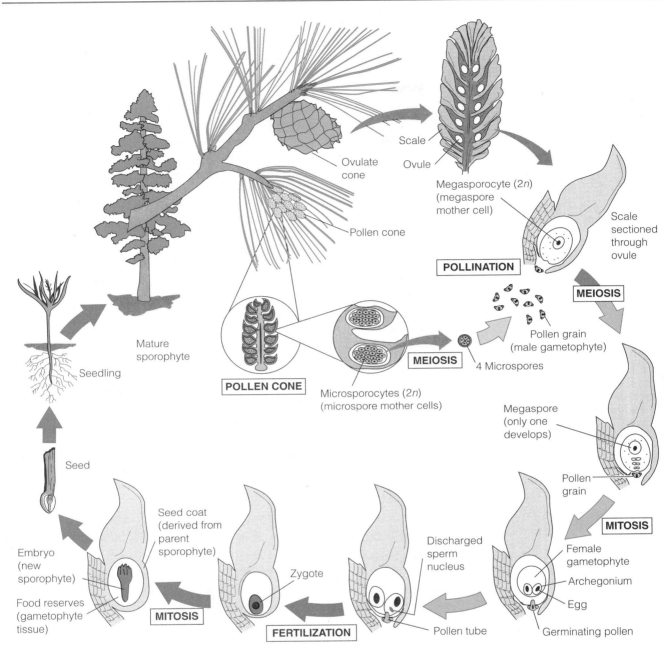

**Figure 16.1.**
**Pine life cycle.** Observe the structures and processes as described in Exercise 16.1. *Using colored pencils, indicate the structures that are haploid or diploid. Circle the terms* mitosis, meiosis, *and* fertilization.

## Procedure

1. Pine sporophyte.

   a. Examine the pine branch and notice the arrangement of leaves in a bundle. A new twig at the end of the branch is in the process of producing new clusters of leaves. Is this plant haploid or diploid?

   *diploid*

b. Examine the small **cones** produced at the end of the pine branch on this specimen or others in lab. Recall that cones contain clusters of sporangia. What important process occurs in the sporangia?

*meiosis*

c. Locate an ovulate cone and a pollen cone. Elongated male **pollen cones** are present only in the spring, producing pollen within overlapping bracts, or scales. The small, more rounded female cones (which look like miniature pine cones) are produced on stem tips in the spring and are called **ovulate cones.** Female cones persist for several years. Observe the overlapping scales, which contain the sporangia.

d. In the margin of your lab manual, sketch observations for future reference.

2. Male gametophyte—development in pollen cones.

a. Examine a longitudinal section of the pollen cone on a prepared slide and identify its parts. Observe that pollen cones are composed of radiating scales, each of which carries two elongated sacs on its lower surface. The sacs are the **microsporangia. Microsporocytes** (microspore mother cells) within microsporangia divide by meiosis. Each produces four haploid **microspores,** which develop into **pollen grains.**

*Pine pollen from fresh male cones in the spring or from dried cones can be used for this slide. Have students make a wet mount.*

b. Observe a slide of pine pollen. If pollen is available, you can make a wet mount. Note the wings on either side of the grain. The pollen grain is the greatly reduced male gametophyte. The outer covering of the pollen is desiccation resistant. Once mature, pollen will be wind dispersed, sifting down into the scales of the female cones.

c. Sketch, in the margin of your lab manual, observations for future reference.

3. Female gametophyte—development in ovulate cones.

a. Examine a longitudinal section of a young ovulate cone on a prepared slide. Note the **ovule** (containing the megasporangium) on the upper surface of the scales. Diploid **megasporocytes** (megasporocyte mother cells) contained inside will produce haploid **megaspores,** the first cells of the gametophyte generation. In the first year of ovulate cone development, pollen sifts into the soft bracts (pollination) and the pollen tube begins to grow, digesting the tissues of the ovule.

b. Observe a second-year cone at your lab bench. During the second year, the ovule develops a multicellular female gametophyte with two **archegonia** in which an **egg** will form. Fertilization will not occur until the second year, when the pollen tube releases a sperm nucleus into the archegonium, where it unites with the egg to form the **zygote.** In each ovule only one of the archegonia and its zygote develops into a **seed.**

c. Observe a mature cone at your lab bench. The development of the embryo sporophyte usually takes another year. The female gametophyte will provide nutritive materials stored in the seed for the early stages of growth. The outer tissues of the ovule will harden to form the **seed coat.**

d. In the margin of your lab manual, sketch observations for future reference.

## Results

1.  Review the structures and processes observed.

2.  Using colored pencils, indicate the structures of the pine life cycle in Figure 16.1 that are haploid or diploid, and circle the processes of mitosis, meiosis, and fertilization.

## Discussion

1.  What is the function of the wings on the pollen grain?

    *wind dispersal*

2.  Why is wind-dispersed pollen an important phenomenon in the evolution of plants?

    *Plants were no longer dependent on water for sexual reproduction.*

3.  Are microspores and megaspores produced by mitosis or meiosis?

    *meiosis*

4.  Describe the structure and function of a seed.

    *A seed contains the embryo sporophyte plant, nutritive tissue, and a hard seed coat. The seed is protected from environmental factors, can remain dormant until conditions are favorable, and can be dispersed away from the parent plant. The embryonic plant has stored resources when germination begins.*

5.  Can you think of at least two ways in which pine seeds are dispersed?

    *Personal observation may help students with this question; wind and squirrels.*

6.  One of the major trends in plant evolution is the reduction in size of the gametophytes. Describe the male and female gametophyte in terms of size and location.

    *The male gametophyte is now the pollen grain with two nuclei; it is wind dispersed. The female gametophyte never leaves the sporophyte; it is composed of a multicellular structure inside the ovule.*

---

### EXERCISE 16.2

# Angiosperms

---

All flowering plants (angiosperms) are classified in the phylum **Anthophyta** (Gk. *anthos,* "flower"). A unique characteristic of angiosperms is the **carpel,** a structure in which **ovules** are enclosed. After fertilization, the ovule develops into a seed (as in the gymnosperms), while the carpel matures into a **fruit** (unique to angiosperms). Other important aspects of angiosperm reproduction include additional reduction of the gametophyte, double fertilization, and an increase in the rapidity of the reproduction process.

The **flowers** of angiosperms are composed of male and female reproductive structures, which are frequently surrounded by attractive or protective leaflike structures collectively known as the **perianth** (Color Plate 43). The flower functions both to protect the developing gametes and to ensure **pollination** and **fertilization.** Although many angiosperm plants are self-fertile, cross-fertilization is important in maintaining genetic diversity. Plants, rooted and stationary, often require transfer agents to complete fertilization. A variety of insects, birds, and mammals transfer pollen from flower to flower. The pollen then germinates into a pollen tube and grows through the female carpel to deliver the sperm to the egg.

Plants must attract pollinators to the flower. What are some features of flowers that attract pollinators? Color and scent are important, as is the shape of the flower. Nectar and pollen provide nutritive rewards for the pollinators as well. The shape and form of some of the flowers are structured to accommodate pollinators of specific size and structure, providing landing platforms, guidelines, and even special mechanisms for the placement of pollen on body parts. While the flower is encouraging the visitation by one type of pollinator, it also may be excluding visitation by others. The more specific the relationship between flower and pollinator, the more probable that the pollen of that species will be successfully transferred. But many successful flowers have no specific adaptations for particular pollinators and are visited by a wide variety of pollinators.

Some plants do not have colorful, showy flowers and are rather inconspicuous, often dull in color, and lacking a perianth. These plants are usually wind-pollinated, producing enormous quantities of pollen and adapted to catch pollen in the wind (Color Plate 44).

The origin and diversification of angiosperms cannot be understood apart from the coevolutionary role of animals in the reproductive process. Colorful petals, strong scents, nectars, food bodies, and unusual perianth shapes all relate to pollinator visitation. Major trends in the evolution of angiosperms involve the development of mechanisms to exploit a wide variety of pollinators (Color Plates 45, 46, and 47).

*Have students begin pollen germination experiments (p. 431, Lab Study C, Procedure, step 3) before beginning Lab Study A.*

In Lab Study A, you will investigate a variety of flowers, observing their shape, structure, and traits that might attract pollinators of various kinds. Following this, in Lab Study B, you will use a key to identify the probable pollinators for some of these flowers. You will follow the life cycle of the lily in Lab Study C and complete the lab by using another key to identify types of fruits and their dispersal mechanisms. *Note: Before beginning Lab Study A, turn to Lab Study C, Procedure step 3, to set up pollination experiments, since the pollen must incubate for 30–60 minutes. Then return to Lab Study A below.*

# Lab Study A. Flower Morphology

## Materials

living flowers provided by the instructor and/or students
stereoscopic microscope

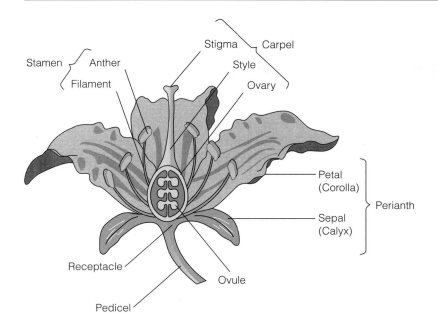

**Figure 16.2.**
**Flower structures.** Determine the structures of flowers in the laboratory by reviewing this general diagram.

## Introduction

Working in teams of two students, you will investigate the structure of the flower (Figure 16.2, Color Plate 43). The instructor will provide a variety of flowers, and you may have brought some with you to lab. You will need to take apart each flower carefully to determine its structure, since it is unlikely that all your flowers will follow the simple diagram used to illustrate the structures. Your observations will be the basis for predicting probable pollinators in Lab Study B.

*Fresh flowers are now available from florists and grocery stores year-round. Use living flowers, not pressed or preserved ones. Ask students to bring in their own, but be prepared for the possibility that carnations and roses have been selected for multiple petals at the expense of reproductive structures, which may be sterile or missing. Daisies and chrysanthemums can be used, but the aster flower is really a composite of many small flowers, which must be pulled apart and viewed under a stereoscopic scope. See the* Preparation Guide *for flower suggestions.*

## Procedure

1. Examine fresh flowers of four different species, preferably with different floral characteristics.

2. Identify the parts of each flower using Figure 16.2 and the list provided following the heading Floral Parts. You may be able to determine the floral traits for large, open flowers by simply observing. However, to really understand the flower structure, you should remove the flower parts stepwise, beginning with the outer sepals and continuing toward the center of the flower. To observe the details of smaller flowers and structures, use the stereoscopic microscope. For example, the ovary is positively identified by the presence of tiny crystal-like ovules, and these are best seen with the stereoscopic scope.

3. In the margin of your lab manual, sketch any flower shapes or structures that you might need to refer to in the future.

4. Record the results of your observations in Table 16.3. You will determine pollinators in Lab Study B.

*Students will passively look at flowers, hoping all the structures are visible by superficial inspection. Some will want to save their flowers rather than pull them apart. Encourage them to dig in and locate the structures. The diversity in the shape and arrangement of structures will amaze them.*

## *Floral Parts*

**Pedicel:** stalk that supports the flower.
**Receptacle:** tip of the pedicel where the flower parts attach.
**Sepal:** outer whorl of bracts, which may be green, brown, or colored like the petals; may appear as small scales or be petal-like.
**Calyx:** all the sepals, collectively.

**Petal:** colored, white, or even greenish whorl of bracts located just inside the sepals.

**Corolla:** all the petals, collectively.

**Stamen:** pollen-bearing structure, composed of filament and anther.

**Filament:** thin stalk that supports the anther.

**Anther:** pollen-producing structure.

**Carpel:** female reproductive structure, composed of the stigma, style, and ovary, often pear-shaped and located in the center of the flower.

**Stigma:** receptive tip of the carpel, often sticky or hairy, where pollen is placed; important to pollen germination.

**Style:** tissue connecting stigma to ovary, often long and narrow, but may be short or absent; pollen must grow through this tissue to fertilize the egg.

**Ovary:** base of carpel; protects ovules inside, matures to form the fruit.

### Results

Summarize your observations of flower structure in Table 16.3.

**Table 16.3**
Flower Morphology and Pollinators

| Features | Plant Names | | | |
|---|---|---|---|---|
| | 1 | 2 | 3 | 4 |
| **Number of petals** | | | | |
| **Number of sepals** | | | | |
| **Parts absent (petals, stamens, etc.)** | | | | |
| **Color** | | | | |
| **Scent (+/−)** | | | | |
| **Nectar (+/−)** | | | | |
| **Shape (including corolla shape: tubular, star, etc.)** | | | | |
| **Special features (landing platform, guidelines, nectar spur, etc.)** | | | | |
| **Predicted pollinator (see Lab Study B)** | | | | |

## Discussion

What structures or characteristics did you observe in your (or other teams') investigations that you predict are important to pollination?

 Student Media Videos—Ch. 30: Flower Blooming (Time Lapse); Flowering Plant Life Cycle

## Lab Study B. Pollinators

### Materials

living flowers provided by the instructor and/or students
stereoscopic microscope

### Introduction

Flowers with inconspicuous sepals and petals are usually pollinated by wind (Color Plate 44). Most showy flowers are pollinated by animals. Some pollinators tend to be attracted to particular floral traits, and, in turn, some groups of plants have coevolved with a particular pollination agent that ensures successful reproduction. Other flowers are generalists, pollinated by a variety of organisms, and still others may be visited by only one specific pollinator (Color Plates 45, 46, and 47). Based on the floral traits that attract common pollinators (bees, flies, butterflies, and hummingbirds), you will predict the probable pollinator for some of your flowers using a **dichotomous key.** (Remember, *dichotomous* refers to the branching pattern and means "divided into two parts.")

In biology, we use a key to systematically separate groups of organisms based on sets of characteristics. Most keys are based on couplets, or pairs of characteristics, from which you must choose one or the other, thus, the term *dichotomous*. For example, the first choice of characteristics in a couplet might be *plants with showy flowers and a scent,* and the other choice in the pair might be *plants with tiny, inconspicuous flowers with no scent.* You must choose one or the other statement. In the next step, you would choose from a second pair of statements listed directly below your first choice. With each choice, you would narrow the group more and more until, as in this case, the pollinator is identified. *Each couplet or pair of statements from which you must choose will be identified by the same letter or number.*

*Students have a hard time following the logic of a taxonomic key. Actually using a key is the best way to explain the process. Have the entire class key the first flower together, reading the key and asking them to make the choices. Go slowly. Students need to see how to locate and choose couplets.*

*A wind-pollinated flower should be available in lab. Remember that these flowers have a greatly reduced perianth or the petals and sepals are missing. See the Preparation Guide for suggestions.*

## Key to Pollination

I. Sepals and petals reduced or inconspicuous; feathery or relatively large stigma; flower with no odor **wind**

II. Sepals and/or petals large, easily identified; stigma not feathery; flower with or without odor

    A. Sepals and petals white or subdued (greenish or burgundy); distinct odor

        1. Odor strong, heavy, sweet **moth**

        2. Odor strong, fermenting or fruitlike; flower parts and pedicel strong **bat**

        3. Odor of sweat, feces, or decaying meat **fly**

    B. Sepals and/or petals colored; odor may or may not be present

        1. Flower shape regular or irregular,* but not tubular

            a. Flower shape irregular; sepals or petals blue, yellow, or orange; petal adapted to serve as a "landing platform"; may have dark lines on petals; sweet, fragrant odor **bee**

            b. Flower shape regular; odor often fruity, spicy, sweet, or carrionlike **beetle**

        2. Flower shape tubular

            a. Strong, sweet odor **butterfly**

            b. Little or no odor; flower usually red **hummingbird**

*A regular flower shape is one that has radial symmetry (like a daisy or carnation), with similar parts (such as petals) having similar size and shape. Irregular flowers have bilateral symmetry.

*This simple key works best when you select flowers whose characteristics will fit the key. Many cultivated flowers do not easily fit these general categories.*

### Procedure

Using the key above, classify the flowers used in Lab Study A based on their floral traits and method of pollination.

### Results

1. Record your results in Table 16.3.
2. If you made sketches of any of your flowers, you may want to indicate the pollinator associated with that flower.

### Discussion

1. Review the Key to Pollination and describe the characteristics of flowers that are adapted for pollination by each of the following agents:

   a. wind

     *inconspicuous sepals and petals, feathery stigma, no odor*

   b. hummingbird

     *red tubular flowers with little or no odor*

   c. bat

     *white flowers that smell like fruit; strong pedicel and flower parts*

2. Discuss with your lab partner other ways in which keys are used in biology. Record your answers in the space provided.

*primarily in the form of taxonomic keys used to identify species of organisms*

Student Media Videos—Ch. 30: Bee Pollinating; Bat Pollinating Agave Plant

## Lab Study C. Angiosperm Life Cycle

### Materials

pollen tube growth medium in dropper bottles
dropper bottle of water
petri dish with filter paper to fit inside
prepared slides of lily anthers and ovary

dissecting probe
brush bristles
compound microscope
flowers for pollen

### Introduction

In this lab study, you will study the life cycle of flowering plants, including the formation of pollen, pollination, fertilization of the egg, and formation of the seed and fruit. You will also investigate the germination of the pollen grain as it grows toward the egg cell.

Refer to Figures 16.2 (flower structures) and 16.3 (angiosperm life cycle) as you complete the exercise.

### Procedure

1. Pollen grain—the male gametophyte.

    a. Examine a prepared slide of a cross section through the **stamens** of *Lilium*. The slide shows six anthers and may include a centrally located ovary that contains ovules.

    b. Observe a single **anther,** which is composed of four **anther sacs** (microsporangia). Note the formation of **microspores** (with a single nucleus) from diploid **microsporocytes** (microspore mother cells). You may also see mature **pollen grains** with two nuclei.

2. Development of the female gametophyte (Color Plate 48).

    a. Examine a prepared slide of the *Lilium* ovary and locate the developing ovules. Each **ovule,** composed of the megasporangium and other tissues, contains a diploid **megasporocyte** (megaspore mother cell), which produces **megaspores** (haploid), only one of which survives. The megaspore will divide three times by mitosis to produce the eight nuclei in the **embryo sac,** which is the greatly reduced female gametophyte. Note that angiosperms do not even produce an archegonium.

*Consider purchasing a set of slides showing all stages of ovule development to use as a microscopy projection demonstration.*

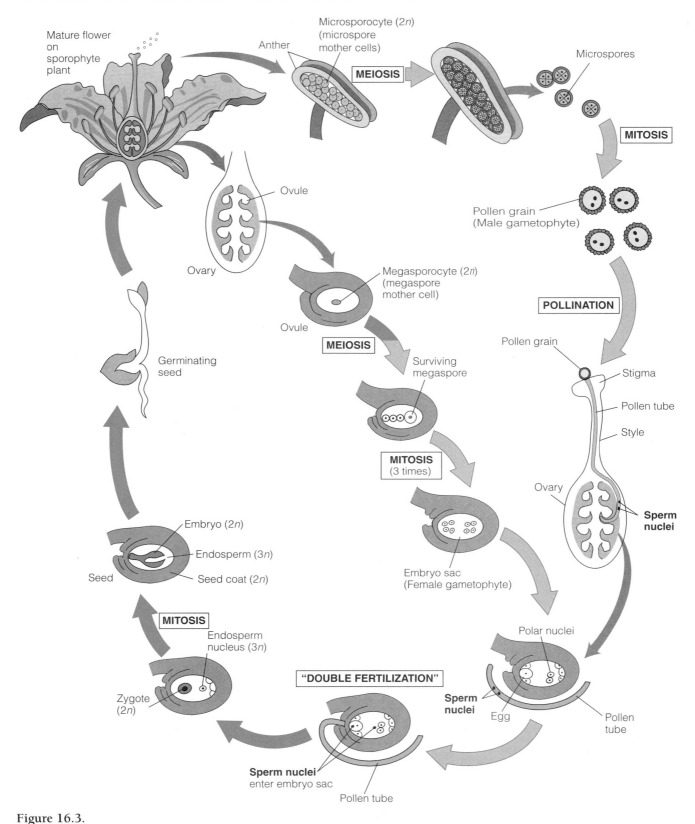

**Figure 16.3.**

**Angiosperm life cycle.** Observe the structures and processes as described in Exercise 16.2. *Using colored pencils, indicate the structures that are haploid or diploid. Circle the terms* mitosis, meiosis, *and* double fertilization.

b. Your slide will not contain all stages of development, and it is almost impossible to find a section that includes all eight nuclei. Locate the three nuclei near the opening to the ovule. One of these is the **egg.** The two nuclei in the center are the **polar nuclei** (or central cell).

3. Pollination and fertilization.

When pollen grains are mature, the anthers split and the pollen is released. When pollen reaches the stigma, it germinates to produce a **pollen tube**, which grows down the style and eventually comes into contact with the opening to the ovule. During this growth, one pollen nucleus divides into two **sperm nuclei.** One sperm nucleus fuses with the egg to form the **zygote,** and the second fuses with the two polar nuclei to form the triploid **endosperm,** which will develop into a rich nutritive material for the support and development of the embryo. The fusion of the two sperm nuclei with nuclei of the embryo sac is referred to as **double fertilization.** Formation of triploid endosperm and double fertilization are unique to angiosperms.

You will examine pollen tube growth by placing pollen in pollen growth medium to stimulate germination. Pollen from some plants germinates easily; for others a very specific chemical environment is required. Work with a partner, following the next steps.

a. Using a dissecting probe, transfer some pollen from the anthers of one of the plants available in the lab to a slide on which there are 2 to 3 drops of pollen tube growth medium and a few brush bristles or grains of sand (to avoid crushing the pollen). Add a coverslip. Alternatively, touch an anther to the drop of medium, then add brush bristles and a coverslip.

b. Examine the pollen under the compound microscope. Observe the shape and surface features of the pollen.

c. Prepare a humidity chamber by placing moistened filter paper in a petri dish. Place the slide in the petri dish, and place it in a warm environment.

d. Examine the pollen after 30 minutes and again after 60 minutes to observe pollen tube growth. The pollen tubes should appear as long, thin tubes extending from the surface or pores in the pollen grain (Color Plate 49).

e. Record your results in Table 16.4 in the Results section. Indicate the plant name and the times when pollen tube germination was observed.

4. Seed and fruit development.

The zygote formed at fertilization undergoes rapid mitotic phyla, forming the embryo. The endosperm also divides; the mature ovule forms a seed. At the same time, the surrounding ovary and other floral tissues are forming the fruit. In Lab Study D, you will investigate the types of fruits and their function in dispersal.

*Set up this activity before beginning Lab Study A. Flower Morphology (p. 424). Pollen germination success is variable and depends in part on the concentration of sucrose. Before lab, try pollen from flowers available. Bridal veil can be kept in the greenhouse year round, and the pollen always germinates. Try spiderwort, snapdragon, lilies, sweet peas, and Fast Plants. See the Prep Guide for other suggestions.*

*Germination should begin in 15 to 30 minutes for most. Some pollen grains will produce a spiral of cytoplasmic material that looks like a tube. However, these probably result from osmotic differences. The spirals do not have any outer wall; a pollen tube should have a distinct boundary. Occasionally, some pollen grains will produce more than one tube.*

## Results

1. Review the structures and processes observed in the angiosperm life cycle, Figure 16.3. Indicate the haploid and diploid structures in the life cycle, using two different colored pencils.

Having trouble with life cycles? Return to Lab Topic 15, Plant Diversity I, and review the generalized life cycle in Figure 15.2. The key to success is to determine where meiosis occurs and to remember the ploidal level for the gametophyte and sporophyte.

2. Sketch observations of slides in the margin of your lab manual for later reference.
3. Record the results of pollen germination studies in Table 16.4. Compare your results with those of other teams who used different plants. This is particularly important if your pollen did not germinate.

**Table 16.4**
Results of Pollen Germination Studies

| Plant Name | 30 min(+/−) | 60 min(+/−) |
|---|---|---|
|  |  |  |

**Discussion**

1. What part of the life cycle is represented by the mature pollen grain?

   *male gametophyte*

2. How does the female gametophyte in angiosperms differ from the female gametophyte in gymnosperms?

   *The angiosperm is reduced to eight nuclei and lacks an archegonium.*

3. Do you think that all pollen germinates indiscriminately on all stigmas? How might pollen germination and growth be controlled?

   *This is a recognition problem, similar to immune function, where the cells of the immune system recognize other cells as being either "self" or "nonself" depending on the presence and absence of chemicals on the cells' surfaces. Germination can be controlled by chemical signals indicating that the pollen and stigma are compatible. Students should see this as a chemical recognition problem.*

# Lab Study D. Fruits and Dispersal

## Materials

variety of fruits provided by the instructor and/or students

## Introduction

The **seed** develops from the ovule, and inside is the embryo and its nutritive tissues. The **fruit** develops from the ovary or from other tissues in the flower. It provides protection for the seeds, and both the seed and the fruit may be involved in dispersal of the sporophyte embryo.

## Procedure

1. Examine the fruits and seeds on demonstration.
2. Use the Key to Fruits on the next page to help you complete Table 16.5. Remember to include the dispersal mechanisms for fruits and their seeds in the table.

## Results

1. Record in Table 16.5 the fruit type for each of the fruits keyed. Share results with other teams so that you have information for all fruits in the lab.

**Table 16.5**
Fruit Types and Dispersal Mechanisms

| Plant Name | Fruit Type | Dispersal Method |
|---|---|---|
| | | |

2. For each fruit, indicate the probable method of dispersal—for example, wind, water, gravity, ingestion by birds, mammals, or insects, or adhesion to fur and socks.
3. For some fruits, the seeds rather than the fruit are adapted for dispersal. In the milkweed, for example, the winged seeds are contained in a dry ovary. Indicate in Table 16.5 if the seeds have structures to enhance dispersal. Recall that seeds are inside fruits. The dandelion "seed" is really a fruit with a fused ovary and seed coat.

Achene

Nut

Capsule

Legume

Follicle

Drupe

Berry

Pome

Aggregate fruit

Multiple fruit

## Key to Fruits

I.  Simple fruits (one ovary)
   A.  Dry fruits (at maturity)
     1.  Fruits with one seed
       a.  Ovary wall and seed coat are fused    **achene***
       b.  Ovary wall hard or woody but can be separated from the seed    **nut**
     2. Fruits with two to many seeds
       a.  Ovary with several cavities (seen when cut in cross section) and several too many seeds    **capsule**
       b.  Ovary with one cavity
       c.  Mature ovary opens along both sides    **legume**
       d.  Mature ovary opens along one side    **follicle**
   B. Fleshy fruits
     1. Ovary with one seed, which is surrounded by a very hard stone (outer covering of the seed is formed from the inner ovary wall)    **drupe**
     2. Ovary with many seeds; does not have a "stone"
       a.  All of mature ovary tissue is soft and fleshy; surrounding flower tissue does not develop into fruit    **berry†**
       b.  Fleshy fruit develops in part from surrounding tissue of the flower (base of sepals and petals); therefore, ovary wall seen as "core" around seeds    **pome**
II. Compound fruits (more than one ovary)
   A. Fruit formed from ovaries of many flowers  **multiple fruit**
   B. Fruit formed from several ovaries in one flower  **aggregate fruit**

*In the grass family, an achene is called a **grain**.
†Berries of some families have special names: citrus family = **hesperidium**; squash family = **pepo**.

*Students often ask about bananas; botanically they are berries. They lack sexual reproduction. All commercial plants are clones of the same stock.*

## Discussion

1.  How might dry fruits be dispersed? Fleshy fruits?

    *Some dry fruits open when mature. Seeds from these might be dispersed by wind or gravity. Other dry fruits may be eaten by animals. Fleshy fruits do not open and may be eaten.*

2.  Describe the characteristics of an achene, drupe, and berry.

    *An achene is a dry one-seeded fruit with the ovary and seed fused.*

    *A drupe is a fleshy one-seeded fruit with a stony covering around the seed.*

    *A berry is a fleshy many-seeded fruit.*

# Reviewing Your Knowledge

1. Complete Table 16.6. Compare mosses, ferns, conifers, and flowering plants relative to sexual life cycles and adaptations to the land environment. Return to Table 16.1 and modify your entries.

2. Identify the function of each of the following structures found in seed plants. Consider their function in the land environment.

   pollen grain: *the male gametophyte; transports sperm to egg without the requirement of water; produced in large numbers with desiccation-resistant coat*

   microsporangium: *sterile jacket of cells protecting spore production—meiosis*

   flower: *attracts pollinator to enhance efficiency of movement of pollen, increasing cross-fertilization and, in some cases, genetic variation*

   carpel: *protects the ovules and developing embryo plants*

**Table 16.6**
Comparison of Important Characteristics of Land Plants

| Features | Moss | Fern | Conifer | Flowering Plant |
|---|---|---|---|---|
| Gametophyte or sporophyte dominant | | | | |
| Water required for fertilization | | | | |
| Vascular tissue (+/−) | | | | |
| Homosporous or heterosporous | | | | |
| Seed (+/−) | | | | |
| Pollen grain (+/−) | | | | |
| Fruit (+/−) | | | | |
| Examples | | | | |

seed: *can be dispersed to new areas; a prepackaged embryonic plant with nutritional resources and a desiccation-resistant covering*

fruit: *enhances dispersal of seeds, protects seeds*

endosperm: *rich nutrient resource available for embryonic plant in developing seed*

3. Plants have evolved a number of characteristics that attract animals and ensure pollination, but what are the benefits to animals in this relationship?

   *Plants provide pollen and nectar, which are energy-rich food sources for animals. Some also produce waxes important to bees.*

4. Why is internal fertilization essential for true terrestrial living?

   *Internal fertilization results in protection of the sperm or sperm nuclei from desiccation; water in the environment is, therefore, not required for reproduction to occur.*

# Applying Your Knowledge

1. Explain how the rise in prominence of one major group (angiosperms, for example) does not necessarily result in the total replacement of a previously dominant group (gymnosperms, for example).

   *As one group colonizes new habitats and displaces existing groups, extinction is not the only outcome; the groups may coexist, the displaced groups being restricted to a limited number of habitats. In this case, gymnosperms, particularly conifers, are dominant in the dry forests of the southeastern United States and the boreal forest in northern climes.*

2. In 1994, naturalists in Australia discovered a new genus and species of conifer, *Wollemia nobilis*, growing in a remote area not far from the city of Sydney. This was the first new conifer discovered since 1948. Wollemi pine (not really a pine) is in a family that had a global distribution 90 million years ago in the Cretaceous period. Scientists used a variety of different evidence to decide where this rare tree should be placed in the "Tree of Life." What evidence would be necessary to determine that this is in the phylum Coniferophyta?

   *Conifers are gymnosperms so pollen and seeds should be present, but not fruits or flowers. Conifers also have male and female cones. Scientists also compared fossilized pollen, leaves, stems, and cones to the same structures in living plants.*

   To support the conservation of this ecosystem and three groups of trees (less than 100 individuals), botanical gardens are propagating these trees for sale. Although seeds have been collected, the commercially available plants are from cuttings and tissue culture. Based on your understanding of the life cycle of conifers, why is it not practical to reproduce Wollemi pines from seed or even to sell the seeds?

   *Conifers are generally long-lived perennials that take many years to mature and reproduce. Also for many conifers the time from pollination to mature seeds is two or more years. Wollemi pines easily sprout new stems from their base, so vegetative cuttings are easy to make and quick to propagate.*

3. Your neighbor's rose garden is being attacked by Japanese beetles, so she dusts her roses with an insecticide. Now, to her dismay, she realizes that the beans and squash plants in her vegetable garden are flowering, but are no longer producing vegetables. She knows beetles feed on leaves of roses and squash plants. What is the problem? Explain to your neighbor the relationship among flowers, fruits (vegetables, in the gardening language), and insects.

   *The vegetables of these plants are really fruits (by definition) that result when flowers are pollinated and the ovary of the flower enlarges. For many plants, insects are needed to pollinate the flowers. The insecticide probably killed the bees as well as the beetles.*

4. Seed plants provide food, medicine, fibers, beverages, building materials, dyes, and psychoactive drugs. Using web resources, your textbook, and library references, describe examples of human uses of plants in Table 16.7. Indicate whether your example is a gymnosperm or angiosperm. Based on your research, what is the relative economic importance of angiosperms and gymnosperms?

**Table 16.7**
Uses of Seed Plants: Angiosperms and Gymnosperms

| Uses of Plants | Example | Angiosperm/ Gymnosperm |
|---|---|---|
| Food | *Wheat, rice, potatoes, cassava, and more* | *Angiosperm* |
| Beverage | *Tea from Camellia, coffee, cocoa* | *Angiosperm* |
| Medicine | *Digitoxin from foxglove, taxol from yew* | *Angiosperm, Gymnosperm* |
| Fibers | *Linen from flax, cotton* | *Angiosperm* |
| Materials | *Pine, oak* | *Gymnosperm, Angiosperm* |
| Dyes | *Blue dye from indigo and woad* | *Angiosperm* |

*Angiosperms are the dominant group of land plants with the greatest diversity. They also provide nearly 100% of the agricultural plant food and the basis for about 25% of the prescription drugs in the U.S. Gymnosperms are important sources of building materials and resins.*

5. Review the introduction to Lab Topic 15 Plant Diversity I, and describe the major trends in the evolution of land plants.

   *Gametophyte is dominant in bryophytes, but the sporophyte is dominant in all vascular plants. The gametophyte becomes more reduced, eventually being reduced to eight nuclei in the embryo sac of flowering plants. Seedless vascular plants and bryophytes require at least a watery film for sperm to fertilize the egg. Seed plants produce pollen, which does not require water. The sporophyte embryo is increasingly protected and retained within the sporophyte tissue of the previous generation. Bryophytes and most seedless vascular plants are homosporous. All seed plants, Selaginella, and a few ferns are heterosporous.*

## Investigative Extensions

Pollen germinates easily in the laboratory for some species and not at all for others. In some species, a biochemical signal is required from the stigma to initiate germination. Think about the advantages to the species if pollen germinates easily or if it requires a biochemical signal. You can investigate the factors that affect pollen germination and pollen tube growth using flowers available in the laboratory that did not germinate using a general growth medium. Develop an investigation based on questions generated by your observations in this lab topic or consider one of the following suggestions.

1. Are factors present in the stigma necessary for pollen germination? Mince a small piece of the stigma in sucrose and then add it to a slide with pollen in pollen growth medium. (The sucrose concentration in the growth medium can also be varied, since this may affect pollen germination). Compare pollen germination using the stigma material from closely related species and then from those distantly related. For example, try different species of the Mustard Family or even different varieties of one species of *Brassica*. Remember that Fast Plants (*B. rapa*) are mustards as is *Arabadopsis,* another plant used in genetic studies.

2. What essential micronutrients are needed for pollen germination? Some species are sensitive to micronutrients in the growth medium, including boron and calcium. Research various growth media and test these with flowers that failed to germinate. Which of the micronutrients in the growth medium is necessary for pollen germination in your plants? Prepare growth medium omitting one of each of the components and observe pollen germination.

*See the* Preparation Guide *for suggestions of plants to investigate.*

3. How do environmental factors, such as temperature and light, affect the rate of pollen tube growth?

## Student Media: BioFlix, Activities, Investigations, and Videos
**www.masteringbiology.com (select Study Area)**

**Activities**—Ch. 29: Terrestrial Adaptations of Plants; Highlights of Plant Phylogeny;
   Ch. 30: Pine Life Cycle; Angiosperm Life Cycle;
   Ch. 38: Seed and Fruit Development; Making Decisions About DNA Technology: Golden Rice
**Investigations**—Ch. 30: How Are Trees Identified by Their Leaves?
   Ch. 38: What Tells Desert Seeds When to Germinate?
**Videos**—Ch. 30: Flower Blooming Time Lapse; Time Lapse of Flowering Plant Life Cycle; Bee Pollinating;
   Bat Pollinating *Agave* Plant;
   Ch. 38: Seed and Fruit Development; Discovery Video: Plant Pollination; Colored Cotton

## References

Berg, L. R. *Introductory Botany: Plants, People and the Environment,* 2nd ed. Belmont, CA: Thomson Brooks/ Cole, 2007.

Levetin, E. and K. McMahon. *Plants and Society,* 4th ed. New York: McGraw Hill Co., 2007.

Mauseth, J. D. *Botany: An Introduction to Plant Biology,* 4th ed. Sudbury, MA: Jones and Bartlett Publishers, 2008.

McLoughlin, S. and V. Vajda. "Ancient Wollemi Pines Resurgent." *American Scientist,* 2005, Vol. 93, pp. 540–547.

Raven, P. H., R. F. Evert, and S. E. Eichhorn. *Biology of Plants*, 7th ed. New York: W. H. Freeman Publishers, 2004.

Rui, M. *The Pollen Tube: A Cellular and Molecular Perspective*, New York: Springer, 2006.

## Websites

American Institute of Biological Science website, *actionbioscience*. Original interview with Pam Soltis, "Flowering Plants: Keys to Earth's Evolution and Human Well-Being." plus extensive website links to plant resources: http://www.actionbioscience.org/genomic/soltis.html

Images, maps and additional links for plants of North America: http://npdc.usda.gov/

Plant Conservation Alliance site with projects, medicinal and other plant uses: http://www.nps.gov/plants/

Society for Economic Botany: http://www.econbot.org/_welcome_/to_seb.php

Tree of Life Project includes phylogeny of living organisms, movies, references, current research, and ideas for independent investigations: http://tolweb.org/tree?group=Spermatopsida& contgroup=Embryophytes

University of California Berkeley Museum of Palentology site with images and resources for both living and fossil seed plants: http://www.ucmp.berkeley.edu/plants/plantae.html

University of Michigan Dearborn site for the uses of plants by Native Americans: http://herb.umd.umich.edu/

## LAB TOPIC 16

# Plant Diversity II: Seed Plants
## Teaching Plan for Laboratories

 If you choose to use this lab but do not use Lab Topic 15, Plant Diversity I, ask students to read the introduction to the first lab. Also, you will have to omit or modify some questions at the end of the lab.

## Main Concepts and Objectives

1. Concept: The most abundant, important, and diverse groups of land plants are the seed plants. Students will identify the phyla of gymnosperms and angiosperms, describe characteristics, and recognize representative members of the taxa. (Exercises 16.1, 16.2)

2. Concept: Flowers have coevolved with insects and other animal agents, ensuring efficient pollen transfer. Students will investigate floral morphology and identify probable pollinators for selected flowers. Students will investigate pollen tube growth. (Exercise 16.2, Lab Study B)

3. Concept: The fruit and seed are important for protection and dispersal of the embryo. Students will determine fruit types and dispersal mechanisms. (Exercise 16.2, Lab Study D)

4. Concept: Variations in basic reproductive cycles (alternation of generations) represent continuing adaptations to the land environment. Students will compare significant

features of the pine and flowering plant life cycles, noting unique features and stating their evolutionary significance. Students will review trends in the evolution of land plants noted in Lab Topic 15 Plant Diversity I. See the Teaching Plan for this lab topic. (Reviewing Your Knowledge)

5. Specific content. Students will be able to describe or define: *gametophyte* (male and female), *sporophyte, microspore, megaspore, egg, pollen grain, ovulate cone, pollen cone, archegonia, seed, stigma, style, ovary, carpel, ovule, anther, filament, stamen, petal, corolla, sepal, calyx, perianth, receptacle, embryo sac, polar nuclei, flower, fruit, double fertilization, endosperm.*

## Order of the Lab

1. Introduce the lab. Review alternation of generations                                          (15 min)

   Review Table 3 and trends in evolution presented in the Teaching Plan for Lab Topic 15. Review innovations of seed plants. (Students should complete Table 1 before coming to lab.)

2. Students study pine life cycle and view demonstrations.                         (30 min)
3. Prepare pollen germination experiments.                                                (10 min)
4. Review angiosperm life cycle and parts of flower.                                 (10 min)
5. Demonstrate the use of the key.                                                               (5 min)
6. Discuss flower morphology and pollinators.                                          (30 min)
7. View slides of life cycle.                                                                        (10 min)
8. Check pollen slides periodically and modify conditions.                      (20 min)
9. Identify fruits and dispersal mechanisms.                                            (30 min)
10. Summarize evolutionary trends and answer discussion questions.      (20 min)

**For a 2-hour lab:** The pollination and fruit keys can be set up as a demonstration with flowers and fruits identified. Omit the demonstration of gymnosperm phyla (Exercise 16.1, Lab Study A) and omit the pollen germination experiment (Exercise 16.2, Lab Study C). Alternately, omit the gymnosperms entirely (Exercise 16.1). It is also possible to divide the material covered in Plant Diversity I and II into three 2-hour lab periods.

## Classroom Management

Each student should have a slide set. Students work on slides individually. All other activities can be done in teams of two. Students will gain much from discussing the life cycles and materials together. Encourage students to view demonstrations in staggered fashion to avoid standing and waiting. Question students about the general life cycle and evidence for trends and adaptations to the land environment as they work through the material. Encourage students to investigate the flowers and fruits. This lab should not be a passive experience for the students. Ask more questions than you answer. Integrate the material from Plant Diversity I. Keep students moving through the material by giving them some time estimates.

## Student Development

Students will lose sight of the trends if they try to memorize terms and structures. Stress the structure/function theme, along with adaptation to the land environment. Students should provide evidence to support trends. Students will develop observational skills and the ability to view slides and investigate how structures relate to plant life cycles. Students will learn to use a key, a basic skill that can be utilized in a variety of contexts. They should be encouraged to practice using the keys. Most of the material in this lab will be new to students. Encourage them to ask questions. Don't worry if you can't assure them that they are correct about the probable pollinators. Ask them how they would test their predictions. Remind them that some flowers are generalists and may have characteristics that could be attractive to a variety of pollinators. Ask questions to prevent passivity in students.

## Discussion and Summary

Review life cycles and summarize trends, asking students to provide examples. Return to Plant Diversity I to discuss trends relative to all land plants. Review trends. Answer questions.

## Evaluation

Test students on all living and pressed plants and prepared slides on the next lab test. Students must also know major concepts and be prepared to provide evidence from the plant labs to support these themes. You may want to ask students to use a key to identify a flower or fruit.

## Lab Topic 17

# Bioinformatics: Molecular Phylogeny of Plants

## Laboratory Objectives

After completing this lab topic, you should be able to:

1. Describe the evolutionary history (phylogeny) of land plants.
2. Describe a phylogenetic tree and explain the relationships depicted in a tree using appropriate terminology.
3. Construct a phylogenetic tree using morphological evidence.
4. Use the tools of bioinformatics to test hypotheses in the form of phylogenetic trees.
5. Describe the steps used in the construction of phylogenetic trees based on molecular data.
6. Analyze results, and write an illustrated report, using critical thinking skills and synthesizing data from a variety of sources.
7. Describe evolutionary relationships, including common ancestry and changes over time in lineages.
8. Discuss the importance of bioinformatics and molecular evidence in evolutionary analysis and in other areas of science.

## Introduction

In Lab Topics 15 and 16 you investigated the diversity of the plant kingdom, comparing representative plants. You studied vegetative and reproductive structures, observed fossils, and investigated moss, fern, pine, and flowering plant life cycles. Land plants have been classified into 10 phyla based on their similarities and differences. Morphological and biochemical features shared by all land plants provide evidence that the closest living relatives to land plants are complex green algae called charophyceans. Advances in biotechnology in the 1970s and 1980s that permitted the analysis and sequencing of DNA provided additional genetic evidence for this relationship. In 1994, the Green Plant Phylogeny Research Coordination Group was formed at the University of California, Berkeley, to facilitate the efforts of various research groups working on the study of genetic relationships between plants and their algal ancestors. This began the "Deep Green" project, a large-scale collaborative study of plant evolution based on examination of nuclear and chloroplast gene sequences of algae and land plants. You can learn more about the Deep Green initiative at http://ucjeps.berkeley.edu/bryolab/GPphylo/proj_summ.php.

*If previously you did not complete Lab Topics 15 and 16, Plant Diversity I and II, then prepare a PowerPoint presentation describing the evolutionary relationships among land plants with examples. Include information on the origin of land plants from algal ancestors (charophytes). Assign students to read the material in appropriate chapters in their text (Campbell Biology, Chapter 29 Plant Diversity I and Chapter 30 Plant Diversity II). Students might also read Chapter 26 Phylogeny and the Tree of Life.*

In this lab topic you will use your previous laboratory experience and knowledge to develop a hypothesis for the evolutionary relationships among land plants and their green algal relatives. You will test your hypothesis by comparing molecular evidence from DNA sequences of a chloroplast gene using the tools of bioinformatics (the discipline connecting computer science and biology). You will construct a tree that represents plant evolutionary history (phylogeny). Before leaving the laboratory, you will prepare a final report incorporating your hypothesis and results.

## Understanding Phylogenetics

**Phylogeny** is the study of evolutionary relationships among groups of living and extinct organisms on Earth. Scientists analyze structural, reproductive, physiological, or molecular changes in specific characters. In particular, they utilize **homologous** characteristics, those features that are similar as a result of shared ancestry (common descent). Molecules, including DNA, can also reveal homology, since these are passed down from one generation to the next.

**Phylogenetic systematics** is the field of biology that examines morphological characteristics, biochemical pathways, and gene sequences to establish relationships among groups of organisms. To show evolutionary relationships, a **phylogenetic tree** (a visual representation of the lineages among organisms) is constructed. All phylogenetic trees are hypotheses that are to be tested, modified, and tested again. One type of phylogenetic tree, known as a **rooted tree,** contains a root, nodes, branches, and clades (Figure 17.1).

The **root** of the tree represents a common ancestor from which all the organisms in the group are derived. Within the tree are several **nodes.** A node represents a branching point for a taxonomic unit (such as group, phylum, genus, or species). **Branches,** the lines that extend from nodes, establish how closely related one group is to the other. The shorter the line, the more closely related the groups. Groups sharing a node share a common ancestor and make up a **clade.** The branching pattern of the tree and the length of the branches reflect the number of changes or differences that have accumulated among the species or groups. In Figure 17.1 wolves, skunks, and *Drosophila* (fruit flies) are all animals, and therefore their **shared ancestral traits** are eukaryotic, multicellular organisms that consume food. But wolves and skunks share some traits that are not shared by *Drosophila*. These shared traits (for example, hair and milk production) are known as

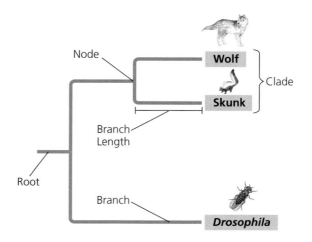

**Figure 17.1.**
Terminology associated with a phylogenetic tree reflecting evolutionary relationships.

**derived traits** and determine the specific clade—mammals—of the organisms that share them.

Phylogenetic trees can be constructed using a variety of different methods to establish the nodes, branches, and topology of the tree. Refer to your textbook (see *Campbell Biology*, 9th edition, Concept 26.3) for further information on methodologies.

# Molecular Phylogenetics

Rapid technological advances in molecular biology have allowed scientists to obtain DNA sequences of genes and entire genomes of a large variety of organisms. **Molecular phylogenetics** is the field that examines evolutionary relationships among groups specifically based on changes occurring in DNA and protein structure. The number of differences in DNA sequences between different groups reflects the accumulation of mutations over the time since they shared a common ancestor. Phylogenetic trees can be constructed to illustrate these differences in DNA sequences for groups of organisms.

Four steps are important in the construction of phylogenetic trees based on molecular data (NCBI, 2004).

1. **Align** similar DNA sequences from different groups to detect similarities and differences in nucleotide bases.

2. **Establish sequence variation** by observing the level of homology or similarity of sequences among groups.

3. **Build a tree** by arranging groups based on the percentage of matching bases for sequences and other factors.

4. **Evaluate the tree**, including the analysis of the resulting tree and comparison with trees constructed with nonmolecular data.

Using DNA sequences in phylogenetics can generate very large data sets. Imagine comparing 10 different species for a gene sequence with 100 nucleotides and the possibility of 4 different nucleotides at each position. To cope with such huge amounts of data, scientists use the tools of bioinformatics to construct molecular phylogenies.

# Using Bioinformatics

**Bioinformatics** is the field of science combining the disciplines of biology and computer science. Bioinformatics uses computer algorithms (a clear set of instructions for solving a problem) to conduct analyses of large biological data sets. Applications include aligning DNA and protein sequences, predicting protein structure, creating restriction enzyme maps for a DNA sequence, and predicting translation products. Many of these computer algorithms are now available for public access through the Internet and have become a staple analysis tool for molecular biologists. One important website is the **National Center for Biotechnology Information (NCBI)** established by the National Library of Medicine at the National Institutes of Health. NCBI manages GenBank®, an international database and collection of all publicly available DNA sequences, which now number over 108 million as of August 2009. NCBI also provides links to tools for molecular data analysis such as BLAST, a program used to find DNA sequences with similarity to an unknown sequence. Another excellent resource for bioinformatics applications is **Biology WorkBench**, a Web interface that allows rapid access to biological databases, such as GenBank®, and analysis tools, such as BLAST and CLUSTALW.

In the following laboratory exercise, you will use bioinformatics tools in Biology WorkBench to construct a phylogenetic tree or gene tree, using DNA sequences of land plants and green algae.

## EXERCISE 17.1
# Establishing Molecular Phylogenetic Relationships among Rubisco Large Subunit Genes

## Materials

*Before the laboratory begins the computers should have two folders on the desktop: one with images of the organisms and the other with nucleotide sequences for rbcL for the 9 algae and plants used in this lab exercise. These sequences have been prepared for use in WorkBench. The sequences are available at www .masteringbiology.com in the Study Area under* Investigating Biology.

*If living examples are not available, place herbarium specimens on demonstration. Images of these organisms should be available in a folder on each computer desktop. They can be accessed at the website listed above. Provide botany and evolution books as references for student use.*

computers with Internet access, one per pair of students
2 computer desktop folders: "images" and "rbcL nucleotide seq" (available at www.masteringbiology.com in the Study Area under *Investigating Biology*).
printer access
living examples of *Arabadopsis, Chara, Equisetum, Lilium, Marchantia, Pinus, Polypodium, Polytrichum,* and *Zamia*, if available
20 sheets of large paper (11″ × 17″)
botany reference books

## Introduction

The ribulose bisphosphate carboxylase (Rubisco) protein is essential to carbon fixation in photosynthesis and is found in green algae and all land plants. Therefore it is an ideal choice to establish phylogenetic relationships among green algae and land plants using nucleotide sequences. The gene sequence for the large subunit (rbcL) of the Rubisco protein has been isolated from a large variety of algae and plants. The DNA sequences of rbcL genes for *Arabadopsis, Chara, Equisetum, Lilium, Marchantia, Pinus, Polypodium, Polytrichum, and Zamia* are available in GenBank.

Working in pairs, you will use your knowledge and laboratory experience from previous lab topics to develop a phylogenetic tree that depicts the evolutionary relationships for the organisms listed above. This tree will serve as your hypothesis, which you will test using molecular data for the rbcL nucleotide sequences. You will access the supercomputer, which will use the tools of bioinformatics to compare the sequences for rbcL genes for each of the above organisms, then perform an alignment of sequences, and construct a phylogenetic tree reflecting the relationships among these genes and groups.

## Developing Hypothesis and Prediction

Examine the organisms selected for this exercise and determine which are charophytes, bryophytes, pterophytes, gymnosperms, or angiosperms. The plants may be on demonstration in the laboratory and images are provided in a folder on the desktop of your computer. Based on your knowledge of plant diversity and the *ancestral* and *derived* characteristics of the major phyla of plants, develop a hypothesis for which organisms will be more closely related to each other. Follow these steps in developing your hypothesis for the phylogenetic relationships of the organisms in the laboratory. Refer to Table 16.6 to review morphological and life cycle characteristics of land plants studied in Lab Topics 15 and 16, Plant Diversity I and II.

- Closely observe the plants on demonstration in the laboratory and the images provided for this lab topic. Compare these with the plant groups in Lab Topics 15 and 16, Plant Diversity I and II (see also Color Plates 28, 30–31, 35–39, and 43). How many different plant groups (taxa) are represented by these plants?

- For these major groups of plants (charophytes, bryophytes, pterophytes, gymnosperms, and angiosperms), what are the ancestral and derived traits?

- Consider how you might arrange these groups. Which groups are more closely related to each other? Which groups of plants are more distantly related to each other?

- Arrange the organisms most closely related to each other into clades.

- Consider which clades might share a common ancestor and group those clades together.

## Hypothesis—Drawing a Phylogenetic Tree

Based on the hypothesized relationships, draw a phylogenetic tree showing appropriate branching to reflect where each organism would be placed relative to the others. The tree you draw is a visual representation of your hypothesis for land plant relationships, which incorporates diverse sources of evidence: morphology, vegetative and reproductive structures, variations in life cycles, and fossils. Refer to the introduction at the beginning of this lab topic to review terminology and concepts. (*Note: This tree will be referred to as the "morphological" tree, even though other evidence was included in its development.*)

*Do not correct any faulty assumptions about the relationships. In particular, students will often place the green algae as the common ancestor on their hypothesized morphological tree. Let them.*

- Make preliminary sketches and a final drawing on the large sheets of paper available in the laboratory.

- Use your laboratory manual (Lab Topics 15 and 16) and your textbook (*Campbell Biology*, 9th edition, Chapters 26, 29, and 30) to guide you in constructing your tree.

- What is the **root** of your tree?

- Which clades should be grouped together?

- Label the **root**, **clades**, and **branches**. (Refer to Figure 17.1.)

## Prediction

You will be testing your hypothesis that the morphological tree accurately represents land plant phylogeny. You will use molecular data from the nucleotide sequences of rbcL. Do you think the molecular evidence will support (be consistent with) your hypothesis or falsify it? Write your prediction below.

*If the morphological tree accurately represents land plant phylogeny, then the molecular evidence and tree will be consistent with the morphological tree.*

## Using Biology WorkBench

Imagine how many comparisons are possible for 9 species, comparing changes at each nucleotide position in rbcL DNA! To make this possible, you

*This exercise uses ClustalW to align the sequences and construct the phylogenetic tree. This program does a multiple sequence alignment and matches up the sequences, so that the similarities and differences are visible. There are other programs as well that use different approaches and can result in generating different trees. Other software can be used to analyze the sequences. For more information on methodologies see* Campbell Biology, *9th edition, Concept 26.3 and the NCBI primer available at http://www.ncbi.nlm.nih.gov/About/primer/phylo.html.*

*The procedure in this laboratory topic uses the current (2010) version of Biology WorkBench v. 3.2, however, a New Generation WorkBench (www.ngwb.org) is being publicly tested and is available for use. Should the New Generation WorkBench replace v. 3.2, then follow the online tutorials for adding nucleotide sequences and generating a phylogenetic tree using ClustalW or other multiple sequence analysis program.*

will use Biology WorkBench with access to the San Diego Supercomputer Center, which will compare and align the nucleotide sequences for all of your organisms. The supercomputer will then construct phylogenetic trees using a mathematical algorithm to select the one phylogeny from those possible that is best explained by the molecular data. *Remember that the more similar two sequences are, the more closely related they are (the more recent their common shared ancestor). The more nucleotide changes that occur, the more time since two groups share a common ancestor, and therefore they are more distantly related.* There are many computer programs that use different algorithms for analyzing these large data sets. In this exercise you will use a multiple sequence analysis program, **ClustalW.** You and your lab partner will do the evaluation and analysis of results, comparing the molecular phylogenetic tree to your hypothesized phylogenetic tree, which was based on morphology and life cycle features.

 While using Biology WorkBench—*do not use the "Back" button* on your Web browser. Use only the "Return" or "Abort" keys on the WorkBench menus. ***Using the "Back" button will overload and crash the supercomputer.***

## Procedure

1. Open a new Word document. Next, open a Web browser and type http://workbench.sdsc.edu/ to enter the Biology WorkBench website. *Note that during this exercise you will move back and forth between these two windows.*

2. To acquire access to Biology WorkBench, you must register your team. Click on the link "register" under first-time users (Figure 17.2). Supply your name (choose one member of the team) and your email address.

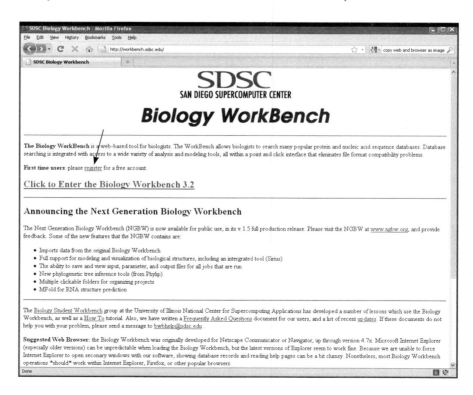

**Figure 17.2.**
**Biology WorkBench home page.**
Register and enter WorkBench.

Select a user ID and a password that is easy to remember; *be sure to write it down*. Click on the "Register" button. Only one member of the team should register.

3. After you register, a new dialog box will appear on your screen asking for your user ID and password; enter the user ID and password you just selected. This will allow you to enter the WorkBench site. Alternatively, if you return to the main page, click on "Enter the Biology WorkBench 3.2" (Figure 17.2), which will prompt you for the user ID and password.

4. When you enter Biology WorkBench, the first page describes the web tools and provides an introduction. Scroll toward the end of the page and select "Session Tools" (Figure 17.3). This will lead you to a new page entitled "Default Session."

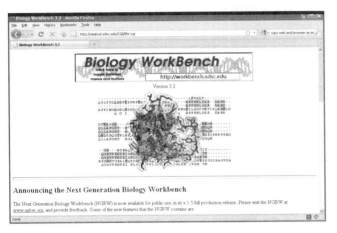
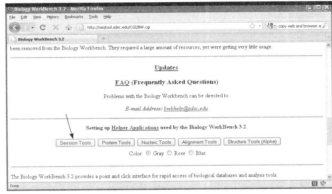

**Figure 17.3.**
**Biology WorkBench.** Selecting "Session Tools."

5. Using the scroll menu in the middle of the page, select "Start New Session" (Figure 17.4). Then click the button "Run" immediately below the scroll menu. A new page will open. Select a name for your session—such as

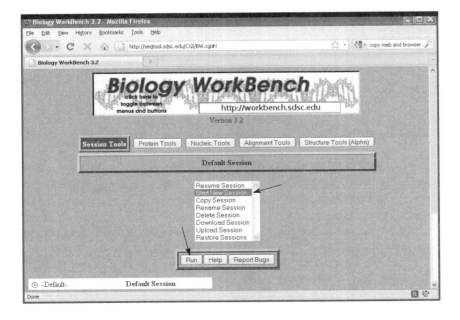

**Figure 17.4.**
**Biology WorkBench.** Starting a new session.

molecular phylogeny lab—and type this in the box labeled "Session Description." Click the button "Start New Session" right below the session description (Figure 17.5).

**Figure 17.5.**
**Biology WorkBench.** Entering a session description.

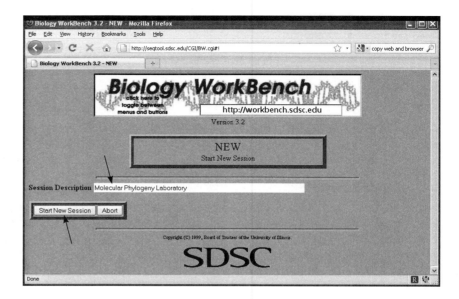

6. A new page will open with your new session listed (Figure 17.6). Next select "Nucleic Tools" from the menu at the top. After you select this option, a new page will appear with a scroll menu in the middle of the page.

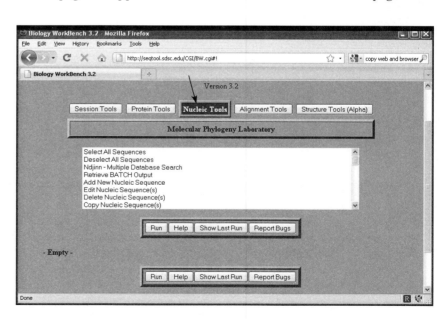

**Figure 17.6.**
**Biology WorkBench.** Selecting "Nucleic Tools."

7. Minimize (but do not close) the Biology WorkBench window. (*If you accidentally close the WorkBench window, follow the steps to log in again.*) Locate the Word file labeled "rbcL nucleotide seq" on your computer Desktop. Open the document, then go to "Edit" and "Select All" to highlight entire text in the document. Right click on the mouse and select "Copy."

8. Minimize the Word document and return to the Biology WorkBench window. Select "Add New Nucleic Sequence" on the scroll menu in the middle of the page (Figure 17.7). Click on "Run."

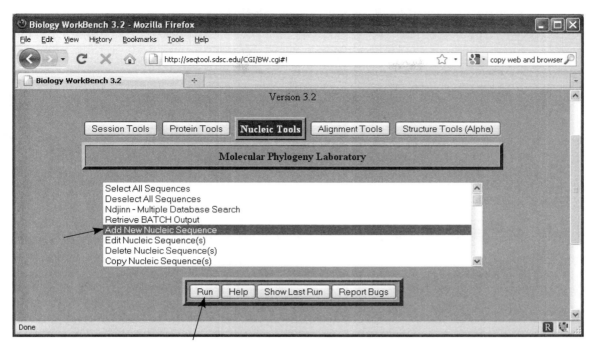

**Figure 17.7.**
**Biology WorkBench.** Adding a new nucleic acid sequence.

A new page requesting input of new sequences will open. Click on the large box labeled "Sequence." Right click on the mouse and choose "Paste" from the menu. You should now see the DNA sequences pasted in WorkBench (Figure 17.8).

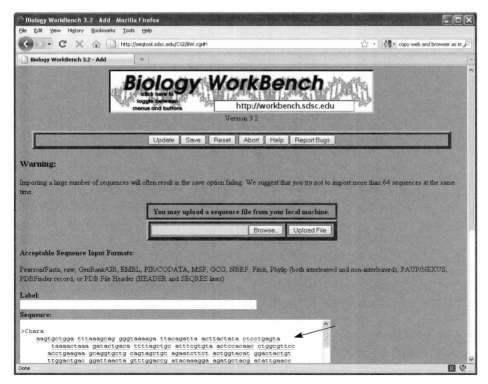

**Figure 17.8.**
**Biology WorkBench.** Pasting the nucleotide sequences.

9. After you have pasted the rbcL sequences into the Sequence box, click "Save" at the bottom of the Web page. A new page will open, containing the main scroll menu. The organism names for your uploaded sequences will appear in the middle of the page as "User Entered" (Figure 17.9). Select one of the uploaded sequences by checking the box on the left of the sequence. Select "View Nucleic Sequence" from the scroll menu (scroll down to find this). Then hit "Run." This allows you to check to be sure that the sequence has been uploaded for the organism listed. (You do not have to check all the organisms.) Next hit "Return" at the bottom of the page to return to the previous page.

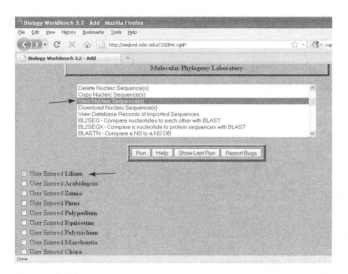
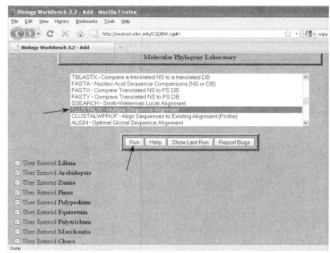

**Figure 17.9.**
**Biology WorkBench.** Entered sequences.

 **Do not use the Back Button!**

10. Now select *all* of the uploaded sequences under the main menu by checking the appropriate boxes. From the scroll menu in the middle of the page, choose "CLUSTALW—Multiple Sequence Alignment" and click "Run" (Figure 17.9). The program will select the sequences and open a new page.

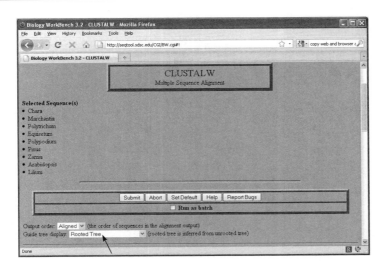

**Figure 17.10.**
**Biology WorkBench.** Options
in ClustalW.

11. In the new ClustalW page, below the selected sequence files, find "Guided Tree Display" (Figure 17.10). Change this to "Rooted Tree." Next click on "Submit" at the bottom of the page. *Be patient* as ClustalW analyzes and processes the data. Do not click on links or arrow keys while the computer is retrieving data. A new page will appear with the results.

12. Scroll to the bottom of the ClustalW results page. Find the phylogenetic tree for the entered sequences. Place the cursor on this image, right click on the mouse, and choose "Copy." *Do not close the browser window!* Return to the open Word document on your desktop and paste the image into the document (alternatively you can paste the image in Paint). Save the document to the Desktop. Keep the Biology WorkBench/ ClustalW browser window open to continue analyzing the results as instructed. *Closing this window will result in the loss of your alignment results.*

13. Print a copy of the molecular tree pasted in the Word document.

14. Return to the Biology WorkBench/ClustalW window. Scroll to the sequence alignments located above the phylogenetic tree. Review the sequence alignments and record any interesting observations about the sequence variability among the rbcL genes of the selected organisms. Continue your analysis in the following Results section.

## Results

1. Record any information about the variation in rcbL sequences you observed in the ClustalW analysis.

2. Review the results. Can you determine the number of nucleotides for this gene? Note: The nucleotide sequence for some of the species may not include the entire rbcL sequence.

3. Observe the alignments and list the plants or algae that have the most similar alignment.

*See Teaching Plan for example of phylogenetic tree generated using rcbL nucleotide sequences in this exercise.*

## Discussion

Discuss your results in the report you prepare in Exercise 17.2. In this report you will analyze your molecular phylogenetic tree and compare your results with the morphological tree that you hypothesized for these nine organisms.

EXERCISE 17.2

# Analyzing Phylogenetic Trees and Reporting Results

## Materials

computer with printer access
computer desktop folder with images
scissors and tape

## Introduction

The supercomputer completed the enormous task of aligning the nucleotide sequences and developing the phylogenetic tree that best fits the data, looking at shared sequences (evidence of common ancestry) and sequence differences (accumulated after divergence from a common ancestor). Now you will compare your results to your hypothesized phylogenetic tree. If you predicted that the molecular data would be consistent (or not) with your hypothesis, what evidence do you have to support or falsify your hypothesis? Once you have completed your analysis, you and your partner will use your results and analysis to construct a report to submit at the end of the laboratory period.

## Procedure

1. Along with your lab partner, examine the observations you made about the sequence alignment results provided by ClustalW. Examine the phylogenetic tree constructed based on rbcL genes of different organisms. Label the **root**, a **branch**, a **node**, and a **clade** within your phylogenetic tree. Paste the images of each organism from the folder on your computer desktop onto your phylogenetic tree. (*Note:* You may cut and paste using the computer or print the images and use scissors and tape!)

*Students will be confronted with two trees that are different. Generally, they assume that the molecular data is more "accurate" rather than trying to understand the differences in the two trees and the limitations of these different sources of evidence. This can be a very productive discussion. The differences reflect the struggle in plant systematics to understand evolutionary history and systematic relationships.*

2. Compare your molecular phylogenetic tree to your hypothesized morphological tree. Observe the similarities and differences in the branching patterns and clades. Compare your results with information provided in your textbook and the websites at the end of the lab topic. Describe any similarities or differences and your thoughts on why there may be differences.

3. The two phylogenetic trees are supported by different types of evidence. What evidence was used to create the phylogenetic tree using bioinformatics in Biology Workbench?

   *nucleotide sequences for the enzyme, rubisco large subunit; one gene*

   What types of evidence support your hypothesized "morphological" tree?

   *morphology, vegetative and reproductive features, life cycles, and fossils*

4. Together with your partner, create and type a one-page report (double spaced, size 12 font), that includes your hypothesis and prediction and a description of your results and analysis. You should attach your morphological tree as part of your hypothesis. Be sure to identify the

phylum for each of the organisms and briefly describe their relationships in the text of your report. Label your hypothesized morphological tree and the molecular tree as figures and refer to them in your report. Include both names of the team members on your report. Turn in your report along with the figures before leaving the laboratory.

## Reviewing Your Knowledge

1. Define and describe the following terms: *phylogeny, phylogenetic tree, phylogenetic systematics, molecular phylogenetics, bioinformatics, sequence alignment, homology.*

2. Draw a simple phylogenetic tree for two sister clades with a common ancestor. Define the following terms and use them to label your diagram: *clade, node, root,* and *branch.*

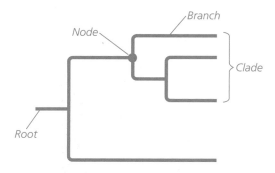

*This is one of several options students might draw.*

3. What is rbcL, and why is it a particularly useful molecule for studying evolutionary relationships in plants and green algae?

*Ribulose bisphosphate carboxylase is the enzyme involved in fixing carbon dioxide in photosynthesis and one of its subunits is coded for by rbcL. This gene is particularly useful for phylogenetic analysis because it is found in all plants and in the green algae that share a common ancestor with land plants. Changes in this gene represent the accumulation of nucleotide changes since groups of plants have diverged from a common ancestor.*

4. What are the limitations in using rcbL to construct a phylogenetic tree?

*Rubisco is an important enzyme in photosynthesis; however, the tree created in this exercise was based on the comparison of a limited number of nucleotides of only one gene. Using additional genes would provide a more robust comparison.*

## Applying Your Knowledge

1. For the following phylogenetic tree, label the boxes on the branches with the derived trait that is shared by the members of each clade. For example, on the lower branch a box has already been labeled for the feature "cuticle," which is shared by all land plants.

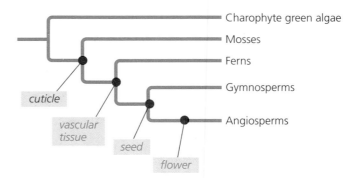

2. Can you suggest reasons why a phylogeny based on molecular evidence and a phylogeny based on morphology and other evidence might not be exactly the same?

*In some instances the morphology may vary greatly between species with very little genetic change; a few mutations produce big effects. This is true for humans and chimpanzees, as well as the silversword plants of Hawaii. In other cases nucleotide changes may have accumulated, but had little effect on the way the organisms look. This is true for some newly proposed species of butterflies in the tropics.*

3. Zoologists worldwide are sequencing a mitochondrial gene CO1 (cytochrome c oxidase subunit), which is found in all animals and appears to be distinctive for each species. The sequence of nucleotides can be used as a universal DNA bar code. By comparing the CO1 DNA sequence for an animal to a growing database of DNA sequences, scientists can accurately identify any animal and also discover species not previously known to science. How might DNA bar coding, which uses molecular biology and bioinformatics, be useful in enforcing international laws for banning the import of endangered species? How might these approaches stimulate the study of biodiversity in remote areas?

*Import officials often lack the expertise to identify endangered species, which may be similar to more common species. With DNA bar coding, species that are similar morphologically can be distinguished using nucleotide sequences. Animal parts that might otherwise be unrecognizable can also be identified. The DNA for animal species can be checked easily using the DNA bar coding. New species would be initially recognized by unique sequences of DNA, thus simplifying the first steps in studying biodiversity.*

4. When scientists compare the phylogenetic trees based on molecular data with existing trees based on morphological characteristics, in over 90% of the cases the molecular data confirms the relationships previously recognized. However, there are some surprises. For example, water lilies and the water lotus were classified as close relatives, but recent molecular analyses have shown that the water lotus is more closely related to the sycamore tree than to water lilies. Scientists have to critically evaluate these conflicts. Do you think that the evidence from molecular analyses should be the deciding factor in resolving these conflicts with other kinds of evidence (morphology, reproduction, biogeography, fossils, etc.)? How do you think scientists should weigh the evidence?

## Investigative Extensions

You may have time to modify your analysis during the laboratory period, or you may be inspired to pursue bioinformatics and molecular phylogeny as an independent investigation.

1. To modify the analysis of the species featured in this laboratory, consider doing the same comparisons (using all species), but select "Unrooted Tree" in the CLUSTALW screen. Then compare your results for rooted and unrooted trees. Go to the NCBI website and scroll to Figure 1 to read about rooted and unrooted trees. http://www.ncbi.nlm.nih.gov/About/primer/phylo.html

2. Do you think the results of your analysis might be different if you selected only some of the nine species used in this lab topic? Return to Figure 17.9 in the Procedure section and select only those sequences that you would like to investigate in a simplified or at least different phylogenetic analysis.

3. A new version of Biology Workbench is available for public testing. You can access the test version: New Generation WorkBench at http://www.ngbw.org/. Follow the online tutorial for uploading the nucleotide sequences and completing a multiple sequence analysis. Once you have completed the tree analysis for the sequences provided, you can make changes to the data set and in the bioinformatics tools used for the analysis. The program provides a diverse set of tools. Consider removing one or more of the plants to observe the tree output.

 ## Student Media: BioFlix, Activities, Investigations, and Videos
www.masteringbiology.com (select Study Area)

**Activities**—Ch. 29: Highlights of Plant Phylogeny
**Investigations**—Ch. 26: How Is Phylogeny Determined by Comparing Proteins?

## References

This lab topic was coauthored with Dr. Nitya Jacob, Associate Professor of Biology, Oxford College.

Cameron, K. M. "Bar Coding for Botany." *Natural History*, 2007, vol. 114(2), pp. 52–57.

Futuyama, D. J. *Evolution*, 2nd ed. Sunderland, MA: Sinauer Associates, Inc., 2009.

Mauseth, J. D. *Botany: An Introduction to Plant Biology*, 4th ed. Sudbury, MA: Jones and Bartlett Publishers, 2008.

National Center for Biotechnology Information. Revised 2004, April. "Just the Facts: A Basic Introduction to the Science Underlying NCBI Resources—Systematics and Molecular Phylogenetics." http://www.ncbi.nlm.nih.gov/About/primer/phylo.html. Accessed March 1, 2010.

Reece, J., et al. *Campbell Biology*, 9th ed. San Francisco, CA: Pearson Education, 2011.

Valentini, A., F. Pompanon, and P. Taberlet. "DNA Barcoding for Ecologists." *Trends in Ecology and Evolution*, 2009, vol. 24, pp. 110-117.

## Websites

Bedrock: Bioinformatics Education Dissemination: Reaching Out, Connecting and Knitting-together. Integrating bioinformatics in undergraduate biology curriculum:
http://www.bioquest.org/bedrock/

Biology WorkBench home page:
http://workbench.sdsc.edu

Consortium for Bar Code of Life (CBOL). See barcode data, read about uses of bar codes:
http://www.barcoding.si.edu/

Green Plant Phylogeny Research Coordination Group home page:
http://ucjeps.berkeley.edu/bryolab/GPphylo/

National Center for Biotechnology Information home page:
http://www.ncbi.nlm.nih.gov/

Tree of Life Web Project site for exploring diversity and phylogeny:
http://www.tolweb.org/tree/home.pages/abouttol.html

University of California Berkley, understanding evolution website with excellent pages on phylogenetics:
http://evolution.berkeley.edu/evolibrary/home.php

# LAB TOPIC 17

# Bioinformatics Molecular Phylogeny of Plants
## Teaching Plan for Laboratories

This lab topic builds on student knowledge and experience after they have completed the two lab topics on plant diversity (Lab Topics 15 and 16). This lab could be used without the prior laboratory experience, but before attending lab, students will have to be prepared through lecture material on the diversity and evolution of land plants that include images of the major groups and a discussion of evolutionary trends and relationships. Another option would be to combine elements of the plant diversity labs into one laboratory period, carefully choosing materials and investigations that will prepare students for developing a phylogenetic tree.

## Main Concepts and Objectives

1. Concept: Evolutionary relationships and phylogenetic trees. (Exercise 17.1)
2. Concept: Bioinformatics, using computers and mathematics to compare and analyze large biological data sets. (Exercise 17.1)
3. Concept: Scientific process—synthesizing diverse data (morphology and life cycles), then developing hypotheses and testing. (Exercise 17.1)
4. Concept: Critical thinking, analysis, and communication. (Exercise 17.2)
5. Specific content. Students will be able to define or describe: *phylogeny, clade, root, branch, node, bioinformatics, algorithm, sequence alignment, phylogenetic tree,* and *homology.*

# Order of the Lab

1. Introduce the concepts of evolution of land plants, phylogenetics, and bioinformatics.                                     (20 min)

2. Develop hypothesis and create a phylogenetic tree using organisms, images, prior knowledge, and other resources.          (30 min)

3. Perform Exercise 17.1. Use WorkBench to align and compare nucleotide sequences, and construct molecular tree.             (30 min)

4. Examine WorkBench results.                                                 (30 min)

5. Analyze molecular tree and compare with morphological phylogenetic tree.                                       (20 min)

6. Construct final report.                                                    (30 min)

**For a 2-hour lab:** Complete the analysis and prepare the report outside of the laboratory period.

# Classroom Management

Students should spend the full 30 minutes developing their hypothesized phylogenetic tree. They will have to think about how to draw the trees and how the shape of the tree reflects evolutionary relationships. Students may need assistance with the concepts of common ancestry, and ancestral and derived traits. Do not correct any faulty assumptions about the relationships. Guide them to sources of information, but do not let them copy a tree from their book or from the Web. Using the living examples and the images provided will help them connect this laboratory with the plant diversity lab topics.

In Exercise 17.2, encourage your students to think carefully about their results and what these mean about evolutionary relationships and the use of molecular evidence in phylogeny. Remind them about controversies discussed in Lab Topic 15 concerning the classification of pterophytes in one or several phyla. Again, this is an opportunity for students to struggle with science and then to think critically in an effort to write a clear report.

# Student Development

Students will have to synthesize their knowledge from lecture and previous laboratories, and then apply that to a group of organisms, some of which they will recognize and others that are unknown. Working cooperatively with a partner they will use this knowledge to create the phylogenetic tree, going from concrete knowledge to a visual model. This lab topic presents an opportunity for students to practice collaborative writing and editing, which is akin to the way most scientists write. Encourage them to edit their work, not to simply finish quickly.

# Discussion and Summary

Encourage student teams to collaborate with other teams in the lab. If time permits, at the end of the laboratory have one or two teams discuss their analysis, presenting summaries of their reports.

# Evaluation

The final report can be evaluated on content, critical thinking, writing, and presentation. Below is a typical phylogenetic tree generated in this laboratory.

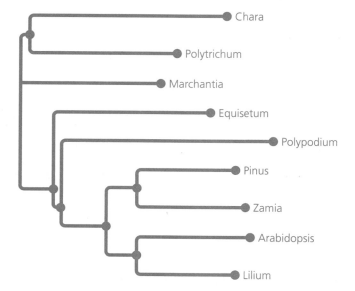

# Lab Topic 18

# Animal Diversity I: Porifera, Cnidaria, Platyhelminthes, Annelida, and Mollusca

## Laboratory Objectives

After completing this lab topic, you should be able to:

1. Compare the anatomy of the representative animals, describing similarities and differences in organs and body form that allow the animal to carry out body functions.
2. Discuss the impact of molecular studies on traditional phylogenetic trees.
3. Discuss the relationship between body form and the lifestyle or niche of the organism.

*For a 2-hour lab: You may choose to extend the two 3-hour labs to three 2-hour labs. If this is not possible, omit the sponge and clamworm.*

## Introduction

Animals are classified in the domain **Eukarya,** kingdom **Animalia** (clade **Metazoa**). They are **multicellular** organisms and are **heterotrophic,** meaning that they obtain food by ingesting other organisms or their by-products. Careful study of comparative anatomy, embryology, and most recently, genetic and molecular data, reveals many similarities in structure and development. Collectively, this evidence implies an ancestral evolutionary relationship among all animals. The ancestor of all animals is thought to have lived somewhere between 675 and 800 million years ago, but most animal fossils range in age from 565 to 550 million years old, with most body forms appearing by the end of the Cambrian period (see Figure 15.3). Scientists recognize over 35 major groups of present-day animals based on differences in body architecture. In this and the following lab topic, you will investigate body form and function in examples of nine major groups of animals. You will use these investigations to ask and answer questions comparing general features of morphology and relating these features to the lifestyle of each animal.

*In the Study Area of masteringbiology.com, see Student Media for suggested activities, investigations, and videos.*

Since the beginning of the scientific study of animals, scientists have attempted to sort and group closely related organisms. Traditionally, taxonomists have divided the metazoa into two major groups: **Parazoa,** which includes the sponges, and **Eumetazoa,** which includes all other animals. This division is made because the body form of sponges is so different from that of other animals that most biologists think that sponges are not closely related to any other animal groups.

Animals in Eumetazoa differ in physical characteristics, such as symmetry which may be **radial** (parts arranged around a central axis), or **bilateral** (right and left halves are mirror images). Other differences include the type of body cavity (coelom) and such basic embryological differences as the number of

461

germ layers present in the embryo and the embryonic development of the digestive tract. Some animals have a saclike body form with only one opening into a digestive cavity. Others have two outer openings, a mouth and an anus, and the digestive tract forms essentially a "tube within a tube." Those animals that are bilaterally symmetrical (clade Bilateria) are divided into two major groups, depending on differences in early development and the origin of the mouth and the anus. An embryonic structure, the blastopore, develops into a mouth in the **protostomes** and into an anus in the **deuterostomes.**

In the study of the protists in Lab Topic 14, you learned that "traditional" phylogenetic trees have been challenged by the results of molecular studies, particularly evidence from analysis of DNA sequences of the gene coding for ribosomal RNA (rRNA). This is true not only for protists, but also for animals. Particularly in the protostomes, molecular studies have led to a regrouping of many traditionally established phylogenetic relationships. For example, for over 200 years zoology publications have assumed that annelids (segmented worms in the phylum Annelida) and arthropods (e.g., insects) are closely related based on their segmented bodies. Zoologists also noted, however, that annelids have developmental patterns similar to several groups that are not segmented. For example, annelids are like molluscs (e.g., clams) in having a developmental stage called the "trochophore larva." Recent molecular evidence helps to clarify this puzzle as it supports the hypothesis that annelids and molluscs are closely related, and separate from arthropods.

Molecular studies have led taxonomists to create two large clades within the protostomes, **Lophotrochozoa** and **Ecdysozoa.** Annelids, molluscs, and several more phyla not studied here are placed in the clade Lophotrochozoa. The name reflects the trochophore larvae found in annelids and molluscs. Also included in this clade are flatworms (phylum Platyhelminthes). Although flatworms lack such characteristics as a body cavity, the "tube-within-a-tube" body plan with mouth and anus, and elaborate internal organs, recent molecular evidence indicates that they should be grouped with annelids and molluscs in the Lophotrochozoa clade. Evidence from ribosomal DNA sequences indicates that roundworms or nematodes (phylum Nematoda), arthropods (phylum Arthropoda), and several other phyla belong in the clade Ecdysozoa. Animals in this clade undergo molting (ecdysis) or the shedding of an outer body cover. In nematodes this covering is called the **cuticle.** In arthropods the covering is the **exoskeleton.**

Another surprising result of rRNA and other molecular evidence is that the nature of the body cavity may not be a characteristic that indicates major phylogenetic branching. A true coelom or pseudocoelom may have been independently gained or lost many times in evolutionary history. In traditional phylogenetic groupings, flatworms and nematodes were considered primitive, neither group having a true coelom. Ribosomal evidence has now moved nematodes to a different position with arthropods in the metazoan tree.

Figure 18.1a is a tree diagram showing organisms studied in Lab Topics 18 and 19 classified in the *traditional* organization of animal phylogeny. This phylogeny is based on *morphology* and *development.* Figure 18.1b is a diagram showing the *new molecular-based* phylogeny. This phylogenetic tree represents a hierarchy of clades nested within larger clades. The smaller clades are more closely related to each other. The order of animals studied in Lab Topics 18 and 19 is based on Figure 18.1b, the molecular-based phylogeny. However, as you study the animals, note those morphological characteristics that were the basis of the traditional system of classification. These characteristics may give evidence of the influence of ecological events in the development of

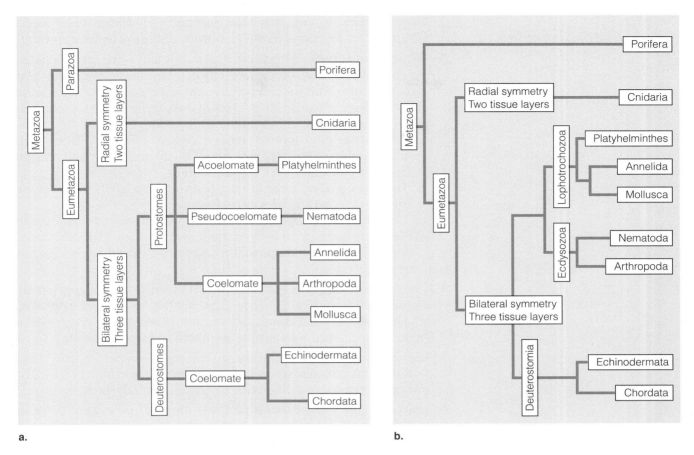

a.                                                                    b.

**Figure 18.1.**
**Phylogenetic organization of animals studied in Lab Topics 18 and 19.** (a) A traditional phylogeny based on morphology and embryology. (b) Proposed phylogeny based on new molecular evidence. Protostomes are assigned to two clades, Lophotrochozoa or Ecdysozoa.

different morphologies. Be ready to discuss how these similarities or differences may have arisen secondarily or through secondary simplifications.

Much work remains to resolve the branching order within the lophotrochozoan and ecdysozoan clades. Scientists are collecting evidence from studies based on mitochondrial DNA sequencing, ribosomal genes, *Hox* genes, and genes coding for various proteins.

The animals you will study in Lab Topic 18 are the sponge, hydra, planarian, clamworm, earthworm, and clam (mussel). In Lab Topic 19, you will study the roundworm, crayfish, grasshopper, sea star, amphioxus (lancelet), and pig. As you study each animal, relate your observations to the unifying themes of this lab: *phylogenetic relationships* and criteria that are the basis for animal classification, the *relationship between form and function,* and the *relationship of the environment and lifestyle to form and function.* The questions at the end of the lab topics will help you do this.

In your comparative study of these organisms, you will investigate 13 characteristics. Before you begin the dissections, become familiar with the following characteristics and their descriptions:

1. *Symmetry.* Is the animal (a) radially symmetrical (parts arranged around a central axis), (b) bilaterally symmetrical (right and left halves are mirror images), or (c) asymmetrical (no apparent symmetry)?

2. *Tissue organization.* Are cells organized into well-defined tissue layers (structural and functional units)? How many distinctive layers are present?

3. *Body cavity.* Is a body cavity present? A body cavity—the space between the gut and body wall—is present only in three-layered organisms, that is, in organisms with the embryonic germ layers ectoderm, mesoderm, and endoderm. There are three types of body forms related to the presence of a body cavity and its type (Figure 18.2).
   a. Acoelomate, three-layered bodies without a body cavity. Tissue from the mesoderm fills the space where a cavity might be; therefore, the tissue layers closely pack on one another.
   b. Pseudocoelomate, three-layered bodies with a cavity between the endoderm (gut) and mesoderm (muscle).
   c. Coelomate, three-layered bodies with the coelom, or cavity, *within* the mesoderm (completely surrounded by mesoderm). In coelomate organisms, mesodermal membranes suspend the gut within the body cavity.

4. *Openings into the digestive tract.* Can you detect where food enters the body and digestive waste exits the body? Some animals have only one opening, which serves as both a mouth and an anus. Others have a body called a "tube within a tube," with an anterior mouth and a posterior anus.

5. *Circulatory system.* Does this animal have open circulation (the blood flows through coelomic spaces in the tissue as well as in blood vessels), or does it have closed circulation (the blood flows entirely through vessels)?

6. *Habitat.* Is the animal terrestrial (lives on land) or aquatic (lives in water)? Aquatic animals may live in marine (sea) or fresh water.

7. *Organs for respiration (gas exchange).* Can you detect the surface where oxygen enters the body and carbon dioxide leaves the body? Many animals use their skin for respiration. Others have special organs, including gills in aquatic organisms and lungs in terrestrial organisms. Insects have a unique system for respiration, using structures called *spiracles* and *tracheae.*

8. *Organs for excretion.* How does the animal rid its body of nitrogenous waste? In many animals, these wastes pass out of the body through the skin by diffusion. In others, there are specialized structures, such as Malpighian tubules, lateral excretory canals, lateral canals with flame cells, structures called *nephridia,* and kidneys.

9. *Type of locomotion.* Does the organism swim, crawl on its belly, walk on legs, burrow in the substrate, or fly? Does it use cellular structures, such as cilia, to glide its body over the substrate?

10. *Support systems.* Is there a skeleton present? Is it an endoskeleton (inside the epidermis or skin of the animal), or is it an exoskeleton (outside the body wall)? Animals with no true skeleton can be supported by water: Fluid within and between cells and in body chambers such as a gastrovascular cavity or coelom provides a "hydrostatic skeleton."

11. *Segmentation.* Can you observe linear repetition of similar body parts? The repetition of similar units, or segments, is called *segmentation.* Segments can be more similar (as in the earthworm) or less similar (as in a lobster). Can you observe any degree of segmentation? Have various segments become modified for different functions?

12. *Appendages.* Are there appendages (organs or parts attached to a trunk or outer body wall)? Are these appendages all along the length of the

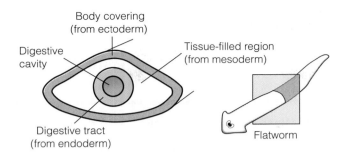

**a. Acoelomate**

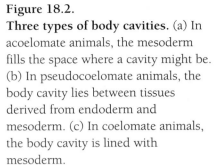

**Figure 18.2.**

**Three types of body cavities.** (a) In acoelomate animals, the mesoderm fills the space where a cavity might be. (b) In pseudocoelomate animals, the body cavity lies between tissues derived from endoderm and mesoderm. (c) In coelomate animals, the body cavity is lined with mesoderm.

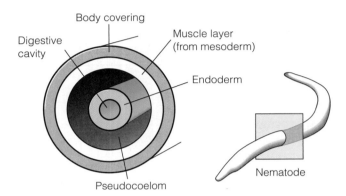

**b. Pseudocoelomate**

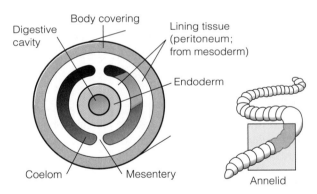

**c. Coelomate**

body, or are they restricted to one area? Are they all similar, or are they modified for different functions?

13. *Type of nervous system.* Do you see a brain and nerve cord? Is there more than one nerve cord? What is the location of the nerve cord(s)? Are sensory organs or structures present? Where and how many? Do you see signs of cephalization (the concentration of sensory equipment at the anterior end)? What purpose do such structures serve (for example, eyes for light detection)?

As you carefully study or dissect each organism, refer to these 13 characteristics, observe the animal, and record your observations in the summary table, Table 19.1, pp. 508–509. You may find it helpful to make sketches of difficult structures or dissections in the margin of your lab manual for future reference.

*Before you begin this study, read Appendix C and become thoroughly familiar with dissection techniques, orientation terms, and planes and sections of the body. Be able to use the terms associated with bilateral symmetry—anterior, posterior, dorsal, ventral, proximal, and distal—as you dissect and describe your animals.*

 **Wear gloves while dissecting preserved animals.**

EXERCISE 18.1
# Phylum Porifera—Sponges (*Scypha*)

## Materials

| | |
|---|---|
| dissecting needle | prepared slide of *Scypha* in |
| compound microscope | longitudinal section |
| stereoscopic microscope | preserved *Scypha* in watch glass |
| preserved and dry bath sponges | |

## Introduction

Within the animal kingdom, sponges are separated from all other animals because of their unique body form. You will observe the unique sponge structure by observing first a preserved specimen and then a prepared slide of a section taken through the longitudinal axis of the marine sponge *Scypha* (Color Plate 50). You will observe other more complex and diverse sponges on demonstration.

*Scypha is the form most often studied in America. A European genus, Grantia, is similar and can be substituted. You might substitute the small, common sponge Leucosolenia. It grows in a branching colony, whereas Scypha vases are unbranched. Also, the body form of Leucosolenia differs from that of Scypha.*

*Using living materials rather than preserved ones throughout both Animal Diversity lab topics will enhance the learning experiences of your students. Check with local businesses (bait suppliers, pet stores, farmers' markets) about the availability of living specimens.*

## Procedure

1. Obtain the preserved sponge *Scypha* and observe its external characteristics using the stereoscopic microscope, comparing your observations with Figure 18.3a.

   a. Note the vaselike shape of the sponge and the **osculum**, a large opening to the body at one end. The end opposite the osculum attaches the animal to the substrate.

   b. Note the invaginations in the body wall, which form numerous folds and channels. You may be able to observe needlelike **spicules** around the osculum and protruding from the surface of the body. These spicules are made of calcium carbonate: They give support and protection to the sponge body and prevent small animals from entering the sponge's internal cavity.

2. Using the compound microscope, examine a prepared slide of a sponge body in longitudinal section and compare it with Figure 18.3b.

   a. Again, locate the osculum. This structure is not a mouth, as its name implies, but an opening used as an outlet for the current of water passing through the body wall and the **central cavity**, or **spongocoel**. The water enters the central cavity from channels and pores in the body. The central cavity is not a digestive tube or body cavity, but is only a channel for water.

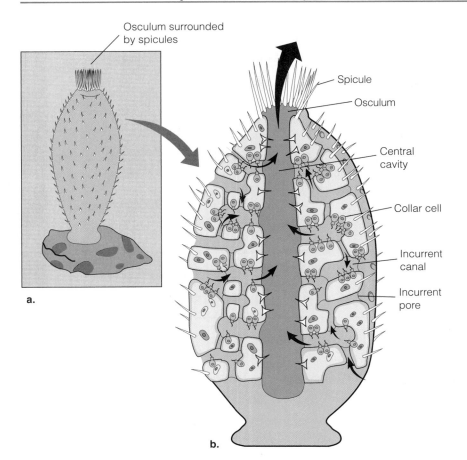

Osculum surrounded by spicules

Spicule

Osculum

Central cavity

Collar cell

Incurrent canal

Incurrent pore

a.

b.

**Figure 18.3.**
**The sponge *Scypha*.** (a) The entire sponge; (b) a longitudinal section through the sponge.

b.  Note the structure of the body wall. Are cells organized into definite tissue layers (well-defined structural and functional units), or are they best described as a loose organization of various cell types? Various cells in the body wall carry out the functions of digestion, contractility, secretion of the spicules, and reproduction (some cells develop into sperm and eggs). One cell type unique to sponges is the **choanocyte,** or **collar cell.** These cells line the central cavity and the channels leading into it. Each collar cell has a flagellum extending from its surface. The collective beating of all flagella moves water through the sponge body. Small food particles taken up and digested by collar cells are one major source of nutrition for the sponge. How would you hypothesize about the movement of oxygen and waste throughout the sponge body and into and out of cells?

*Oxygen and waste exchange takes place by diffusion directly between cells and the surrounding water.*

3.  Observe examples of more complex sponges on demonstration. (See Color Plate 51.) The body of these sponges, sometimes called "bath sponges," contains a complex series of large and small canals and chambers. The same cells that were described in *Scypha* are present in bath sponges, but, in addition to spicules, there is supportive material that consists of a soft proteinaceous substance called **spongin.** These

sponges often grow to fit the shape of the space where they live, and observing them gives you a good clue about the symmetry of the sponge body. How would you describe it?

*Students should conclude that the sponge has no symmetry, neither radial nor bilateral.*

### Results

Complete the summary table, Table 19.1, pp. 508–509, in Lab Topic 19, filling in all information for sponge characteristics in the appropriate row. This information will be used to answer questions in the Applying Your Knowledge section at the end of Lab Topic 19 Animal Diversity II.

---

## EXERCISE 18.2
# Phylum Cnidaria—Hydras (*Hydra*)

### Materials

| | |
|---|---|
| stereoscopic microscope | prepared slide of *Hydra* sections |
| compound microscope | watch glass |
| living *Hydra* culture | depression slide |
| water flea culture | pipettes and bulbs |
| dropper bottles of water, 1% acetic acid, and methylene blue | microscope slide and coverslip |

### Introduction

Cnidarians are a diverse group of organisms, all of which have a **tissue grade** of organization, meaning that tissues, but no complex organs, are present. Included in this group are corals, jellies, sea anemones, and Portuguese men-of-war. Most species are marine; however, there are a few freshwater species. Two body forms are present in the life cycle of many of these animals—an umbrella-like, free-swimming stage, and a cylindrical, attached or stationary form. The stationary forms often grow into colonies of individuals. In this exercise you will observe some of the unique features of this group by observing the solitary, freshwater organism *Hydra* (Color Plate 52).

### Procedure

*Alice Lindahl (Utah State U.) suggests using the colonial brackish water hydroid Cordylophora lacustris in place of hydra. This species is easy to maintain in culture. See the Preparation Guide.*

1. Place several drops of freshwater pond or culture water in a watch glass or depression slide. Use a dropper to obtain a living hydra from the class culture, and place the hydra in the drop of water. Using a stereoscopic microscope, observe the hydra structure and compare it with Figure 18.4a. Note any movement, the symmetry, and any body structures present. Note the **tentacles** that surround the "mouth," the only opening into the central cavity. Tentacles are used in capturing food and in performing a certain type of locomotion, much like a "handspring." To accomplish this motion,

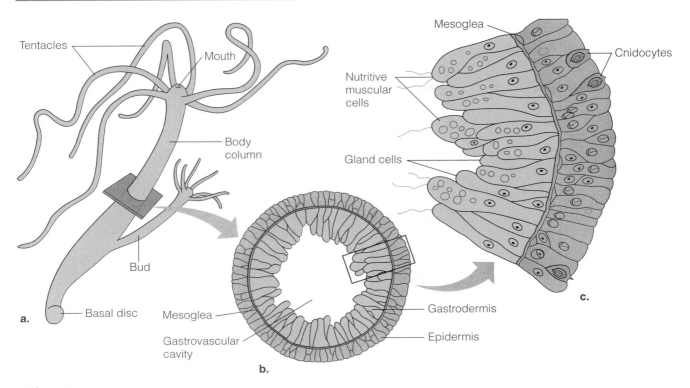

**Figure 18.4.**
*Hydra.* (a) A whole mount of *Hydra*; (b) enlargement showing a cross section through the body wall, revealing two tissue layers; and (c) further enlargement showing details of specialized cells in the body wall, including cnidocytes.

the hydra attaches its tentacles to the substrate and flips the basal portion of its body completely over, reattaching the base to a new position. If water fleas (*Daphnia*) are available, place one or two near the tentacles of the hydra and note the hydra's behavior. Set aside the hydra in the depression slide and return to it in a few moments.

2. Study a prepared slide of *Hydra* sections using the compound microscope and compare your observations with Figure 18.4b and 18.4c.

   Are definite tissue layers present? If so, how many?

   *Yes, two; mesoglea is not cellular.*

   Given what you know of embryology, what embryonic layers would you guess give rise to the tissue layers of this animal's body?

   *ectoderm and endoderm*

3. Not visible with the microscope is a network of nerve cells in the body wall, which serves as the nervous system. There is no concentration of nerve cells into any kind of brain or nerve cord.

4. Observe the central cavity, called a **gastrovascular cavity**. Digestion begins in this water-filled cavity (**extracellular digestion**), but many food particles are drawn into cells in the **gastrodermis** lining the cavity, where **intracellular digestion** occurs.

5. Do you see signs of a skeleton or supportive system? How do you think the body is supported? Are appendages present?

   *There is no supportive skeleton; the animal's body is supported by water in the tissues and in the gastrovascular cavity.*

   *If you define an appendage as any extension from the axial trunk, then the tentacles may be called appendages.*

6. Recalling the whole organism and observing this cross section, are organs for gas exchange present? How is gas exchange accomplished?

   *No organs exist for gas exchange, which takes place between cells and the surrounding water.*

7. Do you see any organs for excretion?

   *There are no special organs for excretion; waste diffuses directly out of cells into the surrounding water.*

8. Are specialized cell types seen in the layers of tissues?

   *a few*

   Cnidarians have a unique cell type called **cnidocytes,** which contain a stinging organelle called a **nematocyst.** When stimulated, the nematocyst will evert from the cnidocyte with explosive force, trapping food or stinging predators. Look for these cells.

9. To better observe cnidocytes and nematocysts, turn your attention again to your living hydra and follow this procedure:

   a. Using a pipette, transfer the hydra to a drop of water on a microscope slide and carefully add a coverslip.

   b. Use your microscope to examine the hydra, first on low, then intermediate, and finally on high powers, focusing primarily on the tentacles. The cnidocytes will appear as swellings. If your microscope is equipped with phase contrast, switch to phase. Alternatively, add a drop of methylene blue to the edge of the coverslip. Locate several cnidocytes with nematocysts coiled inside.

   c. Add a drop of 1% acetic acid to the edge of the coverslip and, watching carefully using intermediate power, observe the rapid discharge of the nematocyst from the cnidocyte.

   d. Using high power, study the discharged nematocysts that will appear as long threads, often with large spines, or barbs, at the base of the thread.

## Results

Complete the summary table, Table 19.1, recording all information for *Hydra* characteristics in the appropriate row. You will use this information to complete Table 19.2 and to answer questions in the Applying Your Knowledge section at the end of Lab Topic 19 Animal Diversity II.

## Discussion

What major differences have you detected between *Scypha* and *Hydra* body forms? List and describe them.

*Scypha—no symmetry, no tissue layers; Hydra —radial symmetry, two tissue layers*

 Student Media Videos—Ch. 33: *Hydra* Building; *Hydra* Eating *Daphnia*; Jelly Swimming; Thimble Jellies

## EXERCISE 18.3
# Phylum Platyhelminthes— Planarians (*Dugesia*)

## Materials

stereoscopic microscope
compound microscope
living planarian
watch glass

prepared slide of whole mount
    of planarian
prepared slide of planarian cross
    sections

## Introduction

The phylum Platyhelminthes (clade Lophotrochozoa) includes planarians, free-living flatworms; that is, they are not parasitic and their body is dorsoventrally flattened. They are found under rocks, leaves, and debris in freshwater ponds and creeks. They move over these surfaces using a combination of muscles in their body wall and cilia on their ventral sides (Color Plate 53).

## Procedure

1. Add a dropperful of pond or culture water to a watch glass. Use a dropper to obtain a living planarian from the class culture. Using your stereoscopic microscope, observe the planarian. Describe its locomotion. Is it directional? What is the position of its head? Does its body appear to contract?

    *Locomotion is directional with the head at the anterior end, slightly raised, usually turning from side to side. The body contracts in rhythmic muscular waves. The gliding motion may lead students to suggest that this comes about by the cilia on its ventral surface.*

As you observe the living planarian, you will see two striking new features with regard to symmetry that you did not see in the two phyla previously studied. What are they?

*(1) bilateral symmetry and (2) anterior/posterior axis with a head present (cephalization)*

*Keep a supply of liver in the freezer and dispense small pieces in watch glasses as needed.*

2. Add a *small* piece of fresh liver to the water near the planarian. The planarian may approach the liver and begin to feed by extending a long tubular **pharynx** out of the **mouth,** a circular opening on the ventral side of the body. If the planarian feeds, it will curve its body over the liver and extend the pharynx, which may be visible in the stereoscopic microscope.

   After observing the planarian's feeding behavior, return it to the culture dish, if possible, without the liver.

3. Using the lowest power on the compound microscope, observe the prepared slide of a whole planarian and compare it with Figure 18.5.

 Do not observe these slides using high power! The high power objective may crack the coverslip, resulting in damage to the lens.

Examine the body for possible digestive tract openings. How many openings to the digestive tract are present?

*one*

Observe again the pharynx and the mouth. The pharynx lies in a **pharyngeal chamber** inside the mouth. The proximal end of the pharynx opens into a dark-colored, branched intestine. If the intestine has been stained on your slide, you will see the branching more easily.

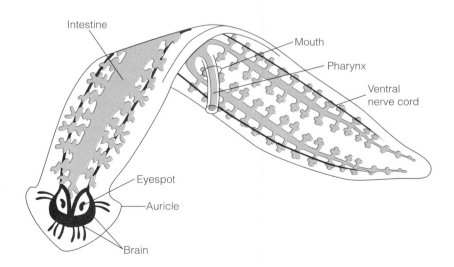

**Figure 18.5.**
**A planarian.** The digestive system consists of a mouth, a pharynx, and a branched intestine. A brain and two ventral nerve cords (plus transverse nerves connecting them, not shown) make up the nervous system.

4. Continue your study of the whole planarian. The anterior blunt end of the animal is the head end. At each side of the head is a projecting **auricle.** It contains a variety of sensory cells, chiefly of touch and chemical sense. Between the two auricles on the dorsal surface are two pigmented **eyespots.** These are pigment cups into which retinal cells extend from the brain, with the photosensitive end of the cells inside the cup. Eyespots are sensitive to light intensities and the direction of a light source but can form no images. Beneath the eyespots are two cerebral ganglia that serve as the **brain.** Two ventral nerve cords extend posteriorly from the brain. These are connected by transverse nerves to form a ladderlike **nervous system.**

5. Study the prepared slide of cross sections of a planarian. You will have several sections on one slide. One section should have been taken at the level of the pharynx and pharyngeal chamber. Do you see a body cavity in any of the sections? (The pharyngeal chamber and spaces in the gut are not a body cavity.) What word describes this body cavity condition (see Figure 18.2a)?

   *acoelomate; no body cavity in flatworms, although they may have had one in their evolutionary history*

   a. How many tissue layers can be detected? Speculate about their embryonic origin.

      *Three. The outer layer is the skin or epidermis derived from ectoderm. There is a layer surrounding the digestive tract derived from endoderm. The space between the two layers is filled with additional tissues from mesoderm.*

      Flatworms are the first group of animals to have three well-defined embryonic tissue layers, enabling them to have a variety of tissues and organs. Reproductive organs and excretory organs consisting of two lateral excretory canals and "flame cells" that move fluid through the canals are derived from the embryonic mesoderm. Respiratory, circulatory, and skeletal systems are lacking.

   b. How do you think the body is supported?

      *by surrounding water and the water in the body tissues*

   c. How does gas exchange take place?

      *by diffusion from water through tissues*

## Results

1. In the space below, diagram the flatworm as seen in a cross section at the level of the pharynx. Label the **epidermis, muscle** derived from **mesoderm**, the lining of the digestive tract derived from **endoderm**, the **pharynx**, and the **pharyngeal chamber**.

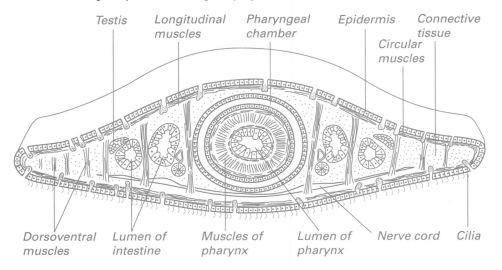

Testis    Longitudinal muscles    Pharyngeal chamber    Epidermis    Connective tissue    Circular muscles

Dorsoventral muscles    Lumen of intestine    Muscles of pharynx    Lumen of pharynx    Nerve cord    Cilia

2. Complete the summary table, Table 19.1, recording all information for planarian characteristics in the appropriate row. You will use this information to complete Table 19.2 and answer questions in the Applying Your Knowledge section at the end of Lab Topic 19 Animal Diversity II.

## Discussion

One of the major differences between Cnidaria and Platyhelminthes is radial versus bilateral symmetry. Discuss the advantage of radial symmetry for sessile (attached) animals and bilateral symmetry for motile animals.

*Radial symmetry allows an animal to interact with its surroundings equally in any direction (which is good if the animal is attached), while bilateral symmetry allows an animal to move head first, with all sensory organs directed into the new environment.*

### EXERCISE 18.4

# Phylum Annelida—Clamworms (*Nereis*) and Earthworms (*Lumbricus terrestris*)

The phylum Annelida (clade Lophotrochozoa) includes a diverse group of organisms inhabiting a variety of environments. Examples range in size from microscopic to several meters in length. Most species are marine, living free in the open ocean or burrowing in ocean bottoms. Others live in fresh water or in soils. One group of annelids, the leeches, are parasitic and live on the blood or tissues of their hosts. In this exercise, you will study the clamworm, a marine annelid, and the earthworm, a terrestrial species. Keep in mind features that are adaptations to marine and terrestrial habitats as you study these organisms.

# Lab Study A. Clamworms (*Nereis*)

## Materials

dissecting tools
dissecting pan
preserved clamworm

disposable gloves
dissecting pins

## Introduction

Species of *Nereis* (clamworms) are commonly found in mud flats and on the ocean floor. These animals burrow in sediments during the day and emerge to feed at night. As you observe the clamworm, note features that are characteristic of all annelids, as well as features that are special adaptations to the marine environment (Color Plate 54).

*You may choose to have students work in pairs for this exercise. One student can dissect Nereis while the other dissects the earthworm. Each then demonstrates their worm to their partner.*

## Procedure

1. Observe the preserved, undissected clamworm and compare it with Figure 18.6. How would you describe the symmetry of this organism?

   *bilateral symmetry*

2. Determine the anterior and posterior ends. At the anterior end, the well-differentiated head bears **sensory appendages.** Locate the mouth, which leads into the digestive tract.

3. A conspicuous new feature of these organisms is the presence of **segmentation,** the division of the body along its length into segments. Posterior to the head region, the segments bear fleshy outgrowths called **parapodia.** Each parapodium contains several terminal bristles called **setae.** In Lab Study B, you will see that the earthworm has setae but does not have parapodia. Suggest functions for parapodia and setae in the marine clamworm.

   *Both of these structures aid in locomotion in the clamworm. The parapodia are paddlelike extensions used for swimming. The setae aid in movement in the burrow. (Students will be asked to speculate about the role of parapodia in gas exchange later.)*

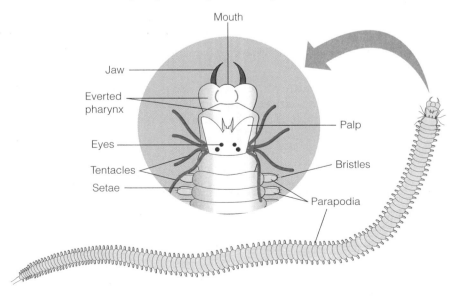

**Figure 18.6.**
**The clamworm, *Nereis*.** The head has sensory appendages, and each segment of the body bears two parapodia with setae.

4. Holding the animal in your hand and using sharp-pointed scissors, make a middorsal incision the full length of the body. Carefully insert the tip of the scissors and lift up with the tips as you cut. Pin the opened body in the dissecting pan but do not put pins through the head region.

5. Locate the **intestine.** Do you see the "tube-within-a-tube" body plan?

    *The tubular digestive tract extending the length of the tubular body appears to be a "tube within a tube."*

6. Two **muscle layers,** one inside the skin and a second lying on the surface of the intestine, may be visible with the stereoscopic microscope. With muscle in these two positions, what kind of coelom does this animal have (see Figure 18.2c)?

    *With muscle lying inside the epidermis (skin) and outside the endodermis (gut), this is a true coelom.*

7. Continuing your observations with the unaided eye and the stereoscopic microscope, look for **blood vessels,** particularly a large vessel lying on the dorsal wall of the digestive tract. This vessel is contractile and propels the blood throughout the body. You should be able to observe smaller lateral blood vessels connecting the dorsal blood vessel with another on the ventral side of the intestine. As you will see, in the earthworm these connecting vessels are slightly enlarged as "hearts" around the anterior portion of the digestive tract (around the esophagus). This is not as obvious in *Nereis.* What is this type of circulatory system, with blood circulating through continuous closed vessels?

    *closed circulation*

8. Gas exchange must take place across wet, thin surfaces. Do you see any organs for gas exchange (gills or lungs, for example)? How do you suspect that gas exchange takes place?

    *There are no lungs or gills. Gas exchange takes place through the skin, and this surface area is expanded in the parapodia.*

9. Do you see any signs of a skeleton? What would serve as support for the body?

    *The body is supported by water in a hydrostatic skeleton.*

10. Clamworms and earthworms have a small bilobed brain (a pair of ganglia) lying on the surface of the digestive tract at the anterior end of the worm. You can see this more easily in an earthworm.

## Lab Study B. Earthworms (*Lumbricus terrestris*)

### Materials

dissecting instruments
compound microscope
stereoscopic microscope

preserved earthworm
prepared slide of cross section
   of earthworm

## Introduction

*Lumbricus* species, commonly called *earthworms,* burrow through soils rich in organic matter. As you observe these animals, note features that are adaptations to the burrowing, terrestrial lifestyle.

## Procedure

1. Obtain a preserved earthworm and identify its anterior end by locating the mouth, which is overhung by a fleshy dorsal protuberance called the **prostomium.** The anus at the posterior end has no such protuberance. Also, a swollen glandular band, the **clitellum** (a structure that secretes a cocoon that holds eggs), is located closer to the mouth than to the anus (Figure 18.7).

   a. Using scissors, make a middorsal incision along the anterior third of the animal, as you did for *Nereis.* You can identify the dorsal surface in a couple of ways. The prostomium is dorsal, and the ventral surface of the worm is usually flattened, especially in the region of the clitellum. Cut to the prostomium. Pin the body open in a dissecting pan near the edge. You may need to cut through the septa that divide the body cavity into segments.

   b. Using a stereoscopic microscope or hand lens, look for the small **brain** just behind the prostomium on the surface of the digestive tract. Note the two nerves that pass from the brain around the pharynx and meet ventrally. These nerve tracts continue posteriorly as a **ventral nerve cord** lying in the floor of the coelom.

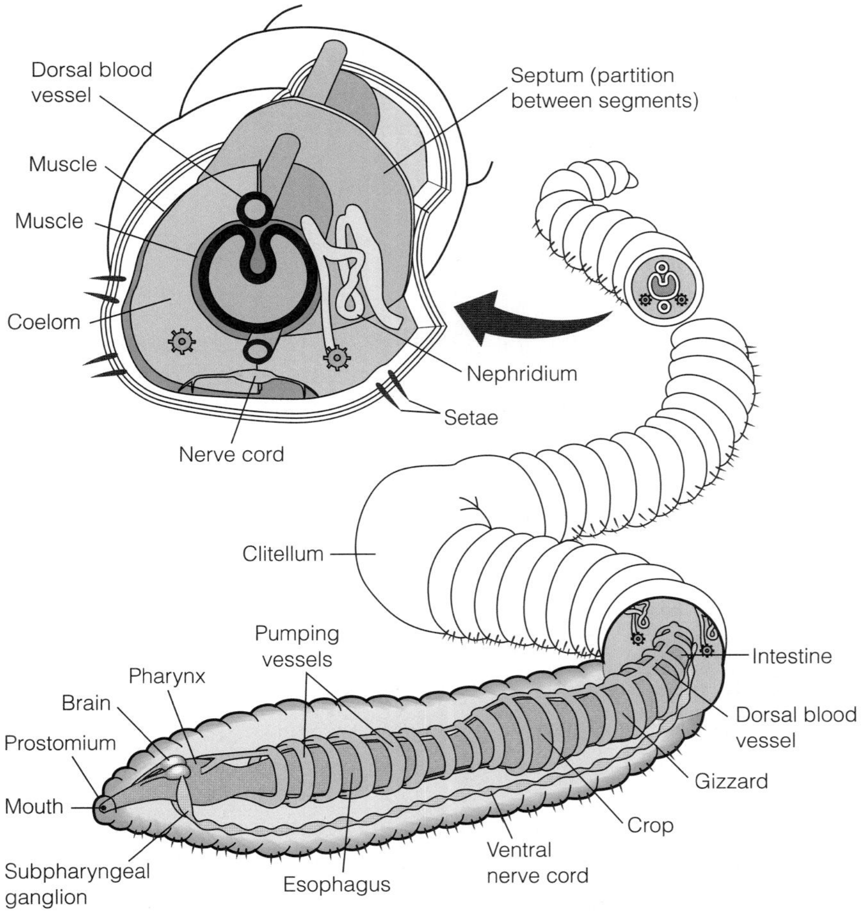

*You may choose to have students first observe behavior, such as locomotion, in a live earthworm, then narcotize the worm by placing it in a 5% ethanol solution for 20 to 30 minutes. Most systems, especially the circulatory system, are more easily studied in living worms.*

**Figure 18.7.**
**The earthworm.** The small brain leads to a ventral nerve cord. A pair of nephridia lie in each segment.

c. Look for the large **blood vessel** on the dorsal wall of the digestive tract. You may be able to see the enlarged lateral blood vessels (**hearts**) around the anterior portion of the digestive tract.

d. Identify (from anterior to posterior) the **pharynx, esophagus, crop** (a soft, swollen region of the digestive tract), **gizzard** (smaller and more rigid than the crop), and **intestine.**

e. Excretion in the clamworm and earthworm is carried out by organs called **nephridia.** A pair of these minute, white, coiled tubes is located in each segment of the worm body. Nephridia are more easily observed in the earthworm than in *Nereis* and should be studied here. To view these organs, cut out an approximately 2-cm-long piece of the worm posterior to the clitellum and cut it open along its dorsal surface. Cut through the septa and pin the piece to the dissecting pan near the edge to facilitate observation with the stereoscopic microscope. The coiled tubules of the nephridia are located in the coelomic cavity, where waste is collected and discharged to the outside through a small pore.

2. Using the compound microscope, observe the prepared slide of a cross section of the earthworm.

a. Locate the **thin cuticle** lying outside of and secreted by the **epidermis.** Recall the habitat of this organism and speculate about the function of the cuticle.

*The cuticle helps to prevent water loss in the terrestrial environment.*

b. Confirm your decision about the type of coelom by locating **muscle layers** inside the epidermis and also lying on the surface of the **intestine** near the body cavity.

*true coelom, or coelomic body cavity*

c. Locate the **ventral nerve cord,** lying in the floor of the coelom, just inside the muscle layer.

## Results

Complete the summary table, Table 19.1, recording all information for clamworm and earthworm characteristics in the appropriate row. You will use this information to complete Table 19.2 and answer questions in the Applying Your Knowledge section at the end of Lab Topic 19 Animal Diversity II.

    Student Media Video—Ch. 33: Earthworm Locomotion

## Discussion

A major new feature observed in the phylum Annelida is the segmented body. Speculate about possible adaptive advantages provided by segmentation.

*Locomotion is improved in earthworms, compared to nematodes, by dividing the hydrostatic skeleton into compartments, facilitating peristaltic contractions. Segmentation also permits the development of segments into parts that perform specific functions for the animal.*

## EXERCISE 18.5
# Phylum Mollusca—Clams

### Materials

dissecting instruments
dissecting pan

preserved clam or mussel
disposable gloves

### Introduction

Second only to the phylum Arthropoda in numbers of species, the phylum Mollusca (clade Lophotrochozoa) includes thousands of species living in many diverse habitats. Most species are marine. Others live in fresh water or on land. Many mollusks are of economic importance, being favorite human foods. Mollusks include such diverse animals as snails, slugs, clams, squids, and octopuses. Although appearing diverse, most of these animals share four characteristic features: (1) a hard external **shell** for protection; (2) a thin structure called the **mantle,** which secretes the shell; (3) a **visceral mass** in which most organs are located; and (4) a muscular **foot** used for locomotion.

*Consider using fresh materials available from grocery stores, fish markets, or farmers' markets.*

In this exercise, you will dissect a clam, a molluscan species with a shell made of two parts called **valves.** Most clams are marine, although many genera live in freshwater lakes and ponds (Color Plate 55).

*You can use freshwater mussels for this dissection. The descriptions will be the same.*

 **Wear gloves while dissecting preserved animals.**

### Procedure

1. Observe the external anatomy of the preserved clam. Certain characteristics will become obvious immediately. Can you determine symmetry, support systems, and the presence or absence of appendages? Are there external signs of segmentation? Record observations in Table 19.1.

2. Before you continue making observations, determine the dorsal, ventral, anterior, posterior, right, and left regions of the animal. Identify the two valves. The valves are held together by a **hinge** near the **umbo,** a hump on the valves. The hinge and the umbo are located **dorsally,** and the valves open **ventrally.** The umbo is displaced **anteriorly.** Hold the clam vertically with the umbo away from your body, and cup one of your hands over each valve. The valve in your right hand is the right valve; the valve in your left hand is the left valve. The two valves are held together by two strong **adductor** muscles inside the shell. Compare your observations with Figure 18.8.

 Be cautious as you open the clam! Hold the clam in the dissecting pan in such a way that the scalpel will be directed toward the bottom of the pan.

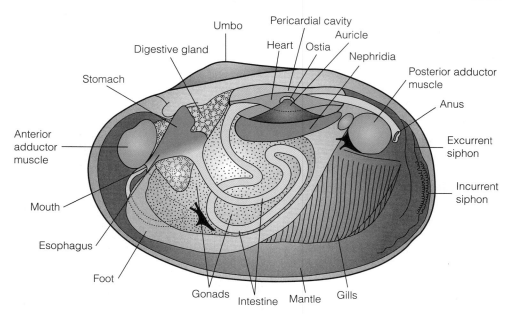

**Figure 18.8.**
**Anatomy of a clam.** The soft body parts are protected by the shell valves. Two adductor muscles hold the valves closed. Most major organs are located in the visceral mass. In this diagram, the left mantle, left pair of gills, and half of the visceral mass have been removed.

3. To study the internal anatomy of the clam, you must open it by prying open the valves. (A wooden peg may have been inserted between the two valves.) Insert the handle of your forceps or scalpel between the valves and twist it to pry the valves farther open. Place your clam in the dissecting pan with the clam's dorsal side supported on the pan bottom. This will allow you to make your cuts with the scalpel blade directed toward the pan bottom and not your hand. Carefully insert a scalpel blade, directed toward the dorsal side of the animal, into the space between the left valve and a flap of tissue lining the valve. The blade edge should be just ventral to (that is, below) the anterior adductor muscle (see Figure 18.8). The flap of tissue is the left **mantle.** Keeping the scalpel blade pressed flat against the left valve, carefully loosen the mantle from the valve and press the blade dorsally. You will feel the tough **anterior adductor muscle.** Cut through this muscle near the valve.

4. Repeat the procedure at the posterior end and cut the posterior adductor muscle. Lay the clam on its right valve and carefully lift the left valve. As you do this, use your scalpel to loosen the mantle from the valve. If you have been successful, you should have the body of the clam lying in the right valve. It should be covered by the mantle. Look for pearls between the mantle and the shell. How do you think pearls are formed?

*A grain of sand or a parasite becomes trapped between the mantle and the shell. The mantle secretes pearly layers around it.*

5. Look at the posterior end of the animal where the left and right mantle come together. Hold the two mantle flaps together and note the two gaps formed. These gaps are called **incurrent** (ventral) and **excurrent** (dorsal) **siphons.** Speculate about the function of these siphons.

*Water, food, and oxygen are carried into the animal through the incurrent siphon. Water and waste pass out of the animal through the excurrent siphon.*

6. Lift the mantle and identify the **visceral mass** and the **muscular foot.**

7. Locate the **gills,** which have a pleated appearance. One function of these structures is obvious, but they have a second function as well. As water comes into the body (how would it get in?), it passes through the gills, and food particles are trapped on the gill surface. The food is then moved anteriorly (toward the mouth) by coordinated ciliary movements.

8. Locate the **mouth** between two flaps of tissue just ventral to the anterior adductor muscle. Look just above the posterior adductor muscle and locate the **anus.** How is it oriented in relation to the excurrent siphon?

   *The anus is just dorsal to the excurrent siphon in such a position that waste can quickly exit the body.*

9. Imagine that this is the first time you have seen a clam. From the observations you have made, what evidence would indicate whether this animal is aquatic or terrestrial?

   *The presence of gills and a mantle forming siphons to move water supports the conclusion that the habitat of this organism is aquatic.*

10. The **heart** of the clam is located in a sinus, or cavity, just inside the hinge, dorsal to the visceral mass (see Figure 18.8). This cavity, called the *pericardial cavity,* is a reduced **true coelom.** The single ventricle of the heart actually surrounds the **intestine** passing through this cavity. Thin auricles, usually torn away during the dissection, empty into the heart via openings called **ostia.** Blood passes from **sinuses** in the body into the auricles. What type of circulatory system is this?

    *This is an open circulatory system: Blood does not circulate through continuous closed channels; blood vessels are not continuous.*

11. Ventral to the heart and embedded in mantle tissue are a pair of greenish brown tissue masses, the **nephridia,** or kidneys. The kidneys remove waste from the pericardial cavity.

12. Open the visceral mass by making an incision with the scalpel, dividing the mass into right and left halves. Begin this incision just above the foot and cut dorsally. You should be able to open the flap produced by this cut and see organs such as the **gonads, digestive gland, intestine,** and **stomach.** Clam chowder is made by chopping up the visceral mass.

13. It is difficult to observe the nervous system in the clam. It consists of three ganglia, one near the mouth, one in the foot, and one below the posterior adductor muscle. These ganglia are connected by nerves.

    Now that you have dissected the clam, you should have concluded that there is no sign of true segmentation. Also, appendages (attached to a trunk or body wall) are absent.

## Results

Complete Summary Table 19.1, recording all information for clam characteristics in the appropriate row. Use this information to complete Table 19.2 and to answer questions in the Applying Your Knowledge section at the end of Lab Topic 19 Animal Diversity II.

## Discussion

List several features of clam anatomy that enable it to survive in a marine environment.

*The heavy shell protects the animal from damaging waves on shores or water pressures in deep seas. The gills allow the animal to filter food and obtain oxygen from water. The muscular foot allows the animal to burrow into sand.*

By the end of today's laboratory period, you should have completed observations of all animals described in Animal Diversity I. The next lab topic, Animal Diversity II, is a continuation of this investigation and will present similar laboratory objectives. In Animal Diversity II, you will continue asking questions and making comparisons as you did in Animal Diversity I. By the end of the two lab topics, you should be able to use what you have learned about the animals to discuss and answer questions about the unifying themes of these laboratory topics.

# Reviewing Your Knowledge

1. What are the characteristics of sponges that have led scientists to classify them in a group separate from the Eumetazoa?

   *The body form of sponges is radically different from that of all other animals. Sponges have no true tissues.*

2. What characteristics are used to separate animals into the two groups, protostomes and deuterostomes?

   *These two groups are separated based on their developmental modes. Students may check their text for differences in cleavage patterns, coelom formation, and formation of the mouth and anus. The blastopore develops into a mouth in protostomes and into an anus in deuterostome.*

3. Molecular evidence has led taxonomists to group bilaterally symmetrical animals in one of three clades, Deuterostomia, Lophotrochozoa, or Ecdysozoa. Give one distinguishing characteristic of each of these groups other than molecular differences.

   *Deuterostomia:   blastopore becomes the anus*

   *Lophotrochozoa:   trochophore larvae in annelids and molluscs*

   *Ecdysozoa:   animals undergo molting, or the shedding of an outer body cover*

# Applying Your Knowledge

A hydra (*Chlorophyra viridissima*) is bright green, and yet it does not synthesize chlorophyll. Think about the structure of the hydra and its feeding and digestive habits. What do you think is the origin of the green pigment in this species?

*Gastrodermal cells in this hydra species bear green algae that give the hydra the green color. This is a case of symbiotic mutualism. The algae are held in vacuoles in the gastrodermal cells, but they are not digested. Students may speculate that gastrodermal cells engulf algae, but do not digest them. This will help them remember that digestion is both extracellular (in the gastrodermal cavity) and intracellular in many cnidarians.*

## Investigative Extensions

1. Earthworms are among the most familiar inhabitants of soil. They play an important role in improving the texture and adding organic matter to soil. You may have read Darwin's estimate that over 50,000 earthworms may inhabit one acre of British farmland. Earthworms are readily available from biological supply houses, or you may collect your own to use in experiments. Following are questions that you might investigate.

   a. Why do earthworms come out of their burrows when it rains? Is it because they may drown in the water in their burrows? Does rain stimulate mating behavior and are the worms coming to the surface to mate? Does the pH of the soil change as it rains, and is the burrow becoming too acidic or alkaline? What *is* the optimum pH range for earthworms? Does rain create conditions more favorable for migration to new habitats?

   b. What effects do chemicals used in agriculture have on earthworm populations? Compare numbers and health of earthworms in containers of soil to which varying amounts of fertilizers, pesticides, or herbicides have been added.

   c. Do earthworms in the soil stimulate plant growth? Compare the biomass of plants grown in containers with and without earthworms present.

2. An amazing diversity of organisms has evolved from the foot-mantle-visceral mass body plan of mollusks. Living terrestrial, freshwater, and marine snails and bivalves are available from biological supply houses, aquarium supply stores, or from ponds or terrestrial sites in your area. Consider the following questions that you might investigate. (A Google search for "snail experiments" yields over 1,140,000 entries, including experiments being performed in the International Space Station.)

   a. What effect does sedimentation have on aquatic snail populations? Consider changes in water chemistry and/or substrate.

   b. What effect does temperature have on the growth and/or reproduction of aquatic or terrestrial snails or slugs? Why would this question be of interest?

   c. Invasive aquatic plants have become a major concern of scientists worldwide. For example, water hyacinth, introduced into ponds in the southern U.S., chokes ponds and waterways, in some cases hindering human and fish navigation. Ponds may become so choked that they dry up, destroying habitat for alligators, turtles, fish, and other native species. Design a greenhouse experiment to test the efficacy of aquatic snails in controlling the growth of invasive aquatic plants.

## Student Media: BioFlix, Activities, Investigations, and Videos
www.masteringbiology.com (select Study Area)

**Activities**—Ch. 32: Animal Phylogenetic Tree;
   Ch. 33: Characteristics of Invertebrates
**Investigations**—Ch. 32: How Do Molecular Data Fit Traditional Phylogenies?
**Videos**—Ch. 33: *Hydra* Budding; *Hydra* Eating *Daphnia*; *Hydra* Releasing Sperm; Jelly Swimming; Thimble Jellies; Earthworm Locomotion; Discovery Videos:
   Ch. 33: Invertebrates

## References

Adoutte, A., G. Balavoine, N. Lartillot, O. Lespinet, B. Prud'homme, and R. de Rosa. "The New Animal Phylogeny: Reliability and Implications." *Proc. Natl. Acad. Sci.* USA, 2000, vol. 97(9), pp. 4453–4456.

Balavoine, G. "Are Platyhelminthes Coelomates Without a Coelom? An Argument Based on the Evolution of Hox Genes." *American Zoologist,* 1989, vol. 38, pp. 843–858.

Erwin, D., J. Valentine, and D. Jablonski. "The Origin of Animal Body Plans." *American Scientist,* 1997, vol. 85, pp. 126–137.

Mallatt, J. and C. Winchell. "Testing the New Animal Phylogeny: First Use of Combined Large-Subunit and Small-Subunit rRNA Sequences to Classify Protostomes." *Molecular Biology and Evolution,* 2002, vol. 19, pp. 289–301.

## Websites

Encyclopedia of Life. A new (2008) online reference source and database for all known species: http://www.eol.org/

The Tree of Life web project provides information on all major groups of organisms, including invertebrates: http://tolweb.org/Animals/2374

University of Michigan Museum of Zoology, Animal Diversity Web. Includes descriptions of many invertebrates and vertebrates, links to insect keys, references: http://animaldiversity.ummz.umich.edu/site/index.html

## LAB TOPICS 18 AND 19

# Animal Diversity I and II
## Teaching Plan for Laboratories

 See pp. 513–517 at the end of Lab Topic 19 for the integrated Teaching Plan for Lab Topics 18 and 19.

If students do not have time to complete the last exercise, they may continue this study at the beginning of lab next week.

The last 10 minutes of this lab period may be used for students to review the animals studied as they complete Table 19.1.

# Animal Diversity II: Nematoda, Arthropoda, Echinodermata, and Chordata

 This lab is a continuation of observations of organisms in the animal kingdom (clade Metazoa) as discussed in Animal Diversity I. Return to Lab Topic 18 and review the objectives of the lab on page 461. Review the descriptions of the 13 characteristics you are investigating in the study and dissection of these animals (pp. 463–465).

*For a 2-hour lab:* Omit the sea star and pig. See Teaching Plan.

In this lab topic you will study examples of two protostome phyla included in the clade **Ecdysozoa,** Nematoda (Exercise 19.1) and Arthropoda (Exercise 19.2). These organisms have coverings on their body surfaces (exoskeletons) that they shed as they grow, a process called *ecdysis*. In Exercises 19.3 and 19.4, you will study two deuterostome phyla, Echinodermata and Chordata.

As you continue your study of representative organisms, continue to record your observations in Table 19.1 at the end of this lab topic. Keep in mind the big themes you are investigating.

*In the Study Area of masteringbiology .com, see Student Media for suggested activities, investigations, and videos.*

1. What clues do similarities and differences among organisms provide about phylogenetic relationships?

2. How is body form related to function?

3. How is body form related to environment and lifestyle?

4. What characteristics can be the criteria for major branching points in producing a phylogenetic tree (representing animal classification)?

## EXERCISE 19.1
## Phylum Nematoda—
## Roundworms (*Ascaris*)

### Materials

dissecting instruments
dissecting pan
dissecting pins
compound microscope
disposable gloves

preserved *Ascaris*
prepared slide of cross section
  of *Ascaris*
hand lens (optional)

*Remind students to refer to Appendix C if they need to check definitions of orientation terms.*

## Introduction

**Roundworms,** or nematodes (clade Ecdysozoa), are among the most abundant and resilient organisms on earth. One species of roundworm, *Caenorhabditis elegans,* has become a model research organism in biology. For example, in 2006, the Nobel Prize in Medicine was awarded to two scientists for their studies of gene regulation using this soil worm as their research organism. NASA recently used *C. elegans* in experiments to test the way weightlessness and space radiation affect an organism's genes. *C. elegans* is a small roundworm—only 1 millimeter in length. *Ascaris,* the roundworm you will study in this exercise, is considerably larger.

Recall that ecdysozoans secrete exoskeletons that must be shed as the animal grows. Nematodes are covered with a proteinaceous **cuticle** that sheds periodically. *Ascaris* lives as a parasite in the intestines of mammals such as horses, pigs, and humans (Color Plate 56). Most often these parasites are introduced into the mammalian body when food contaminated with nematode eggs is eaten. Keep in mind the problem of adaptation to a parasitic lifestyle as you study the structure of this animal.

 **Wear disposable gloves while dissecting preserved animals.**

## Procedure

1. Wearing disposable gloves, obtain a preserved *Ascaris* and determine its sex. Females are generally larger than males. The posterior end of the male is sharply curved.

2. Use a hand lens or a stereoscopic microscope to look at the ends of the worm. A mouth is present at the anterior end. Three "lips" border this opening. A small slitlike **anus** is located ventrally near the posterior end of the animal.

3. Open the animal by making a middorsal incision along the length of the body with a sharp-pointed probe or sharp scissors. Remember that the anus is slightly to the ventral side (Figure 19.1). Be careful not to go too deep. Once the animal is open, pin the free edges of the body wall to the dissecting pan, spreading open the body. Pinning the animal near the edge of the pan will allow you to view it using the stereoscopic microscope. As you study the internal organs, you will note that there is a **body cavity.** This is not a true coelom, however, as you will see shortly when you study microscopic sections. From your observations, you should readily identify such characteristics as symmetry, tissue organization, and digestive tract openings.

   a. The most obvious organs you will see in the dissected worm are **reproductive organs,** which appear as masses of coiled tubules of varying diameters.

   b. Identify the flattened **digestive tract,** or intestine, extending from mouth to anus. This tract has been described as a "tube within a tube," the outer tube being the body wall.

   c. Locate two pale lines running laterally along the length of the body in the body wall. The excretory system consists of two longitudinal tubes lying in these two **lateral lines.**

   d. There are no organs for gas exchange or circulation. Most parasitic roundworms are essentially anaerobic (require no oxygen).

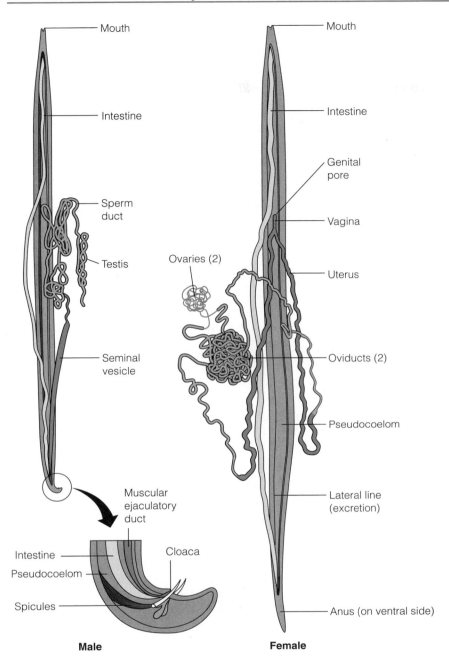

**Figure 19.1.**
**Male and female *Ascaris*.** The digestive tract originates at the mouth and terminates in the anus. Reproductive structures fill the body cavity.

e. How would nourishment be taken into the body and be circulated?

   *Semidigested food is sucked into the mouth, is further digested and absorbed in the intestine, and passes into the pseudocoelom and to all tissues.*

f. The nervous system consists of a ring of nervous tissue around the anterior end of the worm, with one dorsal and one ventral nerve cord. These structures will be more easily observed in the prepared slide.

g. Do you see signs of segmentation in the body wall or in the digestive, reproductive, or excretory systems?

   *No. The body wall is smooth and continuous, and the systems show no signs of segmentation.*

h. Do you see signs of a support system? What do you think supports the body?

*The body is supported by the cuticle and a hydrostatic skeleton.*

4. Using the compound microscope, observe a prepared slide of a cross section through the body of a female worm. Note that the body wall is made up of (from outside inward) **cuticle** (noncellular), **epidermis** (cellular), and **muscle fibers.** The muscle (derived from mesoderm) lies at the outer boundary of the body cavity. Locate the **intestine** (derived from endoderm). Can you detect muscle tissue adjacent to the endodermal layer?

*There is no muscle here, which is characteristic of pseudocoelomates.*

What do we call a coelom that is lined by mesoderm (outside) and endoderm (inside) (see Figure 18.2b)?

*a pseudocoelom*

5. Most of the body cavity is filled with reproductive organs. You should see cross sections of the two large **uteri,** sections of the coiled **oviducts** with small lumens, and many sections of the **ovaries** with no lumen. What do you see inside the uteri?

*eggs*

6. By carefully observing the cross section, you should be able to locate the **lateral lines** for excretion and the dorsal and ventral **nerve cords.**

### Results

1. Sketch the cross section of a female *Ascaris.* Label the **cuticle, epidermis, muscle fibers, intestine, body cavity** (give specific name), **reproductive organs (uterus, oviduct, ovary), lateral lines,** and **dorsal** and **ventral nerve cords.**

*Illustration from Charles F. Lytle and J. E. Wodsedalek,* General Zoology Laboratory Guide, *Complete version 9e (Dubuque, IA: Wm C. Brown, 1987) © 1987 Wm C. Brown Communications, Inc. Reproduced by permission of the McGraw-Hill Companies.*

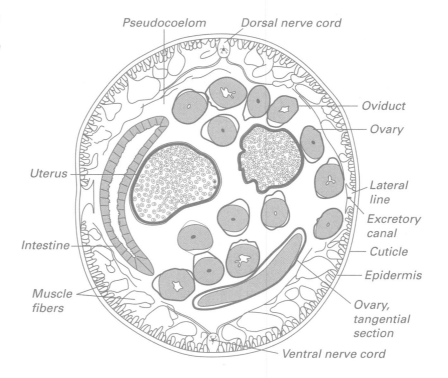

Pseudocoelom · Dorsal nerve cord · Oviduct · Ovary · Uterus · Lateral line · Excretory canal · Cuticle · Epidermis · Intestine · Ovary, tangential section · Muscle fibers · Ventral nerve cord

2.  List some features of *Ascaris* that are possible adaptations to parasitic life.

    *The cuticle protects it from the host's digestive enzymes; sucking lips at mouth to suck up digested food from the host; extensive reproductive system to increase the chances that the offspring will be passed on to another host*

3.  Complete the summary table, Table 19.1, recording all information for roundworm characteristics in the appropriate row. You will use this information to complete Table 19.2 and answer questions in the Applying Your Knowledge section at the end of this lab topic.

 Student Media Video—Ch. 33: *C. elegans* Crawling

## Discussion

1.  Discuss the significance of an animal's having two separate openings to the digestive tract, as seen in *Ascaris*.

    *Students will probably suggest the obvious advantage that food and waste do not mix in the digestive tract. However, point out that from the functional viewpoint, this design allows specialized regions to develop along the length of the gut.*

2.  What are the advantages of a body cavity being present in an animal?

    *A cavity provides space where organs and organ systems can develop. Fluid in the cavity can be used to collect waste and circulate nutrients and may give hydrostatic support.*

# EXERCISE 19.2
# Phylum Arthropoda

Organisms in the phylum Arthropoda (clade Ecdysozoa) have been very successful species. Evidence indicates that arthropods may have lived on Earth half a billion years ago. They can be found in almost every imaginable habitat: marine waters, fresh water, and almost every terrestrial niche. Many species are directly beneficial to humans, serving as a source of food. Others make humans miserable by eating their homes, infesting their domestic animals, eating their food, and biting their bodies. These organisms have an exoskeleton that periodically sheds as they grow. In this exercise, you will observe the morphology of two arthropods: the crayfish (an aquatic arthropod) and the grasshopper (a terrestrial arthropod).

## Lab Study A. Crayfish (*Cambarus*)

### Materials

dissecting instruments
dissecting pan

preserved crayfish
disposable gloves

*Consider using freshly killed crayfish available in some farmers' markets.*

## Introduction

Crayfish live in streams, ponds, and swamps, usually protected under rocks and vegetation. They may walk slowly over the substrate of their habitat, but they can also swim rapidly using their tails. The segmentation seen in annelids is seen also in crayfish and all arthropods; however, you will see that the segments are grouped into functional units (Color Plate 57).

## Procedure

1. Obtain a preserved crayfish, study its external anatomy, and compare your observations with Figure 19.2. Describe the body symmetry, supportive structures, appendages, and segmentation, and state the adaptive advantages of each characteristic.

   a. body symmetry

   *bilaterally symmetrical—promotes cephalization, directed locomotion*

   b. supportive structures

   *exoskeleton—protects soft body parts*

   c. appendages

   *Appendages are segmented, allowing more complex limb movements, and specialized, promoting greater diversity of function.*

   d. segmentation

   *The segmented body promotes greater flexibility and more complex body movements.*

2. Identify the three regions of the crayfish body: the **head, thorax** (fused with the head), and **abdomen.** Note the appendages associated with each region. Speculate about the functions of each of these groups of appendages.

   a. head appendages

   *moving food into the mouth, sensory*

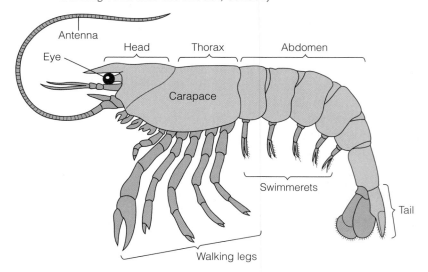

**Figure 19.2.**
**External anatomy of a crayfish.** The body is divided into head, thorax, and abdominal regions. Appendages grouped in a region perform specific functions.

b.  thoracic appendages

*walking, defense, obtaining food*

c.  abdominal appendages

*swimming, reproduction*

3.  Feathery **gills** lie under the lateral extensions of a large, expanded exoskeletal plate called the **carapace** (see Figure 19.2). To expose the gills, use scissors to cut away a portion of the plate on the left side of the animal. What is the function of the gills? Speculate about how this function is performed.

*Gas exchange. Water flowing under the carapace flows through the gills; oxygen leaves the water and enters the gills, while carbon dioxide passes into the water and out of the body.*

4.  Remove the dorsal portion of the carapace to observe other organs in the head and thorax. Compare your observations with Figure 19.3.

a.  Start on each side of the body at the posterior lateral edge of the carapace and make two lateral cuts extending along each side of the thorax and forward over the head, meeting just behind the eyes. This should create a dorsal flap in the carapace.

b.  Carefully insert a needle under this flap and separate the underlying tissues as you lift the flap.

c.  Observe the **heart,** a small, angular structure located just under the carapace near the posterior portion of the thorax. (If you were not successful in leaving the tissues behind as you removed the carapace, you may have removed the heart with the carapace.) Thin

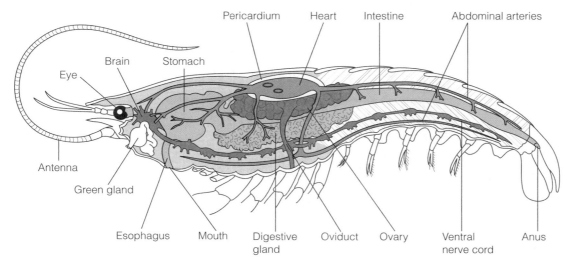

**Figure 19.3.**
**Internal anatomy of the crayfish.** Large digestive glands fill much of the body cavity. The intestine extends from the stomach through the tail to the anus. The green glands lie near the brain in the head.

*Have both plain and injected crayfish available. The heart will be destroyed in the injected animals, but arteries are more obvious. Have students view the heart in the plain crayfish.*

threads leading out from the heart are **arteries.** Look for holes in the heart wall. When blood collects in **sinuses** around the heart, the heart relaxes, and these holes open to allow the heart to fill with blood. The holes then close, and the blood is pumped through the arteries, which distribute it around the body. Blood seeps back to the heart, since no veins are present. What is the name given to this kind of circulation?

*open circulation*

d. Locate the **stomach** in the head region. It is a large, saclike structure. It may be obscured by the large, white **digestive glands** that fill the body cavity inside the body wall. Leading posteriorly from the stomach is the **intestine.** Make longitudinal cuts through the exoskeleton on either side of the dorsal midline of the abdomen. Lift the exoskeleton and trace the intestine to the anus. (When shrimp are "deveined" in preparation for eating, the intestine is removed.) Given all of the organs and tissues around the digestive tract and inside the body wall in the body cavity, what kind of coelom do you think this animal has?

*a true coelom, or eucoelom*

e. Turn your attention to the anterior end of the specimen again. Pull the stomach posteriorly (this will tear the esophagus) and look inside the most anterior portion of the head. Two **green glands** (they do not look green), the animal's excretory organs, are located in this region. These are actually long tubular structures that resemble nephridia but are compacted into a glandular mass. Waste and excess water pass from these glands to the outside of the body through pores at the base of the antennae on the head.

f. Observe the **brain** just anterior to the green glands. It lies in the midline with nerves extending posteriorly, fusing to form a **ventral nerve cord.**

 Student Media Video—Ch. 33: Lobster Mouth Parts

## Results

Complete Table 19.1, recording all information for crayfish characteristics in the appropriate row. Use this information to complete Table 19.2 and answer questions in the Applying Your Knowledge section at the end of this lab topic.

## Discussion

How does the pattern of segmentation differ in the crayfish and the earthworm studied in Lab Topic 18?

*In the earthworm, the segments are uniform, but in the crayfish, segments are grouped into regions that have different functions.*

# Lab Study B. Grasshoppers (*Romalea*)

## Materials

dissecting instruments

dissecting pan

preserved grasshopper

disposable gloves

*Consider using freshly killed grasshoppers or crickets. These can be dissected under water for a better view of the trachea.*

## Introduction

The grasshopper, an insect, is an example of a terrestrial arthropod (Color Plate 58). Insects are the most successful and abundant of all land animals. They are the principal invertebrates in dry environments, and they can survive extreme temperatures. They are the only invertebrates that can fly. As you study the grasshopper, compare the anatomy of this terrestrial animal with that of the aquatic crayfish, just studied. This comparison should suggest ways that terrestrial animals have solved the problems of life out of water.

## Procedure

1. Observe the external anatomy of the grasshopper. Compare your observations with Figure 19.4.

   a. Note the symmetry, supportive structures, appendages, and segmentation of the grasshopper.

   b. Observe the body parts. The body is divided into three regions: the **head,** the **thorax** (to which the legs and wings are attached), and the **abdomen.** Examine the appendages on the head, speculate about their functions, and locate the mouth opening into the digestive tract.

   c. Turning your attention to the abdomen, locate small dots along each side. These dots are **spiracles,** small openings into elastic air tubes, or **tracheae,** that branch to all parts of the body and constitute the respiratory system of the grasshopper. This system of tubes brings oxygen directly to the cells of the body.

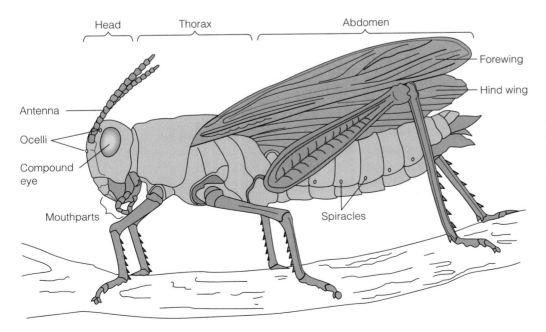

Head    Thorax    Abdomen

Antenna

Ocelli

Compound eye

Mouthparts

Forewing

Hind wing

Spiracles

**Figure 19.4. External anatomy of the grasshopper.** The body is divided into head, thorax, and abdominal regions. Wings and large legs are present. Small openings, called *spiracles,* lead to internal tracheae, allowing air to pass into the body.

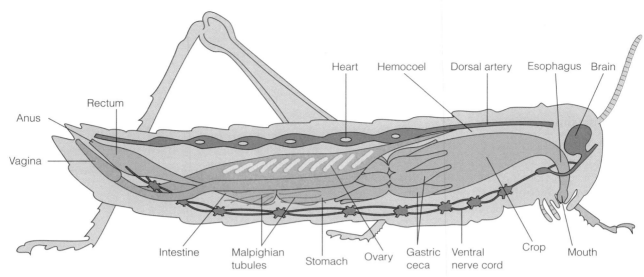

**Figure 19.5.**
**Internal anatomy of the grasshopper.** The digestive tract, extending from mouth to anus, is divided into specialized regions: the esophagus, crop, stomach, intestine, and rectum. Gastric ceca attach at the junction of the crop and the stomach. Malpighian tubules empty excretory waste into the anterior end of the intestine.

2. Remove the exoskeleton. First take off the wings and, starting at the posterior end, use scissors to make two lateral cuts toward the head. Remove the dorsal wall of the exoskeleton and note the segmented pattern in the muscles inside the body wall. Compare your observations with Figure 19.5 as you work.

   a. A space between the body wall and the digestive tract, the **hemocoel** (a true coelom), in life is filled with colorless blood. What type of circulation does the grasshopper have?

      *open*

      The heart of a grasshopper is an elongate, tubular structure lying just inside the middorsal body wall. This probably will not be visible.

   b. Locate the digestive tract and again note the mouth. Along the length of the tract are regions specialized for specific functions. A narrow **esophagus** leading from the mouth expands into a large **crop** used for food storage. The crop empties into the **stomach,** where digestion takes place. Six pairs of fingerlike extensions called **gastric pouches** or **ceca** connect to the digestive tract where the crop and the stomach meet. These pouches secrete digestive enzymes and aid in food absorption. Food passes from the stomach into the **intestine,** then into the **rectum,** and out the **anus.** Distinguish these regions by observing constrictions and swellings along the tube. There is usually a constriction between the stomach and the intestine where the Malpighian tubules (discussed below) attach. The intestine is shorter and usually smaller in diameter than the stomach. *The intestine expands into an enlarged rectum that absorbs excess water from any undigested food, and relatively dry excrement passes out the anus.*

c.  The excretory system is made up of numerous tiny tubules, the **Malpighian tubules,** which empty their products into the anterior end of the intestine. These tubules remove wastes and salts from the blood. Locate these tubules.

d.  Push aside the digestive tract and locate the **ventral nerve cord** lying medially inside the ventral body wall. Ganglia are expanded regions of the ventral nerve cord found in each body segment. Following the nerve cord anteriorly, note that branches from the nerve cord pass around the digestive tract and meet, forming a brain in the head.

## Results

Complete Table 19.1, recording all information for grasshopper characteristics in the appropriate row. Use this information to complete Table 19.2 and answer questions in the Applying Your Knowledge section at the end of this lab topic.

## Discussion

1.  Describe how each of the following external structures helps the grasshopper live successfully in terrestrial environments.

    a.  Exoskeleton

    *prevents desiccation and provides protection to soft tissues*

    b.  Wings

    *allow the insect to extend its range to find favorable habitats and food, and suitable mates*

    c.  Large, jointed legs

    *allow the insect to make rapid, precise movements*

    d.  Spiracles

    *allow air into the body, facilitating internal gas exchange*

2.  Describe how each of the following internal structures helps the grasshopper live successfully in terrestrial environments.

    a.  Tracheae

    *Connected to spiracles, these carry oxygen to all body cells. This system provides an efficient means of respiration with minimum water loss.*

    b.  Malpighian tubules

    *carry waste and water from the coelom into the digestive tract*

    c.  Rectum

    *reabsorbs into the coelom most water in the digestive tract before it exits the body*

EXERCISE 19.3

# Deuterostome—Phylum Echinodermata—Sea Star

*Because of time constraints, we have added the study of the sea star as a demonstration only. If your laboratory schedule allows it, have each student dissect a sea star. Provide more details on external and internal anatomy. Check a zoology text.*

Echinodermata is one of three phyla in the group of animals called deuterostomes. You will study another deuterostome phylum in Exercise 19.4, phylum Chordata. Examples of echinoderms include the sea star, sea urchin, sea cucumber, and sea lily. Some of the most familiar animals in the animal kingdom are in the phylum Chordata—fish, reptiles, amphibians, and mammals. Take a look at a sea star (starfish) in the salt-water aquarium in your lab or in a tidal pool on a rocky shore. What are the most obvious characteristics of this animal? Then imagine a chordate—a fish, dog, or even yourself. You might question why these two phyla are considered closely related phylogenetically. The most obvious difference is a very basic characteristic—the sea star has radial symmetry and most chordates that you imagine have bilateral symmetry. The sea star has no head or other obvious chordate features and it crawls around using hundreds of small suction cups called tube feet. Most chordates show strong cephalization and move using appendages. Your conclusion from the superficial observations might be that these two phyla are not closely related. Your observations are a good example of the difficulty faced by taxonomists when comparing animals based only on the morphology of adults. Taxonomists must collect data from studies of developmental and—as we discovered with the protostomes—molecular similarities before coming to final conclusions.

In this and the following exercise, you will examine an echinoderm, the adult sea star (demonstration only), and two chordates, asking questions about their morphology and adaptation to their habitats. You may not be convinced of their phylogentic relationships, however, until you complete Lab Topic 25, Animal Development, when you will study early development in sea urchins and sea stars. In that lab topic, you will see that chordates and echinoderms have similar early embryonic developmental patterns, including the formation of the mouth and anus and the type of cleavage.

## Materials

whole preserved sea stars on demonstration
several dissected sea stars on demonstration showing the internal contents of the body and the inside surface of oral and aboral halves of the body

## Introduction

The sea star (also called starfish) is classified in the phylum Echinodermata. They are marine animals with an endoskeleton of small, spiny calcareous plates bound together by connective tissue. Their symmetry is radial pentamerous (five-parted). They have no head or brain and few sensory structures. All animals in this phylum have a unique **water-vascular system** that develops from mesoderm and consists of a series of canals carrying water that enters the body through an outer opening, the **madreporite.** The canals are located inside the body and include a ring around the central disk of the body and tubes or canals that extend out into each arm. The canals then terminate in many small structures called **tube feet** visible along the

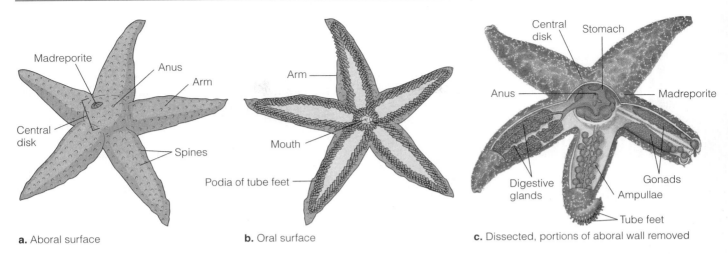

**a.** Aboral surface                    **b.** Oral surface                    **c.** Dissected, portions of aboral wall removed

**Figure 19.6.**
**Anatomy of a Sea Star** (a) Aboral surface. (b) Oral surface. (c) Dissected sea star with portions of the aboral wall removed.

groove on the oral side of each arm. Each tube foot ends blindly in an attachment disk (podium) that extends to the outside of the body. A small canal connects each podium to a small spherical sac (ampulla) inside the body. Using a combination of muscle contraction and adhesive chemicals, a sea star can extend and attach the feet to hard surfaces such as the surface of a clamshell, or rocks on the ocean shore (Color Plate 59).

## Procedure

1. Observe the preserved sea star on demonstration. Locate the **aboral** surface—the "upper" surface away from the mouth (Figure 19.6a). The downside is the **oral** surface where the mouth is located (Figure 19.6b).

2. Count the number of arms that extend out from the **central disk.** Echinoderms are usually pentamerous, meaning that their arms are in multiples of five. Occasionally a sea star with six arms will be found. Arms that are damaged or lost can be regenerated, and an extra arm may regenerate.

3. Observe the animal's aboral surface (Figure 19.6a). Locate the **madreporite,** a small porous plate displaced to one side of the central disk that serves to take water into the water vascular system.

   Notice that the surface of the animal's body is spiny. The spines project from calcareous plates of the **endoskeleton.** The endoskeleton is derived from the embryonic germ layer mesoderm. In life, the entire surface of the body is covered with an **epidermis** derived from ectoderm that may not be visible with the naked eye.

4. Observe the animal's oral surface (Figure 19.6b). The **mouth,** surrounded by spines, is located in the central disk with grooves extending from the mouth out into each arm. **Tube feet** lie along these grooves.

5. Observe the dissected sea star on demonstration (Figure 19.6c). In this dissection the entire aboral surface has been lifted off the body and placed to the side, inside up. This exposes the internal organs. The endoskeleton and its calcareous plates are obvious as viewed from the inside of the body.

6. Inside the body the organs are located in a **true body cavity.** Small delicate projections of the body cavity protrude between the plates of the endoskeleton to the outside of the body. These projections, covered with epidermis, are called **skin gills** or dermal branchiae, and function in the exchange of oxygen and carbon dioxide with the water bathing

*Although Figure 19.6c does not show the inside of the aboral surface, show this in the demonstration dissected animal. Position the aboral portion of the body wall with the inside of the body wall up in the dissecting pan (step 5).*

the animal's body. In addition, nitrogenous waste passes through these skin gills into the surrounding water; these structures thus have both respiratory and excretory functions.

7. The central disk contains the **stomach,** a portion of which can be everted through the mouth on the oral side of the animal. A small anus is located on the aboral body surface, although very little fecal material is ejected here. Most digestion takes place in the stomach, which may be everted into the body of a clam. The digested broth is then sucked up into the sea star body. After feeding, the sea star draws in its stomach by contracting its stomach muscles.

8. Conspicuous organs in the arms of the animal are **gonads** and **digestive glands.** Observe the **ampullae** lying along the grooves extending from the central disk into each arm. Each ampulla connects with a podium of a tube foot. Other systems cannot be easily observed in this preparation. A reduced circulatory system (hemal system) exists, but its function is not well defined. It consists of tissue strands and unlined sinuses. The nervous system includes a nerve ring around the mouth and radial nerves with epidermal nerve networks. There is no central nervous system.

## Results

Complete Table 19.1, recording in the appropriate row all information you have been able to observe. Use this information to complete Table 19.2 and answer questions in the Applying Your Knowledge section at the end of this lab topic.

 **Student Media Video—Ch. 33: Echinoderm Tube Feet**

## Discussion

1. Imagine that you are a zoologist studying sea stars for the first time. What characteristics would you note from the dissection of an adult animal that might give a clue to its phylogenetic relationships—that it belongs with deuterostomes rather than protostomes?

   *There is very little evidence in the adult sea star. One clue is the endoskeleton, similar to chordates. Most protostomes have either no skeleton or an exoskeleton.*

2. What structures have you observed that appear to be unique to echinoderms?

   *The water vascular system is unique.*

3. How would you continue your study to obtain more information that might help in classifying these animals?

   *A study of early development would reveal that these animals show a primitive deuterostome pattern of development. Gastrulation is by invagination. Cleavage, coelom formation, and the formation of the mouth and anus are similar to other deuterostomes. The bipinnaria larva is bilaterally symmetrical and radial symmetry develops secondarily. Point out to students that they will learn more about early development in echinoderms in Lab Topic 25. Students may also suggest molecular studies.*

4.  Given the fact that other deuterostomes are bilaterally symmetrical, what is one explanation for the radial symmetry of most adult echinoderms?

*Radial symmetry probably evolved secondarily in response to their relatively sessile adult lifestyle.*

## EXERCISE 19.4
# Deuterostome—Phylum Chordata

Up to this point, all the animals you have studied in Lab Topics 18 and 19 are commonly called **invertebrates,** a somewhat artificial designation based on the absence of a backbone. The phylum Chordata studied in this exercise includes two subphyla of invertebrates and a third subphylum, Craniata, that includes those animals called **vertebrates.** Animals in this subphylum have an endoskeleton of cartilage or bone and a head and skull. Craniates with a vertebral column are called vertebrates. Chordates inhabit terrestrial and aquatic (freshwater and marine) environments. One group has developed the ability to fly. The body plan of chordates is unique in that these animals demonstrate a complex of four important characteristics at some stage in their development. In this exercise, you will discover these characteristics.

You will study two chordate species: the lancelet, an invertebrate in the subphylum Cephalochordata, and the pig, a vertebrate in the subphylum Craniata. The third subphylum, Urochordata, will not be studied.

## Lab Study A. Lancelets (*Branchiostoma,* formerly *Amphioxus*)

### Materials

| | |
|---|---|
| compound microscope | prepared slide of whole mount of |
| stereoscopic microscope | lancelet |
| preserved lancelet in watch glass | prepared slide of cross section |
| | of lancelet |

### Introduction

Lancelets are marine animals that burrow in sand in tidal flats. They feed with their head end extended from their burrow. They resemble fish superficially, but their head is poorly developed, and they have unique features not found in fish or other vertebrates. They retain the four unique characteristics of chordates throughout their life cycle and are excellent animals to use to demonstrate these features. In this lab study, you will observe preserved lancelets, prepared slides of whole mounts, and cross sections through the body of a lancelet (Color Plate 60).

## Procedure

1. Place a preserved lancelet in water in a watch glass and observe it using the stereoscopic microscope. Handle the specimen with care and *do not dissect it.* Note the fishlike shape of the slender, elongate body. Locate the anterior end by the presence at that end of a noselike **rostrum** extending over the mouth region, surrounded by small tentacles. Notice the lack of a well-defined head. Look for the segmented muscles that surround much of the animal's body. Can you see signs of a tail? If the animal you are studying is mature, you will be able to see two rows of 20 to 25 white gonads on the ventral surface of the body.

2. Return the specimen to the correct container.

3. Observe the whole mount slide of the lancelet and compare your observations with Figure 19.7.

 Use only the lowest power on the compound microscope to study this slide.

a. Scan the entire length of the body wall. Do you see evidence of segmentation in the muscles?

   *Students should notice the muscle segmentations.*

b. Look at the anterior end of the animal. Do you see evidence of a sensory system? Describe what you see.

   *Students may notice the single dark "eyespot" (sensory function not established), and may note the absence of large sensory structures. Sensory cells are concentrated in the rostrum and mouth region, but these are not visible with the microscope. Students may conclude that the tentacles are sensory, and, indeed, more sensory cells may be located on these structures. However, their main function is to prevent large food particles from entering the mouth. Students may ask about the apparent absence of a brain.*

**Figure 19.7.**
**The lancelet, whole mount.** The rostrum extends over the mouth region. The pharynx, including the pharyngeal gill slits, leads to the intestine, which exits the body at the anus. Note that a tail extends beyond the anus. Structures positioned from the dorsal surface of the body inward include a dorsal fin, the nerve cord, and the notochord.

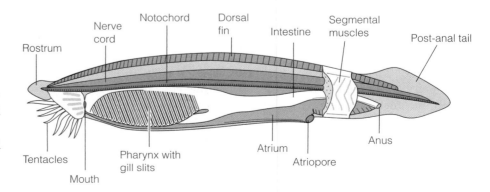

c. Locate the mouth of the animal at the anterior end. See if you can follow a tube from just under the rostrum into a large sac with numerous gill slits. This sac is the **pharynx with gill slits,** a uniquely chordate structure. Water and food pass into the pharynx from the mouth. Food passes posteriorly from the pharynx into the intestine, which ends at the anus on the ventral side of the animal, several millimeters before the end. The extension of the body beyond the anus is called a **postanal tail.** Think of the worms you studied in Lab Topic 18 Animal Diversity I. Where was the anus located in these animals? Was a postanal region present? Explain.

*The anus in worms is at the end of the body. The small overextension in* Ascaris *is not a postanal tail.*

d. Water entering the mouth passes through the gill slits and collects in a chamber, the **atrium,** just inside the body wall. The water ultimately passes out of the body at a ventral pore, the **atripore.** Surprisingly, the gill slits are not the major gas exchange surface in the lancelet body. Because of the great activity of ciliated cells in this region, it is even possible that blood leaving the gill region has less oxygen than that entering the region. The function of gill slits is simply to strain food from the water. The major site for gas exchange is the body surface.

e. Now turn your attention to the dorsal side of the animal. Beginning at the surface of the body and moving inward, identify the listed structures and speculate about the function of each one.

**dorsal fin:**

*locomotion, important in swimming*

**nerve cord:**

*nerve impulse transmission*

**notochord:**

*support; an endoskeleton*

The nerve cord is in a dorsal position. Have you seen only a dorsal nerve cord in any of the animals previously studied?

*no*

The notochord is a cartilage-like rod that lies ventral to the nerve cord and extends the length of the body. Have you seen a notochord in any of the previous animals?

*no*

The lancelet circulatory system is not visible in these preparations, but the animal has **closed circulation** with dorsal and ventral aortae, capillaries, and veins. Excretory organs, or nephridia (not visible here), are located near the true coelom, which surrounds the pharynx.

4. Observe the slide of cross sections taken through the lancelet body. There may be several sections on this slide, taken at several positions along the length of the body. Find the section through the pharynx and compare it with Figure 19.8.

 **Study this slide on the lowest power.**

In cross section, it is much easier to see the structural relationships among the various organs of the lancelet. Identify the following structures and label them on Figure 19.8.

a. *Segmental muscles.* They are located on each side of the body, under the skin.

b. *Dorsal fin.* This projects upward from the most dorsal surface of the body.

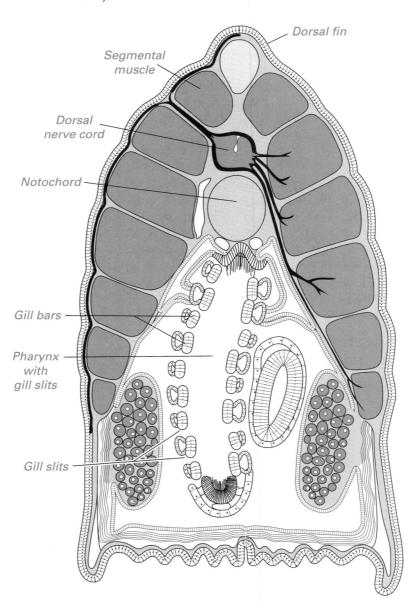

**Figure 19.8.**
Cross section through the pharyngeal region of the lancelet.

c. *Nerve cord.* You may be able to see that the nerve cord contains a small central canal, thus making it hollow. The nerve cord is located in the dorsal region of the body, ventral to the dorsal fin between the lateral bundles of muscle.

d. *Notochord.* This is a large oval structure located just ventral to the nerve cord.

e. *Pharynx with gill slits.* This structure appears as a series of dark triangles arranged in an oval. The triangles are cross sections of **gill bars.** The spaces between the triangles are **gill slits,** through which water passes into the surrounding chamber.

## Results

1. Complete the diagram of the lancelet cross section in Figure 19.8. Label all the structures listed in step 4 of the Procedure section.

2. Complete Table 19.1, recording all information for lancelet characteristics in the appropriate row. Use this information to complete Table 19.2 and answer questions in the Applying Your Knowledge section at the end of this lab topic.

## Discussion

Describe the uniquely chordate features that you have detected in the lancelet that were not present in the animals previously studied.

*(1) a dorsal hollow or tubular nerve cord; (2) a notochord, an endoskeletal rod that supports the body; (3) pharyngeal gill slits, which function in filter feeding or gas exchange in aquatic organisms and develop into other structures in land animals; (4) a post-anal tail containing a posterior extension of the notochord and nerve cord*

# Lab Study B. Fetal Pigs (*Sus scrofa*)

## Materials

preserved fetal pig                    disposable gloves
dissecting pan

## Introduction

The pig is a terrestrial vertebrate. You will study its anatomy in detail in Lab Topics 22, 23, and 24. In this lab study, working with your lab partner, you will observe external features only, observing those characteristics studied in other animals in previous exercises. Compare the organization of the vertebrate body with the animals previously studied. As you dissect the pig in subsequent labs, come back to these questions and answer the ones that cannot be answered in today's lab study.

## Procedure

1. Obtain a preserved fetal pig from the class supply and carry it to your desk in a dissecting pan.

 **Use disposable gloves to handle preserved animals.**

2. With your lab partner, read each of the following questions. Drawing on observations you have made of other animals in the animal diversity lab studies, predict the answer to each question about the fetal pig. Then examine the fetal pig and determine the answer, if possible. Give evidence for your answer based on your observations of the pig, your knowledge of vertebrate anatomy, or your understanding of animal phylogeny.

a. What type of symmetry does the pig body have?

Prediction:

*bilaterally symmetrical (based on phylogenetic trends)*

Evidence:

*right and left halves are mirror images*

b. How many layers of embryonic tissue are present?

Prediction:

*three (based on phylogenetic trends)*

Evidence:

*Skin, gut, muscles, and many organs are present; organs are not possible without three tissue layers.*

c. Are cells organized into distinct tissues?

Prediction:

*yes (based on phylogenetic trends)*

Evidence:

*Organs are composed of tissues; this answer is obvious.*

d. How many digestive tract openings are present? Would you describe this as a "tube within a tube"?

Prediction:

*two (based on phylogenetic trends); yes*

Evidence:

*mouth and anus present*

e. Is the circulatory system open or closed?

Prediction:

*closed (based on phylogenetic trends)*

Evidence:

*arteries, veins, heart present*

f. What is the habitat of the animal?
   Prediction:

   *terrestrial*

   Evidence:

   *no external gills; no appendages for locomotion in water; appendages for movement on land*

g. What are the organs for respiration?
   Prediction:

   *lungs*

   Evidence:

   *no external gills; skin too thick*

h. What are the organs for excretion?
   Prediction:

   *kidney*

   Evidence:

   *only evidence would be student's prior knowledge*

i. What is the method of locomotion?
   Prediction:

   *walking, using appendages*

   Evidence:

   *four walking legs present*

j. Are support systems internal or external?
   Prediction:

   *internal (based on phylogenetic trends)*

   Evidence:

   *no external skeleton; body holds shape*

k. Is the body segmented?
   Prediction:

   *yes (based on phylogenetic trends)*

   Evidence:

   *segmentation of muscle bundles and skeletal components (ribs) will be observed in dissections*

l.  Are appendages present?
    Prediction:

    *yes*

    Evidence:

    *four legs*

m.  What is the position and complexity of the nervous system?
    Prediction:

    *well-developed brain, dorsal nerve cord (based on phylogenetic trends)*

    Evidence:

    *well-defined head, complex body*

### Results

*As students complete the two Animal Diversity lab topics, call on volunteers to report their results recorded in Table 19.1. Then, if time allows, in class or as outside lab assignments, students complete Table 19.2. The instructor then leads a discussion of the "big themes" that have emerged from these observations.*

Complete Table 19.1 on pp. 508–509, recording all information for pig characteristics in the appropriate row. Use this information to complete Table 19.2 and answer questions in the Applying Your Knowledge section that follows.

## Reviewing Your Knowledge—Lab Topics 18 and 19

1.  Complete the summary table, Table 19.1, recording in the appropriate row information about characteristics of all animals studied.
2.  Using Table 19.1, complete Table 19.2. Categorize all animals studied based on the 13 basic characteristics discussed in Lab Topic 18. Use this information to answer questions in the Applying Your Knowledge section that follows.

## Applying Your Knowledge

*See Teaching Plan for answers to Applying Your Knowledge questions.*

1.  Using specific examples from the animals you have studied in Lab Topics 18 and 19, describe ways that organisms have adapted to specific environments.
    a.  Compare organisms adapted to aquatic environments with those from terrestrial environments.
    b.  Compare adaptations of the parasitic *Ascaris* with the earthworm or clamworm, free-living organisms.
2.  In Lab Topic 18 you studied the free-living flatworm, *Planaria*. The phylum Platyhelminthes also includes many examples of parasitic flatworms, for example, tapeworms and trematodes (flukes). Using Web resources, choose an example of a parasitic flatworm and compare morphological differences between this organism and the planarian that reflect specific life-style adaptations.

3. In your studies of animal phyla, you observed segmentation in widely diverse clades, for example, annelids (Lophotrochozoa), arthropods (Ecdysozoa), and chordates (Deuterostomia). How can you explain this in terms of their evolutionary history?

4. Upon superficial examination, the body form of certain present-day animals might be described as simple, yet these animals may have developed specialized structures, perhaps unique to their particular phylum. Illustrate this point using examples from some of the simpler organisms you have dissected.

5. From Lab Topics 18 and 19, one might conclude that certain trends can be detected, trends from ancestral features (those that arose early in the evolution of animals) to more derived traits (those that arose later). However, animals with ancestral characteristics still successfully exist on Earth today. Why is this so? Why have the animals with derived traits not completely replaced the ones with ancestral traits? Use examples from the lab to illustrate your answer.

6. A major theme in biology is the relationship between form and function in organisms. Select one of the major characteristics from Table 19.1, and illustrate the relationship of form and function for this characteristic using examples from the organisms studied.

7. Students in the Coastal Biology Class have found an animal they cannot identify in the lower beach sand of Jekyll Island. What are some questions that they should ask to help them determine the phylum of the animal?

8. As you review the information you have recorded in Tables 19.1 and 19.2, do you see a relationship between *symmetry* and (1) the organization of the nervous system and (2) the number of tissue layers?

**Table 19.1**
Summary Table of Animal Characteristics

| Animal | Symmetry | Tissue Organization | Type of Body Cavity | Digestive Openings | Circulatory System | Habitat | Respiratory Organs |
|--------|----------|---------------------|---------------------|--------------------|--------------------|---------|--------------------|
| **Sponge** | no symmetry`` | loose, aggregate of cells in layers, but no true tissues | none (not applicable) | intracellular digestion | water transported through body | aquatic | cells; directly across membranes |
| **Hydra** | radial | two tissue layers | none (not applicable) | one opening into gastro-vascular cavity | cells come into contact with water | aquatic | cells; directly across membranes |
| **Planarian** | bilateral | three tissue layers | acoelomate | one opening into gastro-vascular cavity | none | aquatic | cells; directly across membranes |
| **Clamworm/ earthworm** | bilateral | three tissue layers | coelomate | two openings: tube within a tube | closed circulation with blood vessels | clamworm —aquatic earthworm —terrestrial | skin, parapodia |
| **Clam** | bilateral | three tissue layers | coelomate | two openings: tube within a tube | open; heart blood vessels, sinuses | aquatic | gills |
| **Roundworm** | bilateral | three tissue layers | pseudocoe-lomate | two openings: tube within a tube | none; trans-port in pseudo-coelom | parasitic | none; essentially anaerobic |
| **Crayfish** | bilateral | three tissue layers | coelomate | two openings: tube within a tube | open; heart, blood vessels, sinuses | aquatic | gills |
| **Grasshopper** | bilateral | three tissue layers | coelomate | two openings: tube within a tube | open; heart, blood vessels, sinuses | terrestrial | spiracles; tracheae |
| **Sea star** | bilateral as larva; radial as adult | three tissue layers | coelomate | two | reduced | marine; bottom dwelling | skin gills; extensions of coelomic cavity |
| **Lancelet** | bilateral | three tissue layers | coelomate | two openings: tube within a tube | closed | aquatic | skin (body surface) |
| **Pig** | bilateral | three tissue layers | coelomate | two openings tube within a tube | closed | terrestrial | lungs |

**Table 19.1**
Summary Table of Animal Characteristics  (*continued*)

| Animal | Excretory System | Locomotion | Support System | Segmentation | Appendages | Nervous System Organization |
|---|---|---|---|---|---|---|
| **Sponge** | *none* | *none; sessile* | *spicules/ spongin* | *no* | *no* | *none* |
| **Hydra** | *none* | *limited locomotion using tentacles* | *water; hydrostatic skeleton* | *no* | *broadly defined, tentacles are appendages* | *no brain; network of nerve cells* |
| **Planarian** | *flame cells; two lateral excretory canals* | *creep over surface, crawling* | *water; hydrostatic skeleton* | *no* | *no* | *brain; two ventral nerve cords; "ladderlike" nervous system* |
| **Clamworm/ earthworm** | *nephridia* | *clamworm— swimming, parapodia; earthworm— setae, crawling from muscle contraction* | *hydrostatic skeleton* | *yes* | *clamworm— parapodia; earthworm— none* | *dorsal brain, ventral nerve cord* |
| **Clam** | *nephridia* | *foot for digging* | *external shell* | *no* | *no* | *three ganglia connected by nerves* |
| **Roundworm** | *two lateral lines* | *live in gut; limited movement* | *hydrostatic skeleton; external cuticle* | *no* | *no* | *dorsal and ventral nerve cord* |
| **Crayfish** | *green glands (resemble nephridia)* | *legs, tail, swimmerets* | *rigid, jointed exoskeleton* | *yes* | *mouth append- ages, walking legs, swimming appendages* | *dorsal brain, ventral nerve cord* |
| **Grasshopper** | *Malpighian tubules* | *wings, legs* | *rigid, jointed exoskeleton* | *yes* | *wings, walking legs, mouth appendages* | *dorsal brain, ventral nerve cord* |
| **Sea star** | *skin gills* | *tube feet* | *endoskeleton of calcareous plates* | *no* | *pentamerous arms extend from central disk* | *nerve ring in central disk; epidermal nerve network* |
| **Lancelet** | *nephridia* | *tail, fin for swimming* | *notochord, an endoskeleton* | *yes (muscles)* | *no* | *dorsal brain and nerve cord* |
| **Pig** | *kidneys* | *legs for walking/ running* | *embryonic notochord; bony endoskeleton* | *yes (muscles, etc.)* | *four legs* | *dorsal brain and nerve cord* |

**Table 19.2**
Comparison of Organisms by Major Features

<table>
<tr><td>

1. **Tissue Organization**
   a. distinct tissues absent:

   *sponge*

   b. distinct tissues present:

   *all other groups*

2. **Symmetry**
   a. radial:

   *Cnidaria; hydra; sea star, but only in adult*

   b. bilateral:

   *all other groups except sponges, which have no symmetry*

3. **Body Cavity**
   a. acoelomate:

   *planarian*

   b. pseudocoelomate:

   *roundworm*

   c. coelomate:

   *all other groups*

4. **Openings to Digestive Tract**
   a. one:

   *hydra, planarian*

   b. two:

   *all other groups*

</td><td>

5. **Circulatory System**
   a. none:

   *sponge, hydra, planarian, roundworm*

   b. open:

   *clam, crayfish, grasshopper*

   c. closed:

   *earthworm, clamworm, lancelet, pig*

6. **Habitat**
   a. aquatic:

   *sponge, hydra, planarian, clamworm, clam, crayfish, sea star, lancelet*

   b. terrestrial:

   *earthworm, grasshopper, pig*

   c. parasitic:

   *roundworm*

7. **Organs for Gas Exchange**
   a. skin:

   *clamworm, earthworm, sea star, lancelet, hydra*

   b. gills:

   *clam, crayfish*

   c. lungs:

   *pig*

   d. spiracles/tracheae:

   *grasshopper*

</td></tr>
</table>

**Table 19.2**
Comparison of Organisms by Major Features  (*continued*)

| | |
|---|---|
| **8. Organs for Excretion**<br>(list organ and animals)<br><br>*flame cells—planarian*<br><br>*nephridia—clamworm, earthworm, clam, lancelet*<br><br>*green glands—crayfish*<br><br>*Malpighian tubules—grasshopper*<br><br>*kidneys—pig* | **11. Segmented Body**<br>a. no:<br><br>*clam, roundworm, planarian*<br><br>b. yes:<br><br>*clamworm, earthworm, crayfish, grasshopper, pig* |
| **9. Type of Locomotion**<br>(list type and animals)<br><br>*none to limited—sponge, hydra*<br><br>*swimming—clamworm, crayfish, lancelet*<br><br>*walking—crayfish, grasshopper, pig*<br><br>*crawling—earthworm, planarian, roundworm* | **12. Appendages**<br>a. yes:<br><br>*(hydra), clamworm, crayfish, grasshopper, pig*<br><br>b. no:<br><br>*all others* |
| **10. Support System**<br>a. external:<br><br>*clam, roundworm (cuticle), crayfish, grasshopper*<br><br>b. internal:<br><br>*sea star, lancelet, pig*<br><br>c. hydrostatic:<br><br>*roundworm, clamworm, earthworm* | **13. Nervous System**<br>a. ventral nerve cord:<br><br>*grasshopper, crayfish*<br><br>b. dorsal nerve cord:<br><br>*lancelet, pig*<br><br>c. other:<br><br>*nerve net—hydra*<br><br>*ladderlike—planarian (two ventral)*<br><br>*dorsal and ventral—roundworm*<br><br>*nerve ring and radial nerves—sea star* |

# Investigative Extensions

1. Scientists worldwide are concerned about reports of climate change. Crayfish are common inhabitants of freshwater streams, ponds, and swamps, all of which may be affected by a warming earth. Design an experiment to test the thermal limits that can be tolerated by crayfish.

   Design similar experiments to test the effects of pesticides, fertilizers, herbicides, petroleum products, or human wastes—all of which may be present in runoff into crayfish habitat from farmlands or urban development.

2. Arthropods are the dominant animals on the earth, both in number of species and number of individuals. Now that you are familiar with the characteristics of insects (terrestrial arthropods), using an entomology (study of insects) text or insect identification key, determine the diversity of arthropods, or specifically insects, found in various habitats. You might sample a given amount of soil taken from several different environments. For example, you might compare arthropod diversity in a deciduous forest with that of a cultivated field with that of a manicured lawn.

You might also investigate arthropod diversity in habitats that differ in moisture. Compare well-drained soil (along a ridge) with saturated soil (along a creek, in a marsh, or in a bog).

3. N. A. Cobb, famous nematologist, writes: "If all the matter in the universe except the nematodes were swept away, our world would still be dimly recognizable . . . We should find its mountains, hills, vales, rivers, lakes, and oceans represented by a film of nematodes" (1915). Nematodes are everywhere and many are readily available for study.

Design an experiment to investigate the diversity of nematodes present in several sources; for example, fresh and rotting fruits, soils collected from different sources or treated with different chemicals, roots of plants—trees or vegetables grown for human consumption. (Many nematodes are plant parasites, and many have been imported on foods or nursery stock.)

Are there nematodes in drinking water? How could you investigate this question? How could you collect the nematodes?

 ## Student Media: BioFlix, Activities, Investigations, and Videos
**www.masteringbiology.com (select Study Area)**

**Activities**—Ch. 32: Animal Phylogenetic Tree;
  Ch. 33: Characteristics of Invertebrates;
  Ch. 34: Characteristics of Chordates;
  Ch. 41: Feeding Mechanisms of Animals

**Investigations**—Ch. 32: How Do Molecular Data Fit Traditional Phylogenies?
  Ch. 33: How Are Insect Species Identified?

**Videos**—Ch. 33: *C. elegans* Crawling; Butterfly Emerging; Bee Pollinating; Lobster Mouth Parts; Echinoderm Tube Feet; Discovery Videos: Ch. 33: Invertebrates

# References

Aguinaldo, A. M., J. M. Turbeville, L. S. Linford, M. C. Rivera, J. R. Garey, R. A. Raff, and J. A. Lake. "Evidence for a Clade of Nematodes, Arthropods, and other Moulting Animals." *Nature*, 29 May 1997 (387), pp. 489–493.

Cobb, N. A. "Nematodes and Their Relationships." Year Book Dept. Agric. 1914, pp. 457–490. Washington, DC: Dept. Agric. 1915.

Hickman, C. P., L. S. Roberts, A. Larson, and H. l'Anson. *Integrated Principles of Zoology,* 12th ed. Boston: McGraw Hill, 2004.

## Websites

Includes descriptions of many invertebrates and vertebrates, links to insect keys, references:
http://animaldiversity.ummz.umich.edu/index.html

Report of an experiment testing nematodes for space experiments:
http://www.nasa.gov/mission_pages/station/science/experiments/ICE-First.html

The Tree of Life web project provides information on all major groups of organisms, including invertebrates:
http://tolweb.org/Bilateria

# LAB TOPIC 18 AND 19

# Animal Diversity I and II
## Teaching Plan for Laboratories

 These lab topics integrate evolutionary concepts with animal diversity. They are not intended to be a comprehensive survey of all animal phyla.

## Main Concepts and Objectives

1. Concept: Understanding the organization and form of an animal's body is essential to discovering how animals carry out body functions. Students will dissect and compare examples of several major animal phyla and determine how those animals carry out body functions. (all exercises)

2. Concept: Meaningful similarities and differences in animal form give clues to phylogenetic relationships. Recent data from molecular studies are confirming some phylogenetic relationships and revising others.

3. Concept: The particular niche (habitat, mode of obtaining nutrition, means of waste disposal, for example) occupied by an organism is directly related to its body form. Students will describe the niche of the organisms being studied based on their observations. (all exercises)

4. In traditional systems of classification, there were seven critical characteristics for major phylogenetic groups:

   **Grade of organization:**   colonial (sponges) or multicellular with true tissues (all other animal groups).

   **Symmetry:**   either radial (Cnidaria) or bilateral (all other animal groups). With the introduction of bilateral symmetry, **cephalization**, or the formation of a head end with a brain, is introduced. Sensory organs become concentrated at the head end, and **directed motion**—that is, motion in one direction—is possible.

   **Number of tissue layers in the embryo of multicellular forms:**   two (Cnidaria) or three (all other animal groups).

   **Number of openings into the gut:**   one (Cnidaria and Platyhelminthes) or two (all other groups).

**Body cavity or coelom:**    (in those animals with three tissue layers in the embryo):

(a)  acoelomate—example, phylum Platyhelminthes.

(b)  pseudocoelomate—example, phylum Nematoda.

(c)  coelomate—most other animal groups.

**Type of development:**    protostomes or deuterostomes.

**Presence of segmentation:**    repetition of similar units or segments.

5. Specific content. Students will be able to define or describe names and descriptions of 13 basic characteristics of animals and all terms in boldface.

## Order of the Lab: Animal Diversity I: Porifera, Cnidaria, Platyhelminthes, Annelida, and Mollusca

1. Discuss techniques and safety precautions used when dissecting organisms.                                                        (10 min)

2. Introduce the main concepts and objectives of the lab. Review and explain the 13 characteristics for comparing organisms (pp. 463–465). Remind students to keep remembering the "big picture" as they do these dissections. Remembering the objectives and the 13 characteristics should help keep them focused                                      (20 min)

3. Point out the Applying Your Knowledge questions at the end of Animal Diversity II. These questions relate to both lab topics. Tell students to fill in Table 19.1 as they observe animals in Lab Topics 18 and 19.                                    (5 min)

4. If a video on animal diversity is available, show it at this time.

5. Students complete Animal Diversity I.                      (2 hr 20 min)

6. Lead class discussion of the correct answers for Table 19.1 to this point.                                        (5 min)

## Order of the Lab: Animal Diversity II: Nematoda, Arthropoda, Echinodermata, and Chordata

1. Review the main concepts and objectives of the lab as stated in Animal Diversity I. Review the 13 basic animal characteristics being investigated.                              (15 min)

2. If a second video on animal diversity is available, show it at this time.

3. Students complete Animal Diversity II.                        (2.5 hr)

4. Summary and assignments. You may ask students to report the information that they recorded in Table 19.1. Selected Applying Your Knowledge questions can be discussed at this time.                                 (15 min)

The animal diversity labs are written to take place in two consecutive 3-hour laboratory periods. We show a 20-minute videotape on animal diversity each week and are able to complete the two lab topics with time to spare. We ask students to work the full 3 hours and to complete the first lab topic during the first lab period. If they do not complete the first lab topic in the first lab period, they can complete Animal Diversity I in the second lab period before proceeding to Animal Diversity II.

**For a 2-hour lab:** You might choose to extend the two labs to three and avoid having to omit some of the animals. If you only have two lab periods allotted to invertebrates, we suggest omitting the sponge and the clamworm in Animal Diversity I and the Echinoderm and the pig in Animal Diversity II. The major themes of the study should not be lost if these animals are omitted.

## Classroom Management

Students should work independently on all dissections, but discussing observations with a lab partner will enhance the learning experience. *Urge students to answer every question in the lab topics and to fill in Table 19.1 very carefully and completely as they dissect.* Your role as instructor is to keep the students focused on the major concepts, the comparison of the animals, and the interpretation of their observations. As you circulate among your students, ask questions that will lead students to compare and interpret what they are seeing. Although you will not be very popular, do not immediately answer students' questions, encouraging them, instead, to think for themselves. Although students must remember names of structures and organs to write intelligent discussions, name memorization is not the primary objective of this study. The dissection of one animal is not to be carried out as a discrete exercise and then forgotten as the student goes on to the next animal. The dissections in both lab topics should be one continuous study.

Encourage your students to make accurate sketches of details of difficult structures in the margins of their lab manual in addition to studying the diagrams in the text. Remind them that the best way to remember and study these animals is by means of a well-labeled drawing that they have personally produced. Making an accurate drawing requires a level of understanding that goes beyond simply identifying structures labeled on a diagram.

## Student Development

These lab topics will improve skills in dissection techniques. Observation skills improve as students are required to make careful observations and use those observations to propose relationships. Students practice communication as they discuss observations, and they practice writing as they answer questions.

## Lab Safety Precautions

Instruct students to:

1. Always dissect preserved animals in a room with adequate ventilation.
2. Wear disposable gloves.
3. Be careful when using sharp dissecting instruments.

 Warn students to be cautious as they open the clam. Hold the clam in the dissecting pan in such a way that the scalpel blade will be directed toward the bottom of the pan.

## Discussion and Summary

Discussion will take place informally as students make their observations and formally as they answer the questions at the end of Animal Diversity II. At the end of the first lab, and then the second, ask students to read the observations that they have recorded in Table 19.1. If there is enough time, you can discuss the questions in the Applying Your

Knowledge section with the entire class rather than have students write out the answers. If you ask students to answer questions orally, make sure that all students participate.

## Evaluation

Students will answer the questions in the exercises and at the end of Lab Topic 19 and will be required to recall their conclusions on a lab test. A portion of the lab test should be in a practical format. Ask students to identify and give the function of selected structures. They should be able to relate structures with similar functions among different animals. The main objectives of the labs should be reflected in the questions on the lab test.

## Notes on Applying Your Knowledge

1a. The most obvious comparisons are the crayfish with the grasshopper and the clamworm with the earthworm.

### Crayfish Compared with Grasshopper

Respiration

*Crayfish:* Gills (require aquatic environment).

*Grasshopper:* Oxygen passes into the body through spiracles that lead to tracheal tubes, which in turn branch out into deep body tissues. Respiratory surfaces are protected from desiccation by this design.

Excretion

*Crayfish:* Green glands for excretion in crayfish are modified nephridia that dump waste and water directly to the outside of the body. Water is not conserved.

*Grasshopper:* Malpighian tubules collect and empty wastes and water into the digestive tract. The water is subsequently conserved as it is reabsorbed in the rectum.

Appendages

*Crayfish:* Appendages for swimming (tail and abdominal appendages).

*Grasshopper:* Lacks tail and abdominal appendages. Has wings and walking legs, adaptations for terrestrial habitation.

### Clamworm Compared with Earthworm

Locomotion

*Clamworm:* Parapodia are used for swimming and creeping over sand when the worm leaves its burrow. The worm moves very rapidly through water using parapodia.

*Earthworm:* Moves by peristalsis-like contractions of its longitudinal and circular muscles. Setae anchor the worm's body in the soil as it crawls. Students may recall that earthworms must live in fairly moist soil, however, since they need moist skin for gas exchange.

1b. Students should mention the protective cuticle on the *Ascaris* body surface, its anaerobic physiology, and its mode of feeding on digested material in the host gut. Whereas the digestive system in *Ascaris* is a simple, straight tube, the digestive system in the earthworm and the clamworm is more specialized, being divided into compartments for special functions. Recall that *Ascaris* is processing semidigested food from the host's digestive system; the earthworm is processing food from the soil passing through its digestive tract. The reproductive system in *Ascaris* is the largest, most conspicuous system in the body, ensuring that there will be an adequate number of eggs to be passed to the next host. In the earthworm and the clamworm, however, the reproductive system is visible only with careful observation. Negative environmental forces are not as significant against these free-living organisms.

2. Students will probably note expanded reproductive systems, structures to attach to a host, and reduced digestive systems in parasites.

3. Segmentation probably evolved independently in these groups that are now classified in three separate clades. Segmentation may have been gained (or lost) several times in the course of evolutionary history.

4. Sponges are described as being "simple," yet they have unique structures such as collar cells, spicules, and spongin. Hydras have cnidocytes and nematocysts, found only in Cnidaria. Flame cells and the extensive gastrovascular system in planarians are unique to Platyhelminthes. Annelids have nephridia. Insects have spiracles and trachea. Students may suggest many other examples.

5. What do ancestral and derived really mean? The word *ancestral* is sometimes used to imply "less fit" or "less developed," meaning that only those organisms that are specialized are successful. Organisms with derived traits are best described as those that have changed considerably compared with the ancestral stock. They should not necessarily be thought of as better or more successful. Success is relative to the environmental niche in which that organism occurs.

   A comparison of monotremes and eutherians (placentals) illustrates this point. Monotremes branched early in the history of mammals and carry on the egg-laying characteristic of their ancestors. The spiny anteater of New Zealand, an egg-laying mammal, is quite successful in its special habitat. It is specialized for this habitat and has successfully survived for millions of years.

   The generally accepted rule is that early branches in an evolutionary tree maintain ancestral traits that are lost or modified in later branching groups.

6. Students might choose to answer this question with any number of examples. Students should choose a characteristic and explain how this allows the animal to function in its particular environment. For example, the student might choose to discuss how segmented animals have more varied specialization along the length of their bodies, bringing about improved locomotion and new means of getting food, allowing them to live in a diversity of habitats.

7. What is the symmetry of the animal? Does it have a shell or an exoskeleton that it sheds? Is it segmented? Does it have appendages, and are they segmented?

8. Only bilaterally symmetrical animals demonstrate cephalization where the brain and sensory structures are concentrated at their anterior end. Also, only bilaterally symmetrical animals have three embryonic tissue layers.

## Lab Topic 20
# Plant Anatomy

## Laboratory Objectives

After completing this lab topic, you should be able to:

1. Identify and describe the structure and function of each plant cell type and tissue type.
2. Describe the organization of tissues and cells in each plant organ.
3. Relate the function of an organ (root, stem, and leaf) to its structure.
4. Describe primary and secondary growth and identify the location of each in the plant.
5. Relate primary and secondary growth to the growth habit (woody or herbaceous).
6. Discuss adaptation of land plants to the terrestrial environment as illustrated by the structure and function of plant anatomy.
7. Apply your knowledge of plants to the kinds of produce you find in the grocery store.

*For a 2-hour lab: Omit Exercise 20.5, Grocery Store Botany, or omit the embedding and study of the living stem. View only the prepared slide. See the Teaching Plan for other suggestions.*

## Introduction

Vascular plants have been successful on land for over 420 million years, and their success is related to their adaptations to the land environment. An aquatic alga lives most often in a continuously homogeneous environment: The requirements for life are everywhere around it, so relatively minor structural adaptations have evolved for functions such as reproduction and attachment. In contrast, the terrestrial habitat, with its extreme environmental conditions, presents numerous challenges for the survival of plants. Consequently, land plants have evolved structural adaptations for functions such as the absorption of underground water and nutrients, the anchoring of the plant in the substrate, the support of aerial parts of the plant, and the transportation of materials throughout the relatively large plant body. In angiosperms, the structural adaptations required for these and other functions are divided among three vegetative plant organs: stems, roots, and leaves. Unlike animal organs, which are often composed of unique cell types (for example, cardiac muscle fibers are found only in the heart, osteocytes only in bone), plant organs have many tissues and cell types in common, but they are organized in different ways. The structural organization of basic tissues and cell types in different plant organs is directly related to their different functions. For example, leaves function as the primary photosynthetic organ and generally have thin, flat blades that maximize light absorption and gas

*In the Study Area of masteringbiology .com, see Student Media for suggested activities, investigations, and videos.*

exchange. Specialized cells of the root epidermis are long extensions that promote one of the root functions, absorption. The interrelationship of structure and function is a major theme in biology, and you will continue to explore it in this lab topic.

Use the figures in this lab topic for orientation and as a study aid. Be certain that you can identify all items by examining the living specimens and microscope slides. These, and not the diagrams, will be used in the laboratory evaluations.

*The instructor may introduce these tissues and cells using microscopy projection of* Cucurbita *stem x.s. and l.s., and* Ranunculus *stem x.s.*

# Summary of Basic Plant Tissue Systems and Cell Types

The plant body is constructed into **tissue systems** based on their shared structural and functional features. There are three tissue systems—**dermal**, **ground**, and **vascular**—that are continuous throughout the organs of roots, stems, and leaves. The plant tissues that actively divide by mitosis are called **meristematic tissues.** These are located in specific regions—for example the root tip. Following is a review of plant tissue systems and the most common types of cells seen in plant organs, as well as their functions. Other specialized cells will be described as they are discussed in lab. Refer back to this summary as you work through the exercises.

### Dermal Tissue System: Epidermis

The **epidermis** forms the outermost layer of cells, usually one cell thick, covering the entire plant body. The epidermal cells are often flattened and rectangular in shape (Figure 20.1a and 1b). Specialized epidermal cells include the **guard cells** of the stomata, hairs called **trichomes**, and unicellular **root hairs**. Most epidermal cells on aboveground structures are covered by a waxy **cuticle**, which prevents water loss. The epidermis provides protection and regulates movement of materials.

### Ground Tissue System: Parenchyma, Collenchyma, and Sclerenchyma

The ground tissue system is distributed throughout the plant beneath the epidermis and surrounding the vascular tissues. Parenchyma, collenchyma, and/or sclerenchyma cells are typically found in ground tissue as seen in the cross section of a pumpkin stem (Figure 20.1c and Color Plate 61).

**Parenchyma** cells are the most common cell in plants and are characteristically thin-walled with large vacuoles. These cells may function in photosynthesis, support, storage of materials, and lateral transport.

**Collenchyma** cells are usually found near the surface of the stem, leaf petioles, and veins. These living cells are similar to parenchyma cells but are characterized by an uneven thickening of cell walls. They provide flexible support to young plant organs.

**Sclerenchyma** cells have thickened cell walls that may contain lignin. They provide strength and support to mature plant structures and may be dead at functional maturity. The most common type of sclerenchyma cells are long, thin **fibers.**

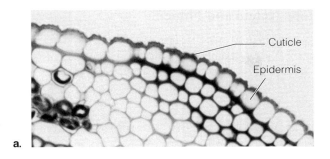

a.

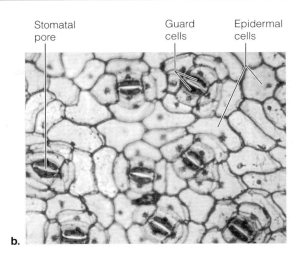

b.

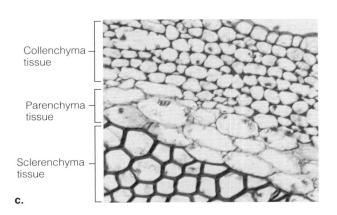

c.

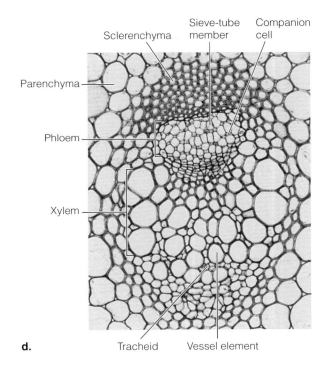

d.

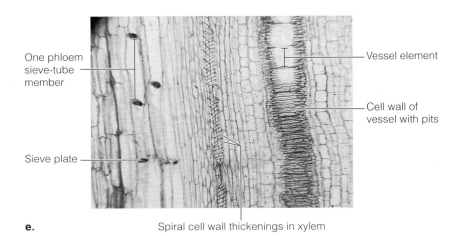

e.

**Figure 20.1.**
**Plant tissue systems and cell types.**
(a) Dermal tissue—a single layer of epidermis covered by waxy cuticle;
(b) Leaf surface showing epidermis with stomata and guard cells;
(c) Ground tissue—cross section of pumpkin stem (Color Plate 61);
(d) Vascular tissue—cross section of a vascular bundle in a buttercup stem;
(e) Long section through xylem and phloem of pumpkin stem.

## Vascular Tissue System: Xylem and Phloem

The vascular tissue system functions in the transport of materials throughout the plant body. Xylem tissue and phloem tissue are complex tissues (composed of several cell types) seen in the cross section of a buttercup stem (Figure 20.1d) and a long section of a pumpkin stem (Figure 20.1e).

**Xylem** cells form a complex vascular tissue that functions in the transport of water and minerals throughout the plant and provides support. **Tracheids** and **vessel elements** are the primary water-conducting cells. Tracheids are long, thin cells with perforated tapered ends. Vessel elements are larger in diameter, open-ended, and joined end to end, forming continuous transport systems referred to as **vessels.** Parenchyma cells are present in the xylem and function in storage and lateral transport. Fibers in the xylem provide additional support.

**Phloem** tissue transports the products of photosynthesis throughout the plant as part of the vascular tissue system. This complex tissue is composed of living, conducting cells called **sieve-tube members,** which lack a nucleus and have **sieve plates** for end walls. The cells are joined end to end throughout the plant. Each sieve-tube member is associated with one or more adjacent **companion cells,** which are thought to regulate sieve-tube member function. Phloem parenchyma cells function in storage and lateral transport, and phloem fibers provide additional support.

## Meristematic Tissue: Primary Meristem, Cambium, and Pericycle

**Primary meristems** consist of small, actively dividing undifferentiated cells located in buds of the shoot and in root tips of plants. These cells produce the primary tissues along the plant axis throughout the life of the plant. You will study primary meristems in the apical bud in Exercise 20.2 (Figure 20.3).

**Pericycle** is a layer of meristematic cells just outside the vascular cylinder in the root. These cells divide to produce lateral branch roots (Exercise 20.3 Lab Study B, Figure 20.6c).

**Vascular cambium** is a lateral meristem also composed of small, actively dividing cells that are located between the xylem and phloem vascular tissue. These cells divide to produce secondary growth, which results in an increase in circumference (Exercise 20.4, Figure 20.10).

**Cork cambium** is a lateral meristem located just inside the cork layer of a woody plant. These cells divide to produce secondary growth (Exercise 20.4, Figure 20.10).

This lab topic begins with a study of the shape and form of the whole plant, then moves to an investigation of the primary plant body derived from apical meristems. You will look at a slide of the apical bud from the tip of the stem. Next you will look at the structure of the three organs of the primary plant body: stems, roots, and leaves. Some plants continue to grow from lateral meristems producing secondary tissues, which you will investigate in stems of a woody plant. The lab topic concludes with an application of your knowledge to plants commonly found in the grocery store.

# EXERCISE 20.1
# Plant Morphology

## Materials

living bean or geranium plant          paper towel
squirt bottle of water

## Introduction

As you begin your investigation of the structure and function of plants, you need an understanding of the general shape and form of the whole plant. In this exercise, you will study a bean or geranium plant, identifying basic features of the three vegetative organs: *roots, stems, and leaves*. In the following exercises, you will investigate the cellular structure of these organs as viewed in cross sections. Refer to the living plant for orientation before you view your slides.

## Procedure

1. Working with another student, examine a living **herbaceous** (nonwoody) plant and identify the following structures in the *shoot* (*stems and leaves*).

   a. **Nodes** are regions of the stem from which leaves, buds, and branches arise and which contain meristematic tissue (areas of cell division).

   b. **Internodes** are the regions of the stem located between the nodes.

   c. **Terminal buds** are located at the tips of stems and branches. They enclose the shoot apical meristem, which gives rise to leaves, buds, and all primary tissue of the stem. Only stems produce buds.

   d. **Axillary**, or **lateral, buds** are located in the leaf axes at nodes; they may give rise to lateral branches.

   e. Leaves consist of flattened **blades** attached at the node of a stem by a stalk, or **petiole**.

2. Observe the *root* structures by gently removing the plant from the pot and loosening the soil from the root structure. You may need to rinse a few roots with water to observe the tiny, active roots. Identify the following structures.

   a. **Primary** and **secondary roots**. The primary root is the first root produced by a plant embryo and may become a long taproot. Secondary roots arise from meristematic tissue deep within the primary root.

   b. Root tips consist of a **root apical meristem** that gives rise to a **root cap** (protective layer of cells covering the root tip) and to all the primary tissues of the root. A short distance from the root tip is a zone of **root hairs** (specialized epidermal cells), the principal site of water and mineral absorption.

## Results

1. Label Figure 20.2.

2. Sketch in the margin of your lab manual any features not included in this diagram that might be needed for future reference. For example, your plant may have small green bracts (leaflike structures) at the base of the petiole. These are called stipules.

**Figure 20.2.**
**A herbaceous plant.** The vegetative plant body consists of roots, stems, and leaves. The buds are located in the axils of the leaves and at the shoot tip. The roots also grow from meristem tissues in the root tip. Label the diagram based on your observations of a living plant and the structures named in Exercise 20.1.

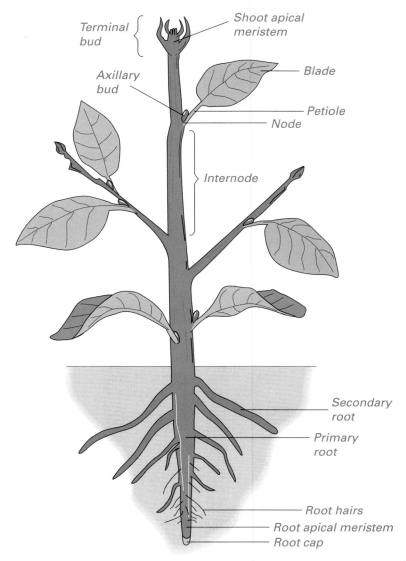

Terminal bud

Shoot apical meristem

Axillary bud

Blade

Petiole

Node

Internode

Secondary root

Primary root

Root hairs

Root apical meristem

Root cap

## Discussion

1. Look at your plant and discuss with your partner the possible functions of each plant organ. Your discussion might include evidence observed in the lab today or prior knowledge. Describe proposed functions (more than one) for each organ.

Stems:

*support for the plant, positioning leaves for exposure to sunlight, transport from roots to leaves and vice versa, photosynthesis for herbaceous plants*

Roots:

*absorption, anchorage, transport, storage*

Leaves:

*photosynthesis, light absorption, transpiration*

2. Imagine that you have cut each organ—roots, stems, and leaves—in cross section. Sketch the overall shape of that cross section in the margin of your lab manual. Remember, you are not predicting the internal structure, just the overall shape.

*One objective of this exercise is to help students visualize the three-dimensional structure of plants before looking at cross sections. Informally check drawings and ask questions.*

---

E X E R C I S E   2 0 . 2
# Plant Primary Growth and Development

---

### Materials

prepared slides of *Coleus* stem (long section)
compound microscope

### Introduction

Plants produce new cells throughout their lifetime as a result of cell divisions in meristems. Tissues produced from apical meristems are called **primary tissues,** and this growth is called **primary growth.** Primary growth occurs along the plant axis at the shoot and the root tip. Certain meristem cells divide in such a way that one cell product becomes a new body cell and the other remains in the meristem. Beyond the zone of active cell division, new cells become enlarged and specialized (differentiated) for specific functions (resulting, for example, in vessels, parenchyma, and epidermis). Using the model plant, *Arabadopsis,* research into the genetic and molecular basis of cell differentiation has rapidly advanced.

In this exercise, you will examine a longitudinal section through the tip of the stem, observing the youngest tissues and meristems at the apex, then moving down the stem, where you will observe more mature cells and tissues.

### Procedure

1. Examine a prepared slide of a longitudinal section through a terminal bud of *Coleus.* Use low power to get an overview of the slide; then increase magnification. Locate the **apical meristem,** a dome of tissue nestled between the **leaf primordia,** young developing leaves. Locate the axillary **bud primordial** between the leaf and the stem.

2. Move the specimen under the microscope so that cells may be viewed at varying distances from the apex. The youngest cells are at the apex of the bud, and cells of increasing maturity and differentiation can be seen as you move away from the apex. Follow the early development of vascular tissue, which differentiates in relation to the development of primordial leaves.
   a. Locate the narrow, dark tracks of **undifferentiated vascular tissue** in the leaf primordia.
   b. Observe changes in cell size and structure of the vascular system as you move away from the apex and end with a distinguishable vessel element of the **xylem,** with its spiral cell wall thickening in the older leaf primordia and stem. You may need to use the highest power on the microscope to locate these spiral cell walls.

*Depending on the prepared slides, students may or may not easily follow the vascular tissue as it elongates and differentiates. Once they see the spiral pattern of secondary cell walls, they can easily follow these changes. Share good slides with others.*

### Results

1. Label Figure 20.3, indicating the structures visible in the young stem tip.
2. Modify the figure or sketch details in the margin of the lab manual for future reference.

### Discussion

1. Describe the changes in cell size and structure in the stem tip. Begin at the youngest cells at the apex and continue to the xylem cells.

   *The size of the cells increases as you move away from the apex. The vascular tissues elongate and develop interesting cell walls.*

2. The meristems of plants continue to grow throughout their lifetime, an example of **indeterminate growth.** Imagine a 200-year-old oak tree, with active meristem producing new buds, leaves, and stems each year. Contrast this with the growth pattern in humans.

   *Humans grow to a certain size and age, after which the only growth is for maintenance and repair, not an increase in size. This is determinate growth.*

---

EXERCISE 20.3

# Cell Structure of Primary Tissues

---

All **herbaceous** (nonwoody) flowering plants produce a complete plant body composed of primary tissue, derived from apical primary meristem. This plant body consists of *organs*—roots, stems, leaves, flowers, fruits, and seeds—and *tissue systems*—**dermal**, **ground**, and **vascular**. In this exercise, you will investigate the cellular structure and organization of plant organs and tissues by examining microscopic slides. You will make your own thin cross sections of stems, and view prepared slides of stems, roots, and leaves. Woody stems will be examined in Exercise 20.4.

## Lab Study A. Stems

### Materials

*We presently recommend the "typical herbaceous dicot stem" slides available from Triarch for their distinct vascular bundles. Medicago slides recommended in previous editions are still usable.*

| | |
|---|---|
| prepared slide of herbaceous dicot stem | warm paraffin |
| dropper bottle of distilled water | living plant for sections |
| small petri dish with 50% ethanol | new single-edged razor blade |
| dropper bottle of 50% glycerine | forceps |
| dropper bottle of 0.2% toluidine blue stain | microscope slides |
| nut-and-bolt microtome | coverslips |
| | compound microscope |
| | dissecting needles |

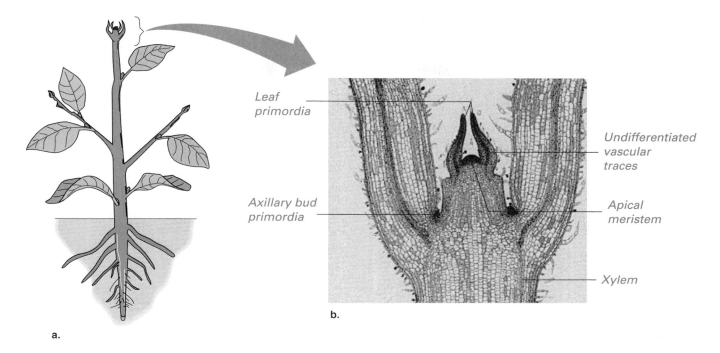

*a.*

*b.*

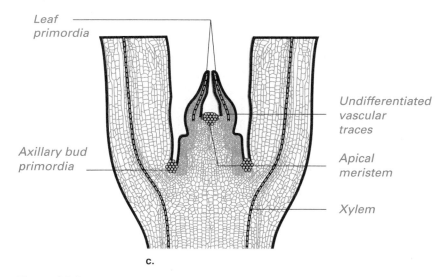

*c.*

**Figure 20.3.**
***Coleus* stem tip.** (a) Diagram of entire plant body. (b) Photomicrograph of a
longitudinal section through the terminal bud. (c) Line diagram of the growing
shoot tip with primordial leaves surrounding the actively dividing apical
meristem. The most immature cells are at the tip of the shoot and increase in
stages of development and differentiation farther down the stem. Label the cells
and structures described in Exercise 20.2.

## Introduction

A stem is usually the main stalk, or axis, of a plant and is the only organ that produces buds and leaves. Stems support leaves and conduct water and inorganic substances from the root to the leaves and carbohydrate products of photosynthesis from the leaves to the roots. Most herbaceous stems are able to photosynthesize. Stems exhibit several interesting adaptations, including water storage in cacti, carbohydrate storage in some food plants, and thorns that reduce herbivory in a variety of plants.

You will view a prepared slide of a cross section of a stem, and, working with another student, you will use a simple microtome—an instrument used for cutting thin sections for microscopic study—to make your own slides. You will embed the stem tissue in paraffin and cut thin sections. You will stain your sections with toluidine blue, which will help you distinguish different cell types. This simple procedure is analogous to the process used to make prepared slides for subsequent lab studies.

 **Read through the procedure and set up the materials before beginning.**

*If young bean plants are not available, Coleus can be substituted. Students should select the youngest stems to avoid secondary growth. They will see multicellular epidermal trichomes and collenchyma cells in the corners.*

*The best way to melt paraffin is in a paraffin oven. Alternatively, the paraffin can be kept warm by placing a 250-ml beaker of paraffin in a 400-ml beaker partially filled with water on a hot plate set on "low" under the hood. Avoid getting water in the paraffin or the sections will not embed. For other suggestions, refer to the Preparation Guide.*

*Caution: remember that paraffin is flammable!*

## Procedure

1. Embed the sections of the stem.

   a. Using a new single-edged razor blade, cut a 0.5 cm section of a young bean stem.

   b. Obtain a nut-and-bolt microtome. The nut should be screwed just into the first threads of the bolt. Using forceps, carefully hold the bean stem upright inside the nut.

   c. Pour the warm paraffin into the nut until full. Continue to hold the top of the stem until the paraffin begins to harden. While the paraffin completely hardens, continue the exercise by examining the prepared slide of the stem.

2. Examine a prepared slide of a cross section through the herbaceous dicot stem (Color Plate 62). As you study the stem tissues and cells, refer to "Summary of Basic Plant Tissue Systems and Cell Types," p. 520, Figure 20.1, and Color Plate 61.

3. Identify the **dermal tissue system,** characterized by a protective cell layer covering the plant. It is composed of the **epidermis** and the **cuticle.** Occasionally, you may also observe multicellular **trichomes** on the outer surface of the plants.

4. Locate the **ground tissue system,** background tissue that fills the spaces between epidermis and vascular tissue. Identify the **cortex region** located between the vascular bundles and the epidermis. It is composed mostly of **parenchyma,** but the outer part may contain **collenchyma** as well.

5. Next find the **pith region,** which occupies the center of the stem, inside the ring of vascular bundles; it is composed of parenchyma. In herbaceous stems, these cells provide support through turgor pressure. This region is also important in storage.

6. Now identify the **vascular system,** a continuous system of xylem and phloem providing transport and support. In your stems and in many stems, the **vascular bundles** (clusters of xylem and phloem) occur in rings that surround the pith; however, in some groups of flowering plants, the vascular tissue is arranged in a complex network.

7. Observe that each bundle consists of phloem tissue toward the outside and xylem tissue toward the inside. A narrow layer of vascular cambium, which may become active in herbaceous stems, is situated between the xylem and the phloem. Take note of the following information as you make your observations.

   **Phloem tissue** is composed of three cell types:

   a. Dead, fibrous, thick-walled **sclerenchyma cells** that provide support for the phloem tissue and appear in a cluster as a **bundle cap.**

   b. **Sieve-tube members,** which are large, living, elongated cells that lack a nucleus at maturity. They become vertically aligned to form sieve tubes, and their cytoplasm is interconnected through sieve plates located at the ends of the cells. Sieve plates are not usually seen in cross sections.

   c. **Companion cells,** which are small, nucleated parenchyma cells connected to sieve-tube cells by means of cytoplasmic strands.

   **Xylem tissue** is made up of two cell types:

   a. **Tracheids,** which are elongated, thick-walled cells with closed, tapered ends. They are dead at functional maturity, and their lumens are interconnected through pits in the cell walls.

   b. **Vessel elements,** which are cylindrical cells that are large in diameter and dead at functional maturity. They become joined end to end, lose their end walls, and form long, vertical vessels.

   **Vascular cambium** is a type of tissue that is located between the xylem and the phloem and which actively divides to give rise to secondary tissues.

8. Complete the Results section on the next page for this slide, then return to step 9 to prepare and observe your own handmade sections of stem preparations.

9. Cut the stem sections in the hardened paraffin.

   a. Support the nut-and-bolt microtome with the bolt head down and, using the razor blade, carefully slice off any excess paraffin extending above the nut. Be careful to slice in a direction away from your body and to keep your fingers away from the edge of the razor blade (Figure 20.4).

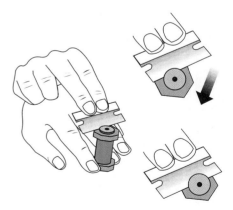

**Figure 20.4.**
**Using the nut-and-bolt microtome.**
A piece of stem is embedded in paraffin in the bolt. As you twist the bolt up, slice thin sections to be stained and viewed. Slide the entire blade through the paraffin to smoothly slice thin sections. Follow the directions in Exercise 20.3, Lab Study A carefully.

 Be careful to keep fingers and knuckles away from the razor blade. Follow directions carefully.

   b. Turn the bolt *just a little,* to extend the stem/paraffin above the edge of the nut.

*Demonstrate the method for slicing. Use the entire blade surface, not a sawing action. Thin slices are best, but students can usually view a part of a thicker section where it tapers to a thin area.*

c. Produce a thin section by slicing off the extension using the full length of the razor blade, beginning at one end of the blade and slicing to the other end of the blade (see Figure 20.4). Use the entire blade surface, not a sawing action.

d. Transfer each section to a small petri dish containing 50% ethanol.

e. Continue to produce thin sections of stem in this manner. The thinnest slices may curl, but this is all right if the stem section remains in the paraffin as you make the transfer. Cell types are easier to identify in very thin sections or in the thin edges of thicker sections.

10. Stain the sections.

a. Leave the sections in 50% ethanol in the petri dish for 5 minutes. The alcohol *fixes*, or preserves, the tissue. Using dissecting needles and forceps, carefully separate the tissue from the surrounding paraffin.

b. Using forceps, move the stem sections, free of the paraffin, to a clean slide.

*Experiment with staining times before lab.*

c. Add several drops of toluidine blue to cover the sections. Allow the sections to stain for 10 to 15 seconds.

d. Carefully draw off the stain by placing a piece of paper towel at the edge of the stain.

e. Rinse the sections by adding several drops of distilled water to cover the sections. Draw off the excess water with a paper towel. Repeat this step until the rinse water no longer looks blue.

f. Add a drop of 50% glycerine to the sections and cover them with a coverslip, being careful not to trap bubbles in the preparation.

g. Observe your sections using a compound microscope. Survey the sections at low or intermediate power, selecting the specimens with the clearest cell structure. You may have to study more than one specimen to see all structures. The thinnest edges of sections will provide the clearest view, particularly of the cells in vascular bundles.

11. Follow steps 3–7 above and identify *all structures and cells*. Incorporate your observations into the Results section (step 4).

### Results

1. Label the stem section in Figure 20.5b and 20.5c.

2. Were any epidermal trichomes present in your stem?

3. Note any features not described in the procedure. Sketch these in the margin of your lab manual for future reference. Return to Procedure step 9 in this lab study and complete the preparation of hand sections of the bean stem.

4. Compare your hand sections with the prepared slide. Modify Figure 20.5 or sketch your hand section in the margin. Is there any evidence of vascular cambium and secondary growth (Exercise 20.4)? Compare your results with those of other students.

 The functions of cells were described in the Summary of Basic Plant Tissue Systems and Cell Types, which appeared near the beginning of this lab topic (Figure 20.1 and Color Plate 61).

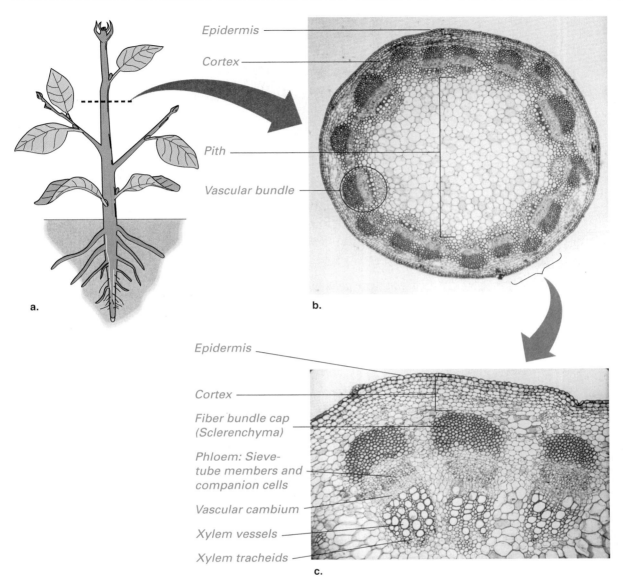

**Figure 20.5.**
**Stem anatomy.** (a) Diagram of whole plant. (b) Photomicrograph of cross section through the stem portion of the plant. (c) Enlargement of one vascular bundle as seen in cross section of the stem (Color Plate 62).

## Discussion

1. Which are larger and more distinct, xylem cells or phloem cells?

   *xylem cells*

2. What types of cells provide support of the stem? Where are these cells located in the stem?

   *Sclerenchyma fibers provide support and are located near phloem in vascular bundles. Collenchyma may also be present under the epidermis, in corners of square stems (if Coleus is used).*

3. For the cells described in your preceding answer, how does their observed structure relate to their function, which is support?

*Fibers have very thick walls (most are dead at maturity), and these cells provide strength. Collenchyma cells have thickenings only in the corners of otherwise thin-walled cells, and they provide flexible, elastic support.*

4. What is the function of xylem? Of phloem?

*Xylem transports water and nutrients. Phloem transports carbohydrates.*

5. The pith and cortex are made up of parenchyma cells. Describe the many functions of these cells. Relate parenchyma cell functions to their observed structure.

*Parenchyma cells function in storage, photosynthesis, and support for nonwoody plants. These cells have large vacuoles for water storage, which provides turgor pressure and support. Carbohydrates are stored in numerous plastids, which may or may not be visible. If chloroplasts are present, then the cells in the stem are photosynthetic.*

6. What differences did you observe in the prepared stem sections and your hand sections? What factors might be responsible for these differences?

*The hand sections may be shaped differently or have different epidermal features, depending on the plants chosen. Some herbaceous stems produce vascular cambium early, and this may be visible in the hand sections. Collenchyma cells may be easier to distinguish in the hand sections of Coleus. Student preparations often provide surprises. Have a copy of a botany book available for their investigations.*

## Lab Study B. Roots

### Materials

prepared slide of buttercup (*Ranunculus*) root (cross section)
demonstration of fibrous roots and taproots
colored pencils
compound microscope

### Introduction

Roots and stems often appear to be similar, except that roots grow in the soil and stems above the ground. However, some stems (rhizomes) grow underground, and some roots (adventitious roots) grow aboveground. Roots and stems may superficially appear similar, but they differ significantly in their functions. One of the major themes of biology is that structure and function are closely related at all levels of the hierarchy of life. Therefore, we would expect that the structure of stems and roots might differ in important ways.

What are the primary functions of stems?

*support of leaves and transport of materials*

Roots have four primary functions:

1. anchorage of the plant in the soil
2. absorption of water and minerals from the soil
3. conduction of water and minerals from the region of absorption to the base of the stem
4. starch storage to varying degrees, depending on the plant

## Hypothesis

The working hypothesis for this investigation is that the *structure* of the plant body is related to particular *functions*.

## Prediction

Based on the hypothesis, make a prediction about the similarity of root and stem structures that you expect to observe (if/then).

*If structure and function of the plant body are related, then the cells present and the organization of the root should differ in important ways from that observed for the stem.*

You will now test your hypothesis and predictions by observing the external structure of roots and their internal cellular structure and organization in a prepared cross section. This activity is an example of collecting evidence from observations rather than conducting a controlled experiment.

## Procedure

1. Examine the external root structure. When a seed germinates, it sends down a **primary root**, or **radicle**, into the soil. This root sends out side branches called lateral roots, and these in turn branch out until a root system is formed.

   If the primary root continues to be the largest and most important part of the root system, the plant is said to have a **taproot** system. If many main roots are formed, the plant has a **fibrous root** system. Most grasses have a fibrous root system, as do trees with roots occurring within 1 m of the soil surface. Carrots, dandelions, and pine trees are examples of plants having taproots.

   a. Observe examples of fibrous roots and taproots on demonstration in the laboratory.

   b. Sketch the two types of roots in the margin of your lab manual.

2. Examine the internal root structure.

   a. Study a slide of a cross section through a buttercup (*Ranunculus*) root. Note that the root lacks a central pith. The vascular tissue is located in the center of the root and is called the **vascular cylinder** (Figure 20.6b).

   b. Look for a cortex. The **cortex** is primarily composed of large parenchyma cells filled with numerous purple-stained organelles. Which of the four functions of roots listed in the introduction to

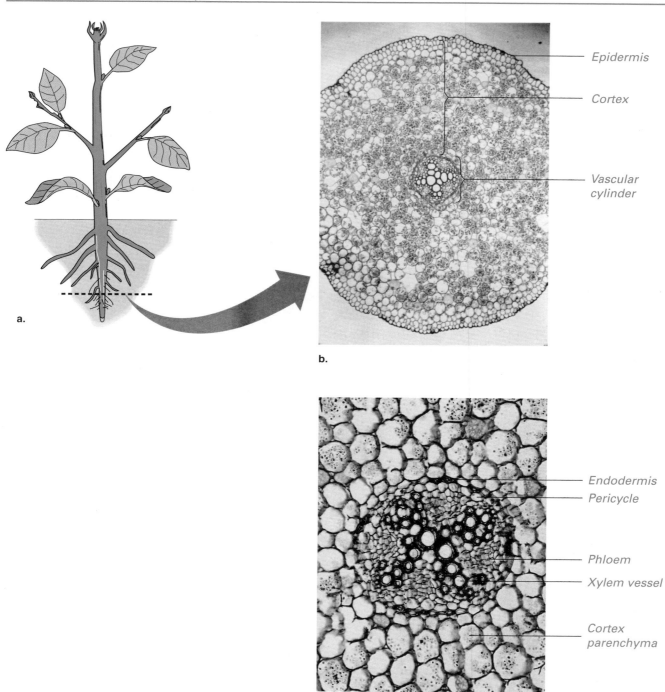

**Figure 20.6.**
**Cross section of the buttercup root.** (a) Whole plant. (b) Photomicrograph of a cross section of a root. (c) Enlargement of the vascular cylinder. Label the root based on your observations of a prepared microscope slide.

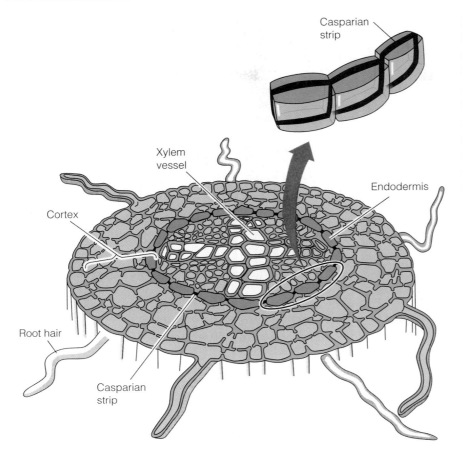

**Figure 20.7.**
**Root endodermis.** The endodermis is composed of cells surrounded by a band containing *suberin*, called the *Casparian strip* (seen in enlargement), that prevents the movement of materials along the cells' walls and intercellular spaces into the vascular cylinder. Materials must cross the cell membrane before entering the vascular tissue.

this lab study do you think is related to these cortical cells and their organelles?

*starch storage*

c. Identify the following tissues and regions and label Figure 20.6b and 20.6c accordingly: **epidermis**, parenchyma of **cortex**, **vascular cylinder**, **xylem**, **phloem**, **endodermis**, and **pericycle**. The endodermis and the pericycle are unique to roots. The endodermis is the innermost cell layer of the cortex. The walls of endodermal cells have a band called the **Casparian strip**—made of **suberin**, a waxy material—that extends completely around each cell, as shown in Figure 20.7. This strip forms a barrier to the passage of anything moving between adjacent cells of the endodermis. All water and dissolved materials absorbed by the epidermal root hairs and transported inward through the cortex must first pass through the living cytoplasm of endodermal cells before entering the vascular tissues. The pericycle is a layer of dividing cells immediately inside the endodermis; it gives rise to lateral roots. Refer to "Summary of Basic Plant Tissue Systems and Cell Types" and Figure 20.1.

*The endodermis is difficult to visualize in three dimensions. Use a model of a root if you have one. If not, use a shoe box with a large rubber band to demonstrate the relationship of the cell and suberin. Remind students that the cells are stacked on top of each other like bricks in a brick tower.*

## Results

1. Review Figure 20.6 and note comparable structures in Figure 20.7.
2. Using a colored pencil, highlight the representations of structures or cells found in the root but not seen in the stem.

## Discussion

1. Suggest the advantage of taproots and of fibrous roots under different environmental conditions.

   *Taproots allow plants to reach water in deep aquifers (some desert plants). Fibrous roots provide an advantage where water is available at the soil surface (many forest trees).*

2. Did your observations support your hypothesis and predictions (p. 533)?

3. Compare the structure and organization of roots and stems. How do these two organs differ?

   *Stems have nodes and produce buds; roots do not.*
   *The tip of the root is always covered with a cap of some kind, and the tip of a stem is not.*
   *The arrangement of internal tissues differs between stems and roots.*
   *Roots have a pericycle and endodermis; stems do not.*
   *Roots have root hairs, and stems do not.*

4. Explain the relationship of structure and function for two structures or cells found only in roots.

   *Root hairs increase the surface area available for absorption.*
   *The endodermis regulates the transport of materials into the vascular cylinder.*
   *The cortex provides an extensive storage area.*
   *The epidermis lacks a cuticle, which would prevent absorption.*

5. Note that the epidermis of the root lacks a cuticle. Can you explain why this might be advantageous?

   *Absorption in active roots is enhanced.*

6. What is the function of the endodermis? Why is the endodermis important to the success of plants in the land environment?

   *The endodermis regulates the transport of materials, which otherwise would move indiscriminately through the plant. Absorption occurs only in the roots; thus, the endodermis provides an important control in the plant's environment.*

 **Student Media Video—Ch. 35: Root Growth in a Radish Seed (time-lapse)**

# Lab Study C. Leaves

## Materials

prepared slide of lilac (*Syringia*) leaf
slides
compound microscope
coverslips

dropper bottles of water
leaves of purple heart (*Setcreasia*)
    kept in saline and DI water

## Introduction

Leaves are organs especially adapted for photosynthesis. The thin blade portion provides a very large surface area for the absorption of light and the uptake of carbon dioxide through stomata. The leaf is basically a layer of parenchyma cells (the **mesophyll**) between two layers of epidermis. The loose arrangement of parenchyma cells within the leaf allows for an extensive surface area for the rapid exchange of gases. Specialized epidermal cells called **guard cells** surround the stomatal opening and allow carbon dioxide uptake and oxygen release, as well as evaporation of water at the leaf surface. Guard cells are photosynthetic (unlike other epidermal cells), and are capable of changing shape in response to complex environmental and physiological factors. Current research indicates that the opening of the stomata is the result of the active uptake of $K^+$ and subsequent changes in turgor pressure in the guard cells.

In this lab study, you will examine the structure of a leaf in cross section. You will observe stomata on the leaf epidermis and will study the activity of guard cells under different conditions.

## Procedure

1.  Before beginning your observations of the leaf cross section, compare the shape of the leaf on your slide with Figure 20.8a and 20.8b.

2.  Observe the internal leaf structure.

    a.  Examine a cross section through a lilac leaf and identify the following cells or structures: **cuticle** (a waxy layer secreted by the epidermis), **epidermis** (upper and lower), parenchyma with chloroplasts (**mesophyll**), **vascular bundle** with **phloem** and **xylem**, and **stomata** with **guard cells** and **substomatal chamber**. Refer to "Summary of Basic Plant Tissue Systems and Cell Types" and Figure 20.1.

    b.  The vascular bundles of the leaf are often called **veins** and can be seen in both cross section and longitudinal sections of the leaf. Observe the structure of cells in the central midvein. Is xylem or phloem on top in the leaf?

    *xylem*

    c.  Observe the distribution of stomata in the upper and lower epidermis. Where are they more abundant?

    *lower epidermis*

    d.  Label the cross section of the leaf in Figure 20.8.

3.  Observe the leaf epidermis and stomata.

    a.  Obtain two *Setcreasia* leaves, one placed in saline for an hour and the other placed in distilled water for an hour.

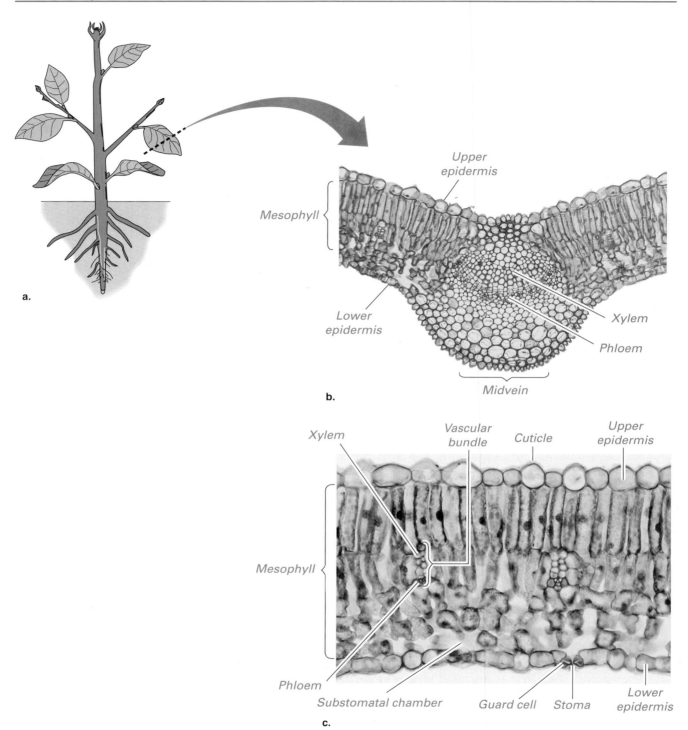

**Figure 20.8.**
**Leaf structure.** (a) Whole plant. (b) Photomicrograph of a leaf cross section through the midvein. (c) Photomicrograph of a leaf cross section adjacent to the midvein.

b. Label two microscope slides, one "saline" and the other "H$_2$O."

c. To remove a small piece of the lower epidermis, fold the leaf in half, with the lower epidermis to the inside. Tear the leaf, pulling one end toward the other, stripping off the lower epidermis (Figure 20.9). If you do this correctly, you will see a thin purple layer of lower epidermis at the torn edge of the leaf.

d. Remove a small section of the epidermis from the leaf in *DI water* and mount it in water on the appropriate slide, being sure that the outside surface of the leaf is facing up. View the slide at low and high power on your microscope, and observe the structure of the stomata. Sketch your observations in the margin of your lab manual.

e. Remove a section of the epidermis from the leaf in *saline* and mount it on the appropriate slide in a drop of the *saline*. Make sure that the outside surface of the leaf is facing up. View the slide with low power on your microscope, and observe the structure of the stomata. Sketch your observations in the margin of your lab manual.

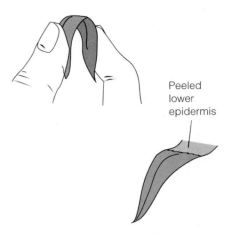

Peeled lower epidermis

**Figure 20.9.**
**Preparation of leaf epidermis peel.**
Bend the leaf in half and peel away the lower epidermis. Remove a small section of lower epidermis and make a wet mount.

## Results

1. Review the leaf cross section in Figure 20.8.
2. Describe the structure of the stomata on leaves kept in DI water.

   *Stomata are open in DI water.*

3. Describe the structure of the stomata on leaves kept in saline.

   *Stomata are closed in saline.*

*Practice making peels before lab. Students tend to break the leaves in half and not strip the lower epidermis as they peel one half from the other. Remind them to label their slides so that they know which leaf is which. The stomata will be difficult to get in focus if the mesophyll is up rather than the lower epidermis.*

## Discussion

1. Describe the functions of leaves.

   *The primary function of leaves is photosynthesis.*

2. Provide evidence from your observations of leaf structure to support the hypothesis that structure and function are related. Be specific in your examples.

   *Leaves are thin and flat, which minimizes the distance for gas exchange and maximizes the quantity of light that can be absorbed.*
   *All mesophyll cells of the leaf are in contact with air spaces, providing good gas exchange.*
   *Stomata are abundant, again for enhanced gas exchange.*

3. Explain the observation that more stomata are found on the lower surface of the leaf than on the upper.

   *Fewer stomata are found on the upper surface, where water loss would be greater owing to evaporation.*

4. Explain the differences observed, if any, between the stomata from leaves kept in DI water and those kept in saline. Utilize your knowledge of osmosis to explain the changes in the guard cells. (In this activity, you stimulated stomatal closure by changes in turgor pressure due to saline rather than $K^+$ transport.)

*Stomata were open in the DI water and closed in saline. In the hypotonic solution of DI water, the water moved into the guard cells, turgor pressure increased and the guard cells expanded, and the stomata opened. In the hypertonic saline solution, water moved out of the guard cells, which became flaccid, and the stomatal opening closed.*

EXERCISE 20.4

# Cell Structure of Tissues Produced by Secondary Growth

## Materials

prepared slides of basswood (*Tilia*) stem
compound microscope

## Introduction

Secondary growth arises from meristematic tissue called cambium. Vascular cambium and cork cambium are two types of cambium. The vascular cambium is a single layer of meristematic cells located between the secondary phloem and secondary xylem. Dividing cambium cells produce a new cell at one time toward the xylem, at another time toward the phloem. Thus, each cambial cell produces files of cells, one toward the inside of the stem, another toward the outside, resulting in an increase in stem circumference. The secondary phloem cells become differentiated into sclerenchyma fiber cells, sieve-tube members, and companion cells. Secondary xylem cells become differentiated into tracheids and vessel elements. Certain cambial cells produce parenchyma ray cells that can extend radially through the xylem and phloem of the stem.

The cork cambium is a type of meristematic tissue that divides, producing cork tissue to the outside of the stem and other cells to the inside. The cork cambium and the secondary tissues derived from it are called periderm. The periderm layer replaces the epidermis and cortex in stems and roots with secondary growth. These layers are continually broken and sloughed off as the woody plant grows and expands in diameter.

## Procedure

1. Examine a cross section of a woody stem (Color Plate 63).

   a. Observe the cork cambium and periderm in the outer layers of the stem. The outer **cork** cells of the periderm have thick walls impregnated with a waxy material called **suberin.** These cells are dead at maturity. The thin layer of nucleated cells that may be visible next to the cork cells is the **cork cambium.** The **periderm** includes the layers of cork and associated cork cambium. The term **bark** is used to describe the periderm and phloem on the outside of woody plants.

b. Observe the cellular nature of the listed tissues or structures, beginning at the periderm and moving inward to the central pith region. **Sclerenchyma fibers** have thick, dark-stained cell walls and are located in bands in the phloem. **Secondary phloem** cells with thin cell walls alternate with the rows of fibers. The **vascular cambium** appears as a thin line of small, actively dividing cells lying between the outer phloem tissue and the extensive secondary xylem. **Secondary xylem** consists of distinctive open cells that extend in layers to the central **pith** region. Lines of parenchyma cells one or two cells thick form **lateral rays** that radiate from the pith through the xylem and expand to a wedge shape in the phloem, forming a **phloem ray.**

2. Note the **annual rings** of xylem, which make up the **wood** of the stem surrounding the pith. Each annual ring of xylem has several rows of **early wood,** thin-walled, large-diameter cells that grew in the spring and, outside of these, a few rows of **late wood,** thick-walled, smaller-diameter cells that grew in the summer, when water is less available.

3. By counting the annual rings of xylem, determine the age of your stem. Note that the phloem region is not involved with determining the age of the tree.

## Results

1. Review Figure 20.10 and Color Plate 63.
2. Sketch in the margin of your lab manual any details not represented in the figure that you might need for future reference.
3. Indicate on your diagram the region where primary tissues can still be found.

   *Primary parenchyma cells are present in the pith.*

## Discussion

1. What has happened to the several years of phloem tissue production?

   *Phloem gets crushed toward the periderm on the outside of the tree.*

2. Based on your observations of the woody stem, does xylem or phloem provide structural support for trees?

   *xylem*

3. What function might the ray parenchyma cells serve?

   *lateral transport of materials and storage*

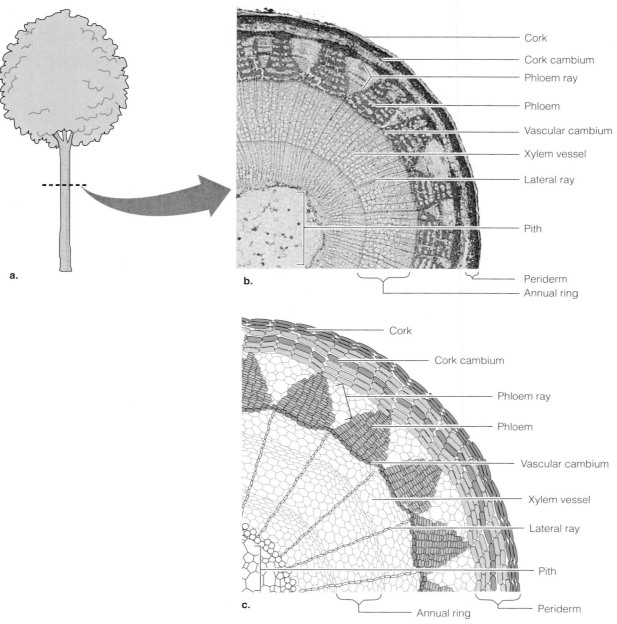

**a.**

**b.**

**c.**

**Figure 20.10.**
**Secondary growth.** (a) Whole woody plant. (b) Photomicrograph of a cross section of a woody stem. (c) Compare the corresponding diagram with your observations of a prepared slide. If necessary, modify the diagram to correspond to your specimen (Color Plate 63).

4. How might the structure of early wood and late wood be related to seasonal conditions and the function of the cells? Think about environmental conditions during the growing season.

*In spring, there is abundant water, and large cells will move more water during periods of active growth. As summer heat arrives and water availability decreases, cell size decreases to conserve water.*

## EXERCISE 20.5
# Grocery Store Botany: Modifications of Plant Organs

## Materials

variety of produce: squash, lettuce, celery, carrot, white potato, sweet potato, asparagus, onion, broccoli, and any other produce you wish to examine

*You may want to ask students to bring in their own grocery store plants for identification. Farmers' markets often have unusual and challenging produce.*

## Introduction

Every day you come into contact with the plant world, particularly in the selection, preparation, and enjoyment of food. Most agricultural food plants have undergone extreme selection for specific features. For example, broccoli, cauliflower, cabbage, and brussels sprouts are all members of the same species that have undergone selection for different features. In this exercise, you will apply your botanical knowledge to the laboratory of the grocery store.

## Procedure

1. Working with another student, examine the numerous examples of root, stem, and leaf modifications on demonstration. (There may be some reproductive structures as well. Refer to Lab Topic 16, Plant Diversity II, if needed.)

2. For each grocery item, determine the type of plant organ, its modification, and its primary function. How will you decide what is a root, stem, or leaf? Review the characteristics of these plant organs and examine your produce carefully.

*For this exercise to be successful, do not tell your students the answers until all students have completed the activity. However, students will usually assume that sweet and white potatoes are both roots. So you might ask them what evidence they are using to make their decision. If they don't know what the "eyes" are, ask if anyone has ever cut and grown potatoes; ask what grew from the eyes. Hopefully, someone will know. You might also display a potato, long left in a cabinet, ready to show stems creeping out of the bud. Ask questions rather than answer them.*

## Results

Complete Table 20.1.

## Discussion

1. What feature of the white potato provided key evidence in deciding the correct plant organ?

   *Buds, or eyes, of the white potato indicate it is a stem.*

2. Based on your knowledge of the root, why do you think roots have been selected so often as food sources?

   *Roots have very large cortex regions with parenchyma for storage. These carbohydrate storage structures are excellent food sources.*

**Table 20.1**
Grocery Store Botany

| Name of Item | Plant Organ (Root, Stem, Leaf, Flower, Fruit) | Function/Features (Storage, Support, Reproduction, Photosynthesis) |
|---|---|---|
| Sweet potato | Root | Storage |
| White potato | Stem (buds present) | Storage |
| Celery | Leaf petiole | Water storage |
| Carrots | Root | Storage |
| Broccoli | Flowers and stems | Reproduction |
| Asparagus | Stems | Photosynthesis |
| Eggplant | Fruit | Reproduction |
| Lettuce | Leaves | Photosynthesis |
| Onion | Leaves | Storage |
| Many other possibilities | | |

*Other suggestions include artichokes (a flower bud), beets, fruits, rhubarb (a petiole), and more.*

# Reviewing Your Knowledge

1. Use Table 20.2 to describe the structure and function of the cell types seen in lab today. Indicate the location of these in the various plant organs examined. Refer to "Summary of Tissue Systems and Cell Types," Figure 20.1, and Color Plates 61, 62, and 63.

2. Some tissues are composed of only one type of cell; others are more complex. List the cell types observed in xylem and in phloem.

Xylem:

*vessel elements, tracheids, parenchyma, fibers*

Phloem:

*sieve-tube members, companion cells, parenchyma, fibers*

**Table 20.2**
Structure and Function of Plant Cells

| Cell Type | Structure | Function | Plant Organ |
|---|---|---|---|
| Epidermis | Flattened rectangular, living, closely packed | Protection, water retention, water absorption | Leaf, stem, root |
| Guard cells | Sausage-shaped, inepidermis, contains chloroplasts | Surrounds stoma, regulates air in/out | Leaf, stem (occasionally) |
| Parenchyma | Thin-walled polyhedral, living | Photosynthesis, storage, healing wounds | Leaf, stem, root |
| Collenchyma | Elongate, unevenly thickened walls | Support | Leaf, primary stem |
| Sclerenchyma | Thick-walled, lignified, usually lacking protoplast | Support | Leaf, stem, root |
| Tracheids | Elongate xylem cells, dead when mature | Support, water transport | Leaf, stem, root |
| Vessels | Barrel-shaped xylem cells, dead when mature | Support, water transport | Leaf, stem, root |
| Sieve tubes | Elongate, without nucleus when functional | Transport | Leaf, stem, root |
| Endodermis | Compact, with waxy Casparian strip | Selective absorption | Root |
| Primary meristems | Dividing cells, small, thin-walled | Gives rise to primary body | Stem tip, root tip |
| Vascular cambium | Dividing cells, small, thin-walled | Gives rise to secondary xylem and phloem | Stem, root |
| Pericycle | Dividing cells, small, thin-walled | Gives rise to lateral roots | Root |
| Periderm | Compact, prismatic, suberized | Water retention | Woody stem (component of bark) |
| Ray parenchyma | Thin-walled | Lateral transport, storage | Secondary xylem, phloem of stem, root |

3. What characteristic of sieve-tube structure provides a clue to the role of companion cells?

   *Sieve tubes are living but lack a nucleus. Cell function must be controlled in some way.*

4. Compare primary and secondary growth. What cells divide to form primary tissue? To form secondary tissue? Can a plant have both primary growth and secondary growth? Explain, providing evidence to support your answer.

   *Primary growth is derived from the division of apical meristems in the root tip and shoot tip. Primary growth results in an increase in length, along the plant axis. Secondary growth is derived from the division of vascular cambium and cork cambium. The cells divide and the diameter is increased. Woody plants, trees, and shrubs continually produce primary tissues as buds grow and develop and new roots are formed. At the same time, year after year, they increase in diameter as secondary xylem and phloem are produced by the vascular cambium. Herbaceous plants have only primary growth.*

## Applying Your Knowledge

1. Cells of the epidermis frequently retain a capability for cell division. Why is this important? (*Hint:* What is their function?)

   *repair damaged surface layer of the plant; to resist pathogens*

2. Why is the endodermis essential in the root but not in the stem?

   *The function of the root is to absorb water into the plant. The endodermis controls which dissolved substances are taken into the root tip and subsequently into the plant's vascular system. The uptake of water is not a function of the stem; therefore, a layer like an endodermis is not advantageous in the stem.*

3. In the summer of 1998, after extremely hot, dry weather, the Georgia corn harvest was expected to be reduced by at least 25%. Using your knowledge of the dual functions of guard cells relative to water retention and uptake of carbon dioxide, explain the reduction in photosynthetic productivity.

   *In response to insufficient supplies of water and physiological stress, guard cells will remain closed during the day. Carbon dioxide uptake will be reduced to low levels and photosynthesis will be limited.*

4. The density of stomata on the lower surface of leaves is affected by genetics and the environment. As the concentration of $CO_2$ decreases, the density of stomata increases. Hypothesize about the trends in stomatal density that might be expected in response to climate change and increased carbon dioxide concentration.

   *The density of stomata on the lower surface of leaves would decrease as the concentration of carbon dioxide increases. Any study would have to include a large number of plants and a range of species.*

5. The belt buckle of a standing 20-year-old man may be a foot higher than it was when he was 10, but a nail driven into a 10-year-old tree will be at the same height 10 years later. Explain.

   *Vertical growth is generally from the tip of the stems, not the base. In 10 years, the tree has increased in diameter (secondary growth), but all vertical growth is at stem tips.*

6. The number of annual rings in trees growing in temperate climates corresponds to the age of the trees. Scientists use tree ring analysis (dendrochronology) to age trees. By comparing patterns of tree ring size in cores from living trees with patterns in older dead trees and even samples of wood from old buildings, scientists have been able to piece together a chronology back thousands of years. Scientists in Europe are working collaboratively to develop a history of climate change for the last 10,000 years using tree ring analysis (dendroclimatology). How would regional climate changes affect the growth of trees? What features of secondary growth in trees would provide evidence for scientists to determine regional climate change? Why do you think this research is being conducted in temperate forests, but not in tropical forests?

   *Local changes in rainfall and temperature affect the growth of trees, and that in turn affects the size of tree rings. Scientists (often using computer analysis) can measure tree ring size and relate that to the master chronology. These can be compared for different species in different regions. Temperate climates have seasonal changes that result in spring growth and winter dormancy, thus providing clear evidence of yearly growth. This is not clear in tropical trees.*

7. The oldest living organisms on Earth are plants. Some bristlecone pines are about 4,600 years old, and a desert creosote bush is known to be 10,000 years old. What special feature of plants provides for this incredible longevity? How do plants differ from animals in their pattern of growth and development?

   *Animals have determinate growth; they reach their maximum size and cease to grow. Plants, however, have living meristems that continue to divide and grow (indeterminate growth).*

8. Many of the structural features studied in this laboratory evolved in response to the environmental challenges of the terrestrial habitat. Complete Table 20.3 naming the cells, tissues, and organs that have allowed vascular plants to adapt to each environmental factor.

**Table 20.3**
Adaptations of Plant Cells and Structures to the Land Environment

| Environmental Factor | Adaptations to Land Environment |
|---|---|
| Desiccation | *Cuticle, stomata, gametangia, sporangia* |
| Transport of materials between plant and environment | *Roots that lack cuticle; root hairs present* |
| Gas exchange | *Stomata with guard cells* |
| Anchorage in substrate | *Fibrous roots, taproots* |
| Transport of materials within plant body | *Stems; vascular tissues: xylem (vessels, tracheids), phloem (sieve tubes)* |
| Structural support in response to gravity | *Sclerenchyma fibers, parenchyma, xylem, tracheids, vessels* |
| Sexual reproduction without water | *Pollen grain, ovule, fruit, flowers, endosperm, seeds, carpel* |
| Dispersal of offspring from immobile parent | *Fruit, seeds* |

# Investigative Extensions

*If you would like to include monocots in this lab study, we encourage this as an investigation that could substitute for the hand sections in Exercise 20.3.*

1. A walk across campus, through the forest or along the path at the botanical garden will reveal an amazing diversity in the structure of roots, stems, and leaves. This outward diversity is accompanied by anatomical diversity in structure that is directly related to the functioning of these organs. Using the nut-and-bolt technique in Exercise 20.3, you can investigate questions relating to structural diversity. For other techniques for studying plant anatomy see http://www.publicbookshelf .com/public_html/Methods_in_Plant_Histology. Develop an investigation based on your observations of plant structures or consider one of the following suggestions.

   a. C3 and C4 plants use different photosynthetic pathways with corresponding differences in leaf anatomy. Using Web and other resources identify C3 and C4 species that are commercially available and compare the leaf anatomy for two related species.

   b. Angiosperms are divided into several groups, including eudicots and monocots. One feature that distinguishes these two groups is the difference in the organization of stem and root tissues. Investigate differences in anatomy for monocots and eudicots.

2. Plants balance the complex problems of gas exchange, temperature regulation, and water loss through stomata opening and closing. Using techniques from this lab topic investigate microenvironmental (at the scale of the plant) factors, such as light, temperature, wind, or soil moisture that might affect stomata function. See Brewer (1992) for a method for making leaf casts using nail polish and then observing these surface replicas under the microscope.

## Student Media: BioFlix, Activities, Investigations, and Videos
www.masteringbiology.com (select Study Area)

**BioFlix**—Ch. 35: Tour of a Plant Cell; Ch. 36: Water Transport in Plants
**Activities**—Ch. 35: Root, Stem, and Leaf Sections; Primary and Secondary Growth;
   Ch. 36: Transport of Xylem Sap; Translocation of Phloem Sap
**Investigations**—Ch. 35: What Are Functions of Monocot Tissues?
**Videos**—Ch. 35: Root Growth in a Radish Seed (Time-lapse);
   Ch. 36: Turgid Elodea

# References

Beck, C. *An Introduction to Plant Structure and Development: Plant Anatomy for the Twenty-First Century.* Cambridge: Cambridge University Press, 2009.

Berg L. *Introductory Botany: Plants, People, and the Environment,* 2nd ed. Belmont, CA: Thomson Brooks/Cole, 2007.

Brewer, C. A. "Responses by Stomata on Leaves to Microenvironmental Conditions," in *Tested Studies for Laboratory Teaching* (volume 12), Proceedings of the 13th Workshop/Conference of the Association for

Biology Laboratory Education (ABLE), 1992, Corer A. Goldman, Editor.

Mauseth, J. D. *Botany: An Introduction to Plant Biology,* 4th ed. Sudbury, MA: Jones and Bartlett Publishers, 2008.

Raven, P. H., R. F. Evert, and S. E. Eichorn. *Biology of Plants,* 7th ed. New York: W.H. Freeman Publishers, 2004.

Zeiger, E., G. D. Farquhar, and I. R. Cowan. *Stomatal Function.* Stanford, CA: Stanford University Press, 1987.

## Websites

American Society of Plant Biologists website with links for undergraduates:
http://www.aspb.org/resourcelinks/ scripts/cats2.cfm? cat=65

Click on General Botany and browse this site for images of plant cells and tissues and plant organ anatomy:
http://botit.botany.wisc.edu

Photographic Atlas of Plant Anatomy by J. Curtis, N. Lersten, and M. Nowak with excellent images:
http:// botweb.uwsp.edu/anatomy/

# LAB TOPIC 20

# Plant Anatomy
## Teaching Plan for Laboratories

## Main Concepts and Objectives

1.  Concept: Plant organs are composed of a few basic cell types and tissues, which are organized in different ways. Students will be able to identify and describe the structure and function of each cell and tissue type. They will describe the structure of each plant organ and the organization of cells, tissues, and regions within these organs. (Exercises 20.1–20.4)

2.  Concept: The structure (organization) of a plant organ is related to the special functions of that organ. Students should be able to compare structure and function for each plant organ. (Exercises 20.1–20.4)

3.  Concept: Plants have indeterminate growth, which results in *primary* and *secondary growth*. Students should be able to define primary and secondary growth, describe the distribution of each in the plant, describe the products of each, and be able to relate this to a herbaceous versus a woody habit. (Exercises 20.1–20.4)

4.  Concept: The anatomy of vascular plants is related to their terrestrial habitat. Students will discuss the adaptive significance of plant organs, tissues, and cells relative to the land environment. (Exercises 20.1–20.4)

5.  Concept: Understanding plants is important in human activities. Students will apply their knowledge to produce in the grocery store. (Exercise 20.5)

6.  Specific content. Students will be able to describe or define all terms in bold.

## Order of the Lab

1.  Introduce objectives and concepts and describe cell types (structure and function). (15 min)

2.  Perform Exercises 20.1 and 20.2. (20 min)

3.  Perform Exercise 20.3, Lab Study A, in this sequence:
    Embed living stem sections.
    Examine prepared stem sections.
    Introduce technique for making hand sections.
    Prepare and view hand sections. (60 min)

4.  Complete Exercise 20.3, leaves and roots. See the following note on classroom management. (20 min)

5. Perform Exercise 20.4.                                                    (15 min)

6. Perform Exercise 20.5, Grocery Store Botany. Review results
   after all students finish.                                                (20 min)

7. Discuss questions/answers.                                                (30 min)

**For a 2-hour lab:** Omit Exercise 20.5, Grocery Store Botany, and have students complete discussion questions outside of lab. Alternately, study only the prepared slide of the cross section of the stem and omit the embedding and study of the living stem. Complete questions outside of lab.

## Classroom Management

Each student should have a slide set and view slides individually. Students work in pairs for plant morphology (Exercise 20.1), to prepare hand sections (Exercise 20.3, Lab Study A), and for Grocery Store Botany (Exercise 20.5).

Three-dimensional models of plant organs are helpful aids for students to understand the structural organization of tissues. Students can become passive during an observational lab. Ask them how structure relates to function, how they will recognize cell types, distinguish plant organs, and so on. Encourage students to share particularly good slides with one another. Do not provide the answers to the Grocery Store Botany exercise until all students have finished this section. Urge them to discuss their ideas with one another.

## Student Development

Students will improve observational skills and develop the ability to envision three-dimensional structure from two-dimensional slides. They will learn to make their own slides. Students will apply knowledge of plant anatomy to daily experiences. Students will continue to learn specific content but also use this information to develop major concepts (relationship of structure and function).

## Lab Safety Precautions

Razor blades are used in Exercise 20.3. Remind students to keep fingers well out of the way of their slicing. Demonstrate the correct technique before students begin.

## Discussion and Summary

Use final questions with tables to review the material in this lab. Discuss the questions at the end of exercises, and help students relate these questions to lab observations.

## Evaluation

Test concepts on the lab practical. This lab provides evidence to support and explain major themes in the course. Students should be able to apply their knowledge to plants in their daily experience.

## Lab Topic 21
# Plant Growth

 This lab topic gives you another opportunity to practice the scientific process introduced in Lab Topic 1. Before going to lab, review scientific investigation in Lab Topic 1 and carefully read Lab Topic 21. Be prepared to use this information to design an experiment for plant growth.

*This lab topic may be used independent of lecture as a research project. If you do this, assign readings to prepare students for lab.*

## Laboratory Objectives

After completing this lab topic, you should be able to:

1. Describe external and internal factors that influence the germination of angiosperm seeds.
2. Explain the effect of auxin on plant growth.
3. Explain the effect of gibberellins on the growth of dwarf corn seedlings.
4. Define and give examples of *phototropism* and *gravitropism*.
5. Design and execute an experiment testing factors that influence seed germination and plant growth.
6. Present the results of the experiment in oral or written form.

*For a 2-hour lab: For the first week of the lab, omit one of the exercises. Students will design an original experiment to investigate the remaining exercise. See the Teaching Plan for additional suggestions.*

*If possible in your laboratory situation, 1 week before this lab, instruct students to read the lab topic and meet with their investigative team to discuss possible independent investigations before coming to the lab. Ask them to develop preliminary questions and hypotheses. Provide students with a list of available materials before lab so that they can submit a request for any additional materials prior to lab.*

## Introduction

In Lab Topic 16, Plant Diversity II, you studied the life cycle of angiosperms. You observed that fertilization of the egg in the female gametophyte results in an embryo protected in a **seed** consisting of the young embryonic sporophyte, food, and a protective seed coat. Seeds develop in the parent plant, and when they are mature, they are separated from the parent and dispersed. Most seeds go through a period of dormancy, but when a dormant seed finds a favorable environment, it will begin to **germinate;** that is, it renews its development and the embryo resumes its growth. Most plants continue to grow as long as they live, a condition known as **indeterminate growth.** Indeterminate growth is possible because of plant tissues called **meristems,** which remain embryonic as long as the plant lives. Seed germination and plant growth are regulated by external factors such as light, temperature, nutrients, and water availability. Plants respond to these environmental stimuli by internal mechanisms regulated by chemical messengers called **hormones.** A hormone, whether it is found in plants or animals, is produced in small quantities in one part of the organism and transported to another site, where it causes some special effect.

*Instructors and/or students can download data tables in Excel format. Go to www .masteringbiology.com. Look in the Study Area under Lab Media.*

In this lab topic, you will work in teams, investigating external stimuli and internal mechanisms that influence the germination of seeds and the growth of plants. You will be led through several brief introductory experiments (Experiment A of Exercises 21.1, 21.2, and 21.3); then your team will propose one or more testable hypotheses based on questions generated during the lab studies. You will then design and carry out an independent investigation based on your hypotheses (Exercise 21.4). You may design an experiment that can be completed in the laboratory period. However, you should plan to make observations of your plants over several days or at the beginning of the next laboratory. Your instructor will tell you if you will be able to return to the lab to make observations or if you should carry your experiment elsewhere for observations. You may be given time to finalize your results and presentation at the beginning of the next laboratory period. Your team should discuss the results and prepare an oral presentation. One member of your team will present your team's results to the class for discussion. This person should be prepared to persuade the class that your experimental design is sound and that your results support your conclusions. If required by the lab instructor, each of you will submit an independent laboratory report describing your experiment and results in the format of a scientific paper (see Appendix A).

The following summarizes the components of each activity in this lab topic.

*   Complete Experiment A in Exercises 21.1, 21.2, and 21.3.
*   Discuss possible questions that your research team might investigate.
*   Choose one interesting question from your list; be certain you can develop a *testable hypothesis.*
*   Design and initiate your open-inquiry investigation (Exercise 21.4).
*   Complete the experiment and report your results during the following laboratory period.

## Procedures for Independent Investigations: Germinating Seeds and Growing Plants

The following sections provide general procedures you will use in designing your open-inquiry investigation (Exercise 21.4). Return to this introductory material when planning your investigation.

### Experimental Plants

You may choose plants used in the lab studies for your independent investigation. These include *Zea mays* (corn), *Phaseolus vulgaris* (pinto bean), *Phaseolus limensis* (lima bean), *Coleus blumei* (a common ornamental annual with colorful variegated leaves), and *Brassica rapa* (related to mustard and cabbage). If you decide to use different plant species, check with your laboratory instructor about the availability of additional plants.

### Germinating Bean and Pea Seeds

Bean and pea seeds can be germinated by first submerging them in a 10% sodium hypochlorite solution for 5 minutes to kill bacteria and fungus spores on their surfaces. Follow this with a distilled water rinse and plant the seeds 1 cm deep in flats of vermiculite, a clay mineral that looks like mica and is frequently used as a starting medium for seeds. Add water or a test solution to the flats daily.

## Growing Wisconsin Fast Plants

The *Brassica rapa* seeds used in this exercise were developed by Dr. Paul Williams of the Department of Plant Pathology, University of Wisconsin, Madison. Dr. Williams used traditional breeding techniques to produce plants, called Wisconsin Fast Plants™, that can complete an entire breeding cycle from seed to seed in 35 days (Figure 21.1). Because of the rapid growth and shortened breeding cycle of these plants, they are excellent investigative tools for use in plant growth experiments.

### A. Seed Germination Exercises

*Brassica rapa* seeds can be germinated by placing them on wet filter paper in the lid of a petri dish. Stand the dish, tilted on its end, in a water reservoir such as the bottom of a 2-L soft-drink bottle (Figure 21.2a). The dish and reservoir should be placed under fluorescent lights. Germination begins within 24 hours, and observations can be made for several days. It is important to keep the filter paper moist by carefully adding water. If you wish to make quantitative measurements of seed germination, tape a transparent grid sheet marked in measured increments to the outside of the petri dish lid. Place the wet filter paper in the lid, as before, and plant the seeds at a particular position in relation to the grid. As the seeds germinate and grow, you can easily use the grid to measure their size (Figure 21.2b).

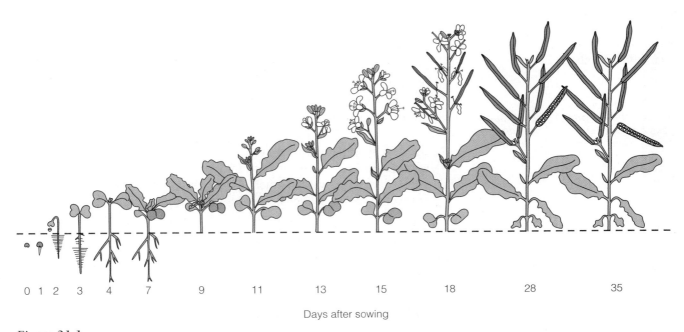

0 1 2   3   4   7      9      11      13      15      18         28         35

Days after sowing

**Figure 21.1.**
**Life cycle of Fast Plant *Brassica rapa*.** These plants can complete their entire life cycle from seed to seed in 35 days.

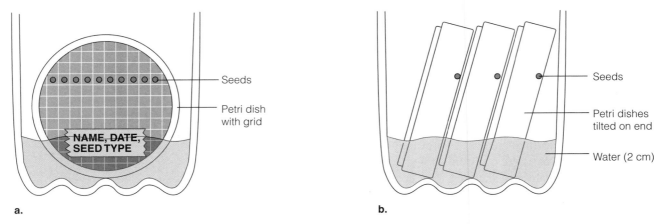

**a.**

**b.**

**Figure 21.2.**

**Germinating** *Brassica rapa* **seeds in a petri dish.** (a) Place the seeds on wet filter paper in the lid of a petri dish. Attach a grid to the outside of the lid for easy seedling measurement. (b) Stand the dishes on end in a water reservoir.

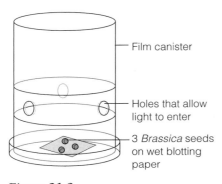

**Figure 21.3.**

**Growing** *Brassica rapa* **seeds in film canisters.** Place seeds on moist blotting paper in the lid. Holes can be punched in the canister to allow light to enter the chamber.

*Instructions for making floral foam disks and wick and grid germination strips are in the Preparation Guide.*

## B. Tropism Studies

1. **Phototropism.** *Brassica rapa* seeds can be germinated in empty 35-mm clear or black film canisters (Figure 21.3). The canisters can be used as is or black canisters can be modified by punching holes in the sides to allow light to enter the chamber. Place small, appropriately sized squares of wet blotting paper or floral foam disks in the lid, and place two or three seeds on the blotting paper or floral foam disks. (Do not use filter paper; it dries out too quickly.) Invert the canister and snap it into the lid. Holes in the canister can be covered with different-colored filters, and the size of the holes can be varied to alter the quality or quantity of light hitting the plants.

2. **Gravitropism.** Seeds and seedlings may be used to demonstrate gravitropism. For investigating gravitropism and *seed germination,* prepare a windowless black film canister with wet blotting paper as described for phototropism. Place three seeds on the blotting paper. Place the chamber horizontally and use a permanent marker to indicate which side is up. The chamber will resemble Figure 21.3, but there will be no holes and the chamber will be placed on its side. (*Hint: Tape a film canister lid to the outside of the canister to keep it from rolling.*) Within 1 or 2 days you will be able to observe the effects of gravitropism.

To observe gravitropism in *3-day-old seedlings,* attach a *wick and grid germination strip* to the inside wall of a windowless black film canister, with the grid strip between the wick and the side of the canister. Add just enough water to cover the bottom of the canister, and have the wick in contact with the water. Attach a young seedling with cotyledons present to the center of the wick strip with the *cotyledons* against the wick and the *hypocotyl* pointing out, into the canister. (See Figure 21.4.)

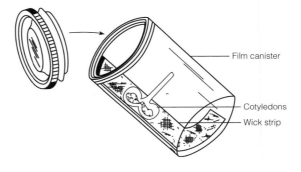

**Figure 21.4.**

**Growing** *Brassica rapa* **seedlings in film canisters.** Attach the seedling to the center of the wick strip with the cotyledons against the wick as shown.

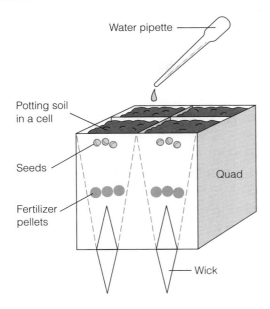

Water pipette

Potting soil
in a cell

Seeds

Quad

Fertilizer
pellets

Wick

**Figure 21.5.**
**Growing *Brassica rapa* plants in quads.** Pull a wick through the hole in each cell. Add potting soil, fertilizer, and seeds. Initially, water using a pipette.

## C. Growing Brassica rapa Seedlings in Quads

Scientists working with Wisconsin Fast Plants suggest germinating seeds in small, commercially available Styrofoam™ containers called *quads*, which contain four cells, or chambers. To germinate seeds in quads (Figure 21.5):

1. Add a wick to each cell to draw water from the source into the soil.
2. Add potting mix until each cell is about half full.
3. Add three fertilizer pellets.
4. Add more soil and press to make a depression.
5. Add two or three seeds to each cell and cover them with potting mix.
6. Carefully water each section using a pipette until water soaks through the potting mix and drips from the wick.
7. Place the quad on the watering tray under fluorescent lights.

After the seeds begin to germinate, you can manipulate the plants in many different ways to investigate plant growth.

*We recommend that you purchase the* Wisconsin Fast Plants Manual, *published by Carolina Biological Supply, and follow the very helpful tips for success in these experiments. Fast Plant prep is not difficult, but it does require attention to details, such as ensuring a continuous supply of water to plants and constant light.*

---

# E X E R C I S E  2 1 . 1
# Factors Influencing Seed Germination

---

Seeds are the means of reproduction, dispersal, and, frequently, survival for a plant. Plants are immobile and can colonize new habitats and escape inhospitable weather only through the dispersal and dormancy of seeds.

*Dormancy* is a special condition of arrested growth in which the seed cannot germinate without special environmental cues. In this exercise, you will observe germinating bean and *Brassica rapa* seeds. The beans have been germinating in a moist environment for several days. The *Brassica rapa* seeds are germinating on wet filter paper in the lid of a petri dish.

# Experiment A. Germinating Bean and *Brassica rapa* Seeds

## Materials

germinating bean seeds
petri dishes with germinating *Brassica rapa* seeds
stereoscopic microscope or hand lens

## Introduction

In this lab study, you will examine seed and seedling morphology. Working individually, you will identify seed parts in two plants, a species of bean and *Brassica rapa*. As you investigate the morphology of seeds, consider the role of each structure in facilitating the function of the seed.

*We have tried germinating lima beans, pinto beans, and mung beans. Of these three, pinto beans germinate earlier and in greater numbers. Begin germinating the seeds of choice in vermiculite about a week before you will use them in lab. For labs taking place throughout the week, we begin a batch 1 week before the first lab, and another batch in 2 or 3 days. This ensures seeds in a usable stage of germination for the entire week. Begin germinating B. rapa seeds 2 to 3 days before lab. See the Preparation Guide for more details.*

## Procedure

1. Examine a germinating bean seed and identify the **seed coat, cotyledons** (seed leaves), and **embryo** consisting of the **radicle** (embryonic root), **hypocotyl** (plant axis below the cotyledons), and **epicotyl** with young leaves (plant axis above the cotyledons).

2. Secure a petri dish with germinating *Brassica rapa* seeds. Carefully examine the seeds using a hand lens or the stereoscopic microscope and identify the **seed coat** (which may have been shed), **cotyledons** (how many?), **hypocotyl** (the portion of the plant below the cotyledons), **primary root**, **root hairs**, and **young shoot**, consisting of the epicotyl and young leaves.

## Results

1. Label the parts of the bean seed in Figure 21.6.
2. Draw and label several germinating *B. rapa* seeds in the margin of your lab manual.

## Discussion

1. What is the function of the seed coat? From what does it develop?

   *It prevents water loss from the seed. It develops from the parent sporophyte tissue in the ovule.*

**Figure 21.6.**
**The structure of a bean seed.** Add the labels: *seed coat, cotyledons, radicle, hypocotyl,* and *epicotyl.*

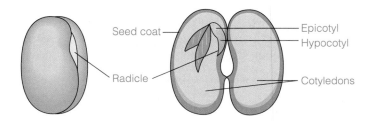

Seed coat — Epicotyl

Hypocotyl

Radicle — Cotyledons

2. What has happened to the endosperm that formed in the embryo sac of the developing ovule observed in Lab Topic 16 Plant Diversity II?

   *It has been consumed as it supplied nutrition to the developing embryo.*

3. How have the cotyledons developed, and from what? (Check your text; see, for example *Campbell Biology* 9th ed. Chapter 38.)

   *Cotyledons, or seed leaves, develop from the zygote. They are a component of the young, developing embryo.*

4. How does the structure of the seed facilitate the dispersal and survival of the plant?

   *The seed separates from the parent plant and is carried to new environments. It provides nourishment and protection for the developing embryo and can withstand adverse environmental conditions. It responds when conditions are favorable for renewed germination and growth of the embryo.*

# Experiment B. Student-Designed Investigation of Seed Germination

## Materials

seeds of *Brassica rapa*

various other seeds—beans, corn, radish, okra, lettuce

35-mm film canisters—clear plastic, black, or black with holes punched in the sides

small squares of blotting paper soaking in water

floral foam disks 28 mm diameter, 2–4 mm thick, soaking in water

red, green, and blue light filter sheets

tape

water baths, oven, refrigerator

grid sheets that fit the lids of petri dishes

wick and grid germination strips

reservoir

forceps

scissors

waterproof pen

agar plates

empty petri dishes

sandpaper

various pH solutions

10% sodium hyperchlorite solution

solutions of gibberellin or other hormones

fertilizer solutions in various concentrations and/or dry fertilizer pellets

hole punch

*Depending on the season and the availability of seeds in stores and gardens, consider lettuce, okra, black locust, mimosa, four-o'clocks, and others.*

*See the Preparation Guide for instructions to prepare wick and grid germination strips.*

## Introduction

If your team chooses to study seed germination for your independent investigation and report, use the available materials. Design a simple experiment to investigate factors involved in the germination of seeds.

*Students should rinse seeds for germination studies in 10% Clorox (sodium hyperchlorite) for 5 minutes to kill fungus and bacteria spores.*

## Procedure

1. Collaborating with your research team, read the following questions, check your text and other sources for supporting information, and choose a question to investigate, using this list or your own imagination.

   a. Do seeds need light to germinate? What effect will germination in total darkness have on the process? Is germination better in alternating light and dark, as in nature?

b. Is germination different in different wavelengths (colors) of light?

c. What effect does scarification (scratching the seed coat) have on germination of various seeds? How does scarification occur in nature?

d. What temperature regimens are optimum for seed germination? Is germination affected by alternating temperatures (as in nature)? Is a constant temperature favorable?

e. Does gibberellin (Exercise 21.3) promote or inhibit seed germination? What about other hormones—for example, cytokinins or abscisic acid? Check your text for other test substances.

f. What effect does fertilizer have on seed germination and seedling growth? Is more always better?

g. What effect will salt solutions or acid solutions have on seed germination? Why would questions such as these be of interest?

h. Does seed size have an effect on germination rates? On seedling size?

i. Some crops are limited in their distribution by cold soils that affect seed germination. Will changes in regional climate affect seed germination in okra, broccoli, or varieties of wheat?

2. Design your experiment, proposing hypotheses, making predictions, and determining procedures as instructed in Exercise 21.4.

*Acid rain and saltwater intrusion into groundwater may be topics of interest in your area.*

---

# EXERCISE 21.2
# Plant Growth Regulators: Auxin

Both plants and animals respond to environmental cues. Animals generally respond rapidly via the nervous system or more slowly via hormones secreted from endocrine glands. Plants lack a nervous system and respond to environmental stimuli via chemical messengers or hormones. Changes in these hormones lead to altered patterns of growth and development. **Auxin** is the name given to a complex of substances that promotes stem growth. The natural auxin, indoleacetic acid (IAA), is a hormone produced in apical meristems. It migrates down the stem from the zone of production. If the growing tip is removed, the stem will not elongate, but if the tip is replaced with a paste containing IAA, elongation will continue. At low concentrations, IAA facilitates cell elongation and promotes growth by breaking linkages among cellulose fibers and loosening the cell wall. In this exercise, you will investigate the role of auxin in stem and root curvature in response to light and gravity.

## Experiment A. Gravitropic and Phototropic Curvature in *Coleus blumei*

### Materials

*Place these plants in the respective conditions about 7 to 10 days before they will be used for this lab. Other rapidly growing plants can be substituted for Coleus. If we have young Coleus plants (or other plants) available, we sometimes transplant them to glass containers with the plants placed near the sides of the container so that students can see the roots.*

(on demonstration)
*Coleus* plant placed on its side
*Coleus* plant in unilateral light
*Coleus* plant in upright position

## Introduction

In this exercise, you will investigate the growth of the stem and root in response to two environmental stimuli, gravity and light. **Gravitropism** (or geotropism) is the response of plant organs to gravity. Different plant organs may show positive or negative gravitropism. **Phototropism** is the response of a plant to light. A plant organ may grow toward a light stimulus (positive phototropism) or away from a light stimulus (negative phototropism). Three *Coleus blumei* plants are on demonstration in the lab room. One was placed on its side several days ago. Another has been growing in unilateral light for several days. The third, the control, was left undisturbed in the greenhouse until lab time.

## Procedure

1. Study gravitropic curvature in *Coleus blumei.*
   a. Carry your lab notebook to the demonstration area and observe the *Coleus* plant placed on its side.
   b. Examine the plant, noting the appearance of different regions of the stems and roots (if visible). What part of each stem has curved? To what degree?
   c. Compare this plant with the control plant left in an upright position.
   d. Describe your observations and answer the questions in the Results section.

2. Study phototropic curvature in *Coleus blumei.*
   a. Examine the plant growing in unilateral light, noting the appearance of different regions of the stems.
   b. Compare the plant to the control plant, which has received light from all directions in the greenhouse.
   c. Record your observations and answer the questions in the Results section.

## Results

1. Describe the appearance of the plant lying on its side. *Sketch* the plant in the margin of your lab manual. How does this compare with the appearance of the control plant?

   *The plant has curved upward near the growing tip of the stem. The stem of the control plant is totally vertical with no curvature in the stem. If plants are in glass containers, or if students are allowed to remove the plant from the pot, they will see that the roots have grown downward.*

2. What is the extent of the response to gravity? How could you measure or quantify the response?

   *The curvature is almost 90°. You could measure this with a protractor. A student might use a marker to make circles around the stem at measured intervals and then determine the position of the marks after bending takes place.*

3. Describe the appearance of the plant in unilateral light. Sketch the plant in the margin of your lab manual. Compare this plant with the control plant.

   *The control plant should have a straight stem, in contrast to the plant in unilateral light, the stem of which grows toward the light.*

4. What is the extent of the response to the light? How could you measure or quantify the response?

*A protractor could be used to measure the angle of growth. Marks could be made on the stem to determine how much the plant grew and how far from the apical meristem the curvature took place.*

5. What part of the plant specifically is being affected? (If you need to, review the anatomy of the apical meristem in Lab Topic 20, Plant Anatomy.) Can you explain why?

*The zone of elongation in the meristematic region is being affected. Auxin causes cells to elongate, not to multiply. Therefore, this zone, and not the zone of division, would be affected.*

## Discussion

1. What type of response would you expect to see if you reoriented the plant lying on its side? Explain.

*Students might suggest that the stem would straighten out. It does not, but a second curve is formed as the tip now grows vertically in relation to the new orientation.*

2. How do plants detect the force of gravity (according to current theory)? Consult your text, if necessary.

*Campbell Biology, 9th ed. (Chapter 39) discusses several hypotheses. One proposes that gravity is detected by the "settling of statoliths, specialized plastids containing dense starch grains, to the lower portions of cells." The accumulation of the statoliths in the lower portion of the cells causes the distribution of calcium, which in turn causes movement of auxin in the root. Research with plants that lack statoliths indicates that their roots still respond to gravity.*

3. Review with your team members the physiological basis for the growth response. How is auxin involved in gravitropism? Where is it produced? Consult your text, if necessary.

*The growth response is due to the asymmetrical distribution of auxin in the stem and the root. The high concentration of auxin in the lower parts of the stem causes those cells to elongate more than the cells in the upper parts of the stem. As the lower cells elongate, the stem curves upward. In contrast, the root curves downward because the higher concentration of auxin inhibits the elongation of root cells in the lower parts of the root. This points out the differential sensitivities of cells to auxin concentrations.*

*Auxin is synthesized in the apical meristem and young leaves of shoots.*

4. Is curvature of the stem in response to gravity and light the result of additional cell division or cell elongation? How do you know, or how could you investigate this?

*As stated previously, curvature of the stem is due to cell elongation, not cell division. We could examine this histologically by sectioning stem tips treated with auxin and comparing the number of cells and the size of these cells with a stem tip that had not been treated with auxin (see the procedure for sectioning stems in Lab Topic 20, Exercise 20.3, Lab Study A).*

5. What is the role of auxin in phototropism? How is directional light detected in the plant? Use your text and discuss these items with members of your research team.

*According to the classical hypothesis explaining phototropism in grass coleoptiles, cells on the darker side of a stem elongate faster than cells on the brighter side because of an asymmetric distribution of auxin from the shoot tip. However, studies of phototropism by organs other than grass coleoptiles provide less support for this idea. For example, illumination from one side does not cause an asymmetrical distribution of auxin in the stems of sunflowers, radishes, and other eudicots. There is, however, an asymmetrical distribution of substances that may act as growth inhibitors, with these substances more concentrated on the lighted side of a stem (Campbell Biology, 9th ed.).*

 Student Media Videos—Ch. 39: Phototropism; Gravitropism

# Experiment B. Student-Designed Investigation of Auxins

## Materials

auxin solutions in various concentrations
auxin in lanolin paste
lanolin with no auxin
scissors
*Coleus* plants
glass containers for planting
*Brassica rapa* seedlings in quads
corn and bean seedlings in pots or flats
lamps
toothpicks
protractor

spray bottles
cotton-tipped applicators
aluminum foil
35-mm black film canisters
floral foam disks, 28 mm diameter, 2–4 mm thick, soaking in water
wick and grid germination strips
forceps
red, green, and blue light filter sheets
*Brassica rapa* seeds
3-day-old *Brassica rapa* seedlings germinated in petri dishes on wet filter paper

## Introduction

Having made observations of gravitropism and phototropism in the preceding experiment, discuss with your team members ways to use *Coleus, Brassica,* or corn or bean seedlings to further investigate these phenomena. If you choose to carry out your independent investigation with this system, the questions provided in the following Procedure section will be appropriate for your study. Using the materials available, design a simple experiment to investigate the role of auxin in plant growth or factors involved in phototropism and gravitropism.

## Procedure

1. Collaborating with your research team, read the following questions, check your text and other sources for supporting information, and choose a question to investigate, using this list or your own imagination.

   a. If only some wavelengths stimulate phototropic response, which ones do and which do not?

b. If the apical meristem is removed, will plants respond to unilateral light?

c. How does the root respond to gravity (if the cotyledons are kept stationary)?

d. If the tip of the root is removed, will roots respond to gravity? (Seedlings can be planted close to the wall in glass containers so that root growth can be viewed, or you may use *Brassica rapa* seeds germinating in dark film canisters.)

e. Can these tropisms be altered by applying auxin paste to the plant?

f. What will happen if the tips of the plants (root or stem) are covered with aluminum foil?

g. How else does auxin affect plants? How does auxin affect apical dominance? Can auxin be used as an herbicide? (In what concentrations? What is the effect on plants?) What concentration of auxin produces the largest roots on cut stems? (What horticultural application would this have?)

h. Will seedlings growing in the dark respond to auxin applied to the side of the stem? At all locations on the stem?

i. The herbicide 2,4-dichlorophenoxyacetic acid (2,4-D), used to control weeds, is described as a synthetic auxin. How does 2,4-D affect plant growth? What plants are affected or unaffected by 2,4-D and at what concentrations?

*If students work with herbicides, be sure they isolate their plants for treatment (preferably outside) and avoid contact with other plants. Designate glassware and spray bottles for herbicide only.*

2. Design your experiment, proposing hypotheses, making predictions, and determining procedures as instructed in Exercise 21.4.

---

## EXERCISE 21.3
# Plant Growth Regulators: Gibberellins

**Gibberellins** are another group of important plant growth hormones found in high concentrations in seeds and present in varying amounts in all plant parts. In some plants, gibberellins produce rapid elongation of stems; in others they produce **bolting**, the rapid elongation of the flower stalk. Produced near the stem apex, gibberellins work by increasing *both* the number of cell divisions and the elongation of cells produced by those divisions. The effects of gibberellins can be induced by artificially applying solutions to plant parts. Not all plants respond to the application of gibberellins, however, and in this exercise, you will investigate the effect of applying gibberellin solutions to normal and to dwarf (mutant) corn seedlings.

## Experiment A. Effects of Gibberellins on Normal and Dwarf Corn Seedlings

### Materials

2 pots each with 4 tall (normal) corn seedlings
2 pots each with 4 dwarf (mutant) corn seedlings
calculator

*Occasionally the dwarf corn seeds that we order do not germinate well or are not available. We recommend that you order early and try germinating a few of the seeds before you begin the germinations for lab. If the corn does not germinate, we substitute dwarf and normal peas. See the* Preparation Guide *for details.*

## Introduction

The seedlings used in this exercise are approximately the same age, but they exist in two phenotypes, tall and dwarf (Figure 21.7). The tall seedlings are wild type, or normal. A genetic mutation produces dwarf plants that lack gibberellins. Each team of students has four pots, each with four corn seedlings. Your instructor has previously treated the plants with either water or a gibberellin solution. In this experiment, you will investigate the effects of these treatments on the corn plants.

## Procedure

1.  Several days ago, your instructor sprayed the plants with either distilled water or a gibberellin solution, as follows:

|              *Control*              |              *Treated*              |
| Pot 1: normal corn, water treatment | Pot 2: normal corn, gibberellin treatment |
| Pot 3: dwarf corn, water treatment  | Pot 4: dwarf corn, gibberellin treatment |

2.  Observe the results of the treatments.
3.  Measure the height of each of your plants and record these data in Table 21.1.

*If using peas, be careful not to break their stems when measuring them.*

**Figure 21.7.**
**Normal corn plants and recessive dwarf mutants.**

**Table 21.1**
Height of Normal and Dwarf Corn Seedlings with and Without
Gibberellin Treatment

| | Normal Control (Pot 1) | Normal Treated (Pot 2) | Dwarf Control (Pot 3) | Dwarf Treated (Pot 4) |
|---|---|---|---|---|
| Plant 1 Height | | | | |
| Plant 2 Height | | | | |
| Plant 3 Height | | | | |
| Plant 4 Height | | | | |
| Mean Height | | | | |

## Results

1. Determine and record the mean height of plants in each category in Table 21.1.

2. Using the mean height for each category of plants, calculate the percentage difference in the mean height of treated normal plants and control normal plants. Then calculate the percentage difference in the mean height of treated dwarf plants and control dwarf plants. Use the given formula for your calculations:

$$\text{Normal \% difference} = \frac{\text{treated} - \text{control}}{\text{control}} \times 100 = \underline{\hspace{1cm}}\%$$

$$\text{Dwarf \% difference} = \frac{\text{treated} - \text{control}}{\text{control}} \times 100 = \underline{\hspace{1cm}}\%$$

*Have each student team record its final results on the board or on an overhead projector acetate. Then students can calculate mean values for the entire class.*

3. Record your data for the average percentage difference in the mean values for both normal and dwarf plants from Table 21.1 on the class master sheet. Then calculate the average percentage differences for the entire class.

|  | *Your Data* | *Class Data* |
|---|---|---|
| Average % difference: normal | _____ | _____ |
| dwarf | _____ | _____ |

## Discussion

1. How do values for percentage difference compare for dwarf versus normal treated and untreated plants?

   *The percentage difference in height is much greater between dwarf treated and dwarf untreated than between normal treated and normal untreated. The percentage difference in height of normal treated and normal untreated should approach zero.*

2. What is the action of gibberellins? Discuss your results with your group, and consult your text or other references in the laboratory.

   *The dwarf condition is due to a mutation that results in no gibberellin synthesis. Adding gibberellins reverses this phenotype, causing the internodes to elongate and the plants to grow to normal phenotype. Adding gibberellins to plants that are normally tall has little effect on size. Apparently, these plants have adequate gibberellins present to attain maximum height.*

## Experiment B. Student-Designed Investigation of Gibberellins

### Materials

normal and dwarf *Brassica rapa* seedlings (*petite* and *rosette* strains)
normal and dwarf corn seedlings
normal and dwarf pea seedlings (Little Marvel peas, *Pisum sativum*)

Solutions of gibberellin
dropper bottles
sprayers
cotton-tipped applicators
supplies to grow *Brassica rapa* seedlings in quads

### Introduction

Having seen the effect of gibberellins on the growth of normal and dwarf corn seedlings in the preceding experiment, discuss with your team members possible questions for further study of this group of hormones. If you choose to carry out your independent investigation with this system, the questions provided in the following Procedure section will be appropriate for your study. Using the materials available, design a simple experiment to investigate the actions of gibberellins in plant growth or seed germination.

### Procedure

1. Collaborating with your research team, read the following questions, check your text and other sources for supporting information, and choose a question to investigate, using this list or your own imagination.

   a. Plant scientists have discovered a mutant strain of *Brassica rapa* in which plants are dwarf. In these plants, the internodes do not elongate, and plants consist of a cluster of leaves spreading close to the

*When purchasing dwarf seeds, order rosette seeds.*

soil. Flowers cluster close to the leaves. What could be the cause of this phenotype?

b. Would other plant hormones produce the same response in dwarf corn seedlings as do gibberellins?

c. In the demonstration experiment, the gibberellin solution was sprayed on all parts of the plant. If the gibberellin solution were added only to specific regions, such as the roots or apical meristem, would the effect be the same?

d. Would the results in the corn experiment differ with different concentrations of gibberellin solution?

e. What effect do gibberellins have on seed germination?

f. What effect do gibberellins have on root growth on cut stems?

g. Is the dwarf condition seen in certain strains of peas (Little Marvel peas) due to the lack of gibberellins?

h. There are two dwarf forms of *Brassica rapa*—*rosette* and *petite*. Is dwarfism in these mutant strains due to insufficient gibberellin or to some other factor?

i. *Brassica rapa* mutant, *tall* (ein/ein genotype) produces an excess of gibberellins. Can the elongated growth in this mutant be inhibited by growth regulator, Cyocel™? Can growth be further stimulated by applying gibberellins? Research the mode of action for Cyocel™.

2. Design your experiment, proposing hypotheses, making predictions, and determining procedures as instructed in Exercise 21.4.

---

## EXERCISE 21.4
# Designing and Performing an Open-Inquiry Investigation

---

### Materials

see each Experiment B materials list in Exercises 21.1, 21.2, and 21.3.

### Introduction

Now that you have completed Experiment A in each of the exercises, your research team should select one factor that affects plant growth to investigate. Return to the investigation of your choice and review the suggestions in Experiment B. Use Lab Topic 1 Scientific Investigation as a reference for designing and performing this independent investigation. You will need to think critically and creatively as you ask questions and formulate your hypothesis. As a team, review and modify the procedures, determine any additional required materials, review the techniques and procedures for growing plants and germinating seeds, and assign tasks to all members of your research team. Your experiments will be successful if you plan carefully, think critically, perform techniques accurately and systematically, and record and report data accurately. The following outline will assist you in designing and performing your original investigation.

You and your lab partner are responsible for the care and maintenance of your plants. Remember to check the water. The success of your investigation depends on the plants' survival.

## Procedure

1. **Decide on one or more questions to investigate.** Suggested questions are included in Experiment B of Exercises 21.1, 21.2, and 21.3 as a starting point. (Refer to Lab Topic 1, Exercise 1.1, Lab Study A. Asking Questions.)
   **Question:**

2. **Formulate a testable hypothesis.** (Refer to Exercise 1.1, Lab Study B. Developing Hypotheses.)
   **Hypothesis:**

3. **Summarize the essential elements of the experiment.** (Use separate paper.)

4. **Predict the results of your experiment based on your hypothesis.** (Refer to Lab Topic 1, Exercise 1.2, Lab Study C. Making Predictions.)
   **Prediction:** (If/then)

5. **Outline the procedures used in the experiment.** (Refer to Lab Topic 1 Exercise 1.2, Lab Study B. Choosing or Designing the Procedure.)
   a. Review and modify the procedures used in Experiment A (see Germinating Seeds, Auxin, or Gibberellin). Review the Procedures for Independent Investigations: Germinating Seeds and Growing Plants in the Introduction section of this lab topic. List each step in your procedure in numerical order.
   b. Critique your procedure: check for replicates, level of treatment, controls, duration of experiment, growing conditions, glassware, and age and size of plants. Review measurements and intervals between measurements. Assign team members to check plants periodically and water if needed.
   c. If your experiment requires materials other than those provided, ask your instructor about their availability. If possible, submit requests in advance.
   d. Create a table for data collection. Using examples in this lab topic as a model or design your own. If computers are available, create your table in Excel. Remember to include space for general observations of plant growth.

6. **Perform the experiment,** making observations and collecting data for analysis.

7. **Record results including observations and data** in your data table. Make notes about experimental conditions and observations. Do not rely on your memory for information that you will need when reporting your results.

8. **Prepare your discussion.** Discuss your results in light of your hypothesis.

   a. Review your prediction. Did your results correspond to the prediction you made? If not, explain how your results are different from your predictions, and why this might have occurred.

   b. Review your hypothesis. Review your results (tables and graphs). Do your results support or falsify your hypothesis? Explain your answer, using your data for support.

   c. If you had problems with the procedure or questionable results, explain how they might have influenced your conclusion.

   d. If you had an opportunity to repeat and expand this experiment to make your results more convincing, what would you do?

   e. Summarize the conclusion you have drawn from your results.

9. **Be prepared to report your results to the class.** Prepare to persuade your fellow scientists that your experimental design is sound and that your results support your conclusions.

10. If your instructor requires it, **submit a written laboratory report** in the form of a scientific paper (see Appendix A). Keep in mind that although you have performed the experiments as a team, you must turn in a lab report of *your original writing*. Your tables and figures may be similar to those of your team members, but your paper must be the product of your own literature search and creative thinking.

*Consider having a symposium day for teams to present their results. A poster session also would allow students to communicate results in ways that are similar to scientific meetings.*

## Reviewing Your Knowledge

1. Having completed this lab topic, you should be able to define and use the following terms, providing examples when appropriate: *seed, seedling, seed coat, cotyledon, endosperm, radicle, hypocotyl, epicotyl, germination, dormancy, phototropism, gravitropism, apical dominance, auxin, gibberellins, bolting.*

2. Students investigating plant growth and the effects of hormones removed the seeds from developing strawberries and compared the size and time of fruit development. The strawberries failed to enlarge and become red and fleshy in plants where the seeds were removed. Research the roles of auxin and gibberellin and determine which hormone is responsible for fruit development.

   *Auxin is produced by seeds as they develop, stimulating the formation of the fleshy strawberry.*

## Applying Your Knowledge

1. Auxin is directly or indirectly responsible for apical dominance in plants. In this phenomenon, the growth of lateral or axillary buds (described in Lab Topic 20, Plant Anatomy) is inhibited by the auxin that moves down the stem from the apical meristem. It has long been the practice of horticulturists to clip off the apical meristems of certain young houseplants. What impact should this practice have on subsequent plant growth and appearance?

   *When the apical meristem is removed, the source of auxin for the stem is removed, thus eliminating the inhibiting influence of the*

*auxin on the lateral buds, which now begin to grow. The plant be-*
*comes more bushy and compact.*

2. What adaptive role would positive phototropism play in a natural ecosystem where plants are crowded?

   *This reduces the effect that shading by nearby plants might have on growing tips and increases the chance that apical meristems of all plants will receive light.*

3. Heliotropism, or solar tracking, is a plant movement whereby some plants orient their leaves or flowers to follow the position of the sun over the course of the day. Based on your investigations, what hormone do you predict might be involved in solar tracking? Briefly describe an experiment to test whether a particular plant species demonstrates solar tracking.

   *The hormone involved is auxin. Plants could be placed in a green-house with an unobstructed view of the sun's path or in a laboratory with a movable light set up. Using replicate potted plants, you could mark the leaves or flowers and periodically during the day measure the angle of movement from a starting point. Movements could also be recorded from a set point using a digital camera.*

4. During the 1960s, a world food shortage was solved by the breeding and cultivation of a race of wheat with higher grain production and short stems that resisted the damage of wind and rain. Norman Borlaug won the Nobel Prize in 1970 for his contributions to this "Green Revolution" that reportedly saved the lives of more than 1 billion people. Based on your knowledge of plant hormones, suggest the hormone that would be deficient in this race of wheat. Explain using the evidence observed in the laboratory.

   *Gibberellin was deficient in these plants and the stems failed to elongate. This is similar to the investigation of corn in which the dwarf plants were treated with gibberellin and then grew to normal heights.*

5. Brassinosteroids are a recently discovered class of plant hormones, which are similar in structure to animal steroids. Initially isolated in *Brassica napus* (thus their name), they now have been identified in a wide range of plants. Brassinosteroids are more potent than auxins and gibberellins requiring much smaller concentrations to effect changes in plant growth. These plant hormones can cause cell elongation among other effects on plant growth. Scientists have long known that stems of normal seedlings grown in the dark elongate rapidly and become light colored, a condition called *etiolation*. Recently scientists have cultured a mutant race of *Arabadopsis*, a model plant for physiology studies, that does not become etiolated in the dark. Utilizing *Arabadopsis* mutant and normal plants, devise an experiment to test the role of brassinosteriods in etiolation. Consider controlled variables, control treatments, replicates, how you will measure plant growth, and the time period for making measurements and observations. State your prediction based on the experiment you design.

   *An example of a possible answer:*

   *Small* Arabadopsis *plants can be grown—both normal and mutant— in the dark. One group of 6 normal and 6 mutant plants would be sprayed with brassinosteriod hormones, while a second group of 6 normal and 6 mutant plants would be sprayed with water. The length of the stem would be measured after 7 days in the dark. If the mutant condition were due to a lack of brassinosteroids, then the mutant plants treated with the hormone would have long stems equal in length to the normal plants. The mutant plants treated with water would not have elongated stems. Their stems would be much shorter than the normal plants treated with water.*

**Student Media: BioFlix, Activities, Investigations, and Videos**
www.masteringbiology.com (select Study Area)

**Activities**—Ch. 35: Primary and Secondary Growth; Ch. 39: Flowering Lab
**Investigations**—Ch. 39: What Plant Hormones Affect Organ Formation?
**Videos**—Ch. 35: Root Growth in a Radish Seed (time-lapse); Ch. 39: Phototropism; Gravitropism

## References

Berg, L. *Introductory Botany: Plants, People, and the Environment,* 2nd ed. Belmont, CA: Thomson Brooks/Cole, 2007.

Mauseth, J. D. *Botany. An Introduction to Plant Biology.*, 4th ed. Sudbury, MA: Jones and Bartlett Publishers, 2009.

Reece, J., et al. *Campbell Biology*, 9th ed. San Francisco, CA: Pearson Education, 2011.

Srivastava, L. M. *Plant Growth and Development: Hormones and Environment*, San Diego, CA: Academic Press, 2002.

Taiz, L. and E. Zeiger. *Plant Physiology,* 4th ed. Sunderland, MA: Sinauer, 2006.

*Wisconsin Fast Plants Manual.* Burlington, NC: Carolina Biological Supply, 1989.

## Websites

American Society of Plant Biologists website with links for undergraduates:
http://www.aspb.org/resourcelinks/scripts/cats2.cfm?cat=65

For information on Wisconsin Fast Plants and detailed instructions for activities:
http://www.fastplants.org/
http://www.fastplants.org/pdf/activities

Information on plant hormones from the University of Bristol UK:
http://www.plant-hormones

North Carolina State University Extension Office site with information on plant propagation:
http://www .ces.ncsu.edu/depts/hort/hil/hil-8702.html

*Plant Physiology* is a journal of the American Society of Plant Biologists. Current research in plant physiology:
http://www.plantphysiol.org/

Planting Science. Botanical Society of America website connecting scientists and students in collaborative research:
http://www.plantingscience.org/index.php

USDA website with agricultural applications and information on plants, growth, and hormones:
http:www.usda.gov/wps/portal/usdahome

# LAB TOPIC 21

# Plant Growth
## Teaching Plan for Laboratories

## Main Concepts and Objectives

1. Concept: scientific process. Students will propose hypotheses, make predictions, and design and carry out an experiment in which they investigate factors influencing seed germination and plant growth. (Exercise 21.1, Experiment B; Exercise 21.2, Experiment B; Exercise 21.3, Experiment B)

2. Concept: seed germination. Students will describe the morphology of seeds and seedlings. Depending on the topic chosen for their independent experiment, students may investigate factors that influence seed germination. (Exercise 21.1)

3. Concept: internal regulators control plant growth. Students will describe gravitropism and phototropism and the role played by auxin in each. Students will also describe genetic causes of dwarfism in corn plants and how the application of gibberellins can reverse this phenotype. Depending on the topic chosen for their independent experiment, students may investigate internal regulators that control plant growth. (Exercises 21.2, 21.3).

4. Specific content. Students will be able to describe or define: morphology of seeds and seedlings; effects of auxin on stems and roots; gravitropism and phototropism; effects of gibberellins on plants; scientific process.

## Order of the Lab

If possible in your laboratory situation, 1 week before this lab, instruct students to read the lab topic and meet with their research team to discuss possible independent investigations before coming to the first plant growth lab. You may choose to ask them to submit a materials list prior to the lab.

This exercise is designed to be completed in two 3-hour laboratory periods.

### Week 1

1. Introduce the main concepts and explain the design of the exercise. Depending on how prepared your students are, you may need to review the role of auxins and gibberellins in plant growth. Give instructions about oral reports and writing assignments.                    (25 min)

2. Study germinating seeds.                    (15 min)

3. Observe and answer questions about gravitropism.                    (20 min)

4. Observe and answer questions about phototropism.                    (20 min)

5. Compare the growth of normal and dwarf corn seedlings treated with a gibberellin solution.                    (30 min)

6. Design and begin an experiment based on questions that arise in the introductory observations.                    (60 min)

### Week 2

1. Students complete observations of experiments and organize data, designing tables and graphs.                    (60 min)

2. A representative from each student research team reports the results of the team's experiment. Classmates critique the experiments and discuss conclusions.                    (120 min)

Consider the following suggestions for completing the lab in 1 week and for 2-hour laboratory periods.

If this lab is to be completed in 1 week, omit the student-designed investigation. Instead, have the students help design one additional investigation that can be completed in lab. Refer to the Experiment Bs for suggestions.

**For a 2-hour lab:** Estimated times for both weeks of this lab are very generous. You may be able to complete the first week's activities in 2 hours if your students come prepared to design their experiment and you do not have to give a lengthy introduction. The first week of this lab can be easily completed in 2 hours if you omit one of the exercises. Have the students design an independent investigation for one of the two exercises they completed in lab. The choice of which experiment to omit will depend on availability of supplies in your area and your budget. See the Preparation Guide for suggested sources of plants and seeds.

## Classroom Management

Students independently observe the germinating seeds. As they perform other introductory experiments (Experiment As), they will work in investigative teams. Experiments on phototropism and gravitropism in *Coleus* are demonstrations. Students carry their lab notebooks to the demonstration table to observe these experiments and record their observations and results. After completing all introductory experiments, students design and initiate their open-inquiry investigations within their investigative team. They may be allowed to return to lab to make observations between the two lab periods, or they may take their experiment home and make observations during the week. The success of this exercise depends on the students' keeping the plants alive. Students may forget to check that plants left in the lab room are surviving and have sufficient water. A student assistant might be assigned the task of checking the plants, especially before a weekend.

## Student Development

In the introductory studies, students will practice asking questions, proposing hypotheses, making predictions, and collecting and interpreting data. As they design and execute their original experiments, they will practice the entire scientific process. If a paper is required, they will practice organizing data in tables and graphs and using results to support conclusions. If oral reports are required, students will practice using persuasion and presenting data.

## Discussion and Summary

Students may discuss results and complete questions in the exercises. A representative from each student team will report results of the team's experiment to the class. Students may write a scientific paper.

## Evaluation

You can evaluate the quality of the students' experiments and also grade them on their oral presentations. You may choose to require a written report in scientific format for evaluation. If you are following the plan to integrate scientific writing, as suggested in this manual, this lab topic should be carried out after students have written separate Introduction, Results, and Discussion sections in previous laboratories. Writing a complete scientific paper for this lab allows students to use the skills that they practiced when they wrote separate sections for grading in previous exercises. You might consider having them submit their proposals as a writing assignment, too. Students sometimes prepare posters as part of their presentations. These can then be posted in the hall to generate student and faculty interest.

# Lab Topic 22
# Vertebrate Anatomy I: The Skin and Digestive System

## Overview of Vertebrate Anatomy Labs (Lab Topics 22, 23, and 24)

In Lab Topics 18 and 19, Animal Diversity I and II, you investigated several major themes in biology as illustrated by biodiversity in the animal kingdom. One of these themes is the relationship between form and function in organ systems. In this and the following two lab topics, you will continue to expand your understanding of this theme as you investigate the relationship between form, or structure, and function in vertebrate organ systems. For these investigations, you will be asked to view prepared slides and isolated adult vertebrate organs, and to dissect a representative vertebrate, the fetal pig. The purpose of these investigations is not to complete a comprehensive study of vertebrate morphology but rather to use several select vertebrate systems to analyze critically the relationship between form and function.

*For a 2-hour lab: Omit the mouth cavity dissection, the intestine slide, and the discussion of answers to questions, Consider expanding the three vertebrate labs into four lab periods. See the Teaching Plan.*

You will explore the listed concepts in the designated exercises.

1. The specialization of cells into tissues with specific functions makes possible the development of functional units, or organs (Exercise 22.1, Histology of the Skin).

2. Multicellular heterotrophic organisms must obtain and process food for body maintenance, growth, and repair (Exercise 22.3, The Digestive System in the Fetal Pig).

3. Because of their size, complexity, and level of activity, vertebrates require a complex system to transport nutrients and oxygen to body tissues and to remove waste from all body tissues (Exercise 23.1, Glands and Respiratory Structures of the Neck and Thoracic Cavity; Exercise 23.2, The Heart and the Pulmonary Blood Circuit; Exercise 24.1, The Excretory System).

4. Reproduction is the ultimate objective of all metabolic processes. Sexual reproduction involves the production of two different gametes, the bringing together of the gametes for fertilization, and limited or extensive care of the new individual (Exercise 24.2, The Reproductive System).

5. Complex animals with many organ systems must coordinate the activities of the diverse parts. Coordination is influenced by the endocrine and nervous systems. Integration via the endocrine system is generally slower and more prolonged than that produced by the nervous system, which may receive stimuli, process information, and elicit a response very quickly (Exercise 24.3, Nervous Tissue, the Reflex Arc, and the Vertebrate Eye).

## Laboratory Objectives

After completing this lab topic, you should be able to:

1. Describe the four main categories of tissues and give examples of each.
2. Identify tissues and structures in mammalian skin.
3. Describe the function of skin. Describe how the morphology of skin makes possible its functions.
4. Identify structures in the fetal pig digestive system.
5. Describe the role played by each digestive structure in the digestion and processing of food.
6. Relate tissue types to organ anatomy.
7. Apply knowledge and understanding acquired in this lab to problems in human physiology.
8. Apply knowledge and understanding acquired in this lab to explain organismal adaptive strategies.

## Introduction to Tissues

All animals are composed of **tissues,** groups of cells that are similar in structure and that perform a common function. During the embryonic development of most animals, the body is composed of three tissue layers: **ectoderm, mesoderm,** and **endoderm.** (Recall from Lab Topic 18 Animal Diversity I, that animals in the phylum Porifera lack true tissue organization and that animals in the phylum Cnidaria have only two tissue layers—ectoderm and endoderm.) It is from these embryonic tissue layers that all other body tissues develop. There are four main categories of tissues: epithelium, connective tissue, muscle, and nervous tissue. Organs are formed from these tissues, and usually all four will be found in a single organ.

Tissues are composed of cells and extracellular substances. The extracellular substances are secreted by the cells. **Epithelial tissue** has cells in close aggregates with little extracellular substance (see Figure 22.1). These cells may be in one continuous layer, or they may be in multiple layers. They generally cover or line an external or internal surface of an organ or cavity. If formed from single layers of cells, the epithelium is called **simple.** If cells are in multiple layers, the epithelium is called **stratified.** If epithelial cells are flat, they are called **squamous.** If they are cube-shaped, they are called **cuboidal.** Tall, prismatic cells are called **columnar.** Thus, epithelium can be stratified squamous (as in skin), simple cuboidal (as in kidney tubules), or in other combinations of characteristics. Epithelial layers may be derived from embryonic ectoderm, mesoderm, or endoderm.

In **connective tissue,** cells are widely separated by extracellular substances consisting of a web of **fibers** embedded in an amorphous **ground substance,** which may be solid, gelatinous, or liquid (Figure 22.2). **Loose fibrous connective tissue,** embedded in a liquid ground substance, binds together other tissues and organs and helps hold organs in place. Adipose tissue is another connective tissue consisting of adipose cells with fibers in a soft, liquid extracellular matrix. Adipose cells store droplets of fat causing the cells to swell and forcing the nuclei to one side. **Bone** and **cartilage** are specialized connective tissues found in the skeleton with, respectively, hard and gelatinous ground substances. In bone, the ground substance is secreted by cells called **osteocytes.** The ground substance in cartilage is secreted by cells

*The term* osteocyte *describes the bone cell in an adult bone. An* osteoblast *is a bone-forming cell.*

**Epithelial tissue**

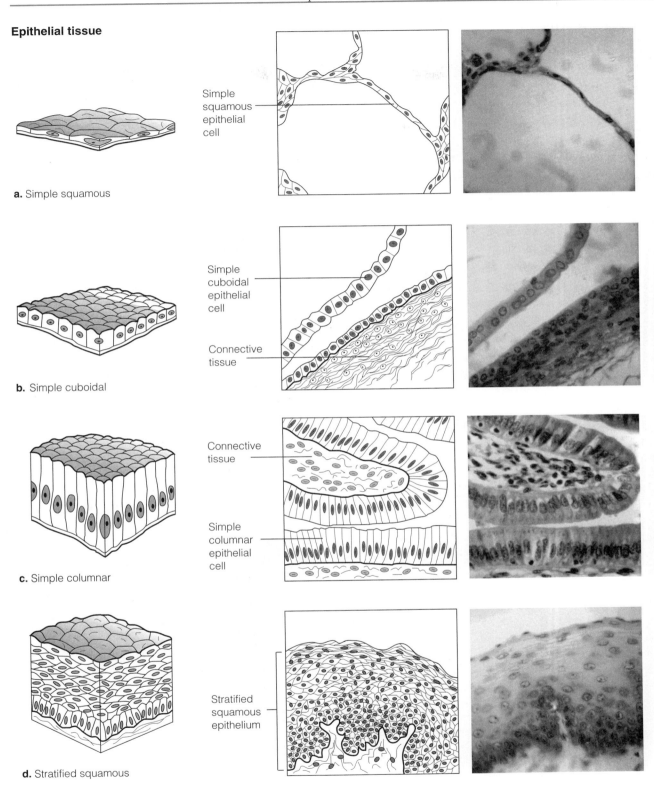

**a.** Simple squamous

Simple
squamous
epithelial
cell

**b.** Simple cuboidal

Simple
cuboidal
epithelial
cell

Connective
tissue

**c.** Simple columnar

Connective
tissue

Simple
columnar
epithelial
cell

**d.** Stratified squamous

Stratified
squamous
epithelium

**Figure 22.1.**
**Epithelial tissue.** Epithelial tissue has closely packed cells with little extracellular matrix. Cells may be (a) squamous (flat), (b) cuboidal (cube-shaped), or (c) columnar (elongated). They may be simple (in single layers) or (d) stratified (in multiple layers).

**Connective tissue**

**a.** Loose fibrous connective tissue

Fiber

Cell

Ground substance

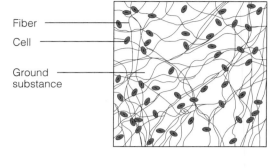
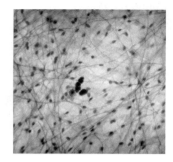

**b.** Adipose tissue

Nucleus of adipose cell

Fat globule

Cytoplasm of adipose cell

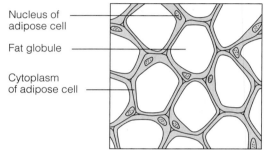
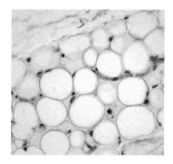

**c.** Bone

Osteocytes

Hard ground substance

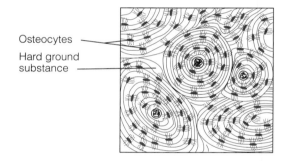
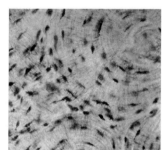

**d.** Cartilage

Gelatinous ground substance

Chondrocytes

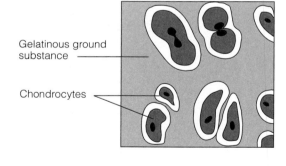
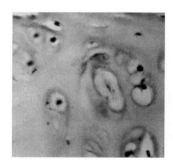

**e.** Blood

Platelet

Erythrocyte

Leukocytes

Liquid ground substance

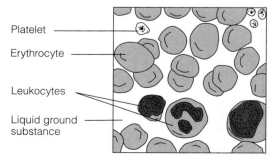
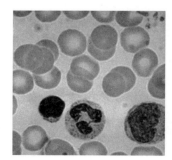

## Muscle tissue

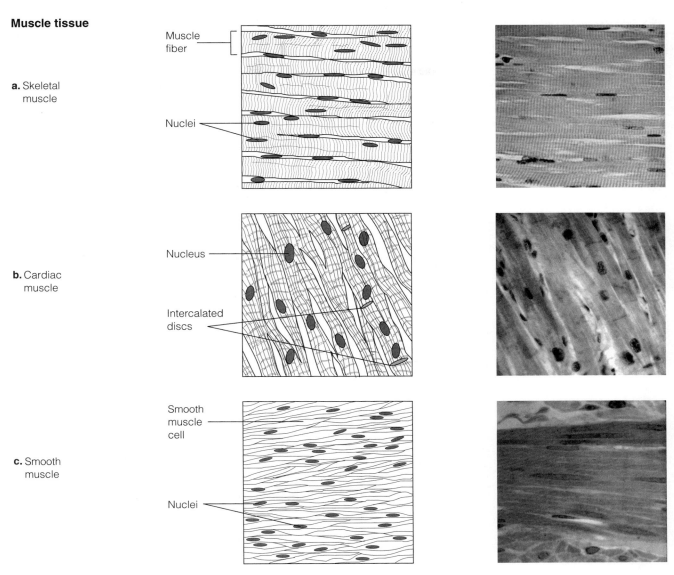

**a.** Skeletal
muscle

Muscle
fiber

Nuclei

**b.** Cardiac
muscle

Nucleus

Intercalated
discs

**c.** Smooth
muscle

Smooth
muscle
cell

Nuclei

**Figure 22.3.**
**Muscle tissue.** Muscle tissue is either striated or smooth. (a) Skeletal muscle
is striated. (b) Cardiac muscle is also striated. (c) Smooth, or visceral, muscle is
not striated.

called **chondrocytes. Blood** is a connective tissue consisting of cellular components called **erythrocytes** (red blood cells), **leukocytes** (white blood cells),
and **platelets** (cell fragments) in a liquid ground substance called **plasma.**
Other connective tissues fill the spaces between various tissues, binding them
together or performing other functions. Connective tissues are derived from
mesoderm.

( ✒ )**Figure 22.2.**
**Connective tissue.** (a) In loose connective tissue, cells are embedded in a liquid
fibrous matrix. (b) Adipose tissue stores fat droplets in adipose cells. (c) In bone,
cells are embedded in a solid fibrous matrix. (d) In cartilage, cells are embedded
in a gelatinous fibrous matrix. (e) In blood, cells are embedded in a liquid matrix.

**Figure 22.4.**
**Nervous tissue.** Neurons and glial cells.

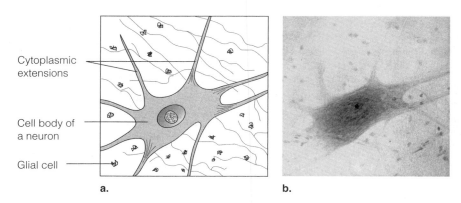

Cytoplasmic extensions

Cell body of a neuron

Glial cell

a.                                    b.

**Muscle tissue** may be **striated,** showing a pattern of alternating light and dark bands, or **smooth,** showing no banding pattern (Figure 22.3). There are two types of striated muscle, skeletal and cardiac. **Skeletal** muscle moves the skeleton and the diaphragm and is made of muscle fibers formed by the fusion of several cells end-to-end. **Cardiac** muscle, found only in the wall of the heart, is also striated, but the cells do not fuse. Cells are attached by **intercalated discs.** Smooth muscle, also called **visceral** muscle, is found in the skin and in the walls of organs such as the stomach, intestine, and uterus. Muscle, like connective tissue, is derived from mesoderm.

**Nervous tissue** is found in the central nervous system (brain and spinal cord) and in the peripheral nervous system consisting of nerves (Figure 22.4). Nervous tissue is found in every organ throughout the body. There are two basic cell types, neurons and glial cells. **Neurons** are capable of responding to physical and chemical stimuli by creating an **impulse,** which is transmitted from one locality to another. **Glial cells** support and protect the neurons. Nervous tissue is derived from ectoderm.

The organs that you will investigate in the following exercises are made up of the four basic tissues. The tissues, each with a specific function, are organized into a functional organ unit.

---

EXERCISE 22.1
# Histology of the Skin

---

### Materials

compound microscope
prepared slide of pig and/or monkey skin

### Introduction

*If possible, have both pig and monkey skin slides available, and have students work in pairs, one student with each. Then students can demonstrate the slides to each other. Some structures are more obvious in the pig, others in the monkey.*

Tissues are structurally arranged to function together in **organs,** which are adapted to perform specific functions. Organs are found in all but the simplest animals. The largest organ of the vertebrate body, the skin, illustrates the organization of tissues, each with a specific function, into a functioning organ (Figure 22.5). The skin protects the body from dehydration and bacterial invasion, assists in regulating body temperature, and receives stimuli from the environment. As you work through this exercise, ask how the unique function

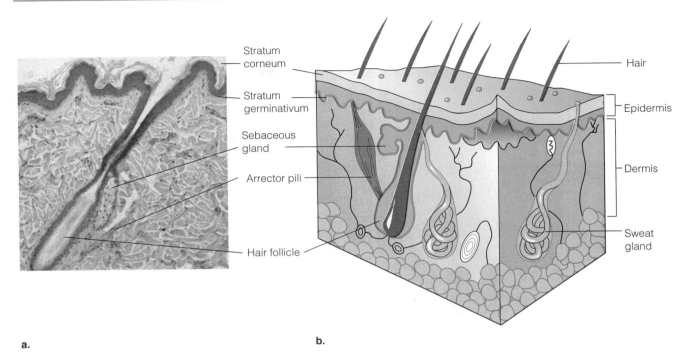

Stratum corneum

Stratum germinativum

Sebaceous gland

Arrector pili

Hair follicle

Hair

Epidermis

Dermis

Sweat gland

a.

b.

**Figure 22.5.**
**Mammalian skin structure.** (a) Photomicrograph of a cross section of the skin.
(b) A diagram detailing the skin structure.

of each tissue produces the functioning whole—the skin. Review information about each observed tissue type in "Introduction to Tissues" (pp. 574–578).

## Procedure

1. Obtain a prepared slide of mammalian skin (pig or monkey). View it using the low and intermediate power objectives on the compound microscope.

2. Identify the two main layers of the skin. The thin outer layer, the **epidermis,** consists of **stratified squamous epithelium** (Figure 22.1d); the thicker inner layer, the **dermis,** consists mainly of **dense connective tissue** and scattered **blood vessels.** The dermis merges into layers of loose connective tissue (Figure 22.2a) and smooth muscle (Figure 22.3c), which are not considered part of the skin.

3. Locate **hair follicles** extending obliquely from the epidermis into the dermis. In the living animal, each follicle contains a hair, but the hair shaft may or may not be visible in every follicle on your slide, depending on the plane of the section through the follicle. The follicle is lined by epithelial cells continuous with the epidermis. Carefully observe several hair follicles. You may be able to find a band of smooth muscle cells attached to the side of a follicle. This muscle, called the **arrector pili,** attaches the hair follicle to the outermost layer of the dermis. When stimulated by cold or fright, it pulls the hair erect, causing "goose bumps." In furry animals this adaptation increases the thickness of the coat to provide additional temperature insulation. Clusters of secretory cells making up **sebaceous glands** (oil glands) are also associated with hair follicles. These are more obvious in monkey skin slides than in pig skin slides.

4. Focus your attention on the epidermis and locate the outermost layer, the **stratum corneum,** a layer of dead, keratinized cells, impermeable to water. This layer is continually exfoliated and replaced. The thickness

of the stratum corneum varies, depending on the location of the skin. This layer is very thick on the soles of feet or palms of human hands.

5. The innermost layer of cells in the epidermis is called the **basal layer** or the **stratum germinativum.** These cells divide mitotically and produce new cells, which, as they mature, are pushed to higher and higher layers of the epidermis, until they fill with keratin and form the stratum corneum. Scattered through the basal layer (though not seen on your slide) are cells called **melanocytes** that produce **melanin,** a pigment that produces brown or black hues in the skin. The melanin is inserted into newly forming epidermal cells as they are pushed outward. Regular exposure to sunlight stimulates melanocytes to produce more melanin, helping protect the body against the potentially harmful effects of sun exposure.

*Sweat glands are more obvious in pig skin slides than in monkey skin slides.*

6. In addition to hair follicles, coiled tubular **sweat glands** lined with **cuboidal epithelium** (Figure 22.1b) extend from the epidermis into the dermis. They appear as circular clusters in cross section and may be easily located in pig skin but are less numerous or absent in furry animals, such as monkeys. The tubular secretory portion is convoluted into a ball, which connects with a narrow unbranched tube leading to the skin surface. It is unlikely that you will see an entire intact sweat gland in one section.

7. Look for connective tissue and blood vessels, which often contain red blood cells in the dermis. Look for adipose tissue (Figure 22.2b) below the dermis.

## Results

In Table 22.1, list the tissues you have identified in the skin and indicate the specific function of each.

**Table 22.1**
Tissues of the Skin and Their Functions

| Tissue | Function |
|---|---|
| Epithelial | Prevents dehydration, secretes sweat and oil, regulates temperature |
| Loose connective | Cushions the epidermis, connects skin with underlying muscle |
| Blood (connective) | Supplies nutrients to skin, regulates temperature |
| Smooth muscle in arrector pili | Pulls hair erect |
| Nervous | Receives stimuli and transmits nervous impulse |
| Adipose | Forms layer of insulation to prevent heat loss |

## Discussion

1. How does the skin prevent dehydration?

   *The dead, keratinized cells of the epidermal stratum corneum form a barrier that is impermeable to water, preventing the loss of water through the epidermis. Water passes through the epidermis via secretion from sweat glands upon nervous stimulation.*

2. How does the skin protect from bacterial invasion?

   *The barrier formed by the stratum corneum prevents the invasion of bacteria.*

3. Discuss how each of the following helps regulate body temperature: blood vessels in the dermis, sweat glands, adipose tissue below the dermis, hair, and hair follicles.

   *Blood vessels may be relaxed, bringing blood and body heat to the surface of the body to be transferred to the environment. Conversely, if these blood vessels are constricted, heat is retained internally. Sweat glands secrete water to the skin surface for evaporative cooling. Hair and the fatty layers below the dermis prevent heat loss. Arrector pili attached to hair follicles may contract and hold the hair erect, increasing the thickness of the fur and its insulating capacity.*

---

## EXERCISE 22.2
# Introduction to the Fetal Pig

---

### Materials

preserved fetal pig
dissecting pan

disposable gloves
preservative

### Introduction

Fetal, or unborn, pigs are obtained from pregnant sows being slaughtered for food. The size of your pig will vary, depending on its stage of gestation, a total period of about 112 to 115 days. After being embalmed in a formaldehyde- or phenol-based solution, pigs are stored in a preservative that usually does not contain formaldehyde, although the smell of formaldehyde may remain. Most preserving solutions are relatively harmless; however, they will dry the skin, and occasionally a student may be allergic to the solutions. For these reasons, you should not handle the pigs with your bare hands, and you should perform your dissections in a well-ventilated room.

---

 **Wear disposable gloves when handling the pig and other preserved animals.**

In this exercise, you will become familiar with the external anatomy of the fetal pig, noting the regions of its body and the surface structures. The skin, just studied in microscopic sections, will be the first organ observed.

Each student, working independently (unless otherwise instructed), will dissect a fetal pig; however, we encourage you to engage in collaborative discussions with your lab partner. Discuss results and conclusions and compare dissections.

   Read carefully the rules and techniques for dissection in Appendix C before you begin your study of the fetal pig.

## Procedure

1. Obtain a fetal pig and place it in a dissecting pan. Add a small amount of preservative to the pan. Do not allow the pig to dry out at any time. However, use the preservative, not water, to moisten the tissue unless otherwise instructed.

   Your pig may have been injected with red and blue latex. The red was injected into arteries through one of the umbilical arteries; the blue was injected into the external jugular vein through an incision in the neck.

2. Using information in Appendix C and collaborating with your lab partner, locate your pig's left and right, dorsal and ventral, and anterior (cranial) and posterior (caudal) regions. Use the terms *proximal* and *distal* to compare positions of several structures.

3. Locate the body regions on your pig. The pig has a **head, neck, trunk,** and **tail** (Figure 22.6a). The trunk is divided into an anterior **thorax,** encased by ribs, and a posterior **abdomen.** The thoracic and abdominal regions of the body house corresponding cavities that are divisions of the body cavity, or coelom. The **thoracic cavity** is in the thorax, and the **abdominal (peritoneal) cavity** is in the abdomen.

4. Examine the head with its concentration of sensory receptors. Identify the mouth; **external nostrils** on the end of the snout; ears, each with an external flap supported by cartilage (Figure 22.2d), the **auricle;** eyes with two eyelids, as in humans, and a third eyelid, the **nictitating membrane,** near the inside corner of each eye.

5. Examine the pig's skin and review its functions. In a fetal pig, an outer embryonic skin, the **epitrichium,** lies over the skin. You may find pieces of this layer on your pig. Is hair present on your pig?

6. Locate the cut **umbilical cord** on the ventral surface of the abdomen. Blood vessels pass from the placenta, attached to the wall of the mother's uterus, to the fetal pig through this cord. If your pig's umbilical cord is collapsed, use scissors to make a fresh transverse cut and examine the end more closely. Identify the cut ends of two round, thick-walled **umbilical arteries;** one larger, flattened **umbilical vein;**

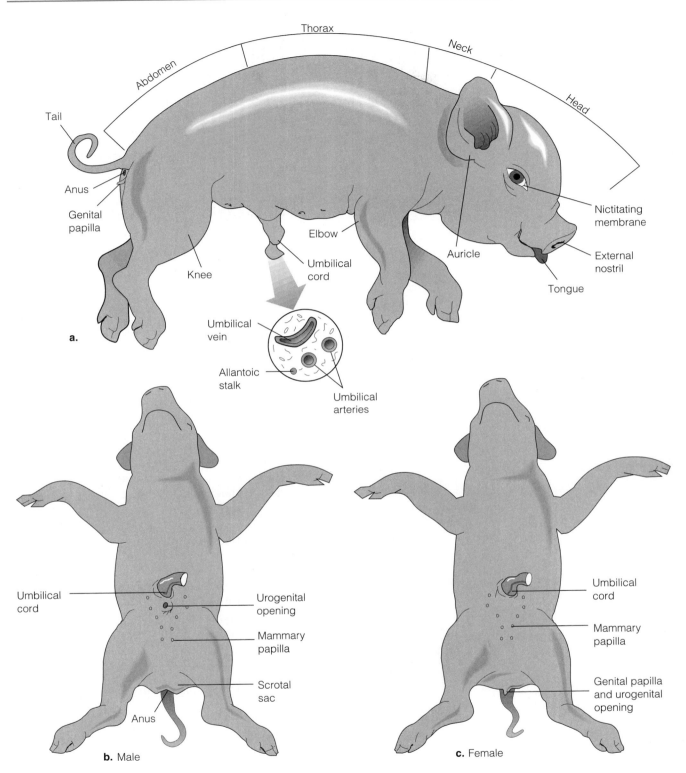

**Figure 22.6.**
**Fetal pig.** (a) Body regions and external structures of the fetal pig with an
enlarged cross section of the umbilical cord. (b) Posterior region of male pig.
(c) Posterior region of female pig.

and one very small, round **allantoic stalk.** You may remember that blood vessels carrying blood away from the heart of the animal are called **arteries.** Blood vessels carrying blood back to the heart are called **veins.** In fetal circulation, the umbilical arteries carry blood from the fetus to the placenta. The umbilical vein carries blood from the placenta to the fetus. The allantois is an extension of the urinary bladder of the fetus into the umbilical cord. Speculate about the nature of blood in the umbilical arteries and the umbilical vein. Explain your conclusions.

a. Which vessel—umbilical artery or umbilical vein—would carry blood *high* in oxygen (oxygen-rich blood)?

*The umbilical vein, carrying blood from the mother to the fetus, would carry oxygen-rich blood.*

b. Which vessel would carry blood *low* in oxygen (oxygen-poor blood)?

*The umbilical arteries, carrying blood from the fetus to the mother, would carry oxygen-poor blood.*

c. Which vessel would carry blood high in nutrients?

*The umbilical vein, carrying blood directly from the placenta.*

d. Which vessel would carry blood high in metabolic waste?

*The umbilical arteries, carrying blood away from the fetus.*

7. Look just caudal to the umbilical cord to determine the sex of your pig. If it is a male (Figure 22.6b), you will see the **urogenital opening** in this position. This opening is located below the tail in the female. Locate the **anus** just ventral to the base of the tail in both sexes. In the male, **scrotal sacs** will be present ventral to the anus and caudal to the hind legs. In the female pig (Figure 22.6c), the **urogenital opening** is located ventral to the anus. Folds, or **labia,** surround this opening, and a small protuberance, the **genital papilla,** is visible just ventral to the urogenital opening. The urogenital opening is a common opening from the urinary and reproductive tracts. Notice that **mammary papillae** are present in pigs of both sexes. Locate a pig of the opposite sex for comparison. Having determined the sex of your pig, you are now ready to begin your dissection.

## Results

1. List structures observed in the fetal pig that are no longer present in the pig after birth.

*epitrichium, umbilical cord with umbilical arteries, vein, and allantoic stalk*

2. Modify Figure 22.6 or make a sketch in the margin of your lab manual with any additional details needed for future reference.

## Discussion

Review the definition of the term *cephalization* defined and observed in some of the animals studied in Lab Topics 18 and 19, Animal Diversity I and II, and describe how the pig demonstrates this phenomenon.

*Cephalization is the evolutionary development of the head, complete with sensory structures, as a dominant body part. The head of the pig is a conspicuous division of the body with sensory structures—eyes, nose, ears.*

---

EXERCISE 22.3
# The Digestive System in the Fetal Pig

## Materials

supplies from Exercise 22.2
dissecting instruments
twine

plastic bag with twist tie and label
stereoscopic microscope or hand lens

## Introduction

Most internal organs, including the entire digestive system, are located in the body cavity, or coelom. A large muscular structure, the **diaphragm**, divides the body cavity into the **thoracic** cavity and the **abdominal (peritoneal)** cavity. The **thoracic** cavity includes two additional cavities, the **pleural** cavity housing the lungs and the **pericardial** cavity housing the heart. **Coelomic epithelial membranes** line these cavities and cover the surface of all organs. Those epithelial membranes *lining the wall* of the cavity are called **parietal** (L., *pariet*, "wall"). Epithelial membranes *covering organs* are called **visceral** (L., *viscera*, "bowels"). Each epithelial membrane lining in the coelom is named according to its cavity and location (lining the wall or covering the organ). Thus, the epithelial membrane covering the lungs would be the **visceral pleura.** The epithelial membrane lining the wall of the pleural cavity would be the **parietal pleura.** Likewise, the lining of the abdominal (peritoneal) cavity is **parietal peritoneum.** The epithelium covering organs in the peritoneal cavity is **visceral peritoneum.** What would be the name of the epithelium covering the heart?

*visceral pericardium*

*Note that human anatomy texts call the abdominal cavity the abdominopelvic cavity and point out that the upper portion is the abdominal cavity with the lower portion being the pelvic cavity.*

Use this convention to complete Table 22.2, naming coelomic epithelial membranes. As you open each body cavity described in these lab topics, return to this table to check your answers.

**Table 22.2**
Divisions of the Body Cavity and Associated Membranes

| Body Cavity | Divisions | Epithelial Membrane Lining the Cavity Wall | Epithelial Membrane Covering the Organs | Organs |
|---|---|---|---|---|
| Thoracic | Pleural | Parietal pleura | Visceral pleura | Lungs |
|  | Pericardial | *Parietal pericardium* | *Visceral pericardium* | *Heart* |
| Abdominal | Abdominal (peritoneal) | *Parietal peritoneum* | Visceral peritoneum | Stomach, pancreas, spleen, liver, gallbladder, intestines |

Recall from studies in Lab Topics 18 and 19, Animal Diversity I and II, that the more complex animals have a tubular digestive system with an anterior opening, the mouth, and a posterior opening, the anus. This pattern in the digestive system allows specialization to take place along the length of the tract, resulting in the development of specific organs that carry out specific functions. As you investigate the digestive system in the fetal pig, ask how each structure or region solves a particular problem in nutrition procurement and processing. Remember that memorizing the names of structures is meaningless unless you understand how the morphology is related to the function of the organ.

**Procedure**

1. Hold the pig ventral side up in the dissecting pan and use twine to tie the two anterior legs together and the two posterior legs together, leaving enough twine between to slip under the dissecting pan. The pig should be positioned "spread-eagle" in the pan, ventral side up.

2. To expose structures in the mouth cavity, use heavy, sharp-pointed scissors to cut at the corners of the mouth along the line of the tongue. Continue to cut until the lower jaw can be lowered, being careful not to cut into the tissues in the roof of the mouth cavity. Continue to cut, pulling down on the lower jaw until your cuts reach the muscle and tissue at the back of the mouth. Cut through this muscle until lowering the jaw exposes the back of the mouth cavity.

   a. Identify structures in the mouth cavity (Figure 22.7). Locate the **teeth**; the **tongue** covered with **papillae**, which house taste buds; and the roof of the mouth, composed of the **hard palate** supported by bone (Figure 22.2c) and the **soft palate.** The hard palate is anterior to the soft palate and is marked by ridges.

   b. Identify structures and openings at the rear of the mouth cavity: the **glottis,** the space in the beginning of the respiratory passageway; the **epiglottis,** a small flap of tissue supported by cartilage (Figure 22.2d) that covers the glottis when swallowing; the **esophagus,** the beginning of the digestive tube (alimentary canal); and the **opening into the nasal chamber.** All these open into the **pharynx,** the chamber located posterior to the mouth.

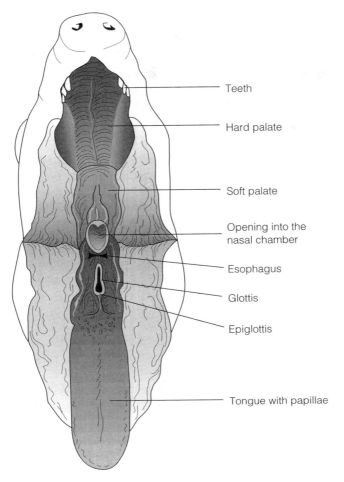

Teeth

Hard palate

Soft palate

Opening into the
nasal chamber

Esophagus

Glottis

Epiglottis

Tongue with papillae

**Figure 22.7.**
**The mouth cavity.**
Openings into the pharynx in the rear
of the mouth cavity lead to the
respiratory system, the digestive
system, and the nasal cavity. The
tongue occupies the floor of the
mouth, and the roof of the mouth
consists of the hard and soft palates.

c.  Probe into the nasal chamber through the opening at the rear of the
soft palate, the esophagus just posterior and ventral to the opening
into the nasal chamber, and the glottis. Notice that the opening into
the esophagus lies dorsal to the opening into the respiratory tract.

3.  Expose the digestive organs in the abdominal region by opening the
posterior portion of the abdominal cavity.

a.  Begin the dissection by using the scalpel to make a shallow midventral
incision from the base of the throat to the umbilical cord (Figure 22.8,
incision 1). Cut lateral incisions (Figure 22.8, incision 2) around each
side of the umbilical cord and continue the two incisions, one to the
medial surface of each leg.

b.  Now use the scissors to cut deep into one of the lateral incisions beside the cord until you penetrate into the abdominal cavity, piercing
the parietal peritoneum. At this point, fluid in the cavity should begin to seep out. Use the scissors to cut through the body wall along
the two lateral incisions to the legs and around the umbilical cord.

c.  Pull lightly on the umbilical cord. If your dissection is correct, you will
see that the umbilical cord and ventral wedge of body wall could be
reflected, or pulled back, toward the tail, except for a blood vessel, the
**umbilical vein.** This vein passes from the umbilical cord anteriorly
toward a large brown organ, the liver. Cut through this vein, leaving a
stub at each end. Tie a small piece of twine around each stub so you
can find them later. This should free the flap of body wall, which may
now be reflected toward the tail, exposing the abdominal organs.

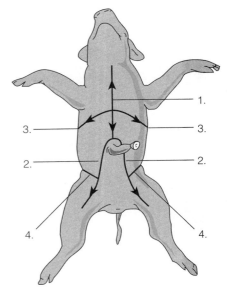

**Figure 22.8.**
**Incisions to open the abdominal**
**cavity.** Incision 1 makes a shallow
midventral incision from the base of the
throat to the umbilical cord. Incision
2 cuts around the umbilical cord to the
medial surface of each leg. Incision
3 cuts through the body wall laterally
just posterior to the diaphragm.
Incision 4 cuts laterally at the posterior
margin of the abdominal cavity.

4. Open the anterior portion of the abdominal cavity.

   a. Cut anteriorly through the body wall along the midventral incision until you reach the diaphragm, separating the thoracic and abdominal cavities.

   b. Make four lateral cuts, two adjacent to the rib cage just posterior to the diaphragm (Figure 22.8, incision 3) and two at the posterior margin of the abdominal cavity (Figure 22.8, incision 4). This will produce two flaps of body wall that can be folded back like the lids of a box.

   c. If your specimen contains coagulated blood or free latex from the injection, pull out the latex and rinse the body cavity under running water.

5. Identify the various structures in the abdominal cavity (Figure 22.9).

   a. Shiny epithelial membranes line the cavity and cover the organs. The **parietal peritoneum** lines the cavity, and the **visceral peritoneum** covers the organs. Speculate about the contents of the space between these two membranes and its function.

   *Coelomic fluid fills this space. This moisture acts as a lubricant, allowing organs some degree of easy movement.*

   b. The **diaphragm** is a large domed, striated (skeletal, Figure 22.3a) muscle forming the transverse cranial wall of the abdominal cavity, separating this from the thoracic cavity. Only mammals have a diaphragm. The contraction and relaxation of this muscle and muscles between the ribs cause the thoracic cavity to expand and contract, changing the pressure in the cavity and lungs, thus facilitating movement of air into and out of the lungs.

   c. The **liver** is the large brown organ that appears to fill the abdominal cavity. Notice that it consists of several lobes. Pull it cranially and locate the **gallbladder,** a small, thin-walled, paddle-shaped sac embedded in the underneath surface near the umbilical vein identified earlier. The liver has many functions, including processing nutrients and detoxifying toxins and drugs. Its main digestive function is the production of **bile,** a substance that emulsifies fats. Bile is stored in the gallbladder until needed.

   d. The **stomach** is a large, saclike organ lying dorsal to the liver (Figure 22.9b). Reflect the liver cranially to get a better view. The larger upper-left portion of the stomach tapers down to a narrower portion to the right. The **esophagus** passes through the diaphragm and enters the upper medial border of the stomach. Locate the **spleen,** a dark organ lying along the **greater curvature** of the stomach (the lateral convex border). The spleen filters blood.

   e. Cut into the stomach along the greater curvature. Rinse out the stomach contents and identify the numerous longitudinal folds on the inside surface called **gastric rugae.** Speculate about the role played by these folds.

   *These folds allow the stomach to expand when filled with food.*

   f. Food enters the stomach from the esophagus through the **cardiac valve,** or **cardiac sphincter.** A sphincter valve is a circular band of muscle that encircles an opening. Looking inside the stomach, locate

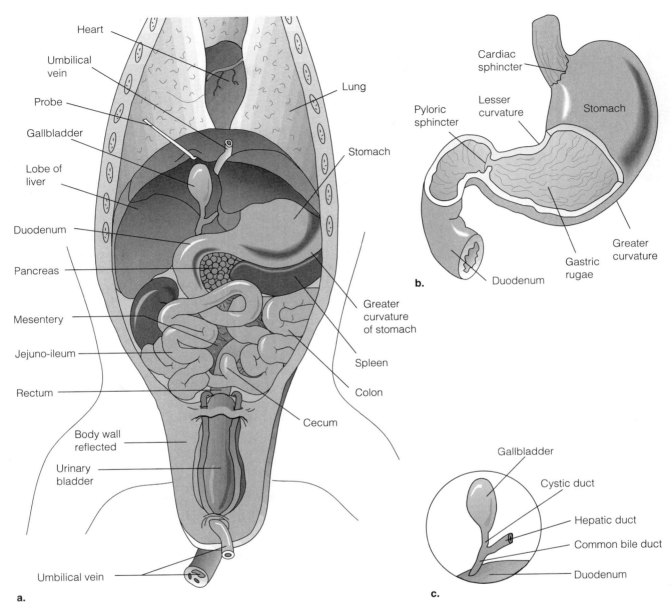

**Figure 22.9.**
**Digestive organs of the abdominal cavity.** (a) The liver is reflected cranially to expose deeper-lying organs. The stomach leads into the duodenum of the small intestine. The jejuno-ileum continues from the duodenum and empties into the colon near the cecum. (b) Cutaway view of stomach showing internal gastric rugae. (c) Enlargement of gallbladder and associated ducts.

this valve between the esophagus and stomach and speculate about its function.

*This valve prevents backflow of stomach contents into the esophagus. It opens only when stimulated by food passing down the esophagus.*

The stomach mechanically churns food and mixes it with water, mucus, hydrochloric acid, and protein-digesting enzymes.

g. From the stomach, food then passes into the **small intestine,** which joins the stomach at its extreme right narrow portion. Locate another valve, the **pyloric sphincter,** which lies between the stomach and small intestine. This sphincter is closed when food is present in the stomach, preventing food from entering the small intestine prematurely.

h. The **duodenum** is the portion of the small intestine connecting with the stomach (Figure 22.9b). The **pancreas**, an irregular, granular-looking gland, lies in a loop of the duodenum. As food passes through the duodenum, enzymes from the pancreas and the duodenal wall are added to it along with bile, which is produced in the liver and stored in the gallbladder. Using your forceps and a blunt probe to pick away surrounding tissue, locate and separate the **common bile duct,** which enters the duodenum. It is formed by the confluence of the **hepatic duct** passing from the liver and the **cystic duct** leading from the gallbladder (Figure 22.9c).

i. The **jejuno-ileum** is the extensive, highly convoluted portion of the small intestine extending from the duodenum to the colon (Figure 22.9a). Whereas the jejunum and the ileum are separate anatomical regions, macroscopically (without a microscope) it is difficult to distinguish between the two in the pig except by position. The duodenum joins the jejunum, which leads to the ileum. The ileum joins the colon.

Most digested food is absorbed into the circulatory system through the walls of the jejuno-ileum. The surface area of the lining of this organ is increased by the presence of microscopic **villi** and **microvilli**, greatly enhancing its absorbing capacity.

Spread apart folds in the jejuno-ileum and notice the thin membrane called **mesentery,** which supports the folds. Do you see blood vessels in this mesentery? Speculate about the relationship between these vessels and food processing.

*Digested food is absorbed from the intestine and passes into this network of blood vessels that converge, ultimately passing to the liver for processing.*

j. Cut out a 1-cm-long segment of the jejuno-ileum and cut it open to expose the inner surface. Use a stereoscopic microscope or hand lens to examine this surface. Can you see villi?

k. Follow the ileum to its junction with the large intestine, or **colon.** The diameter of the colon is slightly greater than that of the small intestine, and it is tightly coiled and held together by mesentery. Look for a small outpocketing or fingerlike projection of the colon at its proximal end. This projection is the **cecum,** which is much larger in herbivores than in carnivores. In animals with a large cecum, it probably assists in digestion and absorption. In humans, a **vermiform** (wormlike) appendix extends from the cecum.

One of the important activities in the colon is the reabsorption of water that has been added, along with enzymes and mucus, to the food mass as it passes down the digestive tract. Water conservation is one of the most critical problems in terrestrial animals.

l. The distal portion of the colon is the **rectum,** which passes deep into the caudal portion of the abdominal cavity and to the outside of the body at the anus. Water reabsorption continues in the rectum.

6. After you complete the dissection, use an indelible pen to prepare two labels with your name, lab day, and room. Tie one label to your pig and place the pig in a plastic bag with the label in view. Add preservative

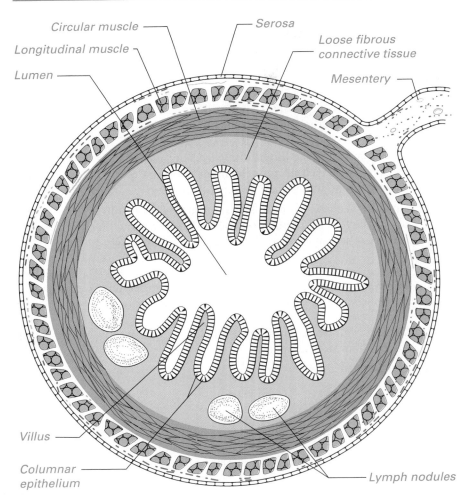

Circular muscle

Longitudinal muscle

Lumen

Serosa

Loose fibrous connective tissue

Mesentery

Villus

Columnar epithelium

Lymph nodules

**Figure 22.10.**
**Cross section through the pig jejuno-ileum.** Label the cells and structures described in Exercise 22.3, Procedure section, step 7.

from the lab stock and securely close the plastic bag. Tie the second label to the outside of the bag.

7. Using the compound microscope, study a prepared microscope slide of a section through a region of the jejuno-ileum (Figure 22.10, Color Plate 64).

a. Use the lowest power on the microscope and scan the section, finding the smooth outer surface and then the **lumen**, or central cavity, located within the intestine. Food passing through the intestine passes through the lumen. Sketch the outline of the section in the margin of your lab manual.

b. The fingerlike **villi** previously observed are now easily discernible, projecting into the lumen of the intestine. Switch to 10× magnification and focus on the simple **columnar epithelial** tissue lining each villus (Figure 22.1c). This tissue functions in the absorption of nutrients into the circulatory system. Capillaries and lacteals (small lymph vessels) are located within each villus. These vessels are not usually visible.

c. Continue your observations, scanning outward toward the surface of the intestine. You will pass through regions with **loose fibrous connective tissue** (Figure 22.1a) containing many blood vessels. You may see large masses of cells that are lymphocytes in **lymph nodules**.

d. On the outer surface of the section, locate a thin layer of simple squamous epithelium (Figure 22.1a), called the **visceral peritoneum,** or **serosa.** Two large bands of smooth muscle lie just inside the visceral peritoneum.

e. Locate the outermost muscle layer, composed of smooth muscle fibers extending longitudinally along the intestine. This muscle layer is called the **longitudinal muscle layer.** Since this is a cross section, the longitudinal nature of the fibers will not be apparent. Look just inside the longitudinal muscle layer to see a wide band of muscle, the **circular muscle layer** (appears as in Figure 22.3c). These muscle fibers encircle the intestine.

Imagine each band of muscle contracting as a unit, and speculate about the function of these muscles. Do you know the special name given to the waves of contraction of these muscles?

*Contraction of these muscles moves food through the small intestine, and the contraction produces rhythmic waves called peristalsis.*

f. Turn to high power and locate nuclei in both the longitudinal and the circular muscle fibers.

g. Label Figure 22.10.

## Results

1. In Lab Topics 18 and 19, you learned that the digestive tract of many invertebrates is described as a "tube-within-a-tube"—a tubular digestive tract within a "tubular" body. You also concluded that the pig digestive tract has this tubular design. Which structures and organs of the pig digestive system develop as tubes or chambers within its tubular digestive tract?

*The mouth, esophagus, stomach, all regions of intestines, caecum, rectum, and anus are all specializations of the "tube-within-a-tube" design.*

2. Which organs in the digestive system lie outside the "digestive tube" but are important in the digestive process?

*liver, gall bladder and its ducts, pancreas*

## Discussion

1. Conservation of water is a critical problem faced by terrestrial organisms. Given the water requirements for digestion, how is the digestive tract anatomy adapted to life on land?

*The addition of water for the digestive process all along the digestive tract could potentially result in the loss of large amounts of water from the body; however, most of this water is reabsorbed in the colon and rectum, so that waste leaving the body is relatively devoid of water.*

2. Speculate about the outcome if food passes too slowly or too rapidly through the colon.

*Most water is not reabsorbed from food passing too rapidly through the colon, resulting in diarrhea. When food passes too slowly, too much water is reabsorbed, resulting in constipation.*

3. Consult a text and describe the process of food absorption in the small intestine, relating it to the structures observed in this exercise.

*Digested food must pass from the lumen of the small intestine into the body by crossing the lining of the digestive tract. Two single layers of cells separate nutrients in the lumen from the bloodstream: the epithelium lining the intestine and the cells in the wall of capillaries in villi. Products of digestion cross this barrier either passively by diffusion or by active transport across epithelial membranes. They then enter the bloodstream or, in the case of some lipids, the lacteals, also located in villi. Villi and microvilli greatly increase the area for absorption.*

 Student Media Video—Ch. 41: Whale Eating a Seal

## Reviewing Your Knowledge

In Table 22.3, beginning with the mouth, list in sequence each region or organ through which food passes in the pig and describe for each the primary digestive functions, the enzymes active in that chamber, and the macromolecules affected, when appropriate. Consult your text if necessary.

**Table 22.3**
Organs of the Pig Digestive Tract and Their Functions

| Region/Organ | Function/Macromolecule Digested | Enzymes |
|---|---|---|
| Mouth | Chews food, digests carbohydrates | Salivary amylase |
| Esophagus | Transports food to stomach | — |
| Stomach | Churns food to liquid, digests protein | Pepsin |
| Duodenum | Digests carbohydrates, proteins, lipids | Trypsin, chymotrypsin, aminopeptidase, carboxypeptidase, pancreatic amylase, lipase |
| Jejuno-ileum | Digestion continues, food is absorbed | — |
| Colon | Water is reabsorbed | — |
| Rectum | Water is reabsorbed, waste carried to outside | — |

# Applying Your Knowledge

1. The Peachtree Road Race is over, and you have just been awarded the coveted T-shirt. Your body is dripping wet and your skin appears bright red. Explain, from a physiological perspective, what is happening to your body.

   *Capillaries in the dermis are dilated, increasing the flow of blood, carrying heat to the surface of your body. This causes the red color. The nervous system is sending signals to your sweat glands, stimulating the secretion of water to the body surface for evaporative cooling.*

2. What happens to your skin when you get a "sunburn"? Describe the process of "getting a tan." Relate these processes to skin histology.

   *The sun penetrates through the epidermis to the dermis. Blood vessels in the dermis dilate, producing the characteristic red color, elevated temperature, and often swelling. This is a first-degree burn. If the sun exposure is more severe and there is damage to the dermis, blisters may appear, which is a second-degree burn. A tan results when sun stimulates melanocytes to produce more melanin, and, as the pigment increases in newly forming epidermal cells, the skin darkens. This process takes 3–4 days to develop and lasts for about 1½ weeks. As the pigmented cells age, they are pushed outward and eventually are sloughed off.*

3. Tattooing is the risky and unregulated procedure of using a needle to deposit pigment in the skin. Where in the skin would it be necessary to deposit the pigment to make the tattoo permanent?

   *The pigment is deposited below the epidermal stratum germinativum in the dermis.*

4. The three most common forms of skin cancer are (1) basal cell carcinoma, (2) squamous cell carcinoma, and (3) malignant melanoma. Using your knowledge of skin histology, predict the skin layers and/or cells that are involved in each of these cancers.

   1. *basal cell carcinoma—the basal layer or stratum germinativum*

   2. *squamous cell carcinoma—the layers of stratified squamous epithelium derived from the basal layer*

   3. *malignant melanoma—the melanocytes*

5. Give examples of structures supported by bone and by cartilage observed in this lab topic. What differences in flexibility and function have you noticed in these structures?

   *Bone: skeleton, hard palate*

   *Cartilage: epiglottis, auricle of ear*

   *Bone is rigid and gives greater support.*

   *Cartilage is found in structures where flexibility is important.*

# Case Study

You have just completed a study of the digestive tract, and now you know the names, structures, and have been introduced to the functions of each part. However, this information alone is just the beginning of understanding the role the digestive system plays in the body. This system functions to provide nourishment for optimum functioning of your tissues and organs. You eat carbohydrates, proteins, and fats and they are digested in the organs of your digestive system. But are all carbohydrates, proteins, and fats equally useful in providing nourishment? Should your diet include more of one component and less of another?

Over 10 years ago the U.S. Department of Agriculture created a "Food Guide Pyramid" with the goal of giving advice on the best diet for healthy living. In recent years, however, many health professionals have criticized this pyramid, pointing out that it was not based on sound scientific research and its design was influenced by the food industry. Recently, new food pyramids have been created with a stronger foundation in scientific evidence linking diet and health.

The "freshman 15" (refers to the number of pounds that many students gain their first year in college) is an indication that many college students are unaware or choose to ignore the wealth of information now available to help plan a healthy diet. This may be due to erratic eating patterns, high stress levels, sedentary lifestyles, or any number of factors of college life. The objective of this study is to encourage you to examine your diet and become familiar with the latest guidelines for healthy living.

1. To begin your study, access "The Nutrition Source" Healthy Eating Pyramid at the Harvard School of Public Health website: http://www.hsph.harvard.edu/nutritionsource then click on the healthy eating pyramid. This website, created by faculty from the Department of Nutrition, presents the latest scientific evidence for the blueprint of a healthy diet. Here you can find an up-to-date "Healthy Eating Pyramid" and links to information about using this guide to design a healthy diet.

2. Next, compile a personal "meal log" for a typical 24-hour period. Create a table and record each food that you consume, keeping a record of numbers of servings and amounts (volume, e.g., ½ cup; or weight, e.g., one 6 oz. chocolate candy bar). Your table might look something like:

| Time | Item | Number and Amounts of Servings | Category in Food Pyramid |
|------|------|-------------------------------|--------------------------|
| 8:00 a.m. | Scrambled egg | 1 large | Fish, poultry, eggs |
| | Whole wheat toast | 2 slices | Wholegrain foods |
| 10:00 a.m. | Banana | 1 | Fruits |
| | Yogurt, plain | ½ cup | Dairy |

3. Working with one or more classmates, evaluate your diets. Compare your diets to the content and proportions of a healthy diet presented in the Healthy Eating Pyramid. Make a list or write a paragraph describing changes you should make to improve your diet.

4. Survey the food available in the food services for your college or university. Based on what you have learned about healthy eating, write a recommended diet for college students from the menu of food available.

## Student Media: BioFlix, Activities, Investigations, and Videos
www.masteringbiology.com (select Study Area)

**BioFlix**—Ch. 41: Membrane Transport
**Activities**—Ch. 40: Overview of Animal Tissues; Epithelial Tissue; Connective Tissue; Muscle Tissue; Nervous Tissue;
  Ch. 41: Feeding Mechanisms of Animals; Digestive System Function; Case Studies of Nutritional Disorders
**Investigations**—Ch. 41: What Role Does Amylase Play in Digestion?
**Videos**—Ch. 40: Shark Eating Seal; Hydra Eating *Daphnia* (time-lapse); Discovery Videos:
  Ch. 40: An Introduction to the Human Body;
  Ch. 41: Nutrition

## References

Fawcett, D. W. and R. P. Jensh. *Bloom and Fawcett: Concise Histology*, 2nd ed. Cambridge, MA: Oxford University Press, 2002. The successor to Bloom and Fawcett's classic *Textbook of Histology.*

Marieb, E. N. and K. Hoehn. *Human Anatomy and Physiology,* 8th ed. Menlo Park, CA: Benjamin Cummings, 2008.

Reece, J., et al. *Campbell Biology*, 9th ed. San Francisco, CA: Pearson Education, 2011.

Walker, W. F., Jr. *Anatomy and Dissection of the Fetal Pig,* 5th ed. New York: W. H. Freeman, 1998.

Willett, W. C. and P. J. Skerrett. *Eat, Drink, and Be Healthy: The Harvard Medical School Guide to Healthy Eating.* New York: Simon & Schuster, 2005.

## Websites

Current information based on scientific evidence for designing a healthy diet:
http://www.hsph.harvard.edu/nutritionsource/
then click on the healthy eating pyramid

Photographs of fetal pig anatomy:
http://biology.uco.edu/AnimalBiology/pigweb/Pig.html

# LAB TOPIC 22

# Vertebrate Anatomy I
# The Skin and Digestive System
## Teaching Plan for Laboratories

## Main Concepts and Objectives

1. Concept: Tissue specialization results in functional organs. Students will describe the structure of the skin, discuss the functions of tissues, and relate the functions of tissues to the function of the skin. (Exercise 22.1)

2. Concept: vertebrate morphology. Students will describe the morphology of the fetal pig as an example of the vertebrate digestive system. (Exercises 22.2, 22.3)

3. Concept: structure/function relationships. Identifying structures and organs in the digestive system and learning their functions will lead students to discuss the organization and function of the digestive system. (Exercise 22.3)

4. Concept: laboratory technique. Students will continue to develop techniques in dissection. (All exercises)

5. Specific content. Students will be able to name, describe, give functions, and/or define: four basic tissue types; tissues and structures in vertebrate skin; structures and organs in the digestive system; orienting terminology; rules for dissection.

## Order of the Lab

| | | |
|---|---|---|
| 1. | Present laboratory objectives and main concepts. | (15 min) |
| 2. | Study a prepared slide of mammalian skin. | (15 min) |
| 3. | Pass out pigs, plastic bags, and tags. | (10 min) |
| 4. | Study external anatomy and determine the sex of the pig. | (10 min) |
| 5. | Tie the pig in the dissecting pan and dissect the mouth cavity. | (30 min) |
| 6. | Cut open the pig. | (20 min) |
| 7. | Dissect the digestive organs. | (30 min) |
| 8. | Study the prepared slide of the intestine. | (20 min) |
| 9. | Discuss answers to questions or make assignments for work outside of class. | (20 min) |

**For a 2-hour lab:** Omit the dissection of the mouth cavity, the slide of the intestine, and the discussion of answers to questions. If you have only 2 hours for a single lab period, you might consider increasing the number of vertebrate anatomy labs to a total of four. The total amount of time estimated for the three labs as written is approximately 9 hours. By omitting two 30-minute activities, you would have four 2-hour labs.

## Classroom Management

Students work independently on all exercises; however, encourage collaboration and comparison of dissections and slide observations. Encourage students to discuss and answer questions presented in the exercises as they dissect. If time permits, call on students to answer questions at the end of the laboratory period. Informally note the quality of your students' dissections, offering suggestions for improvement and ways to perform difficult dissections. Whenever possible, ask students to recall concepts investigated in Lab Topics 18 and 19, Animal Diversity I and II.

## Student Development

Students develop observational skills and skills in dissection. They begin to think critically about the relationship between form and function. They use information learned about pig morphology and apply this knowledge to problems in human physiology.

## Lab Safety Precautions

Instruct students to:

1. Wear gloves when handling preserved animals.
2. Keep the room well ventilated, especially if using pigs that have been fixed in formaldehyde solutions.
3. Be cautious when using sharp laboratory instruments, such as scalpels and dissecting needles.

## Discussion and Summary

Students will answer questions asked in the exercises as they dissect. You may choose to have students discuss the answers to questions at the end of each exercise before they leave the lab or make an assignment to answer questions before the next lab and discuss answers at that time. You could also choose to ask students to submit answers in written form for evaluation.

## Evaluation

Formally evaluate students on laboratory tests with two components: a practical component, where students are asked to identify structures dissected in lab, and a conceptual component, where students are asked to recall the concepts learned and the integrative themes. Students should be evaluated on the quality of their dissection.

## Lab Topic 23

# Vertebrate Anatomy II:
# The Circulatory and Respiratory Systems

## Laboratory Objectives

After completing this lab topic, you should be able to:

1. Identify and describe the function of the main organs and structures in the circulatory system and trace the flow of blood through the pulmonary and systemic circuits.

2. Identify and describe the function of the main organs and structures in the respiratory system and describe the exchange of oxygen and carbon dioxide in the lungs.

3. Describe how the circulatory and respiratory systems work together to bring about the integrated functioning of the body.

4. Apply knowledge and understanding acquired in this lab to problems in human physiology.

5. Apply knowledge and understanding acquired in this lab to explain organismal adaptive strategies.

*For a 2-hour lab:* Limit introductory remarks; omit respiratory system; give questions as take-home assignment. See the Teaching Plan.

## Introduction

In Lab Topic 22, Vertebrate Anatomy I, you learned that nutrients are taken into the digestive tract, where they are processed: chewed, mixed with water and churned to a liquid, mixed with digestive enzymes, and finally digested into the component monomers, or building blocks, from which they were synthesized. For an animal to receive the benefits of these nutrients, these products of digestion must pass across intestinal cells and into the circulatory system to be transported to all the cells of the animal's body. Oxygen is necessary for the release of energy from these digested products. Oxygen from the atmosphere passes into the respiratory system of the animal, where it ultimately crosses cells in the lungs (in a terrestrial vertebrate) or gills (in an aquatic vertebrate) and enters the circulatory system for transport to cells of all organs, to be utilized in nutrient metabolism. Waste products of cellular metabolism—carbon dioxide and urea—are transported from the tissues that produce them via the blood and are eliminated from the body through the lungs of the respiratory system and the kidney of the excretory system, respectively. Thus, the circulatory, respiratory, and excretory systems function collectively, utilizing environmental materials, eliminating wastes, and maintaining a stable internal environment.

In this and the following lab topic, you will investigate the morphology of the circulatory, respiratory, and excretory systems in the fetal pig. As you dissect, relate the structure and specific function of each system to its role in the integrated body.

---

## EXERCISE 23.1

# Glands and Respiratory Structures of the Neck and Thoracic Cavity

---

### Materials

These materials will be used for the entire lab topic.

fetal pig

dissecting pan

dissecting instruments

twine

disposable gloves

plastic bag with twist tie and labels

preservative

### Introduction

To study the glands and respiratory structures of the neck, you must first open the thoracic cavity and then remove the skin and muscles in the neck region. This will expose several major glands that lie in the neck region in close proximity to the respiratory structures.

### Procedure

---

 **Wear disposable gloves when dissecting preserved animals.**

---

1. Begin the dissection by opening the thoracic cavity, which houses the heart and lungs, and making an incision that extends to the jaw.

    a. Use scissors to deepen the superficial incision previously made anterior to the abdominal cavity, and continue deepening this incision to the base of the lower jaw.

    b. Cut through the body wall in the region of the thorax, clipping through the ribs slightly to the right or left of the **sternum** (the flat bone lying midventrally to which ribs attach).

    c. Continue the incision past the rib cage to the base of the lower jaw.

2. Using the blunt probe to separate tissues, carefully remove the skin and muscles in the neck region. You will expose the **thymus gland** on each side of the neck (Figure 23.1). This gland is large in the fetal pig and in young mammals, but regresses with age. It plays an important role in the development of the body's immune system.

3. Push the two thymus masses to the side to expose the **larynx** and **trachea** lying deep in the masses. Recall your knowledge about the **glottis**, observed in the dissection of the mouth in Lab Topic 22. The glottis leads into the larynx, an expanded structure through which air passes from the mouth to the narrower trachea. The larynx houses vocal cords.

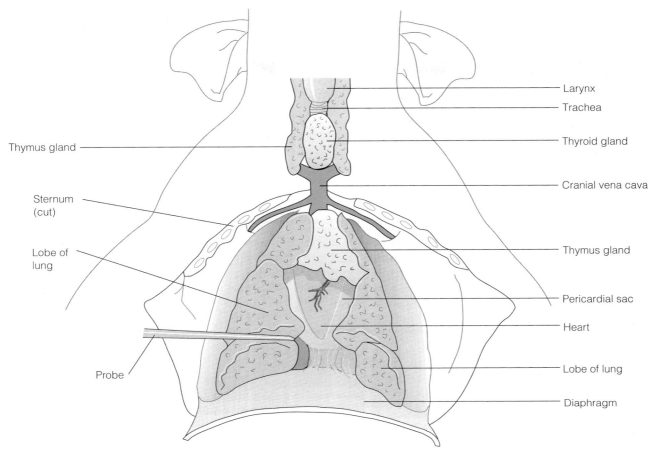

**Figure 23.1.**
**Ventral view of the anterior region of the pig,** showing structures in the neck region and the thoracic cavity. The pericardial sac encloses the heart.

4. A small reddish gland, the **thyroid gland,** covers the trachea. The thyroid gland secretes hormones that influence metabolism. Push this gland aside and observe the rings of cartilage that prevent the collapse of the trachea and allow air to pass to the lungs. Push aside the trachea to observe the dorsally located **esophagus.**

5. Do not continue the dissection of the neck and thoracic regions at this time. To prevent damage to blood vessels, you will complete the dissection of the remainder of the respiratory system (Exercise 23.5) following the dissection of the circulatory system.

# EXERCISE 23.2
# The Heart and the Pulmonary Blood Circuit

The heart and lungs lie in the **pericardial** and **pleural** (Greek for "rib") cavities, respectively, within the thoracic cavity. In your dissection of the heart and blood vessels, you will distinguish the two circulatory pathways found in mammalian circulation: the **pulmonary circuit,** which carries blood from the heart to the lungs in arteries and back to the heart in veins; and the **systemic circuit**, which carries blood from the heart in arteries to all organs *but the lungs* and back to the heart in veins. This exercise investigates circulation in fetal and adult pig hearts and the pathway of blood to the lungs in the pulmonary circuit.

## Materials

isolated adult pig or sheep heart dissected to show chambers and valves, demonstration only
supplies from Exercise 23.1

## Procedure

 Although, generally, veins contain blue latex and arteries contain red latex, the colors can vary and should not be used as guides to distinguish veins from arteries or vessels carrying oxygen-rich blood from vessels carrying oxygen-poor blood.

1. In the fetal pig, expose the heart lying in the **pericardial cavity** between the two pleural cavities. Gently push open the rib cage, using scissors and a probe to cut through muscle and connective tissue. Another lobe of the thymus gland will be seen lying over the **pericardial sac** housing the heart. The wall of the pericardial sac is a tough membrane composed of two fused coelomic epithelial linings, the **parietal pericardium** and the **parietal pleura**.
2. Cut into and push aside the pericardial sac. Carefully dissect away membranes adhering to the heart until you can identify the four chambers of the heart (Figure 23.2). The walls of heart chambers consist of cardiac muscle (Figure 22.3b, p. 577).

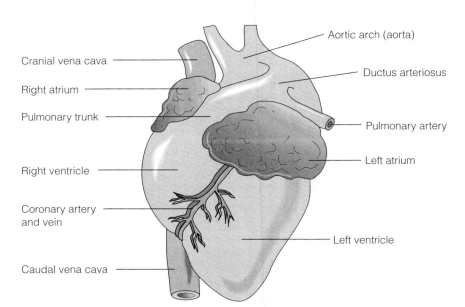

**Figure 23.2.**
**Enlarged ventral view of a fetal heart,** showing the four chambers and the major associated blood vessels. Compare this anatomy with that of an adult heart.

a. The **right atrium** and **left atrium** are small, dark, anteriorly located heart chambers that receive blood from the **venae cavae** and the pulmonary veins, respectively.

b. The **right ventricle** and **left ventricle** are large muscular heart chambers that contract to pump blood. A branch of the **coronary artery** may be seen on the heart surface where the left and right ventricles share a common wall. The coronary artery carries blood to heart tissue.

What is the name of the epithelial lining adhering to the heart surface?

*visceral pericardium*

3. Trace the **pulmonary circuit.** As the heart contracts, blood is forced from the right ventricle into the **pulmonary trunk,** a large vessel lying on the ventral surface of the heart. Another large vessel, the **aorta,** lies just dorsal to the pulmonary trunk.

a. Use forceps to pick away tissue around the pulmonary trunk and trace the pulmonary trunk as it curves cranially, giving off three branches: the right and left **pulmonary arteries** and the **ductus arteriosus.**

b. Identify the ductus arteriosus and the left pulmonary artery (the right pulmonary artery is not readily visible).

The right and left pulmonary arteries are relatively small at this stage of development. They conduct blood to the right and left lungs, respectively. The ductus arteriosus is the short, large-diameter vessel that connects the pulmonary trunk to the aorta. Because the small right and left pulmonary arteries and compact lung tissue present an extremely resistant blood pathway, the greatest volume of blood in the pulmonary trunk will flow through the ductus arteriosus and directly into the aorta and systemic circulation, bypassing the pulmonary arteries and lungs. At the time of the fetus's birth, when air enters the lungs and the tissues expand, blood will more easily flow into the lungs. The ductus arteriosus closes off and eventually becomes a ligament.

4. Observe the isolated **adult pig or sheep heart** on demonstration and locate the dorsal and ventral surfaces (Figure 23.2).

a. Identify the **right atrium** with associated **cranial** and **caudal venae cavae** and the **left atrium** with associated **pulmonary veins.**

b. Locate the **right** and **left ventricles** and the **atrioventricular valves** between the atria and the ventricles.

c. Locate the **pulmonary trunk,** which carries blood from the right ventricle, and the **aorta,** carrying blood from the left ventricle. The first two small branches of the aorta are **coronary arteries.** Locate these vessels and the **coronary veins** lying on the surface of the heart between the left and right ventricles. Coronary arteries and veins form a short circuit servicing heart tissues.

*Owing to their bipedal stance, the cranial vena cava in humans is called the superior vena cava and the caudal vena cava is called the inferior vena cava.*

*If possible, have two hearts with one adult heart opened to show internal valves and chambers and the other showing the external anatomy. The stumps of attached blood vessels are often longer in hearts purchased directly from slaughter houses.*

## Results

Review the heart chambers, blood vessels, and organs in the pathway of the pulmonary circuit in the *adult* heart. To facilitate this review, fill in the blanks in the next paragraph. As you fill in the blanks, draw the pathway of circulation through the heart on Figures 23.2 and 23.3.

**Figure 23.3.**
**Internal structures of the adult heart.** Label the structures as you trace the pathway of blood flow in the pulmonary circuit.

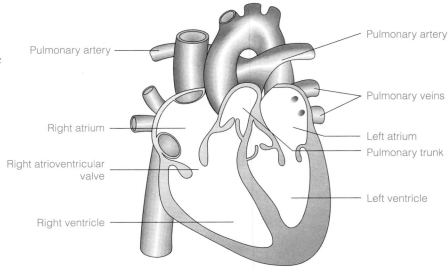

Pulmonary artery

Pulmonary artery

Pulmonary veins

Right atrium

Left atrium
Pulmonary trunk

Right atrioventricular valve

Left ventricle

Right ventricle

Blood entering the heart passes first into the right atrium. From there it flows through the right _atrioventricular valve_ into the right ventricle. When the heart contracts, this blood is forced out of the ventricle into the _pulmonary_ trunk. Branches of this trunk called _pulmonary arteries_ carry blood to the lungs. After birth, the blood will become oxygen-rich in the lungs. Blood from the lungs passes back to the heart through _pulmonary veins_, thus completing the circuit. It enters the left atrium of the heart.

## Discussion

1. Define *artery*. Define *vein*.

   *An artery carries blood away from the heart; a vein carries blood toward the heart.*

2. Why would pulmonary arteries be relatively small at the fetal stage of development?

   *Because the lung tissue is not functional and is compacted, relatively small amounts of blood flow in these vessels. The role of pulmonary arteries at this stage of development is to supply blood for the lung tissue, not to carry blood to the lungs for aeration.*

3. Although a pulmonary circuit exists, the heart in amphibians and most reptiles is made up of only three chambers—two atria and one ventricle. The latter receives blood from both atria. Speculate about possible disadvantages to this circulatory pathway.

   *In some amphibians more than others, there is a mixing of oxygen-rich and oxygen-poor blood in the ventricle.*

EXERCISE 23.3
# The Heart and the Systemic Circuit in the Thorax

Blood returning from the lungs collects in the left atrium and flows into the left ventricle. When the heart contracts, blood is forced out the **aorta,** the origin of which is obscured by the pulmonary trunk. The first branch from the aorta is the small **coronary artery,** previously identified, leading to the heart muscle. The larger volume of blood passes through the aorta to all organs of the body but the lungs. Blood returns to the heart from organs of the body through two large veins, the cranial and caudal venae cavae.

## Procedure

1. Identify the venae cavae and their major branches.

    a. Push the heart to the pigs left to see two large veins entering the right atrium; these are the **cranial** and **caudal venae cavae** (Figure 23.4).

    b. Using the blunt probe to separate the vessels from surrounding tissues, follow the cranial vena cava toward the head and identify the two large **brachiocephalic veins,** which unite in the cranial vena cava.

    c. Identify the three major veins that unite to form each brachiocephalic vein: the **external** and **internal jugulars** that carry blood returning from the head, and the **subclavian vein** that drains blood from the front leg and shoulder. Follow the subclavian vein into the front leg. Probe deep into the muscle covering the underside of the scapula (shoulder blade) and you should see the **subscapular vein,** draining blood from the shoulder region. The **axillary vein** carries blood from the front leg, becoming the subclavian vein at the subscapular branch. Occasionally, the subclavian vein is very short, and the subscapular and axillary veins unite close to the brachiocephalic vein. Another vein that is often injected and prominent in the shoulder area is the **cephalic vein.** This vein lies just beneath the skin on the upper front leg. It typically enters the external jugular near its base.

2. Identify branches of the aorta near the heart (Figure 23.5).

    a. Push the pulmonary trunk ventrally and posteriorly to observe the curve of the aorta, the **aortic arch,** lying behind.

    b. Remove obscuring tissue and expose the first two major branches of the arch, which carry blood anteriorly. It may be necessary to remove the veins to do so. The larger of the branches, the **brachiocephalic trunk,** branches off first. The **left subclavian artery** branches off second.

    c. Identify the three major branches from the brachiocephalic trunk: the **right subclavian artery,** which gives off several branches that serve the right shoulder and limb area, and two **common carotid**

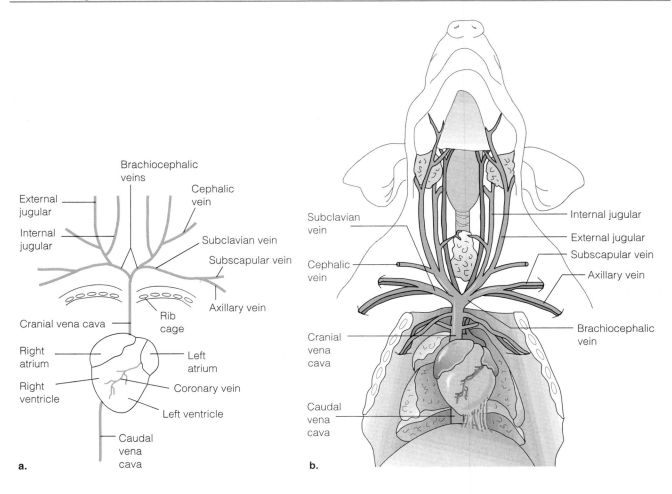

**Figure 23.4.**
**Veins near the heart.** The subclavian vein and the external and internal jugulars carry blood to the brachiocephalic veins, which unite into the cranial vena cava. The caudal vena cava carries blood from the posterior regions of the body.

arteries, which carry blood to the head. The common carotid arteries lie adjacent to the internal jugular veins.

d. Trace the branches of the left subclavian artery into the left shoulder and front leg. The branch that passes deep toward the underside of the scapula is the **subscapular artery.** After the subscapular artery branches off, the left subclavian continues into the front leg as the left **axillary artery.** Additional branches of this artery complex may also be visible.

3. Pull the lungs to the pigs right side and trace the dorsal aorta as it extends posteriorly from the aortic arch along the dorsal thoracic wall. Notice again the **ductus arteriosus** connecting from the pulmonary trunk.

4. Note the small branches of the dorsal aorta carrying blood to the ribs. A large conspicuous vein, the **azygos vein,** lies near this region of the aorta. This vein carries blood from the ribs back to the heart.

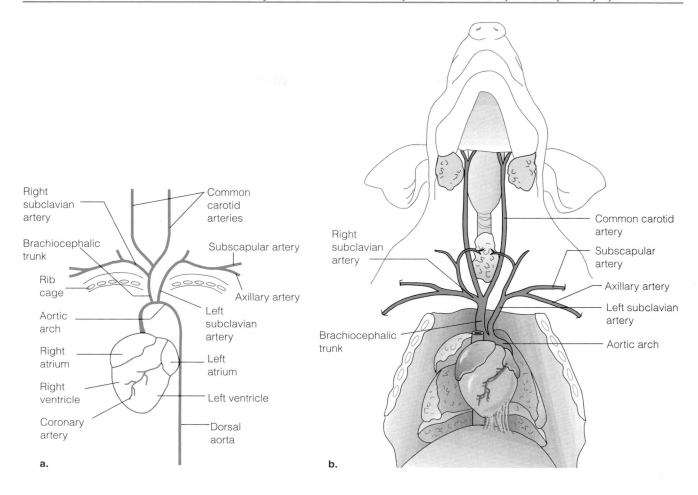

**Figure 23.5.**
**Branches of the aorta.** Branches from the aortic arch carry blood to the head and anterior limbs. The first branch, the brachiocephalic trunk, branches into the right subclavian artery to the right limb and two common carotid arteries to the head. The second branch is the left subclavian to the left limb.

## Results

Modify Figures 23.4 and 23.5 or sketch additional details in the margin of your lab manual to indicate particular features of your pigs circulatory system for future reference.

EXERCISE 23.4
# The Systemic Circuit in the Abdominal Cavity

The dorsal aorta passes into the abdominal cavity, where it branches into arteries supplying the abdominal organs, the legs, and the tail. In fetal circulation, it also branches into two large umbilical arteries to the placenta.

Blood from the legs, tail, and organs collects in veins that ultimately join the caudal vena cava to return to the heart. Blood draining from organs of the digestive system passes through additional vessels in the hepatic portal system before emptying into the caudal vena cava.

## Lab Study A. Major Branches of the Dorsal Aorta and the Caudal Vena Cava

In this lab study, you will identify the major blood vessels branching from the dorsal aorta and those emptying into the caudal vena cava.

### Procedure

1. Identify branches of the dorsal aorta (Figures 23.6 and 23.8).

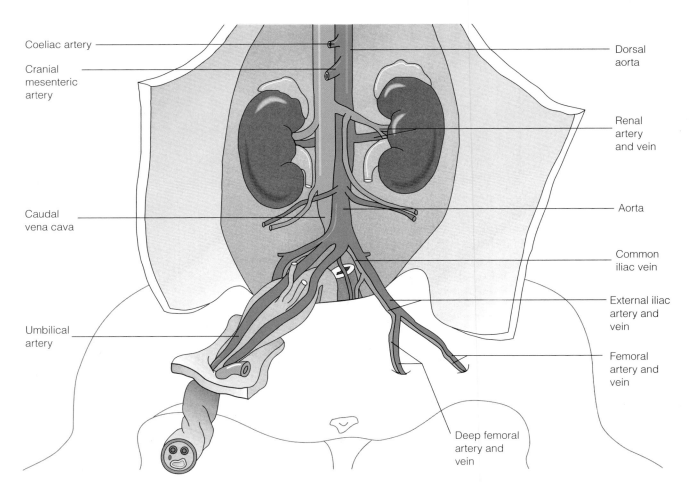

Coeliac artery

Cranial mesenteric artery

Caudal vena cava

Umbilical artery

Dorsal aorta

Renal artery and vein

Aorta

Common iliac vein

External iliac artery and vein

Femoral artery and vein

Deep femoral artery and vein

**Figure 23.6.**
**Branches of the aorta and caudal vena cava in the abdomen.** Branches of the aorta supply blood to the stomach (the coeliac artery), the small intestine (the cranial mesenteric artery), the kidney (renal arteries), the hind limbs (iliac arteries), and the placenta (umbilical arteries). Branches of the caudal vena cava drain blood from the kidney (renal veins) and posterior limbs (common iliac veins).

a. The first large branch of the aorta in the abdominal cavity exits the aorta at approximately the level of the diaphragm. Clip the diaphragm where it joins the body wall, pull all the organs (lungs and digestive organs) to the pig's right, and search for the **coeliac artery**, which carries blood to the stomach and the spleen. You may have to pick away pieces of the diaphragm that are attached to the aorta to see this vessel.

b. Once you have identified the coeliac artery, look for the next branch of the aorta, the **cranial mesenteric artery**, arising slightly caudal to the coeliac artery and carrying blood to the small intestine. The cranial mesenteric artery ultimately branches to the **mesenteric arteries** you observed when you studied the digestive system.

c. Following the dorsal aorta posteriorly, identify the two **renal arteries** leading to the kidneys.

You will observe the posterior branches of the aorta after the dissection of the reproductive system.

d. The dorsal aorta sends branches into the hind legs (the **external iliac arteries**) and to the placenta (the **umbilical arteries**) through the umbilical cord.

e. Separate the muscles of the leg to see that the external iliac artery divides into the **femoral artery** and the **deep femoral artery**. The femoral artery carries blood to the muscles of the lower leg, and the deep femoral artery carries blood to the thigh muscles.

*In the fetal pig, the umbilical arteries appear to branch directly from the caudal end of the aorta. To be precise, the internal iliacs (not identified in this exercise) branch from the aorta, give rise immediately to the umbilical arteries, and continue as very small branches into the pelvic area.*

2. Identify branches of the caudal vena cava.

a. Using Figure 23.6 as a reference, push the digestive organs to the pigs left and trace the caudal vena cava into the abdominal cavity. It lies deep to the membrane lining the wall of the abdominal cavity, the **parietal peritoneum.** Peel off this membrane to see the vena cava, the dorsal aorta, and the kidneys.

b. Identify **renal veins** carrying blood from the kidneys. **Common iliac veins** (to be identified in Lab Topic 24) carry blood from the hind legs, and **hepatic veins** carry blood from the liver to the caudal vena cava. Hepatic veins are presented in Lab Study B.

# Lab Study B. The Hepatic Portal System

In the usual pathway of circulation, blood passes from the heart to arteries, to capillaries in an organ, and to veins leading from the organ back to the heart (Figure 23.7a). In a few rare instances, a second capillary bed is inserted in a second organ in the circulation pathway (Figure 23.7b). When this occurs, the circulatory circuit involved is called a **portal system.** Such a system of portal circulation exists in the digestive system (Figure 23.7c). An understanding of this circulation pathway will increase your understanding of the absorption and processing of nutrients.

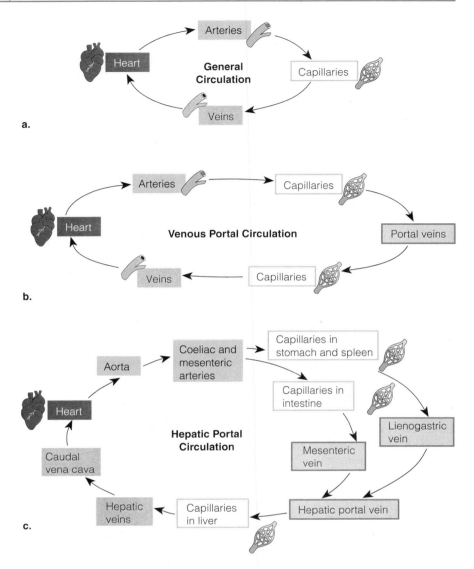

**Figure 23.7.**
**Circulatory pathways.** (a) General circulatory pathway; (b) circulation in a portal system; and (c) circulation in the hepatic portal system. Arteries are depicted in dark color, veins are gray, and portal veins are gray and outlined in color.

You have previously exposed the coeliac and cranial mesenteric arteries, which send branches to the stomach, spleen, and small intestine. These arteries divide into smaller arteries, to arterioles, and, finally, to capillaries, thin-walled vessels that are the site of exchange between blood and the tissues of the organs. Arteries associated with the small intestine are called **mesenteric arteries;** veins leaving the small intestine are called **mesenteric veins,** and they unite to form one large **mesenteric vein.** Veins from the stomach and spleen unite to form the larger **lienogastric vein.** The mesenteric and lienogastric veins unite to form the **hepatic portal vein,** which enters the liver (Figure 23.8). In the fetal pig, small branches of the **umbilical vein** join the hepatic portal vein as it enters the liver. However, the greatest volume of blood in the umbilical vein passes directly through the liver into the caudal vena cava.

In the liver, the hepatic portal vein branches into a second capillary bed, where exchange takes place between blood and liver tissue. These capillaries reunite into **hepatic veins,** which join the caudal vena cava. To identify these vessels, begin by dissecting the veins.

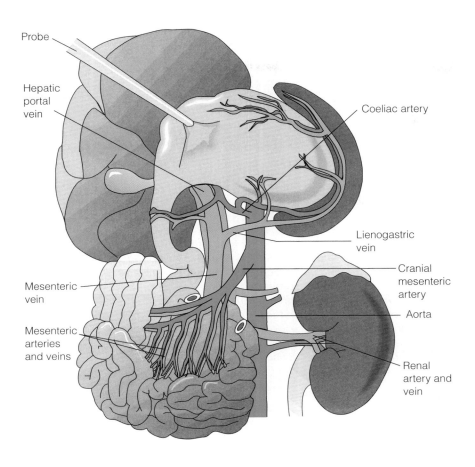

Probe

Hepatic portal vein

Coeliac artery

Lienogastric vein

Cranial mesenteric artery

Mesenteric vein

Aorta

Mesenteric arteries and veins

Renal artery and vein

**Figure 23.8.**
**The hepatic portal system.** Blood from the small intestine passes into the mesenteric vein, which unites with the lienogastric vein to form the hepatic portal vein. This vessel leads to the liver, where it breaks into a capillary bed. Blood leaves the liver through the hepatic veins.

## Procedure

1. Push the stomach and spleen anteriorly and dissect away the pancreas.
2. Use the blunt probe to expose a vein (it will probably not be injected) leading from the mesenteries of the small intestine. This is the **mesenteric vein.** It is joined by a vein leading from the stomach and spleen, the **lienogastric vein.** The two fuse to form the **hepatic portal vein,** which continues to the liver.
3. Review the flow of blood from the mesenteric arteries to the liver.

## Results

Review the blood vessels and organs in the pathway of blood through the hepatic portal system of an adult pig with functioning digestive organs. Fill in the blanks in the next paragraph:

Blood that is poor in nutrients is carried from the aorta to the _cranial mesenteric_ artery to smaller mesenteric arteries, which divide to a capillary bed in the wall of the _small intestine_, where, in the process of absorption, nutrients enter the blood. This nutrient-rich blood now flows into the _mesenteric_ vein, which joins with the _lienogastric_ vein from the spleen and stomach and becomes the _hepatic portal_ vein. This vein now carries blood to the liver, where it breaks into a second capillary bed. Capillaries in the liver converge into the _hepatic_ veins, which empty into the caudal vena cava for transport back to the heart.

### Discussion

Referring to your text, review the function of the liver in nutrient metabolism and relate this to the function of the hepatic portal system. Include information on digestive products, drugs, and toxins.

> _Digested nutrients and other absorbed substances, such as alcohol, aspirin, and other drugs, enter the blood through the capillary beds in the stomach and small intestine and are carried to the liver through the hepatic portal vein, which divides into a second capillary bed in the liver. Here the products of digestion pass into liver cells. Glucose may be converted to glycogen or fat, depending on the needs of the body and the amount in the diet. Amino acids may be converted to glucose or used to synthesize proteins such as plasma proteins. Many drugs and toxins are chemically altered and detoxified in these liver cells._

## EXERCISE 23.5
# Fetal Pig Circulation

_If time is short, omit this exercise._

As you dissected the circulatory system in the fetal pig and observed the adult pig heart, you noted differences between the fetal heart and the adult heart, and you identified blood vessels found in the fetus but not in the adult. In this exercise you will review these vessels and structures, tracing blood flow through the fetal pig.

### Procedure

1. Return the umbilical cord to the position it occupied before you began your dissection. Locate again the umbilical vein as it passes from the umbilical cord toward the liver. You cut through this vein when you opened the abdominal cavity. The umbilical vein carries blood from the umbilical cord into the liver. In the liver, small branches of this vein join the hepatic portal vein, passing blood into the liver tissue. However, the majority of the blood passes through a channel in the liver called the **ductus venosus** into the caudal vena cava. Would blood be _high_ or _low_ in oxygen in the caudal vena cava?

   _high in oxygen_

2. Review the anatomy of the fetal pig heart, and retrace the flow of blood through the heart into the dorsal aorta by way of the **ductus arteriosus**. This represents one pathway of blood through the fetal heart.

3.  A second pathway of blood through the heart is created by a structure in the fetal heart called the **foramen ovale.** To study this pathway, use your scalpel to open the pig heart by cutting it along a frontal plane, dividing it into dorsal and ventral portions. Begin at the caudal end of the heart and carefully slice along the frontal plane, cutting just through the ventricles, keeping the atria intact. Carefully lift the ventricles and look inside the heart for the wall between the two atria. Using your blunt probe, carefully feel along this wall for an opening between the two atria. This hole is the foramen ovale, which makes possible the second pathway of blood through the heart. How would this hole change the flow of blood through the heart?

    *Blood will flow from the right atrium to the left atrium, rather than from the right atrium to the right ventricle.*

    In fact, most blood coming into the heart from the caudal vena cava passes from the right atrium through this hole into the left atrium. After leaving the left atrium, where would blood go next?

    *left ventricle, aortic arch, brachiocephalic trunk to the head*

4.  Follow the dorsal aorta into the abdominal cavity to the umbilical artery branches. These branches pass through the umbilical cord to the placenta. Would blood in these branches be *high* or *low* in oxygen?

    *low in oxygen*

## Results

Trace fetal blood circulation from the umbilical vein to the umbilical artery by filling in the blanks in the next paragraph.

Blood from the umbilical vein passes through the liver in a channel called the _ductus venosus_ and into the _caudal vena cava_, which carries blood into the heart, specifically into the chamber called the _right atrium_. In one circuit of blood flow, blood goes from this chamber into the right ventricle and out the _pulmonary trunk_. A branch from this vessel, the _ductus arteriosus_ (present only in fetal circulation), carries most of this blood into the dorsal aorta. The dorsal aorta passes through the body, giving off branches to all organs of the body. Two large branches located near the tail lead into the umbilical cord and are called the _umbilical arteries_.

An alternate route carries blood from the right atrium through a fetal hole called the _foramen ovale_ into the heart chamber, the _left atrium_. From this chamber, blood next goes into the left ventricle and out the _aortic arch_. Branches of this vessel lead to the head.

## Discussion

1. What is the advantage of the circuit of fetal blood flow through the ductus arteriosus?

   *This circuit allows blood leaving the heart by way of the pulmonary trunk to bypass the nonfunctional lungs and pass directly into systemic circulation.*

2. What is the advantage of fetal blood flow through the foramen ovale?

   *If all of the blood passing through the heart passed through the ductus arteriosus into the aorta, the head would be deprived of the highly oxygenated blood coming from the placenta. This potential problem is overcome, however, by the foramen ovale, which allows blood to pass from the heart into the aortic arch. The first branches of the aortic arch carry the blood to the head.*

# EXERCISE 23.6
# Details of the Respiratory System

*Because students often destroy blood vessels as they expose bronchi leading into the lungs, we have students locate the trachea and bronchi on a sheep pluck (sheep heart with aorta, lungs with trachea) demonstration. This ensures that the blood vessels are intact for them to study for their practical exam. We tease away lung tissue to expose several branches of bronchi.*

You have previously located several of the major structures of the respiratory system (Exercise 23.1). Direct your attention again to the neck region of the pig and complete the study of the respiratory system.

## Procedure

1. Identify again the **larynx** and the **trachea** (Figure 23.9).
2. Follow the trachea caudally to the pleural cavities housing the lungs. The trachea branches into **bronchi** (sing., **bronchus**), which lead into the lobes of the **lungs** (Color Plate 65). It will be necessary to push aside blood vessels to see this. *Take care not to destroy these vessels.*
3. Tease apart lung tissues to observe that the larger bronchi branch into smaller and smaller bronchi. When the tubes are about 1–2 mm in diameter, they are called **bronchioles.** Bronchioles continue to branch and ultimately lead to microscopic **alveoli** (not visible with the unaided eye), thin-walled, blind-ending sacs that are covered with capillaries. It is here that the exchange of oxygen and carbon dioxide takes place between the blood and the atmosphere.
4. Identify the epithelial lining of the pleural cavity. How would this epithelium be named?

   *parietal pleura*

   What is the name of the epithelial lining adhering to the lung surface?

   *visceral pleura*

5. After you complete this lab topic, return your pig to its plastic bag. Check that your labels are intact and that your name, lab room, and lab day are legible. Add preserving solution and securely close the bag.

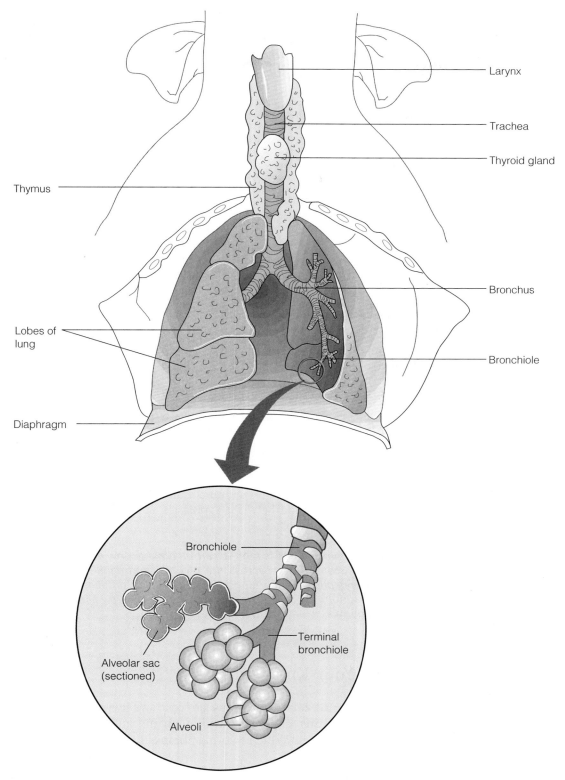

**Figure 23.9.**
**The respiratory system of the fetal pig.** Air passes through a succession of
smaller and more numerous tubes: the larynx, trachea, bronchi, bronchioles, and
ultimately, microscopic alveoli (see enlarged area).

## Results

List, in order, the structures, tubes, and cellular barriers through which air passes as it travels from outside the body to the circulatory system of a pig, a terrestrial vertebrate.

*nose or mouth, glottis, larynx, trachea, bronchi, bronchioles, alveoli, alveolar cells, capillary cells into the blood*

## Discussion

1. In terrestrial vertebrates, what is the advantage of having the surfaces for oxygen and carbon dioxide exchange embedded deep in lung tissue?

   *Barriers for gas exchange must be kept moist at all times. For terrestrial animals, this is a potential problem unless the surface is embedded deep inside the animal, protected by impermeable layers. Housing the lungs inside the thoracic cavity is an adaptation to terrestrial living that prevents dehydration of the membranes and loss of body water.*

2. The capillaries that lie in close contact with alveoli are branches of what blood vessel?

   *the pulmonary artery*

3. The confluence of these capillaries forms what blood vessel?

   *the pulmonary vein*

4. Compare blood composition in adult circulation with reference to oxygen and carbon dioxide between capillaries approaching alveoli and capillaries leaving alveoli.

   *Blood approaching alveoli is high in carbon dioxide, low in oxygen. In alveoli, oxygen diffuses across the alveolar and capillary cells from the atmosphere to the blood. Carbon dioxide diffuses in the reverse direction. Blood leaving alveoli is, therefore, higher in oxygen and lower in carbon dioxide.*

# Reviewing Your Knowledge

1. Return to Table 22.2 in the previous lab topic and review the coelomic cavities, the organs contained within them, and the associated coelomic membranes. Which two membranes fuse to form the pericardial sac?
   *Parietal pericardium and parietal pleura*

2. Complete Table 23.1. List uniquely fetal circulatory structures and vessels, describe the position or location of each, and give the fate of each after birth.

**Table 23.1**

Names, Positions, and Fate of Uniquely Fetal Circulatory Structures

| Fetal Structure or Vessel | Location | Fate After Birth |
|---|---|---|
| Umbilical vein | In the umbilical cord; extends from the placenta to the ductus venosus in the liver | Atrophies and falls away with the umbilical cord |
| Ductus venosus | In the liver; carries blood from the umbilical vein to the caudal vena cava | Becomes a ligament in the liver |
| Foramen ovale | Opening between the right and left atria | Closes as the lungs become functional |
| Ductus arteriosus | From the pulmonary trunk to the dorsal aorta | Closes, atrophies, and becomes a ligament |
| Umbilical artery | In the umbilical cord; carries blood from the dorsal aorta to the placenta | Atrophies and falls away with the umbilical cord |

# Applying Your Knowledge

1. What differences would you expect to see in the appearance of lung tissue in adult and fetal pigs? Explain.

   *The lungs of the fetal pig have not been inflated and appear dense. The adult pig would have inhaled air, and one could predict that the lungs would be spongy and filled with air.*

2. Using the Web, materials provided in the lab, your text, or library materials, answer the following questions related to the effects of smoking on the structure and function of the human respiratory system.

   a. Describe the effects and symptoms of the following diseases linked to cigarette smoking.

   Chronic bronchitis

   *Excessive mucous production in respiratory tubes. Thickening of tube walls. Blocked airflow to regions in lungs. Leads to coughing and difficulty breathing.*

   Emphysema

   *Destroys tissues between alveoli. Alveoli collapse. Inelastic tissue surrounds alveoli. Gas exchange is reduced, and all physical activity becomes difficult.*

   Lung cancer

   *Epithelial cells proliferate, forming a tumor. These cells break through the basement membrane and begin to invade other tissues. Symptoms are a persistent cough, blood in the sputum, difficulty breathing, chest pain, and recurring bronchitis.*

b. Describe smoking-induced changes in the cells and tissues of the lungs and describe the concomitant effects on function.

*Smoking reduces the cilia inside bronchioles, increases mucous secretions that clog airways, and kills cells of the immune system that would otherwise fight foreign invaders. The result is damage to lung tissue, increased fluid in the lungs, and a decrease in the lung area capable of gas exchange.*

c. What effects does smoking during pregnancy have on the fetus?

*Smoking during pregnancy increases the risk of miscarriage, fetal death, low birth weight and resulting illnesses, and sudden infant death syndrome (SIDS).*

3. The trachea is composed of rings of cartilage, while the nearby esophagus is composed of muscle and lacks cartilage. How are these structural differences related to the functions of each?

*The function of the trachea is to transport inhaled and exhaled air to and from the lungs. The rings of cartilage provide support for the trachea, preventing collapse of the airway. The esophagus functions in the transport of food from the mouth to the stomach. The muscular walls move food by swallowing and muscular contractions called peristalsis.*

4. More than any other organ in the body, the liver is most often damaged by excessive intake of substances such as alcohol, anabolic steroids, vitamin A, acetaminophen, and some prescription drugs, a condition called toxic hepatitis. Why would this be?

*Substances absorbed into the blood from the stomach and intestine ultimately pass into the hepatic portal vein that leads directly to the liver. One of the important functions of the liver is to convert toxins into inactive products that can be excreted in urine and bile. When excessive amounts of these substances are ingested, this may go beyond the detoxification capacity of the liver, and the toxins, or their toxic byproducts, damage liver tissue leading to toxic hepatitis.*

5. Scientists have concluded that a four-chambered heart is necessary to support the high metabolic rates seen in "warm-blooded" animals (endotherms)—that is, birds and mammals (see question 3, p. 605). In 2000, scientists at North Carolina State University reported that a 66-million-year-old dinosaur found in South Dakota appeared to contain a fossilized heart with two ventricles, as one would find in a four-chambered heart. What does this discovery suggest about the metabolism of this dinosaur and the position of this species of dinosaur in the evolutionary tree?

*The presence of a four-chambered heart suggests that this species of dinosaur could have had a metabolic rate capable of supporting endothermy. This discovery is another bit of evidence supporting the suggestion that dinosaurs may occupy a position in the evolutionary lineage to birds.*

# Case Studies

Extending their knowledge from this lab topic, students may investigate the following medical cases using various texts, the library, or Web resources. They may submit their findings in written or oral reports.

1. During cardiac development in an embryo or fetus, adequate flow of blood through the heart is necessary for normal development. Use your knowledge of blood flow through the fetal heart to predict which chambers and valves of the heart would not develop if the foramen ovale is absent or closes prematurely. Research identified syndromes that result from this condition.

2. Marilyn was admitted to the emergency room with pain in her chest and left shoulder, extending on to the pit of her stomach. After several tests to rule out other conditions, physicians administered nitroglycerin underneath her tongue and her pain subsided. Did this response give her physician any information to help determine if Marilyn suffered from *angina pectoris* or *myocardial infarction*? How do these two conditions differ? Which of these can be treated with *nitroglycerin*?

3. Bob, a middle-aged male, has been a moderate smoker since his teen years. After many months of procrastination and increasing chest pain, he visited his physician for a physical examination. Tests revealed that he has stage 3B lung cancer. Using Web resources, research the stages of lung cancer. What would be the condition of Bob's lungs in stage 3B? What is the prognosis for his recovery?

*If the foramen ovale closes prematurely, blood flow into the left atrium is reduced, resulting in poor development in the left atrium, the left atrioventricular valve (mitral or bicuspid valve) and the left ventricle. This condition is called hypoplastic left heart syndrome.*

*Angina pectoris is a decrease in oxygen to the heart muscle and myocardial infarction is irreversible injury to heart muscle. Nitroglycerin dilates coronary arteries, increasing blood flow to the heart. This is used to treat angina pectoris, but is ineffective in treating myocardial infarction.*

## Student Media: BioFlix, Activities, Investigations, and Videos
**www.masteringbiology.com (select Study Area)**

**Activities**—Ch. 42: Mammalian Cardiovascular System Structure; Path of Blood Flow in Mammals; Mammalian Cardiovascular System Function; The Human Respiratory System; Transport of Respiratory Gases
**Investigations**—Ch. 42: How Is Cardiovascular Fitness Measured?
  Biology Labs On-Line: Ch. 42: CardioLab; HemoglobinLab
**Videos**—Ch. 42: Discovery Video: Blood

# References

Burggren, W. "And the Beat Goes On—A Brief Guide to the Hearts of Vertebrates." *Natural History,* 2000, vol. 109, pp. 62–65.

Fisher, P. E., D. A. Russell, M. K. Stoskopf, R. E. Barrick, M. Hammer, and A. A. Kuzmitz. "Cardiovascular Evidence for an Intermediate or Higher Metabolic Rate in an Ornithischian Dinosaur." *Science,* 2000, vol. 288, pp. 503–505.

Marieb, E. N. and K. Hoehn. *Human Anatomy and Physiology,* 8th ed. Menlo Park, CA: Benjamin Cummings, 2008.

"What You Need to Know about Cancer." *Scientific American,* 1996, Special Issue, vol. 275.

## Websites

All About Smoking:
http://www.nlm.nih.gov/medlineplus/smoking.html
http://www.cdc.gov/tobacco/quit_smoking/index.htm
http://www.americanheart.org/presenter.jhtml?
identifier= 4545

Descriptions of the stages of lung cancer:
http://www.cancerhelp.org.uK/type/lung-cancer/
treatment/the-number-stages-of-lung-cancer

Excellent photographs of dissected pig including the
heart and blood vessels:
http://biology.uco.edu/AnimalBiology/pigweb/Pig.html

For information on smoking, women's health, and pregnancy, search these terms at the following websites:
http://www.lungusa.org/site/
http://www-med.stanford.edu/medicalreview/smrp
14-16.pdf

Includes photographs of lungs of smokers and non-
smokers and diagrams of microscopic changes in
lungs of smokers:
http://whyquit.com/joel/Joel_02_17_smoke_in_
lung.html

Information on maintaining a healthy heart:
http://www.americanheart.org/presenter.jhtml?identifier
=1200000

# LAB TOPIC 23

# Vertebrate Anatomy II: The Circulatory and Respiratory Systems
## Teaching Plan for Laboratories

## Main Concepts and Objectives

1.  Concept: vertebrate morphology. Students will identify structures in the circulatory and respiratory systems of the fetal pig, a representative vertebrate. (all exercises)

2.  Concept: transport of nutrients and oxygen and waste removal in complex multicel-lular organisms. Students will describe the structure and function of circulatory path-ways of the fetal pig and use this information to discuss the pathway of nutrient transport to all cells in the body and the removal of wastes. (Exercises 23.2–23.4)

3.  Concept: double-circuit pathways of circulation in mammals. Students will trace the flow of blood through the pulmonary and systemic circuits of mammalian circula-tion. (Exercises 23.2–23.4)

4.  Concept: portal circulation. Students will describe the pathway of blood in a portal system. (Exercise 23.4, Lab Study B)

5.  Concept: fetal circulation. Students will trace blood flow through the fetal pig, nothing changes that take place at birth. (Exercise 23.5)

6.  Concept: gas exchange in terrestrial vertebrates. Students will identify structures in the fetal pig respiratory system and describe the movement of air into and out of the lungs. They will explain gas exchange across barriers between air passageways and the circulatory system. (Exercise 23.6)

7. Concept: structure/function relationships. Students will apply their knowledge of the structure of circulatory and respiratory systems to questions associated with human physiology. (Applying Your Knowledge)

8. Specific content. Students will be able to name and describe: organs, vessels, and structures in the circulatory system; structures in the respiratory system; the pulmonary circuit and systemic circuit; the path of blood through the heart; a portal system; the pathway of the hepatic portal system.

## Order of the Lab

1. Introduce the main concepts and objectives, including a brief discussion of the two circuits of circulation in birds and mammals (animals with four-chambered hearts) and an overview of portal circulation. (30 min)

2. Dissect the neck region and the pericardial region; identify parts of fetal and adult hearts. (30 min)

3. Expose and identify branches of the venae cavae. (20 min)

4. Dissect branches of the aorta. (30 min)

5. Expose vessels in the hepatic portal system. (25 min)

6. Dissect structures in the respiratory system. (30 min)

7. Discuss answers to questions or assign as out-of-class work. (15 min)

**For a 2-hour lab:** Omit steps 6 and 7 and abbreviate your introduction.

Because of time constraints in a 3-hour lab, we have abbreviated our discussion of fetal circulation. If time permits, you might choose to project a diagram showing fetal pig circulation and discuss the differences in fetal and adult circulation and changes that take place at birth. Warren F. Walker's *Anatomy and Dissection of the Fetal Pig* (New York: W. H. Freeman, 5th ed., 1998) is a good source of information for your discussion.

For additional guidelines, see the Teaching Plan for Lab Topic 22, Vertebrate Anatomy I.

# Vertebrate Anatomy III: The Excretory, Reproductive, and Nervous Systems

## Laboratory Objectives

After completing this lab topic, you should be able to:

1. Identify and describe the function of all parts of the excretory system of the fetal pig, noting differences between the sexes and noting structures shared with the reproductive system.

2. Identify and describe the function of all parts of the reproductive systems of male and female fetal pigs and trace the pathway of sperm and egg from their origin out of the body.

3. Compare reproductive systems in pigs and humans.

4. Describe the structure of a neuron.

5. Describe the pathway of a simple reflex, relating this to the structure of the spinal cord.

6. Describe the structure of a representative sensory receptor, the eye.

7. Discuss the role played by the nervous and endocrine systems in integrating all vertebrate systems into a functioning whole organism.

*For a 2-hour lab: Omit the studies of the nervous system and the pregnant pig uterus. See the Teaching Plan.*

## Introduction

In Lab Topic 23, Vertebrate Anatomy II, you saw that, functionally, the excretory system is closely related to the circulatory and respiratory systems. Developmentally, however, the excretory system shares many embryonic and some adult structures with the reproductive system. In the first two exercises of this lab topic, you will investigate form and functional relationships in the excretory and reproductive systems. In the last exercise of this lab topic, you will study the nervous system, which keeps all organ systems functioning appropriately and in harmony.

The action and interaction of organ systems must be precisely timed to meet specific needs within the animal. Two systems in the body, the nervous system and the endocrine system, coordinate the activities of all organ systems. The nervous system consists of a **sensory component,** made up of **sensory receptors** that detect such stimuli as light, sound, touch, and the concentration of oxygen in the blood, and **sensory nerves,** which carry the data to the **central nervous system.** The central nervous system consists of the brain and spinal cord. It integrates information from all stimuli, external and internal, and, when appropriate, sends signals to the motor system. The **motor system** carries impulses along motor nerves to **effectors** such as glands, muscles, and other organs, bringing about the appropriate response. The nervous system provides rapid, precise, and complex control of body activities.

The endocrine system consists of endocrine glands, which respond to stimuli by secreting hormones into the blood to be transported to target tissues in the body. The target tissues then bring about the response. You have already observed several endocrine glands, including the thymus, thyroid, and pancreas. In this lab topic, you will study additional endocrine glands: ovaries and testes. Control mediated by hormones in the endocrine system is slower and less precise than nervous system control. The interaction of the nervous and endocrine systems brings about the coordination of physiological processes and the maintenance of internal **homeostasis,** the steady state condition in the vertebrate body.

EXERCISE 24.1
# The Excretory System

## Materials

preserved fetal pig
preserved adult kidney on demonstration
dissecting instruments

dissecting pan
disposable gloves

## Introduction

Several important functions are performed by the vertebrate excretory system, including **osmoregulation,** the control of tissue water balance, and the elimination of excess salts and urea, a waste product of the metabolism of amino acids. In terrestrial animals, including most mammals, water conservation is an important function of the excretory system. Studying this system in the pig will reveal the organs and structures involved in producing and eliminating metabolic waste with minimal water loss.

 Wear disposable gloves when dissecting preserved animals.

## Procedure

1. Locate the blood vessels serving the kidneys, exposed in the dissection of the circulatory system. The arteries branch from the dorsal aorta caudal to the cranial mesenteric artery. Blood enters the kidney through the **renal artery** and exits through the **renal vein.** Identify these vessels and the **kidneys** lying deep to the **parietal peritoneum** lining the abdominal cavity. This membrane was observed and removed in Lab Topic 23, Vertebrate Anatomy II.

2. Dissect the left kidney as follows. Leaving the kidney in the body and attached to all blood vessels and tubes, make a frontal section along the outer periphery, dividing it into dorsal and ventral portions (Figure 24.1a). First, observe the **renal cortex, renal medulla, renal pyramids, renal pelvis,**

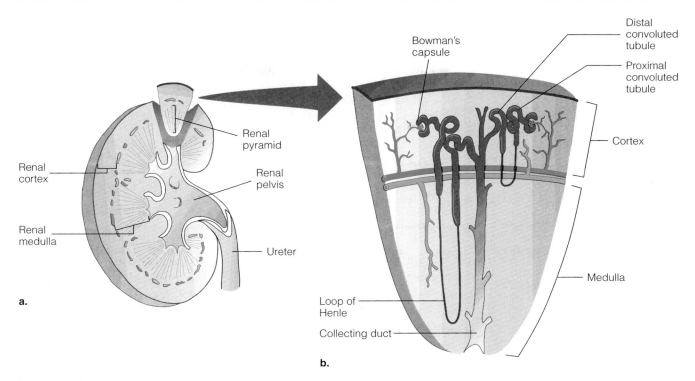

**Figure 24.1.**
**Structure of the kidney.** (a) The kidney consists of three major regions: the cortex, the medulla, and the pelvis. Renal pyramids make up the medulla, and the pelvis is continuous with the ureter. (b) An enlarged wedge of the kidney, including the cortical region over one pyramid. Nephrons—consisting of Bowman's capsule, a proximal convoluted tubule, the loop of Henle, a distal convoluted tubule, and a collecting duct—extend over the cortical and medullary regions. Waste carried in the collecting duct ultimately passes into the pelvis and ureter.

and **ureter** in your fetal pig. Then observe these parts in the adult kidney on demonstration.

Each kidney is made up of microscopic tubules, blood vessels, and thousands of nephrons (over 1 million in humans). A nephron (not visible in your dissections) consists of Bowman's capsule, a proximal convoluted tubule, the loop of Henle, a distal convoluted tubule, and a collecting duct (Figure 24.1b). Cuboidal epithelial cells line most regions of the nephron (see Figure 22.1b, p. 575). Bowman's capsule, a cup-shaped swelling at the end of the nephron, surrounds a ball of capillaries, the **glomerulus.** Blood is filtered as water and waste pass from the glomerulus into Bowman's capsule. (For details of nephron function, see your text.) Bowman's capsule, proximal and distal convoluted tubules, and associated blood vessels lie in the *renal cortex.* Loops of Henle and collecting ducts extend into *renal pyramids,* which make up the *renal medulla.* Both the loop of Henle and the collecting duct play a role in producing a concentrated urine, a significant adaptation for terrestrial vertebrates. The hypertonic urine passes into the collecting ducts, which ultimately empty into the renal pelvis, an expanded portion of the ureter into the kidney.

*In addition to dissecting the fetal pig kidney, have students observe structures in preserved adult sectioned kidneys on demonstration. Injected specimens can be purchased from biological supply houses.*

*We place a microscope slide of kidney cortex on demonstration and ask students to identify glomeruli, Bowman's capsules, and cuboidal epithelium in the kidney tubules.*

3. Using Figure 24.2 as a reference, follow the ureter as it exits the kidney at its medial border and turns to run caudally beside the dorsal aorta. The ureter then enters the **urinary bladder**. Also locate the ureter draining the right kidney and trace it to the urinary bladder. In the fetal pig, the urinary bladder is an elongate structure lying between the two **umbilical arteries** identified in Lab Topic 23 Vertebrate Anatomy II. It narrows into the small **allantoic stalk** identified in the study of the umbilical cord in Lab Topic 22 Vertebrate Anatomy I.

 Do not damage reproductive organs as you expose the structures of the excretory system.

4. Pull on the umbilical cord, extending the urinary bladder, and locate a single tube, the **urethra,** exiting the urinary bladder near the attachments of the ureters. At this stage, you will see only the end of the urethra near the entrance of the ureters. In male pigs (see Figure 24.2a), the urethra leads into the **penis.** This will be visible only after you have dissected the reproductive structures. In female pigs (Figure 24.3a), the urethra joins the **vagina,** forming a chamber, the **vaginal vestibule.** You will identify these structures after exposing the reproductive structures.

In male humans, the urethra is a tube in the penis. In female humans, the urethra does not join the vagina but empties to the outside of the body through a separate opening. The urethra becomes functional after birth when the umbilical cord and allantois wither and fall away. Waste stored in the bladder passes into the urethra, where it is carried to the outside of the body.

## Results

Describe the pathway of metabolic waste from the aorta to the outside of the body in the fetal pig.

*Urea and other wastes pass into the kidney via the renal artery, where they are removed from the blood in the nephron. The collecting duct of the nephron empties into the renal pelvis, and waste passes from here into the ureter, to the urinary bladder, to the allantoic stalk, and to the placenta.*

## Discussion

How does the elimination of metabolic waste in the pig change after birth?

*After birth, as the umbilical cord withers and ceases to function, waste passes from the bladder to the urethra and to the outside of the body via the penis or vaginal vestibule.*

EXERCISE 24.2
# The Reproductive System

## Materials

items from Exercise 24.1

## Introduction

Reproduction is perhaps the ultimate adaptive activity of all organisms. It is the means of transmitting genetic information from generation to generation. Less complex animals may reproduce sexually or asexually, but in general, vertebrates reproduce sexually. Sexual reproduction promotes genetic variation, which is important for species to adapt to changing environments. For evolution to occur, heritable variation must exist in populations. Although mutation is the source of variation, sexual reproduction promotes new and diverse combinations of genetic information. Ultimately, all sexual reproduction involves the production of gametes and the bringing together of gametes to enable fertilization to take place.

## Lab Study A. Male Reproductive System

The male reproductive system consists of gonads, ducts, and glands. Testes, the male gonads, produce sperm and secrete testosterone and other male sex hormones. Sperm pass from the testes into the epididymis, where they mature and are stored. When ejaculation takes place, sperm pass from the epididymis through the ductus deferens—also called the *vas deferens*—to the urethra. The urethra leads to the penis, which carries the sperm to the outside of the body. As sperm pass through the male tract, secretions from the seminal vesicles, the prostate gland, and the bulbourethral glands are added, producing semen, a fluid containing sperm, fructose, amino acids, mucus, and other substances that produce a favorable environment for sperm survival and motility.

*Traditionally, the tube leading from the epididymis to the urethra is called the vas deferens in humans and the ductus deferens in the pig.*

## Procedure

You will dissect the reproductive system of only one sex. However, you should observe the dissection of a pig of the opposite sex and be able to identify and describe various structures of both sexes.

1. Expose the structures of the male reproductive system (Figure 24.2a). The penis is located in the flap of ventral body wall caudal to the umbilical cord. To prevent damage to this structure, locate it before you make an incision. Hold the flap between your fingers and feel for the cordlike penis just below the skin. Once you locate the penis, using scissors, begin at the urogenital opening, the **preputial orifice** (identified in Lab Topic 22 Vertebrate Anatomy I), and make a longitudinal incision, extending caudally, just through the skin. Push aside the skin and use the

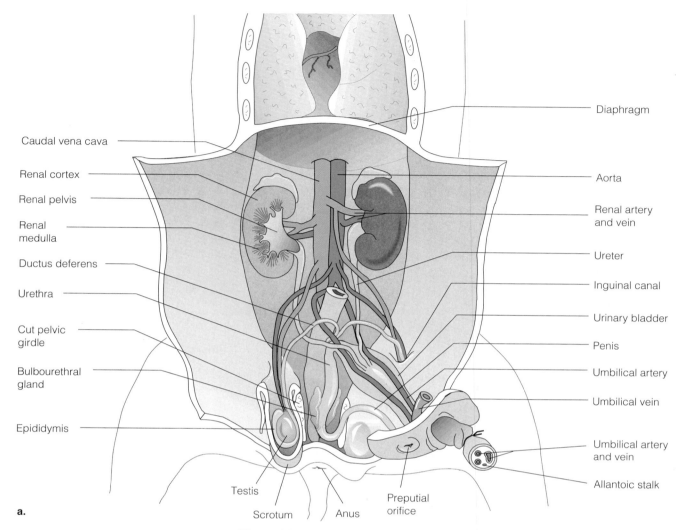

Caudal vena cava

Renal cortex

Renal pelvis

Renal medulla

Ductus deferens

Urethra

Cut pelvic girdle

Bulbourethral gland

Epididymis

Diaphragm

Aorta

Renal artery and vein

Ureter

Inguinal canal

Urinary bladder

Penis

Umbilical artery

Umbilical vein

Umbilical artery and vein

Allantoic stalk

Testis

Scrotum

Anus

Preputial orifice

**a.**

**Figure 24.2a.**

**Organs of the excretory and reproductive systems in the male fetal pig.**
The ureters enter the urinary bladder between the umbilical arteries. The urethra exits the urinary bladder and leads to the penis. The penis leads to the preputial orifice. The testes lie in pouches in the scrotum. Sperm are produced in the testes, stored in the epididymis on the testis surface, and pass to the ductus deferens, which leads to the urethra.

probe to locate and expose the long penis from the orifice caudally until it turns dorsally to meet the urethra (see Figure 24.2b).

2. Next, begin to expose the testis, epididymis, and ductus deferens. To do this, locate the **ureters** (identified in Exercise 24.1) and observe the right and left **ductus deferentia** (sing., deferens), which loop over the ureters. Follow a ductus deferens outward and caudally to the **inguinal canal** leading into the **scrotum**. Use scissors to cut carefully along the canal to expose the **testis** lying in a membranous sac. Remove this sac and identify the various structures.

a. Identify the **testis**, a bean-shaped gonad. The testes first develop in the abdominal cavity and descend before birth into the scrotal sacs.

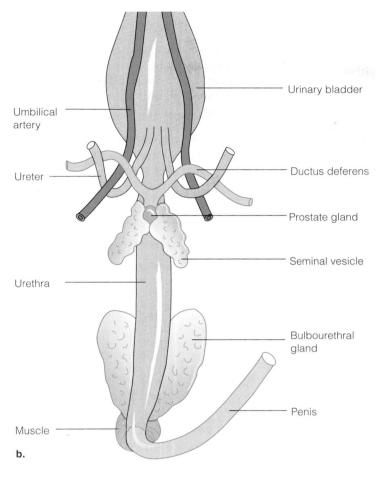

Urinary bladder

Umbilical
artery

Ureter

Ductus deferens

Prostate gland

Seminal vesicle

Urethra

Bulbourethral
gland

Penis

Muscle

**b.**

**Figure 24.2b.**
**Enlarged dorsal view of male excretory and reproductive structures in the fetal pig.** Seminal vesicles lie near the junction of the urethra and ductus deferens. Bulbourethral glands lie on either side of the urethra.

b. Identify the **epididymis,** a convoluted duct that originates at the cranial end of the testis, extends caudally along one side, then turns and continues cranially as the ductus deferens.

c. Identify the **ductus deferens,** a duct that leads away from the epididymis back into the abdominal cavity, where it loops over the ureter and enters the urethra. Also locate the ductus deferens from the other testis.

3. Turn your attention again to the area where the penis turns dorsally to meet the urethra. Push the penis to one side and probe through the muscle between the legs to locate the pubic symphysis, the portion of the pelvic girdle that fuses in a position ventral to several of the reproductive structures and the rectum. *Being careful not to go too deep or to cut the penis,* use heavy scissors to cut the pubic symphysis from posterior to anterior beginning at the bend in the penis. Press the hind limbs apart and trim the ends of the symphysis. Use the probe to remove connective tissue, and expose the **urethra,** which continues anteriorly from the bend of the penis. The urethra continues into the **urinary bladder** lying between the umbilical arteries. Identify the two large **bulbourethral glands** lying on either side of the urethra anterior to its junction with the penis (see Figure 24.2b).

4. Pull on the umbilical cord, reflecting the urethra, and locate a pair of glands, the **seminal vesicles,** that lie on the dorsal surface of the urethra near the junction of the ductus deferens and the urethra. The **prostate gland** lies between the lobes of the seminal vesicles, but because of its immature stage of development, it is difficult to identify.

At this time, complete the study of the branches of the dorsal aorta (Exercise 23.4, Lab Study A). Identify the **umbilical arteries** and the **external iliac arteries** to the legs and their branches, the **femoral** and **deep femoral arteries**. Also identify the **deep femoral, femoral,** and **common iliac veins,** which drain the legs and empty into the **caudal vena cava.**

*You may choose to have students locate the brachial plexus, the network of nerves from the spinal cord to the upper limbs, before they store their pigs. This will give them an idea of the appearance of nerves as they begin the next exercise.*

5. After you conclude the study of the male pig, find someone with a female pig, and demonstrate the systems to each other.
6. Place your pig in its plastic bag, make sure the labels are legible, add preservative, secure the bag, and store it.

## Results

In Table 24.1, list the organs and ducts through which sperm pass from their origin to the outside of the body. Describe what takes place in each organ or duct, and note glandular secretions when appropriate. Refer to your text if needed.

## Discussion

1. Vasectomy is the most common form of human male sterilization used for birth control. Describe this process.

   *A small section of each vas deferens near the epididymis is surgically removed. The cut ends of the ducts are then tied. Remember that the vas deferens is called the ductus deferens in the fetal pig.*

**Table 24.1**
Pathway of Sperm

| Organ/Duct | Activity and Glandular Secretion |
|---|---|
| Testis | Sperm and testosterone production |
| Epididymis | Storage and maturation of sperm |
| Ductus deferens | Transport of sperm, addition of seminal fluid |
| Urethra | Transport of sperm and urine, prostatic fluid, and fluid from bulbourethral glands |
| Penis | Transport of semen to female reproductive system |

2. What structures identified are common to both reproductive and excretory systems?

*urethra, penis*

3. The testes develop inside the abdominal cavity and descend through the inguinal canal into the scrotum before birth. Explain the significance of the external scrotum and external testes in mammals. Refer to your text if needed.

*Normal sperm production is temperature-sensitive and cannot occur at internal body temperature. The temperature in the scrotal sac is cooler. To adjust for extreme cold, smooth muscular contractions pull the testes closer to the body, where their temperature is increased.*

## Lab Study B. Female Reproductive System

The female reproductive system consists of the ovaries (female gonads), short uterine tubes (called *fallopian tubes,* or *oviducts* in humans), the uterus, the vagina, and the vaginal vestibule. The vaginal vestibule is present in the pig but not in the human. In the pig, the uterus consists of a uterine body and two uterine horns in which embryonic pigs develop. In the human female, the uterus does not have uterine horns but consists of a dome-shaped portion, the fundus, which protrudes above the entrance of the fallopian tubes, and an enlarged main portion, the body of the uterus, where embryos develop.

*Traditionally, the tubes that carry eggs from the ovary to the uterus are called fallopian tubes, or oviducts, in humans, but are called uterine tubes in the pig.*

### Procedure

1. To study the female reproductive system (Figure 24.3a), use scissors and make a median longitudinal incision, cutting through the skin posterior to the umbilical cord. Push aside skin and muscles and probe in the midline to locate the pubic symphysis, the portion of the pelvic girdle that fuses in a position ventral to many of the female reproductive structures and the rectum. Being careful not to go too deep, use heavy scissors to cut through the muscles and the symphysis. Press apart the hind limbs and trim away the cut ends of the symphysis.

2. Begin observations by locating the **ovaries** in the abdominal cavity just caudal to the kidneys (Figure 24.3a). They are a pair of small, bean-shaped organs, one caudal to each kidney. (When the testes of the male first develop, they are located in approximately the same position in the abdominal cavity as the ovaries; however, the testes later descend, becoming supported in the scrotal sacs.) A small convoluted tube, the **uterine tube,** can be observed at the border of the ovary.

3. The reproductive structures form a long, continuous tract. Follow a uterine tube from one ovary into the associated **horn of the uterus.** Left and right horns join to form the **body of the uterus.** The body of the uterus lies dorsal to the urethra. Push the urethra aside and use the probe to separate the urethra from the uterus. Notice that the urethra and the reproductive structures meet.

4. The body of the uterus leads into the **cervix** of the uterus, which leads into the **vagina.** To conclusively identify these regions, you must open the uterus. Without disturbing the junction of the urethra and the

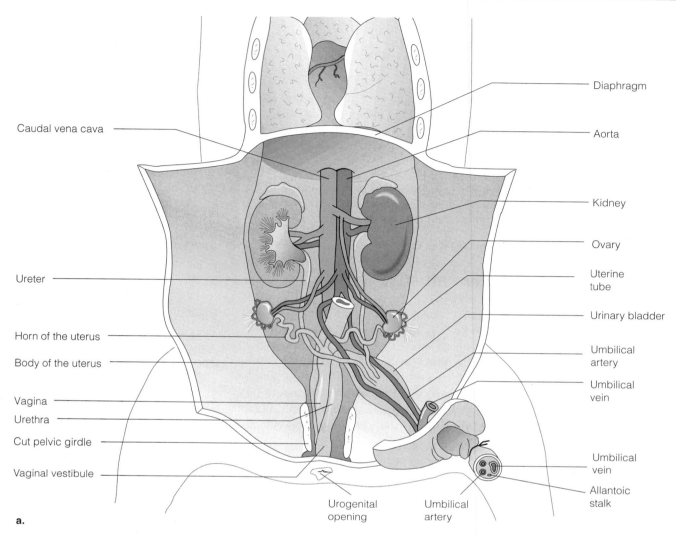

Caudal vena cava

Ureter

Horn of the uterus

Body of the uterus

Vagina

Urethra

Cut pelvic girdle

Vaginal vestibule

a.

Diaphragm

Aorta

Kidney

Ovary

Uterine tube

Urinary bladder

Umbilical artery

Umbilical vein

Umbilical vein

Allantoic stalk

Urogenital opening

Umbilical artery

**Figure 24.3a.**
**Organs of the excretory and reproductive systems in the female fetal pig.**
The ureters enter the urinary bladder. The urethra exits the urinary bladder and joins the vagina, forming the vaginal vestibule.

reproductive structures, use scissors to make a longitudinal, lateral incision in the reproductive structures and push back the sides, exposing the interior. Your dissection should resemble Figure 24.3b. Now you should be able to identify all parts of the uterus, the vagina, and the opening of the urethra into the reproductive tract. Identify the cervix, easily identified by the presence of internal ridges. The vagina, which joins the cervix, does not have these ridges. The vagina joins the urethra to form a common chamber, the **vaginal vestibule,** leading to the outside of the body. The outer opening is the **urogenital opening,** ventral to the anus (identified in Lab Topic 22 Vertebrate Anatomy I).

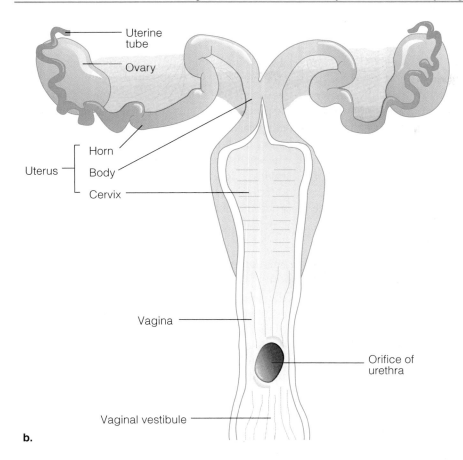

**b.**

**Figure 24.3b.**
**Enlarged view of the female reproductive system in a fetal pig.**
The cervix and vagina have been opened to show the ridges in the cervix, which are absent in the vagina. The vaginal vestibule is the common chamber formed by the confluence of the vagina and the urethra.

 At this time, complete the study of the branches of the dorsal aorta (Exercise 23.4, Lab Study A). Identify the umbilical **arteries** and the **external iliac arteries** to the legs and their branches, the **femoral** and **deep femoral** arteries. Also identify the **deep femoral, femoral,** and **common iliac veins,** which drain the legs and empty into the **caudal vena cava.**

5. After you conclude your study of the female pig, find someone with a male pig, and demonstrate the systems to each other.
6. Place your pig in its plastic bag, make sure your labels are legible, add preservative, secure the bag, and store it.

## Results

Describe the pathway of an egg from the ovary to the outside of the body in a fetal pig, naming regions of organs when appropriate.

*ovary, uterine tube, horns of the uterus, body of the uterus, cervix, vagina, vaginal vestibule*

# Lab Study C. The Pregnant Pig Uterus

*Dissect one or two pregnant uteri exposing one fetal pig still encased in the chorionic vesicle. Open another chorionic vesicle to expose the fetal pig. The amnion is easily torn, so take care to keep this intact. The instructor may choose to demonstrate this to groups of students.*

On demonstration is an isolated pregnant pig uterus, which should include uterine horns and the body of the uterus. Ovaries and uterine tubes may be attached. Fetal pigs are located in the uterine horns. Each fetal pig is attached to the mother pig by means of the **placenta**, a structure consisting of tissue from the inner lining of the uterus (maternal tissue) and the **chorionic vesicle** (embryonic tissue). These tissues are convoluted, creating interdigitating folds that increase the surfaces where the exchange of nutrients, oxygen, and wastes takes place between mother and fetus. Remember that blood does not flow directly between the mother and fetus, but the close proximity of the tissues allows diffusion across the placental membranes.

## Procedure

1. Observe the uterus with one uterine horn partially opened (Figure 24.4). One or more fetal pigs should be visible.

2. If it is not already dissected, using scissors, carefully cut into the **chorionic vesicle**, a saclike structure surrounding each fetal pig. Note that the chorionic vesicle is composed of two fused membranes, the outer **chorion** and the inner **allantois.** Blood vessels are visible lying within the thin allantois. In Lab Topic 22 Vertebrate Anatomy I, you identified the allantoic stalk, a small tube in the umbilical cord extending between the fetal pig's urinary bladder and the allantois. Speculate about the function of the allantois. The blood vessels are branches of which vessels?

   *It helps rid the embryo of nitrogenous wastes.*

   *umbilical arteries and veins*

3. Observe the **amnion**, a very thin, fluid-filled sac around the fetus. What function do you think this membrane performs?

   *It contains amniotic fluid that serves to cushion and protect the developing fetus.*

**Figure 24.4.**
**Section of uterine horn from an adult pig with two fetuses.** Two saclike structures, an amnion and a chorionic vesicle, surround the fetus on the left. The chorionic vesicle around the other fetus has been opened and the amnion removed.

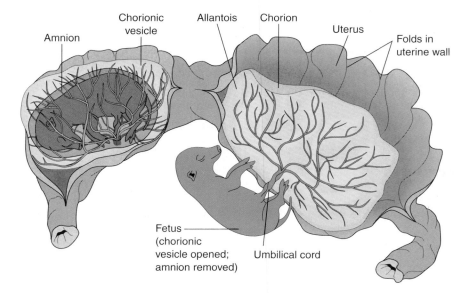

Amnion — Chorionic vesicle — Allantois — Chorion — Uterus — Folds in uterine wall

Fetus (chorionic vesicle opened; amnion removed) — Umbilical cord

4. Open the amnion and see the **umbilical cord** attaching each fetus to the fetal membranes.

## Results

Beginning with those membranes closest to the fetal pig, list in order all embryonic and maternal membranes and tissues associated with the fetal pig.

*Adjacent to the pig are embryonic membranes, the amnion, allantois, and chorion (allantois and chorion comprise the chorionic vesicle); and outside these lie the maternal tissues of the uterine wall.*

## Discussion

1. Using your text if necessary, compare the female reproductive organs in a human and an adult pig with respect to the oviduct and uterus in the human and uterine tube, horns of the uterus, and body of the uterus in the pig. Speculate about the adaptive advantage of the differences.

*The human has two oviducts that lead to the body of the uterus. Fertilization takes place in the oviduct, and usually only one or two embryos implant in the body of the uterus. Uterine horns are not found in the human. In the pig, fertilization takes place in the uterine tubes and fertilized eggs pass into the horns of the uterus where they implant. Six to 12 fetuses develop equally spaced along the two horns but not in the body of the uterus as in humans. The adaptive advantage of the horns of the uterus is the increased surface provided for development of a litter of pigs.*

2. Describe differences in the arrangement of the vagina and urethra in the fetal pig and human.

*In the human female, there is no common duct, the vaginal vestibule, as in the pig; instead, the urethra and the vagina exit the body through separate openings.*

3. Tubal ligation is a common form of human female sterilization. Describe this process.

*Oviducts are cut and tied, preventing fertilization of the egg.*

---

## EXERCISE 24.3
# Nervous Tissue, the Reflex Arc, and the Vertebrate Eye

---

In this exercise, you will study several components of the nervous system: the structure of neurons, the pathway of a reflex arc as it relates to the structure of the spinal cord, and the structure of a sensory receptor, the vertebrate eye.

# Lab Study A. Nervous Tissue and the Structure of the Neuron

## Materials

compound microscope
prepared slide of nervous tissue

## Introduction

To understand the function of nervous tissue, review the structure of the **neuron,** the functional cell of nervous tissue. Neuron structure facilitates nervous impulse transmission. Each neuron has three parts: a **cell body,** which contains cytoplasm and the nucleus; **dendrites,** extensions from the cell body that receive nervous impulses from other neurons and transmit those signals toward the cell body; and an **axon,** an extension that transmits nervous impulses away from the cell body to the next neuron or sometimes to a muscle fiber (Figure 24.5). Neurons are found in the brain and the spinal cord and in nervous tissue throughout the body. You will study the structure of nervous tissue and neurons in the spinal cord.

## Procedure

1. Using the intermediate objective on your compound microscope, scan the prepared slide of nervous tissue provided. This is a smear preparation of tissue taken from the spinal cord. You will see hundreds of small, dark dots, which are the nuclei of **glial cells.** Glial cells are nonconducting cells that support and protect neurons.

2. Look for large angular **cell bodies** of **motor neurons** scattered among the fibers and glial cells. Study one of these cell bodies on high power and locate the **nucleus,** usually with a prominent **nucleolus.** Try to identify the two types of **processes** extending from the cell body: the **axon** and **dendrites.** Although it is difficult to be certain, you may be able to differentiate between the single, broader axon and one or more slender dendrites extending from the cell body.

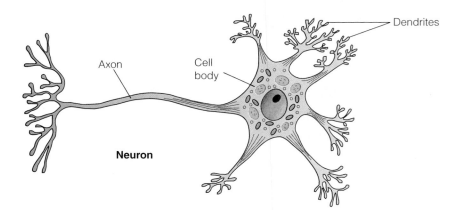

**Figure 24.5.**
**Structure of a neuron.** Dendrites and an axon, cytoplasmic processes, extend from the cell body.

# Lab Study B. The Reflex Arc
# and Structure of the Spinal Cord

## Materials

stereoscopic microscope
compound microscope
prepared slide of a cross section of spinal cord

## Introduction

By studying the anatomy of the spinal cord, you will be able to better envision the interaction of the three components of the nervous system: the sensory component with sensory receptors and sensory nerves; the central nervous system, consisting of the brain and the spinal cord; and the motor system, consisting of motor nerves and effectors. Each of these components plays a role in a simple reflex, a rapid, involuntary response to a particular stimulus. One example is the knee-jerk reflex.

## Procedure

1. Using the stereoscopic microscope, examine a prepared slide of a spinal cord cross section taken at a level that shows **dorsal** and **ventral roots.** The roots are collections of processes of neurons in spinal nerves.

2. Identify the dorsal and ventral surfaces of the spinal cord by locating the **ventral fissure** (Figure 24.6). Recall from Lab Topic 19 Animal Diversity II, that vertebrates have a tubular nervous system. This is supported by the presence of the **central canal,** a small channel in the center of the cord.

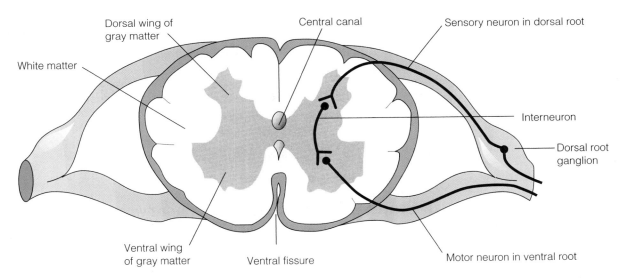

**Figure 24.6.**
**Cross section of the spinal cord at the level of dorsal and ventral roots.**
Sensory neurons enter the dorsal wing of the gray matter, and the cell bodies of motor neurons lie in the ventral portion of the gray matter. Interneurons may be present in simple reflex arcs. A simple reflex arc may include two neurons—one motor neuron and one sensory neuron—or three neurons if an interneuron is present.

3. Locate **gray** and **white matter.** In the spinal cord, white matter lies outside the butterfly-shaped gray matter. In sections through the spinal cord at the level where dorsal and ventral roots enter and exit the cord, you will be able to identify the **dorsal root ganglion,** in which cell bodies of sensory neurons lie. Look for the neuronal processes from these cell bodies. These processes continue into the tip of the dorsal "wing" of the gray matter. Sensory neurons receive impulses directly from the environment or from a specific sensory receptor.

4. Locate cell bodies of **motor neurons** in the ventral "wing" of the gray matter. These are best studied using lower powers on the compound microscope. Many of these cell bodies contain conspicuous nuclei and nucleoli. Whereas the simplest reflex involves only one sensory and one motor neuron, most reflexes involve many **interneurons,** lying between sensory and motor neurons. Motor neurons carry impulses to muscles or glands and bring about a **response.**

5. Careful observations may reveal axons from motor neuron cell bodies coursing through the white matter and into the ventral root of the spinal nerve.

## Results

Using information from the study of the spinal cord, list in sequence the structures or neurons involved in the simplest reflex.

*sensory receptor, sensory neuron, motor neuron, muscle*

## Discussion

Most reflexes involve a specific sensory receptor (such as the eye, ear, or pain or touch receptors), several sensory neurons, several interneurons, several motor neurons, and effectors (muscles or glands). Propose a reflex arc that would result from your touching a hot plate in the lab.

*Sensory receptors in the hand would be stimulated, and impulses would rush to the spinal cord through sensory neurons and pass to interneurons. From certain interneurons, the impulse would be passed to motor neurons leading to muscles (effectors) in your arm, and you would quickly pull your hand away. Simultaneously, interneurons would carry the impulses from sensory neurons to the brain, where you would become conscious of the pain. Perhaps motor impulses would pass to your vocal cords and you would yell.*

# Lab Study C. Dissection of a Sensory Receptor, the Eye

## Materials

preserved cow or sheep eye        dissecting pan
dissecting instruments        disposable gloves

## Introduction

The goal of this study is not to perform a comprehensive study of eye structure but rather to identify those structures in the eye that enable it to receive stimuli and transmit the signals in sensory nerves to the central nervous system for processing. After processing, the signals are sent through motor nerves to the effector, bringing about the response.

The vertebrate eye is a complex sensory organ containing nervous tissue capable of being stimulated by light to produce nervous impulses. Sensory neurons carry these impulses to the brain, where they are interpreted, resulting in the perception of sight. The light-sensitive, or photoreceptor, cells in the eye are called *rods* and *cones*. They are the sensory part of a multilayered tissue, the retina. Other structures in the eye protect the retina and regulate the amount and quality of light stimulating the photoreceptor cells.

As you dissect an isolated eyeball from a cow or sheep, determine the contribution of each structure to the production of sight.

## Procedure

1. Examine the isolated eye and notice that it is covered with fatty tissue and muscle bands except in the region of the **cornea,** a tough, transparent layer that allows light to enter the eye (Figure 24.7a).

2. Search through the fatty tissue on the eye sphere approximately opposite the cornea and locate the round stub of the **optic nerve** exiting the eyeball (Figure 24.7b). The optic nerve contains axons of sensory cells that transmit sensations from the eye to the brain.

3. Use forceps and scissors to trim away all fat and muscle on the eye surface, taking care to leave the optic nerve undisturbed.

4. Once the fat is removed, you will see that the cornea is the anterior portion of the tough, outer layer of the eyeball, the **fibrous tunic.** The posterior portion of this layer is the white **sclera** (Figure 24.7b). The fibrous tunic protects the internal eye structures.

5. Use scissors to cut the eye in half, making an equatorial incision and separating the anterior hemisphere (with the cornea) from the posterior hemisphere (bearing the optic nerve). Open the eye by placing it, nerve down, in the dissecting tray and lifting off the front hemisphere. Place this hemisphere in the tray with the cornea down. Your dissection should look like Figure 24.7c.

*You may choose to have students carry out this portion of the dissection with the eyeball submerged in water in a finger bowl to lessen the chance of damage to internal structures.*

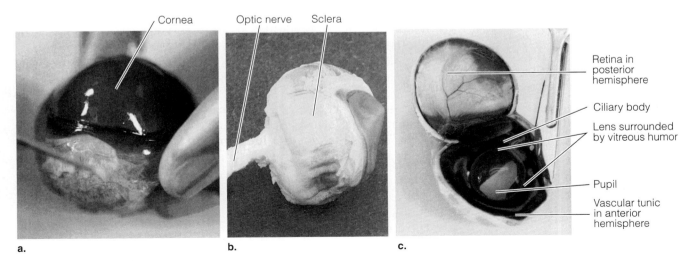

Cornea  Optic nerve  Sclera

Retina in posterior hemisphere

Ciliary body

Lens surrounded by vitreous humor

Pupil

Vascular tunic in anterior hemisphere

a.  b.  c.

**Figure 24.7.**

**Cow and sheep eyes.** (a) Isolated cow eye, (b) sheep eye with fatty tissue removed, and (c) cow eye opened to show internal structures.

6. Identify the various structures in the anterior hemisphere.

   a. Identify the **lens,** a hard, oval-shaped structure that focuses the light on the retina (Figure 24.7c). In life, this is transparent. Surrounding and attached to the lens may be a jellylike clear substance, the vitreous humor, described in step 8.

   b. Identify the **ciliary body,** a dark, ridged, muscular structure surrounding and attached to the lens by thin ligaments. The ciliary body is a component of the second tunic of the eye, the darkly pigmented **vascular tunic.** Contraction of muscle fibers in the ciliary body changes the shape of the lens. What role does this process play in eye function?

   *Changing the shape of the lens focuses the light on the retina.*

7. Carefully remove the lens (and vitreous humor, if attached) and observe that the ciliary body merges anteriorly into another component of the vascular tunic, the **iris.** The iris surrounds an opening, the **pupil.** In the cow or sheep eye, the pupil is more irregular in shape than in the human eye. The pupil allows light to pass through the vascular tunic to the lens. The iris is a sphincter muscle. What is its function?

   *The iris controls the amount of light entering the eye.*

8. Turn your attention to the posterior hemisphere of the eye. If dissected as described, this hemisphere should hold the **vitreous humor,** which holds the retina in place and is the major internal support of the eye.

9. Using forceps, carefully remove the vitreous humor and identify the pale, delicate **retina,** the third tunic of the eye. The retina contains microscopic rods and cones. What is the function of the retina?

   *Rods and cones in the retina are modified neurons that are receptors of light stimuli (photoreceptors). After being stimulated by light, they transmit impulses that are ultimately interpreted by the brain as vision.*

10. The retina lies on the inside surface of the pigmented **choroid layer,** another component of the vascular tunic. The choroid layer absorbs extraneous light rays passing through the retina. Gently push the retina aside and notice that it appears to be attached to the choroid layer in only one spot. This point of attachment is actually where processes from sensory neurons exit the retina as fibers of the **optic nerve.** You may notice a semicircular area of rainbow-colored tissue in the choroid layer. This is the **tapetum lucidum,** a tissue found in the choroid of some animals (but not humans) that enhances vision in limited light.

## Results

Examining your dissected eye, list in sequence all tissues and structures in the eye through which light passes to create an image, beginning outside the eye through to the brain.

*cornea, pupil surrounded by iris, lens surrounded by ciliary body, vitreous humor, retina, optic nerve, brain*

## Discussion

1. Using your text or library sources, describe each listed functional impairment of the eye.

   myopia (nearsightedness):

   *elongation of the eyeball that causes light waves to focus at a point in the vitreous humor in front of the retina*

   hyperopia (farsightedness):

   *The eyeball is too short, and the light waves focus behind the retina.*

   astigmatism:

   *An irregular curvature of the cornea or lens distorts the refraction of light waves.*

   cataracts:

   *a clouding of the lens due to a chemical change of the protein in the lens*

   Which of the above impairments is (are) most likely to occur as a result of aging?

   *The elderly occasionally develop cataracts.*

# Reviewing Your Knowledge

Complete Table 24.2, naming the three tunics of the eye and their subdivisions, if appropriate. Give the function of each subdivision.

**Table 24.2**
Eye Tunics and Their Functions

| Tunic | Subdivision | Function |
|---|---|---|
| *Fibrous* | *Sclera* <br> *Cornea* | *Protection* <br><br> *Protection; allows light to pass* |
| *Vascular* | *Choroid layer* <br><br> *Ciliary body* <br><br> *Iris* | *Absorbs extraneous light* <br><br> *Suspends and changes shape of lens* <br><br> *Controls amount of light* |
| *Retina* | — | *Photoreceptor layer* <br><br> *Contains rods and cones* |

# Applying Your Knowledge

1. How would hypertrophy (swelling) of the prostate gland (often a symptom of prostate cancer) affect functioning of the excretory system?

   *The location of the prostate gland is near the junction of the urinary bladder and the urethra. Swelling of this gland can constrict the urethra, making it difficult to void, leading to kidney damage.*

2. Normally a fatty encasement surrounds the kidney helping to maintain its normal position in the body. In cases of extreme emaciation in humans—for example, as in anorexia—the kidneys may drop to a lower position. Consider the ducts associated with the kidney and propose one side effect to the kidney that could result from severe weight loss.

   *The ureter may become kinked and prevent normal drainage of urine from the kidney. Urine backs up into the kidney causing damage to kidney tubules.*

3. A person who has lost a limb may experience phantom pain, feeling pain in the part of the body that is gone. Suggest an explanation for this phenomenon.

   *Severed sensory neurons heal and function in the remaining portion of the limb, sending impulses to the brain. For reasons unknown, the brain interprets these impulses as pain.*

4. Both the eye and a camera focus an image using a lens, but the mechanisms differ. How does the eye lens focus light on the retina? How is this different in a camera?

   *In the eye, the ciliary body focuses the light on the retina by changing the shape of the lens. The camera lens being glass, its shape does not change, so moving the lens closer to or farther from the film focuses the image.*

# Case Studies

Extending their knowledge from this lab topic, students may investigate the following medical cases using various texts, the library, or Web resources. They may submit their finding in written or oral reports.

1. Consult a text and write a definition of *homeostasis*. It is said that disease is the failure to maintain homeostatic conditions in the body. Investigate disorders or diseases that may result when homeostasis is disrupted owing to problems in the respiratory, digestive, circulatory, nervous, or excretory systems.

2. Of the couples in the U.S., 10–15% experience infertility, defined as the inability to get pregnant after a year of frequent, unprotected sex. Kim and Joseph Lang have been married 5 years. Two years ago they decided to have a baby, but have been unable to conceive. Laboratory tests show that Kim is ovulating normally and has normal hormone cycles. Now that you know more about male and female reproduction systems, use the Web and other resources to investigate possible causes for their infertility. Consider potential obstructions in structures of both male and female systems, the effects of sexually transmitted diseases, and causes of abnormal pregnancies.

3. Kate has peritonitis, which she was told resulted from a urinary tract infection. What is peritonitis? Why would this result from a urinary tract infection in females but not males? Discuss the occurrence of this disease in relation to the anatomy of female and male reproductive tracts.

4. Victoria, a 45-year-old woman, has noticed that she has difficulty focusing on close objects and must hold a book at arms length when reading. What could be causing this condition, known as *presbyopia*? Relate the changes in her vision to your understanding of the structure and function of the parts of the eye.

*Guide your students to investigate peritonitis and then imagine how the infection could have spread to the peritoneum in the female. They should discover why males would not have the same problem.*

 ## Student Media: BioFlix, Activities, Investigations, and Videos
www.masteringbiology.com (select Study Area)

**BioFlix**—Ch. 48: How Synapses Work
**Activities**—Ch. 44: Structure of the Human Excretory System; Nephron Function; Control of Water Reabsorption;
   Ch. 46: Reproductive System of the Human Female; Reproductive System of the Human Male;
   Ch. 48: Neuron Structure;
   Ch. 49: Neuron Structure;
   Ch. 50: Structure and Function of the Eye
**Investigations**—Ch. 46: What Might Obstruct the Male Urethra?
**Videos**—Ch. 46: Ultrasound of Human Fetus 1; Ultrasound of Human Fetus 2; Discovery Videos:
   Ch. 49: Teen Brains

# References

Fawcett, D. W. and R. P. Jensh. *Bloom and Fawcett: Concise Histology*, 2nd ed. Cambridge, MA: Oxford University Press, 2002. The successor to Bloom and Fawcett's classic *Textbook of Histology*.

Marieb, E. N. and K. Hoehn. *Human Anatomy and Physiology*, 8th ed. Menlo Park, CA: Benjamin Cummings, 2008.

Werbin, F. and B. Rosha. "The Movies In Our Eyes." *Scientific American*, April, 2007.

# Websites

A Google search for "reflex arc" will yield many informative websites:
http://www.google.com

Photographs of fetal pig anatomy:
http://biology.uco.edu/AnimalBiology/pigweb/Pig.html

Vertebrate eye anatomy:
http://www.macula.org/anatomy/eyeframe.html

Website for the National Eye Institute with information on eye anatomy, diseases, and disorders:
http://www.nei.nih.gov/health/

# LAB TOPIC 24

# Vertebrate Anatomy III: The Excretory, Reproductive, and Nervous Systems
## Teaching Plan for Laboratories

## Main Concepts and Objectives

1. Concept: homeostasis. Students will identify and describe structures in the excretory system of the fetal pig and discuss the role of that system in maintaining optimum conditions in the body by the removal of waste and excesses in the blood. (Exercise 24.1)

2. Concept: reproduction. Students will identify structures in the reproductive systems of male and female fetal pigs. As structures are identified, students will describe functions. (Exercise 24.2)

3. Concept: coordination of activities in the vertebrate body. Students will describe the structure of neurons and the spinal cord, relating this information to the pathway of a simple reflex. (Exercise 24.3, Lab Study A, Lab Study B)

4. Concept: the role of sensory receptors in detecting the internal and external environment of an organism. Students will describe the structure of the vertebrate eye and its role in receiving and transmitting light stimuli. (Exercise 24.3, Lab Study C)

5. Specific content. Students will be able to describe or define: regions of the kidney, organs and structures of the excretory system, organs and structures of the male and female reproductive systems, the pathway of sperm, the parts of the pregnant pig uterus, the pathway of the reflex arc, structure of the spinal cord.

## Order of the Lab

1. Introduce the main concepts and objectives.                                    (15 min)
2. Dissect the excretory system.                                                  (20 min)
3. Dissect the reproductive system.                                               (50 min)
4. View the reproductive system in a pig of the opposite sex.                      (15 min)
5. Observe the demonstration of the pregnant pig uterus.                           (10 min)
6. Introduce the parts and functions of neurons.                                    (5 min)
7. View the prepared slides of neural tissue and a crosssection
   of spinal cord.                                                                (20 min)
8. Discuss the reflex arc and answers to questions.                               (15 min)
9. Dissect the vertebrate eye.                                                     (30 min)
10. Final instructions. Tell students how to dispose of fetal pig carcass.          (5 min)

**For a 2-hour lab:** Omit the entire nervous system and the demonstration of the pregnant pig uterus. See the Teaching Plan for Lab Topic 22 Vertebrate Anatomy I, for suggestions about converting the three anatomy labs to four 2-hour labs.

For additional guidelines, see the Teaching Plan for Lab Topic 22 Vertebrate Anatomy I.

## Lab Topic 25
# Animal Development

## Laboratory Objectives

After completing this lab topic, you should be able to:

1. Describe early development in echinoderms (sea urchin and sea star), amphibian (frog or salamander), fish (zebrafish), and bird (chicken).

2. List the events in early development that are common to all organisms.

3. Compare early development in the organisms studied, speculating about factors causing differences.

4. Relate the events of early development in vertebrates to the formation of a dorsal nerve cord.

5. Discuss the effects of large amounts of yolk on the events of early development.

*For a 2-hour lab: Omit the echinoderm development video, Fertilization in Living Sea Urchins (Exercise 25.1, Lab Study A), and the amphibian exercise (Exercise 25.2).*

## Introduction

The development of a multicellular organism involves many stages in a long process beginning with the production and fusion of male and female gametes, continuing with the development of a multicellular embryo, the emergence of larval or juvenile stages, growth and maturation to sexual maturity, and the process of aging, and eventually ending with the death of the organism. A range of biological processes functions in development, including **cell division; differentiation,** where cells, tissues, and organs become specialized for a particular function; and **morphogenesis,** the development of the animal's shape, or body form, and organization.

Early developmental studies focused on the description of form, or **morphology,** of animals as they grow. Developmental biologists asked questions about the process of development and the forces involved in morphogenesis. Model organisms used in these studies included the sea urchin and sea star (echinoderms), frogs and salamanders (amphibians), and the chick. More recently, the zebrafish and other organisms have become important subjects for developmental studies. Currently, developmental biologists use these same animals to ask questions about the genetic control of development and the processes involved in activating different genes in different cells. Before these questions can be pursued, however, it is important to have a basic understanding of morphology in early development.

In this lab topic, you will use several of the model organisms of classical and current research to investigate major early developmental events common to most animals. These events include **gametogenesis,** the production of gametes; **fertilization,** the union of male and female gametes; **cleavage** and **blastulation,** the production of a multicellular blastula; **gastrulation,** the

*In the Study Area of masteringbiology.com, see Student Media for suggested activities, investigations, and videos.*

**Figure 25.1.**
**Egg types based on amount and distribution of yolk.** (a) Isolecithal eggs have small amounts of evenly distributed yolk. (b) Telolecithal eggs have large amounts of yolk concentrated at one end.

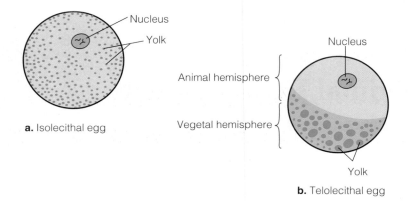

**a.** Isolecithal egg

**b.** Telolecithal egg

formation of three primary germ layers—ectoderm, mesoderm, and endoderm; **neurulation,** the formation of the nervous system in chordates; and **organogenesis,** the development of organs from the three primary germ layers. Although you may observe all of these developmental stages, you will study primarily cleavage and blastulation, gastrulation, neurulation, and organogenesis.

## Overview of Stages in Early Development

### Stage 1: Preparation of the Egg, Fertilization, and Cleavage

Development begins as sperm and egg prepare for fertilization. Sperm develop a flagellum, which propels the cell containing the haploid genetic complement of the paternal parent to the egg, which contains the haploid maternal genetic complement. As an egg matures, sperm-binding receptors develop on its surface. These can only bind to matching receptors in the sperm, insuring fertilization of egg and sperm of the same species. Within the egg, food reserves called **yolk** accumulate. These reserves are mainly protein and fat and will be utilized by the early embryo.

When egg and sperm unite, their nuclei, each containing a haploid set of chromosomes, combine to form one diploid cell, the **zygote.** The mitotic cell divisions of cleavage rapidly convert the zygote to a multicellular ball, or disc, called the **blastula.** The cells of the blastula are called **blastomeres.** A cavity, the **blastocoel,** forms within the ball of cells. The blastocoel is centrally located in embryos developing from isolecithal eggs. In embryos developing from telolecithal eggs, the blastocoel is associated with the dividing cells in the animal hemisphere.

### Egg Types

Because early events in development are affected by the amount of yolk present in the egg, the classification of eggs is based on the amount and distribution of yolk. Eggs with small amounts of evenly distributed yolk are called **isolecithal** eggs (Figure 25.1a). Eggs containing large amounts of

**Figure 25.2.** (☞)
**Cleavage types based on amount and distribution of yolk.**
(a) In isolecithal eggs, cleavage is holoblastic, and the blastocoel is centrally located. (b) In moderately telolecithal eggs, cleavage is holoblastic, and the blastocoel develops in the animal hemisphere. (c) In strongly telolecithal eggs, cleavage is meroblastic. Only the active cytoplasm divides, producing a cap of cells, the blastoderm. The blastocoel forms within the blastoderm.

**Holoblastic cleavage**

**Meroblastic cleavage**

**a.** Isolecithal egg

Cleavage plane I

Zygote

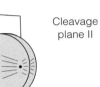

Cleavage plane II

2-cell stage

Cleavage plane III

4-cell stage

8-cell stage

16-cell stage

Blastocoel

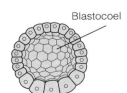

**b.** Moderately telolecithal egg

Zygote

2-cell stage

4-cell stage

8-cell stage

16-cell stage

**c.** Strongly telolecithal egg

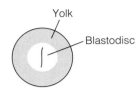

Yolk

Blastodisc

2-cell stage

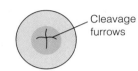

Cleavage furrows

4-cell stage

8-cell stage

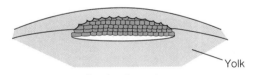

Blastoderm

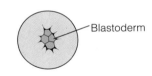

Yolk

Section through blastoderm and yolk

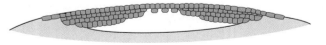

Blastocoel

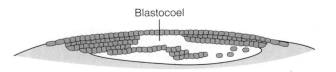

yolk concentrated at one end are called **telolecithal** eggs (Figure 25.1b). Some species are moderately telolecithal, whereas others are strongly telolecithal. Eggs may also be classified as **centrolecithal** (yolk in the center of the egg) or **alecithal** (no significant yolk reserves). Neither of these conditions will be studied in this lab topic.

In strongly telolecithal eggs, the nucleus is surrounded by **active cytoplasm**, which is relatively devoid of yolk. This nuclear-cytoplasmic region is called the **blastodisc**. The blastodisc is displaced toward the pole of the egg where polar bodies budded from the cell in meiosis. This pole is designated the **animal pole**. The half of the egg associated with the animal pole is the **animal hemisphere**. In these eggs, the yolk is concentrated in the other half of the egg, designated the **vegetal hemisphere**. The pole of this hemisphere is the **vegetal pole**.

### Cleavage Types

Although the end result of cleavage, the formation of the blastula, is the same in all organisms, the pattern of cleavage can differ. One factor that influences the pattern of cleavage is the amount of yolk present. In total, or **holoblastic**, cleavage, cell divisions pass through the entire fertilized egg. This type of cleavage takes place in isolecithal eggs, where the impact of yolk is minimal (Figure 25.2a). In these eggs, the blastocoel forms in the center of the blastula. In moderately telolecithal eggs, the yolk will retard cytoplasmic divisions and affect the size of cells. However, if the entire egg is cleaved, cleavage is considered holoblastic (Figure 25.2b). In this case, the blastocoel develops in the animal hemisphere. Cells in this hemisphere will be smaller and have less yolk than cells in the vegetal hemisphere.

In a strongly telolecithal egg, only the active cytoplasm is divided during cleavage. This process is called **meroblastic** cleavage, and it produces a cap of cells called a **blastoderm** at the animal pole (Figure 25.2c). In meroblastic embryos, the blastocoel forms within the cell layers of the blastoderm.

### Stage 2: Gastrulation

Gastrulation transforms the blastula, the hollow ball of cells (in holoblastic cleavage) or cap of cells (in meroblastic cleavage), into a **gastrula** made up of three embryonic, or germ, layers: endoderm, ectoderm, and mesoderm (Figure 25.3). Whereas cleavage is characterized by cell division, gastrulation is characterized by cell movement. Surface cells migrate into the inte-

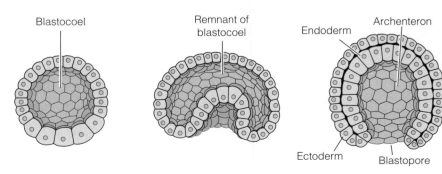

**Figure 25.3.**
**Gastrulation.** The blastula is converted to a three-layer embryo. Ectoderm and endoderm germ layers form first. Mesoderm forms later between the ectoderm and endoderm.

**a.** Blastula          **b.** Early gastrula          **c.** Late gastrula

rior of the embryo forming a new internal cavity, the **archenteron.** This new cavity is lined by the **endoderm,** the embryonic germ layer that ultimately forms the digestive tract. The archenteron opens to the outside through the **blastopore,** which in deuterostomes becomes the anus. In protostomes, the blastopore becomes the mouth. The cells that remain on the surface of the embryo become the **ectoderm.** A third layer of cells, the **mesoderm,** develops between ectoderm and endoderm.

## Stage 3: Neurulation

Late in gastrulation, neurulation, the formation of a dorsal, hollow neural tube, begins (Figure 25.4). In this *strictly chordate* process, certain ectodermal cells flatten into an elongated **neural plate** extending from the dorsal edge of the blastopore to the anterior end of the embryo. The center of the plate sinks, forming a **neural groove.** The edges of the plate become elevated to form **neural folds,** which approach each other, touch, and eventually fuse, forming the hollow **neural tube.** The anterior end of the tube develops into the brain, while the posterior end develops into the nerve (spinal) cord.

## Stage 4: Organogenesis

After the germ layers and nervous system have been established, organogenesis, the formation of rudimentary organs and organ systems, takes place. Ectoderm, the source of the neural tube in chordates, also forms skin and associated glands. In chordates, somites and the notochord (see Lab Topic 19, Animal Diversity II) develop early from mesodermal cells. Later, muscles, the skeleton, gonads, the excretory system, and the circulatory system develop from mesoderm. Nonchordate animals lack somites and the notochord, but their muscles and organs of the excretory, circulatory, and reproductive systems develop from mesoderm. The endoderm develops into the lining of the digestive tract and such associated organs as the liver, pancreas, and lungs.

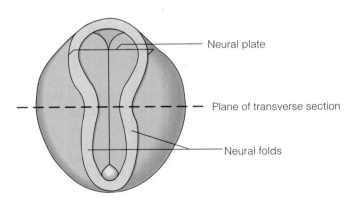

**a.** Dorsal view of frog embryo

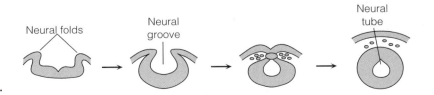

**b.**

**Figure 25.4.**
**Neurulation.** (a) Dorsal view of the entire frog embryo, showing the ectodermal neural plate with edges elevated, forming the neural folds. (b) Seen in transverse section, the neural folds meet and fuse, forming the neural tube.

Today's lab will be a comparative study of early development in five organisms including the sea urchin and the sea star, both echinoderm invertebrates; the salamander, an amphibious vertebrate; the fish, an aquatic vertebrate; and the chick, a terrestrial vertebrate.

---

EXERCISE 25.1

# Development in Echinoderms: Sea Urchin and Sea Star

---

Sea urchins and sea stars (starfish) are classified in the phylum Echinodermata, the invertebrate group that is phylogenetically closer to chordates than any other. Echinoderms release large numbers of gametes into the sea, and fertilization is external. Early development leads to a larval stage that is free-swimming and free-feeding. In this exercise, you will observe fertilization and early development in living sea urchins, and then you will observe a prepared slide of sea star development.

## Lab Study A: Fertilization in Living Sea Urchins

### Materials

clean slides and coverslips
sand or glass chips in a small petri dish
transfer pipettes, one labeled "egg," the other labeled "sperm," cut to make a slightly larger bore
small clean test tube labeled "egg" containing a suspension of living eggs from a sea urchin, e.g., *Lytechinus* sp. or *Arbacia* sp.
small clean test tube labeled "sperm" containing a suspension of living sperm from a sea urchin, e.g., *Lytechinus* sp. or *Arbacia* sp.
moisture chamber made from a petri dish containing a piece of moist (not wet) filter paper

### Introduction

*Although activities and videos that show sea urchin fertilization and early development are readily available, we recommend that students perform this study with living organisms to create immediate interest before viewing a video.*

Before your laboratory began, your instructor collected sperm and eggs from living sea urchins by injecting a KCl solution into the body cavity of the urchin. An injection of this solution causes the urchin to extrude gametes from its genital pore located on its upper (aboral—opposite the mouth) surface. It is not possible to determine the sex of a sea urchin from its external anatomy. However, it is possible to determine its sex once it begins to extrude its gametes. Whereas eggs are colored (e.g., light orange in *Lytechinus*), sperm in all species are whitish. Each female sea urchin can spawn over a million eggs and each male a billion sperm. Your instructor collected living eggs and sperm in sea water.

An unfertilized echinoderm egg is surrounded by a protective *vitelline layer* just outside the plasma membrane and a **jelly coat** that slowly dissolves in sea water. When eggs and sperm come into contact, fertilization takes place and a halo—the **fertilization envelope**—forms around the fertilized egg. This envelope helps prevent multiple fertilizations, or *polyspermy*. Two sequential processes prevent polyspermy, the *fast block* and the *slow block*. When a sperm first fuses with an egg, the permeability of the egg plasma membrane immediately changes, allowing an influx of sodium ions. The sodium ions change the electric potential across the cell membrane and, within a second or so, create a *fast block to polyspermy*. A second block to multiple fertilization takes about 20 to 30 seconds. This *slow block to polyspermy* involves the fusion of egg cytoplasmic vesicles with the egg plasma membrane. These vesicles lie in the cortex, or outer portion of the egg cytoplasm, and are called *cortical granules*. When they fuse with the egg membrane, their contents are expelled to the egg surface. Enzymes from the vesicles break bonds between the vitelline layer and the egg plasma membrane, and water flows between the two layers. The vitelline layer rises up from the egg membrane and becomes the fertilization envelope. Some time after the egg and sperm cells fuse, the egg and sperm nuclei, called **pronuclei,** move toward the center of the egg, where they fuse and almost immediately begin to prepare for the first mitotic division of cleavage.

In this lab study you will observe fertilization and early development. In Lab Study B, you will learn more details about the process and early stages of development in the sea star.

## Procedure

1. Place a generous drop of egg suspension on a clean microscope slide. Add a few grains of sand or glass chips and cover with a coverslip. Place the slide on your compound microscope stage and observe the eggs using first 4 × and then 10 × objectives.

2. Take a drop of sperm suspension and add this to the edge of the coverslip.

3. Working quickly, note the time and immediately begin observing the slide using first the 10 × and then the high power objectives. Use phase-contrast microscopy on both powers if your microscope has this capability. Look for sperm swimming toward the eggs.

4. Carefully observe the eggs, watching for the formation of fertilization envelopes (a clear halo surrounding the egg), indicating that fertilization has taken place.

5. After observing fertilization for several minutes, place your slide in a moisture chamber made from a petri dish.

6. At 30-minute intervals until the end of lab, remove your slide from the petri dish, carefully dry off the bottom of the slide, and use the compound microscope to observe developing embryos. After you complete your observations, return the slide to the moisture chamber each time.

## Results

1. Describe the events that take place as sperm and eggs are mixed. How long did it take for fertilization envelopes to become visible?

2. Are membranes present around all eggs? If not, can you estimate the percent of eggs that have been fertilized?

3. Describe the activity of the sperm. Do active sperm bounce off the egg, or do they become stuck to the jelly coat?

4. Did you observe cleavage? How long after you first observed fertilization? Was the cleavage pattern *holoblastic* or *meroblastic* (Figure 25.2)?

*In a 3-hour lab period, students will be able to see first and second cleavages, resulting in two-celled and four-celled embryos. The first cleavage division takes place approximately 90 minutes after fertilization. Some students will see four-celled stages after 2 hours. Cleavage will be holoblastic.*

5. At the end of the laboratory period, at what stage of development are most of your embryos?

*Most embryos will be in early cleavage, in the four- to eight-cell stage.*

## Discussion

1. What role might the jelly coat surrounding the egg perform in the process of fertilization?

*As the jelly coat slowly dissolves in sea water, it releases molecules that trigger the acrosomal reaction allowing sperm to penetrate the jelly coat, and the fertilization process begins.*

2. What would you predict would happen if sperm were exposed to a solution containing only egg jelly coat? Would these sperm still be capable of fertilization?

*No. Once the acrosomal reaction takes place, if no eggs are present the sperm clump together and lose their ability to fertilize eggs.*

 Student Media Video—Ch. 47: Sea Urchin Embryonic Development

## Lab Study B: Development in the Sea Star

### Materials

compound microscope
prepared slide of whole sea star embryos in different stages of development

### Introduction

In this lab study, you will observe a prepared slide containing an assortment of whole sea star embryos in various stages of development. You will identify each developmental stage and determine the type of egg and cleavage pattern of the sea star.

### Procedure

1. View the prepared slide of sea star embryos using low and intermediate powers on the compound microscope.

 Use only low and intermediate powers. Using the high power objective to view this slide will destroy the slide!

2. Find examples of all stages of development. When you find a good example of each of the stages described, make a careful drawing of that stage in the appropriate square in Figure 25.5. (See Color Plate 66 for photomicrographs of various developmental stages in an echinoderm.)

## Unfertilized Egg

By the time sea star eggs leave the body of the female, meiosis I and II are completed. The nucleus, called the *germinal vesicle,* is conspicuous because the nuclear envelope is intact. A nucleolus is usually distinct. The plasma membrane surrounding the egg cytoplasm closely adheres to a thin external layer known as the *vitelline layer* (not visible at this magnification). Species-specific sperm receptors extend from the egg plasma membrane into the vitelline layer.

## Fertilized Egg

The fertilized egg, or zygote, has no visible nuclear envelope giving this cell a uniform appearance. Look on the zygote surface for a **fertilization envelope,** most easily seen using phase-contrast microscopy. In Lab Study A you learned that this envelope forms as a result of sperm–egg fusion. The presence of the fertilization envelope and the absence of the visible nuclear envelope will help distinguish fertilized and unfertilized eggs.

## Early Cleavage

As cleavage begins, the zygote divides by mitosis and cytokinesis and continues to divide as this single cell is converted into a multicellular embryo. The $G_1$ and $G_2$ phases of the cell cycle (see Lab Topic 7 Mitosis and Meiosis) are essentially skipped in these mitotic events. Find two-, four-, and eight-cell stages. Is the entire zygote involved in early cleavage?

*yes*

The fertilization envelope remains intact around the embryo until the gastrula stage. What is happening to the size of the cells as cleavage takes place and cell numbers increase?

*Without the $G_1$ and $G_2$ phases of the cell cycle, as the number of cells increases, the size of cells decreases.*

## Late Cleavage

As cleavage continues, a cavity, the **blastocoel,** forms in the center of the cell cluster. The end product of cleavage will be a hollow ball of cells, the **blastula.** Locate and study a blastula. How does the size of individual cells compare with the size of the fertilized egg?

*Help students distinguish zygote and late blastula stages by suggesting that they focus very carefully on the edge of the embryo. If the embryo is at the blastula stage, the blastocoel and a thick blastula wall will be visible.*

*Cells in the blastula are much smaller than the fertilized egg.*

How does the overall size of the blastula compare with that of the fertilized egg?

*No growth takes place in this stage of development, and the blastula is only slightly larger than the fertilized egg. In fact, some students will mistake a blastula for a fertilized egg.*

### Early Gastrulation

Gastrulation converts the blastula into the gastrula, an embryo composed of three primary germ layers. The early gastrula can be recognized by a small bubble of cells protruding into the blastocoel. These cells push into the blastocoel through a region on the embryo surface called the **blastopore.** As cells continue to *invaginate*, or move inward, a tube called the **archenteron** forms. The archenteron eventually becomes the adult gut. Which embryonic germ layer lines the archenteron?

*endoderm*

### Middle Gastrulation

The archenteron continues to grow across the blastocoel. It takes on a bulb-like appearance as the advancing portion swells.

### Late Gastrulation

Cells at the leading edge of the advancing archenteron extend pseudopodia that attach to a specific region across the blastocoel. These cells continue to pull the archenteron across the blastocoel. As the tip of the archenteron approaches the opposite wall of the embryo, it bends to one side and fuses with surface cells. The site of fusion will eventually become the mouth of the embryo. What will be formed from the blastopore at the opposite end of the archenteron? (Recall from Lab Topic 19 Animal Diversity II that echinoderms are deuterostomes.)

*anus*

What is the germ layer of cells on the surface of the embryo called?

*ectoderm*

The amoeboid cells that attach the archenteron to the embryo wall are called *mesenchyme cells.* These cells later detach from the archenteron, proliferate, and form a layer of cells within the old blastocoel, now divided by the archenteron. This layer of cells will become the mesodermal germ layer.

### Bipinnaria Larval Stage

The archenteron of the gastrula differentiates into a broad **esophagus** leading from the **mouth** to a large oval **stomach** and on to a small, tubular **intestine.** All these structures will be visible in the bilaterally symmetric bipinnaria larva. Locate these structures in larvae on your slide and in Color Plate 67. The larva is now self-feeding and begins to grow. It will later be transformed into the radially symmetric adult sea star.

### Results

Draw stages of sea star development in the appropriate boxes in Figure 25.5.

| | |
|---|---|
| **a.** Unfertilized egg | **b.** Fertilized egg |
| **c.** Early cleavage | **d.** Late cleavage |
| **e.** Early gastrulation | **f.** Middle gastrulation |
| **g.** Late gastrulation | **h.** Bipinnaria larva |

**Figure 25.5.**
**Early stages of development in the sea star.**

## Discussion

1. What is the advantage of species-specific sperm receptors in the plasma membrane and vitelline layer of the egg?

   *Their presence ensures that eggs will be fertilized only by sperm of the same animal species.*

2. What type of egg does the sea star have? What evidence have you observed that supports your answer?

   *Isolecithal egg; the uniform appearance (no large obvious yolk platelets), the uniformity of the cleavage pattern, and the central blastocoel all support this answer.*

3. Describe the pattern of cleavage seen in the sea star and give the name for this type of cleavage.

   *Cell divisions pass through the entire fertilized egg. This type of cleavage is holoblastic.*

---

### EXERCISE 25.2
# Development in an Amphibian

---

### Materials

video or film of early development in a salamander or some other amphibian

*See the* Preparation Guide *for information about an excellent film on salamander development. See also www.masteringbiology.com Activities and Videos listed in Student Media at the end of this lab topic.*

### Introduction

Amphibians are vertebrates that lay jelly-coated eggs in water or in moist areas on land. Common examples include frogs and salamanders. For most species, fertilization is external, with the male depositing sperm over the eggs after the female releases them. Internal fertilization takes place in some amphibians, however, in which cases the young are born in advanced developmental stages. Early development is similar in all species. After fertilization, the zygote begins cleavage followed by gastrulation, neurulation, and organogenesis.

In this exercise, you will study early development in an amphibian by observing a video or film presentation of some species such as the salamander *Triturus alpestris*. The film or video shows dramatic time-lapse photography of cleavage, gastrulation, and neurulation.

*These terms are defined in* Campbell Biology, *9th ed., Chapter 47.*

### Procedure

1. Before viewing the film or video, complete Table 25.1 by defining terms commonly used when describing early embryos. (You may need to refer to your text.)

**Table 25.1**
Common Terms Used in Embryology

| Term | Definition |
|------|-----------|
| Animal pole | *The pole where polar bodies are given off in meiosis. In telolecithal eggs, the pole with the least amount of yolk.* |
| Animal hemisphere | *The half of a telolecithal egg with the lesser amount of yolk* |
| Equator | *Imaginary plane midway between animal and vegetal hemispheres* |
| Vegetal hemisphere | *The half of a telolecithal egg with the greater amount of yolk* |
| Vegetal pole | *The pole with the most yolk in a telolecithal egg* |

2. Read the questions in the Results section, view the film or video, and then answer the questions.

## Results

1. Would you describe the amphibian egg as isolecithal, moderately telolecithal, strongly telolecithal, or alecithal?

   *moderately telolecithal*

2. Describe the cleavage pattern. Is it holoblastic or meroblastic? Are cleavages synchronous or irregular? Can you detect any particular pattern in the cleavage? Where is the second cleavage plane in relation to the first?

   *holoblastic; synchronous and regular at first, less so later; both cleavages pass from animal pole to vegetal pole, and they are at right angles to each other*

3. Does the size of the embryo change as cleavage progresses?

   *no, not noticeably*

4. Visually follow surface cells during gastrulation. Do they all move at the same rate? Describe gastrulation, comparing the process with that in the sea star. Notice the position of the blastopore and the yolk plug located in the blastopore.

   *Cells do not move at the same rate. The blastopore does not form directly at the vegetal pole as in the sea star but is displaced to one side. Surface cells involute through the blastopore, and a yolk plug forms in the blastopore of the salamander.*

5. During neurulation, do the neural ridges (folds) meet and fuse simultaneously along the entire length of the neural tube or do they close like a zipper?

*close like a zipper*

## Discussion

1. Name at least two major differences in early development between the salamander and the sea star and describe factors responsible for these differences.

   *1. In the salamander there is unequal cleavage of cells and displacement of the blastopore to the side, owing to the greater amount of yolk in the salamander as compared with the sea star.*

   *2. Neurulation in the salamander forms a dorsal hollow nerve cord, which does not develop in the sea star.*

2. Compare the video of amphibian development with Figure 25.6. Label the following in the appropriate figure: **animal pole**, **vegetal pole**, **blastocoel**, **archenteron**, **yolk plug**, **neural plate**, and **neural folds**.

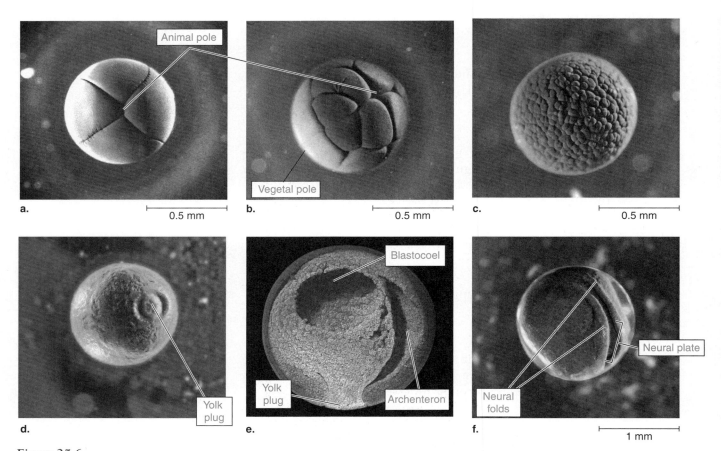

**Figure 25.6.**
**Early amphibian development.** (a) and (b) Early cleavage. (c) Surface view of the hollow blastula in late cleavage. (d) Surface view of a gastrula. (e) Cross section of a gastrula. (f) Surface view of a neurula.

Student Media Video—Ch. 47: Frog Embryo Development

# EXERCISE 25.3
# Development in the Zebrafish

## Materials

small petri dishes
depression slides
embryo-rearing solution
clean toothpicks
fish embryos in various developmental stages (some on ice)
stereoscopic microscope
compound microscope

## Introduction

In this portion of the laboratory, you will observe living embryos in early developmental stages using a freshwater fish commonly known as the zebrafish, or zebra danio (*Danio rerio*). This fish is a native of streams in India. Male and female fish are similar in appearance, but the male is generally smaller, with a streamlined body shape. The female is larger and broader than the male, especially when carrying eggs. (See Color Plate 68.)

In nature, zebrafish are stimulated to reproduce when days consist of approximately 16 hours of light and 8 hours of dark. This photoperiod corresponds to favorable weather and food supplies for developing embryos. By artificially creating this photoperiod in the laboratory, we can produce conditions that stimulate the zebrafish to spawn. After only 2 or 3 days on a cycle of 16 hours of light and 8 hours of dark, female fish will lay eggs, and male fish will deposit sperm for external fertilization.

The embryos you will observe today were collected from zebrafish on the artificial schedule. Newly spawned embryos were placed in embryo-rearing solution in petri dishes. Some embryos were maintained at room temperature, while others were placed on ice to retard development. The petri dishes were labeled to indicate the approximate stage of development. Because some embryos have been kept on ice, and because not all female fish lay their eggs at exactly the same time, a variety of early developmental stages should be available for your study. Neurulation and organogenesis stages are available from yesterday's spawning.

The approximate schedule of development at 25°C is described in Table 25.2. Use this schedule to predict the approximate stage of development for eggs collected at 8 A.M. and maintained at room temperature (approximately 23°C).

*Zebrafish are rapidly becoming a model system for research in genetics and development (in much the same manner that fruit flies and chickens have been popular research organisms in the past). By introducing your students to the organism at this level, you will be planting seeds of interest that may be reinforced in upper division courses and independent research for your majors. Zebrafish can be purchased from a tropical fish supplier and easily reared in a freshwater aquarium. See the Preparation Guide for additional information.*

*To retard development of the embryos, collect them and place them in rearing solution in petri dishes on paper towels in a metal pan. Place the pan on ice in an ice chest. If petri dishes are placed directly on the metal, the embryos die.*

*Some females lay later than others, so to obtain more embryos, collect them two or three times, at 1-hour intervals, after spawning first takes place.*

**Table 25.2**
Developmental Schedule for Zebrafish at 25°C

| Hour | Time | Comments |
|---|---|---|
| 0 | 8 A.M. | Lights on. Fish stimulated to spawn. Fish begin to dart back and forth, depositing eggs and sperm (spawning) close to the bottom of the aquarium. Fertilization takes place, forming the zygote. Cleavage begins in 35 minutes. |
| 1 | 9 A.M. | Cleavage continues. Some embryos are collected and placed on ice to slow development. |
| 2 | 10 A.M. | Embryos are in midblastula stage (approximately 64 cells). |
| 4 | 12 noon | Late blastula. |
| 5 | 1 P.M. | Early gastrula. |
| 6 | 2 P.M. | Midgastrula. |
| 12 | 8 P.M. | Gastrulation completed; neurulation taking place. |
| 18 | 2 A.M. | Neurulation completed; organogenesis beginning. |
| 24 | 8 A.M. | Second day begins; organogenesis continues. |
| 96 | 8 A.M. | Day 4 begins; embryos hatch. |

## Procedure

1. Obtain a petri dish with an embryo in embryo-rearing solution from the lab supply. View it using the stereoscopic microscope. Gently roll the embryo using a toothpick to see the embryo from several angles.

2. Read the description (following) of each stage of development, and determine the stage of the embryo. Remember that these are living embryos, and the stage of development may have changed from that indicated on the petri dish label.

3. Using a pipette, carefully transfer the embryo and rearing solution to a depression slide and view it on the lowest power of the compound microscope.

4. Remembering that these are living embryos, watch carefully to observe cells dividing. You may observe a two-cell embryo changing to a four-cell embryo or a late blastula beginning gastrulation. With careful and patient observations, you may be fortunate enough to see the developmental stages unfolding.

5. Make drawings in the margin of your lab manual so that you will be able to refer to them later.

6. Using the pipette, return the embryo to the petri dish. Then return the petri dish to the lab supply.

7. Obtain a petri dish with an embryo in a different stage of development and repeat steps 1 to 6. Continue your observations until you have seen all listed stages of development.

## Unfertilized Egg

Because most eggs are immediately fertilized after spawning, you may not find any unfertilized eggs. The unfertilized egg is about 0.5 mm in diameter. It is spherical, with yolk granules evenly distributed. There is a membrane, the **chorion**, around the egg.

## Fertilized Egg (Zygote)

A thin **fertilization membrane** forms after fertilization. The yolk granules condense, and the active cytoplasm migrates to the **animal pole**, where it becomes the **blastodisc**, or germinal disc. The blastodisc is visible as a bulge in the otherwise spherical cell. The future embryo will develop from the blastodisc, and the yolk will serve as the nutrient supply.

## Cleavage

Cleavage takes place in the blastodisc. Cells resulting from cleavage are called **blastomeres.** The blastodisc divides into 2, then 4, 8, and 16 cells of equal size.

## Blastula

By the 64-cell stage, the blastula appears as a high dome of cells at the animal pole (Figure 25.7a). This dome of cells is called the **blastoderm.** As cleavage continues, the cells become more compact, and the interface between the blastoderm and the yolk flattens out (Figure 25.7b).

## Gastrula

Gastrulation in the zebrafish is strikingly different from the process in the sea star and the salamander, because cleavage is restricted to the blastodisc. The surface cells of the blastula spread over the entire yolk mass toward the vegetal pole in a process called **epiboly.** The surface cells produce the ectoderm. When the advancing blastoderm cells have covered approximately one-half of the yolk, a type of cell movement called **involution** begins when cells move into the interior of the embryo in a ring at the edge of the advancing blastoderm. These involuting cells will eventually form mesoderm and endoderm. Figures 25.7c and 25.7d illustrate early and late gastrulation.

## Neurulation

The antero/posterior, dorso/ventral embryonic axes are first obvious during neurulation. As neurulation takes place, ectodermal cells form the neural plate and eventually the neural tube. In the whole embryo, the region of the neural tube appears as a ridge on the dorsal surface. With careful investigation, you can discern that one end of the neural tube is enlarged. This enlarged portion is the **brain** developing at the anterior end of the embryo. The posterior end of the neural tube develops into the **nerve (spinal) cord.** As development progresses, a supportive rod, the **notochord**, forms from mesoderm beneath the neural tube. The notochord is later replaced by vertebrae. Blocks of tissue called **somites** form from mesoderm along each side of the nerve cord (Figure 25.7e).

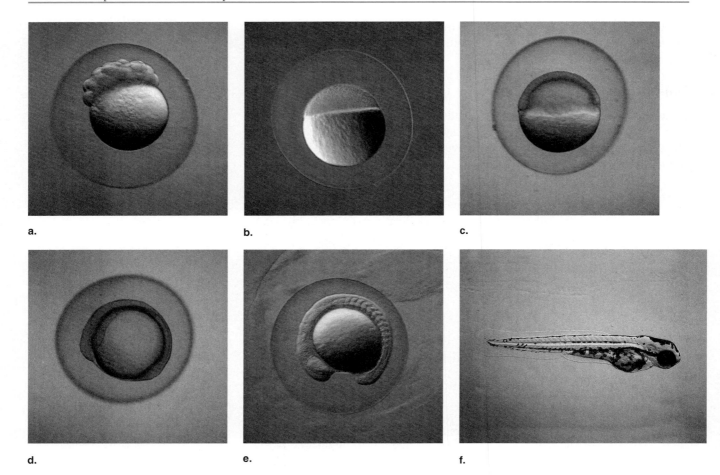

a.         b.         c.

d.         e.         f.

**Figure 25.7.**
**Development in the zebrafish.** (a) The 64-cell blastula. (b) Late blastula in side view. (c) The early gastrula. Surface cells move over the yolk mass. When they cover approximately one-half of the yolk, they begin to involute at the blastopore. (d) The late gastrula and early neurulation. (e) Neurulation and organogenesis. Ectodermal cells form the neural tube, which eventually develops into the anterior brain and posterior spinal cord. (f) Newly hatched embryo.

## Organogenesis

In organogenesis, formation of rudimentary organs and organ systems takes place. In the fish, the brain, eyes, somites, spinal cord, and tail bud are visible. The rhythmic beating of the heart and circulating blood are seen.

Among other things, somites develop into skeletal muscles, and older embryos will be actively twisting and turning in the fertilization membrane. Developing **pigmentation** will be visible in the skin and eye.

## Hatching

As organogenesis continues, embryos grow and develop, and after 4 days, the eggs hatch (Figure 25.7f).

## Results

In the margin of your lab manual, draw the stages of development. Include at least two stages in early cleavage (2-cell, 4-cell, 8-cell, 16-cell) and label the animal pole, vegetal pole, blastomere, and fertilization membrane.

## Discussion

1. What type of egg does the zebrafish have (Figure 25.2)?

   *strongly telolecithal*

2. Is cleavage in the zebrafish holoblastic or meroblastic?

   *meroblastic*

3. As the region of active cytoplasm (the blastodisc) undergoes cleavage, it becomes the blastoderm. Compare the size of the blastodisc and the blastoderm, and the size of cells as cleavage takes place.

   *The overall size does not change noticeably during cleavage. Cells become smaller and smaller with each division.*

---

EXERCISE 25.4

# Development in a Bird: The Chicken

---

## Materials

compound microscope
stereoscopic microscope
unincubated egg (demonstration)
prepared slide of 16-hour chick
prepared slide of 24-hour chick
living egg incubated 48 hours
living egg incubated 96 hours

finger bowl
warm 0.9% NaCl
flat-tipped forceps
sharp-pointed scissors
watch glass
disposable pipette
pipette bulb

## Introduction

Immature eggs, or **oocytes,** develop within follicles in the single ovary of the adult female bird. (Two ovaries begin to develop in birds, but the second ovary degenerates.) In the sexually mature bird, hormonal stimulation brings about **ovulation,** the release of oocytes into a single oviduct. An oocyte consists of active cytoplasm, called the **blastodisc,** or germinal disc, floating on a huge amount of food reserve, the yolk, surrounded by a plasma membrane. At the time of ovulation, chromosomes in the large oocyte nucleus have just completed the first maturation division of meiosis (meiosis I). At ovulation in chickens, the oocyte nucleus measures approximately 0.5 mm and the oocyte measures approximately 35 mm in diameter.

Fertilization is internal in birds. If sperm are present in the oviduct at ovulation, they will penetrate each oocyte (one per oocyte), stimulating the completion of meiosis in the oocyte nucleus. The sperm nucleus and the now mature egg nucleus fuse, producing the zygote nucleus, which begins to divide by mitosis followed by cytoplasmic cleavage. As this developing embryo continues its passage down the oviduct, albumin, shell membranes, and finally a calcareous shell are deposited on its surface. In chickens, passage down the oviduct takes about 25 hours. This means that a freshly laid chicken egg, if it has been fertilized, has completed about 25 hours of development. The cleaved blastodisc is now called the **blastoderm**, or **blastula**. Development continues in the blastoderm, giving rise to all parts of the embryo, with yolk containing carbohydrates, proteins, lipids, and vitamins serving as the food reserves.

In this exercise, you will observe an unincubated egg and incubated eggs in several stages of development. As you study the embryos, identify developing structures and compare bird development with that of the sea star, salamander, and zebrafish.

### Procedure

1. Refer to Figure 25.8 and observe the unincubated chicken egg on demonstration.

   Most eggs sold for human consumption are purchased from commercial egg farms where hens are not allowed contact with roosters. The egg you are studying may or may not have been fertilized.

   a. Observe the broken calcareous **shell** with outer and inner **shell membranes** just inside. The shell and membranes are porous, allowing air to pass through to the embryo inside. You have probably noticed an air chamber at one end of a hard-boiled egg, between the two membranes.

   b. Observe the watery, proteinaceous **egg albumin** (egg white) and the yellow yolk. The layers of albumin closest to the yolk are more viscous and stringy than the outer albumin. As the yolky egg passes down the oviduct, it rotates, twisting the stringy albumin into two whitish strands on either end of the yolk. Called **chalaza**, these strands suspend the yolk in the albumin.

   c. Locate the small whitish disc lying on top of the yolk. This is larger in a fertilized egg because of the development that has taken place. Cleavage takes place in this disc of active cytoplasm, called the *blastodisc*, before cleavage begins and the *blastoderm* after cleavage

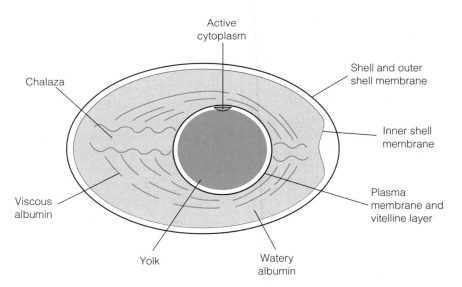

**Figure 25.8.**
**The unincubated chicken egg.** The chalaza suspends the yolk in the albumin.

has begun. If you are studying a fertilized egg, cleavage is completed. Cleavage is restricted to the blastoderm; the yolk does not divide.

As cleavage takes place, the blastoderm, now a mass of cells, becomes elevated above the yolk. Subsequent horizontal cleavages create three or four cell layers in the blastoderm, and a space, the blastocoel, forms within these layers.

2. Study the prepared slide of the 16-hour egg (the **gastrula**). Refer to Figure 25.9a, a surface view of the embryo.

 Use only low and intermediate powers when viewing this slide. The high power objective will break the slide!

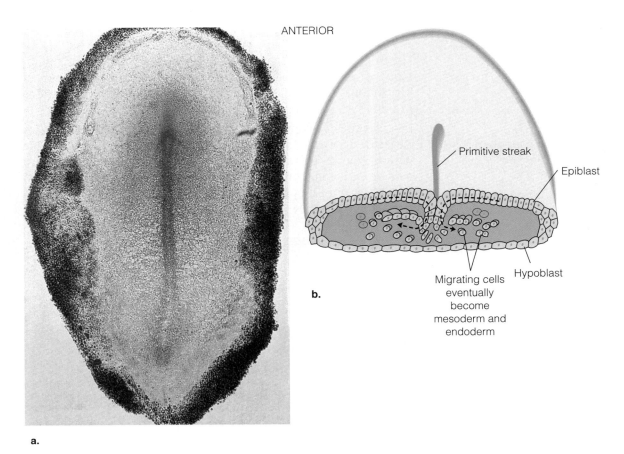

**a.**

**Figure 25.9.**
**Chick gastrulation.** (a) Surface view of chick blastoderm after 16–19 hours of incubation, the gastrula stage. The primitive streak is visible. (b) Cross section of blastoderm after 16 hours of incubation. Cells turn in at the primitive streak, initiating the formation of mesodermal and endodermal tissues.

a. Using low and then intermediate powers on the compound microscope, view a prepared slide of the whole chick embryo after about 16–19 hours of incubation. At this stage, cells in the blastoderm have separated into an upper layer, the *epiblast*, and an inner layer, the *hypoblast*, lying next to the yolk. These layers are visible only in sections of the embryo.

b. Locate a dark, longitudinal thickening, the **primitive streak.** During gastrulation, cells in the epiblast migrate toward the primitive streak and then turn under, through the primitive streak. By 18 hours, they have spread out and initiated the formation of mesodermal and endodermal tissues, the latter eventually replacing the hypoblast (Figure 25.9b). Epiblast cells that remain on the surface of the developing embryo become the ectoderm.

3. Study the prepared slide of the 24-hour chick (**neurulation**). Refer to Figure 25.10.

  Use only low and intermediate powers to view this slide. The high power objective will break the slide!

a. Use the low and then intermediate powers on the compound microscope to view the prepared slide of a chick after at least 24 hours of incubation. At this stage, the neural tube is forming anterior to the primitive streak in a process similar to neurulation in fish.

b. Look for the developing longitudinal ectodermal **neural tube** with elevated edges called **neural folds** and a depressed center called the **neural groove.** The margins of the neural folds, which appear as a pair of dark longitudinal bands, become elevated and approach each other until they touch and eventually fuse. This fusion of the folds completes the formation of the neural tube. The anterior end

*Labels on your slides may not be precisely accurate. A slide with five somites, as in our diagram, may be labeled 24 hours, but probably was incubated 27 to 30 hours. The point is that neurulation starts after about 24 hours.*

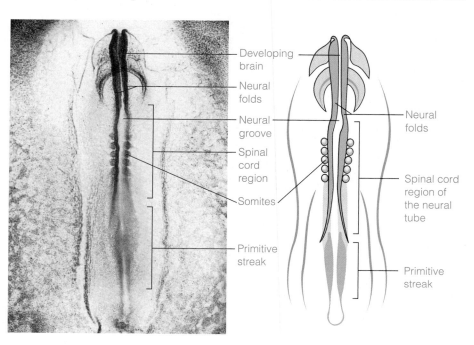

**Figure 25.10.**
**The chick after 24 hours of incubation (neurulation).** Edges of the ectodermal neural plate elevate, forming neural folds. The depressed center is the neural groove. The neural folds eventually fuse, forming the neural tube.

of the neural tube becomes the brain; the posterior end becomes the spinal, or nerve, cord.

c. Observe **somites,** blocks of tissue lying on either side of the **spinal cord region.** Somites develop into body musculature and several other mesodermal organs. To determine the number of hours your chick was incubated, count the number of somites. A chick with five somites was incubated approximately 27 to 30 hours.

d. Label the following on the photo and diagram of Figure 25.10: *neural folds, neural groove, developing brain, spinal cord region* of the *neural tube, somites* and *primitive streak.*

---

 Study the next two stages of development using living chick embryos and the materials listed at the beginning of the exercise. Work in pairs. One student will open the 48-hour chick; the other will open the 96-hour chick. Collaborate as you observe both eggs.

---

4. Prepare each egg to study the 48- and 96-hour chicks (**organogenesis**).

   a. Pour warm (heated to about 38°C) 0.9% NaCl solution into a clean finger bowl.

   b. Obtain an egg and carry it to your desk, keeping it oriented with the "top" up as it has been in the incubator. Crack the "down" side of the egg on the edge of the finger bowl. Hold the egg, cracked side down, in the NaCl solution, and carefully open the shell, allowing the egg to slide out of the shell.

   c. Using the stereoscopic microscope, observe the embryo on the surface of the yolk. The embryo should either be on top of the yolk or it should float to the top. The embryo is in the center of the vascularized blastoderm.

   d. Remove the finger bowl from the microscope and, using broad-tipped forceps and sharp scissors, remove the embryo from the yolk surface by carefully snipping outside the vascularized region of the embryo. Use the forceps to hold the blastoderm as you cut. Do not let go or you may lose the embryo in the mass of yolk.

   e. Hold a watch glass under the NaCl solution and carefully pull the blastoderm away from the surface of the yolk into the watch glass. Carefully lift the watch glass and embryo out of the solution and pipette away excess solution until only a small amount remains.

   f. Wipe the bottom of the watch glass, place on the stereoscopic microscope, and observe.

5. Study the 48-hour chick (early organogenesis). Refer to Figure 25.11a.

   a. Identify structures in the circulatory system. In the living embryo, the **heart** is already beating at this stage, pumping blood through the **vitelline blood vessels,** which emerge from the embryo and carry food materials from the yolk mass to the embryo. If the heart is still beating, you should be able to see blood passing from the **atrium** into the **ventricle.** The atrium lies behind the ventricle, which is a larger, U-shaped chamber.

   b. Identify structures in the nervous system. The anterior part of the neural tube has formed the **brain. Eyes** are already partially formed.

*The "down" side should be the broad surface, not the end of the egg. If you incubate eggs on one end, have students carefully turn the egg and crack the side.*

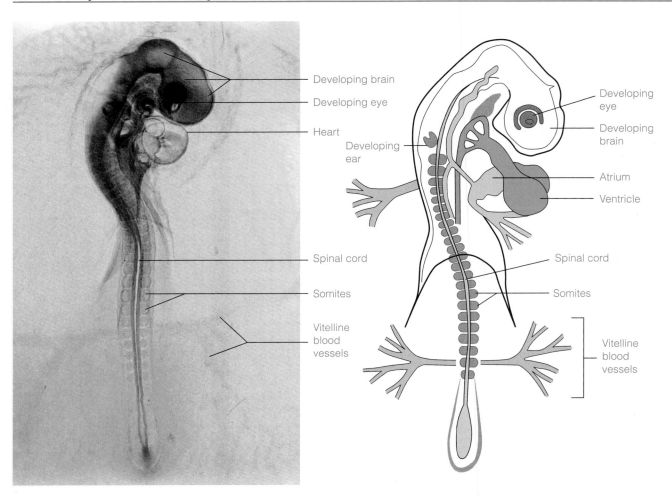

**Figure 25.11.**
**The chick after 48 hours of incubation.** Identify the heart (atrium and ventricle), vitelline blood vessels, the brain, an eye, the spinal cord, and somites.

You may be able to locate the developing ear. Follow the tube posteriorly to the **spinal cord.**

   c. Count the number of **somites** lying on either side of the **spinal cord.** A chick with 23 to 24 somites has been incubated about 48 hours.

   d. Label Figure 25.11.

   e. Swap chick embryos with your lab partner.

6. Study the 96-hour chick (later organogenesis). Refer to Figure 25.12 and Color Plate 69.

   a. Notice that the 96-hour chick has a strong **cervical flexure,** bending the body into a C configuration. Several organs are noticeably larger than in the 48-hour chick.

   b. Locate the developing **brain, eyes,** and **ears.**

   c. Identify the conspicuous **heart.**

   d. Locate one of the two **anterior limb buds** just behind the heart. Anterior limb buds develop into wings.

   e. Locate the **posterior limb buds** near the tail. These limb buds grow into legs.

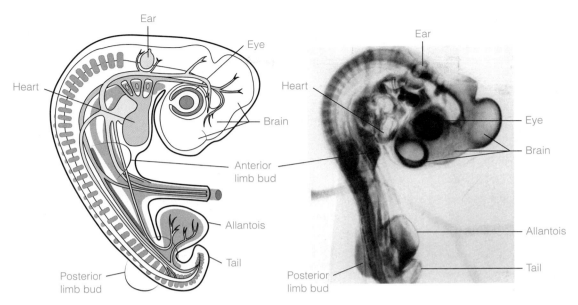

**Figure 25.12.**
**The chick after 96 hours of incubation.** A strong cervical flexure has developed. Limb buds, allantois, ears, and tail are visible.

   f. Identify one of four **extraembryonic membranes**, the **allantois**, which protrudes outward from the hindgut near the posterior limb bud (see Figure 25.12, Color Plate 69, and Figure 25.13). Be sure you can distinguish between limb bud and allantois. Extraembryonic membranes are derived from embryonic tissue, but are found outside the embryo proper. These membranes are important adaptations for land-dwelling organisms like reptiles, birds, and mammals. They help solve problems such as desiccation, gas exchange, and waste removal in the embryo. The large saclike allantois will continue to grow until it lies close inside the porous shell. It functions to bring oxygen to the embryo, carry away carbon dioxide, and store liquid wastes.

   g. Look for a second extraembryonic membrane, the **amnion**, a thin, transparent membrane that encloses the embryo in a fluid-filled sac.

   h. Using the broad-tipped forceps, carefully lift the embryo to observe the yolk stalk and **yolk sac**, a third extraembryonic membrane that surrounds the yolk. The fourth extraembryonic membrane, the **chorion**, is not easily observed.

*Students often have trouble locating the amnion and the yolk stalk.*

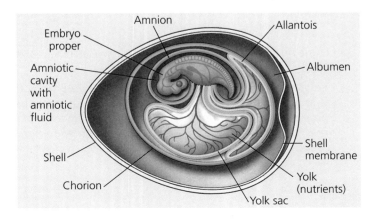

**Figure 25.13.**
**Extraembryonic membranes in a chick between 96 and 120 hours of incubation.** The allantois protrudes from the gut near the posterior limb bud. The amnion surrounds the embryo proper, creating a fluid-filled amniotic cavity; the yolk sac surrounds the yolk; and the chorion grows until it eventually fuses with the shell membrane.

    i.  Label Figure 25.12.

    j.  Swap chick embryos with your lab partner.

### Results

Label Figures 25.10, 25.11, and 25.12 and make additional sketches in the margin of your lab manual for future reference.

### Discussion

1. What type of egg is the chicken egg (Figure 25.2)?

   *strongly telolecithal*

2. This egg undergoes what type of cleavage?

   *meroblastic*

3. Collaborating with your lab partner, describe major differences between the 48-hour and 96-hour chicken embryos.

   *In the 96-hour embryo, the limb buds and allantois have developed. Other organs (such as the eye) are bigger and better developed, the cervical flexure is greater, and the blastoderm is noticeably larger.*

# Reviewing Your Knowledge

1. Having completed this lab topic, define and describe the following terms, giving examples when appropriate: *cleavage, blastula, gastrula, involution, archenteron, blastocoel, blastopore, isolecithal, telolecithal, alecithal, meroblastic, holoblastic, bipinnaria, epiboly, neurulation, organogenesis, animal pole, vegetal pole, blastodisc, blastoderm, ectoderm, mesoderm, endoderm, neural tube, neural groove,* and *neural fold.*

2. List the embryonic germ layers and the organs derived from each.

   *ectoderm—skin and associated glands, neural tube*

   *mesoderm—muscles, skeleton, gonads, excretory system, circulatory system*

   *endoderm—lining of the digestive tract and associated organs*

3. Define *primitive streak* and *epiblast,* two structures identified in the gastrula of the chick.

   *The epiblast is the outer layer of cells of the embryo at the gastrula stage and this layer gives rise to all three embryonic germ layers. Some epiblast cells migrate through the primitive streak to the interior of the embryo to form the mesoderm and endoderm. Cells that remain on the surface become the ectoderm.*

4. Differentiate between epiboly and involution.

*epiboly—when a layer of surface cells spreads over the entire embryo, as in gastrulation in the fish*

*involution—when a sheet of cells enters the interior of the embryo, as in gastrulation in the amphibian and the fish*

5. Review the stages of early development in animals. How are these stages similar in the animals you have studied? How do they differ? Complete Table 25.3.

**Table 25.3**
Comparison of Stages of Early Development in the Sea Urchin, Sea Star, Salamander, Fish, and Chick

| Organism | Cleavage | Gastrulation | Neurulation | Organogenesis |
|---|---|---|---|---|
| Sea urchin and sea star | *Holoblastic, little yolk Blastocoel centrally located.* | *3 germ layers formed; cells invaginate at blastopore located at the vegetal pole* | *None* | *Esophagus, stomach, intestines form in larva* |
| Salamander | *Holoblastic, moderate yolk Blastocoel displaced to animal hemisphere* | *3 germ layers formed; blastopore displaced to one side; cells roll over the lip of the blastopore into the interior (involution)* | *Similar in all vertebrates; a dorsal neural tube forms; the anterior portion becomes the brain; the posterior, the spinal cord* | *Germ layers develop into all organs* |
| Fish | *Meroblastic, blastoderm formed* | *3 germ layers formed; blastoderm spreads over yolk mass (epiboly) and then cells move inward (involution)* | *Similar in all vertebrates; a dorsal neural tube forms; the anterior portion becomes the brain; the posterior, the spinal cord* | *Organs form; movement visible as muscles form; pigment visible* |
| Chick | *Meroblastic, blastoderm formed* | *3 germ layers formed; epiblast cells migrate into the interior at the primitive streak* | *Similar in all vertebrates; a dorsal neural tube forms; the anterior portion becomes the brain; the posterior, the spinal cord* | *All organ systems and most body structures begin to develop. You can ask students to list examples.* |

# Applying Your Knowledge

1. What is yolk? What is its function in development? Do mammals have yolk? Explain.

   *Yolk is the most usual form of food storage in the egg. It is not a definite chemical substance but may be made up of proteins, phospholipids, and neutral fats.*

   *The function of yolk is to provide nourishment for the early embryo until it develops into a form that can obtain food from outside sources.*

   *The amount of yolk in mammals varies. Monotremes, egg-laying mammals, have large amounts of yolk. Marsupials have a small amount of yolk, but it is not used to nourish the embryo because it is ejected at the beginning of cleavage. Placental mammals have practically no yolk in their eggs from the beginning. However, all mammals do have a yolk sac similar to that seen in birds.*

2. Giving examples from organisms studied, explain how the amount of yolk affects cleavage and gastrulation.

   *In the salamander, the yolk at the vegetal pole retards formation of cleavage furrows and causes cells to be larger in the vegetal hemisphere. These differences in cleavage cause a displacement of the blastopore to the side of the embryo.*

   *In the fish and chick, yolk is so abundant that cleavage is restricted to a small disc of cytoplasm at the animal pole of the cell. In the fish, gastrulation takes place around the yolk mass. In the chick, a primitive streak forms in the blastoderm. Invagination takes place here.*

3. How do the differences in development lead to adaptations to the particular lifestyles of these organisms? (Some are aquatic, others terrestrial, and so on.)

   *Development in the sea urchin and sea star progresses rapidly to a motile, self-feeding larval stage. This is necessary, since there is little food stored in the embryo. The motile aquatic larva also facilitates rapid dispersion of the embryos.*

   *Large supplies of yolk allow for development that is independent of the parent.*

   *Extraembryonic layers, such as albumin, shell, and shell membranes, permit development away from water (see question 5).*

4. Sea urchins live in marine communities that include a variety of species living in close proximity. Most of these species, including sea urchins, have external fertilization, a reproductive strategy that potentially has a low probability of sperm and egg contact. List several adaptations of sea urchins that help ensure successful reproduction.

   *1. Huge numbers of gametes are produced and released simultaneously.*

   *2. Products from the soluble egg jelly coat may increase sperm mobility.*

   *3. The egg vitelline layer and plasma membrane have species-specific sperm receptors.*

5. Eggs of reptiles, birds, and a few primitive mammals are laid on land, usually away from water and independent sources of food. What features of the land egg (chick) allow this type of early development?

   *abundant food supply—yolk; watery environment—albumin; hard, porous shell and two shell membranes that prevent water loss while allowing exchange of $O_2$ and $CO_2$; presence of the allantois to bring $O_2$ to the embryo and carry $CO_2$ away*

6. The extraembryonic membranes in mammals (including humans) are homologous to those of birds and develop in a similar way. In what ways are their functions the same, and how are they different? Use your text, if needed.

   **Chorion**

   *Functions in gas exchange in both birds and mammals. In the chick this membrane adheres to the shell membrane, allowing the exchange of $CO_2$ and $O_2$ between the egg and the environment.*

   **Amnion**

   *Encloses the embryo in a fluid-filled sac in both birds and mammals, providing protection and allowing unencumbered development in the embryo.*

   **Yolk sac**

   *In the chick this encloses the yolk, the nutrients stored in the egg. In mammals this contains no yolk, but, being homologous, is given the same name. It serves as a site for early formation of blood cells in the mammalian embryo.*

   **Allantois**

   *In the chick this is a disposal sac for metabolic waste contained in the egg. It plays a similar function in mammals where this membrane is incorporated into the umbilical cord. Students identified the allantoic stalk, a continuation of the pig's urinary bladder, in Lab Topic 22.*

7. Using your text or other resources, describe possible differences in the arrangement of the chorion and amnion in the formation of identical twins in humans.

   *(From Campbell Biology, 9e, Ch. 47) When embryonic cells become separated forming identical twins, the timing of the separation determines the nature of their extraembryonic membranes. If cells in the embryo separate early, then each embryo has its own chorion and amnion. If the separation occurs a little later, then the two embryos share a chorion, but have separate amnions. Rarely, the cells separate even later, and the two embryos share a common chorion and amnion.*

8. Using information from the Web or other resources, describe the birth defect spina bifida. Which stage of embryonic development studied in this lab topic would be related to this condition?

   *Spina bifida is a developmental birth defect caused by the incomplete closure of the embryonic neural tube. This condition would be related to abnormalities in the process of neurulation.*

## Investigative Extensions

Sea urchin, fish, and chicken eggs collected for this lab topic provide an opportunity for student independent projects. Living amphibian eggs also may be available from colleagues or, in season, from streams or ponds in

*If you perform the exercises in this lab topic using living materials, you will have an excess of sea urchin and fish eggs. In addition, amphibian eggs may be available from colleagues conducting amphibian research. Take this opportunity to encourage students to perform independent projects. Have students come to lab with a question in mind. Before they leave lab, they can set up their experiment and then return on several consecutive days to collect data. Students will then process their data and submit their results in the form of an oral report, a poster, or a scientific paper.*

your area. Recent concern about declining populations of sea urchins (for example, *Diadema antillarum,* the black sea urchin once abundant throughout the Caribbean) and the decline of healthy populations of amphibians in North America are just two examples where environmental factors may be having an impact on animal development.

Suggestions follow for possible questions to investigate.

1. What effects, if any, and at what concentration do specific pollutants-pesticides, herbicides (e.g., atrazine), fertilizers—have on the following?

   a. Fertilization rates in sea urchins, amphibians, or fish

   b. Early development in sea urchins, amphibians, or fish (Recently atrazine, a common herbicide, has been implicated in contributing to the decline and development of abnormalities in amphibian populations.)

2. What effect, if any, does acid rain–induced pH change have on the following?

   a. Fertilization rates in sea urchins, amphibians, or fish

   b. Early development in these organisms

3. Does UV radiation have an effect on the following?

   a. The rate of fertilization in sea urchins

   b. Early development in amphibians or fish

4. In sea urchins, will the addition of a calcium solution bring about the formation of the fertilization envelope in the absence of sperm?

## Student Media: BioFlix, Activities, Investigations, and Videos
www.masteringbiology.com (select Study Area)

**Activities**—Ch. 47: Sea Urchin Development; Frog Development
**Investigations**—Ch. 47: What Determines Cell Differentiation in the Sea Urchin?
**Videos**—Ch. 47: Sea Urchin Embryonic Development; Frog Embryo Development; Discovery Videos: Ch. 20: Cloning

## References

Beams, H. W. and R. G. Kessel. "Cytokinesis: A Comparative Study of Cytoplasmic Division in Animal Cells." *American Scientist,* 1976, vol. 64, pp. 279–290. Scanning electron micrographs of development in zebrafish.

Gilbert, S. F. *Developmental Biology,* 8th ed. Sunderland, MA: Sinauer Associates, 2006.

Hisaoka, K. K. and C. F. Firlit. "Further Studies on the Embryonic Development of the Zebra Fish, *Brachydanio rerio* (Hamilton-Buchanan)." *Journal of Morphology,* 1960, vol. 107, pp. 205–226.

Patten, B. M. *Early Embryology of the Chick,* 5th ed. New York, NY: McGraw-Hill, 1971.

Reece, J. et al. *Campbell Biology,* 9th ed. San Francisco, CA: Pearson Education, 2011.

Westerfield, M. (ed.). *The Zebrafish Book; A Guide for the Laboratory Use of Zebrafish (Brachydanio rerio).* Eugene, OR: University of Oregon Press, 1989. Also available online. See below.

Exercise 25.3, Development in the Zebrafish, is based on an exercise by John Pilger, Professor of Biology, Agnes Scott College, Decatur, GA, and a workshop presented by Robert R. Cohen, Professor of Biology, Metropolitan State College, Denver, at a meeting of the American Society of Zoologists. Used by permission.

## Websites

Developmental biology web links:
http://www.uoguelph.ca/zoology/devobio/dbindex.htm

The Fish Net-Zebrafish Information Network:
http://zfin.org/index.html

Sea urchin fertilization animations developed by teachers and Stanford University researchers. Includes

instructions on injecting sea urchins to stimulate egg and sperm release:
http://www.stanford.edu/group/Urchin

*The Zebrafish Book* online:
http://zfin.org/zf_info/zfbook/zfbk.html

*The Zebrafish Science Monitor.* Past issues available at:
http://zfin.org/zf_info/monitor/mon.html

## LAB TOPIC 25

# Animal Development
## Teaching Plan for Laboratories

## Main Concepts and Objectives

1. Concept: the orderly process of early development in animals. Students will observe and describe stages of early development in the sea urchin, sea star, salamander, fish, and chick. (Exercises 25.1–25.4)

2. Concept: the unity and diversity in patterns of animal development. Students will compare development in the animals being studied. They will describe similarities and variations in early developmental stages. (Exercises 25.1–25.4)

3. Concept: neurulation. Students will describe variations in early animal development due to the development of a dorsal central nervous system, comparing the invertebrates (sea urchin and sea star) with the three vertebrates. (Exercises 25.1–25.4)

4. Concept: the amount of yolk and its position in an egg influence early development. Students will compare development in eggs with differing amounts of yolk. (Exercises 25.1–25.4)

5. Specific content. Students will be able to describe or define: isolecithal, telolecithal alecithal eggs; holoblastic and meroblastic cleavage; extra embryonic membranes; and terminology: *ectoderm, mesoderm, endoderm, blastopore, blastocoel, blastodisc, blastoderm, neural groove, neural plate, neural folds, neural tube, archenteron, epiboly, invagination, primitive streak, epiblast, hypoblast, somites.*

## Order of the Lab

1. Introduce concepts and objectives.                                    (15 min)
2. Observe fertilization in living sea urchins.                          (15 min)
3. Show the video of echinoderm development.                            (15 min)
4. Observe a slide of sea star embryos.                                   (20 min)
5. Introduce salamander development. Students read questions and prepare to view the film.    (15 min)
6. Show the film of salamander development.                               (15 min)
7. Observe living fish embryos.                                           (30 min)
8. Observe slides and living chick embryos.                              (40 min)
9. Observe again the living sea urchin slides to determine the stage of development embryos achieved by the end of the lab period.    (10 min)

**For a 2-hour lab:** Omit the video on echinoderm development, the discussion of salamander development, and the film on salamander development. The video and film could be shown in the lecture portion of the course with concomitant discussion. You may choose to omit either the sea urchin fertilization or the zebrafish development study. Remember, however, that study of living organisms captures students' interest and can lead to additional investigations.

## Classroom Management

Students work independently with sea urchin fertilization and sea star slides. With zebrafish embryos, ideally, each team of four students will have at least one of each stage of early development and several embryos in organogenesis and hatched stages. The number available may vary, however. In the event that students must share stages, embryos will be in labeled petri dishes on the demonstration table. Students should observe a stage and return it to the demo table for the next student to observe, progressing through all stages. With chick development, each pair of students will have one 16-hour and one 24-hour slide. Each pair of students will have two eggs, one incubated for 48 hours and one incubated for 96 hours. Each student opens an egg and removes the blastoderm. Students then study their egg and their partner's egg.

## Student Development

Students will improve observational skills and practice laboratory skills such as microscopy and the manipulation of living chick embryos.

## Discussion and Summary

Observations during lab should stimulate discussion. If time permits, conclude the lab by summarizing the major concepts, asking students to contribute ideas from observations made in the lab. Students will summarize what they have observed as they answer the questions in the laboratory exercises.

## Evaluation

Note individual students' laboratory skills and participation for subjective evaluation. Test students' understanding of concepts using Applying Your Knowledge questions. Evaluate retention of concepts on a laboratory test. If possible, use the discussion format on the test, asking students to make comparisons and discuss, for example, the significance of differences in developmental stages. Put some questions in practical format, asking students to identify developmental stages. We sometimes ask these questions using images.

# Lab Topic 26
# Animal Behavior

## Laboratory Objectives

After completing this lab topic, you should be able to:

1. Define *ethology.*
2. Define and give an example of *taxis, kinesis, agonistic behavior,* and *reproductive behavior.*
3. State the possible adaptive significance of each of these behaviors.
4. Propose hypotheses, make predictions, design experiments to test hypotheses, collect and process data, and discuss results.
5. Present the results of your experiments in a scientific paper.

*For a 2-hour lab:* Omit two of the organisms or expand to two 2-hour labs. See Teaching Plan.

*In the Study Area of masteringbiology.com, see Student Media for suggested activities, investigations, and videos. Although the videos may not directly relate to the laboratory exercises, they will stimulate interest and possible ideas for student-designed research.*

## Introduction

**Behavior,** broadly defined, is the sum of the responses of an organism to stimuli in its environment. In other words, behavior is what organisms do. **Ethology** is the study of animal behavior in the context of the evolution, ecology, social organization, and sensory abilities of an animal (Gould, 1982). Ethologists concentrate on developing accurate descriptions of animal behavior by carefully observing and experimentally analyzing overt behavior patterns and by studying the physiology of behavior.

Explaining a particular behavior in the broad, multivariable context of evolution or ecology can become a complex undertaking. It is often necessary, therefore, to study behavior in animals that have a limited range of behaviors and for which more is known about their evolution, ecology, and sensory abilities. Understanding simple and isolated behaviors is important in unraveling more complex behaviors.

There are two basic categories of behavior—**learned** (depends on experiences) and **innate** (inherited—exhibited by nearly all individuals of a species). Experimental evidence suggests that the basis of both lies in the animal's genes. As with all genetically controlled features of an organism, behavior is subject to evolutionary adaptation. As you study animal behavioral activities in this lab topic, think in terms of both **proximate causes,** the immediate physiological events that led to the behavior, and **ultimate causes,** the adaptive value and evolutionary origin of the behavior. To illustrate, a fiddler crab will respond to human intrusion into its feeding area by running into its burrow. The proximate cause of this behavior might be the vibration caused by footsteps stimulating sensory receptors and triggering nervous impulses. The nervous impulses control muscle contractions in the crab's legs. Ultimate causes are the adaptive value of retreating from predators to avoid being eaten.

Note that you will be asking **causal** questions in your investigations. It is inappropriate to ask **anthropomorphic** questions—that is, questions that ascribe human attributes to the animal. Consider, for example, a behavior that places an animal in its best environment. An anthropomorphic explanation for this behavior would be that the animal makes a conscious choice of its environment. There is no way for us to come to this conclusion scientifically. The causal explanation would be that the animal is equipped with a sensory system that responds to environmental stimuli until the favorable environment is reached.

Ethologists have categorized behavioral patterns based on the particular consequence of that behavior for the organism. **Orientation behaviors** place the animal in its most favorable environment. Two categories of orientation behaviors are **taxis** (plural, **taxes**) and **kinesis.** *A taxis is movement directly toward or away from a stimulus.* When the response is toward a stimulus, it is said to be *positive;* when it is away from the stimulus, it is *negative.* Prefixes such as *photo, chemo,* and *thermo* can be added to the term to describe the nature of the stimulus. For example, an animal that responds to light may demonstrate positive phototaxis and is described as being *positively phototactic.*

*A kinesis differs from a taxis in that it is undirected, or random, movement.* A stimulus initiates the movement but does not necessarily orient the movement. The intensity of the stimulus determines the rate, or velocity, of movement in response to that stimulus. If a bright light is shined on an animal and the animal responds by moving directly away from it, the behavior is a taxis. But if the bright light initiates random movement or stimulates an increase in the rate of turning with no particular orientation involved, the behavior is a kinesis. The terms *positive* and *negative* and the prefixes mentioned earlier are also appropriately used with *kinesis.* An increase in activity is a positive response; a decrease in activity is a negative response.

Another complex of behaviors observed in some animals is **agonistic behavior.** In this case, the animal is in a conflict situation where there may be a threat or approach, then an attack or withdrawal. Agonistic behaviors in the form of force are called **aggression;** those of retreat or avoidance are called **submission.** Often the agonistic behavior is simply a display that makes the organism look big or threatening. It rarely leads to death and is thought to help maintain territory so that the dominant organism has greater access to resources such as space, food, and mates.

**Reproductive behavior** can involve a complex sequence of activities, sometimes spectacular, that facilitate *finding, courting,* and *mating* with a member of the same species. It is an adaptive advantage that reproductive behaviors are species-specific. Can you suggest reasons why?

For the first hour of lab, you will perform Experiment A in each of the exercises that follow, briefly investigating the four behaviors just discussed: *taxis in brine shrimp, kinesis in pill bugs, agonistic behavior in Siamese fighting fish,* and *reproductive behavior (courting and mating) in fruit flies.* After completing every Experiment A, your team will choose one of the systems discussed and perform Experiment B in that exercise. *In Experiment B, you will propose one or more testable hypotheses and design a simple experiment to test your hypotheses.* Then you will spend the remainder of the laboratory period carrying out your experiments.

Near the end of the laboratory period, several of you may be asked to present your team's results to the class for discussion. One part of the scientific process involves persuading your colleagues that your experimental design is sound and that your results support your conclusions (either negating or supporting your hypothesis). Be prepared to describe your results in a brief presentation in which you will use your experimental evidence to persuade the other students in your class.

You may be required to submit a laboratory report describing your experiment and results in the format of a scientific paper (see Appendix A). You should discuss results and come to conclusions with your team members; however, you must turn in an originally written lab report. Your Materials and Methods section and your tables and figures may be similar, but your Introduction, Results, and Discussion sections must be the product of your own library and online research, creative thinking, evaluating, and writing.

Remember, first complete Experiment A in each exercise. Then discuss with your research team a possible question for your original experiment, choosing one of the animals investigated in Experiment A as your experimental organism. Be certain you can pose an interesting question from which you can develop a testable hypothesis. Then turn to Experiment B in the exercise for your chosen organism and design and execute an experiment.

---

## EXERCISE 26.1
# Taxis in Brine Shrimp

---

Brine shrimp (*Artemia salina*) are small crustaceans that live in salt lakes and swim upside down using 11 pairs of appendages. Their sensory structures include two large compound eyes and two pairs of short antennae (Figure 26.1). They are a favorite fish food and can be purchased in pet stores.

**Figure 26.1.**
**Brine shrimp (*Artemia salina*) magnified about 20×.** A type of fairy shrimp, brine shrimp live in inland salt lakes such as the Great Salt Lake in Utah.

# Experiment A. Brine Shrimp Behavior in Environments with Few Stimuli

## Materials

brine shrimp
2 large test tubes
black construction paper

1 small finger bowl
salt water
dropper

## Introduction

In this experiment, you will place brine shrimp in a test tube of salt water similar to the water of their normal environment. You will not feed them or disturb them in any way. You will observe their behavior in this relatively stimulus-free environment. Notice their positions in the test tube. Are they in groups, or are they solitary? Are they near the top or near the bottom? You should make careful observations of their behavior, asking questions about possible stimuli that might initiate taxes in these animals.

## Hypothesis

Hypothesize about the behavior of brine shrimp in an environment with few stimuli.

> *In an environment with few stimuli, the shrimp will swim through-out the environment.*

## Prediction

Predict the result of your experiment based on the hypothesis (if/then).

> *If in an environment with few stimuli, shrimp swim throughout their environment, then they will be randomly distributed in the test tube.*

## Procedure

1. Place six brine shrimp in a test tube filled two-thirds with salt water. Rest the test tube in the finger bowl in such a way that you can easily see all six shrimp. You may need to use black construction paper as a background.

2. Describe the behavior of the brine shrimp in the Results section; for example, are they randomly distributed throughout the test tube or do they collect in one area?

3. Record your observations in the Results section.

## Results

1. By describing the behavior of brine shrimp in an environment with relatively few stimuli, which component of experimental design are you establishing?

   *the control experiment*

2. Describe the behavior of the brine shrimp.

   *They should be randomly distributed and swimming randomly.*

## Discussion

On separate paper, list four stimuli that might initiate taxes in brine shrimp and predict the response of the animal to each. What possible adaptive advantage could this behavior provide?

*Students may list such stimuli as light, gravity, cold, heat, and food types. Encourage students to discuss ideas with their teammates.*

# Experiment B. Student-Designed Investigation of Brine Shrimp Behavior

## Materials

supplies from Experiment A
piece of black cloth
lamp

dropper bottles—solutions
  of sugar, egg albumin, acid,
  and base

## Introduction

If your team chooses to perform your original experiment investigating taxes in brine shrimp, return to this experiment after you have completed all the introductory investigations (Experiment A of each exercise). Using the materials available and collaborating with other members of your research team, design a simple experiment to investigate taxes in brine shrimp.

## Hypothesis

State the hypothesis that you will investigate.

## Prediction

Predict the results of your experiment based on your hypothesis (if/then).

## Procedure

 Allow a conditioning period of several minutes after the shrimp have been disturbed or stimulated. If you add something to the water in one experiment, begin additional experiments with fresh water and shrimp.

1. On separate paper, list in numerical order each step of your procedure. Remember to include the number of repetitions, levels of treatment, the duration of each stimulus, and other time intervals when appropriate (see Lab Topic 1).
2. If you have an idea for an experiment that requires materials other than those available, ask your laboratory instructor. If possible, additional supplies will be made available.
3. Quantify your data whenever possible (count, weigh, measure, time).

### Results

On separate paper, record your data and describe your results. You should design at least one table and figure.

### Discussion

1. Among members of your team, discuss your results in light of your hypothesis. If possible, come to conclusions about the behaviors you have been investigating. Record your conclusions on a separate paper.
2. You may be asked to report the results of your experiments to the class.

---

EXERCISE 26.2

# Kinesis in Pill Bugs

*If you know the species of terrestrial isopod you are using, tell the students to include the scientific name in their independent experiment. Armadillium sp. (pill bug is the common name) is the most common, but you may have Porcellio sp. (commonly called wood louse and sow bug).*

Kinesis can be studied using a crustacean in the order Isopoda (called *isopods*). These animals are called by many common names, including *pill bugs, sow bugs, wood lice,* and *roly-polies* (Figure 26.2). Although most crustaceans are aquatic, pill bugs are truly terrestrial, and much of their behavior is involved with their need to avoid desiccation. They are easily collected in warm weather under flowerpots, in leaf litter, or in woodpiles. Some species respond to mechanical stimuli by rolling up into a ball.

## Experiment A. Pill Bug Behavior in Moist and Dry Environments

### Materials

pill bugs                              filter paper
2 large petri dishes              squirt bottle of water

**Figure 26.2.**
**Pill bugs measure up to about 15 mm in length.** These terrestrial isopods are also called *sow bugs*, *wood lice*, and *roly-polies*.

## Introduction

In this experiment, you will investigate pill bug behavior in moist and dry environments by observing the degree of their activity, that is, the number of times they circle and turn. As you observe their behavior, ask questions about possible stimuli that might modify this behavior.

## Hypothesis

Hypothesize about the degree of activity of pill bugs in moist and dry environments.

> *Pill bugs will be less active in moist environments than in dry environments.*

## Prediction

Predict the results of the experiment based on your hypothesis (if/then).

> *If pill bugs are less active in moist environments than in dry, then they will circle and turn more frequently in the dry environment than in the moist one.*

## Procedure

1. Prepare two large petri dishes, one with wet filter paper, the other with dry filter paper.
2. Place five pill bugs in each dish.
3. Place the dishes in a dark spot, such as a drawer, for 5 minutes.
4. After 5 minutes, carefully observe the pill bugs in the petri dishes. Before you open the drawer or uncover the petri dishes, assign each of the following procedures to a member of your team.
   a. Count the number of pill bugs moving in each dish.
   b. Choose one moving pill bug in each dish and determine the rate of locomotion by counting revolutions per minute (rpm) around the petri dish.

c. Determine the rate of turning by counting turns (reversal of direction) per minute for one pill bug in each dish.

## Results

Record your results in Table 26.1.

### Table 26.1
Kinesis in Pill Bugs: Response to Wet and Dry Environments

| Environmental Condition | Number Moving | Rate of Locomotion (rpm) | Rate of Turning (turns/min) |
|---|---|---|---|
| Moist | | | |
| Dry | | | |

## Discussion

1. Kinetic response to varying moisture in the environment is called *hygrokinesis*. What other environmental factors might influence the behavior of pill bugs?

   *light, temperature, gravity, contact with other pill bugs of same sex and opposite sex, contact with dead bugs*

2. On separate paper, list four factors that might initiate kineses in pill bugs and predict their response to each. What possible adaptive advantage could this behavior provide?

   *See answers in question 1, preceding. Encourage collaboration among students.*

3. Some terrestrial arthropods display a behavior where they avoid decaying organisms of their own species, but not other species. Can you suggest an adaptive advantage for this behavior? How would you design an experiment to test this?

   *Students should discuss how to design an experiment to test this, what factors they should control, etc.*

# Experiment B. Student-Designed Investigation of Pill Bug Behavior

## Materials

supplies from Experiment A
white enamel pan
wax pencils
beaker of water

construction paper
manila folder
large pieces of black cloth

## Introduction

If your team chooses to perform your original experiment investigating kineses in pill bugs, return to this experiment after you have completed all the introductory investigations (Experiment A of each exercise). Using the materials available and collaborating with other members of your research team, design a simple experiment to investigate kineses in pill bugs.

### Hypothesis

State the hypothesis that you will investigate.

### Prediction

Predict the results of the experiment based on your hypothesis (if/then).

### Procedure

1. On separate paper, list in numerical order each step of your procedure. Remember to include the number of repetitions, the levels of treatment, the duration of stimulus, and other time intervals where appropriate (see Lab Topic 1).

2. If you have an idea for an experiment that requires materials other than those available, ask your laboratory instructor. If possible, additional supplies will be made available.

3. Quantify your data whenever possible (count, weigh, measure, time).

### Results

On separate paper, record your data and describe your results. You should design at least one table and figure.

### Discussion

1. Among members of your team, discuss your results in light of your hypothesis. If possible, come to conclusions about the behaviors you have been investigating. Record your conclusions on a separate paper.

2. You may be asked to report the results of your experiments to the class.

---

EXERCISE 26.3

# Agonistic Display in Male Siamese Fighting Fish

The innate agonistic behavior of the male Siamese fighting fish (*Betta splendens*) has been widely studied (Simpson, 1968; Thompson, 1969). The sight of another male *Betta* or even its own reflection in a mirror will stimulate a ritualized series of responses toward the intruder. If two fish are placed in the same aquarium, their agonistic behavior usually continues until one fish is defeated or subordinated (see Color Plate 70).

## Experiment A. Display Behavior in Male Siamese Fighting Fish

### Materials

male Siamese fighting fish in a 1- to 2-L flat-sided fishbowl
mirror
watch (must record seconds, one per student)

### Introduction

The purpose of this experiment is to describe the ritualized agonistic display of a male Siamese fighting fish after being stimulated by its own reflection in a mirror. Before you begin the experiment, become familiar with the fish's anatomy, identifying its dorsal, caudal (tail), anal, pectoral, and ventral (pelvic) fins, and its gill cover (Figure 26.3 and Color Plate 70).

When you begin the experiment, you will be looking for several possible responses: frontal approach (facing intruder), broadside display, undulating movements, increased swimming speed, and enhanced coloration in tail, fin, or body. You should note the duration and intensity of elevation for each fin and changes in the gill cover.

### Hypothesis

Hypothesize about the response of the fish to its image in the mirror.

*The fish will behave as if it is seeing an intruder.*

### Prediction

Predict the result of the experiment based on your hypothesis (if/then).

*If the fish behaves as if it is seeing an intruder, then it will elevate its fins, extend its gill cover, expand its tail, and attempt to attack the mirror.*

### Procedure

1. Plan your strategy.

   a. Be ready with your pencil and paper to record your observations. Behaviors can happen very quickly. Your entire team should observe and record them.

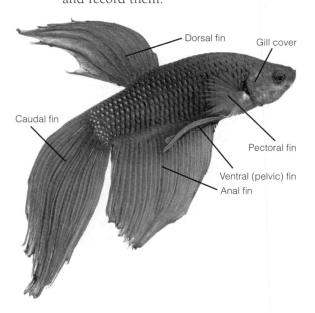

**Figure 26.3.**
**Male Siamese fighting fish**
(*Betta splendens*).

b. Each team member should be responsible for timing the beginning and duration of particular responses (listed in the Introduction). For example, one or two team members will observe the response in one or two specific fins and record their results in Table 26.2 of the Results section. Another team member will observe more general responses, such as "increased swimming speed" or "broadside display for 60 seconds."

2. Place the mirror against the fishbowl.

3. As the fish reacts to its reflection, record your observations in the table or on separate paper.

4. Compare collective results.

5. In the Results section, using the time you present the mirror as time zero, make a *sequential* list of the recognizable responses involved in the display. Be as quantitative as possible; for example, you might record "gill cover extended 90° for 30 seconds," or "broadside display for 60 seconds."

6. Note in the Results section those responses that take place simultaneously.

## Results

1. Record your data for each response in Table 26.2. Consider the time when you place the mirror next to the fishbowl as time zero. Then note how many seconds lapse before the specific response and its intensity. Use these symbols as estimates of response intensity: - (none), + (slight response), + + (moderate response), + + + (intense response).

## Table 26.2
Behaviors, Response Times, and Response Intensity for Siamese Fighting Fish Presented with a Mirror Stimulus

| Response | Time Before Response (seconds) | Intensity - (none), + (slight), + + (moderate), + + + (intense) |
|---|---|---|
| Fish notices mirror | | |
| Approaches mirror | | |
| Dorsal fin flare | | |
| Caudal fin (tail) flare | | |
| Anal fin flare | | |
| Ventral fin flare | | |
| Pectoral fin flare | | |
| Gill cover flare | | |

2. Record your sequential list.

3. Record the responses that take place simultaneously.

**Discussion**

1. Collaborating with your teammates, write a descriptive paragraph, as quantitative and detailed as possible, describing the agonistic display elicited in the Siamese fighting fish in response to its reflection.

   *Dorsal and anal fins elevate, and their color becomes greatly enhanced. Previously flattened gill covers are extended. Tail is expanded and waved. The fish then proceeds to alternate frontal approach with broadside display. This will vary somewhat from situation to situation. Some fish are not as aggressive as others, and they also become conditioned.*

   *Additional notes: There are conflicting reports in the literature about the role of colors in releasing the display. Display is not all or none; it can be partial. Display can be released by fish of other species and even by the female Betta.*

2. Consult your text for the definition of the behavior called *fixed action pattern*. Do you conclude that agonistic behavior in Siamese fighting fish is an example of this type of behavior? Why or why not?

   *Fixed action pattern (FAP) is defined as a sequence of unlearned acts that, once initiated, is usually carried to completion. In a well-controlled experiment, the behavior in the Siamese fighting fish appears to be such a behavior.*

3. What is the obvious adaptive advantage of complex agonistic displays that are not followed by damaging fights? Are there advantages that are not so obvious?

   *Intruder is chased away without harm to either. Other advantages: defends territory, nest; ensures food or mate access.*

4. Name several other animals that demonstrate a strong display that is seldom followed by a damaging fight.

   *many birds, fish species, squirrels, fiddler crabs*

5. Name several animals that do engage in damaging fights.

   *dogs, cats, humans, swans, bees, lizards, elephants, rodents, bighorn sheep*

# Experiment B. Student-Designed Investigation of Siamese Fighting Fish Behavior

## Materials

supplies from Experiment A
colored pencils and index cards or brightly colored paper
scissors

wooden applicator sticks
transparent tape
fish of different species in fishbowls
female Siamese fighting fish

## Introduction

If your team chooses to perform your original experiment investigating agonistic behavior in Siamese fighting fish, return to this experiment after you have completed all the introductory investigations (Experiment A of each exercise). Using the materials available and collaborating among your research team, design a simple experiment to investigate this behavior.

Discuss with your team members possible investigations that might be carried out. Several questions follow that might give you ideas.

1. What is the simplest stimulus that will initiate the response? Is color important? Size? Movement?
2. Is the behavior "released" by a specific stimulus or by a complex of all the stimuli?
3. Will another species of fish initiate the response?
4. Will a female *Betta* fish initiate the response, and, if so, how does the response compare with the response to a fish of a different species?
5. Is the response all or none—that is, are there partial displays with different stimuli?
6. Does the fish become "conditioned"—that is, after repeated identical stimuli, does the duration of the display change, or does the display cease?
7. Could chemical stimulation contribute to the response? (Transfer water from one fishbowl to another.)

### Hypothesis

After your team has decided on one or more questions to investigate, formulate a testable hypothesis.

### Prediction

Predict the results of the experiment based on your hypothesis (if/then).

### Procedure

1. On separate paper, list in numerical order each step of your procedure. Remember to include the number of repetitions, the levels of treatment, the duration of a stimulus, and other time intervals where appropriate (see Lab Topic 1).
2. If you have an idea for an experiment that requires materials other than those available, ask your laboratory instructor. If possible, additional supplies will be made available.
3. Quantify your data whenever possible (count, weigh, measure, time).

### Results

On separate paper, design a table and then record your data. Describe your results. Create a figure to illustrate some aspect of your results. Your Results section should include at least one table and one figure.

### Discussion

1. Among members of your team, discuss your results in light of your hypothesis. If possible, come to conclusions about the behaviors you have been investigating. Record your conclusions on a separate paper.
2. You may be asked to report the results of your experiments to the class.

---

### EXERCISE 26.4
# Reproductive Behavior in Fruit Flies

---

Reproductive behavior in some animals involves a complex sequence of activities that may include courtship and mating. Spieth (1952, described in Marler, 1968) has classified courtship and mating behavior of the fruit fly *Drosophila melanogaster* as being a complex of at least 14 behaviors.

Courtship behavior in these animals is an example of a *stimulus-response chain*. The response to a stimulus becomes the stimulus for the next behavior. If there is no response, then the behavior changes. Described below are 10 of the most common and easily recognized of these behaviors. Read the list carefully and become familiar with the behaviors you will be required to recognize. Six of the behaviors are seen in males, four in females. The behavior sequence begins as the male orients his body toward the female (Figure 26.4a).

## Male Behaviors

1. *Tapping.* The forelegs are extended to strike or tap the female (tactile communication) (Figure 26.4b).
2. *Waving.* The wing is extended and held 90° from the body, then relaxed without vibration (Figure 26.4c).
3. *Wing vibration.* The male extends one or both wings from the resting position and moves them rapidly up and down (producing a courtship song—auditory communication) (Figure 26.4c).
4. *Licking.* The male licks the female's genitalia (on the rear of her abdomen) (Figure 26.4d).
5. *Circling.* The male postures and then circles the female, usually when she is nonreceptive.
6. *Stamping.* The male stamps forefeet as in tapping but does not strike the female.

**a.** Orienting          **b.** Tapping          **c.** Waving and wing vibration

**d.** Licking          **e.** Attempting copulation          **f.** Copulation

**Figure 26.4.**
**Mating behavior in fruit fly, *Drosophila melanogaster.***

*Female Behaviors*

1. *Extruding.* A temporary, tubelike structure is extended from the female's genitalia.
2. *Decamping.* A nonreceptive female runs, jumps, or flies away from the courting male.
3. *Depressing.* A nonreceptive female prevents access to her genitalia by depressing her wings and curling the tip of her abdomen down.
4. *Ignoring.* A nonreceptive female ignores the male.

If the behavior display is successful, the flies will copulate (Figure 26.4e, f).

# Experiment A. Reproductive Behavior in *Drosophila melanogaster*

## Materials

stereoscopic microscope
fly vials with 2 or 3 virgin female *D. melanogaster* flies
fly vials with 2 or 3 male *D. melanogaster* flies

## Introduction

In this experiment, you will place virgin female *D. melanogaster* flies in the same vial with male flies and observe the behavior of each sex. Working with another student, discuss the behaviors described in the introduction to this exercise and plan the strategy for your experiment. Identify mating behaviors of *D. melanogaster* and record their sequence and duration (when appropriate). As you observe the behavior of the flies, discuss possible original experiments investigating courtship and mating behavior in flies.

## Hypothesis

Hypothesize about the presence of flies of the opposite sex in the same vial.

> *The presence of flies of the opposite sex will stimulate male and female flies to display courtship behaviors and mate.*

## Prediction

Predict the results of the experiment based on your hypothesis (if/then).

> *If the flies display courtship behaviors, then the male will display wing vibration, will wave his wings, and will tap and lick the female, while the female will display receptive behavior, extruding a tubelike structure from her genitalia. Then they will mate.*

*Place virgin females and males in separate vials with medium 3 to 4 days before this experiment. They must have time to recover from the anesthesia. They will mate very quickly under these conditions.*

## Procedure

1. Set up the stereoscopic microscope.
2. Have paper and pencil ready. The behaviors can happen very rapidly. One person should call out observations while the other person records.

3. Obtain one vial containing virgin females and one vial containing males, and gently tap the male flies into the vial containing females.

4. Observe first with the naked eye. Once flies have encountered each other, use the stereoscopic microscope to make observations.

## Results

1. Describe, in sequence, the response of the male to the female and the female to the male. Quantify your observations. To do this, you may consider counting the number of times a behavior takes place and timing the duration of behaviors.

2. Describe rejection if this takes place.

3. In the margin of your lab manual, note any behaviors that can be analyzed quantitatively.

## Discussion

Speculate about the adaptive advantage of elaborate courtship behaviors in animals.

> *"[It] assures each animal not only that the other is not a threat, but also that the other animal's species, sex, and physiological condition are all correct." (Campbell and Reece, 2008)*

# Experiment B. Student-Designed Investigation of Reproductive Behavior in Fruit Flies

## Materials

supplies from Experiment A
fly vials with 2 or 3 virgin females of an alternate fly species (other than *D. melanogaster*)

fly vials with 2 or 3 males of the alternate species (other

## Introduction

If your team chooses to perform your original experiment investigating courtship and mating behavior in fruit flies, continue with this experiment after you have completed all the introductory investigations (Experiment A of each exercise). Using the materials available and collaborating with your research partner, design a simple experiment to investigate this behavior. Several questions follow that might provide ideas.

1. Will courtship and mating in another species be identical to that in *D. melanogaster*?
2. Will males placed in the same vial demonstrate courtship behaviors?
3. Will males respond to dead females?
4. What is the response of a male *D. melanogaster* to females of a different species?
5. Do males compete?

Quantify your observations. To do this, you may consider counting the number of times a behavior takes place or timing the duration of behaviors.

### Hypothesis

After your team has decided on one or more questions to investigate, formulate a testable hypothesis.

### Prediction

Predict the results of your experiment based on your hypothesis (if/then).

### Procedure

1. On a separate paper, list in numerical order each step of your procedure. Remember to include the number of repetitions, the levels of treatment, the duration of stimulus, and other time intervals where appropriate (see Lab Topic 1).
2. If you have an idea for an experiment that requires materials other than those available, ask your laboratory instructor. If possible, additional supplies will be made available.
3. Quantify your data whenever possible (count, weigh, measure, time).

### Results

On a separate paper, record your data and describe your results. You should design at least one table and figure.

### Discussion

1. Within your team, discuss your results in light of your hypothesis. If possible, come to conclusions about the behaviors you have been investigating. Record your conclusions on a separate paper.
2. You may be asked to report the results of your experiments to the class.

## Reviewing Your Knowledge

Define, compare, and give examples for each item in the following pairs.

1. Learned behavior—innate behavior
2. Proximate cause of behavior—ultimate cause of behavior

3.  Causal explanation for a behavior—anthropomorphic explanation for a behavior

4.  Taxis—kinesis

Define the following terms.

1.  Fixed action pattern

2.  Stimulus-response chain

# Applying Your Knowledge

1.  Adult male European robins have red feathers on their breasts. A male robin will display aggressive behavior and attack another male robin that invades his territory during mating season. Immature male robins with all brown feathers do not elicit this behavior in the adult robin. How could you explain this behavior? Design an experiment to test your explanation (hypothesis).

    *The red feathers on the breast of an adult robin are a "releaser" for the aggressive behavior. To test this, you could dye the feathers of immature males or present a tuft of brown and then red feathers on a stick to the male robin to see if the red color is the stimulus for the behavior.*

2.  Bees, ants, and other social insects are able to detect and will remove dead organisms from their colony. Scientists have analyzed decaying insects and found they give off compounds—mostly oleic acid and linoleic acid—as they decay. You know that social insects display some of the most complex behaviors in response to environmental stimuli of any other animal groups. In lab today, you learned that it is possible to explain behaviors based on anthropomorphic causes, but that scientists base explanations on proximate and ultimate causes (see Lab Topic Introduction). Using bees as the experimental animals, in the space below, propose explanations (hypotheses) based on these three perspectives—anthropomorphic, proximate, and ultimate—and then propose a research project for those hypotheses that can be tested scientifically.

    a.  Anthropomorphic causes:

      *Bees are very intelligent animals and they prefer a clean hive with no clutter. (Students will propose an explanation that reflects human behavior.)*

      Possible research project:

      *This is not a testable hypothesis (see Lab Topic 1). It is difficult to quantify "intelligent" and preference behaviors in a scientific research project.*

    b.  Proximate causes:

      *(Proximate causes are the physiological events leading to the behavior.) Students may hypothesize that the products of decay (oleic acid and linoleic acid) are chemical stimuli and activate neurons in the central nervous system and ultimately muscles in the bee's appendages. The compounds (as bees come into contact with the dead bees) stimulate the removal behavior.*

      Possible research project:
      *Some ideas and variations:*

      •   *Apply the products of crushed bees onto inanimate objects (cotton balls, glass beads, etc.) and observe the bees' behavior.*

- *Apply oleic acid or linoleic acid to inanimate objects.*
- *Vary the concentration of these substances to see the lowest concentration that will induce a reaction.*

c. Ultimate causes:

*(Ultimate causes tell the adaptive value and the evolutionary origin of the behavior.) The adaptive value of this behavior is that dead organisms are removed from the hive and disease does not spread to the whole colony.*

*Research at this level might compare this behavior in other social insects or in insects that do not display social behavior. Consider if this behavior has led to greater success in survival of social insects versus nonsocial insects.*

Could an experiment similar to this be used to investigate ancestors of insects and isopods and to answer questions about when these groups diverged?

*Testing a variety of arthropods and comparing positions on the phylogenetic tree, determine which organisms demonstrate this behavior. For example, if the terrestrial isopods studied in this become more active in the presence of the decay products (indicating a negative stimulus), then this behavior could be either a convergence between isopods and insects or a common response that goes back to the evolutionary ancestor of both groups.*

## Student Media: BioFlix, Activities, Investigations, and Videos
**www.masteringbiology.com (select Study Area)**

**Activities**—Ch. 51: Honeybee Waggle Dance Video
**Investigations**—Ch. 51: How Can Pillbug Responses to Environments Be Tested?
  LabBench: Animal Behavior
**Videos**—Ch. 51: Ducklings; Chimp Cracking Nut; Snake Ritual Wrestling; Albatross Courtship Ritual; Blue-footed Boobies Courtship Ritual; Chimp Agonistic Behavior; Wolves Agonistic Behavior; Giraffe Courtship Ritual

## References

Alcock, J. *Animal Behavior: An Evolutionary Approach,* 9th ed. Sunderland, MA: Sinauer Associates, 2009.

Gould, J. L. *Ethology: The Mechanisms and Evolution of Behavior.* New York, NY: W. W. Norton & Company, 1982.

Greenspan, R. J. "Understanding the Genetic Construction of Behavior." *Scientific American,* 1995, vol. 272(4) p. 72.

Johnson, R. N. *Aggression in Man and Animals.* Philadelphia, PA: Saunders College Publishing, 1972.

Marler, P. "Mating Behavior of *Drosophila*." In *Animal Behavior in Laboratory and Field,* editor A. W. Stokes. San Francisco, CA: Freeman, 1968.

Reebs, S. "Death Whiff." *Natural History,* October 2009. (pill bugs' response to their dead)

Raham, G. "Pill Bug Biology." *The American Biology Teacher,* 1986, vol. 48(1).

Reece, J. et al. *Campbell Biology,* 9th ed. San Francisco, CA: Pearson Education, 2011.

Simpson, M. J. A. "The Threat Display of the Siamese Fighting Fish, *Betta splendens*." *Animal Behavior Monograph,* 1968, vol. 1, p. 1.

Thompson, T. "Aggressive Behavior of Siamese Fighting Fish," in *Aggressive Behavior,* editors S. Garattini and E. B. Sigg. Proceedings of the International Symposium on the Biology of Aggressive Behavior. New York, NY: Wiley, 1969.

Waterman, M., and E. Stanley. *Biological Inquiry: A Workbook of Investigative Cases.* San Francisco, CA: Pearson Benjamin Cummings, 2011. In this supplement to Campbell Biology, 9th ed., see "Back to the Bay," a case study applying principles of animal behavior to an environmental problem.

## Websites

Animal Behavior Society Website. The education section describes experiments submitted by college professors suitable for independent projects. Select Education and Behavior.
http://animalbehaviorsociety.org/

J. Kimball's online discussion of innate behavior: http://users.rcn.com/jkimball.ma.ultranet/BiologyPages/I/InnateBehavior.html

Overview of topics in animal behavior: http://cas.bellarmine.edu/tietjen/animal_behavior_and_ethology.htm

# LAB TOPIC 26

# Animal Behavior
## Teaching Plan for Laboratories

## Main Concepts and Objectives

1.  Concept: behavior classification. Students will define and give examples of *taxis, kinesis, agonistic behavior,* and *reproductive behavior.* (Lab Study A of each exercise)

2.  Concept: genetic control of behavior. Students will understand that behavior is controlled by genes and is subject to evolutionary adaptation. Students will discuss the adaptive values of the behaviors studied. (Lab Study A of each exercise)

3.  Concept: scientific study of behavior. Students will design and execute behavioral experiments using one of the four organisms studied. (Lab Study B of one exercise)

4.  Concept: Ethology is a science. Students will learn to ask causality questions rather than anthropomorphic questions. (Introduction)

5.  Concept: scientific persuasion. Students will give brief presentations of their experiments to the class and may be required to submit a laboratory report on their experiment in the format of a scientific paper. (Lab Study B of one exercise)

6.  Specific content. Students will be able to describe and define different classes of behaviors: orientation behaviors (taxis, kinesis), agonistic behaviors, reproductive behaviors, and learned and innate behaviors. They should be able to explain differences in causality and anthropomorphic explanations of behavior.

## Order of the Lab

1.  Introduce the concepts, objectives, and procedures. (10 min)

2.  Students perform each introductory investigation, and each team decides which organism it will study in an original experiment. (60 min)

3.  Students design and execute an original experiment using one of the behavioral systems. (90 min)

4.  Students practice communicating results and persuading colleagues in oral reports. (15 min)

5. Give instructions about preparing a written report practicing scientific writing. (Appendix A)                                    (5 min)

**For a 2-hour lab:** Omit two of the organisms, depending on availability in your particular area. The fruit fly experiment requires the greatest preparation, so you may choose to omit this one. Have students perform Experiment A for only two organisms for about 30 minutes and then design and perform their original experiment for about 1.5 hours. Omit the oral reports.

**Alternate 2-hour plan:** If your schedule will allow you to use two lab periods for the study of behavior, have students perform every Experiment A and design their original experiment during the first lab. Students then perform their original experiment in the second lab. The last hour of the second lab can be reserved for student reports. All teams will be able to report with this arrangement.

## Classroom Management

All students carry out all four of the introductory investigations (Experiment A of each exercise) as teams. Each team of four students (two if they choose to investigate flies) will then design one original experiment using the supplies available. If students request additional supplies, attempt to accommodate them. Encourage creative ideas. We have found that this laboratory works best if you instruct students to meet with their research teams before lab day and discuss the original experiment they would like to perform. We ask them to come to lab with a couple of hypotheses and procedures already worked out; then they choose from these after they have performed the introductory investigations (Experiment A of each exercise).

Ask for volunteers or select three or four students to report on their findings. Encourage questions and discussion of results by other students, as would take place in a scientific meeting. Give final instructions about the written scientific paper. The Materials and Methods and Results sections can be a collaborative effort among team members. The Introduction, Discussion, and Conclusions sections must be the original work of each student.

## Student Development

The most important aspects of student development in this exercise will be laboratory skills, creative thinking, and critical thinking as students plan their own experiments and evaluate those of class members. Students will practice scientific processes: proposing hypotheses, making predictions, testing hypotheses, recording and processing data, presenting results to colleagues, and writing scientifically.

## Discussion and Summary

Students ask questions of other students and discuss results in the oral reports at the conclusion of the lab. Students develop their ideas in the Discussion and Conclusions sections of their lab report.

## Evaluation

If time permits, have a student from each group report on the results of the group's experiment. If time is short, have only one report on each organism. Keep notes on the quality of the reports and the experiments for a subjective evaluation of the student's performance. Make an effort to have as many different students as possible report throughout the term. Grade the written laboratory reports. Mastery of main concepts (descriptions of kinds of behaviors) can be tested on a laboratory exam. Actually, we do not test students on these concepts in lab; the topic is covered on tests in the lecture portion of the course.

## Lab Topic 27
# Ecology I: Terrestrial Ecology

## Laboratory Objectives

After completing this lab topic, you should be able to:

1. Describe the trophic levels of an ecosystem and provide examples from field experience.
2. Describe the environmental factors that are important components of the ecosystem.
3. Calculate density, frequency, dominance, and species diversity.
4. Describe the relative importance of particular species in the ecosystem as determined by their density, size, or role in the ecosystem.
5. Construct an illustrative model of the ecosystem components.

*For a 2-hour lab:* Select portions of the lab for field studies. Provide a description of the study site before lab. Complete sampling and have instructor pool the data. In a second lab period, view demonstrations and complete data analysis. See Teaching Plan for other suggestions.

*Instructors and/or students can download data tables in Excel format. Go to www .masteringbiology.com. Look in the Study Area under Lab Media.*

## Introduction

Organisms are influenced by their physical environment and by interactions with other living organisms. The study of the relationships of organisms with their environment (both physical and biological) is called **ecology.** Ecological investigations may address questions at several hierarchical levels: **individuals, populations** (organisms of the same species that share a common gene pool and occur in the same area), **communities** (populations of different species that inhabit the same area), and **ecosystems** (the community of plants and animals plus the physical environment). A physiological ecologist studies the effects of the environment on individual organisms. A population ecologist might be interested in questions about the reproductive biology of a population of endangered plant species. The community ecologist might investigate the sequence of species composition changes in a forest following a disturbance. And the ecosystem ecologist might question how biological diversity or the interactions among members of different trophic levels (feeding levels) are influenced by environmental conditions.

An ecosystem can be divided into **biotic** components (living organisms) and **abiotic** components (physical features). In a forest ecosystem, the abiotic components include the climatic factors, soil type, water availability, and landscape features. The biotic components of the forest include trees, shrubs, wildflowers, squirrels, foxes, caterpillars, eagles, spiders, millipedes, fungi, and bacteria on the forest floor and in the soil (Figure 27.1). The biotic components can be further characterized based on **trophic structure,** the ecological role of organisms in the food chain. Plants and some protists are categorized as **primary producers** (autotrophic organisms), capable of

*This laboratory is designed to encourage field experiences for students in introductory biology. Particular exercises can be selected by the instructor depending on the ecosystem, number of students, and time available. Suggestions are provided in the Teaching Plan for solving problems often viewed as reasons for omitting field labs from the laboratory program.*

*Information about developing a case study concerning local environmental issues is included in the Teaching Plan.*

**Figure 27.1.**
**Forest ecosystem.** A simplified forest system showing trophic (feeding) relationships among primary producers, consumers, detritivores, and decomposers.

transforming light energy into chemical energy stored in carbohydrates through the process of photosynthesis. The amount of energy available for all other trophic levels is dependent on the photosynthetic ability of the primary producers. **Consumers** are animals and heterotrophic protists, which literally consume the primary producers or each other or both. They may be divided into **primary consumers** (**herbivores**, which consume plants) and **secondary** and **tertiary consumers** (**carnivores**, which eat other consumers). Rarely do ecosystems support additional levels of consumers. In general, only 10–20% of the energy available at one trophic level is transferred to the next trophic level. The number of organisms that can be supported at subsequent higher trophic levels is limited by the available energy. **Detritivores** obtain their nutrients and energy from dead organisms and waste materials, and **decomposers** (fungi and bacteria) absorb nutrients from nonliving organic material.

With the assistance of your lab partner, match each of the following trophic levels with an example of an organism from the forest system just described. Remember that the trophic level reflects who is eating whom.

Primary producers: *trees, shrubs, wildflowers*

Primary consumers: *squirrels, foxes, caterpillars*

Secondary consumers: *foxes, eagles*

Tertiary consumers: *eagles*

Detritivores: *millipedes*

Decomposers: *fungi*

Note that some organisms fit into more than one trophic level—because, like many humans, some are omnivores, eating whatever is available.

*Students may not know what these animals eat, so you may have to provide some information. Ask students first.*

*The Discovery Video Ch. 52: Trees, introduces components of a forest ecosystem. See www.masteringbiology.com.*

## Designing and Organizing Your Investigation

In this lab topic, you will investigate the structure and function of a local ecosystem. The exercise is designed for a forest ecosystem but can be adapted easily for use in grasslands or even a weedy urban lot. (An outline for adapting the lab for use in a weedy field is included at the end of the lab topic.) The study site has been selected in advance by your instructor, who may have prepared an introductory description for you.

*This lab topic can be adapted to practically any local terrestrial ecosystem, from a weedy corner lot to rock outcrops to open prairie. You may want to prepare a brief description of the ecosystem for the students to read before they go outdoors.*

This lab topic provides rich resources and methodologies to pursue questions about the trophic structure of a forest ecosystem, or to perform a comparative study of two sites, for example, with different records of disturbance (cutting, herbicides, or invasive species). As another option, you might choose to study biological diversity for only one component of the ecosystem, for example, the litter layer. *As a class discuss the possible questions that you will pursue. Record a brief description of your study site, and then state the question.*

*Study Site*

*Question*

*Modify the responsibilities of students within a group if you are only doing selected field studies.*

*Using the organization suggested in this table, you could sample two or three plots, or you may choose to sample more areas and not include all components of the ecosystem.*

**Table 27.1**
Suggested Organization of Student Teams

| Exercise | Sampling | No. of Students |
|---|---|---|
| **Exercise 27.1** | **Biotic Components** | |
| A | Trees | 2 |
| B | Shrubs, saplings, and vines | 2 |
| C | Seedlings and herbaceous vegetation | 2* |
| D | Macroinvertebrates | 2* |
| E | Microinvertebrates | 2** |
| F | Microorganisms | 2* |
| G | Other forest animals | All |
| **Exercise 27.2** | **Abiotic Components** | 2** |

*Two students can complete Field Studies C, D, and E.
**Two students can complete both Exercise 27.2 and Field Study F in Exercise 27.1.

### Organization of Teams

Based on your question, select the components of the ecosystem that you will be studying. Next determine the number of replicate plots; this will be the number of teams. Each team, composed of six to eight students, will sample one plot, thus providing three or four replicate samples, depending on the number of teams. Within teams, each student will have specific assigned responsibilities, as suggested in Table 27.1 and fully described in the following exercises.

Following the field sampling, you will share results with other teams, make calculations, pool and analyze data, and develop a model of the ecosystem. Students should read assignments for all field studies.

# EXERCISE 27.1
# Biotic Components

As you begin your observations of the forest, note the vertical layers of the forest from the forest floor up to the tallest and largest trees, which form a canopy. The forest vegetation can be subdivided arbitrarily into categories according to the structural pattern of the forest (Figures 27.2 and 27.7): forest **trees** (woody plants with a diameter at breast height (DBH) of >10 cm); **shrubs, saplings,** and **vines** (DBH 2.5–10 cm); and **seedlings** and **herbaceous plants** (DBH < 2.5 cm). (Note that DBH is measured at a height of 1.5 m from the base of the tree.) **Microorganisms** and **small animals** can be sampled from the forest floor, **litter** (fallen leaves, for example), and **soil**, with **large animals** observed directly or indirectly by animal signs (nest sites, feces, tracks).

In your student teams, you will sample part of the study site using circular plots to estimate the abundance of plants and animals in each of the categories

**Figure 27.2.**
**Vertical stratification of the forest.**
The forest can be divided into layers:
the trees (DBH >10 cm); shrubs,
saplings, and vines (DBH 2.5–10 cm);
and seedlings and herbaceous plants
(DBH <2.5 cm). See also Figure 27.7.

Trees
DBH >10 cm

Shrubs, vines,
and saplings
DBH 2.5–10 cm

Seedlings
and herbs
DBH <2.5 cm

**Table 27.2**
Sampling Design for Determining the Biotic
Components of a Forest Ecosystem

| Organisms | Plot Size |
|---|---|
| Trees (DBH >10 cm) | 100 m$^2$ |
| Shrubs, saplings, and vines (DBH 2.5–10 cm) | 50 m$^2$ |
| Seedlings and herbs (DBH <2.5 cm) | 0.50 m$^2$ |
| Litter macroinvertebrates | 0.50 m$^2$ |
| Microinvertebrates from litter and soil samples | — |
| Microorganisms from litter and soil samples | — |
| Large animals observed by all | — |

described above. The size of the circular plots varies according to the size and abundance of the organisms (Table 27.2). Thus, trees are sampled using the largest plot, while the herbs of the forest floor are sampled in small plots. Imagine the difficulty of sampling trees in plots of only 1 m² or counting all the nonwoody plants in a plot with a diameter of over 10 m²!

 Avoid contact with poison ivy. Your instructor will identify the plant for you. After completing the fieldwork, thoroughly wash all areas of exposed skin with soap and cold water. Avoid contact with your clothing if you think you have come into contact with poison ivy. Notify your instructor if you are allergic to bee stings or other plants and animals that you might encounter outdoors.

POISON IVY

# Field Study A. Trees

## Materials

*Refer to the* Preparation Guide *for construction of the simple, inexpensive, and easy-to-store sampling materials used with circular plots.*

| | |
|---|---|
| 3 lines, 5.64 m long, with a clip on one end and a stake on the other end of each | DBH measuring tape |
| | index cards |
| | permanent marker |
| center post | plastic bags and rubber bands |
| mallet | |

## Introduction

Trees form the uppermost layer, or **canopy,** of the forest. They influence the physical environment, such as the quality and quantity of light reaching all other vegetation, water availability, and temperature. A forest description is usually based on the largest and most abundant species of trees.

## Procedure

*Plan to go to the field site in advance and flag plot center points or even install the center posts for sample plots. If these are to be permanent plots that will be sampled at some future date, consider using weather-resistant metal fence posts with a more permanent installation.*

1. Determine the sample plot location based on recommendations from your instructor.
2. Locate the center of the sample plot and hammer the center post into the ground at that location.
3. Clip one end of a line to the center post. Extend the line, keeping it straight and taut, and hammer the stake into position (Figure 27.3, line 1).
4. Attach and secure lines 2 and 3 at an angle forming two wedge-shaped sections A and B (see Figure 27.3). Each wedge should be equal to approximately one-fifth of the plot.
5. Identify and measure DBH (>10 cm) for all trees in sections A and B (see Figure 27.3). The number of DBH measurements will also provide a record of abundance. If a tree is on the outer perimeter of the plot, it should be counted in the plot if at least 50% of the tree is within the plot.

 Diameter at breast height (DBH) is measured 1.5 m above the base of the tree. Measure or estimate this height on your body to ensure that you determine DBH in a consistent fashion. If DBH measuring tapes are not available, measure trees with a circumference of >32 cm.

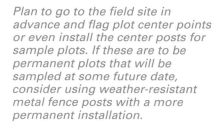

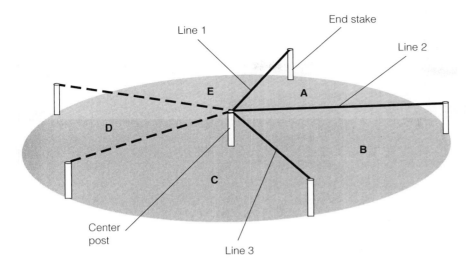

**Figure 27.3.**
**Establishing and sampling a circular plot.** Locate the center post and attach three lines (1, 2, 3) to form two wedge-shaped sections, A and B. After sampling A and B, move line 2 to form section C. After sampling C, move line 3 to form sections D and E.

6. If you cannot identify a tree, assign a number to that type of tree. Collect a leaf sample, if possible, for later identification. Label the sample by attaching a numbered card or by writing the number on the leaf blade. If leaves are not available, write a brief description.

7. Before moving any lines, check with the shrub, sapling, and vine group to be sure they have completed their sampling. Their plot is smaller and lies within the boundary of the tree plot.

8. Do not move line 1. Detach line 2 and move it into position to form the boundary between sections C and D (the first dotted line in Figure 27.3).

9. Identify and measure DBH for all trees in section C (see Figure 27.3).

10. Detach line 3 and move it into position to form the boundary between sections D and E (the second dotted line in Figure 27.3).

11. Identify and measure DBH for all trees in sections D and E (see Figure 27.3).

*Identification of organisms is always a problem in a field lab when you cannot spend time teaching the students the identity of all the plants and animals. One inexpensive and efficient method is to collect examples of leaves from the most commonly observed plants in the sample sites. Use a photocopier to make black-and-white outline copies of each plant, and label each sheet. Make an identification booklet that is specific for your study site. Members of your department or a local landscaper or forester can help with identifications. Any unknowns can be numbered and bagged for later identification, although often simply identifying species by number is adequate as long as they are not the most abundant species.*

## Results

1. Record your data in Table 27.3 on the next page.

2. In Exercise 27.3, Data Analysis, you will calculate:

    DBH (if measurements were taken of circumference rather than DBH)

    Density and relative density

    Frequency and relative frequency

    Basal area, dominance, and relative dominance

    Importance value

# Field Study B. Shrubs, Saplings, and Vines

## Materials

circular plot established for trees in Field Study A; the radius lines should be flagged conspicuously at 4 m for the SSV plot

plastic bags and rubber bands
calipers
permanent marker
index cards

**Table 27.3**
Sampling Results for **Trees** (DBH >10 cm)

Locality: _____  Plot ID #: _____  Plot size: _____  Date: _____

Students: _____  Instructor: _____

| SPECIES: | Circumference (cm) | DBH (cm) = Circumference/π | Basal Area (cm²) = 0.7854 (DBH)² | | | | |
|---|---|---|---|---|---|---|---|
| | | | | | | | |
| | | | | | | | |
| | | | | | | | |
| | | | | | | | |

| SPECIES: | Circumference (cm) | DBH (cm) = Circumference/π | Basal Area (cm²) = 0.7854 (DBH)² | | | | |
|---|---|---|---|---|---|---|---|
| | | | | | | | |
| | | | | | | | |
| | | | | | | | |
| | | | | | | | |

| SPECIES: | Circumference (cm) | DBH (cm) = Circumference/π | Basal Area (cm²) = 0.7854 (DBH)² | | | | |
|---|---|---|---|---|---|---|---|
| | | | | | | | |
| | | | | | | | |
| | | | | | | | |
| | | | | | | | |

| SPECIES: | Circumference (cm) | DBH (cm) = Circumference/π | Basal Area (cm²) = 0.7854 (DBH)² | | | | |
|---|---|---|---|---|---|---|---|
| | | | | | | | |
| | | | | | | | |
| | | | | | | | |
| | | | | | | | |

| SPECIES: | Circumference (cm) | DBH (cm) = Circumference/π | Basal Area (cm²) = 0.7854 (DBH)² | | | | |
|---|---|---|---|---|---|---|---|
| | | | | | | | |
| | | | | | | | |
| | | | | | | | |
| | | | | | | | |

| SPECIES: | Circumference (cm) | DBH (cm) = Circumference/π | Basal Area (cm²) = 0.7854 (DBH)² | | | | |
|---|---|---|---|---|---|---|---|
| | | | | | | | |
| | | | | | | | |
| | | | | | | | |
| | | | | | | | |

## Introduction

Lower vertical layers of the forest are inhabited by young trees of the types seen in the upper levels, trees and shrubs unique to this layer, and some vines. The shrub, sapling, and vine (SSV) plot is located within the tree plot (Figure 27.4). Therefore, each line should be boldly marked to indicate the radius for 50 m$^2$.

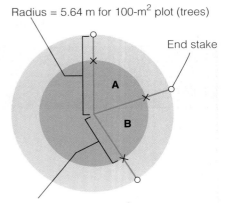

Radius = 5.64 m for 100-m$^2$ plot (trees)

End stake

A

B

Radius = 4 m for 50-m$^2$ plot (SSV)

 Coordinate your sampling with the tree group, Field Study A.

## Procedure

1. Locate the circular plot established by your team members sampling trees. Restrict your sampling to the smaller plot located within the tree plot. The radius of the smaller plot should be marked clearly with flagging or tape along each line. Sample vines and other woody plants with a DBH of 2.5–10 cm within the SSV plot.

2. Begin sampling in section A (see Figure 27.3). Identify and measure each individual in the SSV category for all sections of the plot. If available, use calipers, rather than a meter tape, to determine DBH.

**Figure 27.4.**
**Placement of circular plot for shrubs, saplings, and vines.** A circular plot with a smaller diameter is marked within the boundary of the large tree plot. Radius lines are clearly marked to indicate the outer perimeter for the plot (Xs).

 Diameter at breast height (DBH) is measured 1.5 m above the base of the woody plant. Measure or estimate this height on your body to ensure that you determine DBH in a consistent fashion. If DBH measuring tapes are not available, measure plants with a circumference of 8–32 cm.

3. If you cannot identify a woody plant, assign a number to that type of plant. Collect a leaf sample, if possible, for later identification. Label the sample by attaching a numbered card or writing the number on the leaf blade. If leaves are not available, write a brief description.

*Refer to the Teaching Plan for suggestions on plant identification.*

## Results

1. Record your data in Table 27.4 on the next page.
2. In Exercise 27.3, Data Analysis, you will calculate:
   DBH (if measurements were taken of circumference rather than DBH)
   Density and relative density
   Frequency and relative frequency
   Basal area, dominance, and relative dominance
   Importance value

**Table 27.4**
Sampling Results for **Shrubs**, **Saplings**, and **Vines** (DBH 2.5–10 cm)

| Locality: _____ | Plot ID #: _____ | Plot size: _____ | Date: _____ |
| Students: _____ | | Instructor: _____ | |

SPECIES:

| Circumference (cm) | DBH (cm) = $\frac{\text{Circumference}}{\pi}$ | Basal Area (cm²) = 0.7854 (DBH)² |
|---|---|---|
| | | |
| | | |
| | | |
| | | |
| | | |
| | | |

SPECIES:

| Circumference (cm) | DBH (cm) = $\frac{\text{Circumference}}{\pi}$ | Basal Area (cm²) = 0.7854 (DBH)² |
|---|---|---|
| | | |
| | | |
| | | |
| | | |
| | | |
| | | |

SPECIES:

| Circumference (cm) | DBH (cm) = $\frac{\text{Circumference}}{\pi}$ | Basal Area (cm²) = 0.7854 (DBH)² |
|---|---|---|
| | | |
| | | |
| | | |
| | | |
| | | |
| | | |

SPECIES:

| Circumference (cm) | DBH (cm) = $\frac{\text{Circumference}}{\pi}$ | Basal Area (cm²) = 0.7854 (DBH)² |
|---|---|---|
| | | |
| | | |
| | | |
| | | |
| | | |
| | | |

SPECIES:

| Circumference (cm) | DBH (cm) = $\frac{\text{Circumference}}{\pi}$ | Basal Area (cm²) = 0.7854 (DBH)² |
|---|---|---|
| | | |
| | | |
| | | |
| | | |
| | | |
| | | |

SPECIES:

| Circumference (cm) | DBH (cm) = $\frac{\text{Circumference}}{\pi}$ | Basal Area (cm²) = 0.7854 (DBH)² |
|---|---|---|
| | | |
| | | |
| | | |
| | | |
| | | |
| | | |

# Field Study C. Seedlings and Herbaceous Vegetation

## Materials

circular 0.50-m$^2$ plot
compass
index cards

plastic bags and rubber bands
permanent marker

## Introduction

Tree seedlings and herbaceous plants appear near the forest floor. These plants may be difficult to count, and, in some cases, it may be almost impossible to determine what is actually one individual. Because of this, the abundance of these plants is estimated based on the percent cover.

The seedling and herb plot is not located within the larger plots but should be placed near the tree plot at a predetermined position.

## Procedure

1. Place a 0.50-m$^2$ plot just outside the tree plot. Determine the exact location for the plot before beginning. For example, it might always be 1 m away from the tree plot at a compass heading of due north.

2. Identify and estimate the abundance of each species by estimating the percent of the plot area covered by the species. Sketching the plot may help determine cover and ensure consistency in your estimates.

3. If you cannot identify a plant, assign a number to that type of plant. Collect a leaf or flower sample if possible for later identification. Label the sample by attaching a numbered card or by writing the number on the leaf blade. If leaves are not available, write a brief description.

*Refer to the Teaching Plan for suggestions on plant identification.*

## Results

1. Record results in Table 27.5 on the next page.

2. In Exercise 27.3, Data Analysis, you will calculate average percent cover, frequency, and relative frequency.

# Field Study D. Macroinvertebrates

## Materials

circular 0.50-m$^2$ plot
thin plastic sheet
forceps
dissecting probes

vials with 70% alcohol
labeling tape
permanent marker
compass

## Introduction

Large invertebrates—for example, grasshoppers, ants, and spiders—can be surveyed by sifting through the dead leaves or litter on the forest floor. These animals may fit into any of the consumer trophic levels. For example, grasshoppers are primary consumers, spiders are secondary consumers, and ants are omnivores.

**Table 27.5**

Sampling Results for **Seedlings** and **Herbaceous Vegetation** (DBH <2.5 cm)

| Locality: _____ Plot ID #: _____ Plot size: _____ Date: _____ | | | |
|---|---|---|---|
| Students: _____     Instructor: _____ | | | |
| Species | Percent Cover | Species | Percent Cover |
|  |  |  |  |
|  |  |  |  |
|  |  |  |  |
|  |  |  |  |
|  |  |  |  |
|  |  |  |  |
|  |  |  |  |
|  |  |  |  |
|  |  |  |  |
|  |  |  |  |
|  |  |  |  |
|  |  |  |  |
|  |  |  |  |
|  |  |  |  |

*Refer to the* Preparation Guide *for information on the construction of simple circular plots.*

### Procedure

1. Carefully place a 0.50-m² circular plot on the ground in an undisturbed spot outside and near the large vegetation plot. This should be a preselected spot; for example, it might always be 1 m away from the large plot at a compass heading of due south. Avoid disturbing the animals present. Quickly cover the plot with plastic to discourage "escapees."

2. Carefully remove all animals from the vegetation and litter at ground level.

3. Sort animals into similar groups in the field. Place the organisms into vials of 70% alcohol according to their group. If you cannot identify an animal, assign a number to that type of animal. Preserve the animal in alcohol and label the sample using tape and a permanent marker.

### Results

1. Record results in Table 27.6.

2. In Exercise 27.3, Data Analysis, you will calculate density for the groups.

**Table 27.6**
Sampling Results for **Consumers, Detritivores,** and **Decomposers**

| Locality: _____ | | Plot ID #: _____ Date: _____ |
| --- | --- | --- |
| Students: _____ | | Instructor: _____ |
| **Macroinvertebrates** | | **Microinvertebrates** |
| **Species/Group** | **No. of Individuals** | **Groups Observed** |
| | | |
| | | |
| | | |
| | | |
| | | |
| | | |
| | | |
| | | |
| | | |

Microorganisms:

Vertebrates:

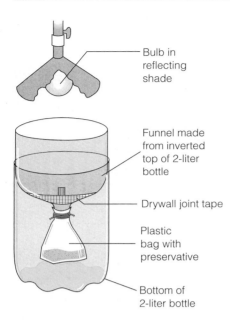

- Bulb in reflecting shade
- Funnel made from inverted top of 2-liter bottle
- Drywall joint tape
- Plastic bag with preservative
- Bottom of 2-liter bottle

**Figure 27.5.**
**Berlese-Tullgren funnel used for the extraction of microinvertebrates from litter and soil.** The samples are placed in a funnel with the narrow opening covered with two crossed pieces of drywall tape, adhesive side down. The light and heat drive the animals down into the small sampling bag containing preservative. (Setup modified from Kingsolver, 2006.)

*You may choose to prepare the extractions in advance and provide these for your students or to have the students set up the extractions and view the organisms either before or during the next lab period. These "hidden" organisms are the biggest surprise to students. They think of the soil and litter as dead, or abiotic. There is merit in having them actually do the extractions.*

# Field Study E. Microinvertebrates

## Materials

circular 0.50-m² plot
index cards
self-sealing plastic bags
Berlese-Tullgren funnels

light sources
soil sample collected in
  Exercise 27.2, Field Study A

## Introduction

A variety of invertebrates, including mites and springtails, inhabit the litter layer and the upper layers of the soil. Many of these animals are not easily seen with the naked eye and would be impossible to collect by inspecting samples. These consumers and detritivores can be sampled by collecting soil and litter and placing these in Berlese-Tullgren funnels (Figure 27.5). The animals move away from the light and heat source into a small sampling vial. The animals can then be identified in the laboratory using a dissecting microscope. You may be asked to collect samples for extraction later in the lab, or your instructor may have already collected the samples and started the extractions.

## Procedure

1. Place the 0.50-m² plot in an undisturbed location. Carefully remove half the litter (discard sticks) and place it into the bag provided. Include a card with the plot number in your sample bag. Close the bag securely.

2. Upon return to the lab, extract the microinvertebrates using a Berlese-Tullgren funnel. A soil sample will also be collected by the soil group (see Exercise 27.2, Field Study A), and microinvertebrates will be extracted using the same procedure.

   a. Place the soil or litter sample on the mesh tape in a large funnel under a lightbulb. The soil samples can be wrapped in a layer of cheesecloth to prevent soil particles from sifting into the collection bags.

   b. The light and heat force the animals down the funnel into the collecting bag containing 70% alcohol. (See Figure 27.5 and *Preparation Guide* for extraction funnel setup.)

3. After 24 to 48 hours, the organisms can be sorted and identified using the dissecting scope. Refer to the illustrated key to common microinvertebrates in Figure 27.6.

## Results

1. Record in Table 27.6 the microinvertebrates present.

2. Sketch the most common organisms in the margin of your lab manual if they are not represented in the illustrated key.

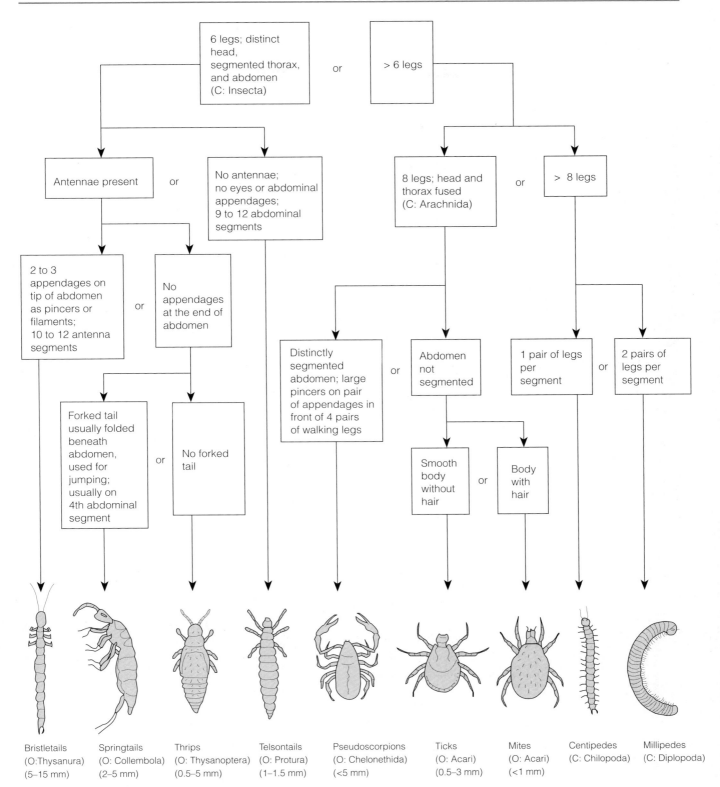

**Figure 27.6.**
**Picture key to microinvertebrates** commonly found in litter and soil samples.
C: class; O: order.

# Field Study F. Microorganisms

## Materials

petri dishes of nutrient agar
soil sample collected in Exercise 27.2, Field Study A
litter sample collected in Exercise 27.1, Field Study E

## Introduction

Microorganisms in the litter and soil are important decomposers, the primary functions of which are soil building and nutrient recycling. Your instructor may have collected the samples in advance and begun the cultures. If not, you will need to obtain a small portion of the soil sample from the students sampling the abiotic components and some litter from the microinvertebrate group. You will prepare cultures for later evaluation.

## Procedure

*These plates can be prepared in advance by the instructor.*

1. Upon returning to the lab, place small samples of leaf litter and soil on separate agar plates and incubate them at room temperature for several days until bacterial and fungal growth are abundant. Refrigerate them to prevent further growth.

2. Observe the diversity of colonies present after 2 or 3 days.

## Results

Record the presence of different bacterial colony types and fungi in Table 27.6. Refer to Lab Topic 13 Bacteriology for help with describing organisms.

# Field Study G. Other Forest Animals

## Materials

binoculars
field guides

## Introduction

You will not trap forest vertebrates in this field study, but all students will need to make observations of these larger but often secretive animals. Observe the activities of these animals as well as their calls, tracks, nest sites, and feces. Think about their role in the ecosystem.

## Procedure

1. After completing your field sampling, survey the surrounding forest for birds, mammals, reptiles, and amphibians.

2. Record observations of animals sighted, plus any tracks, nests, burrows, carcasses, feces, owl pellets, songs, or calls. Field guides are provided to assist with identification.

## Results

Record your observations in Table 27.6.

# EXERCISE 27.2
# Abiotic Components

Abiotic components include the physical features of the environment that influence the biotic components and in turn may be influenced by the organisms inhabiting the area. You will record the climatic conditions both for the local environment and for the smaller scale microenvironment within the ecosystem. Two students in each team will measure and record all abiotic features.

## Field Study A. Soil

### Materials

soil thermometer
soil auger
self-sealing plastic bags
index cards
permanent marker

trowel
diagram of soil profile
soil map of the area
soil test kit

### Introduction

The soil contains biotic and abiotic components of the ecosystem. You will determine the soil type and collect samples to be used in extracting microinvertebrates (Exercise 27.1, Field Study E) and microorganisms (Exercise 27.1, Field Study F).

### Procedure

1. Read about soil temperature (Field Study A) and air temperature (Field Study B); then place thermometers in appropriate locations to equilibrate while you begin your other work.

2. Remove the vegetation from a small undisturbed area near the tree plot. Place the soil thermometer in the soil and allow it to equilibrate for 10 to 15 minutes before recording the temperature.

3. Carefully remove the leaf litter. Using the soil auger, remove two soil cores approximately 30–45 cm long.

4. Sketch the profile of the soil layers in the margin of your lab manual. (See an example of a soil profile in your materials.) Note changes in color and texture. How much sand or clay or dark organic matter appears to be present?

5. Place each soil sample in a separate plastic (self-sealing) bag. Follow the directions provided with the soil test kit. Determine the soil pH and any other characteristics suggested by your instructor for one sample. Instructions are included with the soil test kit. Save the other sample to share with your team.

6. Remove a third sample of soil to be used for determining the presence and types of microinvertebrates and microorganisms. Collect this sample by using a trowel to dig some of the upper loose dirt. Seal this sample in a third plastic bag. Label it "organisms" with the plot number. The organisms will be extracted as described in Exercise 27.1, Field Studies E and F.

*To begin, students should equilibrate all thermometers. Instructions for air temperature are in Field Study B, point 2a in the Procedure section.*

*Students can simply test the soil pH by making a wet slurry with water mixed with the soil sample. The pH can be determined using pH test papers. Suggest that students test the pH at intervals along the soil core. If you purchase soil test kits, additional characteristics can be determined. See suggestions for soil kits in the* Preparation Guide.

7. Study the soil association map of this general area and determine the type of soil typical of this region.

### Results

1. Record the soil temperature and pH in Table 27.7.
2. Sketch the soil profile in the margin of your lab manual.
3. Record in Table 27.7 the soil type and other pertinent information from soil maps.

## Field Study B. Climatology

### Materials

*The local weather station may have the NOAA data sheets for your area. These can also be obtained through the library from government documents. Try the NOAA website: http://www.erl.noaa.gov.*

NOAA climatological data sheets
sling psychrometer (with directions)
min/max thermometer
  (with directions)
gun-style infrared thermometers

light meter
thermometers
wind anemometer
  (with directions)

### Introduction

The climatic conditions of the environment can be viewed on the local level or on a smaller scale (microclimate) within the forest ecosystem. Not only does the microclimate affect the organisms that are present, but the process is mutual: Organisms will influence the microclimatic conditions as well. For example, the air temperature above bare ground and open grassland may differ significantly owing to differences in water loss and the shading effect of the vegetation.

### Procedure

1. *Local climate.* For general climate patterns in your area, consult the local climatological data sheets provided. You should record in Table 27.7 the average annual rainfall for this month, temperature averages, and the relative humidity at 1300 hours (1 P.M.).

2. *Microclimate.* Variation in microclimate due to such factors as elevation, slope, and shade can result in temperatures, humidities, and light intensities quite different from those of surrounding areas only a few meters away. For example, the microclimate conditions near ground level will be different from those a few meters above the surface. Measure the following microclimate conditions and record your results in Table 27.7.

   a. *Air temperature.* Record the temperature at the soil surface and 2 m or so above the ground for a sunny spot and a shady spot. Allow 5 to 10 minutes for the thermometer to equilibrate at each position. Record the minimum and maximum temperature for the previous 24 hours if a min/max thermometer was left in the forest overnight.

*Have students measure the light intensity of full sun before entering the forest.*

   b. *Light.* Using a light meter, measure the light intensity in the forest and in full sunlight. Calculate the percentage of full sun that reaches the forest floor.

   c. *Humidity.* Humidity refers to the amount of water vapor in the air. It has important biological effects on respiration, transpiration, and evaporation. **Relative humidity** (the most common measurement)

**Table 27.7**
Characteristics of the Physical Environment

| Locality: _____   Plot ID #: _____   Date: _____ |
| --- |
| Students: _____   Instructor: _____ |

**Soil**

   Temperature (°C):

   pH:

   Other:

   Description:

**Local Climate**

   Annual rainfall:

   Annual average temperature:

   Relative humidity:

**Microclimate**

   Temperature

      Soil surface: sun: _____ shade: _____

      2 m above surface: sun: _____ shade: _____

      Min _____

      Max _____

   Light

      Forest _____

      Full sun _____

      % full sun _____

   Relative humidity _____

   Wind speed _____

**Topography**

**Disturbance**

**Comments**

is the actual amount of water vapor in the air divided by the total possible water vapor (or saturation vapor) in the air at its temperature. Refer to the directions provided and measure relative humidity with a **sling psychrometer.** Do not hit anyone with the psychrometer!

d. *Wind.* Using the **wind anemometer,** measure the wind speed in the sampling area. What factors can you suggest that affect wind speed?

### Results

Record in Table 27.7 the results of your measurements for air temperature, light, humidity, and wind.

## Field Study C. Topographic Features

### Materials

field notebook
pencil
USGS topographic map for study site

### Introduction

Before completing your study of the abiotic components, you should stand back and observe the general features of the landscape that are difficult to measure but that may influence other physical factors and the organisms as well.

### Procedure

Observe the topography of the sampling site. Is the area sloping, eroded, on a hillside, or cut by a stream or ditch? Record in Table 27.7 indications of past or present disturbance—for example, scarring from past fires, cut stumps, old fences, or terracing.

### Results

1. Record the results of your observations in Table 27.7.
2. In the margin of your lab manual, sketch the topographic features that appear to influence your sample site.

EXERCISE 27.3
# Data Analysis

## Materials

summary of all student data
calculator

## Introduction

Students will need to pool data with team members and other teams. You may determine the abundance (density or cover), distribution (frequency), overall size (dominance), and importance (the sum of relative density, frequency, and dominance) for each species in the sampling categories (trees, SSV, and so on).

**Density,** the number of individuals per unit area, provides a summary of abundance by species. (**Percent cover** is the measure of abundance used for seedlings and herbaceous plants.) However, the density of a species does not necessarily reflect the distribution of the species on the landscape. **Frequency** provides information on the distribution of a species and is calculated as the percent of plots sampled that have at least one individual of the species present. **Dominance** is a measure of the influence of a species based on the size of individuals. (Dominance is determined from the area, called basal area, calculated from DBH measurements.) Adding the relative values for each of these measures provides an estimate of overall **importance,** which includes abundance, distribution, and size. The equations for calculating these parameters are provided in Table 27.8 on the next page.

*Sample calculations are provided in the Teaching Plan at the end of the lab topic. The calculations assume that more than one plot has been sampled. The total area sampled is the sum of the areas for all plots. If four 100-m² plots were sampled, the total area sampled would be 400 m².*

## Procedure

### Trees, Shrubs, Saplings, and Vines

1. If circumference was measured for trees and shrubs, saplings, and vines, then calculate DBH for individuals in each sample plot and record the results in Tables 27.3 and 27.4, respectively.

2. Calculate basal area for each tree and shrub, sapling, and vine. Record your results in Tables 27.3 and 27.4, respectively.

3. Pool the data for all sample plots and record for each species the total number of individuals, number of plots in which a species was present, and total basal area. Summarize tree data in Table 27.9, SSV data in Table 27.10.

4. Calculate density. Total the densities for all species and calculate the relative density of each species.

5. Calculate frequency. Total the frequencies for all species and calculate the relative frequency of each species.

6. Calculate dominance. Total the dominance for all species and calculate the relative dominance of each species.

7. Calculate importance value by totaling the relative density, relative frequency, and relative dominance for each species.

$$A = \pi \left(\frac{d}{2}\right)^2$$

$$= \pi \frac{d^2}{4}$$

$$= \frac{3.1416}{4} d^2$$

$$= 0.7852 \, d^2$$

**Table 27.8**
Calculating Density, Frequency, Dominance, and Importance Values for the Biotic Components of the Ecosystem

| | |
|---|---|
| **DBH** $= \dfrac{\text{circumference}}{\pi}$ | **Basal area** $= 0.7854 \, (\text{DBH})^2$ |
| **Dominance** $= \dfrac{\text{total basal area}}{\text{total area sampled}}$ | |
| **Relative dominance** $= \dfrac{\text{dominance for a species}}{\text{total dominance for all species}} \times 100$ | |
| **Density** $= \dfrac{\text{no. of individuals}}{\text{total area sampled}}$ | |
| **Relative density** $= \dfrac{\text{dominance for a species}}{\text{total density for all species~norm}} \times 100$ | |
| **Frequency** $= \dfrac{\text{number of plots in which species recorded}}{\text{total number of plots sampled}}$ | |
| **Relative frequency** $= \dfrac{\text{frequency for a species}}{\text{total frequency for all species}} \times 100$ | |
| **Average percent cover** $= \dfrac{\text{total percent cover}}{\text{total number of plots sampled}}$ | |
| Importance **Value** = relative density + relative dominance + relative frequency | |

### Seedlings and Herbs

1. Pool the data for all sample plots and record in Table 27.11 the percent cover for each species and the number of plots in which a species was present.
2. Calculate average percent cover.
3. Calculate frequency. Sum the frequencies for all species and calculate the relative frequency of each species.

### Macroinvertebrates

1. Pool the data for all sample plots and record in Table 27.12 the total number of individuals for each species and the number of plots in which a species was present.
2. Calculate density. Sum the densities for all species and calculate the relative density of each species.
3. Calculate frequency. Sum the frequencies for all species and calculate the relative frequency of each species.

### Abiotic Components

Pool the data for all sample plots and record it in Table 27.13.

## Results

1. Record all summary results for *trees* in Table 27.9 and those for *shrubs, saplings,* and *vines* in Table 27.10.

2. List in the spaces provided the three most important tree species and the three most important shrub, sapling, and vine species.

   *Trees*:

   *Shrubs, saplings, and vines*:

3. Record the summary results of *seedlings* and *herbs* in Table 27.11.

4. List in the space provided the three most common species based on their abundance.

   *Seedlings and herbs*:

5. Record summary results for *macroinvertebrates* in Table 27.12.

6. List in the space provided the three most common macroinvertebrate species based on abundance. Indicate, if possible, the appropriate trophic level: *primary* (1°) or *secondary* (2°) *consumers,* or *detritivores* (D) by placing the appropriate letter or number by each. To make these determinations, observe mouthparts or other body structures. Consult reference books or handouts provided in the laboratory or from the library. See "Who Eats What" (Hogan, 1994).

   *Macroinvertebrates*:

**Table 27.9**
Summary of Results for **Trees**

| Locality: _____ | Size class: Trees | Date: _____ |
| No. of plots sampled: _____ | Total area sampled: _____ | |

| Species | Total No. of Individuals | Density | Relative Density | No. of Plots Present | Frequency | Relative Frequency | Total Basal Area | Dominance | Relative Dominance | Importance Value |
|---------|--------------------------|---------|------------------|----------------------|-----------|--------------------|-----------------|-----------|---------------------|------------------|
|         |                          |         |                  |                      |           |                    |                 |           |                     |                  |
| Totals  |                          |         | 100              |                      |           | 100                |                 |           | 100                 |                  |

**Table 27.10**
Summary of Results for **Shrubs, Saplings,** and **Vines**

| Locality: | Size class: SSV | Date: |
| --- | --- | --- |
| No. of plots sampled: | Total area sampled: | |

| Species | Total No. of Individuals | Density | Relative Density | No. of Plots Present | Frequency | Relative Frequency | Total Basal Area | Dominance | Relative Dominance | Importance Value |
| --- | --- | --- | --- | --- | --- | --- | --- | --- | --- | --- |
| | | | | | | | | | | |
| | | | | | | | | | | |
| | | | | | | | | | | |
| | | | | | | | | | | |
| | | | | | | | | | | |
| | | | | | | | | | | |
| | | | | | | | | | | |
| | | | | | | | | | | |
| Totals | | | 100 | | | 100 | | | 100 | |

**Table 27.11**
Summary of Results for **Seedlings** and **Herbaceous Vegetation**

Locality: _____   Size class: S/Herb  Date:_____

No. of plots sampled: _____   Total area sampled: _____

| Species | Average Percent Cover | No. of Plots Present | Frequency | Relative Frequency |
|---------|-----------------------|----------------------|-----------|--------------------|
|  |  |  |  |  |
|  |  |  |  |  |
|  |  |  |  |  |
|  |  |  |  |  |
|  |  |  |  |  |
|  |  |  |  |  |
| **Totals** |  |  |  | 100 |

**Table 27.12**
Summary of Results for **Macroinvertebrates**

Locality: _____   Size class: Macroinvert.  Date:_____

No. of plots sampled: _____   Total area sampled: _____

| Species | Total No. of Individuals | Density | Relative Density | No. of Plots Present | Frequency | Relative Frequency |
|---------|--------------------------|---------|------------------|----------------------|-----------|--------------------|
|  |  |  |  |  |  |  |
|  |  |  |  |  |  |  |
|  |  |  |  |  |  |  |
|  |  |  |  |  |  |  |
|  |  |  |  |  |  |  |
|  |  |  |  |  |  |  |
| **Totals** |  |  | 100 |  |  | 100 |

7. List in the space below the five most common *microinvertebrates* observed in all plots. Indicate, if possible, the appropriate trophic level: *primary* (1°), *secondary* (2°), or *tertiary* (3°) *consumers*, or *detritivores* (D) by placing the appropriate letter or number by each. To determine trophic level, observe mouthparts and other body structures. Consult reference books or handouts provided in the laboratory or from the library. See "Who Eats What" (Hogan, 1994).

   *Microinvertebrates*:

8. What types of microorganisms grew in your agar plates? Did you observe fungal hyphae (filaments)? Did you observe bacterial colonies? Describe them briefly. (Refer to Lab Topic 13 Bacteriology.)

9. Record the summary results for all *abiotic* components in Table 27.13.

10. List the abiotic features that appear to be important influences in the ecosystem.

11. Describe any summary information not included in the tables.

### Table 27.13
Summary of Results for **Physical Environment**

| | Plot | | | | |
|---|---|---|---|---|---|
| | 1 | 2 | 3 | 4 | 5 |
| **Soil** | | | | | |
| **Description** | | | | | |
| **pH** | | | | | |
| **Temperature** | | | | | |
| **Climate** | | | | | |
| **Annual rainfall** | | | | | |
| **Annual avg. temp.** | | | | | |
| **Microclimate** | | | | | |
| **Temperature** | | | | | |
| **°C soil surface** | | | | | |
| **°C 2 m above** | | | | | |
| **°C min** | | | | | |
| **°C max** | | | | | |
| **Light intensity** | | | | | |
| **Forest** | | | | | |
| **Full sun** | | | | | |
| **% full sun** | | | | | |
| **Relative humidity** | | | | | |
| **Wind speed** | | | | | |
| **Topography** | | | | | |
| **Comments** | | | | | |

**Figure 27.7.**
**Profile diagram of a typical forest system.** Using your results, label the
important species in each of the vertical layers of the forest, from trees to herbs
on the forest floor.

## Discussion

1. Prepare ecosystem profiles. Using the results from your study of the forest, label the profile diagram provided in Figure 27.7. Label the trees, shrubs, saplings, vines, and herbs to illustrate the species composition of each vertical layer of the forest ecosystem. If you are not investigating a forest system, construct a profile diagram that will illustrate the specific composition and patterns of your study site. Refer to Figure 27.7.

2. Incorporating all components of the ecosystem, complete the model of the ecosystem in Figure 27.8, indicating the trophic levels and interactions observed in the forest. Provide examples for each trophic level, from producers to top carnivores to decomposers.

3. Write a one-page discussion of your results. Characterize the forest by vegetation type. Describe trophic levels and provide examples, using your model as an illustration. Propose features of the physical environment that appear to influence or be influenced by the biotic community.

*If this lab falls at the end of the term, you may choose not to assign a paper, but students' preparation of a one-page discussion section will help them pull together information from many sources and experiences. See the Teaching Plan for suggestions for organizing and evaluating this exercise.*

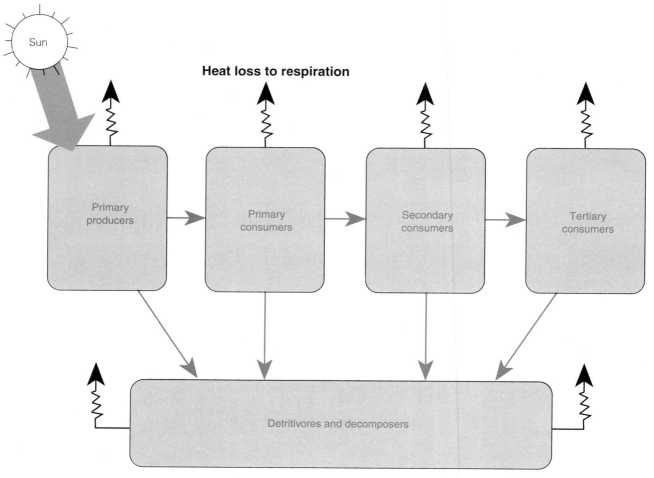

**Figure 27.8.**
**Compartmental model for trophic relationships in an ecosystem.** Heat losses to respiration are represented by zigzag arrows. Label the trophic levels and add arrows to represent energy flow from trophic level to trophic level. For each trophic level, list examples from organisms observed in this laboratory.

# Reviewing Your Knowledge

1. What are the factors that determine the importance of species in the tree category?

   *The tree category will usually have fewer but larger individuals com-pared with other categories; therefore, relative dominance deter-mines importance in the tree category.*

2. Are the same species important in the tree and the shrub, sapling, and vine categories?

   *If yes, then the community is self-replacing and probably relatively stable. If no, then the community may be early successional, and the pioneer species will be replaced by more tolerant species that are forming the understory of saplings. Some species (for example, dogwood) are restricted to the understory and would not be present in the tree class.*

3.  What are the advantages and disadvantages of using cover as a measure of abundance for herbaceous plants?

    *Cover is a better measure when plants occur in high densities or are connected as vegetative clones. The disadvantage is that cover is less accurate, and estimates can vary from investigator to investigator.*

4.  Which group of consumers, primary or secondary, had the highest density?

    *primary consumers*

5.  What are the top carnivores in this ecosystem?

    Explain the relatively low density of carnivores (for example, hawks, owls, or snakes) in this ecosystem.

    *Primary consumers and detritivores are generally more abundant. Students will observe few carnivores, particularly vertebrates. Only 10% on average of the energy of one trophic level will be passed on to the next trophic level. Therefore, there is limited energy available to support tertiary or higher consumers.*

6.  Do your results from this ecosystem analysis adequately represent the forest ecosystem you studied? Explain.

## Applying Your Knowledge

1.  To investigate the structure and function of ecosystems, ecologists may construct a microcosm using organisms and materials from the ecosystem. Properly constructed, these model systems should be self-sustaining. If you remove the primary producers from the microcosm, would you predict that your model would continue to be self-sustaining? Explain.

    *Ecosystems are maintained by a constant input of solar energy, which is converted to chemical energy by producers carrying out photosynthesis. No producers means no energy for the ecosystem.*

    If you remove the decomposers and detritivores, would the microcosm be self-sustaining? Explain.

    *This system would not remain for long. Organic matter from death, loss, and excretion would accumulate in the ecosystem. Many nutrients would be tied up and unavailable to the community. The recycling of nutrients by decomposers is necessary for continued growth and reproduction of organisms in the ecosystem.*

2. Using your knowledge of ecosystem structure and function, compare the trophic structure of a desert to that of a temperate hardwood forest. Include the relative number of organisms and energy availability for the different trophic levels.

   *Few plants can survive desert conditions. Therefore, in the desert the producer level will have lower density and less available energy compared to that of the forest. Less energy at the producer level means fewer consumers can be supported at each subsequent level.*

3. Sudden oak death, the rapid decline of red oaks and tan oaks, was first described for central California forests in 1994. The disease is caused by the pathogen *Phytopthora ramorum,* which infects and kills tan oak seedlings and trees. However, mature red oaks are more susceptible, developing large lesions on their stem, with clusters of trees rapidly dying. The death of saplings could have very different effects on the forest community compared with the death of mature trees. How might forest structure and regeneration be affected by the loss of mature oaks that dominate the canopy? How might the forest ecosystem be affected by the loss of young saplings?

   *The loss of canopy species would result in rapid and dramatic changes. Large openings in the canopy would change temperature and moisture conditions, while increasing light and space. Trees, saplings, and shrubs capable of rapid growth in response to changing environmental conditions would increase in the openings, changing the species composition of the forest. Red oaks would decline rapidly with little recruitment from saplings.*

   *The loss of saplings would result in gradual changes over a longer period of time because the dominant canopy would not be affected dramatically. However, the forest understory would be more open, and no new tan oaks would be recruited into the canopy. As older trees die, they would be replaced by other species.*

4. Wolves did not inhabit Yellowstone National Park from the 1930s until 1995, when 14 gray wolves were introduced by the National Park Service and U.S. Fish and Wildlife Service. In 2004, there were 16 packs with about 10 wolves per pack. Some of the changes in the ecosystem were expected, for example the decline of elk, a major food source for the wolves. However, changes in other animals and vegetation over 10 years provide a study of trophic structure and ecological connections. Between 1930 and 1990, there was no growth of young aspen, cottonwood, and willows along creeks, but now young saplings are thriving. No beavers were seen in these areas, but they have now returned. Draw a simple model of interactions among these components in the Yellowstone ecosystem, including wolves, elk, aspen, and beavers.

Explain how the introduction of a top carnivore could affect vegetation in an ecosystem.

*Aspen saplings* → *Elk* → *Wolves*

*Beavers*

*As carnivores, the wolves changed the number of elk and where they feed in the park. The result has been an increase in primary producers: aspen, willow, and cottonwood. The increase in aspen and willow saplings has encouraged the return of beaver to the park by providing them with food and shelter.*

5. Design a sampling regimen to answer the following question: Are there differences in the abundance and species composition of microinvertebrates found in the tree litter from forests of different ages? Include the types of plots, organisms, and physical features to be measured and the selection of sample sites.

*Sample the tree category in 100-m² circular plots to determine forest type and size of trees, since size is roughly related to age. For each tree plot, sample one or two 0.5-m² plots for microinvertebrates, extracting the organisms using Berlese-Tullgren funnels. Physical factors that might be sampled include soil pH, relative humidity, and percentage receiving full sun.*

## OPTIONAL EXERCISE 27.1
# Biotic Components: Weedy Fields

If a forest system is not accessible for study, small and more numerous plots can be used to study weedy fields, which are ubiquitous in both urban and rural environments. The scale for environmental factors can also be reduced and will require sampling at intervals that correspond to the smaller plot sizes. If the weedy field has woody vegetation, then refer to the plot sizes recommended earlier for the forest ecosystem. If only a few woody plants are present, consider recording the presence of these in the general description of the study site, but do not sample them. If the field has only seedlings or nonwoody (herbaceous) vegetation (or both), then use 0.50-m² plots to sample the vegetation and invertebrates.

*This is only one option for using this lab topic. See the Teaching Plan for suggestions on ways to use only parts of the lab topic and to organize students and select study sites.*

It may be necessary to modify your field studies to correspond to the ecosystem being studied.

Students should work in teams of four: two for seedlings, herbaceous plants, and macroinvertebrates, and two for microinvertebrates and abiotic factors (optional). A class of 24 could sample 6 to 12 plots.

*See Exercise 27.1, Field Study A, for the instructor's note regarding identification of organisms.*

### *Suggested Outline for Weedy Fields*

Exercise 27.1, Biotic Components
Field Study C. Seedlings and Herbaceous Plants
     D. Macroinvertebrates
     E. Microinvertebrates
     F. Microorganisms
     G. Other Forest Animals

Exercise 27.2, Abiotic Components (optional)
Field Study A. Soil
     B. Climatology

Exercise 27.3, Data Analysis

## Student Media: BioFlix, Activities, Investigations, and Videos
www.masteringbiology.com (select Study Area)

**Activities**—Ch 52: Terrestrial Biomes; Ch. 54: Interspecific Interactions; Food Webs; Primary Succession
**Investigations**—Ch. 30: How Are Trees Identified by Their Leaves? Ch. 52: How Do Abiotic Factors Affect Distribution of Organisms? Ch. 54: How Are Impacts on Community Diversity Measured?
**Videos**—Discovery Videos: Ch. 52: Trees

## References

Boyce, R. "Life Under Your Feet: Measuring Soil Invertebrate Diversity," *Teaching Issues and Experiments in Ecology,* Volume 3, Ecological Society of America, 2005. Downloadable soil invertebrate key accessed at http://tiee.ecoed.net/vol/v3/experiments/soil/downloads.html.

Brower, J. E. and J. H. Zar. *Field and Laboratory Methods for General Ecology.* Dubuque, IA: William C. Brown, 1972. This edition is more comprehensive than later editions.

Cain, M., W. E. Bowman, and S. D. Hacker. *Ecology.* Sunderland, MA: Sinauer Associates, 2008.

Cox, G. W. *Laboratory Manual of General Ecology,* 8th ed. Columbus, OH: McGraw Hill, 2001.

Gurevitch, J., S. M. Seheiner, and G. A. Fox. *Ecology of Plants.* Sunderland, MA: Sinauer Associates, Inc., 2002.

Hogan, K. "Who Eats What." *Eco-Inquiry,* Appendix A. Dubuque, IA: Kendall-Hunt Publishing, 1994, pp. 355–382.

Kingsolver, R. W. *Ecology on Campus.* San Francisco, CA: Benjamin Cummings, 2006.

*National Audubon Society Regional Guides.* This series provides an overview of ecosystems and natural areas by region: Pacific Northwest, Southeastern, Southwestern, New England, Rocky Mountain, California, Florida, etc.

Rizzo, D. M. and M. Garbelotto. "Sudden Oak Death: Endangering California and Oregon Forest Ecosystems." *Frontiers in Ecology and the Environment,* 2003, vol. 1(5), pp. 197–204.

Robbins, J. "Lessons from the Wolf." *Scientific American.* 290(6):76–81, 2004.

Vodopich, D. *Ecology Laboratory Manual.* Dubuque, IA: McGraw Hill, 2009.

Waterman, M. and E. Stanley. *Biological Inquiry: A Workbook of Investigative Cases.* San Francisco, CA: Pearson Benjamin Cummings, 2011.

## Websites

EPA Ecosystems web page, Terrestrial and Aquatic Ecosystems, Watersheds in your area of U.S.: http://www.epa.gov/ebtpages/ecosystems.html

Finding and ordering USGS topographic maps: http://topomaps.usgs.gov/ordering_maps.html

*Soil Biology Primer* [online]. Available: http://soils.usda.gov/sqi/concepts/soil_biology/soil_food_web.html

Tree of Life Project includes current information on relationships of all major groups of living organisms: http://tolweb.org/tree/phylogeny.html

The world's biomes (and interesting information on all organisms): http://www.ucmp.berkeley.edu/exhibits/biomes/index.php

## LAB TOPIC 27

# Ecology I: Terrestrial Ecology
## Teaching Plan for Laboratories

## Introduction

The laboratory experience is greatly enhanced when we expose students to the vast array of biological phenomena and approaches encountered when exploring the living world. Field experience enhances students' observational skills and their understanding of the hierarchy of life. Three problems often prevent instructors from doing a field ecology laboratory topic: (1) location of a suitable field study area, (2) transportation of students to the study area, and (3) identification of local plants and animals.

### Location of a Field Study Area

This lab is designed for a forest ecosystem (a complex ecosystem). However, by omitting the larger vegetation, it is possible to apply this lab to a variety of study areas. The option is provided in the exercise for a weedy lot or field, which can be located in either urban or rural areas.

Ask an ecologist, botanist, or local naturalist to assist you in locating a suitable site. Be on the lookout for a field on or near your campus that might not be mowed for a substantial period of time before the lab (the longer, the better). Solicit the assistance of the grounds/maintenance crews in establishing a study area with a limited mowing schedule.

Students appreciate participating in a continuing investigation or research project. This is possible in ecology if the study area will be available year after year for continued sampling. A number of long-term projects can be initiated. Each year, a portion of the study area can be sampled by students and data accumulated for several years. Changes in species composition (succession) can be studied by comparing data over 5 or more years. If written and photographic records of natural and human disturbance are maintained, students can study the response of an ecosystem to these factors.

For multiple sections, the study area should be large enough to allow teams of students to sample plots along lines or transects through the study area. Avoid having students sample the same plots on consecutive days. Alternatively, two sites that differ for a key factor (length of time since mowing or other disturbance) could be chosen and the results compared.

### Transportation

Ideally, a study area would be near campus, so that the second problem, transportation, could be minimized. See suggestions for transportation of students in the section titled Field Laboratory Management.

### Identification of Local Plants and Animals

Few biologists are experts in every area of biology, including the identification of plants and animals. Locate a botanist, ecologist, or naturalist to assist with plant identification. If your institution offers courses in invertebrate or plant identification, solicit the assistance of students from those courses. They will be pleased to share their expertise, as will graduate students and colleagues. The Prep Guide includes a bibliography of field guides for many regions of the United States, plus suggestions for local sources, such as the Forest Service and county Extension Service.

In the field, identification can be greatly simplified by creating a field guide specific for your study area. One inexpensive and efficient method of doing this is to collect examples of leaves from the most commonly observed plants in the sample sites. Use a photocopier to make black-and-white outline copies of each plant, and label each sheet. Make an identification booklet specific for your study site. Members of your department or a local landscaper or forester can help with identifications if necessary. Any unknowns can be numbered and bagged for later identification, although simply identifying species by number is adequate as long as they are not the most abundant species.

## Main Concepts and Objectives

1. Concept: sampling and measurement of biotic components of an ecosystem. Students will sample and measure vegetation, consumers, and decomposers. They will calculate density, cover, frequency, dominance, and importance value for species in the ecosystem studied. (Exercise 27.1 and Exercise 27.3)

2. Concept: measurement of abiotic components of an ecosystem. Students will measure physical environmental factors that appear to influence the abundance and distribution of organisms. (Exercise 27.2 and Exercise 27.3)

3. Concept: ecosystem structure and function. Students will describe and provide examples of the trophic levels in the ecosystem. (All exercises)

4. Concept: models. Students will use a profile diagram and compartmental model to describe the ecosystem studied. (Exercise 27.3)

5. Concept: discussion of results. Students will discuss their results in written form. (Exercise 27.3)

6. Specific content. Students will be able to define and describe: *ecosystem, trophic levels (primary producer, consumer, detritivore, decomposer), herbivore, carnivore, omnivore, abiotic and biotic components, sample plot, density, cover, frequency, dominance, importance value, profile diagram, compartmental model.*

## Order of the Lab

*Before lab,* select a field site and prepare a short description of the site or provide questions to guide students in writing their general description of the field site. Locate and mark the center of the sample plots. The center post can be installed, especially if these are permanent plots. Prepare a map or arrange for transportation. Prepare a local field guide by collecting and identifying common plants. See the preceding notes on identification.

1. Assemble at the field site and introduce the
   study area and problem. (15–30 min)

2. Organize teams and describe the tasks
   and materials for each field study. (15 min)

3. Sample the study area. (60 min)

4. Pool data and discuss calculations. (30 min)

5. Some students will return to the lab and set up
   Berlese-Tullgren funnels for extracting microorganisms.
   (This may be done by the instructor.) (15–30 min)

At this point, you can assign students to complete calculations outside of lab or you can use all or part of a second lab period to complete calculations and develop models.

6. Observe microinvertebrates and microorganisms. (30 min)

7. Complete calculations and analyze data. (45 min)

8. Construct a profile diagram and model of the ecosystem. (15 min)

9. Assignment: Prepare written discussion of results; or, if you are using a second lab period, have groups report on their results.

**For a single 3-hour lab:** To complete the lab in one lab period, complete the extraction of microinvertebrates (Exercise 27.1, Field Study E) and the culture of microorganisms (Exercise 27.1, Field Study F) before lab. These materials, along with hand lenses or a field scope, can be taken to the sample site as a demonstration, or have students view the demonstration during the week following lab. After step 4 in Order of the Lab, describe the calculations and discuss the profile diagram and compartmental model. Even without having completed the calculations, students will have a feel for the forest and can discuss these models. Students complete Exercise 27.3, Data Analysis, on their own.

**For a 2-hour lab:** Students will need a written description of the study site before lab. Complete the sampling, but do not pool data. Collect data sheets from student groups and provide a master copy for students the following week. The instructor will extract microinvertebrates. The second week, students view demonstrations of organisms and complete calculations. Models and preparation of the discussion are assigned.

**For night lab:** This laboratory is difficult to modify for use in night laboratories. Consider having students read the introduction and perhaps explore a local forest system through slides, illustrating the trophic structure with real examples. Plan a field trip outside of class to a local natural area. Follow this with a computer simulation exercise. See Lab Topic 28 Ecology II: Computer Simulations of a Pond Ecosystem.

See Optional Exercise 27.1 for alternate plans for this lab topic. Some require less time than others, depending on which field studies are included.

## Field Laboratory Management

Students should be organized into teams of six to eight students. Each team will sample one plot, thus providing three or four replicate samples, depending on the number of teams. Within teams, each student will have specific responsibilities:

Exercise 27.1, Biotic Components

Field Study A—two students

Field Study B—two students

Field Studies C, D, E—two students

Field Study F—students in Exercise 27.2 complete this field study

Field Study G—all students

Exercise 27.2, Abiotic Components—two students

Following the field sampling, students will share results with other teams, make calculations, pool and analyze data, and develop a model of the ecosystem. Students should read assignments for all groups.

Each team will have a box containing all sampling equipment and instructions needed by students. These can be grouped into smaller boxes or bags according to the field study. (See the Prep Guide.)

Transportation: The logistics of getting students, particularly large numbers of students, into the field can be daunting. The simplest solution for small numbers of students is to transport them in vans or buses. In general, the best solution is to locate a study site within reasonable walking distance, provide the students with a map, and have them meet you there at lab time. Again, the choice of a study area is crucial. Although the lab is written for a forest ecosystem, the basic framework and exercises are adaptable to the ubiquitous weedy field, which, if you are fortunate, will be within walking distance for students.

## Student Development

Students develop field skills, including observation of diversity, pattern of vegetation, and structure of ecosystems. Students practice organizational skills and teamwork. They also practice quantitative skills and the ability to analyze results. Students apply data to models. Students continue to develop skills of persuasion in discussing results.

## Lab Safety Precautions

Instruct students to take appropriate precautions.

1.  Have students advise the instructor of any allergic reactions, particularly to biting and stinging insects.
2.  Have access to a first aid kit adapted for field excursions.
3.  Check for poisonous and toxic plants, such as poison ivy or pokeweed.
4.  Wash skin that has been exposed to poison ivy with soap and cold water and avoid contact with clothing that has been in contact with the plants.
5.  If you live in an area affected by Lyme disease, wear long pants, carry insect repellent, and look for and remove ticks following the field lab.

## Discussion and Summary

Students should pool data and summarize results. Encourage students to think; more than one ecosystem model can be proposed. Students will write a discussion section of a scientific paper.

## Evaluation

Note the effort and contribution of students in the lab and in the field for a subjective grade. Students can be graded on written reports. Students can be tested on their knowledge of ecosystem structure and function, including organisms observed and trophic relationships.

## Alternate Plans for Investigations

This laboratory topic is designed to cover both biotic and abiotic components of an ecosystem and all trophic levels. The alternate study area, a weedy field, does not include the woody vegetation and calls for less time and effort from each team. For any study area, you might choose to sample only certain parts of the ecosystem, depending on time, the number of students, and other resources. The laboratory might focus on the macroinvertebrates and microinvertebrates found in a mowed field. The vegetation could be of limited interest, but depending on the physical environment and past disturbance, the consumers might be of interest.

Examples of plans using portions of the lab topic:

1. Biotic component only: Exercise 27.1, Field Studies A–F; Exercise 27.3 (Data Analysis)

2. Plants, soil, and climate: Exercise 27.1, Field Studies A, B, C; Exercise 27.2, Field Studies A, B; Exercise 27.3 (Data Analysis)

3. Plants and climate: Exercise 27.1, Field Studies A, B, C; Exercise 27.2, Field Study B; Exercise 27.3 (Data Analysis)

4. Herbaceous plants, soil, and climate: Exercise 27.1, Field Study C; Exercise 27.2, Field Studies A, B; Exercise 27.3 (Data Analysis)

5. Microinvertebrates: Exercise 27.1, Field Study E; Exercise 27.2, Field Study A; Exercise 27.3. Compare microinvertebrates from different study sites, or following different treatments (mowing, pesticides, irrigation).

6. For field studies A and B, omit DBH, basal area, and dominance, and use only density and frequency in the data analysis.

## Case Studies

Case studies utilize stories or scenarios to engage students in learning through critical thinking, problem solving, and research. Investigative case studies can connect problem solving in lecture and laboratory. Students are given a scenario that describes a situation or problem. They analyze the problem, develop questions, and research the problem. These case studies can be used to develop laboratory and field investigations and to prepare oral or written presentations. Students generally work in groups or teams dividing the workload, sharing ideas, and learning independently. There are many environmental case studies available on the Web that can be adapted for use by students, depending on interest, local environmental issues, or lecture topics. We suggest that you select a case study that relates to biotic or abiotic components of an ecosystem similar to one that you and your students may access for study using Lab Topic 27 Ecology I: Terrestrial Ecology. The case study would provide the "hook" for the laboratory project. Using selected field studies from this lab topic, students could then pursue questions raised in the case study.

One suggestion for using Lab Topic 27 with investigative case-based learning follows. For example, students might be interested in the effects of herbicide and pesticide treatment used by the college grounds crew on landscaped areas or athletic fields. Using a case to present the problem or even question prompts, students might hypothesize that soil organisms would be negatively affected by herbicide treatment. They could design a comparative study of pesticide treated and untreated fields, sampling macroinvertebrates (Field Study D), microinvertebrates (Field Study E), and microorganisms (Field Study F). (Refer to Lab Topic 1 Scientific Investigation for assistance in developing hypotheses, designing an experiment, stating predictions, and collecting and analyzing data.) See Lab Topic 28 Ecology II: Computer Simulations of a Pond Ecosystem, *Exercise 28.2, Experiment D, Computer Simulation of an Invasive Species* for an example of an investigative case study.

You may want to develop a case study based on environmental issues in your community. Using information from professionals, newspaper clippings, or government reports, you can develop a case. See the references below for ideas and suggestions for how to write and teach with case studies.

http://bioquest.org/BQLibrary/library_result.php This site provides background information on investigative case-based learning and some examples of case studies. Select and download module, ICBL.

http://bioquest.org/lifelines/index.html Margaret Waterman and Ethel Stanley have developed investigative cases for the BioQUEST Curriculum Consortium. This site provides references, background information, resources, and examples from Lifelines, intensive workshops for 2-year college faculty to develop case studies.

A supplement for Reece, J. et al., *Campbell Biology*, 9th ed. is Waterman, M. and E. Stanley. *Biological Inquiry: A Workbook of Investigative Cases*. San Francisco, CA: Pearson Benjamin Cummings, 2011.

# Data Analysis

Sample calculations for student results follow (Tables IA 27.1 and 27.2).

## Table IA 27.1
Sample Calculations for Student Results Corresponding to Table 27.3

Locality: _____ Plot ID #: _____ Plot size: ___100 m²___ Date: _____

Students: ____Sample Data_____ Instructor: _____

| **Species:** White Oak | | | **Species:** Pignut Hickory | | | **Species:** Tulip Poplar | | |
|---|---|---|---|---|---|---|---|---|
| Circumference (cm) | DBH (cm) $= \frac{\text{Circumference}}{\pi}$ | Basal Area (cm²) $= 0.7854\ (\text{DBH})^2$ | Circumference (cm) | DBH (cm) $= \frac{\text{Circumference}}{\pi}$ | Basal Area (cm²) $= 0.7854\ (\text{DBH})^2$ | Circumference (cm) | DBH (cm) $= \frac{\text{Circumference}}{\pi}$ | Basal Area (cm²) $= 0.7854\ (\text{DBH})^2$ |
| | 20 | 314.16 | | 10.5 | 86.59 | | 75 | 4417.88 |
| | 10 | 78.54 | | 20 | 314.16 | | 10.5 | 86.59 |
| | 75 | 4417.88 | | 15 | 176.72 | | 10.5 | 86.59 |
| | 37 | 1075.21 | | | | | | |
| | | | | | | | | |
| | | | | | | | | |

## Table IA 27.2
Sample Calculations for Student Results Corresponding to Table 27.9

Locality: ___Sample Data_____ Size class: Trees   Date: _____

No. of plots sampled: _____2_____   Total area sampled: __2 ×100 m² = 200 m²__

| Species | Total No. of Individuals | Density | Relative Density | No. of Plots Present | Frequency | Relative Frequency | Total Basal Area | Dominance | Relative Dominance | Importance Value |
|---|---|---|---|---|---|---|---|---|---|---|
| White Oak | 6 | $\frac{6}{200} = 0.03\ \frac{}{m^2}$ | 31.6 | 2 | 2/2 = 1.00 | 40 | 10,129.87 | 0.00506 | 52.65 | 124.25 |
| Pignut Hickory | 10 | $\frac{10}{200} = 0.05\ \frac{}{m^2}$ | 52.6 | 2 | 2/2 = 1.00 | 40 | 1,732.41 | 0.00087 | 9.05 | 101.65 |
| Tulip Poplar | 3 | $\frac{3}{200} = 0.015\ \frac{}{m^2}$ | 15.8 | 1 | 1/2 = 0.5 | 20 | 7,335.14 | 0.00368 | 38.30 | 74.1 |
| Totals | | 0.095 | 100 | | 2.50 | 100 | | 0.00961 | 100 | |

## Lab Topic 28
# Ecology II: Computer Simulations of a Pond Ecosystem

## Laboratory Objectives

After completing this lab topic, you should be able to:

1. Develop a computer model to investigate a pond ecosystem.
2. Calibrate the model using information from field investigations.
3. Determine the steady-state values for the model ecosystem.
4. Answer questions and test hypotheses using a computer model.
5. Evaluate the effects of disturbance on the model ecosystem.
6. Apply the results of computer simulations to predict the results in real ecosystems.

*For a 2-hour lab: Omit one or two simulations. Omit student reports and have students submit tables and graphs for one or more simulations.*

## Introduction

Ecosystems are difficult to investigate using the experimental method because they are so large and complex. Scientists are nevertheless studying ecosystem structure and function in a selected number of systems across the United States as part of the Long Term Ecological Research Network, effectively implementing experiments at the landscape level in places such as the Okefenokee Swamp in Georgia, Konza Prairie in Kansas, the hardwood forest of Coweeta Hydrological Station in North Carolina, North Temperate Lakes in Wisconsin, the Santa Barbara Coastal Preserve in California, and the urban ecosystem of Baltimore, Maryland. However, to understand these complex systems and to make predictions concerning their responses to disturbance (both natural and human), ecologists often depend on computer models that correspond to the ecosystem of interest. Information obtained in field investigations is used to determine the structure of the model, the appropriate interactions among components, and the actual values used in the model.

You have already worked with models in other laboratory topics, including the bead models of cellular reproduction and population genetics and the diagrammatic models in terrestrial ecology. Models generally provide a simplified view of the complex phenomenon of interest. In this lab topic, you will actually construct a computer model of an aquatic ecosystem, the pond. As you develop your model, you can simulate a variety of conditions, including the effects of disturbance and fishing.

If you have completed Lab Topic 27 Ecology I: Terrestrial Ecology, you should be able to construct a compartmental model of a forest ecosystem and develop a general model based on the trophic levels present in ecosystems.

**Table 28.1**
Definitions and Examples of Ecological Terms

| Terms | Definitions | Examples |
|---|---|---|
| Primary producer | | |
| Consumer | | |
| Trophic level | | |
| Ecosystem | | |
| Biotic component | | |
| Abiotic component | | |
| Food chain | | |

Before continuing this lab, complete Table 28.1, providing definitions and examples. Refer to Lab Topic 27 and your textbook if necessary.

## The Pond Ecosystem

*One way to increase the interest of your students is to use local examples. Since water quality is a concern in most communities, you should be able to obtain information about local ponds or lakes from government agencies responsible for natural resources. Fishery biologists are particularly knowledgeable and helpful. You might even include a short videotape, a list of fish native to your ponds and lakes, or other information concerning water resources. See the References for a video suggestion and the EPA website for aquatic ecosystems and watersheds.*

The primary producers in a pond are generally the algae floating in the surface layers and some plants growing at the pond edge, or margin. These autotrophic organisms convert the energy of sunlight into chemical energy stored in organic compounds. The producers, in turn, are consumed by primary consumers (herbivores such as aquatic invertebrates), which are ingested by secondary consumers (carnivores such as sunfish) and even tertiary consumers (carnivores such as bass) (Figure 28.1). The movement of energy through these systems is called **energy flow** (energy does not cycle), and at each trophic level, the amount of available energy is reduced. All organisms expend some energy in the activities of life, and this energy is dissipated as heat. The efficiency of transfer from one level to another is only about 10%.

The rate at which light energy is converted to chemical energy is called **primary productivity.** Productivity is measured as the amount of **biomass,** or organic matter, added to the system per unit area per unit time (for example, $kg/m^2/yr$). In tropical rain forests, for example, the primary productivity is 2.2 $kg/m^2/yr$; in temperate grassland, 0.6 $kg/m^2/yr$; and in lakes and streams, 0.25 $kg/m^2/yr$ (*Campbell Biology*, 9th ed.). **Secondary productivity** is the rate at which consumers and detritivores accumulate new biomass from organic matter that was consumed. In the computer model of the pond, the changes in each trophic level will be measured as biomass.

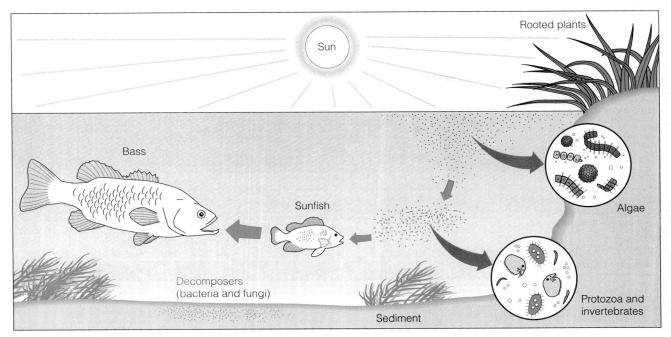

**Figure 28.1.**
**The food web of a pond ecosystem.** The primary producers (algae) are consumed by primary consumers (protozoa and invertebrates), which in turn are consumed by secondary consumers (sunfish and bass). In this pond, bass also may be tertiary consumers.

Refer to the food web in Figure 28.1, and in the margin of your lab manual, sketch a compartmental model of a pond ecosystem. (See also the example of a compartmental model in Figure 27.8.) Indicate connections between compartments (inputs and outputs).

## EXERCISE 28.1
# Computer Model of the Pond Ecosystem

The computer model of an ecosystem must be based on observations of the natural ecosystem in the field. Your compartmental model, which includes biotic components and interactions, represents the first step in constructing a model. You must also determine the biomass associated with each compartment based on data collected from field investigations. These data are used to **calibrate** the model—that is, to set the starting values for each component.

In this exercise, you will construct and calibrate a computer model for a pond ecosystem. You will run an initial simulation to determine the **steady-state,** or stable, values for your pond, then compare your results with those expected from a field study. In the process, you will practice changing components of the model, graphing, and printing your results. Once you have determined that your model adequately represents a natural ecosystem, you can begin to ask questions, formulate hypotheses and predictions, and test these using the model ecosystem. This process will continue as you adjust the model, analyze your results, and simulate conditions.

# Experiment A. Constructing and Calibrating the Model

## Materials

See BioQUEST manual for system requirements.

computer software (Environmental Decision Making)
computer
printer

## Introduction

Working in groups of three, using the computer software Environmental Decision Making, written by E. C. Odum, H. T. Odum, and N. S. Peterson for the BioQUEST project, you will develop a compartmental model for a pond ecosystem. You will calibrate the model and run the simulation to determine the steady-state values, the values at which the model stabilizes.

*The 2002 version of Environmental Decision Making is used in this lab topic. Environmental Decision Making is now available for free download at http://bioquest.org/ BQLibrary/library_result.php Both the CD and the online versions include user notes to assist with installation, opening the model, and running the simulations. Note that the Extend LT application must be downloaded to create and use the models in Environmental Decision Making.*

*Other software is available to construct a model, but it generally does not incorporate all the features of this program. See the References section for other models that are available. Note that we have not used these models with our students.*

The 1-hectare (= 2.47 acres) pond that you are modeling is inhabited by a variety of **pond life**, including such small organisms as algae, micro- and macroinvertebrates, plants, and animals and microbes that inhabit the pond bottom. Two species of fish are present: **sunfish**, which feed primarily off the pond life, and **bass**, which eat the sunfish. Each component of the ecosystem is represented by an icon (Figure 28.2).

## Procedure

Environmental Decision Making is available in both Mac and Windows™ versions. Depending on the version you are using, the screens may be slightly different.

1. **Open the program** by double-clicking on the icon Environmental Decision Making (EDM), then double-click on *Pond Worksheet.* (Double-click on *Pondwork* in the Windows version.) Click on the *Extend* screen. You should see an empty window with the plotter icon in the corner with the label "Quantity." Note: You can also open the model file from the Extend LT application by clicking on Open in the File menu.

*Students should have their own new disks or USB flash drives, so that they can save their models and simulation graphs for review at a later date.*

2. **Construct the model** to incorporate pond life, sunfish, and bass.

   a. Choose BioQUEST Library from the Library menu, and then choose BIOQULIB.LIX from the Library menu. Select *Sunlight* from the submenu.

   b. When you select an icon and hold down the mouse button, the pointer changes to a hand that allows you to move the icon. Move the sunlight icon to the left of the window.

   c. Choose *Pond Life* from the BioQUEST Library submenu and position it to the right of the sunlight icon.

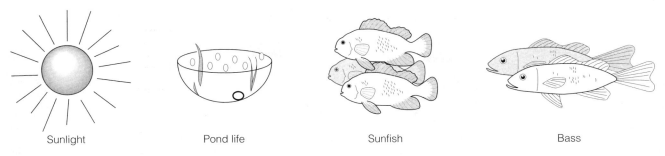

Figure 28.2.
**Icons for the components of the model ecosystem.**

d.  To connect the sunlight and pond life components, draw a line from one icon to another, connecting the small boxes on each icon. To do this, move the pointer over the small dark box by the sunlight icon. The pointer changes to a pen. Drag the pen from the dark box of the sunlight icon to the open box of pond life (Figure 28.3). Release the button. The solid line should connect the dark box of the sunlight icon (indicating flows out of this component) with the open box of the pond life icon (indicating flows into the second component). For example, sunlight provides the energy input to the ecosystem through pond life. Therefore, the dark box of sunlight is connected to the open box of pond life.

e.  Connect the pond life to the plotter (Figure 28.4) by drawing a line connecting the upper dark box of the pond life icon to the first open box on the plotter. The plotter can connect up to four components.

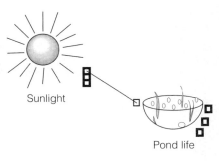

Figure 28.3.
**Drawing lines with the mouse to connect icons.** Position the mouse over the dark box, click and hold. A pen will appear, and you can draw a connecting line.

 To erase an icon, move the pointer to the icon, and a hand should appear. Select the icon and press *Delete*.

To erase a line, move the pointer to the line and click it to select it. Press *Delete*.

f.  Choose *Sunfish* from the BioQUEST Library and position it to the right of the pond life icon. Connect the open box of the sunfish icon to a dark box of the pond life icon. Connect the upper dark box of the sunfish to the second box of the plotter.

g.  Choose *Bass* from the BioQUEST Library and position the icon to the right of the sunfish icon. Connect one of the open boxes of the bass icon to one of the dark boxes of the sunfish icon. Connect the

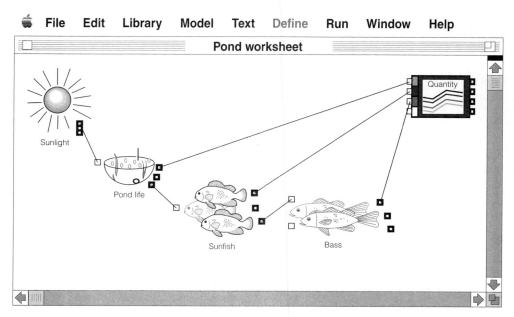

**Figure 28.4.**
**An example of the model ecosystem.**

upper dark box of the bass icon to the third box of the plotter. Beginning on the left side of the screen, you should have sunlight connected to pond life; pond life should in turn be connected to sunfish, which consume the pond life. Finally, the bass should be connected to the sunfish. All the organisms should be connected to the plotter (Figure 28.4). What trophic level is represented by the bass?

*in this model, tertiary consumers*

3. **Calibrate the model** by determining the starting values for the components of the ecosystem: sunlight, pond life, sunfish, and bass.

   a. Refer to the map of solar radiation for the United States (Figure 28.5). Determine the solar energy for your location in kilocalories per meter squared. If your location is not included on the map, ask your instructor for appropriate values to use.

   b. Double-click on the sunlight icon. A *dialog box* should appear on the screen. Calibrate the model by entering the appropriate energy values in the box and click on OK.

   c. Double-click on the pond life icon. Enter *1,000 kg/ha (kilograms per hectare) for your starting value.* Click on OK.

   d. Double-click on the sunfish icon and enter the total biomass of sunfish in kilograms in the dialog box. An appropriate starting value for sunfish might be 500 sunfish for a 1-ha pond. (*Assume 10 sunfish per kilogram, or 50 kg/ha.*) Click on OK.

   e. Double-click on the bass icon and enter the total biomass of bass in kilograms in the dialog box. *The starting value for bass might be 10 large bass, 1 kg each, or 10 kg/ha.* Click on OK.

   f. Record the starting values for all components in Table 28.2.

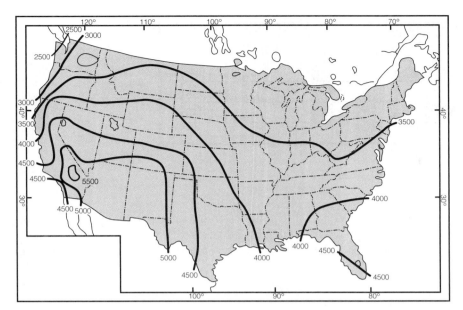

**Figure 28.5.**
**Solar radiation for regions of the United States.** Select the kilocalories of solar radiation for your location.

4. To **calibrate the plotter**, choose Simulation Setup from the Run menu. In the time dialog box that appears, change the time to end the simulation to 2 years (730 days). Enter *730* in the first box and click on OK.

5. **Run the simulation**: Choose Run Simulation from the Run menu. The simulation should run and the simulation graph, similar to the screen shown in Figure 28.8, should be plotted.

   a. Check the lines on the graph. There should be three, representing pond life, sunfish, and bass. If not, the icons on the model are not correctly attached to the plotter or each other. Return to the model by closing the Pond System window. Correct the model and run the simulation again.

   b. Check the axes of the graph. You can modify the graph and table, if needed, to accommodate all the components of the ecosystem and to scale axes appropriately. To change the values on the axes, click on the lower or upper value for the axis. Type in the new number and press *Enter*. To modify the labels, click in the space above the axis and type in the appropriate label. Press *Enter*.

   c. Label the axes for your graph. The controls for changing the features of the graph appear in a bar at the top of the screen. You may want to experiment with these in a later simulation. (Refer to Figure 28.6.) Click on the icon in the first box of the display bar. A palette showing the variables in the graph will appear. *Pond life* should appear in the first box, *Sunfish kg/ha* in the second, and *Bass kg/ha* in the third (Figure 28.7). Click and type in the box to change the variables. To read the small values for the sunfish and bass, you will use a different scale on the right axis (Y2). If the right axis is not selected, click on

Sets graph variables

**Figure 28.6.**
**Display bar for graph.** Select the graph icon to adjust the axes and labels for the simulation graph.

**Figure 28.7.**

**Example of palette for modifying graph variables.** Modify the axes and labels by entering information in this palette.

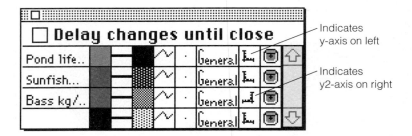

Indicates y-axis on left

Indicates y2-axis on right

the graph icon (eighth column) on the same row for Sunfish and Bass. The graph icons should now be reversed, indicating that the right axis is selected. Click on the Close box to make the variable palette disappear and return to the graph.

d. The graph will run again with the new graph features. Make additional changes as needed.

### Results

1. On the simulation graph, note the point at which biomass levels off for the ecosystem components. These various points indicate the steady state values for the energy levels used in your model, showing that the system has stabilized. *To read values from the graph, move the cursor to the point of interest (for example, the steady-state point) and read the corresponding biomass in the table at the bottom of the graph (Figure 28.8).* The top row of the table should show the time and biomass at the position on the graph indicated by the cursor line. By moving along the graph, you can follow the values on the top row, indicating increases, decreases, and relative stability. Other values and times are listed in the table.

2. Record the steady-state values and the time it took to reach steady state in Table 28.2.

**Figure 28.8.**

**Example of reading values from a simulation graph.** To determine the values at any point on the graph, move the cursor to the position. Read the values on the top row of the table below the graph. The cursor line is at 5,225 kg/ha of pond life on day 30.

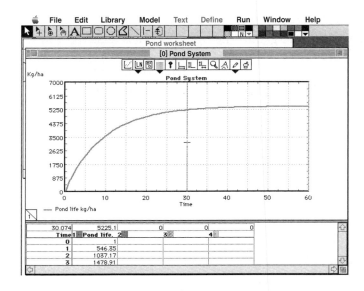

**Table 28.2**
Starting and Steady-State Values for Computer Simulation, Experiment A

| Component | Starting Value | Steady-State Value |
|---|---|---|
| Sunlight | *4,000* | *4,000* |
| Pond life | *1,000* | *~6,000* |
| Sunfish | *50* | *~100* |
| Bass | *10* | *~30* |
| **Time to steady state:** | | |

*Suggested starting values and resulting steady-state values.*

3. Choose *Print Plotter* from the File menu. If you want to print only the graph, then click on the *Top Plot Only* button. If *Plot Data Tables* is selected (a ✓ appears in the box), then click on this box to remove the table option. Click on *Print.*

4. Close the simulation graph window.

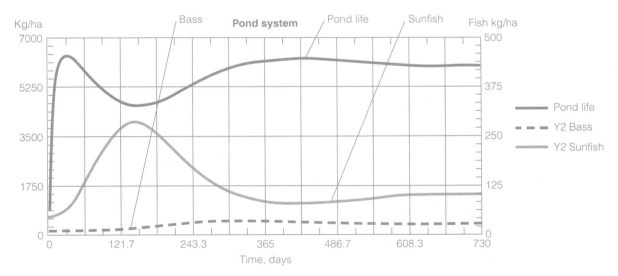

**Figure IA 28.1.**
**Simulation graph for the model ecosystem.** Sunlight is 4,000 kcal.

## Discussion

1. Describe the changes in the biomass of each component of the ecosystem over time.

   Pond life:

Sunfish:

Bass:

2. Did the highest and lowest values for the three components occur at the same time? Explain why or why not.

   *There should be some lag between decreases in one component and the next in the trophic structure. A loss initially in pond life would be expected to result in an increase in sunfish followed by a decrease.*

3. What do you predict would happen to the steady-state values if you increased or decreased the amount of sunlight available?

   *The steady-state values are dependent on the amount of energy available to the ecosystem from solar radiation. New steady-state values would occur with a change in energy input.*

# Experiment B. Steady-State Model

## Materials

computer software (Environmental Decision Making)
computer
printer

## Introduction

Your initial calibration was based on data from field observations and investigations. Having run the model, you should have estimates for the steady-state values. In this simulation, you will calibrate and run the model using the steady-state values from Experiment A.

## Hypothesis

State a hypothesis for changes in the components of the ecosystem at steady state.

   *At steady state the biomass for each of the pond components should remain at the same values.*

## Prediction

Based on your hypothesis, predict the appearance of the simulation.

   *If the components of the ecosystem are at steady state, each component will be represented on the simulation graph by a straight line.*

## Procedure

1. Calibrate the model. (Refer to Experiment A, Procedure step 3.) Double-click on each of the icons and set the starting value for each component at the steady-state value based on results in Table 28.2.
2. Record the new starting values in Table 28.3.
3. Run the simulation.

### Table 28.3
Starting and Steady-State Values for Computer Simulation, Experiment B

| Component | Starting Value | New Steady-State Value |
| --- | --- | --- |
| Sunlight | | |
| Pond life | | |
| Sunfish | | |
| Bass | | |
| Time to steady state: | | |

## Results

1. Compare the graph of this simulation with the results from the first simulation (Experiment A). If the starting values for this experiment are not similar to the ending steady-state values from the first simulation, check the starting value for each component. Rerun the simulation if necessary.

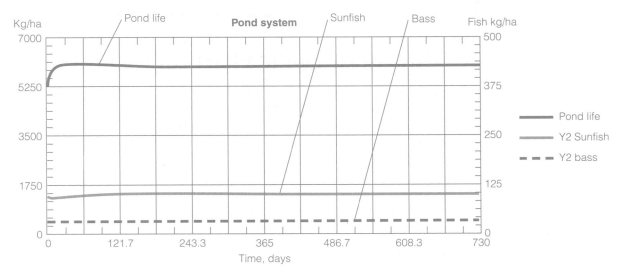

**Figure IA 28.2.**
**Simulation graph for steady state.** Sunlight is 4,000 kcal.

2. Record your results in Table 28.3.
3. Print the graph. Choose *Print* from the File menu.
4. Label each line on your graph.

### Discussion

1. Did your results match your predictions? If not, explain.

2. Were the values you set for this simulation actually at steady state? If not, consider using new steady-state values from this simulation and repeat the simulation.

3. The steady-state value for each component represents the biomass that can be sustained by this ecosystem. This is referred to as the **carrying capacity** for each component of the ecosystem. At carrying capacity, how many bass can be sustained by this pond system? (*Hint:* Refer to Experiment A; The values used in the model are biomass.)

   *Students should convert biomass to numbers based on information in Experiment A; 1 kg per adult bass. For this example, 30 kg/ha would mean 30 bass.*

4. Does the biomass increase or decrease as you move from producers to the secondary and tertiary consumers? Would you ever expect the pond biomass to be the same in all the components? Explain.

   *The biomass decreases as you move up the food chain. The biomass will never stabilize at equal values. Consumption is never 100%. At each level, energy is dissipated as heat, and decomposers receive part of the biomass.*

---

### EXERCISE 28.2

# Investigating the Effects of Disturbance on the Pond Ecosystem

### Materials

computer software (Environmental Decision Making)
computer
printer
CD or USB flash drive (provided by students)

### Introduction

In this exercise, you will discuss questions that interest your group and design experiments to determine the effects of disturbance on the components of the pond ecosystem. You will run a series of three simulations.

In groups of three, discuss with your teammates possible options for *simulating disturbance of the pond ecosystem.* Using Environmental Decision Making, you can change a number of factors. Consider the following options as you discuss your ideas.

1. What happens to biomass and steady-state values for a pond receiving more or less sunlight (see Figure 28.5)?
2. What would happen to each trophic level if more (or fewer) sunfish (or bass) inhabited the pond?
3. What changes would occur to the biomass of the sunfish and bass if another fish (e.g., gar) were added to the pond? (Gar eat sunfish, but not bass.)
4. What changes would you expect in each trophic level if sunfish (or bass) were eliminated due to a massive fish kill?

The questions you pose should have a biological basis; that is, they should be meaningful questions based on your understanding of ecosystems, particularly pond ecosystems. Just because you can change the model does not mean the question is appropriate. If necessary, consult your textbook or additional readings supplied by your instructor.

List the questions of interest to your group.

**Questions**

You will run three computer simulations of your own design. You may choose to pursue one or more questions, depending on your results. With a computer model, you can make any number of changes easily, but, depending on what you do, you may have difficulty in stating predictions and interpreting your results. *It is best to begin with one change at a time, increasing the level of complexity in subsequent simulations.*

In the second simulation (Experiment B), you will investigate the effects of fishing on the components of the pond ecosystem. In Experiment C, you are presented with a case study about the introduction of an exotic fish species. Your team will develop a model to investigate the effects of this invasive species on the pond trophic structure.

Be prepared to describe one of your questions, explain the model, and present your results at the end of the lab period.

# Experiment A. Open-Inquiry
# Computer Simulation of Disturbance

Choose one question that you want to pursue. Formulate a hypothesis and make a prediction for your computer simulation.

## Question

## Hypothesis

Formulate a hypothesis about the effects of disturbance on the pond ecosystem.

## Prediction

State a prediction based on your hypothesis. (This is an if/then statement that predicts the results of changes in the steady-state model in Exercise 28.1.)

Test your prediction using the computer simulation model.

## Procedure

1. In Table 28.4 on the next page, describe the disturbance problem you have chosen to simulate. Record the selected starting values for your model and indicate, by circling, the values that differ from the steady-state model.

2. Note any other changes in the model—for example, changes in interactions, solar radiation, or addition or removal of components.

3. Calibrate the model based on your changes. Modify the model according to your problem. You should be able to make changes in the model based on your experience in Exercise 28.1, Experiment A, Procedure step 2. For additional help, refer to the EDM manual provided with the software.

4. Run the simulation until steady-state values are reached. At steady state, the biomass for each component should change only slightly or not at all.

5. Modify the graph features if necessary.

**Problem:**

**Table 28.4**
Starting and Steady-State Values for Exercise 28.2, Disturbance Model

| Component | Starting Value | Steady-State Value |
|---|---|---|
| Sunlight | | |
| Pond life | | |
| Sunfish | | |
| Bass | | |
| Time to steady state: | | |

## Results

1. Describe the problem and then record the new steady-state values and the time to steady state in Table 28.4.
2. Print the simulation graph.
3. If instructed to do so, save your model on a disk or flash drive. Refer to the following instructions for assistance.

### To Save a Model

a. Insert your flash drive or new formatted disk. (Ask your instructor for assistance if your disk is not formatted.)

b. Click on *File* and pull down to *Save Model As*. Enter the name of the file, which might be, for example, your name and the experiment number (Stef1.A). Select the formatted disk. (Ask for assistance if the disk does not appear on the screen.) Click on *Save*. The file is saved on your disk.

c. If the file has been saved previously, click on *File* and *Save*. The current version of the file will be saved with the existing name. Save your files to a flash drive or disk for future use. You may choose to save each model for Experiments A to C with a different name. Ask your instructor.

*If students will be working with their models outside the lab period (see Teaching Plan), they should save their models as they work. If this lab is completed during the lab period, you may wish to have students omit this step; or you may want your students to save their models and turn in their disks for evaluation.*

**Discussion**

1. Describe the interactions among the trophic levels in the pond ecosystem.

2. Do your results match your predictions? Discuss any differences between results and predictions.

3. How did the disturbance simulated in this experiment affect the pond ecosystem? How might these results be applied to a natural ecosystem?

## Experiment B. Open-Inquiry Simulation of Fishing Activities

Although we often think of our human activities as being separate from natural ecosystems, we can have a dramatic effect on these systems. Ponds support not only carnivores such as bass, but also humans, who fish selectively. In this simulation, you will add fishing as a component to the model. Your objective might be to allow fishing at levels that will maintain a stable pond ecosystem. You might choose to challenge the system by increasing fishing on one or more species. You do not have to begin with the original steady-state model but can choose to add fishing to one of your previous models.

Consider what type of fish will be caught—sunfish, bass, or gar. Is it possible to catch more than one type of fish? Does fishing compound disturbance factors that you may have introduced in previous simulations? State the question of interest below.

## Question

## Hypothesis

State a hypothesis about the effect of fishing on the pond ecosystem.

## Prediction

State a prediction based on your hypothesis.

Test your prediction using the computer simulation model.

## Procedure

1. In Table 28.5, describe the problem you have chosen to simulate. Record the selected starting values for your model.
2. Record in the space provided the proposed changes to the model. What fish will be caught? Are any other interactions being changed?

**Problem:**

**Table 28.5**
Starting and Steady-State Values for Exercise 28.2, Fishing Model

| Component | Starting Value | Steady-State Value |
|---|---|---|
| Sunlight | | |
| Pond life | | |
| Sunfish | | |
| Bass | | |
| Fishing | | |
| Time to steady state: | | |

3. Modify the model. Select *Fishing* from the Library. Connect the open box for fishing to the dark box for the fish of choice. Connect the dark box for catch to the fourth plotter line.

4. Calibrate the model based on your changes. The level for fishing is set at 1 hour of fishing per day. Begin with this level. You may adjust the model according to your hypothesis.

5. Run the simulation until steady-state values are reached. At steady state, the biomass for each component should change only slightly or not at all.

6. You will need to modify the graph features. *Fishing* should now be displayed on the right (Y2) axis. Select the first box on the display bar over the graph. In the fourth row, type in *Fishing kg/ha/d* and select the right axis by clicking in the eighth column. Return to the graph and adjust the right axis values as needed. The right axis label is now *Catch/day* (for fishing) and *kg/ha* (for fish). Refer to the Procedure section for the first simulation for additional assistance. (See also Figures 28.7 and 28.8.)

## Results

1. Record the new steady-state values and the time to steady state in Table 28.5.

2. Print the simulation graph.

3. If instructed to do so, save your model on a disk. Refer to the instructions in Experiment A for assistance.

## Discussion

1. Describe the interactions among the trophic levels in the pond ecosystem.

2. Do your results match your predictions? Discuss any differences between results and predictions.

3. Discuss the effects of fishing on the pond ecosystem. How might these results be applied to a natural ecosystem?

## Experiment C. Computer Simulation— Invasive Species Case Study

### Problem

Jake, a fish farmer, counted the number of dead striped bass in his net. Twenty more today! He skillfully cut one open, exposing intestines and the mottled mass of muscles riddled with worms. "I'm not sure I can even harvest enough stripers to cover my expenses, much less make a profit. At this rate, it looks like the worm has turned. They're eatin' my fish instead of the other way 'round." Bud, the fishery's technician, suggested, "Can't you just kill those worms? You know, worm the fish." Jake countered, "I can't kill the worms because they reproduce in the snails. I can't kill the snails, because every chemical I could use would also kill the fish and contaminate my fish farm." "How about a biological control, like we use grass carp to keep the plants down in the pond?" suggested Bud. Jake thought about this idea. He knew that black carp would eat the snails. No snails, no worms. But he remembered that black carp is considered an invasive species, when it escapes from ponds into streams and lakes. Maybe he should find out more before investing in snail-crunching black carp. Jake is alarmed after finding a video on the web about Asian carp (http://www.invasivespeciesinfo.gov/aquatics/asiancarp.shtml) and reading a news item on black carp.

*See Teaching Plan for suggestions on teaching with case studies. Also see Waterman, M. and E. Stanley. Biological Inquiry: A Workbook of Investigative Cases. San Francisco, CA: Pearson Benjamin Cummings, 2011.*

### News Focus from *Science:*

"Will Black Carp Be the Next Zebra Mussel?" by Dan Ferber, vol. 292, pp. 203–204, April 13, 2001.

*There's good reason to worry about black carp, says ichthyologist Jim Williams of the U.S. Geological Survey (USGS) Caribbean Research Center in Gainesville, Florida. In a detailed 1996 risk assessment of the fish, Williams and USGS colleague Leo Nico concluded that black carp would survive and reproduce in U.S. rivers, consuming native mollusks and competing with native mollusk-eating fish such as redear sunfish and freshwater drum.*

## Additional Information

If Jake adds black carp to his pond and they escape to a natural pond, there could be problems. The structure and function of the pond ecosystem would change. Black carp will eat the snails (part of the pond life in the computer model). The bass prefer to eat the sunfish, because carp are too bony. Therefore, black carp would be competing with the sunfish for food, but with little or no predation by bass. (Assume no bass predation.)

## Questions

What might be the consequences if black carp were to escape to a nearby natural pond? What do you think would happen to the pond life, sunfish, and bass in a natural pond?

## Hypothesis

Formulate a hypothesis about the effects of introducing black carp to the pond ecosystem.

Test your hypothesis using the computer simulation model.

## Prediction

State a prediction based on your hypothesis. (This is an if/then statement that predicts the results of changes in the model.)

Test your prediction using the computer simulation model.

## Procedure

1. In Table 28.6, describe the components of the model you will simulate. Record the selected starting values for your model. The starting value for black carp should be small relative to sunfish. For example, if sunfish is 30 kg/ha, then black carp might be 10 kg/ha.
2. Note below any other changes to the model, such as changes in interactions or additions or removal of components.

   *Black carp should be connected to pond life, but not to bass; bass do not feed on carp.*

3. Calibrate the model based on your changes. Modify the model according to your problem. Select a new fish icon from the Library menu to represent the black carp. (Black carp does not appear as an icon, so you should select the gar icon or add another sunfish icon.) Connect the dark box of pond life to the open box of the new fish icon. Connect the new icon to the plotter. Double-click on the new icon to set the starting value for

black carp (Table 28.6). Remember, the starting value for carp should be less than for sunfish. (You may want to check the starting values for all components at this time.) The carp should not be connected to bass.

4. Run the simulation until steady-state values are reached.

5. You will need to modify the graph features. Select the first box on the display bar over the graph. In the fourth row, type in *Black carp kg/ha* and select the right axis by clicking in the eighth column. Return to the graph and adjust the right axis as needed. Refer to the Procedure section of the first simulation for additional assistance. (See also Figures 28.7 and 28.8.)

## Results

1. Describe the problem and record the new steady-state values and the time to steady state in Table 28.6.

2. Print the simulation graph.

3. If instructed to do so, save your model on a disk. Refer to the instructions in Experiment A for assistance.

## Problem:

**Table 28.6**
Starting and Steady-State Values for Exercise 28.2, Invasive Species Model

| Component | Starting Value | Steady-State Value |
|---|---|---|
| Sunlight | | |
| Pond life | | |
| Sunfish | | |
| Bass | | |
| Black carp | | |
| Time to steady state: | | |

## Discussion

1. Describe the interactions among the trophic levels in the pond ecosystem.

2. Do your results match your predictions? Discuss any differences between results and predictions.

3. Do you think that Jake should add black carp to his pond? Why or why not? Remember that Jake wants to increase his yield of striped bass (after all, he is a fish farmer). Should the federal government ban the sale of black carp, given your prediction of what will happen if black carp invade natural aquatic ecosystems?

*Adding black carp to the fish farm pond would be good for production of striped bass, since the harvest would increase with the decrease of snails and parasitic worms. However, the value to Jake is small compared to the large problem of invasive exotic species and their effect on native species. Competition with sunfish could not only mean the loss of primary and secondary consumers, but could also result in the crash of bass, if bass do not switch to carp or another fish as a food source.*

## Reviewing Your Knowledge

1. Define *model* and provide examples from several areas in biology. Consider areas of biology other than those studied in the laboratory.

*A model represents a structure, mechanism, or process, usually in a simplified form that allows the study of systems that would otherwise be too small, too large, or too complex to investigate. Examples are the double-helix model constructed by Watson and Crick, models of diffusion and osmosis, and models used to show stages of development.*

2. Critique the computer model used in this lab topic. In what ways is it an appropriate model of the pond ecosystem? In what ways does it fail to model the pond adequately?

*The basic components and interactions are present and simplified. The model is dynamic and can be manipulated to represent real ecosystems. However, the pond life compartment contains a mixture of trophic levels. Also, the rate of consumption cannot be altered in this model.*

3. Even when you adjust the starting values for components in the model, the steady-state values remain the same. However, if you change the energy from the sun, new steady-state values will emerge. Explain.

*The initial input of energy from the sun determines the amount of energy available in the ecosystem. At the set rates of consumption in the model, the same steady-state values will emerge unless you change the energy available to the system.*

4. Which disturbance factors had the greatest effect on the pond ecosystem? How are you measuring the effect?

# Applying Your Knowledge

1. In 1997, nonnative and invasive Asian swamp eels were collected in Florida for the first time at two sites near Tampa and Miami. By 2008, the eel had also been found in a lake in New Jersey. These fish are extremely adaptable to a wide range of freshwater habitats, from wetlands to streams and ponds. They are predators that feed on worms, insects, crayfish, frogs, and other fishes, including bluegill and bass. Swamp eels have the ability to gulp air, which allows them to survive in only a few inches of water and to move over land to a nearby body of water. Scientists are tracking their movements and increasing numbers in the Southeast. In one pond, several species of fish have been completely eliminated.

   Based on your understanding of the pond ecosystem, predict the effect of introducing swamp eels on the following components of the pond.

   Bluegill:

   *Bluegill and aquatic insects would be wiped out by the swamp eel.*

   Bass:

   *Bass would be eliminated by the combination of swamp eel predation and the loss of its prey species, bluegill.*

   Pond life:

   *Pond life would flourish. Algae might overrun the pond.*

2. Using your knowledge of ecosystem structure and function, propose a plan of action for eliminating the swamp eels (described in question 1 above) from the pond before they eliminate the other organisms. You cannot use toxins, since the local anglers fish in this pond.

   *Adding another predator, such as an otter or more water snakes, that would preferentially feed on the eels might help. However, if you choose another invasive species, or one that will not eat the eels, the problem may get worse. You might interest a local Asian market in collecting eels for sale, since they are considered a delicacy in Asian countries.*

3. Global warming (increase in global mean temperatures by 0.4°C over the last 150 years) has been greater in the Arctic with twice the rate and a decline in minimum sea ice cover by 45,000 km$^2$/yr (about twice the size of New Jersey) over the past two to three decades. Changes in the ecological systems in the Arctic are rapid and complex with serious effects on plants, animals, and their relationships. During the Fourth International Polar Year, ecological studies revealed the complex dynamics of climate change (Post et al., 2009).

   Over the last decade, Arctic plants have tracked warm temperatures in early spring by growing and flowering earlier. However, caribou continue to calve at the same time of year, which means that calves are born after the peak of plant resources. This has contributed to reduced reproduction and survival of caribou calves. In another example, the loss of snow cover in the Arctic has resulted in the crash of small rodent (lemmings and voles) population cycles and these populations remain low. Small rodents are the food source for predators snowy owls and Arctic foxes.

   For these two examples, state the *trophic level* for each organism: flowering plants, caribou, foxes, owls, lemmings, and voles.

   Sketch the changes in caribou populations and flowering plant populations given this mismatch.

*Caribou reproduction and survival of calves will continue to decline, moving caribou to the brink of extinction given other arctic changes. On the other hand, plant populations will thrive in the longer growing season and with reduced herbivory.*

Sketch the changes in plant, lemming, and snowy owl populations that you might expect as the lemmings crash and then remain low (no rebound).

*As the lemming populations decline, the snowy owls will decrease due to loss of food. The plants will increase in productivity. However, there will not be a rebound in the lemming population, because the populations are no longer cyclic, having been reduced to such low numbers. Both primary and secondary consumers may crash.*

4. A local television news bulletin urges you not to eat fish caught in nearby Lake Ketchum because water levels of the pollutant PCB have reached 0.0001 part per million (ppm). With such a small reading, why should you be concerned?

   *The PCBs will increase in the food chain as they are consumed by each trophic level. Concentration of PCBs will reach toxic levels in the bass, which consume the sunfish, which consume the pond life, which absorb the PCBs from the water.*

5. Eliminating top predators from an ecosystem can have effects that "cascade" down the trophic levels with serious results for the survival and interactions of many species. Over the last 35 years large predatory sharks, including sandbar, blacktip, bull, dusky, tiger, and scalloped hammerhead sharks, have been effectively eliminated from the world's oceans in the hunt for fins and meat. Over the same time period, northwestern Atlantic marine systems suffered severe losses of the economically important bay scallops, as well as clams and oysters. The populations of rays and smaller sharks (blacknose, bonnethead, finetooth sharks, and cownose, lesser devil, spiny butterfly, smooth butterfly and spotted eagle rays, plus several skates) have increased over the same time period, with cownose ray populations increasing by 20-fold (Meyers et al., 2007).

   Given the changes in population sizes for these three groups of consumers (great sharks, smaller rays and sharks, and bivalves), sketch a model of trophic relationships that would best explain the data. What do you think ecologists mean by the "cascade effect"?

   *The great sharks are top carnivores (consumers) that feed on the smaller sharks and rays. As the great sharks were eliminated, their primary prey, the lesser sharks and rays, increased in population size. The increase in numbers of smaller sharks and rays (particularly the cownose ray) has meant increased predation on bay scallops and other bivalves. The "cascade effect" refers to the series of changes in predator/prey interactions, when the top predators either increase or decrease. The consequences cascade down the food web.*

## Investigative Extensions

Environmental Decision Making has two other ecosystem models, *Grasslands* as well as *Forestry* and *Logging*, and an option for creating your own model using general symbols. These programs can be used to pursue additional topics, including ecosystem dynamics and management.

## Student Media: BioFlix, Activities, Investigations, and Videos
www.masteringbiology.com (select Study Area)

**Activities**—Ch. 54: Food Webs; Ch. 55: Pyramids of Production; Energy Flow and Chemical Cycling
**Investigations**—Ch. 54: How Are Impacts on Community Diversity Measured? Ch. 55: How Do Temperature and Light Affect Primary Production?

## References

Cox, G. W. *Alien Species in North America and Hawaii.* Washington, DC: Island Press, 1999.

Mack, R. N., D. Simberloff, W. M. Lonsdale, H. Evans, M. Clout, and F. Bazzaz. "Biotic Invasions: Causes, Epidemiology, Global Consequences and Control." *Issues in Ecology,* 2000, No. 5, Spring.

Myers, R. A., J. K. Baum, T. D. Shepard, S. D. Powers, and C. H. Peterson. "Cascading Effects of the Loss of Apex Predatory Sharks from a Coastal Ocean." *Science,* 2007, vol. 315, pp. 1846–1850.

Odum, E. C., H. T. Odum, and N. S. Peterson. *Environmental Decision Making, The BioQUEST Library Volume VI.* San Diego, CA: Academic Press, 2002. Lab written to correspond to software by permission of publisher.

Odum, H. T. and E. C. Odum. *Computer Minimodels and Simulation Exercises,* Gainesville, FL: Center for Wetlands, Phelps Laboratory, University of Florida, 1989. Examples using BASIC with program listings for Apple II, PC, and Macintosh.

Post, E. et al. "Ecological Dynamics Across the Arctic Associated with Recent Climate Change." *Science,* 2009, vol. 325, pp. 1355–1358.

Reece, J. et al. *Campbell Biology,* 9th ed. San Francisco, CA: Pearson Education, 2011.

*SOS for America's Streams* (video). Gaithersburg, MD: Izaak Walton League of America, 1990. See the Preparation Guide for ordering information.

Waterman, M. and E. Stanley. *Biological Inquiry: A Workbook of Investigative Cases.* San Francisco, CA: Pearson Benjamin Cummings, 2011.

## Websites

The BioQUEST Curriculum Consortium:
http://bioquest.org

BioQUEST Library Online. Environmental Decision Making is now available for free download:
http://bioquest.org/BQLibrary/library_result.php

EPA Ecosystems Web page with terrestrial and aquatic ecosystems and information on watersheds across the U.S.:
http://www.epa.gov/ebtpages/ecosystems.html
http://www.epa.gov/watershed/

USDA Invasive Species Information Center:
http://www.invasivespeciesinfo.gov/

USDA National Invasive Species Information Center. Information on Asian carp, including informative and entertaining video:
http://www.invasivespeciesinfo.gov/aquatics/asiancarp.shtml

U.S. Long Term Ecological Research Network with links to 26 sites:
http://www.lternet.edu/sites/

## LAB TOPIC 28

# Ecology II: Computer Simulations of a Pond Ecosystem
## Teaching Plan for Laboratories

### Main Concepts and Objectives

1. Concept: ecosystem structure and function. Students will develop a computer model of a pond ecosystem based on their knowledge of the trophic structure of an ecosystem. They will use their knowledge to develop questions of biological interest, and they will investigate the response of the system to disturbance. (All exercises)

2. Concept: computer models. Students will construct a computer model of a pond ecosystem, then modify the model to test hypotheses and predictions. They will consider how the results would apply to natural ecosystems. (All exercises)

3. Concept: steady-state model. Students will calibrate their model to establish steady-state values for the components of their system based on the available solar energy. They will compare additional simulations to the steady-state model. (All exercises)

4. Concept: computer experience. Students will gain experience with the way computers contribute to the scientific process through quick and easy manipulations of systems, testing of hypotheses, and graphical presentation of data. (All exercises)

5. Concept: scientific processes. Students will develop questions, state hypotheses and predictions, then test these with the computer model. Students will analyze simulation graphs, make adjustments, and construct additional models. (All exercises)

6. Specific content. Students will be able to define and describe: *models, trophic levels, primary productivity, secondary productivity, biomass, steady state, calibration, invasive species.*

### Order of the Lab

1. Review ecosystem structure and discuss models.                                    (10 min)
2. Describe the computer facilities and guidelines. Review the fundamentals of turning on the computer and opening the program. Introduce the software package.                        (15 min)
3. Organize student groups.                                                          (5 min)
4. Run simulations for Exercise 28.1.                                                (20 min)
5. Discuss Exercise 28.2.                                                            (10 min)
6. Design and run simulations.                                                      (1–1.5 hr)
7. Student reports on simulations.                                                   (20 min)

**For a 2-hour lab:** Omit one of the simulations in Exercise 28.2. Omit the student reports and have students submit their tables and simulation graphs for one or more simulations.

### Case Studies

Case studies utilize stories or scenarios to engage students in learning through critical thinking, problem solving, and research. Investigative case studies can connect problem solving in lecture and laboratory. Students are given a scenario that describes a situation or problem. There are many environmental case studies available on the Web. These can be adapted for use by students, depending on interest, local environmental issues, or lecture topics. We have included an investigative case study for one of the simulations in this lab, Lab Topic 28 Ecology II: Computer Simulations of a Pond Ecosystem, *Exercise 28.2, Experiment C, Computer Simulation of an Invasive Species.*

See the references below for ideas and suggestions for how to write case studies and use them to teach biology.

http://bioquest.org/BQLibrary_result.php This site provides background information on investigative case-based learning and some examples of case studies. Select and download module ICBL.

http://bioquest.org/lifelines/index.html Margaret Waterman and Ethel Stanley have developed investigative cases for the BioQUEST Curriculum Consortium.

See Waterman, M. and E. Stanley. *Biological Inquiry: A Workbook of Investigative Cases.* San Francisco, CA: Pearson Benjamin Cummings, 2011.

## Classroom Management

Ideally, this lab topic is taught in a laboratory with computers available for every three or four students or in a computer lab with similar facilities. This lab topic can be organized so that the actual simulations are completed by students working outside of lab. In that case, students need an introduction to the equipment, access to computer facilities, guidelines for your computer lab, and a demonstration of the software. The laboratory can be divided into smaller groups of 6 to 8 students who come to the lab for instruction at staggered times over the course of the laboratory period. If you have a projector, you can introduce the software to an entire laboratory section.

## Student Development

Students will apply their knowledge of ecosystems from Lab Topic 27 to a pond ecosystem. They will apply their knowledge to the model and will also consider how the results of their investigations would apply to natural systems. Students will practice their skills of graphical interpretation.

## Discussion and Summary

Students describe one or more simulations and report results to the class or they can submit simulation tables and graphs for evaluation.

## Evaluation

Students can submit simulation tables and graphs. Students can submit their simulations for evaluation. Concepts can be tested on a lab test. Consider presenting graphs for interpretation or posing a question that requires students to make predictions about the results of computer simulations.

# Appendix A
# Scientific Writing and Communication

For the scientific enterprise to be successful, scientists must clearly communicate their work. Scientific findings are never kept secret. Instead, scientists share their ideas and results with other scientists, encouraging critical review and alternative interpretations from colleagues and the entire scientific community. Communication, both oral and written, occurs at every step along the research path. While working on projects, scientists present their preliminary results for comments from their co-workers at laboratory group meetings and in written research reports. At a later stage, scientists report the results of their research activities as a poster or oral presentation at a scientific meeting. Then the final report is prepared in a rather standard scientific paper format and submitted for publication in an appropriate scientific journal. At each stage in this process, scientists encourage and require critical review of their work and ideas by their peers. The final publication in a peer-reviewed journal generally promotes additional research and establishes this contribution to current knowledge.

*One of the objectives of every lab topic in this manual is to develop your writing skills.* You will generate and write hypotheses, observations, answers to questions, and more, as one way of learning biology. Also, you will practice writing in a scientific paper format and style to communicate the results of your investigations. The scientific process is reflected in the design of a scientific paper and the format you will use for your laboratory papers.

*See the Preparation Guide Lab Topic 1, Scientific Investigation for a rubric to use for scientific writing.*

A scientific paper usually includes the following parts: a **Title** (statement of the question or problem), an **Abstract** (short summary of the paper), an **Introduction** (background and significance of the problem), a **Materials and Methods** section (report of exactly what you did), a **Results** section (presentation of data), a **Discussion** section (interpretation and discussion of results), and **References Cited** (books and periodicals used). A **Conclusion** (concise restatement of conclusions) and **Acknowledgments** (recognition of assistance) may also be included.

**Writing Program.** *We propose that you practice writing throughout the biology laboratory program by submitting individual sections of a scientific paper.* Your instructor will determine which sections you will write for a given lab topic and will evaluate each of these sections, pointing out areas of weakness and suggesting improvements. By the time you have completed these assignments, you will have submitted the equivalent of one scientific paper. Having practiced writing each section of a scientific paper in the first half of the laboratory program, *you will then write one or two complete laboratory papers in scientific paper format during the second half of the laboratory program,* reporting the results of experiments, preferably those that you and your research team have designed and performed.

767

# Successful Scientific Writing

The following notes for success apply to writing throughout all sections of a scientific paper.

- Your writing should be **clear and concise** to communicate effectively with your audience. Write objectively in direct and informative sentences that explain what you mean with a minimum of words. **Avoid adjectives and adverbs** that can contribute to flowery language and have limited use in describing your work.

- **Write in short and logical, but not choppy, sentences.** Avoid run-on sentences and use grammatically correct English. Avoid long introductions. (See Appendix 4, "Sentences Requiring Revision," in Knisely (2009) for basic rules of writing and practice in editing for common errors.)

- **Write for your audience**: other student-scientists and your professor. Scientists read scientific papers to be current in the field, provide background information on a topic, and to develop or improve methodologies. Scientific writing must be factual and informative, rather than entertaining.

- **Support your writing with evidence.** Your interpretation of your work in the Discussion section will be based on the results of your research presented in the Results section. In the Introduction and Discussion sections you will provide support for your ideas by describing and referencing work of other scientists.

- **Locate sources related to your work early** in the research project. Read the sources carefully to understand the work. Then take notes in your own words. Make a complete record of the reference citation information to use in the References section of your paper.

- **Avoid quotations in scientific writing.** Unlike writing in other disciplines, scientific writing seldom includes direct quotes. You should summarize and explain the work of others in your own words. Then provide a reference citation to the work. *Do not use footnotes.*

- **Do not plagiarize.** When you restate or paraphrase the words of others, you must credit that work by including the source in a reference citation in the text. Simply changing a word or two in the sentence is still plagiarism. See Pechenik (2010) and Knisely (2009) for specific examples and suggestions for avoiding plagiarism.

- **Use the past tense** in the Abstract, Materials and Methods, and Results sections. Also use the past tense in the Introduction and Discussion sections when referring to *your* work. **Use the present tense** when relating the background information as you refer to other investigators' published work. Previously published research is considered established in the present body of knowledge.

- **Use the active voice** when possible. Doing so makes the paper easier to read and more understandable. However, in the Materials and Methods section you may use the passive voice so that the focus of your writing is the methodology, rather than the investigator.

- When referring to the scientific name of an organism, **the genus and species should be in italics or underlined.** The first letter of the genus is capitalized, but the species is written in lowercase letters, for example, *Drosophila melanogaster.*

- **Pay particular attention to the rules for numbers.** Use metric units for all measurements. Use numerals when reporting measurements, percentages, decimals, and magnifications. When beginning a sentence, write the number as a word. Numbers of ten or less that are not measurements are written out. Numbers greater than ten are given as numerals. Decimal numbers less than one should have a zero in the one position (e.g., 0.153; not .153).

- **Include a heading for each section of the scientific paper** (except the title page), placing the heading of the section against the left margin on a separate line. Each section does not begin a new page but continues in order.

- **Revise, revise, and then revise!** For suggestions and examples of how to revise your work, see Chapter 5, "Revision," in Knisely (2009) and Chapter 6, "Revising," in Pechenik (2010).

- **A few points to remember.** Note the word *data* is plural. Remember the results cannot "prove" the hypothesis, but rather they may "support" or "falsify" the hypothesis.

- **Carefully proofread** your work even if your word processor has checked for grammatical and spelling errors. These programs cannot distinguish between "your" and "you're," for example.

- **Save a copy** of your work on a disk or USB flash drive, and **print a copy** of your paper before turning in the original.

- **Begin writing early, leaving time for literature search, analysis, revision and final proofreading.** Writing takes time and thought.

## Locating Appropriate References

References are important to provide a context and evidence from the work of other scientists to support your explanations and interpretations. Scientific papers in general rely on **primary references,** *reports of original research that present the work of scientists in such a way that it can be repeated. Primary references are journal articles that have been reviewed by other scientists and the journal editor.* In addition to articles in journals (for example, *American Journal of Botany, Cell, Ecology,* and *Science*), primary references include conference papers, dissertations, and technical reports. Many scientific journals are available in a full-text version online; these are still primary references. (Websites are not primary references, because they are not required to participate in the peer review process.)

As you begin your study and literature review, it may be helpful to start with secondary references. Textbooks, review articles, and articles from popular science magazines are **secondary references,** *which generally provide a summary and interpretation of research* (for example, *Annual Review of Genetics, Science News,* and *Scientific American*). To develop fundamental concepts and terminology, consult books and review articles. For example, if your project is on plant hormones, you might consult a plant physiology textbook or search in the *Annual Review of Plant Physiology*. In general these secondary references do not provide the primary evidence needed for your scientific paper. They may include references to primary articles that will be useful. (Also see the References section for each Lab Topic in this manual.)

Given the enormous number of scientific papers and journals available, searching for primary references on your research topic may seem to be a daunting task. The key to success is to use effective **search strategies** with **online databases and search engines** that specifically target scientific literature (for example, PubMed, Google Scholar, Biosis Previews, Science Direct, JSTOR, and Web of Science). You can compare these online resources and read about search strategies in Pechenik (2010) and Knisely (2009). Most databases and search engines provide a tutorial for searching under "Help" or "Advanced Searches."

*As you read scientific papers, take notes by summarizing the work in your own words.* Do not record quotes, since scientific writing does not use direct quotations. Record the complete citation information for any references at the time you read the article. Any information from these sources must be acknowledged with a citation in the text and inclusion in the References Cited section. *Refer to the citation format in the References Cited section of this appendix.* Your instructor will indicate the number of primary references required for your paper. Gillen (2007) and Knisely (2009) provide useful suggestions for how to read and evaluate scientific papers.

### Plagiarism

**Plagiarism is the use of someone else's words or ideas without acknowledging the source or author.** Plagiarism is a serious offense and one that can and must be avoided. Many instructors are using Internet and software resources to detect plagiarism. The penalties for plagiarism are severe, ranging from a failing grade on the work to expulsion. All sources of information must be acknowledged even if you write about the work in your own words. Although you will work collaboratively on your laboratory investigations, *students must write and submit their papers independently*. Because performing the experiment will be a collaborative effort, you and your teammates will share the results of your investigation. The Introduction, Discussion, and References Cited (or References) sections must be the product of your own personal literature research and creative thinking. If you are not certain about the level of independence and what constitutes plagiarism in this laboratory program, ask your instructor to clarify the class policy. *In the most extreme case of plagiarism, a student presents another student's report as his or her own. However, representing another person's ideas as your own without giving that person credit is also plagiarism and is a serious offense.* To learn more about and to practice identifying plagiarism, see Frick's (2004) website.

*Each instructor may have his or her own criteria for what constitutes plagiarism. We have a handout for our students explaining our criteria. We also explicitly explain the consequences of plagiarism in this handout.*

*We provide students with a checklist covering the writing and formatting of a scientific paper. Students use this as a guide and attach it to their completed papers. The instructor can also use the checklist in evaluating the papers. See Preparation Guide Lab Topic 1, Scientific Investigation, for sample checklist.*

# Writing a Scientific Paper

The sections of a scientific paper and particular material to be covered in each section are described after the following table in order of appearance in a scientific paper. However, most scientists do not follow that sequence in the actual writing of the paper, but rather begin with the methodology. (See table, "Plan for Writing a Scientific Paper.")

| | Plan for Writing a Scientific Paper | |
|---|---|---|
| **Order of Writing** | **Section** | **Notes** |
| **Begin 1st** | Materials and Methods | The first draft of this section can be written before all the results are completed. Remember to review and carefully edit after completing all work. (See Materials and Methods, pp. 772–773.) |
| **2nd** | Results | Construct the **tables and figures** (see Lab Topic 1 Scientific Investigation). **Compose the text** for the Results section based on the tables and figures. (See Results, p. 773.) |
| **3rd** | Literature Search— Continued | Consult references for background information and interpretation of results. **Locate and review primary and secondary references** for use in the Discussion and Introduction sections. (See References Cited, pp. 774–777.) |
| **4th** | Discussion and References | Write the **Discussion** section and begin the **References Cited** section. (See Discussion, pp. 773–774 and References Cited, pp. 774–777.) Most scientists prefer to write the Discussion before the Introduction. Both sections require background information and a clear understanding of the results of the work. Remember to carefully check and revise your Discussion, if you write it first. |
| **5th** | Introduction and References | Develop the **Introduction** section and complete the **References Cited** section. (See Introduction, p. 772 and References Cited, pp. 774–777.) |
| **6th** | Abstract and Title | **Write the Title and Abstract**. (See pp. 771–772.) The abstract is a short summary of all components of the paper. The title is a succinct statement of the objective of the research. |
| **7th** | Revise and Proofread | Leave the paper for a short time and then reread carefully. **Revise** to be clear, concise, and well written. Review your use of evidence and evaluate your sources. Revise accordingly. **Proofread carefully** for errors. Review a checklist, if available, before preparing the final version of the paper. Check for all formatting rules and suggestions. |

## Title Page and Title

The **title page** is the first page of the paper and includes the **title** of the paper, **your name**, the **course title**, your **lab section**, your **instructor's name**, and the **due date** for the paper. *The title should be as short as possible and long as necessary to describe the objective and significance of your research topic.* For example, if you are asking a question about the inheritance patterns of the gene for aldehyde oxidase production in *Drosophila melanogaster,* a possible title might be "Inheritance of the Gene for Aldehyde Oxidase in *Drosophila melanogaster.*" Something like "Inheritance in Fruit Flies" is too general, and "A Study of the Inheritance of the Enzyme Aldehyde Oxidase in the Fruit Fly *Drosophila melanogaster*" is too wordy. The words "A Study of the" are superfluous, and "Enzyme" and "Fruit Fly" are redundant. The suffix *-ase* indicates that aldehyde oxidase is an enzyme, and most scientists know that *Drosophila melanogaster* is the scientific name of a common fruit fly species. However, it is appropriate to include in the title both common and scientific names of lesser known species.

**The format for the title page.** Place the title about 7 cm from the top of the title page. Place "by" and your name in the center of the page, and place the course name, lab section, instructor's name, and due date, each on a separate centered line, at the bottom of the page. Leave about 5 cm below this information.

## Abstract

The **abstract** is placed at the beginning of the second page of the paper, after the title page. *The abstract concisely summarizes the question being investigated in the paper, the methods used in the experiment, the results, and the conclusions drawn.* The reader should be able to determine the major topics in the paper without reading the entire paper. The abstract should be no more than 250 words, and fewer if possible. Compose the abstract after the paper is completed.

## Introduction

*The* **introduction** *has two functions: (1) to provide the context for your investigation and (2) to state the question asked and the hypothesis tested in the study.* Begin the introduction by providing background information that will enable the reader to understand the objective of the study and the significance of the problem, relating the problem to the larger issues in the field. Include only information that directly prepares the reader to understand the question investigated. Most ideas in the introduction will come from outside sources, such as scientific journals or books dealing with the topic you are investigating. All sources of information must be referenced and included in the References Cited (or References) section of the paper, but the introduction must be in your own words. Refer to the references when appropriate. Unless otherwise instructed, place the author of the reference cited and the year of publication in parentheses at the end of the sentence or paragraph relating the idea; for example, (Finnerty, 1992). Additional information on citing references is provided on pp. 774–775, References Cited. *Do not use citation forms utilized in other disciplines. Do not use footnotes and avoid the use of direct quotes.*

As you describe your investigation, include only the question and hypothesis that you finally investigated. Briefly describe the experiment performed and the outcome predicted for the experiment. Although these items are usually presented after the background information near the end of the introduction, you should have each clearly in mind before you begin writing the introduction. It is a good idea to write down each item (question, hypothesis, prediction) before you begin to write your introduction.

## Materials and Methods

*The* **Materials and Methods** *section describes your experiment in such a way that it can be repeated. This section should be a narrative description that integrates the materials with the procedures used in the investigation.* The Materials and Methods section is often the best place to begin writing your paper. The writing is straightforward and concise, and you will be reminded of the details of the work. The following notes apply to writing the Materials and Methods section.

- Write the Materials and Methods section concisely in paragraph form in the past tense. *Do not list the materials and do not list the steps of the procedure.*

- Be sure to include levels of treatment, numbers of replications, and controls.

- If you are working with living organisms, include the scientific name and the sex of the organism.

- Describe any statistical analyses or computer software used.

- Determine which details are essential for another investigator to repeat the experiment. For example, if in your experiment you incubated potato pieces in different concentrations of sucrose solution, it would not be necessary to explain that the pieces were incubated in plastic cups labeled with a wax marking pencil or to provide the numbers of the cups. In this case, the molarity of the sucrose solutions, the size of the potato pieces and how they were obtained, and the amount of incubation solution are the important items to include.

- Do not include failed attempts unless other investigators may try the technique used. Do not try to justify your procedures in this section.

- Results should not be included in the Materials and Methods.

## Results

The **Results** section is the central section of a scientific paper. It consists of at least four components: *(1) one or two sentences reminding the reader about the nature of the research, (2) one or more paragraphs that describe the results, (3) figures (graphs, diagrams, pictures), and (4) tables.* The data included in tables and graphs should be summarized and emphasized in a narrative paragraph written in past tense. Draw the reader's attention to the results that are important. Describe trends in your data and provide evidence to support your claims.

*Encourage students to return to Lab Topic 1 for specific instructions for preparing tables and figures.*

Before writing the Results section, prepare the tables and figures. Remember to number figures and tables consecutively throughout the paper (*see Lab Topic 1 Scientific Investigation for instructions on creating figures and tables and their presentation*). Refer to figures and tables within the paragraph as you describe your results, using the word Figure or Table, followed by its number; for example, (Figure 1). If possible, place each figure or table at the end of the paragraph in which it is cited.

If you have performed a statistical analysis of your data, such as chi-square, include the results in this section.

Report your data as accurately as possible. Do not report what you expected to happen in the experiment nor whether your data supported your hypothesis. Do not discuss the meaning of your results in this section. Do not critique the results. Any data you plan to include in the Discussion section must be presented in the Results. Conversely, do not include data in the Results that you do not mention in the Discussion.

Write the Results section before attempting the Discussion section. This will ensure that the results of your investigation are clearly organized, logically presented, and thoroughly understood before they are discussed.

## Discussion

*In the **Discussion** section, you will analyze and interpret the results of your experiment.* Simply restating the results is not interpretation. The Discussion must provide a context for understanding the significance of the results. Explain why you observed these results and how these results contribute to our knowledge. Your results either will support or confirm your hypothesis or will negate, refute, or contradict your hypothesis; but the word *prove* is not appropriate in scientific writing. If your results do not support your hypothesis, you must still state why you think this occurred. Review the related work of other scientists and provide a summary of their research. Use this evidence to support your interpretation of the results of your study. State your conclusions in this section.

Complete the Results sections before you begin writing the Discussion. The figures and tables in the Results section will be particularly important as you begin to think about your discussion. The tables allow you to present your results clearly to the reader, and graphs allow you to visualize the effects that the independent variable has had on the dependent variables in your experiment. Studying these data will be one of the first steps in interpreting your results. As you study your data in the Results section, write down relationships and integrate these relationships into a rough draft of your discussion.

*The following steps may be helpful as you begin to outline your discussion and before you write the narrative.*

- Restate your question, hypothesis, and prediction.
- Write down the specific data, including results of statistical tests.
- State whether your results did or did not confirm your prediction and support or negate your hypothesis.
- Write down what you know about the biology involved in your experiment. How do your results fit in with what you know? What is the significance of your results?
- How do your results support or conflict with previous work? Summarize the findings and include references to this work.
- Clearly state your conclusions.
- List weaknesses you have identified in your experimental design that affected your results. The weaknesses of the experiment should not, however, dominate the Discussion. *Include one or two sentences only if these problems affected the results.* Remember the focus of the Discussion is to convey the significance of the results.
- You are now ready to write the narrative for the Discussion. Integrate all of the above information into several simple, clear, concise paragraphs. Discuss the results; do not simply restate the data. Refer to other work to support your ideas.

## References Cited (or References)

*We provide students with the reference section from a scientific paper that illustrates the correct form for citing journal articles, books, chapters in books, and other publications.*

A **References Cited** section lists only those references cited in the paper. A References section (bibliography), on the other hand, is a more inclusive list of all references used in producing the paper, including books and papers used to obtain background knowledge that may not be cited in the paper. Most references will be cited in the Introduction and Discussion sections of your paper. References may also be included in the Materials and Methods when acknowledging the source for a procedure or method of analysis. For your paper you should have a References Cited section that includes only those references cited in the paper. For additional assistance, see "Locating Appropriate References" on p. 769.

## Constructing the References Cited Section

The **format for the References Cited section** differs slightly from one scientific journal to the next. How does an author know which format to use? Every scientific journal provides "Instructions to Authors" that describe specific requirements for this important section and all other aspects of the paper. You may use the format used in this lab manual and provided in the examples below, select the format in a scientific journal provided by your instructor, or use another accepted format for listing your references. Your instructor may provide additional instructions. Be sure to read the references that you cite in your paper.

**Journal article, one author:**

Gould, S. J. "Is a New and General Theory of Evolution Emerging?" *Paleobiology*, 1980, vol. 6, pp. 119–130.

**Journal article, two or more authors:**

Greider, C. W. and E. H. Blackburn. "Identification of a Specific Telomere Terminal Transferase Activity in *Tetrahymena* Extracts." *Science*, 1985, vol. 43, pp. 405–413.

**Book:**

Darwin, C. R. *On the Origin of Species.* London: John Murray, 1859.

**Chapter or article in an edited book:**

Funk, D. J. "Investigating Ecological Speciation," in *Speciation and Patterns of Diversity*, eds. R. K. Butlin, J. R. Bridle, and D. Schluter. Cambridge, UK: Cambridge University Press, 2009, pp. 195–218.

**Government publication:**

Office of Technology Assessment. *Harmful Non-indigenous Species in the United States.* Publication no. OTA-F-565. Washington, D.C.: U.S. Government Printing Office, 1993.

## Citing References in Text

Scientific writing uses a different format than you may have used in other disciplines for acknowledging sources of information in the body of the paper. After referring to and describing the work of another scientist in your own words, place the author's name and the date in parentheses at the end of the passage. This date and year citation format is not the only one used in scientific writing, but it is the one used in this laboratory manual and many scientific journals. *Do not use citation formats from other disciplines.*

**Example for one author:**

The innate agonistic behavior of the male Siamese fighting fish has been widely studied (Simpson, 1968).

**For two authors use both names:**

Telomere terminal transferase was identified as the enzyme that adds nucleotide repeats at the telomeres (ends of chromosomes) during eukaryotic replication (Grieder and Blackburn, 1985).

**For three or more authors use first author plus et al.:**

Taxol, a bioactive compound extracted from the Pacific yew, inhibits cell replication at the metaphase/anaphase stage. The blocking of mitosis occurs in tissue culture cells as a result of stabilizing the microtubules so that the chromosomes cannot move to the poles of the cell (Jordan et al., 1993).

# Information Sources and the Internet

The Internet can provide access to online search engines and databases including *Biological Abstracts, Current Contents, Medline,* and *Annual Reviews* among many others. These search tools provide access to a wide range of published papers, some of which may be available online as full text journals. For suggestions and examples of how to locate sources using the Internet, see Harnack and Kleppinger (2003) and Knisely (2009). Scientific papers published in professional journals have gone through an extensive review process by other scientists in the same field. Most scientific articles have been revised based on comments by the reviewers and the editors. Sources of information that lack this critical review process do not have the same validity and authority.

The Internet is an exciting, immediate, and easily accessible source of information. However, unlike traditional bibliographic resources in the sciences, the Internet includes websites with material that has not been critically reviewed. Your instructor may prefer that you use the Internet only for locating peer-reviewed primary references or as a starting point to promote your interest and ideas. You may not be allowed to use Internet sources at all. Consult your instructor concerning use of Internet information.

If you do use the Internet to locate information, you should be prepared to evaluate these sites critically. Remember always to record the online address (URL) for any site you use as a reference. Tate and Alexander (1996) suggest the following five criteria for evaluating Internet sources.

1. **Author and Publisher of Source.** Determine the author and publisher for the Internet site. What is the professional affiliation of the author? Are phone numbers and addresses included? Is there a link to the publisher's home page? Does the author list his or her qualifications? If the material is copyrighted, who owns the copyright?

2. **Accuracy.** Look for indications of professional standards for writing, citations, figures, and tables. Are there typographical, spelling, and grammatical errors? Are sources of information cited? Are the data presented or simply summarized?

3. **Objectivity.** Is the site provided as a public service, free of advertising? If advertising is present, is it clearly separate from the information? Does the site present only the view of the publisher or advertiser?

4. **Timeliness.** Determine the date of the site and whether it is regularly revised. How long has the site existed? When was it last updated? Are figures and tables dated? Some Internet sites disappear overnight. Always record the date that you visited the site and retrieved information.

5. **Coverage.** Is the information offered in a complete form or as an abstract or summary of information published elsewhere? Is the site under construction? When was the site last revised?

Below find a model format and examples for citing Internet sources in the References Cited section of your paper. Other formats may be suggested by your instructor or librarian.

**Model:**

Author's last name and initials. Date of Internet publication. Document title. <URL> Date of access.

**Examples:**

**Professional site:**

[CBE] Council of Biology Editors. 2010. CBE home page. http://www
.councilscienceeditors.org. Accessed May 15, 2010.

**e-journal:**

Browning T. 1997. Embedded visuals: Student design in Web spaces.
*Kairos: A Journal for Teachers of Writing in Webbed Environments* 3(1).
http://english.ttu.edu/kairos/2.1/features/browning/bridge.html. Accessed May 15, 2010.

**Government publication:**

Food and Drug Administration, 1996, Sep. "Outsmarting Poison Ivy
and Its Cousins." *FDA Consumer Magazine.* http://www.fda.gov/fdac/
features/796_ivy.html. Accessed May 15, 2010.

**Note:** If using a Web search engine like Google Scholar, *do not include the entire
search path.* Cite only the actual Web address or URL in the reference citation.

# Oral and Poster Presentations

## Oral Presentations

Scientists regularly present their work at scientific meetings in oral paper
presentations. They can share preliminary findings and receive critical questions and comments from others in their field that are essential to scientific
research. In this laboratory program you may be asked to give an oral report
on your experiments in the laboratory or in a research symposium. For your
10- to 15-minute presentation, you need to capture the interest of your audience and make a convincing presentation of your results and conclusions.
Following are suggestions to help you prepare and present oral reports.

- Include the components of a scientific paper, with an Introduction, Meth-
  ods, Results, Discussion, and Conclusions.
- In the introduction, provide background information, state your hy-
  pothesis, give a brief description of your experiment, and state your
  predicted results.
- Briefly describe the experimental design and essential procedures.
- Emphasize your results with bold and clearly labeled figures and tables
  that have visual impact.
- Discuss and interpret your results providing supporting evidence from
  primary references.
- State your conclusions.
- Include references to sources of information and images.
- Be prepared to answer questions.

Pay attention to your presentation delivery. Use simple, bold visual aids
that are easily visible from the back of the room. For information on
preparing PowerPoint slides, see Knisely (2009) and www.Dartmouth
.edu/~biomed/new.htmld/ppt_resources.shtml.

*Below are suggestions for effective presentations.*

- Slides should have a small number of bullets, not extensive text.
- Use a simple and consistent template or theme for slides.
- Minimize distracting transitions and sounds.

- Do not read the slides.
- Look at your audience (not the screen). Speak slowly and clearly, projecting your voice so all members of the audience can hear you.
- Keep your objective in mind—to clearly communicate your ideas.

A checklist and other suggestions for preparing successful oral presentations may be found in McMillan (2006). A copy of an evaluation form for oral presentations can be downloaded at http://www.sinauer.com/knisely/.

## Poster Presentations

Another form of scientific communication that has become popular in recent years is the **scientific poster**, a document usually created on a single, large sheet. If a large printer is not available, a poster may be created on a multiple-paneled mat board. Most scientific societies are now organizing poster sessions at their annual meetings, often featuring undergraduate and graduate student research. Your instructor may ask you to model this form of communication by preparing a poster about your student-designed investigations or extended research projects as part of a poster symposium.

In many situations, posters may be a more effective method of presentation than writing a scientific paper or giving an oral report. A poster may be prominently displayed for a wider audience for an extended period. The presenters are then available at designated times to explain the research and engage in meaningful discussion.

*Suggestions for preparing and presenting a poster follow.*

- Present the essentials of your research with minimal text and maximum visual impact. Studies show that you have only 11 seconds to grab the attention of interested persons.
- Include the sections of a scientific paper—Title, Introduction, Experiment Summary (with brief methods), Results, Discussion, and References.
- Use a large typeface for headings and the content in abbreviated format under each heading. Use fonts that are easily readable from 3 to 6 feet. Title fonts should be 72 point; section headings 28 point; and text should be no smaller than 24 point (Knisely, 2009).
- Include simple, bold, and clearly labeled figures and tables.
- Before creating your poster, sketch the layout on paper, blocking out sections in two or three columns. Consider the location of text, images, tables, and figures.
- Consider the use of fonts, color, and images to highlight your poster and direct the viewer's eye.

**You can design your poster in an electronic form using PowerPoint.** This allows you to edit the text, images, and overall design. You can import sections from electronic files and insert digital images. You can effectively collaborate with your research team as you modify electronic versions of your poster. If your institution or community has a printing service, the electronic file can be sent to them for printing on large poster-sized paper. For detailed instructions see Chapter 7 in Knisely (2009).

Your instructor may give specific guidelines for your poster, or you may follow guidelines from other sources. There are several excellent websites with instructions for preparing posters. One of the most complete and well-designed sites is http://www.swarthmore.edu/NatSci/cpurrin1/posteradvice.htm. This website includes a poster template, suggested layouts, and examples of good and mediocre posters.

## References

The following sources are recommended to give additional help and examples in scientific writing.

Frick, T. 2004. "Understanding Plagiarism." Indiana School of Education. Available at: https://www.indiana.edu/~tedfrick/plagiarism/

Gillen, C. *Reading Primary Literature: A Practical Guide to Evaluating Research Articles in Biology.* San Francisco, CA: Benjamin/Cummings, 2007.

Harnack, A. and E. Kleppinger. *Online! A Reference Guide to Using Internet Sources,* 3rd ed. Boston: Bedford/St. Martin's, 2003.

Knisely, K. *A Student Handbook for Writing in Biology,* 3rd ed. Sunderland, MA: Sinauer Associates, 2009.

McMillan, V. E. *Writing Papers in the Biological Sciences,* 4th ed. Boston: Bedford/St.Martin's, 2006.

Pechenik, J. A. *A Short Guide to Writing about Biology,* 7th ed., New York, NY: Addison Wesley, 2010.

Style Manual Committee, Council of Biology Editors. *Scientific Style and Format: The CBE Manual for Authors, Editors and Publishers,* 6th ed. Cambridge, MA: Cambridge Univ. Press, 1994.

Tate, M. and J. Alexander. "Teaching Critical Evaluation Skills for World Wide Web Resources." *Computers in Libraries,* Nov/Dec 1996, pp. 49–55.

## Websites

Evaluation forms for posters and oral presentations: http://www.sinauer.com/knisely/

How to cite Internet sources: http://www.bedfordstmartins.com/online/cite8.html

How to evaluate Web sources: http://library.berkeley.edu/TeachingLib/Guides/Internet/Evaluate.html

How to prepare and present effective PowerPoint presentations: http://www.dartmouth.edu/~biomed/new.htmld/ppt_resources.shtml

Purrington, C.B. 2006. Advice on designing scientific posters: http://www.swarthmore.edu/NatSci/cpurrin1/posteradvice.htm

### Note to Instructors: Scientific Writing Program

Examples of Individual Sections of a Scientific Paper for Suggested Lab Topics:

   Title Page and Materials and Methods for Lab Topic 1 Scientific Investigation

   Discussion for Lab Topic 3 Diffusion and Osmosis

   Results for Lab Topic 4 Enzymes

   Results plus Discussion for Lab Topic 5 Cellular Respiration and Fermentation

   Introduction and References Cited for Lab Topic 6 Photosynthesis

*You may choose to change these assignments, depending on your particular laboratory schedule.*

*We suggest that students be given the opportunity to resubmit each section after you have given suggestions for improvement.*

*We suggest that students write a complete scientific paper on an experiment of their own design from Lab Topics 21, 26, or 28. Lab Topics 5 and 8 also could be used for a complete scientific paper, but they may come too early in the lab program.*

# Appendix B
# Chi-Square Test

Chi-square is a statistical test commonly used to compare observed data with data we would expect to obtain according to a specific scientific hypothesis. For example, if according to Mendel's laws, you expect 10 of 20 offspring from a cross to be male and the actual observed number is 8 males out of 20 offspring, then you might want to know about the "goodness of fit" between the observed and the expected. Were the deviations (differences between observed and expected) the result of chance, or were they due to other factors? How much deviation can occur before the investigator must conclude that something other than chance is at work, causing the observed to differ from the expected? The chi-square test can help in making that decision. The chi-square test is always testing what scientists call the **null hypothesis**, which states that there is no significant difference between the expected and the observed result.

The formula for calculating chi-square ($\chi^2$) is:

$$\chi^2 = \Sigma(o - e)^2/e$$

That is, chi-square is the sum of the squared difference between observed ($o$) and expected ($e$) data (or the deviation, $d$), divided by the expected data in all possible categories.

For example, suppose that a cross between two pea plants yields a population of 880 plants, 639 with green seeds and 241 with yellow seeds. You are asked to propose the genotypes of the parents. Your scientific hypothesis is that the allele for green is dominant to the allele for yellow and that the parent plants were both heterozygous for this trait. If your scientific hypothesis is true, then the predicted ratio of offspring from this cross would be 3:1 (based on Mendel's laws), as predicted from the results of the Punnett square (Figure B.1). The related null hypothesis is that there is no significant difference between your observed pea offspring and offspring produced according to Mendel's laws. To determine if this null hypothesis is rejected or not rejected, a $\chi^2$ value is computed.

To calculate $\chi^2$, first determine the number expected in each category. If the ratio is 3:1 and the total number of observed individuals is 880, then the expected numerical values should be 660 green and 220 yellow ($^3/_4 \times 880 = 660$; $^1/_4 \times 880 = 220$).

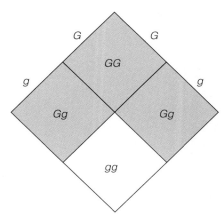

**Figure B.1.**
**Punnett square.** Predicted offspring from cross between green- and yellow-seeded plants. Green ($G$) is dominant (3/4 green; 1/4 yellow).

 Chi-square analysis requires that you use numerical values, not percentages or ratios.

Then calculate $\chi^2$ using the formula, as shown in Table B.1. Note that we get a value of 2.673 for $\chi^2$. But what does this number mean? Here's how to interpret the $\chi^2$ value:

1. Determine **degrees of freedom** (df). Degrees of freedom can be calculated as the number of categories in the problem minus 1. For example, if we had four categories, then the degrees of freedom would be 3. In our example, there are two categories (green and yellow); therefore, there is 1 degree of freedom.

2. Determine a relative standard to serve as the basis for rejecting the hypothesis. Scientists allow some level of error in their decision making for testing their hypotheses. The relative standard commonly used in biological research is $p < 0.05$. The **p value** is the *probability* of rejecting the null hypothesis (that there is no difference between observed and expected), when the null hypothesis is true. In other words, the p value gives an approximate value for the error of falsely stating that there is a significant difference between your observed numbers and the expected numbers, when there is *not* a significant difference. *When we pick $p < 0.05$, we state that there is less than a 5% chance of error of stating that there is a difference when, in fact, there is no significant difference.* Although scientists and statisticians sometimes select a lower significance value, in this manual we will assume a value of 0.05.

3. Refer to a chi-square distribution table (Table B.2). Using the appropriate degrees of freedom, locate the value corresponding to the p value of 0.05, the error probability that you selected. If your $\chi^2$ value is *greater* than the value corresponding to the p value of 0.05, then you reject the null hypothesis. You conclude that the observed numbers are significantly different from the expected. In this example your calculated value, $\chi^2 = 2.673$, is not larger than the table $\chi^2$ value of 3.84 (df = 1, $p = 0.05$). *Therefore, your observed distribution of plants with green and yellow seeds is not significantly different from the distribution that would be expected under Mendel's laws. Any minor differences between your offspring distribution and the expected Mendelian distribution can be attributed to chance or sampling error.*

### Step-by-Step Procedure for Testing Your Hypothesis and Calculating Chi-Square

1. State the hypothesis being tested and the predicted results.

2. Gather the data by conducting the relevant experiment (or, if working genetics problems, use the data provided in the problem).

3. Determine the expected numbers for each observational class. Remember to use numbers, not percentages.

 Chi-square should *not* be calculated if the expected value in any category is less than 5.

4. Calculate $\chi^2$ using the formula. Complete all calculations to three significant digits.

**Table B.1**
Calculating Chi-Square

|  | Green | Yellow |
|---|---|---|
| Observed ($o$) | 639 | 241 |
| Expected ($e$) | 660 | 220 |
| Deviation ($o - e$) | $-21$ | 21 |
| Deviation$^2$ ($d^2$) | 441 | 441 |
| $d^2/e$ | 0.668 | 2.005 |

$\chi^2 = \Sigma\, d^2/e = 2.673$

**Table B.2**
Chi-Square Distribution

| Degrees of Freedom (df) | Probability ($p$) | | | | | | | | | | |
|---|---|---|---|---|---|---|---|---|---|---|---|
|  | 0.95 | 0.90 | 0.80 | 0.70 | 0.50 | 0.30 | 0.20 | 0.10 | 0.05 | 0.01 | 0.001 |
| 1 | 0.004 | 0.02 | 0.06 | 0.15 | 0.46 | 1.07 | 1.64 | 2.71 | 3.84 | 6.64 | 10.83 |
| 2 | 0.10 | 0.21 | 0.45 | 0.71 | 1.39 | 2.41 | 3.22 | 4.60 | 5.99 | 9.21 | 13.82 |
| 3 | 0.35 | 0.58 | 1.01 | 1.42 | 2.37 | 3.66 | 4.64 | 6.25 | 7.82 | 11.34 | 16.27 |
| 4 | 0.71 | 1.06 | 1.65 | 2.20 | 3.36 | 4.88 | 5.99 | 7.78 | 9.49 | 13.28 | 18.47 |
| 5 | 1.14 | 1.61 | 2.34 | 3.00 | 4.35 | 6.06 | 7.29 | 9.24 | 11.07 | 15.09 | 20.52 |
| 6 | 1.63 | 2.20 | 3.07 | 3.83 | 5.35 | 7.23 | 8.56 | 10.64 | 12.59 | 16.81 | 22.46 |
| 7 | 2.17 | 2.83 | 3.82 | 4.67 | 6.35 | 8.38 | 9.80 | 12.02 | 14.07 | 18.48 | 24.32 |
| 8 | 2.73 | 3.49 | 4.59 | 5.53 | 7.34 | 9.52 | 11.03 | 13.36 | 15.51 | 20.09 | 26.12 |
| 9 | 3.32 | 4.17 | 5.38 | 6.39 | 8.34 | 10.66 | 12.24 | 14.68 | 16.92 | 21.67 | 27.88 |
| 10 | 3.94 | 4.86 | 6.18 | 7.27 | 9.34 | 11.78 | 13.44 | 15.99 | 18.31 | 23.21 | 29.59 |
|  | Nonsignificant | | | | | | | | Significant | | |

*Source:* R. A. Fisher and F. Yates, *Statistical Tables for Biological, Agricultural, and Medical Research*, 6th ed., Table IV, Longman Group UK Ltd., 1974.

5. Use the chi-square distribution table to determine the significance of the value.

   a. Determine the degrees of freedom, one less than the number of categories. Locate that value in the appropriate column.

   b. Locate the $\chi^2$ value for your significance level ($p = 0.05$ or less).

   c. Compare this $\chi^2$ value (from the table) with your calculated $\chi^2$.

6. State your conclusion in terms of your hypothesis.

   a. If your calculated $\chi^2$ value is greater than the $\chi^2$ value for your particular degrees of freedom and $p$ value (0.05), then *reject the null hypothesis* of no difference between expected and observed results. *You can conclude that there is a significant difference between your observed distribution and the theoretical expected distribution* (for example, under Mendel's laws).

   b. If your calculated $\chi^2$ value is less than the $\chi^2$ value for your particular degrees of freedom and $p$ value (0.05), then *fail to reject the null hypothesis* of no difference between observed and expected results. *You can conclude that there does not seem to be a significant difference between your observed distribution and the theoretical expected distribution* (for example, under Mendel's laws). *You can conclude that any differences between your observed results and the expected results can be attributed to chance or sampling error.* (Note: It is incorrect to say that you "accept" the null hypothesis. Statisticians either "reject" or "fail to reject" the null hypothesis.)

The chi-square test will be used to test for the goodness of fit between observed and expected data from several laboratory investigations in this lab manual.

# References

Gould, J. L. and G. F. Gould, *Biostats Basics: A Student Handbook.* New York: W. H. Freeman, 2002.

Moore, D. *Basic Practice of Statistics*, 4th ed. New York, NY: W.H. Freeman, 2007.

Motulsky, M. *Intuitive Biostatistics.* New York: Oxford University Press, 1995.

Triola, M. *Elementary Statistics Update*, 9th ed. San Francisco, CA: Addison-Wesley, 2005.

# Websites

Discussion and example of chi-square test: http://www.enviroliteracy.org/pdf/materials/1210.pdf

Electronic Statistics Textbook. Tulsa, OK: StatSoft. A relatively easy-to-understand free online statistics textbook: http://www.statsoft.com/textbook/

# Appendix C
# Terminology and Techniques for Dissection

## Orientation Terminology

The terms defined below are used with bilaterally symmetrical animals, both invertebrates and vertebrates (Figures C.1 and C.2). Use these terms as you describe animals studied in this laboratory manual. Note that texts may use different terminology for those animals called bipeds (for example, humans) and those called quadrupeds (for example, the fetal pig). Terminology used exclusively with bipeds is not included in this appendix.

**Right/left:** always refer to the animal's right or left, not yours.

**Anterior, cranial:** toward the head.

**Posterior, caudal:** toward the tail.

**Dorsal:** backside; from the Latin *dorsum,* meaning "back."

**Ventral:** bellyside; from the Latin *venter,* meaning "belly."

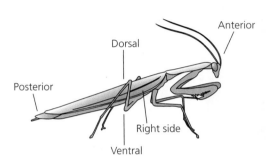

**Figure C.1.**
Orientation terminology in a bilaterally symmetrical invertebrate, the praying mantis.

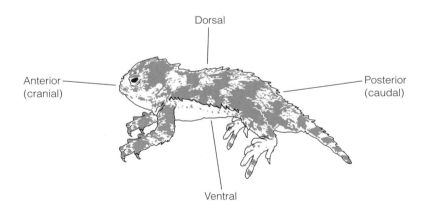

**Figure C.2.**
Orientation terminology in a quadruped vertebrate, a horned lizard.

## Terms Relating to Position in the Body

**Proximal:** near the trunk, attached portion, or point of reference, for example: "The pig's elbow is *proximal* to its wrist."

**Distal:** farther from the trunk, attached portion, or point of reference, for example: "The toes are *distal* to the ankle."

**Superficial:** lying on top or near the body surface.

**Deep:** lying under or below.

## Planes and Sections

A **section** is a cut through a structure. A **plane** is an imaginary line through which a section can be cut. Anatomists generally refer to three planes or sections (Figure C.3).

- **Sagittal section:** divides the body into left and right portions or halves. This is a longitudinal or lengthwise section from anterior to posterior.
- **Frontal section:** A longitudinal or lengthwise section from anterior to posterior, this divides the body into dorsal and ventral portions or halves.
- **Transverse section:** Also called a **cross section**, this divides the body into anterior and posterior portions or cuts a structure across its smallest diameter.

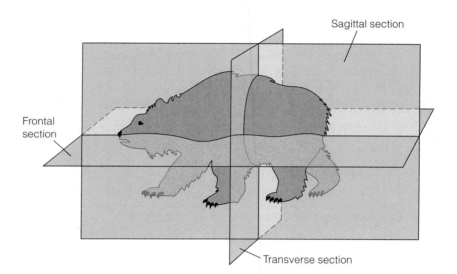

Sagittal section

Frontal section

Transverse section

**Figure C.3.**
Sections of a bilaterally symmetrical animal.

## Dissection Techniques

When studying the anatomy of an organism, the term **dissection** is perhaps a misnomer. *Dissection* literally means to cut apart piece by piece. In lab, however, it is usually more appropriate to expose structures rather than dissect them. Initial incisions do require that you cut into the body, but after body cavities are opened, you will usually only separate and expose body parts, using dissection rarely. Accordingly, you will use the scalpel when you make initial incisions into the body wall of large animals, but seldom when studying small animals or organs of large animals.

Scissors are used to deepen initial cuts made by the scalpel in large animals and to cut into the bodies of smaller animals. When using scissors, direct the tips upward to prevent gouging deeper organs. Once the animal's body is open, use forceps and the blunt probe to carefully separate organs and to pick away connective tissue obstructing and binding organs and ducts. Needle probes are only minimally useful. Never cut away an organ or cut through a blood vessel, nerve, or duct unless given specific instructions to do so.

Producing a good dissection takes time and cannot be rushed. As you study the anatomy of animals, your goal should be to expose all parts so that they can be easily studied and demonstrated to your lab partner or instructor.

## Appendix D

# Calculations and Answers to Problems for Lab Topic 12

## Calculations for Answers in Table 12.7

| Loci | Heterozygotes Observed | Heterozygotes Expected |
|------|------------------------|------------------------|
| ADH-1 | 13 | 12* |

To calculate, expected heterozygosity = $1 - [(f1)^2 + (f2)^2]$

$$f1 = 0.580 \quad (F)$$
$$f2 = 0.420 \quad (S)$$
$$1 - [(0.58)^2 + (0.42)^2] =$$
$$1 - [0.336 + 0.176] =$$
$$1 - [0.512] =$$
$$\text{expected heterozygosity} = 0.488$$

This is *heterozygosity*. To calculate *expected heterozygotes*, multiply this figure by the number of individuals in the total population; here it is *25:*

$$0.488 \times 25 = 12 = \text{expected number of heterozygotes*}$$

Repeat these calculations for ADH-2, PGI-2, and PGM-2.

To calculate the chi-square for the ADH-1 figures:

| | Heterozygotes | Homozygotes | Total |
|---|---|---|---|
| Observed (*o*) | 13 | 12 | 25 |
| Expected (*e*) | 12 | 13 | 25 |
| $\dfrac{(o-e)^2}{e}$ | 0.083 | 0.077 | |

$$\chi^2 = \Sigma(o - e)^2/e = 0.083 + 0.077 = 0.160$$

degree of freedom = 1
(2 classes = heterozygotes and homozygotes)

Refer to the chi-square table in Appendix B. Note that the probability is between 0.5 and 0.7, which is not significant. Therefore, the observed and expected are *not* significantly different. Repeat these calculations for all other loci.

## Answers to Question 2, Questions for Review

### Formulas

$$\text{Allelic frequency} = \frac{\text{number of one allele}}{\text{number of } all \text{ alleles}}$$

$$\text{Genotypic frequency} = \frac{\text{number of one genotype}}{\text{number of } all \text{ genotypes}}$$

## Question 2a

To calculate the genotypic frequency for 1/1, divide the number of 1/1 (634) by the total number of individuals (1,110):

$$\text{genotypic frequency for 1/1} = \frac{634}{1,110} = 0.571$$

$$\text{for 1/2, } \frac{391}{1,110} = 0.352; \text{ for 2/2, } \frac{85}{1,110} = 0.077$$

$$\text{Allelic frequency for allele 1} = \frac{2(634) + 391}{2,220} = \frac{1,659}{2,220} = 0.747$$

$$\text{Allelic frequency for allele 2} = \frac{2(85) + 391}{2,220} = \frac{561}{2,220} = 0.253$$

No. of allele 1 = 1659
No. of allele 2 =  561
Total = 2,220

## Question 2b

$$\text{Allelic frequency for allele 1} = \frac{2(9) + 135}{438} = \frac{153}{438} = 0.349$$

$$\text{Allelic frequency for allele 3} = \frac{135 + 2(75)}{438} = \frac{285}{438} = 0.651$$

## Question 2c

### Expected Frequency of Heterozygotes in Question 2a

$$
\begin{aligned}
\text{Expected frequency} &= 1 - [(f1)^2 + (f2)^2] \\
&= 1 - [(0.747)^2 + (0.253)^2] \\
&= 1 - [0.558 + 0.064] \\
&= 1 - 0.622 \\
&= 0.378
\end{aligned}
$$

$0.378 \times$ total no. of individuals = expected no. of heterozygotes
$0.378 \times 1110$ = 420 heterozygotes expected
391 heterozygotes observed

$1110 - 420$ = 690 homozygotes expected
719 homozygotes observed

**Chi-Square**

|  | Heterozygotes | Homozygotes | Total |
|---|---|---|---|
| Observed (o) | 391 | 719 | 1110 |
| Expected (e) | 420 | 690 | 1110 |
| $(o - e)^2/e$ | 2.002 | 1.219 |  |
| $\chi^2 = \Sigma(o - e)^2/e = 2.002 + 1.219 = 3.221$ |  |  |  |

There is no significant difference between $o$ and $e$.

### Expected Frequency of Heterozygotes in Question 2b

$$
\begin{aligned}
\text{Expected frequency} &= 1 - [(f1)^2 + (f3)^2] \\
&= 1 - [(0.349)^2 + (0.651)^2] \\
&= 1 - [0.122 + 0.424] \\
&= 1 - 0.546 \\
&= 0.454
\end{aligned}
$$

$0.454 \times$ total no. of individuals = expected no. of heterozygotes
$0.454 \times 219$ = 99 heterozygotes expected
135 heterozygotes observed

$219 - 99$ = 120 homozygotes expected
84 homozygotes observed

**Chi-Square**

|  | Heterozygotes | Homozygotes | Total |
|---|---|---|---|
| Observed (o) | 135 | 84 | 219 |
| Expected (e) | 99 | 120 | 219 |
| $(o - e)^2/e$ | 13.091 | 10.800 |  |
| $\chi^2 = \Sigma(o - e)^2/e = 13.091 + 10.800 = 23.891$ |  |  |  |

There is no significant difference between $o$ and $e$.

# Answers to Applying Your Knowledge

## Question 1

**Genotypic frequencies** $= \dfrac{\text{number of genotypes observed}}{\text{total genotypes}}$

Belgians    $M = \dfrac{896}{3,100}$    $MN = \dfrac{1,559}{3,100}$    $N = \dfrac{645}{3,100}$

$M = 0.289$    $MN = 0.503$    $N = 0.208$

English    $M = \dfrac{121}{422}$    $MN = \dfrac{200}{422}$    $N = \dfrac{101}{422}$

$M = 0.287$    $MN = 0.474$    $N = 0.239$

Continue calculations for other populations.

**Allelic frequencies** $= \dfrac{\text{number of times each allele found}}{\text{total alleles}}$

Belgians    $M = \dfrac{2(896) + 1,559}{6,200}$    $N = \dfrac{1,559 + 2(645)}{6,200}$

$M = 0.540$    $N = 0.459$

English    $M = \dfrac{442}{844}$    $N = \dfrac{402}{844}$

$M = 0.524$    $N = 0.476$

Continue calculations for other populations.

## Calculations of Expected Number of Heterozygotes from Allelic Frequencies

$f$M = frequency of allele M
$f$N = frequency of allele N

Belgian expected heterozygotes $= 1\,[(f\text{M})^2 + (f\text{N})^2]$
$= 1 - [(0.540)^2 + (0.459)^2]$
$= 1 - (0.292 + 0.211)$
$= 1 - 0.503$
$= 0.497$
$0.497 \times \text{total } (3,100) = 1,540$

English expected heterozygotes $= 1\,[(f\text{M})^2 + (f\text{N})^2]$
$= 1 - [(0.524)^2 + (0.476)^2]$
$= 1 - [0.275 + 0.227]$
$= 1 - 0.502$
$= 0.498$
$0.498 \times \text{total } (422) = 210$

Continue calculations for other populations.

## Summary

| Population | Genotypic Frequencies | | | Allelic Frequencies | | Heterozygotes | | Homozygotes | |
|---|---|---|---|---|---|---|---|---|---|
| | M | MN | N | M | N | Obs. | Exp. | Obs. | Exp. |
| Belgians | 0.289 | 0.503 | 0.208 | 0.540 | 0.459 | 1,559 | 1,540 | 1,541 | 1,560 |
| English | 0.287 | 0.474 | 0.239 | 0.524 | 0.476 | 200 | 210 | 222 | 212 |
| Egyptians | 0.279 | 0.488 | 0.233 | 0.523 | 0.477 | 245 | 250 | 257 | 252 |
| Ainu | 0.179 | 0.502 | 0.319 | 0.430 | 0.570 | 253 | 247 | 251 | 257 |
| Fijians | 0.110 | 0.445 | 0.445 | 0.332 | 0.668 | 89 | 89 | 111 | 111 |
| Papuans | 0.070 | 0.240 | 0.690 | 0.190 | 0.810 | 48 | 62 | 152 | 138 |

## Chi-Square Calculations

| | | Heterozygotes | Homozygotes | |
|---|---|---|---|---|
| Belgians | $o$ | 1,559 | 1,541 | |
| | $e$ | 1,540 | 1,560 | $\chi^2 = 0.465$ |
| | $(o - e)^2/e$ | 0.234 | 0.231 | |
| English | $o$ | 200 | 222 | |
| | $e$ | 210 | 212 | $\chi^2 = 0.947$ |
| | $(o - e)^2/e$ | 0.476 | 0.471 | |

Continue calculations for other populations.

# COLOR PLATES

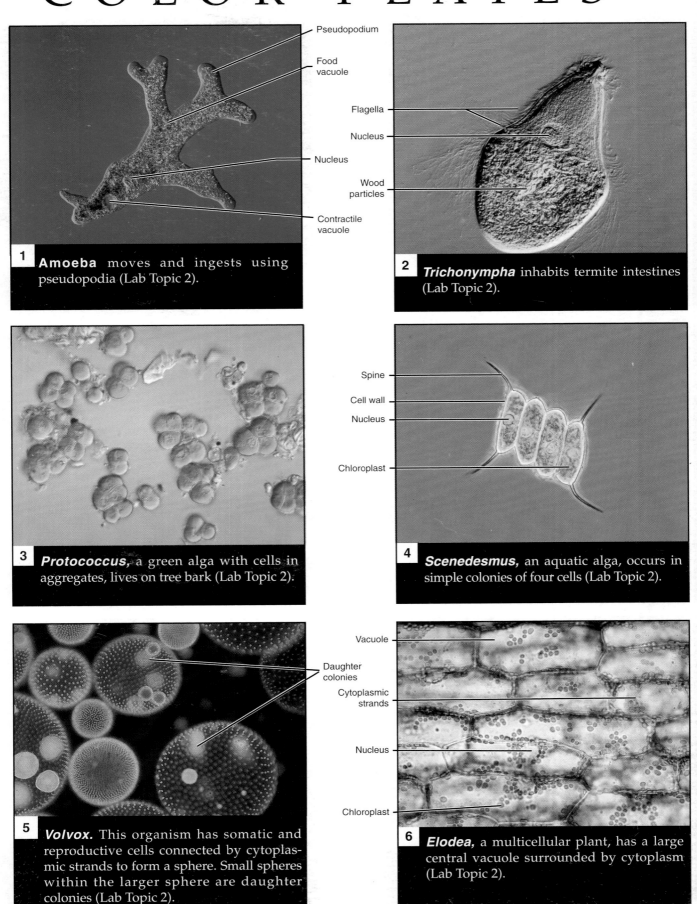

Pseudopodium

Food vacuole

Nucleus

Contractile vacuole

**1** **Amoeba** moves and ingests using pseudopodia (Lab Topic 2).

Flagella

Nucleus

Wood particles

**2** **Trichonympha** inhabits termite intestines (Lab Topic 2).

**3** **Protococcus,** a green alga with cells in aggregates, lives on tree bark (Lab Topic 2).

Spine

Cell wall

Nucleus

Chloroplast

**4** **Scenedesmus,** an aquatic alga, occurs in simple colonies of four cells (Lab Topic 2).

Daughter colonies

**5** **Volvox.** This organism has somatic and reproductive cells connected by cytoplasmic strands to form a sphere. Small spheres within the larger sphere are daughter colonies (Lab Topic 2).

Vacuole

Cytoplasmic strands

Nucleus

Chloroplast

**6** **Elodea,** a multicellular plant, has a large central vacuole surrounded by cytoplasm (Lab Topic 2).

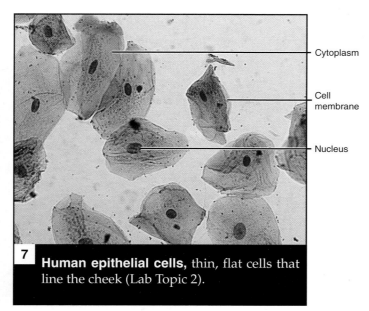

**7** **Human epithelial cells,** thin, flat cells that line the cheek (Lab Topic 2).

Cytoplasm

Cell membrane

Nucleus

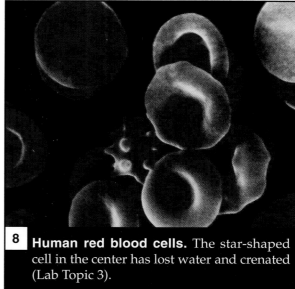

**8** **Human red blood cells.** The star-shaped cell in the center has lost water and crenated (Lab Topic 3).

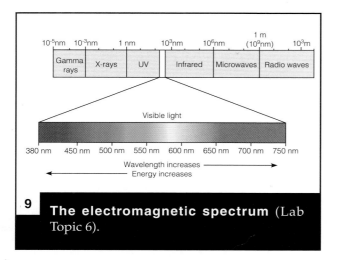

**9** **The electromagnetic spectrum** (Lab Topic 6).

**See color plates 11 and 12 on next page**

**10** ***Coleus*** **leaves with green and pink pigments** (Lab Topic 6).

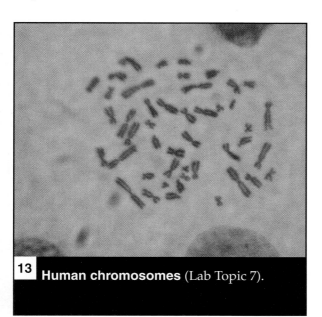

**13** **Human chromosomes** (Lab Topic 7).

**14** Two mating types of ***Sordaria fimicola*** growing on agar (Lab Topic 7).

## 11a–f    Mitosis in plant cells, onion root tip (Lab Topic 7).

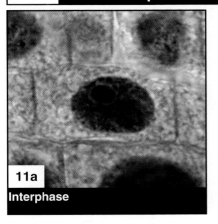

**11a** Interphase

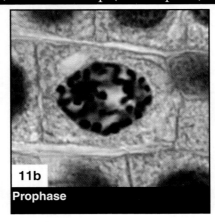

**11b** Prophase

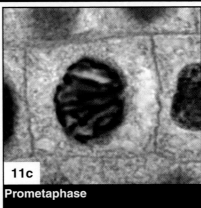

**11c** Prometaphase

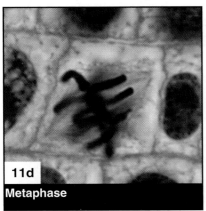

**11d** Metaphase

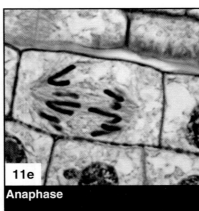

**11e** Anaphase

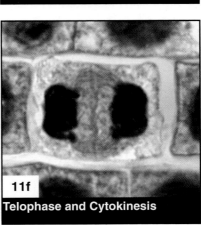

**11f** Telophase and Cytokinesis

## 12a–e    Mitosis in animal cells, whitefish blastula (Lab Topic 7).

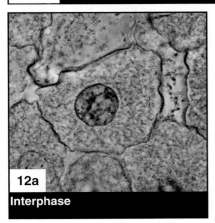

**12a** Interphase

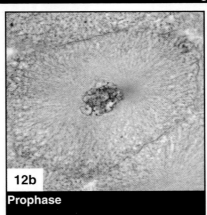

**12b** Prophase

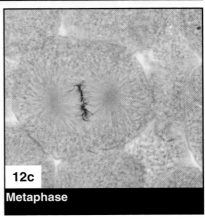

**12c** Metaphase

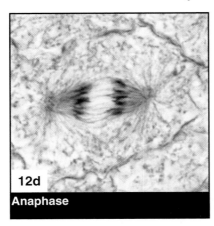

**12d** Anaphase

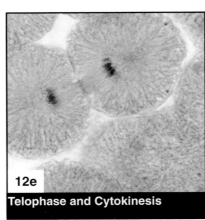

**12e** Telophase and Cytokinesis

**15** Yellow-green mutant *Brassica rapa* seedlings in the left half of the quad (wild-type with anthocyanin in the right half) (Lab Topic 8).

Wild-type

Yellow-green

**16** *Talinum,* a succulent commonly found in shallow soils on rock outcrops (Lab Topic 12).

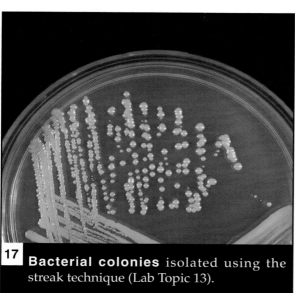

**17** **Bacterial colonies** isolated using the streak technique (Lab Topic 13).

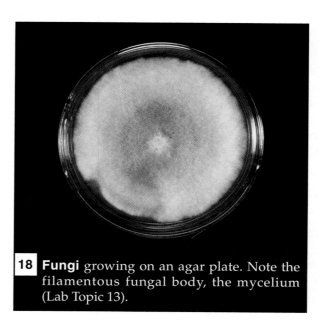

**18** **Fungi** growing on an agar plate. Note the filamentous fungal body, the mycelium (Lab Topic 13).

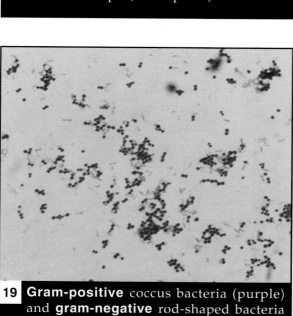

**19** **Gram-positive** coccus bacteria (purple) and **gram-negative** rod-shaped bacteria (pink) (Lab Topic 13).

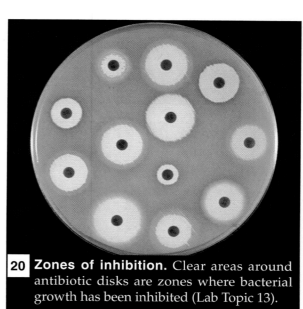

**20** **Zones of inhibition.** Clear areas around antibiotic disks are zones where bacterial growth has been inhibited (Lab Topic 13).

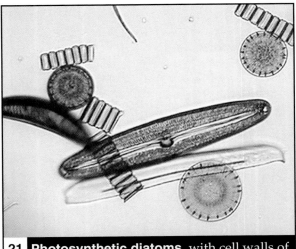

**21** **Photosynthetic diatoms,** with cell walls of silica, are found in two forms, pennate and centric (Lab Topic 14).

**22** The kelp *Macrocystis,* a brown alga (Lab Topic 14).

**23** **Radiolarians,** protozoa that move using pseudopodia, have skeletons of silicon dioxide (Lab Topic 14).

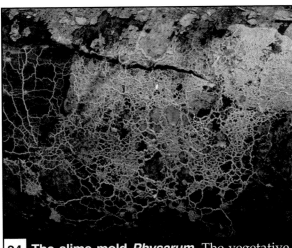

**24** **The slime mold *Physarum.*** The vegetative plasmodium is a multinucleate mass of protoplasm (Lab Topic 14).

**25** **Fruiting bodies of a slime mold** are shown magnified about 50 times (Lab Topic 14).

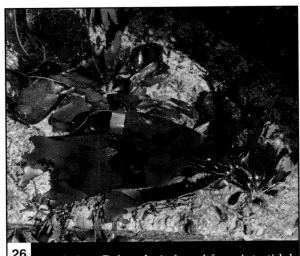

**26** A red alga, ***Palmaria,*** is found from intertidal to deep marine waters (Lab Topic 14).

**27** *Ulva,* sea lettuce, is an edible green alga (Lab Topic 14).

**28** The multicellular green algae, *Chara,* and land plants may share a common ancestor (Lab Topic 14).

**29** Each ascocarp (cup) bears microscopic asci in these **cup fungi** (Lab Topic 14).

**30** In a **moss,** spores develop in sporangia at the end of the sporophyte growing out of the leafy gametophyte (Lab Topic 15).

Gemmae cup

**31** **Liverworts.** Gemmae cups on the surface of this bryophyte function in asexual reproduction (Lab Topic 15).

**32** *Lycopodium.* This club moss is a living member of the division Lycophyta (Lab Topic 15).

**33** *Selaginella* often live in moist habitats, but this species may survive in a desert (Lab Topic 15).

**34** Sporangia, small spherical structures, are seen on stems of the **whisk fern,** *Psilotum* (Lab Topic 15).

**35** **Horsetail stems** are seen here with strobili containing sporangia (Lab Topic 15).

**36** **Fern fronds** are the sporophyte stage of the life cycle. The fiddleheads in the inset are young fronds ready to unfurl (Lab Topic 15).

**37** **Fern frond** with clusters of sporangia in sori. The sorus in the inset has an indusium covering the sporangia (Lab Topic 15).

**38** **Coniferophyta.** First, second, and third-year cones of bristlecone pine. Some bristlecone pines are more than 4,000 years old (Lab Topic 16).

**39** **Cycadophyta.** Pollen is produced in male cones, seeds in female cones in cycads (Lab Topic 16).

**40** **Ginkgophyta.** These ginkgo leaves are shown with ovules (Lab Topic 16).

**41** **Ginkgophyta.** Male strobili, pollen-producing structures, cluster at the base of leaves (Lab Topic 16).

**42** **Gnetophyta.** Mormon tea occurs in the deserts of North and Central America. Inset, enlarged male cones. (Lab Topic 16).

Stamen

Pistil

**43** Close-up of a **tulip** showing reproductive structures stamen and pistil (Lab Topic 16).

**44** **Wind pollinated flowers** are inconspicuous and produce enormous quantities of pollen (Lab Topic 16).

**45** **Flowers pollinated by bees** may be irregular in shape (Lab Topic 16).

**46** **Hummingbirds** pollinate red tubular flowers (Lab Topic 16).

**47** **Bats** pollinate night-blooming flowers with pale sepals and petals (Lab Topic 16).

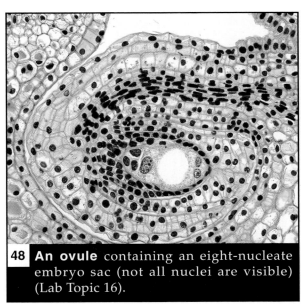

**48** **An ovule** containing an eight-nucleate embryo sac (not all nuclei are visible) (Lab Topic 16).

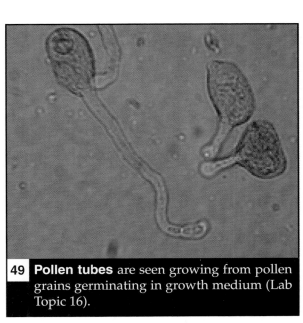

**49** **Pollen tubes** are seen growing from pollen grains germinating in growth medium (Lab Topic 16).

**50** **Sponge.** Needlelike spicules of calcium carbonate protrude from the osculum and the surface of the sponge body (Lab Topic 18).

**51** **Sponges** have a body form unlike any other animal (Lab Topic 18).

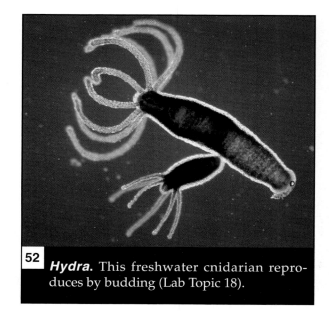

**52** *Hydra.* This freshwater cnidarian reproduces by budding (Lab Topic 18).

**53** *Dugesia,* a freshwater planarian with two pigmented eyespots between the two auricles on its anterior end (Lab Topic 18).

**54** *Nereis.* This segmented clamworm is an annelid that bears fleshy appendages called parapodia (Lab Topic 18).

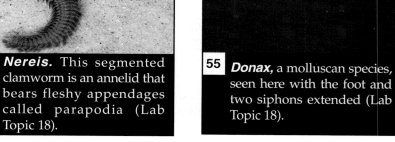

**55** *Donax,* a molluscan species, seen here with the foot and two siphons extended (Lab Topic 18).

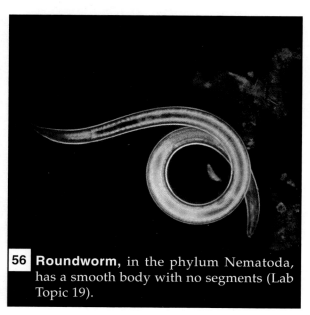

**56** **Roundworm,** in the phylum Nematoda, has a smooth body with no segments (Lab Topic 19).

**57** *Cambarus.* The freshwater crayfish is a member of the phylum Arthropoda (Lab Topic 19).

**58** The segmented body and jointed appendages are visible in this **grasshopper,** a terrestrial anthropod (Lab Topic 19).

**59** **Sea star.** The radial symmetry of echinoderm adults is evident in this deuterostome (Lab Topic 19).

**60** *Branchiostoma.* The lancelet is a small chordate that lives in coastal waters (Lab Topic 19).

Sclerenchyma

Parenchyma

Collenchyma

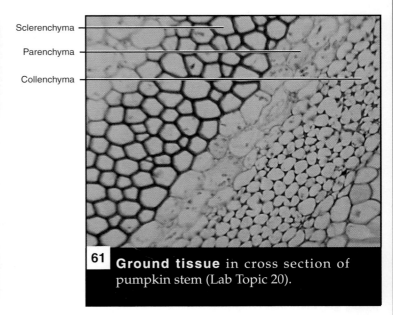

**61** **Ground tissue** in cross section of pumpkin stem (Lab Topic 20).

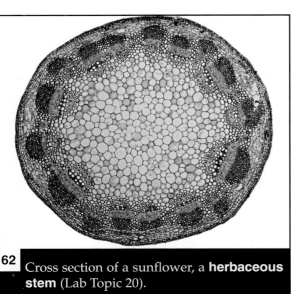

**62** Cross section of a sunflower, a **herbaceous stem** (Lab Topic 20).

**63** This **woody stem,** seen in cross section, is more than 2 years old (Lab Topic 20).

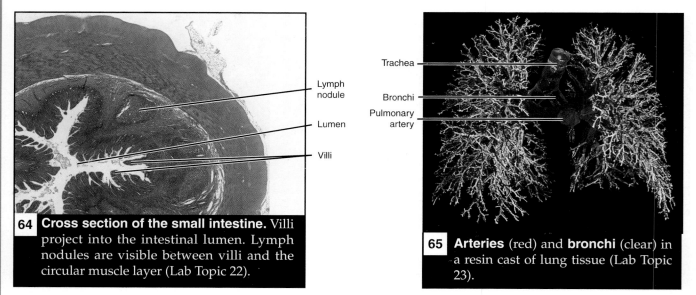

**64** **Cross section of the small intestine.** Villi project into the intestinal lumen. Lymph nodules are visible between villi and the circular muscle layer (Lab Topic 22).

Lymph nodule
Lumen
Villi

**65** **Arteries** (red) and **bronchi** (clear) in a resin cast of lung tissue (Lab Topic 23).

Trachea
Bronchi
Pulmonary artery

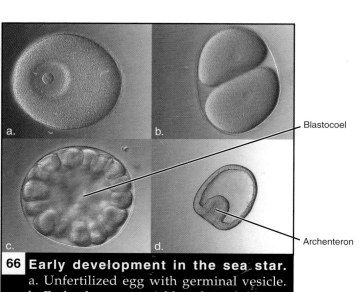
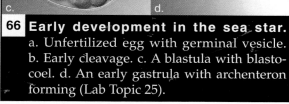

Blastocoel
Archenteron

**66** **Early development in the sea star.** a. Unfertilized egg with germinal vesicle. b. Early cleavage. c. A blastula with blastocoel. d. An early gastrula with archenteron forming (Lab Topic 25).

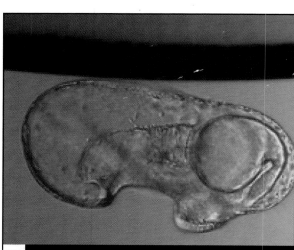

**67** **Sea star bipinnaria larva.** Identify mouth, esophagus, stomach, intestine, and anus (Lab Topic 25).

**68** **Zebrafish** are important organisms for developmental studies (Lab Topic 25).

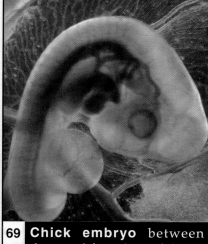

**69** **Chick embryo** between three and four days of development (Lab Topic 25).

**70** *Betta splendens.* The male Siamese fighting fish displays agonistic behavior (Lab Topic 26).